AF528762

The Most Successful Way of JEE Preparation

DPP

Daily Practice Problems

JEE Main & Advanced

Conic Section, Vector & 3D Geometry

MATHEMATICS Vol-5

The Most Successful Way of JEE Preparation

DPP

Daily Practice Problems

JEE Main & Advanced

Conic Section, Vector & 3D Geometry

MATHEMATICS Vol-5

Amit M Agarwal

arihant
ARIHANT PRAKASHAN (SERIES), MEERUT

DPP
Daily Practice Problems

arihant

ARIHANT PRAKASHAN (SERIES), MEERUT

All Rights Reserved

© Author

No part of this publication may be re-produced, stored in a retrieval system or by any means, electronic mechanical, photocopying, recording, scanning, web or otherwise without the written permission of the publisher. Arihant has obtained all the information in this book from the sources believed to be reliable and true. However, Arihant or its editors or authors or illustrators don"t take any responsibility for the absolute accuracy of any information published, and the damages or loss suffered thereupon.

All disputes subject to Meerut (UP) jurisdiction only.

Administrative & Production Offices

Corporate Office 'Ramchhaya' 4577/15, Agarwal Road, Darya Ganj, New Delhi -110002
Tele: 011- 47630600, 43518550; Fax: 011- 23280316

Head Office Kalindi, TP Nagar, Meerut (UP) - 250002
Tel: 0121-2401479, 2512970, 4004199; Fax: 0121-2401648

Sales & Support Offices

Agra, Ahmedabad, Bengaluru, Bareilly, Chennai, Delhi, Guwahati, Hyderabad, Jaipur, Jhansi, Kolkata, Lucknow, Meerut, Nagpur & Pune

ISBN : 978-93-13193-53-1

Published by Arihant Publications (I) Ltd.

Production Team

Publishing Managers	Mahendra Singh Rawat & Keshav Mohan	*Inner Designer*	Mazher Chaudhary
Project Head	Sorabh Chaudhary	*Page Layouting*	Sunil Sharma, Hemant
Cover Designer	Shanu Mahad	*Proof Reading*	Himanshu

For further information about the products from Arihant, log on to www.arihantbooks.com or email to info@arihantbooks.com

/arihantpub /@arihantpub Arihant Publications /arihantpub

ABOUT DPP

Daily Practice Problems

Indian Institutes of Technology better known as IITs have always been the dream destination of each science group students who want to pursue their career as Engineers. These are the institutes which have got very high value not even in India but in the whole world. To get the admission in such institutes one need to face entrance Examinations; JEE Main & JEE Advanced, out of which JEE Main is considered as the preliminary exam & JEE Advanced is the final exam to get into the IITs. Approximately per year 12 lakh students enroll themselves for JEE Main and only the top 2.25 lakh in JEE Main merit can attempt for JEE Advanced, the fact clearly express the degree of toughness of JEE Main & JEE Advanced.

To have success in such an exam one needs to be very planned & systematic in its preparation. A very popular wise saying 'Rome was not built in a day' suggests that to achieve a big target like IITs we need to work continuously. For continuous work, one need to have proper planning that how much work he/she has to do on regular basis, i.e., on daily basis. Each student preparing for JEE has many books but unfortunately none of them gives the students a plan for daily studies, which is the most effective success key for JEE.

Keeping the daily practice needs of the students in mind we have come up with DPP Series (Daily Practice Problems) having approximately 26 books. The primary focus of this series is on achieving success through practice & proper planning. For that we have divided the whole syllabus of JEE Main & JEE Advanced; Physics, Chemistry & Mathematics, into different volumes (around 8 volumes of Mathematics & 9 volumes each of Physics & Chemistry), each volume has Daily Practice Problems Sheets having specific questions on various Topics of the individual chapter of the volume ensuring the complete Practice & Assessment of the topic.

DPP
Daily Practice Problems

In this way we have tried to give you an effective daily planning for JEE keeping in mind the daily hours that a student can devote to his practice, you are just required to practice one sheet of each subject on daily basis. This habit of daily practice will keep you away from exam stress and thoughts like; how much I have prepared/am I preparing properly or not/ do I need to study more on daily basis or what I am doing is ok, etc., which regularly tease your mind and can give you panic at any stage of your preparation.

Some special features that we have tried to incorporate in our Daily Practice Problems (DPP) series are discussed below

1. Micro Level coverage of all the topics of a Chapter/ Unit given in JEE Main & Advanced Syllabus.
2. Each DPP has various types of questions; Subjective Questions, Single/ Multiple Correct Options, Statements type, Comprehension based, Integer Value Answer type, Matching, etc., making you practice & ready for all the question formats of JEE.
3. New & innovative questions in each DPP challenging & improving your problem solving skills.
4. Along with Topical Coverage, Revisal DPPs for JEE Main & JEE Advanced with each Chapter, covering multidisciplinary questions proving comprehensive practice.
5. JEE Main & JEE Advanced Archive (collection of last 13 years' exams questions) with each chapter.
6. Complete Solutions for each DPP.

We hope that DPP series will prove to be very beneficial for the students preparing for JEE, by giving them daily practice in a systematic & planned manner. The suggestions/criticism from the students, teachers & parents about DPP series will be welcomed whole heartedly and we promise to update this series accordingly from time to time.

DPP
Daily Practice Problems

DAILY TASKS...

1 PARABOLA

2 ELLIPSE

DPP
Daily Practice Problems

3 HYPERBOLA

4 VECTORS

5 3D GEOMETRY

1. Parabola

DPP

DAILY PRACTICE PROBLEMS

DPP-1 Standard Equation of Conics

Subjective Questions

Working Space

Direction (Q. Nos. 1-8) *These questions are subjective in nature, need to be solved completely on notebook.*

1 Determine the name of the conic represented by each of the following equations:

i. $\sqrt{px} + \sqrt{qy} = 1$, where $p, q \in R$, $p, q > 0$ ii. $2xy + 4x - 6y + 17 = 0$

iii. $13x^2 - 18xy + 37y^2 + 2x + 14y - 2 = 0$ iv. $x = t^2 + t + 1$, $y = t^2 - t + 1$

2 Find the coordinates of the foci, equations of the directrix and lengths of the latusrectum.

i. $y^2 = 8x$ ii. $x^2 = 6y$

iii. $y^2 = -12x$ iv. $x^2 = -16y$

3 Find the vertex, axis, directrix, latusrectum of the following parabolas. Also, draw their rough sketches.

i. $y^2 - 8y - x + 19 = 0$ ii. $4y^2 + 12x - 20y + 67 = 0$

iii. $y = x^2 - 2x + 3$ iv. $x^2 + 2y - 3x + 5 = 0$

4 Find the equation of parabola whose focus is at $(-1, -2)$ and the directrix $x - 2y + 3 = 0$.

5 Find the equation of parabola with vertex $(2, -3)$ and focus $(0, 5)$.

6 Find the equation of parabola whose focus is $(1, -1)$ and whose vertex is $(2, 1)$.

7 Find the equation of parabola whose latusrectum is 4 units, axis is the line $3x + 4y - 4 = 0$ and the tangent at the vertex is the line $4x - 3y + 7 = 0$.

8 Find the locus of middle points of all chords of the parabola $y^2 = 4ax$ which are drawn through the origin.

Only One Option Correct Type

Direction (Q. Nos. 9-16) *This section contains 8 multiple choice questions. Each question has four choices (a), (b), (c) and (d), out of which ONLY ONE is correct.*

9 The locus of point of trisection of all the double ordinates of the parabola $y^2 = lx$ is a parabola, whose latusrectum is

a. $l/9$ b. $2l/9$

c. $4l/9$ d. $l/36$

10 A variable circle is drawn to touch the line $3x - 4y = 10$ and also the circle $x^2 + y^2 = 1$ externally, then the locus of its centre is a

a. straight line b. circle

c. pair of real, distinct straight lines d. parabola

Working Space

11 The equation of line $y = m(x - a)$ meets the parabola $y^2 = 4ax$ in two distinct real points, if

a. $m \in R$
b. $m \in [-1, 1]$
c. $m \in (-\infty, 1] \cup [1, \infty)$
d. $m \in R - \{0\}$

12 The equation of circle drawn with the focus of the parabola $(x - 1)^2 - 8y = 0$ as its centre and touching the parabola at its vertex, is

a. $x^2 + y^2 - 4y = 0$
b. $x^2 + y^2 - 4y + 1 = 0$
c. $x^2 + y^2 - 2x - 4y = 0$
d. $x^2 + y^2 - 2x - 4y + 1 = 0$

13 Which one of the following equations represented parametrically, represents equation to a parabolic profile?

a. $x = 3\cos t,\ y = 4\sin t$
b. $x^2 - 2 = -2\cos t,\ y = 4\cos^2 \frac{t}{2}$
c. $\sqrt{x} = \tan t,\ \sqrt{y} = \sec t$
d. $x = \sqrt{1 - \sin t},\ y = \sin \frac{t}{2} + \cos \frac{t}{2}$

14 The length of intercept on Y-axis cut off by the parabola $y^2 - 5y = 3x - 6$ is

a. 1
b. 2
c. 3
d. 5

15 If the line $x - 1 = 0$ is the directrix of the parabola $y^2 - kx + 8 = 0$, then one of the values of k is

a. 1/8
b. 8
c. 4
d. 1/4

16 The length of latusrectum of the parabola $25[(x - 2)^2 + (y - 3)^2] = (3x - 4y + 7)^2$ is

a. 4
b. 2
c. 1/5
d. 2/5

One or More than One Options Correct Type

Direction (Q. Nos. 17-20) This section contains 4 multiple choice questions. Each question has four choices (a), (b), (c) and (d), out of which ONE or MORE THAN ONE are correct.

17 The locus of mid-point of the focal radii of a variable point moving on the parabola $y^2 = 8x$ is a parabola whose

a. latusrectum is half the latusrectum of original parabola
b. vertex is (1, 0)
c. directrix is Y-axis
d. focus has the coordinates (2, 0)

18 P is the point on the parabola $y^2 = 4ax (a > 0)$ whose vertex is A. PA is produced to meet the directrix in D and M is the foot of perpendicular from P on the directrix. If a circle is described on MD as a diameter, then its intersect the X-axis at a point whose coordinates are

a. $(-3a, 0)$
b. $(-a, 0)$
c. $(-2a, 0)$
d. $(a, 0)$

19 A variable circle is described to pass through the point (1, 0) and tangent to the curve $y = \tan(\tan^{-1} x)$. The locus of centre of the circle is a parabola whose

a. length of the latusrectum is $2\sqrt{2}$
b. axis of symmetry has the equation $x + y = 1$
c. vertex has the coordinates (3/4, 1/4)
d. directrix is $x - y = 0$

Working Space

20 Let A and B be two distinct points on the parabola $y^2 = 4x$. If axis of the parabola touches a circle of radius r having AB as its diameter, then the slope of the line joining A and B can be **[IIT JEE 2010]**

a. $-1/r$ b. $1/r$
c. $2/r$ d. $-2/r$

Matching List Type

Direction (Q. No. 21) *Choices for the correct combination of elements from Column I and Column II are given as options (a), (b), (c) and (d), out of which one is correct.*

21 Match the Column I with Column II and mark the correct option from the codes given below.

	Column I		Column II
i.	$xy + a^2 = a(x + y)$	p.	Ellipse
ii.	$2x^2 - 72xy + 23y^2 - 4x - 28y - 48 = 0$	q.	Hyperbola
iii.	$6x^2 - 5xy - 6y^2 + 14x + 5y + 4 = 0$	r.	Parabola
iv.	$14x^2 - 4xy + 11y^2 - 44x - 58y + 71 = 0$	s.	Pair of straight lines
v.	$4x^2 - 4xy + y^2 - 12x + 6y + 9 = 0$		

Codes

	i	ii	iii	iv	v
a.	s	q	s	p	s
b.	p	r	q	r	s
c.	q	p	r	s	q
d.	r	s	p	q	r

Comprehension Type

Direction (Q. Nos. 22-25) *This section contains a paragraph, describing theory, experiments, data, etc. Four questions related to the paragraph have been given. Each question has only one correct answer among the four given options (a), (b), (c) and (d).*

Passage

The equation of conic given by
$S : 9x^2 - 24xy + 16y^2 - 20x - 15y - 60 = 0.$

22 The curve S represents a/an

a. circle b. parabola
c. ellipse d. hyperbola

23 The axis of conic is

a. $3x - 4y = 0$ b. $3x + 4y = 0$
c. $3x - 4y + 12 = 0$ d. $4x + 3y + 12 = 0$

24 The vertex of conic is

a. $\left(\frac{48}{25}, \frac{36}{25}\right)$ b. $\left(-\frac{43}{25}, -\frac{129}{100}\right)$
c. $\left(-\frac{48}{25}, -\frac{36}{25}\right)$ d. $\left(\frac{43}{25}, \frac{129}{100}\right)$

25 The directrix of curve is

a. $4x + 3y + \frac{53}{4} = 0$ b. $4x - 3y + 12 = 0$
c. $4x + 3y + 12 = 0$ d. $3x - 4y = 0$

DPP-2 Equation of Chord, Focal Chord and Its Properties

Only One Option Correct Type

Working Space

Direction *(Q. Nos. 1-19) This section contains 19 multiple choice questions. Each question has four choices (a), (b), (c) and (d), out of which ONLY ONE is correct.*

1 The focal chord of parabola $y^2 = 4ax$ makes an angle α with the X-axis is of length

a. $4a\,\text{cosec}^2\alpha$ b. $4\,\text{cosec}^2\alpha$

c. $8a\,\text{cosec}^2\alpha$ d. None of these

2 If $(-2a, a+1)$ lies in the interior (smaller region) bounded by the circle $x^2 + y^2 = 4$ and the parabola $y^2 = 4x$, then

a. $a \in (-1, 5 - 2\sqrt{6})$

b. $a \in (1, 5 + 2\sqrt{6})$

c. $a \in (-1, -5 + 2\sqrt{6})$

d. $a \in (-1, -5 + 4\sqrt{6})$

3 The circle described on any focal chord of a parabola as diameter touches

a. tangent at vertex b. axis of parabola

c. the double ordinate d. the directrix

4 The locus of centre of the circle described on any focal chord of a parabola $y^2 = 4ax$, is

a. $y^2 = 2a(x - a)$ b. $y^2 = 2ax$

c. $y^2 = 4a(x - a)$ d. $y^2 = 2a(x + a)$

5 If a chord of $y^2 = 4ax$ makes an angle α, $\alpha \in (0, \pi/4]$ with the positive direction of X-axis, then the minimum length of this focal chord is

a. $2\sqrt{2}\,a$ units b. $4\sqrt{2}\,a$ units

c. $8\,a$ units d. $16\,a$ units

6 If $(2, -8)$ is one end of a focal chord of the parabola $y^2 = 32x$, then the other end of the focal chord is

a. (32,32) b. (32, – 32)

c. (–2, 8) d. (2, 8)

7 The maximum number of common chords of a parabola and a circle can be equal to

a. 2 b. 4

c. 6 d. 8

Working Space

8 Let A and B be two points on a parabola $y^2 = x$ with vertex V such that VA is perpendicular to VB and θ is the angle between the chord VA and the axis of the parabola. The value of $\frac{|VA|}{|VB|}$ is

a. $\tan\theta$ b. $\tan^3\theta$
c. $\cot^2\theta$ d. $\cot^3\theta$

9 The length of focal chord of the parabola $y^2 = 4ax$ at a distance b from the vertex is c, then

a. $2a^2 = bc$ b. $a^3 = b^2c$
c. $ac = b^2$ d. $b^2c = 4a^3$

10 The straight line joining any point P on the parabola $y^2 = 4ax$ to the vertex and perpendicular from the focus to the tangent at P, intersect at R, then the equation of locus of R is

a. $x^2 + y^2 - ax = 0$
b. $2x^2 + y^2 - 2ax = 0$
c. $2x^2 + y^2 - ax = 0$
d. $2x^2 + y^2 - 2ay = 0$

11 The locus of feet of perpendiculars drawn from vertex of the parabola $y^2 = 4ax$ upon all such chords of the parabola which subtends a right angle at the vertex, is

a. $x^2 + y^2 - 4ax = 0$ b. $x^2 + y^2 - 2ax = 0$
c. $x^2 + y^2 + 2ax = 0$ d. $x^2 + y^2 + 4ax = 0$

12 The locus of trisection point of any arbitrary double ordinate of the parabola $x^2 = 4by$ is

a. $9x^2 = by$ b. $3x^2 = 2by$
c. $9x^2 = 4by$ d. $9x^2 = 2by$

13 The point on $y^2 = 4ax$ nearest to the focus has its abscissa equal to

a. $-a$ b. a
c. $a/2$ d. 0

14 The length of chord of the parabola $x^2 = 4ay$ passing through the vertex and having slope $\tan\alpha$, is

a. $4a \operatorname{cosec}\alpha \cdot \cot\alpha$
b. $4a \tan\alpha \cdot \sec\alpha$
c. $4a \cos\alpha \cdot \cot\alpha$
d. $4a \sin\alpha \cdot \tan\alpha$

15 If the distance of two points P and Q from the focus of a parabola $y^2 = 4ax$ are 4 and 9 respectively, then the distance of the point of intersection of tangents at P and Q from the focus, is

a. 8 b. 6
c. 5 d. 12

16 The intercept of the focal chord of parabola $y^2 = 4ax$ are b and k, then the value of k will be

a. $\frac{ab}{b-a}$ b. $\frac{b}{b-a}$
c. $\frac{a}{b-a}$ d. $\frac{ab}{a-b}$

Working Space

17 If $a \neq 0$ and the line $2bx + 3cy + 4d = 0$ passes through the point of intersection of parabolas $y^2 = 4ax$ and $x^2 = 4ay$, then

a. $d^2 + (2b - 3c)^2 = 0$
b. $d^2 + (3b - 2c)^2 = 0$
c. $d^2 + (2b + 3c)^2 = 0$
d. $d^2 + (3b + 2c)^2 = 0$

18 The locus of vertices of the family parabola $y = \frac{a^3x^2}{3} + \frac{a^2x}{2} - 2a$ is

a. $xy = 3/4$
b. $xy = 35/16$
c. $xy = 64/105$
d. $xy = 105/64$

19 Consider a circle with its centre lying on the focus of the parabola $y^2 = 2px$ such that it touches the directrix of the parabola. Then, a point of intersection of the circle and parabola is

a. $(p/2, p)$ or $(p/2, -p)$
b. $(p/2, -p/2)$
c. $(-p/2, p)$
d. $(-p/2, -p/2)$

Comprehension Type

Direction (Q. Nos. 20-25) *This section contains 2 paragraphs, each describing theory, experiments, data, etc. Six questions related to the paragraphs have been given. Each question has only one correct answer among the four given options (a), (b), (c) and (d).*

Passage I

Two perpendicular chords are drawn from the origin O to the parabola $y = x^2 - x$, which meet the parabola at P and Q, then rectangle $POQR$ is completed.

20 The locus of R is

a. $y = 2x^2 + 3x + 4$
b. $y = x^2 - 3x + 4$
c. $2y^2 + 3y + 4 = x$
d. $3y^2 - 3y + 4 = x$

21 The axis of locus of R is

a. $x = 3/2$
b. $x = -3/2$
c. $y = 3/2$
d. $y = -3/2$

22 The length of latusrectum is

a. 1
b. 1/2
c. 3/4
d. 3/2

Passage II

Through the vertex O of a parabola $y^2 = 4x$, chords OP and OQ are drawn at right angles to one another.

23 For all positions of P, PQ cuts the axis of parabola at fixed point

a. $(-4, 0)$
b. $(-1, 0)$
c. $(4, 0)$
d. $(1, 0)$

24 The locus of middle point of PQ is

a. $y^2 = 2x$
b. $y^2 = 2(x - 2)$
c. $y^2 = 2(x - 4)$
d. $y^2 = 4(x - 2)$

25 The directrix of locus of the middle point of PQ is

a. $x = 0$
b. $x = 2$
c. $x = 9/2$
d. $x = 7/2$

One Integer Value Correct Type

Working Space

Direction (Q. Nos. 26-31) *This section contains 6 questions. When worked out will result in an integer from 0 to 9 (both inclusive).*

26 Chords of a parabola $y^2 = 4ax$, intersect the axis in a common point P which is on the same side of the vertex as the focus. If the distance of P from the vertex be equal to the length of the latusrectum of the parabola. If the chords subtends the same angle θ at the vertex, then $(2\sin^2\theta + 4\cos^2\theta)$ is

27 Let l_1, l_2 be the lengths of perpendicular chords of the parabola $y^2 = 4ax$ drawn through the vertex and $(l_1 \, l_2)^{4/3} = 4a^2\lambda \cdot (l_1^{2/3} + l_2^{2/3})$, then λ is

28 A ΔABC of area Δ is inscribed in the parabola $y^2 = 4ax$, such that the vertex A lies at the vertex of the parabola and side BC is a focal chord. If the distances of B and C from the axis of the parabola is $\frac{\lambda\Delta}{a}$, then the value of λ is

29 Let PQ be a variable focal chord of the parabola $y^2 = 4ax$ whose vertex is A. If the locus of the centroid of ΔAPQ is a parabola, whose length of latusrectum is $\frac{4a}{\lambda}$, then the value of λ is

30 If p be the perpendicular distance of focal chord PQ of length l from the vertex A of the parabola $y^2 = 4x$. If the length $l = \lambda / p^2$. Then, the value of λ is

31 From a variable point $P(t^2, 2t)$ $(1 \le t \le 3)$ on the parabola $y^2 = 4x$, perpendicular PM is drawn to the tangent at the vertex of parabola. Now, from the mid-point Q of PM, perpendicular QL is drawn to the focal chord of parabola through P. If the maximum length of QL is λ/μ, then the value of $(\lambda - 2\mu)$ is

DPP-3 Equation of Tangent and Its Properties

Only One Option Correct Type

Working Space

Direction *(Q. Nos. 1-20) This section contains 20 multiple choice questions. Each question has four choices (a), (b), (c) and (d), out of which ONLY ONE is correct.*

1 If $y = 2x - 3$ is a tangent to the parabola $y^2 = 4a\left(x - \frac{1}{3}\right)$, then a is equal to

a. 1 b. -1
c. $14/3$ d. $-14/3$

2 The equation of tangent to the parabola $y = (x - 3)^2$ parallel to the chord joining the points (3, 0) and (4, 1), is

a. $2x - 2y + 6 = 0$
b. $2x - 2y - 6 = 0$
c. $4x - 4y + 11 = 0$
d. $4x - 4y = 13$

3 The combined pair of tangents from $(-2, 0)$ to the parabola $y^2 = 8x$ is

a. $y = \pm (x + 2)$ b. $x = \pm (y - 2)$
c. $y = \pm (x - 2)$ d. $x = \pm (y + 2)$

4 The angle between the tangents drawn from a point $(-a, 2a)$ to $y^2 = 4ax$ is

a. $\pi/4$ b. $\pi/2$
c. $\pi/3$ d. $\pi/6$

5 The line $4x - 7y + 10 = 0$ intersects the parabola, $y^2 = 4x$ at the points A and B. The coordinates of point of intersection of the tangents drawn at the points A and B are

a. $(7/5, 5/2)$ b. $(-5/2, -7/2)$
c. $(5/2, 7/2)$ d. $(-7/2, -5/2)$

6 An equation of tangent common to the parabolas $y^2 = 4x$ and $x^2 = 4y$ is

a. $x - y + 1 = 0$ b. $x + y - 1 = 0$
c. $x + y + 1 = 0$ d. $y = 0$

7 If on a given base, a triangle be described such that the sum of the tangents of the base angles is constant, then the locus of the vertex is

a. a circle b. a parabola
c. an ellipse d. a hyperbola

8 PN is an ordinate of the parabola $y^2 = 4ax$. A straight line is drawn parallel to the axis to bisect NP and meets the curve in Q. NQ meets the tangent at the vertex A in a point T such that $AT = k \cdot NP$, then the value of k is

a. $3/2$ b. $2/3$
c. 1 d. None of these

Working Space

9 The equation of common tangent to the curves $y^2 = 8x$ and $xy - 1$is

a. $3y = 9x + 2$
b. $y = 2x + 1$
c. $2y = x + 8$
d. $y = x + 2$

10 The equation of common tangent touching the circle $(x - 3)^2 + y^2 = 9$and the parabola $y^2 = 4x$ above the X-axis, is

a. $\sqrt{3}y = 3x + 1$
b. $\sqrt{3}y = -(x + 3)$
c. $\sqrt{3}y = x + 3$
d. None of these

11 AB, AC are tangents to a parabola $y^2 = 4ax$. p_1, p_2 and p_3 are the lengths of the perpendiculars from A, B and C respectively on any tangent to the curve, then p_2, p_1, p_3 are in

a. AP
b. GP
c. HP
d. None of these

12 T is a point on the tangent to a parabola $y^2 = 4ax$ at its point P. TL and TN are the perpendiculars on the focal radius SP and the directrix of the parabola respectively, then

a. $SC = 2(TN)$
b. $3(SL) = 2(TN)$
c. $SL = TN$
d. $2(SL) = 3(TN)$

13 The locus of foot of the perpendiculars drawn from the vertex on a variable tangent to the parabola $y^2 = 4ax$ is

a. $x(x^2 + y^2) + ay^2 = 0$
b. $y(x^2 + y^2) + ax^2 = 0$
c. $x(x^2 - y^2) + ay^2 = 0$
d. None of the above

14 Let $y^2 = 4ax$ be a parabola and $x^2 + y^2 + 2bx = 0$ be a circle. If parabola and circle touch each other externally, then

a. $ab > 0$
b. $a > 0, b < 0$
c. $a < 0, b > 0$
d. None of these

15 The angle between the tangents drawn from the origin to the parabola $(x - a)^2 = -4a(y + a)$ is

a. $\pi/3$
b. $\pi/6$
c. $\pi/4$
d. $\pi/2$

16 AP and BP are tangents to the parabola, $y^2 = 4x$ at A and B. If the chord AB passes through a fixed point $(-1, -1)$, then the equation of locus of P is

a. $y = 2(x - 1)$
b. $y^2 = 2(x - 1)$
c. $y = 2(x + 1)$
d. $y^2 = 2(x + 1)$

17 The angle between the tangents drawn from the point (1, 4) to the parabola $y^2 = 4x$ is

a. $\pi/2$
b. $\pi/3$
c. $\pi/4$
d. $\pi/6$

18 Tangent to the curve $y = x^2 + 6$ at a point $P(1, 7)$ touches the circle $x^2 + y^2 + 16x + 12y + c = 0$ at a point Q. Then, the coordinates of Q are

a. $(-6, -11)$
b. $(-9, -13)$
c. $(-10, -15)$
d. $(-6, -7)$

Working Space

19 At any point P on the parabola $y^2 - 2y - 4x + 5 = 0$ a tangent is drawn which meets the directrix at Q. Then, the locus of R which divides QP externally in the ratio $\frac{1}{2} : 1$ is

a. $(x+1)(y-1)^2 = 4$
b. $(x+1)(y-1)^2 + 4 = 0$
c. $(x-1)(y+1)^2 = 4$
d. $(x-1)(y+1)^2 + 4 = 0$

20 The angle between a pair of tangents drawn from a point P to the parabola $y^2 = 4ax$ is 45°, then the locus of point P is a/an

a. circle
b. parabola
c. ellipse
d. hyperbola

One or More than One Options Correct Type

Direction (Q. Nos. 21-25) *This section contains 5 multiple choice questions. Each question has four choices (a), (b), (c) and (d), out of which ONE or MORE THAN ONE are correct.*

21 The tangents at the extremities of a focal chord of a parabola

a. are perpendicular
b. are parallel
c. intersect on the directrix
d. intersect at the vertex

22 The equation of a tangent to the parabola $y^2 = 12x$ which make an angle 45° with the $y = 3x + 77$, is

a. $2x - 4y + 3 = 0$
b. $x - 2y + 12 = 0$
c. $4x + 2y + 3 = 0$
d. $2x + y - 12 = 0$

23 The focus of the parabola is (1, 1) and the tangent at the vertex has the equation $x + y = 1$. Then,

a. the equation of parabola is $(y - x)^2 = 2(x + y - 1)$
b. the equation of parabola is $(x - y)^2 = 4(x + y - 1)$
c. the coordinates of vertex are (1/2, 1/2)
d. the length of latusrectum is $2\sqrt{2}$

24 P is a point on the parabola $y^2 = 4x$, where abscissa and ordinate are equal. Then, the equation of circle passing through the focus and touching the parabola at P is

a. $x^2 + y^2 - 13x + 2y + 12 = 0$
b. $x^2 + y^2 - 3x - 18y + 2 = 0$
c. $x^2 + y^2 + 13x - 2y - 12 = 0$
d. $x^2 + y^2 - x = 0$

25 The equation of common tangent(s) to the parabola $y = x^2$ and $y = -x^2 + 4x - 4$ is/are

a. $y = 4(x - 1)$
b. $y = 0$
c. $y = -4(x - 1)$
d. $y = -30x - 50$

Working Space

Comprehension Type

Direction (Q. Nos. 26-31) *This section contains 2 paragraphs, each describing theory, experiments, data, etc. Six questions related to the paragraphs have been given. Each question has only one correct answer among the four given options (a), (b), (c) and (d).*

Passage I

Consider $y = x$ is tangent to the parabola $y = ax^2 + c$.

26 If $a = 2$, then the value of c is

a. 1/8
b. −1/2
c. 1/2
d. 1

27 If (1, 1) is point of contact, then a is equal to

a. 1/2
b. 1/3
c. 1/4
d. 1/6

28 If $c = 2$, then the point of contact is

a. (2, 2)
b. (4, 4)
c. (6, 6)
d. (3, 3)

Passage II

Consider the circle $x^2 + y^2 = 9$ and the parabola $y^2 = 8x$. They intersect at P and Q in the first and fourth quadrants, respectively. Tangents to the circle at P and Q intersect the X-axis at R and tangents to the parabola at P and Q intersect the X-axis at S.

29 The ratio of the areas of ΔPQS and ΔPQR is

a. $1:\sqrt{2}$
b. 1 : 2
c. 1 : 4
d. 1 : 8

30 The radius of circumcircle of the ΔPRS is

a. 5
b. $3\sqrt{3}$
c. $3\sqrt{2}$
d. $2\sqrt{3}$

31 The radius of incircle of the ΔPQR is

a. 4
b. 3
c. 8/3
d. 2

One Integer Value Correct Type

Direction (Q. Nos. 32-35) *This section contains 4 questions. When worked out will result in an integer from 0 to 9 (both inclusive).*

32 A chord of the parabola $y^2 = 4ax$ subtends a right angle at the vertex. If the locus of the point of intersection of tangents at its extremities is $x = \lambda a$, then the value of $|\lambda|$ is

33 The orthocentre of any triangle formed by three tangents to a parabola $y^2 = 4x$, lies on $x = \lambda$, then the value of $|\lambda|$ is

34 Two straight lines are perpendicular to each other. One of them touches the parabola $y^2 = 4a(x + a)$ and the other touches the parabola $y^2 = 4b(x + b)$. If the point of intersection of the lines lie on $x + \lambda a + \mu b = 0$, then the value of $(\lambda + \mu)$ is

35 Two equal parabolas $y^2 = 4ax$ and $x^2 = 4ay$ have the same vertex and their axes are at right angles. If the length of their common tangent is $(\lambda\sqrt{2}a)$, then the value of λ is

DPP-4 Equation of Normal and Its Properties

Only One Option Correct Type

Direction (Q. Nos. 1-23) This section contains 23 multiple choice questions. Each question has four choices (a), (b), (c) and (d), out of which ONLY ONE is correct.

Working Space

1 If $y = 2x + (c - 4)$ is a normal to the parabola $y^2 = 4x$, then the value of c is

a. 8 b. −8
c. 12 d. −12

2 The normal drawn at a point $(at_1^2, -2at_1)$ of the parabola $y^2 = 4ax$ meets it again in the point $(at_2^2, 2at_2)$, then

a. $t_2 = t_1 + \frac{2}{t_1}$ b. $t_2 = t_1 - \frac{2}{t_1}$
c. $t_2 = -t_1 + \frac{2}{t_1}$ d. None of these

3 If $x + y = k$ is the normal to $y^2 = 12x$, then k is equal to

a. 3 b. 9
c. −9 d. −3

4 The equation of normal to the parabola $y^2 = 4ax$ at its point $(am^2, 2am)$ is

a. $y = mx + 2am + am^3$ b. $y = -mx + 2am + am^3$
c. $y = mx - 2am - am^3$ d. None of these

5 At what point on the parabola $y^2 = 4x$, the normal makes equal angles with the axes?

a. (4, 4) b. (9, 6) c. (4, − 1) d. (1, 2)

6 The line $2x + y + \lambda = 0$ is a normal to the parabola $y^2 = -8x$, then λ is equal to

a. 12 b. −12 c. 24 d. −24

7 The normal chord of a parabola $y^2 = 4ax$ at the point $P(x_1, x_1)$ subtends a right angle at the

a. focus b. vertex
c. end of the latusrectum d. None of these

8 The distance between a tangent to the parabola $y^2 = 4ax\,(a > 0)$ and the parallel normal with gradient 1, is

a. $4a$ units b. $2\sqrt{2}a$ units c. $2a$ units d. $\sqrt{2}a$ units

9 The normal at the point $P(ap^2, 2ap)$ meets the parabola $y^2 = 4ax$ again at $Q(aq^2, 2aq)$ such that the line joining the origin to P and Q are at the right angle. Then, p^2 is equal to

a. 1 b. 2 c. $\sqrt{2}$ d. 4

Working Space

10 If a line $x - y = 1$ cuts the parabola $y^2 = 4x$ at the points A and B. Normals at A and B intersect at C, other than above two meet the parabola in D, then D is equal to

a. (1, 1) b. (2, 2) c. (4, –4) d. (6, 6)

11 If the tangents and normals at the extremities of a focal chord of a parabola intersect at (x_1, y_1) and (x_2, y_2) respectively, then

a. $x_1 = x_2$ b. $x_1 = y_2$ c. $x_2 = y_1$ d. $y_1 = y_2$

12 The locus of the point of intersection of normals at the ends of parallel chords of gradient m of the parabola $y^2 = 4ax$, is

a. $2xm^2 - ym^3 = 4a(2+m^2)$
b. $2xm^2 + ym^3 = 4a(2+m^2)$
c. $2xm + ym^2 = 4a(2+m)$
d. $2xm^2 + ym^3 = 4a(2-m^2)$

13 The subtangent, ordinate and subnormal to the parabola $y^2 = 4ax$ at a point (different from origin) are in

a. AP
b. GP
c. HP
d. None of these

14 A normal chord of the parabola subtending a right angle at the vertex makes an acute angle θ with X-axis, then θ is

a. $\tan^{-1}\sqrt{2}$
b. $\tan^{-1}2$
c. $\tan^{-1}\dfrac{\sqrt{2}}{3}$
d. None of these

15 The number of distinct normals of a parabola passing through the focus of parabola is

a. 1
b. 2
c. 3
d. None of these

16 Normals are drawn from the point P with slopes m_1, m_2, m_3 to the parabola $y^2 = 4x$. If locus of P with $m_2 m_3 = \alpha$ is a part of the parabola itself. Then, the value of α is

a. 1 b. 2 c. 3 d. 4

17 The ordinates of points P and Q on the parabola $y^2 = 12x$ are in the ratio 1 : 2. Then, the locus of the point of intersection of the normals to the parabola at P and Q is

a. $x = 3\left[7\left(\dfrac{y}{18}\right)^{2/3} + 2\right]$
b. $x = \left(\dfrac{y}{18}\right)^{2/3} + 2$
c. $x = 3\left[7\left(\dfrac{y}{18}\right)^{2/3} + 1\right]$
d. None of these

18 A normal is drawn to parabola $y^2 = 4ax$ at any point other than the vertex. If it cuts the parabola again at a point whose distance from the vertex is not less than

a. $4a$
b. $4\sqrt{6}a$
c. $2\sqrt{6}a$
d. $3\sqrt{6}a$

19 If the normals from any point on the parabola $x^2 = 4y$ cut the line $y = 2$ in points whose abscissa are in AP, then the slopes of tangents at the 3 conormal points are in

a. AP
b. GP
c. HP
d. None of these

Working Space

20 The two parabolas $y^2 = 4ax$ and $y^2 = 4c(x - b)$ cannot have a common normal other than the axes, unless

a. $\frac{b}{a-c} > 2$ b. $\frac{a}{b-c} > 2$ c. $\frac{b}{a-c} < 2$ d. $\frac{a}{b-c} < 2$

21 The length of intercept on the normal at the point $(at^2, 2at)$ of the parabola $y^2 = 4ax$ made by circle which is described on a focal distance of the given point as diameter is

a. $a\sqrt{1+t}$ b. $a\sqrt{1+2t}$
c. $a\sqrt{1+t^2}$ d. $a\sqrt{1+t^4}$

22 The tangent and normal at $P(t)$, for all real positive t, to the parabola $y^2 = 4ax$ meet the axis of the parabola in T and G respectively, then the angle at which the tangent at P to the parabola is inclined to the tangent at P to the circle passing through the points P, T and G, is

a. $\cot^{-1} t$ b. $\cot^{-1} t^2$
c. $\tan^{-1} t$ d. $\tan^{-1} t^2$

23 PQ is a normal chord of the parabola $y^2 = 4ax$ at P, A being the vertex of the parabola. Through P a line is drawn parallel to AQ meeting the X-axis in R. Then, the length of AR is equal to

a. the length of latusrectum
b. the focal distance of point P
c. the twice the focal distance of point P
d. the distance of the point P from the directrix

Comprehension Type

Direction *(Q. Nos. 24-31) This section contains 4 paragraphs, each describing theory, experiments, data, etc. Eight questions related to the paragraphs have been given. Each question has only one correct answer among the four given options (a), (b), (c) and (d).*

Passage I

Consider the parabola $y^2 = 8x$.

24 Area of the figure formed by the tangents and normals drawn at the extremities of its latusrectum, is

a. 8 sq units b. 16 sq units
c. 32 sq units d. 64 sq units

25 Distance between the tangent to the parabola and a parallel normal inclined at 30° with the X-axis, is

a. 16/3 units b. $16\sqrt{3}/9$ units
c. 2/3 units d. $16/\sqrt{3}$ units

Passage II

The locus of the point of intersection of two mutually perpendicular normals to the parabola $y^2 = 4ax$.

26 The abscissae of the point of intersection of perpendicular normals can never be less than

a. a b. $2a$
c. $3a$ d. $4a$

27 The ordinate when the abscissae is smallest, is

a. 1 b. 2
c. 3 d. 0

Working Space

Passage III

If three normals from a point to the parabola $y^2 = 4ax$ cuts the X-axis in points whose distances from the vertex are in AP.

28 The slope of one of the normal is

a. $\left(\frac{k}{a}\right)^{1/3}$ b. $\left(\frac{2k}{a}\right)^{2/3}$

c. $\left(\frac{2k}{a}\right)^{1/3}$ d. $\left(\frac{k}{a}\right)^{2/3}$

29 The point lies on the curve

a. $27ay^2 = 2(x-2a)^3$ b. $27y^2 = 2(x-2a)^3$

c. $27ay^2 = 2(x+2a)^3$ d. $27y^2 = 2(x+2a)^3$

Passage IV

The locus of a point P in each of the following, if the slopes of the three normals drawn from P to the parabola $y^2 = 4ax$.

30 Connected by the relation $\tan^{-1} m_1^2 + \tan^{-1} m_2^2 + \tan^{-1} m_3^2 = \alpha$ is

a. $x^2 \tan\alpha + 2ax(1 - 2\tan\alpha) - y^2 + a^2(3\tan\alpha - 4) = 0$

b. $x^2 \cot\alpha + 2ax(1 - 2\cot\alpha) - y^2 = a^2(3\cot\alpha - 4)$

c. $x^2 \tan\alpha + 2ax(1 - 2\tan\alpha) - y^2 = a^2(3\tan\alpha - 4)$

d. None of the above

31 Connected by the relation
$\tan^{-1} m_1^2 + \tan^{-1} m_2^2 + \tan^{-1} m_3^2 = \tan^{-1} m_1 + \tan^{-1} m_2 + \tan^{-1} m_3$ is

a. $y^2 - 2ax + 3ay + xy + 4a^2 = 0$

b. $y^2 - 2ax - 3ay - xy + 4a^2 = 0$

c. $y^2 - 2ax + 3ay - xy + 4a^2 = 0$

d. None of the above

One Integer Value Correct Type

Direction (*Q. Nos.* 32-35) *This section contains* 4 *questions. When worked out will result in an integer from* 0 *to* 9 (*both inclusive*).

32 Three normals are drawn from $(2k, 0)$ to the parabola $y^2 = 4x$ $(k > 1)$. One normal is always the X-axis. Then, the value of $(2k)$ for which other two normals are perpendicular to each other, is

33 If the locus of the point of intersection of those normals to the parabola $x^2 = 8y$ which are at right angles to each other is $x^2 + 2(4 - y) + \lambda = 0$. Then, the value of λ is

34 The locus of the points such that the sum of the angles which three normals drawn from them to a parabola make with the axis of the parabola $y^2 = 4ax$ is a constant λ, is $2x + (\cot\lambda)\, y - \mu a = 0$. Then, the value of μ is

35 A normal drawn at a point P on the parabola $y^2 = 4ax$ meets the curve again at Q. If the least distance of the point Q from the axis of the parabola is $(\lambda\sqrt{2}a)$, then the value of λ is

DPP-5 Equation of $S'=T$, $T=0$ and $SS'=T^2$

Only One Option Correct Type

Working Space

Direction (Q. Nos. 1-10) *This section contains 10 multiple choice questions. Each question has four choices (a), (b), (c) and (d), out of which ONLY ONE is correct.*

1 The locus of the point of intersection of tangents to the parabola $y^2 = 4ax$ which are perpendicular to each other, is

a. $x = a$ b. $y = a$

c. $x = -a$ d. $y = -a$

2 The locus of the point of intersection of two tangents to the parabola $y^2 = 4ax$ such that they include an angle of 45°, is

a. $y^2 - 4ax = (a + x)^2$

b. $y^2 - 4ax = (a - x)^2$

c. $y^2 + 4ax = (a + x)^2$

d. $y^2 + 4ax = (a - x)^2$

3 The equation of chord of contact of tangents drawn from a point (3, 4) to the parabola $y^2 = 2x$, is

a. $x + 4y + 3 = 0$ b. $x - 4y - 3 = 0$

c. $x + y - 7 = 0$ d. $x - 4y + 3 = 0$

4 Tangents are drawn at the point, where the line $lx + my + n = 0$ is intersected by the parabola $y^2 = 4ax$, then the point of intersection of tangents is

a. $\left(\frac{n}{l}, -\frac{2am}{l}\right)$ b. $\left(-\frac{n}{l}, \frac{2am}{l}\right)$

c. $\left(-\frac{n}{l}, -\frac{2am}{l}\right)$ d. None of these

5 The locus of the point of intersection of tangents drawn at the extremities of a normal chord of the parabola $y^2 = 4ax$, is

a. $y^2(x + 2a) + 4a^3 = 0$

b. $y^2(x - 2a) + 4a^3 = 0$

c. $y^2(x + 2a) - 4a^3 = 0$

d. $y^2(x - 2a) - 4a^3 = 0$

6 The equation of chord of $y^2 = 8x$ which is bisected at $(2, -3)$, is

a. $4x + 3y - 1 = 0$ b. $x + y + 1 = 0$

c. $2x + y - 1 = 0$ d. $4x + 3y + 1 = 0$

7 The locus of the middle points of chords of the parabola $y^2 = 4ax$ which pass through the focus, is

a. $y^2 + 2ax + 2a^2 = 0$ b. $y^2 - 2ax + 2a^2 = 0$

c. $y^2 - 2ax - 2a^2 = 0$ d. None of these

Working Space

8 The polar of the point (2, 1) with respect to the parabola $y^2 = 4ax$ is

a. $y = 3x + 6$ b. $y = 3x - 6$
c. $y = x + 1$ d. None of these

9 The pole of the line $lx + my + n = 0$ with respect to the parabola $y^2 = 4ax$ is

a. $\left(\frac{n}{l}, -\frac{2am}{l}\right)$ b. $\left(\frac{n}{l}, -\frac{2am}{l}\right)$
c. $\left(\frac{n}{l}, -\frac{2am}{l}\right)$ d. None of these

10 The locus of the poles of focal chords of the parabola $y^2 = 4ax$ is

a. the tangent at vertex b. $x = -2a$
c. the directrix d. $x = y + a$

Comprehension Type

Direction (Q. Nos. 11-15) *This section contains 2 paragraphs, each describing theory, experiments, data, etc. Five questions related to the paragraphs have been given. Each question has only one correct answer among the four given options (a), (b), (c) and (d).*

Passage I

Let normal chord cuts the curve $y^2 = 4ax$ at P and Q. Also, $R(h, k)$ be the mid-point of PQ.

11 The slope of the normal chord PQ can be

a. $\frac{2a}{h}$ b. $\frac{ka}{h}$
c. $\frac{2a}{k}$ d. $\frac{ha}{k}$

12 The locus of the point R is

a. $y^2(y^2 - 2ax) + 4a^2(y^2 + 2a^2) = 0$
b. $y^2(y^2 - 2ax) = y^2 + 2a^2$
c. $y^2(y^2 - 2ax) - 4a^2(y^2 + 2a^2) = 0$
d. $y^2(y^2 - 2ax) + (y^2 + 2a^2) = 0$

Passage II

Let the two curves $C_1 : x^2 + y^2 - \frac{a^2}{2} = 0$ and $C_2 : y^2 - 4ax = 0$ intersect at two real and distinct points.

13 The equation of common tangents to C_1 and C_2 is/are

a. $y = x \pm a$ b. $y = \pm x + a$
c. $y = \pm(x + a)$ d. None of these

14 The chord of contact of tangents from the point of intersection of common tangents to the curve C_1 and C_2 are

a. $x = -\frac{a}{2}, x = a$ b. $x = \frac{a}{2}, y = a$
c. $x = -\frac{a}{2}, y = 2a$ d. $x = \frac{a}{2}, y = 2a$

Working Space

15 The area of quadrilateral formed by the common tangents, the chord of contact of circle and the chord of contact of parabola, is

a. $\frac{5}{4}a^2$ sq units
b. $\frac{15}{7}a^2$ sq units
c. $\frac{4}{5}a^2$ sq units
d. $\frac{15}{4}a^2$ sq units

One Integer Value Correct Type

Direction (Q. Nos. 16-20) This section contains 5 questions. When worked out will result in an integer from 0 to 9 (both inclusive).

16 The length of chord of contact of tangents drawn from $(-1, -1)$ to $y^2 = 4x$ is

17 The area of triangle formed by the tangents drawn from $(-3/4, -1)$ to $y^2 = 4x$ is

18 The mid-point of the chord intercepted on the line $4x - 3y + 4 = 0$ by the parabola $y^2 = 8x$ is (λ, μ), then the value of $(4\lambda + \mu)$ is

19 The polar of a point with respect to $y^2 = 4x$ touches $x^2 = 4y$. If the locus of this point is $xy = \lambda$, then the value of $|\lambda|$ is

20 The locus of the poles of tangents to the parabola $y^2 = 4ax$ with respect to the parabola $y^2 = 4bx$ is parabola $y^2 = \frac{\lambda b^2 x}{a}$, then the value of λ is

SET - 1

Revisal Problems
for JEE Main

Only One Option Correct Type

Working Space

Direction *(Q. Nos. 1-43) This section contains 43 multiple choice questions. Each question has four choices (a), (b), (c) and (d), out of which ONLY ONE option is correct.*

1 The equation $\lambda x^2 + 4xy + y^2 + \lambda x + 3y + 2 = 0$ represents a parabola, if

a. $\lambda = -4$
b. $\lambda = 4$
c. $\lambda = 0$
d. None of these

2 If the line $x - 1 = 0$ is the directrix of the parabola $y^2 - kx + 8 = 0$, then one of the value of k is

a. 1/8
b. 8
c. 4
d. 1/4

3 The equation of the directrix of the parabola $y^2 + 4y + 4x + 2 = 0$ is

a. $x = -1$
b. $x = 1$
c. $x = -3/2$
d. $x = 3/2$

4 The length of latusrectum of the parabola $2\{(x-a)^2 + (y-a)^2\} = (x+y)^2$ is

a. $2a$
b. $2\sqrt{2}a$
c. $4a$
d. $\sqrt{2}a$

5 AB is a chord of the parabola $y^2 = 4ax$ with vertex at A. BC is drawn perpendicular to AB meeting the axis at C. The projection of BC on the axis of the parabola is

a. a
b. $2a$
c. $4a$
d. $8a$

6 The parametric coordinates of any point on the parabola whose focus is (0, 1) and the directrix is $x + 2 = 0$, are

a. $(t^2 + 1, 2t - 1)$
b. $(t^2 + 1, 2t + 1)$
c. $(t^2, 2t)$
d. $(t^2 - 1, 2t + 1)$

7 Let $y = f(x)$ be a parabola, having its axis parallel to Y-axis, which is touched by the line $y = x$ at $x = 1$, then

a. $f'(0) = f'(1)$
b. $2f(0) = 1 - f'(0)$
c. $f'(1) = 1$
d. $f(0) + f'(0) + f''(0) = 1$

8 Let A, B and C be three points taken on the parabola $y^2 = 4ax$ with coordinates $(at_i^2, 2at_i)$; $i = 1, 2, 3$; where t_1, t_2 and t_3 are in AP. If AA', BB' and CC' are focal chords and coordinates of A', B' and C' are $(at_i'^2, 2at_i')$; $i = 1, 2, 3$; then t_1', t_2', t_3' are in

a. AP
b. GP
c. HP
d. None of these

9 Vertex of the parabola $x^2 + y^2 - 2xy - 4x - 4y + 4 = 0$ is at

a. (1, 1)
b. (−1, −1)
c. (1/2, 1/2)
d. None of these

10 The parabola $y^2 = \lambda x$ and $25\{(x-3)^2 + (y+2)^2\} = (3x - 4y - 2)^2$ are equal, if λ is equal to

a. 1
b. 2
c. 4
d. 6

11 The circle on focal radii of a parabola as diameter touches

a. the axis
b. the tangent at the vertex
c. the directrix
d. None of these

Working Space

12 The radius of the circle whose centre is (–4, 0) and which cuts the parabola $y^2 = 8x$ at A and B such that its common chord AB subtends a right angle at the vertex of parabola, is

a. 4 units
b. 3 units
c. $\sqrt{18}$ units
d. 5 units

13 Two common tangents to the circle $x^2 + y^2 = 2a^2$ and parabola $y^2 = 8ax$ are

a. $x = \pm (y + 2a)$
b. $y = \pm (x + 2a)$
c. $x = \pm (y + a)$
d. $y = \pm (x + a)$

14 It the tangent at the point $P(2, 4)$ to the parabola $y^2 = 8x$ meets $y^2 = 8x + 5$ at Q and R, then the mid-point of QR is

a. (2, 4)
b. (4, 2)
c. (7, 9)
d. None of these

15 If m be the slope of common tangents to $y = x^2 - x + 1$ and $y = x^2 - 3x + 10$, then m is equal to

a. 16
b. 7
c. 9
d. None of these

16 The angle between the tangents drawn from the point (3, 4) to the parabola $y^2 - 2y + 4x = 0$ is

a. $\tan^{-1}\left(\frac{8\sqrt{5}}{7}\right)$
b. $\tan^{-1}\left(\frac{12}{\sqrt{5}}\right)$
c. $\tan^{-1}\left(\frac{\sqrt{5}}{7}\right)$
d. None of these

17 If $y + b = m_1(x + a)$ and $y + b = m_2(x + a)$ are two tangents to the parabola $y^2 = 4ax$, then

a. $m_1 + m_2 = 0$
b. $m_1 m_2 = 1$
c. $m_1 m_2 = -1$
d. None of these

18 Radius of the largest circle which passes through the focus of the parabola $y^2 = 4x$ and contained in it, is

a. 8
b. 4
c. 2
d. 5

19 The tangent at the point $P(x_1, y_1)$ to the parabola $y^2 = 4ax$ meets the parabola $y^2 = 4a(x + b)$ at Q and R, then mid-point of QR is

a. (x_1, y_1)
b. $(x_1 + b, y_1)$
c. $(x_1 + b, y_1 + b)$
d. None of these

20 If the point $(a, 2a)$ is an interior point of the region bounded by the parabola $y^2 = 16x$ and the double ordinate through the focus, then

a. $a \in (-\infty, -4)$
b. $a \in (0, 4)$
c. $a \in (0, 2)$
d. $a \in (4, \infty)$

21 Minimum distance between the parabolas $y^2 - 4x - 8y + 40 = 0$ and $x^2 - 8x - 4y + 40 = 0$ is

a. 0
b. $\sqrt{3}$
c. $2\sqrt{2}$
d. $\sqrt{2}$

22 The locus of the point of intersection of tangents to the parabola $y^2 = 4(x + 1)$ and $y^2 = 8(x + 2)$, which are perpendicular to each other, is

a. $x + 7 = 0$
b. $x - y = 4$
c. $x + 3 = 0$
d. $y - x = 12$

23 Two parabolas $y^2 = 4a(x - \lambda_1)$ and $x^2 = 4a(y - \lambda_2)$ always touch each other, λ_1 and λ_2 being variable parameters. Then,their points of contact lie on a

a. straight line
b. circle
c. parabola
d. hyperbola

Working Space

24 The area of trapezium whose vertices lie on the parabola $y^2 = 4x$ and its diagonals pass through (1, 0) and having length 25/4 units each, is

a. $\frac{75}{4}$ sq units
b. $\frac{625}{16}$ sq units
c. $\frac{25}{4}$ sq units
d. $\frac{25}{8}$ sq units

25 The angle between the tangents drawn from the point (1, 4) to the parabola $y^2 = 4x$ is

a. $\pi/6$
b. $\pi/4$
c. $\pi/3$
d. $\pi/2$

26 Two perpendicular tangents *PA* and *PB* are drawn to $y^2 = 4ax$, the minimum length of *AB* is equal to

a. a
b. $2a$
c. $4a$
d. $8a$

27 The triangle formed by tangents to the parabola $y^2 = 4ax$ at the ends of the latusrectum and the double ordinate through the focus, is

a. equilateral
b. isosceles
c. dependent on the values of *a* for its classification
d. None of the above

28 The set of values of m for which a chord of slope m of the circle $x^2 + y^2 = 4$ touches the parabola $y^2 = 4x$, is

a. $\left(-\infty, -\sqrt{\frac{\sqrt{2}-1}{2}}\right) \cup \left(\sqrt{\frac{\sqrt{2}-1}{2}}, \infty\right)$
b. $(-\infty, -1) \cup (1, \infty)$
c. $(-1, 1)$
d. R

29 The chords of contact of the pairs of tangents drawn from each point on the line $2x + y = 4$ to the parabola $y^2 = -4x$ pass through the point

a. $(2, -1)$
b. $(1/2, 1/4)$
c. $(-1/2, -1/4)$
d. $(-2, 1)$

30 Two parabolas with the same axis, focus of each being exterior to the other and the latusrectum being $4a$ and $4b$. The locus of the middle points of the intercepts between the parabolas made on the lines parallel to the common axis is a

a. straight line, if $a > b$
b. parabola, if $a \neq b$
c. parabola for all a, b
d. ellipse, if $b > a$

31 The slope of a chord of the parabola $y^2 = 4ax$ which is normal at one end and which subtends a right angle at the origin, is

a. $1/\sqrt{2}$
b. $\sqrt{2}$
c. 2
d. None of these

32 The inclination of the normal with positive direction of X-axis, drawn at another end of a normal to the parabola $y^2 = 4x$ is always

a. $60°$
b. $< 60°$
c. $> 60°$
d. $< 45°$

33 If the tangents drawn from a point to the parabola $y^2 = 4ax$ are normals to the parabola $x^2 = 4by$, then

a. $a^2 \geq b^2$
b. $a^2 \geq 4b^2$
c. $a^2 \geq 8b^2$
d. $8a^2 \geq b^2$

34 The number of distinct normals that can be drawn $(-2, 1)$ to the parabola $y^2 - 4x - 2y - 3 = 0$ is

a. 1
b. 2
c. 3
d. 0

Working Space

35 Let P, Q, R be three points on parabola $y^2 = 4x$ and normal P and R meet at Q, then the locus of the mid-point of the chord PR is a parabola whose vertex is at

a. (2, 0)
b. (0, −2)
c. (−2, 0)
d. None of these

36 The circle $x^2 + y^2 - 2x - 6y + 2 = 0$ intersects the parabola $y^2 = 8x$ orthogonally at the point P. The equation of the tangent to parabola at P can be

a. $x - y - 4 = 0$
b. $2x + y - 2 = 0$
c. $x + y - 4 = 0$
d. $2x - y + 1 = 0$

37 If (x_1, y_1), (x_2, y_2) and (x_3, y_3) are the feet of the three normals drawn from a point to the parabola $y^2 = 4ax$, then $\frac{x_1 - x_2}{y_3} + \frac{x_2 - x_3}{y_1} + \frac{x_3 - x_1}{y_2}$ is equal to

a. $4a$
b. $2a$
c. a
d. 0

38 If normal at the ends of the double ordinate $x = 4$ of parabola $y^2 = 4x$ meets the curve again at P and P' respectively, then PP' is

a. 6 units
b. 12 units
c. 10 units
d. None of these

39 The set of points on the axis of the parabola $y^2 = 4ax$, from which three distinct normals can be drawn to the parabola $y^2 = 4ax$, is

a. $\{(x, 0) : x < 4a\}$
b. $\{(x, 0) : x > 2a\}$
c. $\{x : x > a\}$
d. $\{x : a < x < 2a\}$

40 The set of values of h for which the number of distinct common normals of $(x - 2)^2 = 4(y - 3)$ and $x^2 + y^2 - 2x - hy - c = 0$, $(c > 0)$ is 3, then

a. $(2, \infty)$
b. $(4, \infty)$
c. $(2, 4)$
d. $(10, \infty)$

41 If (h, k) is a point on the axis of the parabola $2(x - 1)^2 + (y - 1)^2 = (x + y + 2)^2$ from where three distinct normals may be drawn, then

a. $h > 2$
b. $h < 4$
c. $h > 8$
d. $h < 8$

42 The set of points on the axis of the parabola $y^2 = 4x + 8$ from which 3 normals to the parabola are real and distinct, is

a. $\{(x, 0) : x \le -2\}$
b. $\{(x, 0) : x > -2\}$
c. $\{(0, y) : y > -2\}$
d. None of these

43 All the three normals drawn from a point (h, k) to the parabola $y^2 = 4ax$ are real and distinct, if

a. $h > 2a$ and $27ak^2 < 4(h - 2a)^3$
b. $h > 2a$ and $27ak^2 > 4(h - 2a)^3$
c. $h > a$ and $27ak^2 > 4(h - 2a)^3$
d. $h > a$ and $27ak^2 < 4(h - 2a)^3$

Statement Type

Direction (Q. Nos. 44-50) ***This section is based on Statement I and Statement II. Select the correct answer from the codes given below.***

a. Both Statement I and Statement II are correct and Statement II is the correct explanation of Statement I
b. Both Statement I and Statement II are correct but Statement II is not the correct explanation of Statement I
c. Statement I is correct but Statement II is incorrect
d. Statement II is correct but Statement I is incorrect

44 **Statement I** The normal at any point of a parabola is equally inclined to the focal distance of the point and the axis of the parabola.

Statement II The normal at any point P of a parabola, bisects the external angle between the focal distance of point and the perpendicular on the directrix from the point P.

45 **Statement I** From any point P on the straight line $x = 1$ a tangent PQ is drawn to the parabola $y^2 - 8x + 24 = 0$, then the abscissa of N, where N is foot of the perpendicular drawn from the point (5, 0) to PQ is 3.

Statement II The locus of foot perpendicular from the focus of the parabola to any of the tangent, is the tangent at vertex.

46 **Statement I** Locus of the point moving such that its distance from the line $2x + y - 3 = 0$ is equal to its distance from the point $(5/4, 1/2)$ is a parabola.

Statement II Parabola is a conic section for which eccentricity is 1.

47 **Statement I** The point of intersection of the tangents at three distinct points A, B, C on the parabola $y^2 = 4x$ can be collinear.

Statement II If a line L neither touches nor intersect the parabola $y^2 = 4x$, then from every point of the line two tangents can be drawn.

48 **Statement I** If $a^2 > 8b^2$, then a point can be found that the two tangents from it to the parabola $y^2 = 4ax$ are normals to the parabola $x^2 = 4by$.

Statement II The equations $yt = x + at^2$ and $x + sy = 2bs + bs^3$, where s and t are parameters representing same lines for two pair of s and t, if $a^2 > 8b^2$.

49 **Statement I** The normals at point (4, 4) and $(1/4, -1)$ of the parabola $y^2 = 4x$ are perpendicular.

Statement II The tangents to parabola at the end of a focal chord are perpendicular.

50 **Statement I** A ray of light from the origin after reflection at the point $P(x, y)$ of any curve becomes parallel to X-axis. The equation of curve may be $y^2 = 2x + 1$.

Statement II A ray of light parallel to axes after reflection from parabolic mirror always through the focus.

Working Space

SET - 2

Revisal Problems
for JEE Advanced

Only One Option Correct Type

Working Space

Direction *(Q. Nos. 1-29) This section contains 29 multiple choice questions. Each question has four choices (a), (b), (c) and (d), out of which ONLY ONE option is correct.*

1 If the points (2, 3) and (3, 2) on a parabola are equidistant from the focus, then the slope of its tangent at vertex is

a. parallel to $y = x$
b. parallel to $y = -x$
c. parallel to X-axis
d. parallel to Y-axis

2 The equation of circle of minimum radius which touches both the parabolas $y = x^2 + 2x + 4$ and $x = y^2 + 2y + 4$ is

a. $4x^2 + 4y^2 - 11x - 11y - 31 = 0$
b. $4x^2 + 4y^2 - 11x - 11y - 13 = 0$
c. $4x^2 + 4y^2 - 11x - 11y + 11 = 0$
d. $4x^2 + 4y^2 - 11x - 11y - 6 = 0$

3 If the parabola $y = ax^2 + bx + c$ has vertex (4, 2) and $a \in [1, 3]$, then difference between the extreme values of abc is equal to

a. 3600
b. 144
c. 3456
d. Can't found

4 Two parabolas have the same focus namely at the point $(3, -28)$. If their directrices are the X-axis and Y-axis, respectively and slope of common chord is t, then the value of $|t|$ is

a. 1
b. 2
c. 3
d. 4

5 Let $P \neq (0, 0)$ be a point on the parabola $y = x^2$. The normal to parabola at P meets the parabola at another point Q. The coordinates of this point P, so that the area bounded by the normal line and the parabola is minimum, is

a. (1/4, 1/16)
b. (1/2, 1/4)
c. (1/3, 1/9)
d. (2/3, 4/9)

6 The area of triangle formed by the tangents at points (4, 6), (10, 8) and (2, 4) on the parabola $y^2 - 2x = 8y - 20$ is

a. 4 sq units
b. 2 sq units
c. 1 sq unit
d. 8 sq units

7 The line $x - b + \lambda y = 0$ cuts the parabola $y^2 = 4ax\ (a > 0)$ at $P(t_1)$ and $Q(t_2)$. If $b \in [2a, 4a]$, then the range of $t_1 t_2$, where $\lambda \in R$ is

a. $[-4, -2]$
b. $[2, 4]$
c. $[4, 16]$
d. $[-16, -4]$

8 Consider one side AB of a square $ABCD$ on the line $y = 2x - 17$ and the other two vertices C, D lies on the parabola $y = x^2$, then the maximum possible area of the square $ABCD$ (in sq units) is

a. 160
b. 1280
c. 640
d. 320

9 The normals are drawn from $(2k, 0)$ to the curve $y^2 = 4x$. One normal is X-axis. Then, the value of k for which other two normals are perpendicular to each other, is

a. 3/2
b. 2
c. 5/2
d. 7/2

Working Space

10 If on a given base, a triangle be described such that the sum of the tangents of the base angles is a constant, then the locus of the vertex is

a. a circle
b. a parabola
c. an ellipse
d. a hyperbola

11 Let a variable circle is drawn so that it always touches a fixed line and also a given circle, the line not passing through the centre of the circle. The locus of centre of the variable circle is

a. a parabola
b. a circle
c. an ellipse
d. a hyperbola

12 Two unequal parabolas have the same common axis which the X-axis and have the same vertex which is the origin with their concavities in opposite direction. If a variable line parallel to the common axis meet the parabolas in P and P', then the locus of middle point of PP' is

a. a parabola
b. a circle
c. an ellipse
d. a hyperbola

13 All points on the curve $y^2 = 4a\left(x + a\sin\frac{x}{a}\right)$ at which the tangent is parallel to X-axis lie on

a. a circle
b. a parabola
c. an ellipse
d. a line

14 A variable circle is described to pass through (1, 0) and touch the line $y = x$. The locus of the centre of the circle is a parabola, whose length of latusrectum is

a. 2
b. $1/\sqrt{2}$
c. $\sqrt{2}$
d. 1

15 Angle between the parabolas $y^2 = 4b(x - 2a + b)$ and $x^2 + 4a(y - 2b - a) = 0$ at the common end of their latusrectum, is

a. $\tan^{-1}(1)$
b. $\cot^{-1}(1) + \cot^{-1}(2)$
c. $\cot^{-1}(1) + \cot^{-1}(1/2) + \cot^{-1}(1/3)$
d. None of the above

16 A point P on a parabola $y^2 = 4x$, the foot of the perpendicular from it upon the directrix and the focus are the vertices of an equilateral triangle, then the area of the triangle is

a. $4\sqrt{3}$ sq units
b. $2\sqrt{3}$ sq units
c. 4 sq units
d. 2 sq units

17 The locus of the centre of a circle, which intercepts a chord of given length $2a$ on the X-axis and passes through a given point on the Y-axis distant b from the origin, is the curve

a. $x^2 - 2yb + b^2 = a^2$
b. $x^2 - 2ay + b^2 = a^2$
c. $y^2 - 2ax + a^2 = b^2$
d. $y^2 - 2bx + b^2 = a^2$

18 A variable parabola is drawn to pass through A and B, the ends of a diameter of a given circle with centre at the origin and radius c and to have as directrix a tangent to a concentric circle of radius $a\,(a > c)$; the axes being AB and a perpendicular diameter, then the locus of focus of parabola is

a. a circle
b. a parabola
c. an ellipse
d. a hyperbola

19 $ABCD$ and $EFGC$ are squares and the curve $y = k\sqrt{x}$ passes through the origin D and the points B and F. The ratio $\frac{FG}{BC}$ is

a. $\frac{\sqrt{5}+1}{2}$
b. $\frac{\sqrt{3}+1}{2}$
c. $\frac{\sqrt{5}+1}{4}$
d. $\frac{\sqrt{3}+1}{4}$

Working Space

20 Let A and B be two points on a parabola $y^2 = x$ with vertex V such that VA is perpendicular to VB and θ is the angle between the chord VA and the axis of the parabola. The value of $\left|\frac{VA}{VB}\right|$ is

a. $\tan\theta$
b. $\tan^3\theta$
c. $\cot^2\theta$
d. $\cot^3\theta$

21 The locus of feet of perpendiculars drawn from the vertex of the parabola $y^2 = 4ax$ upon all such chords of the parabola which subtend a right angle at the vertex, is

a. $x^2 + y^2 - 4ax = 0$
b. $x^2 + y^2 - 2ax = 0$
c. $x^2 + y^2 + 2ax = 0$
d. $x^2 + y^2 + 4ax = 0$

22 C is the centre of the circle with centre (0, 1) and radius unity. P is the parabola $y = ax^2$. The set of values of a for which they meet at a point other than the origin is

a. $a > 0$
b. $a \in \left(0, \frac{1}{2}\right)$
c. $\left(\frac{1}{4}, \frac{1}{2}\right)$
d. $\left(\frac{1}{2}, \infty\right)$

23 PQ is a normal chord of the parabola $y^2 = 4ax$ at P, A being the vertex of the parabola. Through P a line is drawn parallel to AQ meeting the X-axis at R. Then, the length AR is equal to

a. twice the focal distance of the point P
b. length of latusrectum
c. focal distance of the point P
d. the distance of the point P from the directrix

24 TP and TQ are tangents to the parabola $y^2 = 4ax$ at P and Q. If the chord PQ passes through the fixed point $(-a, b)$, then the locus of T is

a. $ay = 2b(x - b)$
b. $bx = 2a(y - a)$
c. $by = 2a(x - a)$
d. $ax = 2b(y - b)$

25 Through the vertex O of the parabola $y^2 = 4ax$, two chords OP and OQ are drawn and the circles on OP and OQ as diameters intersect in R. If θ_1, θ_2 and ϕ are the angles made with the axes by the tangents P and Q on the parabola and by OR, then the value of $\cot\theta_1 + \cot\theta_2$ is

a. $-2\tan\phi$
b. $-2\tan(\pi - \phi)$
c. 0
d. $2\cot\phi$

26 A pair of tangents to a parabola are equally inclined to a straight line whose inclination to the axis is α. The locus of their point of intersection is

a. a circle
b. a straight line
c. a parabola
d. an ellipse

27 Normals are drawn at points A, B and C on the parabola $y^2 = 4x$ which intersect at $P(h, k)$. The locus of the point $P(h, k)$, if the line joining the feet of two of them is 2, is

a. $x + y = 1$
b. $x - y = 3$
c. $y^2 = 2(x - 1)$
d. $y^2 = 2\left(x - \frac{1}{2}\right)$

28 Tangents are drawn from the point $(-1, 2)$ on the parabola $y^2 = 4x$. The length these tangents will intercept on the line $x = 2$ is

a. 6
b. $6\sqrt{2}$
c. $2\sqrt{6}$
d. None of these

29 An equation of the line that passes through $(10, -1)$ and is perpendicular to $y = \frac{x^2}{4} - 2$, is

a. $4x + y = 39$
b. $2x + y = 19$
c. $x + y = 9$
d. $x + 2y = 8$

Matching List Type

Working Space

Direction (Q. Nos. 30-32) *Choices for the correct combination of elements from Column I and Column II are given as options (a), (b), (c) and (d), out of which one is correct.*

30 AB is a chord of the parabola $y^2 = 4x$. Such that the normal's at A and B intersect at the point $C(9, 6)$. Match the Column I with Column II and mark the correct option from the codes given below.

	Column I		Column II
i.	The length of AB is	p.	20
ii.	The area of ΔABC is	q.	$4/\sqrt{13}$
iii.	The distance from vertex of parabola to the chord AB is	r.	$\sqrt{13}$
iv.	The area bounded by the coordinate axes and the line through AB is	s.	4/3

Codes

	i	ii	iii	iv		i	ii	iii	iv
a.	r	p	q	s	b.	p	q	r	s
c.	q	r	s	p	d.	s	s	p	q

31 Consider the curves $y^2 = 4|x|$ and $x^2 = 4|y|$. Match the Column I with Column II and mark the correct option from the codes given below.

	Column I		Column II
i.	The number of distinct points of intersection of the two curves is	p.	4
ii.	Radius of the largest circle inscribed in one loop formed by the two curves is	q.	8
iii.	$x^2 + y^2 = 8$ intersects the curve in ...... points.	r.	5
iv.	$x^2 + y^2 = 64$ intersects the curve in ...... points.	s.	$2\sqrt{10} - 4\sqrt{2}$

Codes

	i	ii	iii	iv
a.	p	q	r	s
b.	s	p	q	r
c.	r	s	q	q
d.	q	r	s	p

32 Match the Column I with Column II and mark the correct option from the codes given below.

	Column I		Column II
i.	The number of points on the curve $y = \|\|1 - e^x\| - 2\|$ from which mutually perpendicular tangent can be drawn to the parabola $x^2 + 4y = 0$, is	p.	2
ii.	Locus of vertex of parabola whose focus is (1, 2) and latusrectum is of 12 units is $(x-1)^2 + (y-2)^2 = a^2$, then a is	q.	4
iii.	A line drawn through the focus F and parallel to tangent at $P(1, 2\sqrt{2})$ on the parabola $y^2 = 8x$ cut the line $y = 2\sqrt{2}$ at Q, then PQ is	r.	3
iv.	A variable parabola touches the X and Y-axes at (1, 0) and (0, 1), then the radius of locus of focus of parabola is	s.	0
		t.	5

Codes

	i	ii	iii	iv
a.	p	r	r	s
b.	t	r	q	p
c.	s	p	t	r
d.	r	t	p	q

Comprehension Type

Working Space

Direction *(Q. Nos. 33-56) This section contains 8 paragraphs, each describing theory, experiments, data, etc. Twenty four questions related to the paragraphs have been given. Each question has only one correct answer among the four given options (a), (b), (c) and (d).*

Passage I

Tangents are drawn to the parabola $y^2 = 4x$ from the point $P(6, 5)$ to touch the parabola at Q and R. C_1 is a circle which touches the parabola at Q and C_2 is a circle which touches the parabola at R. Both the circles C_1 and C_2 pass through the focus of the parabola.

33 The area of ΔPQR is

a. 1/2 sq unit
b. 1 sq unit
c. 2 sq units
d. 1/4 sq unit

34 The radius of circle C_2 is

a. $5\sqrt{5}$ units
b. $5\sqrt{10}$ units
c. $10\sqrt{2}$ units
d. $\sqrt{210}$ units

35 The common chord of the circle C_1 and C_2 passes through the

a. incentre
b. circumcentre
c. centroid
d. orthocentre of the ΔPQR

Passage II

Tangents are drawn to the parabola $y^2 = 4x$ at the point P which is the upper end of latusrectum.

36 Image of the parabola $y^2 = 4x$ in the tangent line at the point P is

a. $(x+4)^2 = 16y$
b. $(x+2)^2 = 8(y-2)$
c. $(x+1)^2 = 4(y-1)$
d. $(x-2)^2 = 2(y-2)$

37 The radius of the circle touching the parabola $y^2 = 4x$ at the point P and passing through its focus is

a. 1
b. $\sqrt{2}$
c. $\sqrt{3}$
d. 2

38 The area enclosed by the tangent line at P, X-axis and the parabola is

a. 2/3
b. 4/3
c. 14/3
d. None of these

Passage III

$y = f(x)$ is a parabola of the form $y = x^2 + ax + 1$. Its tangent at the point of intersection of Y-axis and the parabola also touch the circle $x^2 + y^2 = r^2$. It is known that no points of the parabola is below X-axis.

39 The radius of the circle where a attains its maximum value, is

a. $1/\sqrt{10}$
b. $1/\sqrt{5}$
c. 1
d. $\sqrt{5}$

40 The slope of the tangent when the radius of the circle is maximum, is

a. 0
b. 1
c. -1
d. not defined

41 The minimum area bounded by the tangent and the coordinate axes, is

a. 1/4 sq unit
b. 1/3 sq unit
c. 1/2 sq unit
d. 1 sq unit

Passage IV

Let $C: y = x^2 - 3$, $D: y = kx^2$, $L_1: x = a\,(a \neq 0)$ and $L_2: x = 1$.

42 If the parabolas C and D intersect at a point A on the line L_1, then the tangent line L at A to the parabola D is

a. $2(a^2-3)x - ay + a^3 - 3a = 0$
b. $2(a^2-3)x - ay + a^3 + 3a = 0$
c. $(a^2-3)x - 2ay - 2a^3 + 2a = 0$
d. $2(a^2-3)x - ay - a^3 + 3a = 0$

Working Space

43 If the line L meets the parabola C at point B on the line L_2, other than A, then a is

a. -3 b. -2
c. 2 d. 1

44 If $a > 0$, the angle subtended by the chord AB at the vertex of the parabola C is

a. $\tan^{-1}(5/7)$ b. $\tan^{-1}(1/2)$
c. $\tan^{-1}(2)$ d. $\tan^{-1}(1/8)$

Passage V

Let t be a positive number. Draw two tangents from the point $(t, -1)$ to the parabola $y = x^2$. If the area bounded by the tangent lines and the parabola is denoted by $S(t)$.

45 $S(t)$ is given by

a. $\frac{2}{3}(1+t^2)^{8/3}$ b. $\frac{1}{3}(1+t^2)^{3/2}$
c. $\frac{3}{4}(1+t^2)^{3/2}$ d. $\frac{2}{3}(1-t^2)^{2/3}$

46 $\frac{S(t)}{\sqrt{t}}$ is minimum, then t is equal to

a. $1/\sqrt{5}$ b. $\sqrt{5}$
c. $2/\sqrt{5}$ d. $1/\sqrt{3}$

47 The minimum value of $\frac{S(t)}{\sqrt{t}}$ is

a. $4(5)^{1/4}\sqrt{6}$ b. $4(5^{-5/4})\sqrt{6}$
c. $\frac{16}{9\sqrt{3}}$ d. $4(5^{3/4})\sqrt{6}$

Passage VI

Consider two parabolas $C_1 : (x-1)^2 + f(k)$ and $C_2 : x^2 - 2x + 4y - 7 = 0$, where $f(k) = |k^2 - 3k + 2| + \frac{1}{4}, k \in R$.

48 If C_1 and C_2 does not intersect, then $|k^2 - 3k + 2|$ is greater than

a. 1/4 b. 3/4
c. 7/4 d. 1

49 The distance between the foci of C_1 and C_2 for which $f(k)$ is least, is

a. 3/2 b. 1/4
c. 3/4 d. 1/2

50 The minimum value of the function $g(k) = \frac{f^5(k)+1}{f^2(k)}$ is

Passage VII

Consider a parabola P which touches $y = 0$ at $(1, 0)$ and $x = 0$ at $(0, 2)$.

51 The vertex at P is

a. (16/25, 2/25) b. (4/5, 2/5)
c. (12/25, 6/25) d. None of these

52 Latusrectum of P is

a. $\frac{9}{5\sqrt{5}}$ b. $\frac{16\sqrt{5}}{25}$
c. $\frac{16}{25}$ d. $\frac{8}{25}$

53 The equation of directrix is

a. $2x - y = 0$ b. $x + 2y = 0$
c. $x + y = 0$ d. $x + 3y = 0$

Working Space

Passage VIII

A point P is moving in XY-plane such that if it lies inside the triangle formed by $|x| = 3$ and $y = \pm x$, then its distance from the origin is always r, otherwise it always lies on the curve $y^2 = 4|x|$.

54 The range of r such that there exist exactly eight points on locus P from which if we draw two tangents on the parabolic part, then they are at right angle, is

a. $r > 1$ b. $1 < r < \sqrt{2}$ c. $1 < r \le \sqrt{2}$ d. $r \ge 1$

55 If $r = 1$, then the radius of the largest circle which touches the locus of P at three points is given by

a. $(x-5)^2 + y^2 = 16$ b. $(x+5)^2 + y^2 = 0$ c. $x^2 + (y-5)^2 = 16$ d. $\left(x - \frac{1}{2}\right)^2 + y^2 = \frac{1}{4}$

56 If $r = 1$, then the number of points on the given locus which are situated nearest to $x^2 + y^2 - 10|x| + 21 = 0$ is

a. 2 b. 4 c. 6 d. 3

One Integer Value Correct Type

Direction *(Q. Nos. 57-65) This section contains 9 questions. When worked out will result in an integer from 0 to 9 (both inclusive).*

57 If the parabolas $y^2 = \lambda x$ and $25[(x-3)^2 + (y+2)^2] = (3x - 4y - 2)^2$ are equal, then the value of λ is

58 PQ is a focal chord of $y^2 = 4ax$, PP' and QQ' are normal chord at P and Q respectively. Then, the value of $P'Q'/PQ$ is

59 The locus of point from which three normals drawn to the curve $y^2 + 2y - 4x + 5 = 0$ are such that two of them are perpendicular is $\left(y - \frac{\mu}{2}\right)^2 - (x - \lambda) = 0$, then the value of $(\lambda - 2\mu)$ is

60 From a point T is tangent drawn to touch the parabola $y^2 = 16x$ at $P(16, 16)$. If S is the focus of the parabola and $\angle TPS = \cot^{-1}(k)$, then the value of k is

61 Radius of the largest circle which passes through the focus of the parabola $y^2 = 4x$ and contained in it, is

62 Normals are drawn from the point P with slopes $m_1 \cdot m_2 = \alpha$ in a part of the parabola itself, then the value of α is

63 If ends of latusrectum of parabola are (2, 6) and (6, 2) and the equation of the possible directrix is $x + y = \lambda_i$, where $i = 1, 2$. Then, the value of AM of λ_1 and λ_2 is

64 The tangent at the point $P(1, 1)$ on the curve $y = x^2 + bx - 6$ meets the positive X and Y-axes at A and B respectively and O is the origin. If b_1, b_2, b_3 are the values of b for which the area of the ΔOAB is 2 sq units with $b_1 < b_2 < b_3$, then $|b_1| + |b_2 + b_3|$ is

65 The figure shown a right angle with its hypotenuse OB along the Y-axis and its vertex A on the parabola $y = x^2$. Let h represents the length of the hypotenuse which depends on the x-coordinate of the point A. Then, $\lim_{x \to 0}(t^2 + 1)$ is

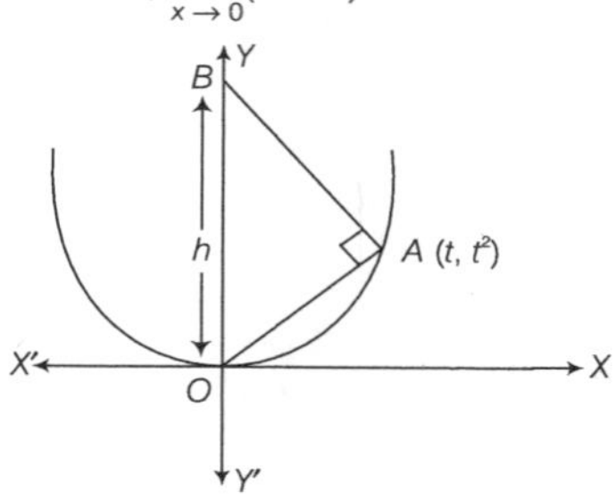

JEE Main & AIEEE Archive

(Compilation of Last 13 Years' Questions)

1 Let O be the vertex and Q be any point on the parabola $x^2 = 8y$. If the point P divides the line segment OQ internally in the ratio $1:3$, then the locus of P is **(JEE Main 2015)**

a. $x^2 = y$ b. $y^2 = x$

c. $y^2 = 2x$ d. $x^2 = 2y$

2 The slope of the line touching both the parabolas $y^2 = 4x$ and $x^2 = -32$ is **(JEE Main 2014)**

a. $\frac{1}{2}$ b. $\frac{3}{2}$

c. $\frac{1}{8}$ d. $\frac{2}{3}$

3 Given, A circle $2x^2 + 2y^2 = 5$ and a parabola $y^2 = 4\sqrt{5}x$.

Statement I An equation of a common tangent to these curves is $y = x + \sqrt{5}$.

Statement II If the line $y = mx + \frac{\sqrt{5}}{m}$, $(m \neq 0)$ is the common tangent, then m satisfies $m^4 - 3m^2 + 2 = 0$. **(JEE Main 2013)**

a. Both Statement I and Statement II are correct and Statement II is the correct explanation of Statement I

b. Both Statement I and Statement II are correct but Statement II is not the correct explanation of Statement I

c. Statement I is correct but Statement II is incorrect

d. Statement II is correct but Statement I is incorrect

4 **Statement I** An equation of a common tangent to the parabola $y^2 = 16\sqrt{3}x$ and the ellipse $2x^2 + y^2 = 4$ is $y = 2x + 2\sqrt{3}$.

Statement II If the line $y = mx + \frac{4\sqrt{3}}{m}$, $(m \neq 0)$ is a common tangent to the parabola $y^2 = 16\sqrt{3}x$ and the ellipse $2x^2 + y^2 = 4$, then m satisfies $m^4 + 2m^2 = 24$. **(AIEEE 2012)**

a. Both Statement I and Statement II are correct and Statement II is the correct explanation of Statement I

b. Both Statement I and Statement II are correct but Statement II is not the correct explanation of Statement I

c. Statement I is correct but Statement II is incorrect

d. Statement II is correct but Statement I is incorrect

5 If two tangents drawn from a point P to the parabola $y^2 = 4x$ are at right angles, then the locus of P is **(AIEEE 2010)**

a. $x = 1$ b. $2x + 1 = 0$

c. $x = -1$ d. $2x - 1 = 0$

6 A parabola has the origin as its focus and the line $x = 2$ as the directrix. Then, the vertex of the parabola is at **(AIEEE 2008)**

a. (2, 0) b. (0, 2)

c. (1, 0) d. (0, 1)

7 The equation of a tangent to the parabola $y^2 = 8x$ is $y = x + 2$. The point on this line from which the other tangent to the parabola is perpendicular to the given tangent is **(AIEEE 2007)**

a. (−1, 1) b. (0, 2)

c. (2, 4) d. (−2, 0)

8 The locus of the vertices of the family of parabolas $y = \frac{a^3x^2}{3} + \frac{a^2x}{2} - 2a$ is **(AIEEE 2006)**

a. $xy = \frac{3}{4}$ b. $xy = \frac{35}{16}$

c. $xy = \frac{64}{105}$ d. $xy = \frac{105}{64}$

9 If $a \neq 0$ and the line $2bx + 3cy + 4d = 0$ passes through the points of intersection of the parabolas $y^2 = 4ax$ and $x^2 = 4ay$, then **(AIEEE 2004)**

a. $d^2 + (2b + 3c)^2 = 0$

b. $d^2 + (3b + 2c)^2 = 0$

c. $d^2 + (2b - 3c)^2 = 0$

d. $d^2 + (3b - 2c)^2 = 0$

10 The normal at the point $(bt_1^2, 2bt_1)$ on a parabola meets the parabola again in the point $(bt_2^2, 2bt_2)$, then **(AIEEE 2003)**

a. $t_2 = -t_1 - \frac{2}{t_1}$ b. $t_2 = -t_1 + \frac{2}{t_1}$

c. $t_2 = t_1 - \frac{2}{t_1}$ d. $t_2 = t_1 + \frac{2}{t_1}$

11 The equation of directrix of the parabola $y^2 + 4y + 4x + 2 = 0$ is **(AIEEE 2002)**

a. $x = -1$ b. $x = 1$

c. $x = -3/2$ d. $x = 3/2$

JEE Advanced & IIT JEE Archive

(Compilation of Last 10 Years' Questions)

1 If the normals of the parabola $y^2 = 4x$ drawn at the end points of its latusrectum are tangents to the circle $(x-3)^2 + (y+2)^2 = r^2$, then the value of r^2 is

(JEE Adv 2015, Integer Type)

2 Let P and Q be distinct points on the parabola $y^2 = 2x$ such that a circle with PQ as diameter passes through the vertex O of the parabola. If P lies in the first quadrant and the area of ΔOPQ is $3\sqrt{2}$, then which of the following is/are the coordinates of P? **(JEE Adv. 2015, One or More than One Options Correct Type))**

a. $(4, 2\sqrt{2})$
b. $(9, 3\sqrt{2})$
c. $\left(\frac{1}{4}, \frac{1}{\sqrt{2}}\right)$
d. $(1, \sqrt{2})$

3 Normals at P, Q, R are drawn to $y^2 = 4x$ which intersect at $(3, 0)$. Then, match the Column I with Column II and mark the correct option from the codes given below.

(2012, Matching Type)

	Column I		Column II
i.	Area of ΔPQR	p.	2
ii.	Radius of circumcircle of ΔPQR	q.	$\frac{5}{2}$
iii.	Centroid of ΔPQR	r.	$\left(\frac{5}{2}, 0\right)$
iv.	Circumcentre of ΔPQR	s.	$\left(\frac{2}{3}, 0\right)$

Codes

	i	ii	iii	iv
a.	p	q	r	s
b.	q	r	p	s
c.	r	s	q	p
d.	p	q	s	r

4 **Given** A circle $2x^2 + 2y^2 = 5$ and a parabola $y^2 = 4\sqrt{5}x$.

Statement I An equation of a common tangent to these curves is $y = x + \sqrt{5}$.

Statement II If the line, $y = mx + \frac{\sqrt{5}}{m}$ $(m \neq 0)$ is the common tangent, then m satisfies $m^4 - 3m^2 + 2 = 0$. **(2012, Statement Type)**

a. Both Statement I and Statement II are correct and Statement II is the correct explanation of Statement I
b. Both Statement I and Statement II are correct but Statement II is not the correct explanation of Statement I
c. Statement I is correct but Statement II is incorrect
d. Statement II is correct but Statement I is incorrect

5 Let (x, y) be any point on the parabola $y^2 = 4x$. Let P be the point that divides the line segment from $(0, 0)$ to (x, y) in the ratio 1 : 3. Then, the locus of P is

(2011, Only One Option Correct Type)

a. $x^2 = y$
b. $y^2 = 2x$
c. $y^2 = x$
d. $x^2 = 2y$

6 Let L be a normal to the parabola $y^2 = 4x$. If L passes through the point $(9, 6)$, then L is given by **(2011, One or More than One Options Correct Type)**

a. $y - x + 3 = 0$
b. $y + 3x - 33 = 0$
c. $y + x - 15 = 0$
d. $y - 2x + 12 = 0$

7 Let S be the focus of the parabola $y^2 = 8x$ and PQ be the common chord of the circle $x^2 + y^2 - 2x - 4y = 0$ and the given parabola. The area of the ΔOPS is

(2011, Integer Type)

8 Consider the parabola $y^2 = 8x$. Let Δ_1 be the area of the triangle formed by the end points of its latusrectum and the point $P\left(\frac{1}{2}, 2\right)$ on the parabola and Δ_2 be the area of triangle formed by drawing tangents at P and at the end points of the latusrectum. Then, $\frac{\Delta_1}{\Delta_2}$ is

(2011, Integer Type)

Passage for Q. Nos. (9 and 10)

Let PQ be a focal chord of the parabola $y^2 = 4ax$. The tangents to the parabola at P and Q meet at a point lying on the line $y = 2x + a, a > 0$. **(2010, Compreshension Type)**

9 Length of chord PQ is

a. $7a$ b. $5a$
c. $2a$ d. $3a$

10 If chord PQ subtends an angle θ at the vertex of $y^2 = 4ax$, then $\tan\theta$ is equal to

a. $\frac{2}{3}\sqrt{7}$ b. $\frac{-2}{3}\sqrt{7}$
c. $\frac{2}{3}\sqrt{5}$ d. $\frac{-2}{3}\sqrt{5}$

11 Let A and B be two distinct points on the parabola $y^2 = 4x$. If the axis of the parabola touches a circle of radius r having AB as its diameter, then the slope of the line joining A and B can be **(2010, One or More than One Options Correct Type)**

a. $-\frac{1}{r}$ b. $\frac{1}{r}$
c. $\frac{2}{r}$ d. $-\frac{2}{r}$

12 The tangent PT and the normal PN to the parabola $y^2 = 4ax$ at a point P on it meet its axis at points T and N, respectively. The locus of the centroid of the ΔPTN is a parabola whose **(2009, One or More than One Options Correct Type)**

a. vertex is $\left(\frac{2a}{3}, 0\right)$ b. directrix is $x = 0$
c. latusrectum is $\frac{2a}{3}$ d. focus is $(a, 0)$

13 Let $P(x_1, y_1)$ and $Q(x_2, y_2)$, $y_1 < 0$, $y_2 < 0$, be the end points of the latusrectum of the ellipse $x^2 + 4y^2 = 4$. The equations of parabolas with latusrectum PQ are

(2008, One or More than One Options Correct Type)

a. $x^2 + 2\sqrt{3}y = 3 + \sqrt{3}$ b. $x^2 - 2\sqrt{3}y = 3 + \sqrt{3}$
c. $x^2 + 2\sqrt{3}y = 3 - \sqrt{3}$ d. $x^2 - 2\sqrt{3}y = 3 - \sqrt{3}$

14 Equation of common tangent of $y = x^2$, $y = -x^2 + 4x - 4$ is

(2006, One or More than One Options Correct Type)

a. $y = 4(x - 1)$ b. $y = 0$
c. $y = -4(x - 1)$ d. $y = -30x - 50$

15 Axis of a parabola is $y = x$ and vertex and focus are at a distance $\sqrt{2}$ and $2\sqrt{2}$ respectively from the origin. Then, the equation of parabola is

(2006, Only One Option Correct Type)

a. $(x - y)^2 = 8(x + y - 2)$ b. $(x + y)^2 = 2(x + y - 2)$
c. $(x - y)^2 = 4(x + y - 2)$ d. $(x + y)^2 = 2(x - y + 2)$

16 The slope of the line touching both the parabolas $y^2 = 4x$ and $x^2 = -32y$ is

(2011, Only One Option Correct Type)

a. $\frac{1}{2}$ b. $\frac{3}{2}$ c. $\frac{1}{8}$ d. $\frac{2}{3}$

2. Ellipse

DPP
DAILY PRACTICE PROBLEMS

DPP-1 Equation of Ellipse

Subjective Questions

Working Space

Direction (Q. Nos. 1-3) *These questions are subjective in nature, need to be solved completely on notebook.*

1 For the following ellipses, find the lengths of major axis, minor axis, coordinates of foci, vertices and eccentricity.

i. $16x^2 + 25y^2 = 400$

ii. $3x^2 + 2y^2 = 6$

iii. $x^2 + 4y - 2x = 0$

2 Find the eccentricity, vertices, foci, directrix, length of latusrectum and equation of latusrectum for $x^2 + 4y^2 + 2x + 16y + 13 = 0$.

3 Find the equation of ellipse, whose axes are along the coordinate axes.

i. Vertices $(\pm 5, 0)$ and foci $(\pm 4, 0)$

ii. Vertices $(0, \pm 10)$ and $e = 4/5$

iii. Foci $(0, \pm 4)$ and $e = 4/5$

iv. Foci $(\pm 2, 0)$ and $e = 1/2$

Only One Option Correct Type

Direction (Q. Nos. 4-14) *This section contains 11 multiple choice questions. Each question has four choices (a), (b), (c) and (d), out of which ONLY ONE option is correct.*

4 The eccentricity of ellipse whose axes are parallel to the coordinate axes having its centre at the point $(2, -3)$, one focus at $(3, -3)$ and one vertex is $(4, -3)$, is

a. $\sqrt{3}/4$ b. $1/\sqrt{3}$

c. $1/2$ d. $1/\sqrt{2}$

5 The eccentricity of ellipse $4(x - 2y + 1)^2 + 9(2x + y + 2)^2 = 180$ is

a. $\sqrt{5}/3$ b. $\sqrt{3}/4$

c. $1/2$ d. $1/4$

6 Let E be the ellipse $\frac{x^2}{9} + \frac{y^2}{4} = 1$ and C be the circle $x^2 + y^2 = 9$. Let P and Q be the points $(1, 2)$ and $(2, 1)$, respectively. Then,

a. Q lies inside C but outside E

b. Q lies outside both C and E

c. P lies inside both C and E

d. P lies inside C but outside E

7 The eccentricity of ellipse $(x - 3)^2 + (y - 4)^2 = y^2/9$ is

a. $\frac{\sqrt{3}}{2}$ b. $\frac{1}{3}$

c. $\frac{1}{3\sqrt{2}}$ d. $\frac{1}{\sqrt{3}}$

Working Space

8 The equation $2x^2 + 3y^2 - 8x - 18y + 35 = k$ represents

a. no locus, if $k > 0$
b. an ellipse, if $k < 0$
c. a point, if $k = 0$
d. a hyperbola, if $k > 0$

9 If the ellipse $\frac{(x-h)^2}{M} + \frac{(y-k)^2}{N} = 1$ has major axis on the line $y = 2$, minor axis on the line $x = -1$. Major axis has length 10 and minor axis has length 4, then the values of h, k, N, M respectively are

a. −1, 2, 5, 2
b. −1, 2, 10, 4
c. 1 − 2, 25, 4
d. −1, 2, 25, 4

10 A circle has the same centre as an ellipse and passes through the foci F_1 and F_2 of the ellipse, such that the two curves intersect in four points. Let P be any one of their point of intersection. If the major axis of the ellipse is 17 and the area of $\Delta PF_1F_2 = 30$, then the distance between the foci is

a. 11
b. 12
c. 13
d. None of these

11 The latusrectum of a conic section is the width of the function through the focus. The positive difference between the length of the latusrectum of $3y = x^2 + 4x - 9$ and $x^2 + 4y^2 - 6x + 16y = 24$ is

a. 1/2
b. 2
c. 3/2
d. 5/2

12 Imagine that you have two thumb tacks placed at two points, A and B. If the ends of a fixed length of string are fastened to the thumb tacks and the string is drawn tant with a pencil, the path traced by the pencil will be an ellipse. The best way to maximise the area surrounded by the ellipse with a fixed length of string occurs, when

I. the two points A and B have the maximum distance between them.
II. two points A and B coincide.
III. A and B are placed vertically.
IV. the area is always same regardless of the location of A and B.

a. Only I
b. Only II
c. Only III
d. Only IV

13 Let $S(5, 12)$ and $S'(-12, 5)$ be the foci of an ellipse passing through the origin. The eccentricity of ellipse is

a. 1/2
b. $1/\sqrt{3}$
c. $1/\sqrt{2}$
d. 2/3

14 Consider $\frac{x^2}{\tan^2 \alpha} + \frac{y^2}{\sec^2 \alpha} = 1$, where $x \in \left(0, \frac{\pi}{2}\right)$, then which of the following remains constant as α varies?

a. Degree of flatness
b. Ordinate of vertex
c. Coordinates of foci
d. Length of the latusrectum

Working Space

Matching List Type

Direction (Q. No. 15) *Choices for the correct combination of elements from Column I and Column II are given as options (a), (b), (c) and (d), out of which one is correct.*

15 Match the Column I with Column II and mark the correct option from the codes given below.

	Column I		Column II
i.	The eccentricity of ellipse which meets the straight line $2x - 3y = 6$ on the X-axis and the straight line $4x + 5y = 20$ on the Y-axis and whose principal axes lie along the coordinate axes, is	p.	$1/2$
ii.	A bar of length 20 units moves with its ends on two fixed straight lines at right angles. A point P marked on the bar at a distance of 8 units from one end describes a conic whose eccentricity is	q.	$1/\sqrt{2}$
iii.	If one extremity of the minor axis of the ellipse $\frac{x^2}{a^2} + \frac{y^2}{b^2} = 1$ and the foci form an equilateral triangle, then its eccentricity is	r.	$\sqrt{5}/3$
iv.	There are exactly two points on the ellipse $x^2/a^2 + y^2/b^2 = 1$, whose distance from the centre of the ellipse are greatest and equal to $\sqrt{\frac{a^2 + 2b^2}{2}}$. Eccentricity of ellipse is	s.	$\sqrt{7}/4$

Codes

	i	ii	iii	iv
a.	s	r	p	q
b.	p	s	q	r
c.	r	q	s	p
d.	q	p	r	s

Comprehension Type

Direction (Q. Nos. 16-19) *This section contains 2 paragraphs, each describing theory, experiments, data, etc. Four questions related to the paragraphs have been given. Each question has only one correct answer among the four given options (a), (b), (c) and (d).*

Passaage I

The equation of ellipse whose axes are of lengths 6 units and $2\sqrt{6}$ units and their equations are $x - 3y + 3 = 0$ and $3x + y - 1 = 0$ respectively, is
$a(x - 3y + 3)^2 + b(3x + y - 1)^2 = 180$.

16 The value of a is

a. 2
b. 3
c. 4
d. 5

17 The value of b is

a. 2
b. 3
c. 4
d. 5

Working Space

Passage II

A straight rod of given length a slides between the two fixed bars which subtends an angle of 90°.

18 The locus of a point on the rod which divides it in a given ratio λ, is

a. $x^2 + \frac{y^2}{\lambda^2} = \frac{a^2}{(\lambda+1)^2}$

b. $\frac{x^2}{\lambda^2} + y^2 = \frac{a^2}{(\lambda+1)^2}$

c. $\frac{x^2}{(\lambda+1)^2} + \frac{y^2}{\lambda^2} = a^2$

d. $\frac{x^2}{\lambda^2} + \frac{y^2}{(\lambda+1)^2} = a^2$

19 If $\lambda = 1/2$, then the eccentricity is

a. 1/2

b. $1/\sqrt{2}$

c. $\sqrt{3}/2$

d. $1/\sqrt{3}$

One Integer Value Correct Type

Direction *(Q. Nos. 20-25) This section contains 6 questions. When worked out will result in an integer from 0 to 9 (both inclusive).*

20 Point O is the centre of ellipse with major axis AB and minor axis CD. Point F is one focus of ellipse. If $OF = 6$ and the diameter of the inscribed circle of ΔOCF is 2, then $\frac{(AB)(CD)}{13}$ is

21 The equation $\frac{x^2}{8-t} + \frac{y^2}{t-4} = 1$ will represent an ellipse, if $t \in (4, 8) - \{\lambda\}$. Then, the value of λ is

22 The equation of ellipse whose focus is the point $(-1, 1)$ whose directrix is the straight line $x - y + 3 = 0$ and eccentricity is 1/2, is $\lambda x^2 + 7y^2 + 2xy + 10x - 10y + \mu = 0$, then the value of $|\lambda - \mu|$ is

23 A man running on the race course notes that the sum of distance of two flag-posts from him is always 10 m and the distance between the flag-posts is 8 m. If the area of path he encloses (in sq m) is $(3\lambda\pi)$, then the value of λ is

24 The length of latusrectum of the ellipse $(10x-5)^2 + (10y-5)^2 = (3x+4y-1)^2$ is λ, then the value of (2λ) is

25 Let P be a variable point on the ellipse $\frac{x^2}{4} + \frac{y^2}{1} = 1$ with foci S_1 and S_2. If A is the area of ΔPS_1S_2, then the value of A^2 is

DPP-2 Equation of Chord, Focal Chord and Its Properties

Only One Option Correct Type

Working Space

Direction (Q. Nos. 1-9) *This section contains 9 multiple choice questions. Each question has four choices (a), (b), (c) and (d), out of which ONLY ONE option is correct.*

1 If Y-axis is the directrix of the ellipse with eccentricity $e = 1/2$ and corresponding focus is at (3, 0), then the equation to its auxiliary circle is

a. $x^2 + y^2 - 8x + 12 = 0$

b. $x^2 + y^2 - 8x - 12 = 0$

c. $x^2 + y^2 - 8x + 9 = 0$

d. $x^2 + y^2 = 4$

2 If α and β are the eccentric angles of the extremities of a focal chord of an standard ellipse, then the eccentricity of ellipse is

a. $\dfrac{\cos\alpha + \cos\beta}{\cos(\alpha + \beta)}$

b. $\dfrac{\sin\alpha - \sin\beta}{\sin(\alpha - \beta)}$

c. $\dfrac{\cos\alpha - \cos\beta}{\cos(\alpha - \beta)}$

d. $\dfrac{\sin\alpha + \sin\beta}{\sin(\alpha + \beta)}$

3 An ellipse is inscribed in a circle and a point with in the circle is chosen at random. If the probability that this point lies outside the ellipse is 2/3, then the eccentricity of ellipse is

a. $\dfrac{2\sqrt{2}}{3}$

b. $\dfrac{\sqrt{5}}{3}$

c. $\dfrac{8}{9}$

d. $\dfrac{2}{3}$

4 The locus of the middle points of chords of an ellipse $\dfrac{x^2}{16} + \dfrac{y^2}{25} = 1$, passing through $P(0, 5)$ is another ellipse E. The coordinates of foci of the ellipse E is

a. $\left(0, \dfrac{3}{5}\right)$ and $\left(0, -\dfrac{3}{5}\right)$

b. $(0, -4)$ and $(0, 1)$

c. $(0, 4)$ and $(0, 1)$

d. $\left(0, \dfrac{11}{2}\right)$ and $\left(0, -\dfrac{1}{2}\right)$

5 The eccentric angle of the extremities of latusrecta of the ellipse $\dfrac{x^2}{a^2} + \dfrac{y^2}{b^2} = 1$ is

a. $\tan^{-1}\left(\pm\dfrac{b}{ae}\right)$

b. $\tan^{-1}\left(\pm\dfrac{a}{be}\right)$

c. $2\tan^{-1}\left(\pm\dfrac{b}{ae}\right)$

d. None of these

6 The line $lx + my + n = 0$ will cut the ellipse $\frac{x^2}{a^2} + \frac{y^2}{b^2} = 1$ in the points whose eccentric angle differ by $\pi/2$, if

a. $a^2m^2 + b^2l^2 = n^2$
b. $a^2l^2 + b^2m^2 = 2n^2$
c. $a^2l^2 + b^2m^2 = n^2$
d. $a^2m^2 + b^2l^2 = 2n^2$

7 Let α and β be the eccentric angles of the extremities of a focal chord of an ellipse of eccentricity e, then $\tan\frac{\alpha}{2} \cdot \tan\frac{\beta}{2}$ is

a. $-\left(\frac{1 \mp e}{1 \mp e}\right)$
b. $\left(\frac{1 \mp e}{1 \mp e}\right)$
c. $-\left(\frac{1 \mp e}{1 \pm e}\right)$
d. None of these

8 If any two chords can be drawn through two points on the major axis of the ellipse $\frac{x^2}{a^2} + \frac{y^2}{b^2} = 1$ equidistant from the centre, then the value of $\tan\frac{\alpha}{2} \cdot \tan\frac{\beta}{2} \cdot \tan\frac{\gamma}{2} \tan\frac{\delta}{2}$ is

(where $\alpha, \beta, \gamma, \delta$ are the eccentric angles of extremities of the chords)

a. 0
b. 1
c. −1
d. 1/2

9 If PQ is focal chord of ellipse $\frac{x^2}{25} + \frac{y^2}{16} = 1$, which passes through $S(3, 0)$ and $PS = 2$, then the length of chord PQ is

a. 8
b. 6
c. 10
d. 4

Matching List Type

Direction (Q. Nos. 10 and 11) *Choices for the correct combination of elements from Column I and Column II are given as options (a), (b), (c) and (d), out of which one is correct.*

10 Match the Column I with Column II and mark the correct option from the codes given below.

	Column I		Column II
i.	A stick of length 10 m slides m coordinate axes, then the locus of a point dividing this stick reckoning from X-axis in the ratio 6 : 4 is a curve whose eccentricity is e, then $9e$ is	p.	$\sqrt{6}$
ii.	AA' is major axis of an ellipse $3x^2 + 2y^2 + 6x - 4y - 1 = 0$ and P is a variable point on it, then the greatest area of $\Delta APA'$ is	q.	$2\sqrt{7}$
iii.	The distance between the foci of the curve represented by the equation $x = 1 + 4\cos\theta$ and $y = 2 + 3\sin\theta$ is	r.	$3\sqrt{5}$

Codes

	i	ii	iii
a.	r	p	q
b.	p	q	r
c.	q	r	p
d.	q	p	q

Working Space

Working Space

11 Let $C_1 : 9x^2 + 8y^2 - 36x - 16y - 28 = 0$ and $C_2 = 4x^2 + 9y^2 + 8x - 36y + 4 = 0$. Match the Column I with Column II and mark the correct option from the codes given below.

	Column I		Column II
i.	The minimum distance of (2, 6) from C_1 is	p.	1
ii.	The maximum distance of (2, 6) from C_1 is	q.	2
iii.	The minimum distance of (1, 2) from C_2 is	r.	5
iv.	The maximum distance of (1, 2) from C_2 is	s.	8

Codes

	i	ii	iii	iv
a.	q	s	p	r
b.	p	q	r	s
c.	s	p	q	r
d.	r	r	p	q

Comprehension Type

Direction *(Q. Nos. 12-14) This section contains a paragraph, describing theory, experiments, data, etc. Three questions related to the paragraph have been given. Each question has only one correct answer among the four given options (a), (b), (c) and (d).*

Passage

If d is a diameter of an ellipse whose principal axes have lengths $2a$ and $2b$. The inclination of the diameter to the major axis is θ.

12 If d^2 is the arithmetic mean of $(2a)^2$ and $(2b)^2$, then $\tan\theta$ is

a. $\pm\sqrt{\frac{a}{b}}$ b. $\pm\frac{b}{a}$

c. $\pm\frac{\pi}{4}$ d. $\pm\frac{\pi}{3}$

13 If d^2 is the geometric mean of $(2a)^2$ and $(2b)^2$, then $\tan\theta$ is

a. $\pm\sqrt{\frac{b}{a}}$ b. $\pm\frac{b}{a}$

c. $\pm\frac{\pi}{4}$ d. $\pm\frac{\pi}{3}$

14 If d^2 is the harmonic mean of $(2a)^2$ and $(2b)^2$, then θ equals

a. $\pm\sqrt{\frac{a}{b}}$ b. $\pm\frac{b}{a}$

c. $\pm\frac{\pi}{4}$ d. $\pm\frac{\pi}{3}$

One Integer Value Correct Type

Direction *(Q. No. 15) This section contains a question. When worked out will result in an integer from 0 to 9 (both inclusive).*

15 If equation of the curves on reflection of the ellipse $\frac{(x-4)^2}{16} + \frac{(y-3)^2}{9} = 1$ about the line $x - y - 2 = 0$ is $16x^2 + 9y^2 + k_1x - 36y + k_2 = 0$, then the units' digit of $(k_1 + k_2)$ is

DPP-3 Tangent and Its Properties

Only One Option Correct Type

Working Space

Direction (Q. Nos. 1-20) *This section contains 20 multiple choice questions. Each question has four choices (a), (b), (c) and (d), out of which ONLY ONE option is correct.*

1 The condition that $lx + my + n = 0$ to touch the ellipse $\frac{x^2}{a^2} + \frac{y^2}{b^2} = 1$ is

a. $a^2l^2 + b^2m^2 = n^2$
b. $a^2m^2 + b^2l^2 = n^2$
c. $a^2l^2 + b^2m^2 = 2n^2$
d. $a^2m^2 + b^2l^2 = 2n^2$

2 The condition for the line $x\cos\alpha + y\sin\alpha = p$ to be a tangent to the ellipse $\frac{x^2}{a^2} + \frac{y^2}{b^2} = 1$ is

a. $p^2 = a^2\sin^2\alpha + b^2\cos^2\alpha$
b. $p^2 = a^2\cos^2\alpha + b^2\sin^2\alpha$
c. $p^2 = 2a^2\sin^2\alpha + 2b^2\cos^2\alpha$
d. None of the above

3 The value of λ, for which $y = x + \lambda$ touches the ellipse $9x^2 + 16y^2 = 144$ is

a. ± 2
b. ± 3
c. ± 4
d. ± 5

4 The equations of tangents to the ellipse $4x^2 + 3y^2 = 5$, which are inclined at an angle of 60° to the X-axis is

a. $y = \sqrt{3}x \pm \sqrt{\frac{65}{12}}$
b. $y = \sqrt{3}x \pm 5$
c. $y = \frac{1}{\sqrt{3}}x \pm \sqrt{\frac{65}{12}}$
d. $y = \frac{1}{\sqrt{3}}x \pm 5$

5 The equations of tangents to the ellipse $4x^2 + 3y^2 = 5$, which are parallel to $y = 3x - 7$ is

a. $y = 3x \pm \sqrt{155}$
b. $y = 3x \pm \sqrt{\frac{155}{12}}$
c. $y = 3x + 1$
d. $y = 3x - 1$

6 The equations of tangents to the ellipse $3x^2 + 4y^2 = 12$, which are perpendicular to $y + 2x = 4$ is

a. $2y = x \pm 3$
b. $2y = x \pm 1$
c. $2y = x \pm 4$
d. None of these

7 The equation of tangent to the ellipse $4x^2 + 9y^2 = 36$ at $(3, -2)$ is

a. $2x - 3y = 6$
b. $3x - 2y = 13$
c. $x + y = 1$
d. $x - y = 5$

Working Space

8 The equation of tangents to the ellipse $x^2 + 4y^2 = 25$ at the points whose ordinate is 2, is

a. $3x - 8y - 25 = 0$ b. $3x + 8y = -25$
c. $3x + 8y = 25$ d. $3x + 8y = -25$

9 The straight line $x + 2y - 4 = 0$ touches the ellipse $3x^2 + 4y^2 = 12$ at

a. (1, 3) b. (1/2, 3/2)
c. $(0, \sqrt{3})$ d. (1, 3/ 2)

10 The product of the length of perpendicular from the foci upon any tangent to the ellipse $\frac{x^2}{a^2} + \frac{y^2}{b^2} = 1$ is

a. a^2 b. $2a^2$
c. b^2 d. $2b^2$

11 The equations of tangents to the ellipse $9x^2 + 16y^2 = 144$ which passes through the point (2, 3) is

a. $y = 3, y = -x + 5$ b. $x = 2, y = x - 5$
c. $y = 3/2, y = -x + 5$ d. $x = 2, y = x + 5$

12 The tangent at the extremities of latusrectum of an ellipse intersect on the

a. tangent at vertex b. directrix
c. $x = \frac{2a}{e}$ d. $x = \frac{4a}{e}$

13 The locus of the mid-points of the portion of tangents to the ellipse $\frac{x^2}{a^2} + \frac{y^2}{b^2} = 1$ intercepted between the axes is

a. $a^2x^2 + b^2y^2 = x^2y^2$ b. $a^2x^2 + b^2y^2 = 4x^2y^2$
c. $a^2y^2 + b^2x^2 = 4x^2y^2$ d. $a^2y^2 + b^2x^2 = x^2y^2$

14 Let d be the perpendicular distance from the centre of ellipse $\frac{x^2}{a^2} + \frac{y^2}{b^2} = 1$ to the tangent drawn at a point P on the ellipse. If F_1 and F_2 are two foci of the ellipse, where $(PF_1 - PF_2)^2 = \lambda a^2\left(1 - \frac{b^2}{d^2}\right)$, then λ is equal to

a. 2 b. 3
c. 4 d. 5

15 The point of intersection of tangents at two points on the ellipse $\frac{x^2}{a^2} + \frac{y^2}{b^2} = 1$, whose eccentric angles differ by a right angle lies on the ellipse, is

a. $\frac{x^2}{a^2} + \frac{y^2}{b^2} = 4$ b. $\frac{x^2}{a^2} + \frac{y^2}{b^2} = 3$
c. $\frac{x^2}{a^2} + \frac{y^2}{b^2} = 2$ d. None of these

16 The locus of foot of the perpendicular on a varying tangent to the ellipse $\frac{x^2}{a^2} + \frac{y^2}{b^2} = 1$, from either of its foci is

a. $x^2 + y^2 = a^2 + b^2$ b. $x^2 + y^2 = a^2$
c. $x^2 + y^2 = b^2$ d. $x^2 + y^2 = 2a^2$

Working Space

17 The radius of the largest circle with centre (1, 0) that can be inscribed in the ellipse $x^2 + 4y^2 = 16$ is

a. $\frac{11}{3}$
b. $\frac{11}{\sqrt{3}}$
c. $\frac{\sqrt{11}}{3}$
d. $\sqrt{\frac{11}{3}}$

18 An ellipse of semi-axis *a*, *b* slides between two perpendicular lines, then the locus of its foci is, (the two lines being taken as the axes of coordinates)

a. $(x^2 + y^2)(x^2y^2 + b^2) = 4a^2x^2y^2$
b. $(x^2 + y^2)(x^2y^2 + a^2) = 4b^2x^2y^2$
c. $(x^2 - y^2)(x^2y^2 + b^2) = 4a^2x^2y^2$
d. $(x^2 - y^2)(x^2y^2 + a^2) = 4a^2x^2y^2$

19 A tangent to the ellipse $x^2 + 4y^2 = 4$ meets the ellipse $x^2 + 2y^2 = 6$ at *P* and *Q*, then the tangent at *P* and *Q* of the ellipse $x^2 + 2y^2 = 6$ subtends an angle

a. 60°
b. 30°
c. 45°
d. 90°

20 The tangents drawn to the ellipse $\frac{x^2}{a} + \frac{y^2}{b} = a + b$ at the points, where it cut by the tangent to $\frac{x^2}{a^2} + \frac{y^2}{b^2} = 1$ subtends an angle of

a. 60°
b. 30°
c. 45°
d. 90°

Comprehension Type

Direction *(Q. Nos. 21-28) This section contains 2 paragraphs, each describing theory, experiments, data, etc. Eight questions related to the paragraphs have been given. Each question has only one correct answer among the four given options (a), (b), (c) and (d).*

Passage I

Consider an ellipse $\frac{x^2}{9} + \frac{y^2}{4} = 1$ and a parabola $y^2 = 2x$, they intersect at *P* and *Q* in the first and fourth quadrants respectively. Tangents to the ellipse at *P* and *Q* intersect the *X*-axis at *R* and tangents to the parabola at *P* and *Q* intersect the *X*-axis at *S*.

21 The ratio of the areas of ΔPQS and ΔPQR is

a. 1 : 3
b. 1 : 2
c. 2 : 3
d. 3 : 4

22 The area of quadrilateral *PQRS* is

a. $\frac{3\sqrt{15}}{2}$ sq units
b. $\frac{15\sqrt{3}}{2}$ sq units
c. $\frac{5\sqrt{3}}{2}$ sq units
d. $\frac{5\sqrt{15}}{2}$ sq units

23 The equation of circle touching the parabola at upper end of its latusrectum and passing through its vertex, is

a. $2x^2 + 2y^2 - x - 2y = 0$
b. $2x^2 + 2y^2 + 4x - \frac{9}{2}y = 0$
c. $2x^2 + 2y^2 + x - 3y = 0$
d. $2x^2 + 2y^2 - 7x + y = 0$

Working Space

Passage II

Let the two foci of an ellipse be $(-1, 0)$ and $(3, 4)$ and the foot of perpendicular from focus $(3, 4)$ upon a tangent to the ellipse be $(4, 6)$.

24 The foot of perpendicular from the focus $(-1, 0)$ upon the same tangent to the ellipse is

a. $\left(\frac{12}{5}, \frac{34}{5}\right)$
b. $\left(\frac{7}{3}, \frac{11}{3}\right)$
c. $\left(2, \frac{17}{4}\right)$
d. $(-1, 2)$

25 The equation of auxiliary circle of the ellipse is

a. $x^2 + y^2 - 2x - 4y - 5 = 0$
b. $x^2 + y^2 - 2x - 4y - 20 = 0$
c. $x^2 + y^2 + 2x + 4y - 20 = 0$
d. $x^2 + y^2 + 2x + 4y - 5 = 0$

26 The length of semi-minor axis of the ellipse is

a. 1
b. $2\sqrt{2}$
c. $\sqrt{17}$
d. $\sqrt{19}$

27 The equations of directrices of the ellipse are

a. $x - y + 2 = 0, x - y - 5 = 0$
b. $x + y + 1 = 0, x + y + \frac{17}{2} = 0$
c. $x - y + \frac{3}{2} = 0, x - y - \frac{5}{2} = 0$
d. $x + y - \frac{31}{2} = 0, x + y + \frac{19}{2} = 0$

28 The point of contact of the tangent with the ellipse is

a. $\left(\frac{40}{11}, \frac{68}{11}\right)$
b. $\left(\frac{4}{7}, \frac{8}{7}\right)$
c. $\left(\frac{8}{5}, \frac{17}{5}\right)$
d. $\left(\frac{41}{13}, \frac{83}{13}\right)$

One Integer Value Correct Type

Direction (Q. Nos. 29 and 30) *This section contains 2 questions. When worked out will result in an integer from 0 to 9 (both inclusive).*

29 $x - 2y + 4 = 0$ is a common tangent to $y^2 = 4x$ and $\frac{x^2}{4} + \frac{y^2}{b^2} = 1$. Then, the value of b^2 is

30 The slope of common tangent to the ellipses $\frac{x^2}{6} + \frac{y^2}{1} = 1$ and $\frac{x^2}{5} + \frac{y^2}{6} = 1$ is m, then the value of m^2 is

DPP-4 Normal and Its Properties

Only One Option Correct Type

Direction (Q. Nos. 1-12) *This section contains 12 multiple choice questions. Each question has four choices (a), (b), (c) and (d), out of which ONLY ONE option is correct.*

Working Space

1 The equation of normal to the ellipse $\frac{x^2}{a^2}+\frac{y^2}{b^2}=1$ at the positive end of latusrectum is

a. $x+ey+e^2a=0$
b. $x-ey-e^3a=0$
c. $x-ey-e^2a=0$
d. None of these

2 The eccentric angle of the point, where the line $5x-3y=8\sqrt{2}$ is a normal to the ellipse $\frac{x^2}{25}+\frac{y^2}{9}=1$ is

a. $3\pi/4$
b. $\pi/4$
c. $\pi/6$
d. $\tan^{-1}2$

3 PQ is a double ordinate of the ellipse $x^2+9y^2=9$, the normal at P meets the diametre through Q at R, then the locus of mid-point of PR is

a. a circle
b. a parabola
c. an ellipse
d. a hyperbola

4 The area of rectangle formed by the perpendiculars from the centre of the standard ellipse to the tangent and normal at its point whose eccentric angle is $(\pi/4)$, is

a. $\frac{(a^2-b^2)ab}{a^2+b^2}$
b. $\frac{(a^2+b^2)ab}{a^2-b^2}$
c. $\frac{a^2-b^2}{ab(a^2+b^2)}$
d. $\frac{a^2+b^2}{(a^2-b^2)ab}$

5 The normal at a variable point P on an ellipse $\frac{x^2}{a^2}+\frac{y^2}{b^2}=1$ of eccentricity e meets the axes of the ellipse in Q and R, then the locus of mid-point of QR is a conic with an eccentricity e' such that

a. e' is independent of e
b. $e'=1$
c. $e'=e$
d. $e'=1/e$

6 The length of normal (terminated by the major axis) at a point of the ellipse $\frac{x^2}{a^2}+\frac{y^2}{b^2}=1$ is

a. $\frac{b}{a}(r+r_1)$
b. $\frac{b}{a}|r-r_1|$
c. $\frac{b}{a}\sqrt{rr_1}$
d. independent of r, r_1

(where, r and r_1 are the focal distance of the points.)

Working Space

7 The normal at a point P on the ellipse $x^2 + 4y^2 = 16$ meets the X-axis at Q. If M is the mid-point of the segment PQ, then the locus of M intersects the latusrectum of the given ellipse at the points

a. $\left(\pm\frac{3\sqrt{5}}{2}, \pm\frac{2}{7}\right)$
b. $\left(\pm\frac{3\sqrt{5}}{2}, \pm\frac{\sqrt{19}}{4}\right)$
c. $\left(\pm 2\sqrt{3}, \pm\frac{1}{7}\right)$
d. $\left(\pm 2\sqrt{3}, \pm\frac{4\sqrt{3}}{7}\right)$

8 If the normal at an end of a latusrectum of an ellipse passes through one extremity of the minor axis, then the eccentricity of ellipse is given by

a. $e^4 - e^2 + 1 = 0$
b. $e^4 + e^2 - 1 = 0$
c. $e^2 + e - 1 = 0$
d. None of these

9 The line $lx + my + n = 0$ is normal to the ellipse $\frac{x^2}{a^2} + \frac{y^2}{b^2} = 1$, if $n^2\left(\frac{a^2}{l^2} + \frac{b^2}{m^2}\right)$ is equal to

a. $(a^2 + b^2)$
b. $(a^2 + b^2)^2$
c. $(a^2 - b^2)$
d. $(a^2 - b^2)^2$

10 Any ordinate MP of an ellipse meets the auxiliary circle in θ. Then, the locus of the point of intersection of the normals at P and Q is

a. $x^2 + y^2 = a^2 + b^2$
b. $x^2 + y^2 = (a + b)^2$
c. $x^2 + y^2 = a^2 - b^2$
d. $x^2 + y^2 = (a - b)^2$

11 If ω is one of the angles between the normals to the ellipse $\frac{x^2}{a^2} + \frac{y^2}{b^2} = 1$ at the points whose eccentric angles are θ and $\frac{\pi}{2} + \theta$, then $\frac{2\cot\omega}{\sin 2\theta}$ is

a. $\frac{e^2}{1-e^2}$
b. $\frac{e}{1-e}$
c. $\frac{e^2}{\sqrt{1-e^2}}$
d. $\frac{e^2}{1-e}$

12 If the tangent drawn at point $(t^2, 2t)$ on the parabola $y^2 = 4x$ is same as the normal drawn at a point $(\sqrt{5}\cos\theta, 2\sin\theta)$ on the ellipse $4x^2 + 5y^2 = 20$, then the value of t is

a. $\sqrt{5}$
b. $-\sqrt{5}$
c. $1/\sqrt{5}$
d. $-2/\sqrt{5}$

Comprehension Type

Direction (Q. Nos. 13-17) *This section contains 2 paragraphs, each describing theory, experiments, data, etc. Five questions related to the paragraphs have been given. Each question has only one correct answer among the four given options (a), (b), (c) and (d).*

Passage I

Consider the ellipse $E : \frac{x^2}{4} + \frac{y^2}{1} = 1$, S and S' are foci and P is any point on E such that $SP \cdot S'P$ be maximum, among the two values of P with lesser ordinate be name it as A and other as B.

13 Possible coordinates of P are

a. $(0, \pm 1)$
b. $(\pm 2, 0)$
c. $(\pm\sqrt{2}, 0)$
d. None of these

Working Space

14 Ray of light is passing through point A it get reflected from curve $\frac{-x^2}{a^2} + \frac{y^2}{b^2} = 1$ such that reflected ray appear to pass through B, curve X is passing through the point $D(2, 1)$, then the curve X is

a. $\frac{-x^2}{2(\sqrt{2}-1)} + \frac{y^2}{(\sqrt{2}-1)^2} = 1$

b. $\frac{x^2}{2(\sqrt{2}-1)^2} - \frac{y^2}{(\sqrt{2}-1)} = 1$

c. $\frac{x^2}{2} - \frac{y^2}{3} = 1$

d. $\frac{x^2}{(\sqrt{2}-1)^2} + \frac{y^2}{2(\sqrt{2}+1)} = 1$

15 From the point D normal be drawn to curve X cut axis at point E, then the area of ΔBDE is

a. $(2\sqrt{2}-2)$ sq units

b. $(\sqrt{2}+1)$ sq units

c. $2(\sqrt{2}+1)$ sq units

d. $2(\sqrt{2}+3)$ sq units

Passage II

To prove geometrical properties of ellipse, we may take its standard equation as $\frac{x^2}{a^2} + \frac{y^2}{b^2} = 1$, we will assume that $a > b > 0$ and foci S, S' lie on positive and negative side of X-axis, respectively.

16 If the normal at any point P meets major and minor axis at G and G', respectively and OF is the perpendicular drawn from centre O to this normal. Then, $PF \cdot PG$ must be equal to

a. b^2

b. a^2

c. ab

d. a^2b^2

17 If P is any point on the ellipse and G is same as above, then $S'G/SG$ must be

a. $\frac{OS'}{OS}$

b. $\frac{G'S}{OG}$

c. $\frac{S'P}{SP}$

d. $\frac{a}{b}$

One Integer Value Correct Type

Direction *(Q. Nos. 18-20) This section contains 3 questions. When worked out will result in an integer from 0 to 9 (both inclusive).*

18 The shortest distance from the ellipse $x^2 + 2y^2 = 2$ and the circle $x^2 + y^2 - 3x - 2\sqrt{2}y + 4 = 0$ is $\frac{\sqrt{k}-1}{2}$, then the value of k is

19 If the normal at the point $P(\theta)$ to the ellipse $\frac{x^2}{14} + \frac{y^2}{5} = 1$ intersect it again at the point $Q(2\theta)$, then the value of $3\cos\theta + 2$ is

20 The number of points on the ellipse $\frac{x^2}{50} + \frac{y^2}{20} = 1$ from which pair of perpendicular tangents are drawn to the ellipse $\frac{x^2}{16} + \frac{y^2}{9} = 1$ is

DPP-5 Equation of $S'=T$, $T=0$ and $SS'=T^2$, Diameter and Conjugate Diameter

Only One Option Correct Type

Working Space

Direction (Q. Nos. 1-12) *This section contains 12 multiple choice questions. Each question has four choices (a), (b), (c) and (d), out of which ONLY ONE option is correct.*

1 The equation of chord of contact of tangents drawn from (3, 2) to the ellipse $x^2 + 4y^2 = 9$ is

a. $3x + 4y = 9$
b. $3x + 8y = 6$
c. $3x + 8y = 9$
d. None of these

2 Tangents are drawn to the ellipse $\frac{h^2}{a^4} + \frac{k^2}{b^4} = 1$ at points, where it is intersected by the line $lx + my + n = 0$. Then, the point of intersection of tangents is

a. $\left(-\frac{a^2 l}{n}, -\frac{b^2 m}{n}\right)$
b. $\left(\frac{a^2 l}{n}, \frac{b^2 m}{n}\right)$
c. $\left(\frac{a^2 l}{n}, -\frac{b^2 m}{n}\right)$
d. $\left(-\frac{a^2 l}{n}, \frac{b^2 m}{n}\right)$

3 The chord of contact of tangents drawn from the point (h, k) to the ellipse $\frac{x^2}{a^2} + \frac{y^2}{b^2} = 1$ will subtend a right angle at the centre, if $\frac{h^2}{a^4} + \frac{k^2}{b^2}$ is

a. $a^2 + b^2$
b. $\frac{1}{a^2} + \frac{1}{b^2}$
c. $\frac{a^2 + b^2}{ab}$
d. None of these

4 The middle point of chord intercepted on the line $2x - y + 3 = 0$ by the ellipse $\frac{x^2}{10} + \frac{y^2}{6} = 1$ is

a. $\left(\frac{30}{23}, -\frac{9}{23}\right)$
b. $\left(-\frac{30}{23}, \frac{9}{23}\right)$
c. $\left(-\frac{30}{23}, -\frac{9}{23}\right)$
d. $\left(\frac{30}{23}, \frac{9}{23}\right)$

5 The locus of mid-point of focal chords of the ellipse $\frac{x^2}{a^2} + \frac{y^2}{b^2} = 1$ is

a. $\frac{xe}{a} = \frac{x^2}{a^2} + \frac{y^2}{b^2}$
b. $\frac{xe}{a} = \frac{x^2}{a^2} - \frac{y^2}{b^2}$
c. $\frac{xe}{a} = \frac{x^2}{b^2} + \frac{y^2}{a^2}$
d. $\frac{ye}{a} = \frac{x^2}{a^2} - \frac{y^2}{b^2}$

Working Space

6 The locus of the point of intersection of tangents at the end-points of conjugate diameter of the ellipse $\frac{x^2}{a^2}+\frac{y^2}{b^2}=1$, is

a. $\frac{x^2}{a^2}+\frac{y^2}{b^2}=4$ b. $\frac{x^2}{a^2}+\frac{y^2}{b^2}=3$

c. $\frac{x^2}{a^2}+\frac{y^2}{b^2}=2$ d. None of these

7 *CP* and *CD* are conjugate diameters of the ellipse $\frac{x^2}{a^2}+\frac{y^2}{b^2}=1$. Then, the locus of mid-point of *PD* is

a. $\frac{x^2}{a^2}+\frac{y^2}{b^2}=\frac{1}{2}$ b. $\frac{x^2}{a^2}+\frac{y^2}{b^2}=1$

c. $\frac{x^2}{a^2}+\frac{y^2}{b^2}=\frac{3}{2}$ d. $\frac{x^2}{a^2}+\frac{y^2}{b^2}=2$

8 *CP* and *CD* are conjugate semi-diameters of the ellipse $\frac{x^2}{a^2}+\frac{y^2}{b^2}=1$and the circles with *CP* and *CD* as diameters intersect at *R*. Then, *R* lies on

a. $2(x^2+y^2)=a^2x^2+b^2y^2$

b. $2(x^2+y^2)^2=a^2+b^2$

c. $2(x^2+y^2)^2=a^2x^2+b^2y^2$

d. None of the abvoe

9 The locus of pole of tangents to the ellipse $\frac{x^2}{a^2}+\frac{y^2}{b^2}=1$with respect to $y^2=4ax$, is

a. $b^2y^2=4a^2(x^2-a^2)$ b. $b^2x^2=4a^2(y^2-b^2)$

c. $a^2x^2=4b^2(y^2-b^2)$ d. $a^2x^2=4a^2(y^2-b^2)$

10 The locus of poles of normal chord of the ellipse $\frac{x^2}{a^2}+\frac{y^2}{b^2}=1$ is

a. $\frac{a^6}{x^2}+\frac{b^6}{y^2}=a^2-b^2$ b. $\frac{a^6}{x^2}-\frac{b^6}{y^2}=(a^2-b^2)^2$

c. $\frac{a^6}{x^2}+\frac{b^6}{y^2}=(a^2-b^2)^2$ d. $\frac{a^6}{x^2}-\frac{b^6}{y^2}=(a^2+b^2)^2$

11 The equation of chord $\frac{x^2}{36}+\frac{y^2}{9}=1$, which is bisected at (2, 1) is

a. $x+2y-4=0$ b. $x-2y+4=0$

c. $x+2y+4=0$ d. $-x+2y+4=0$

12 The locus of point the chord of contact of tangents from which to the ellipse $\frac{x^2}{a^2}+\frac{y^2}{b^2}=1$touches the circle $x^2+y^2=c^2$, is

a. $b^4c^2x^2+a^4c^2y^2=a^4b^4$

b. $b^2c^2x^2+a^2c^2y^2=1$

c. $b^4c^2x^2+a^4c^2y^2=1$

d. None of the above

Working Space

Matching List Type

Direction (Q. No. 13) *Choices for the correct combination of elements from Column I and Column II are given as options (a), (b), (c) and (d), out of which one is correct.*

13 Match the Column I with Column II and mark the correct option from the codes given below.

	Column I		Column II
i.	If P is a point on the ellipse $\frac{x^2}{16}+\frac{y^2}{20}=1$ whose foci are SS', then $S'P + SP$ is	p.	4/5
ii.	The eccentricity of ellipse $2x^2+3y^2-4x-12y+13=0$ is	q.	$4\sqrt{5}$
iii.	Tangents are drawn from the points on the line $x-y-5=0$ to $x^2+4y^2=4$, then all the chords of contact pass through a fixed point whose eccentricity is	r.	8
iv.	The sum of distance of any point on the ellipse $3x^2+4y^2=12$ from its directrix is	s.	$\frac{1}{\sqrt{3}}$

Codes

	i	ii	iii	iv
a.	q	s	p	r
b.	p	q	r	s
c.	r	p	s	q
d.	s	r	q	p

Comprehension Type

Direction (Q. Nos. 14 and 15) *This section contains a paragraph, describing theory, experiments, data, etc. Two questions related to the paragraph have been given. Each question has only one correct answer among the four given options (a), (b), (c) and (d).*

Passage

If $S=0$ is a parabola or ellipse, then the equation of chord of conic with mid-point $P(x_1, y_1)$ is $S_1 = S_{11}$ and $S_{11} < 0$. For the ellipse $\frac{x^2}{16}+\frac{y^2}{4}=1$ and parabola $y^2=4x$.

14 If $Q(a, 0)$ is a point, where $a > 4$, is such that three distinct chords through Q of the ellipse are drawn and each of which is bisected by the parabola, then

a. $a \in (4, 10+2\sqrt{5})$ b. $a \in (16, 8+4\sqrt{5})$
c. $a \in (4, 20)$ d. $a \in (4, \infty)$

15 The length of common chord of ellipse and parabola is

a. $2\sqrt{5-\sqrt{2}}$ b. $4\sqrt{2-\sqrt{3}}$
c. $8\sqrt{\sqrt{5}-2}$ d. $2\sqrt{5}$

SET - 1

Revisal Problems for JEE Main

Only One Option Correct Type

Working Space

Direction *(Q. Nos. 1-40) This section contains 40 multiple choice questions. Each question has four choices (a), (b), (c) and (d), out of which ONLY ONE option is correct.*

1 If A and B are two points and P is a variable point such that $PA + PB = 4$, then locus of P is

a. a parabola
b. an ellipse
c. a hyperbola
d. None of these

2 If a bar of given length moves with its extremities on two fixed straight lines at right angles, then the locus of any point on the bar describes a/an

a. circle
b. parabola
c. ellipse
d. hyperbola

3 The eccentricity of an ellipse with centre at the origin and axes along the coordinate axes is 1/2. If one of the directrices is $x = 4$, then the ellipse is

a. $4x^2 + 3y^2 = 1$
b. $3x^2 + 4y^2 = 12$
c. $4x^2 + 3y^2 = 12$
d. $3x^2 + 4y^2 = 1$

4 If the focal distance of an end of the minor axis of an ellipse (referred to its axes as the X and Y-axes respectively) is k and the distance between its foci is $2h$, then its equation is

a. $\frac{x^2}{k^2} + \frac{y^2}{h^2} = 1$
b. $\frac{x^2}{k^2} + \frac{y^2}{k^2 - h^2} = 1$
c. $\frac{x^2}{k^2} + \frac{y^2}{h^2 - k^2} = 1$
d. $\frac{x^2}{k^2} + \frac{y^2}{k^2 + h^2} = 1$

5 The equation $\frac{x^2}{10 - a} + \frac{y^2}{4 - a} = 1$ represents an ellipse, if

a. $a < 4$
b. $a > 4$
c. $4 < a < 10$
d. $a > 10$

6 A set of points is such that each point is three times as far away from the Y-axis as it is from the point (4, 0), then the locus of points is

a. a hyperbola
b. a parabola
c. an ellipse
d. a circle

7 A man running round a race course notes that sum of the distances of two flag-posts from him is always 10 m and the distance between the flag-posts is 8 m. The area of the path he encloses (in sq m) is

a. 15π
b. 12π
c. 18π
d. 8π

8 PSQ is the focal chord of ellipse $\frac{x^2}{a^2} + \frac{y^2}{b^2} = 1, a > b$, then the harmonic mean of SP and SQ is

a. $\frac{b^2}{a}$
b. $\frac{a^2}{b}$
c. $\frac{2b^2}{a}$
d. $\frac{2a^2}{b}$

9 If PSQ is focal chord of the ellipse $16x^2 + 25y^2 = 400$ such that $SP = 8$, then SQ is

a. 1
b. 2
c. 3
d. 4

Working Space

10 If S and S' are two focii of ellipse $16x^2 + 25y^2 = 400$ and PSQ is a focal chord such that $SP = 16$, then $S'Q$ is equal to

a. 44/9
b. 54/9
c. 64/9
d. 74/9

11 The focus of an ellipse is $(-1, -1)$ and the corresponding directrix is $x - y + 3 = 0$. If the eccentricity of ellipse is 1/2, then the coordinates of centre of the ellipse is

a. $\left(\frac{1}{2}, \frac{3}{2}\right)$
b. $\left(-\frac{1}{2}, \frac{3}{2}\right)$
c. $\left(-\frac{1}{2}, -\frac{3}{2}\right)$
d. None of thses

12 If C is the centre and A, B are two points on the conic $4x^2 + 9y^2 - 8x - 36y + 4 = 0$ such that $\angle ACB = 90°$, then $\frac{1}{CA^2} + \frac{1}{CB^2}$ is

a. 13/36
b. 36/13
c. 16/33
d. 33/16

13 Let P be a variable point on the ellipse $\frac{x^2}{a^2} + \frac{y^2}{b^2} = 1$ with foci F_1 and F_2. If A is the area of ΔPF_1F_2, then the maximum value of A is

a. $2abe$
b. abe
c. $\frac{1}{2}abe$
d. None of these

14 If p and p' denotes the length of perpendiculars from a focus and centre of an ellipse with semi-major axis of length a respectively, on a tangent to the ellipse and r denotes the focal distance of the point, then

a. $ap = rp'$
b. $rp = ap'$
c. $ap = rp' + 1$
d. $ap' + rp = 1$

15 If $P(x, y)$, $F_1(3, 0)$, $F_2(-3, 0)$ and $16x^2 + 25y^2 = 400$, then $PF_1 + PF_2$ equals

a. 8
b. 6
c. 10
d. 12

16 The number of values of c such that the straight line $y = 4x + c$ touches the curve $\frac{x^2}{4} + y^2 = 1$, is

a. 0
b. 1
c. 2
d. 4

17 The locus of the point of intersection of tangents to an ellipse at two points, the sum of whose eccentric angle is constant, is a/an

a. parabola
b. circle
c. ellipse
d. straight line

18 If $\tan\theta_1 \cdot \tan\theta_2 = -\frac{a^2}{b^2}$, then the chord joining two points Q_1 and Q_2 on the ellipse $\frac{x^2}{a^2} + \frac{y^2}{b^2} = 1$ will subtend a right angle at

a. focus
b. centre
c. end of the major axis
d. end of the minor axis

19 The tangent at a point $P(a\cos\theta, b\sin\theta)$ on the ellipse $\frac{x^2}{a^2} + \frac{y^2}{b^2} = 1$ meets the auxiliary circle in two points. The chord joining them subtends a right angle at the centre. Then, the eccentricity of the ellipse is

a. $(1 + \sin^2\theta)^{-1/2}$
b. $(1 + \cos^2\theta)^{-1/2}$
c. $(1 + \sin^2\theta)$
d. $(1 + \cos^2\theta)^{1/2}$

Working Space

20 If F_1 and F_2 be the feet of perpendicular from the foci S_1 and S_2 of an ellipse $\frac{x^2}{5}+\frac{y^2}{3}=1$ on the tangent at any point P on the ellipse, then $(S_1F_1)\cdot(S_2F_2)$ is

a. 2 b. 3
c. 4 d. 5

21 The area of rectangle formed by the perpendiculars from the centre of the ellipse to the tangent and normal at the point whose eccentric angle is $\pi/4$, is

a. $\left(\frac{a^2-b^2}{a^2+b^2}\right)ab$ b. $\left(\frac{a^2+b^2}{a^2-b^2}\right)ab$
c. $\frac{1}{ab}\left(\frac{a^2-b^2}{a^2+b^2}\right)$ d. $\frac{1}{ab}\left(\frac{a^2+b^2}{a^2-b^2}\right)$

22 On the ellipse $4x^2+9y^2=1$, the point at which the tangents are parallel to the line $8x=9y$ are

a. $\left(\frac{2}{5},\frac{1}{5}\right)$ b. $\left(\pm\frac{2}{5},\pm\frac{1}{5}\right)$
c. $\left(-\frac{2}{5},-\frac{1}{5}\right)$ d. None of these

23 On the ellipse $\frac{x^2}{4}+\frac{y^2}{9}=1$, one of the points of which normals are parallel to the line $2x-y=1$, is

a. $\left(\frac{9}{\sqrt{10}},\frac{2}{\sqrt{10}}\right)$ b. $\left(-\frac{9}{\sqrt{10}},\frac{2}{\sqrt{10}}\right)$
c. $\left(\frac{2}{\sqrt{10}},\frac{9}{\sqrt{10}}\right)$ d. None of these

24 The tangent is drawn to the ellipse $\frac{x^2}{27}+\frac{y^2}{1}=1$ at $(3\sqrt{3}\cos\theta, \sin\theta)$, where $\theta\in(0,\pi/2)$. Then, the value of θ such that sum of intercepts on axes made by the tangent is minimum, is

a. $\pi/3$ b. $\pi/6$
c. $\pi/8$ d. $\pi/4$

25 A point on the ellipse $\frac{x^2}{16}+\frac{y^2}{9}=1$ at a distance equal to the mean of lengths of the semi-major and semi-minor axes from the centre is

a. $\left(\frac{2\sqrt{91}}{7},\frac{3\sqrt{105}}{14}\right)$ b. $\left(\frac{2\sqrt{91}}{7},-\frac{3\sqrt{91}}{14}\right)$
c. $\left(-\frac{2\sqrt{105}}{7},-\frac{3\sqrt{91}}{14}\right)$ d. None of these

26 A tangent to the ellipse $4x^2+9y^2=36$ is cut by the tangent at the extremities of the major axis at T and T'. The circle on TT' as diameter passes through the point

a. $(-\sqrt{5},0)$ b. $(\sqrt{5},1)$
c. $(0,0)$ d. $(3,2)$

27 The area of quadrilateral formed by the tangents at the end-points of latusrectum to the ellipse $\frac{x^2}{9}+\frac{y^2}{5}=1$ is

a. $\frac{27}{4}$ sq units b. 9 sq units
c. $\frac{27}{2}$ sq units d. 27 sq units

28 If $\alpha-\beta=$ constant, then the locus of points of intersection of tangents at $P(a\cos\alpha, b\sin\alpha)$ and $Q(a\cos\beta, b\sin\beta)$ to the ellipse $\frac{x^2}{a^2}+\frac{y^2}{b^2}=1$ is

a. a circle b. a straight line
c. an ellipse d. None of these

Working Space

29 Let S and S' be two foci of ellipse $\frac{x^2}{a^2}+\frac{y^2}{b^2}=1$. If a circle described on SS' as diameter intersects the ellipse in real and distinct points, then the eccentricity e of ellipse satisfies

a. $e=\frac{1}{\sqrt{2}}$
b. $e\in\left(\frac{1}{\sqrt{2}},1\right)$
c. $e\in\left(0,\frac{1}{\sqrt{2}}\right)$
d. None of these

30 The locus of foot of perpendicular from the foci upon any tangent to the ellipse $\frac{x^2}{a^2}+\frac{y^2}{b^2}=1$ is

a. $x^2+y^2=b^2$
b. $x^2+y^2=a^2$
c. $x^2+y^2=a^2+b^2$
d. None of these

31 The locus of point of intersection of tangents to the ellipse $\frac{x^2}{a^2}+\frac{y^2}{b^2}=1$, which make complementary angles with X-axis, is

a. $x^2+y^2=a^2+b^2$
b. $x^2+y^2=a^2-b^2$
c. $x^2-y^2=a^2+b^2$
d. $x^2-y^2=a^2-b^2$

32 The locus of mid-points of the chords $2x+3y+\lambda=0$ of the ellipse $x^2+4y^2=1$, is (λ being parameter)

a. $8x-3y=0$
b. $8x+3y=0$
c. $3x-8y=0$
d. None of these

33 The locus of point of intersection of perpendicular tangents to $\frac{x^2}{a^2}+\frac{y^2}{b^2}=1$ and $\frac{x^2}{a+\lambda}+\frac{y^2}{b+\lambda}=1$ is

a. $x^2+y^2=a^2+\lambda$
b. $x^2+y^2=b^2+\lambda$
c. $x^2+y^2=a^2+b^2+\lambda$
d. $x^2+y^2=a^2+b^2$

34 $S(3, 4)$ and $S'(9, 12)$ are two focii of an ellipse.If the foot of perpendicular from S on a tangent to the ellipse has the coordinates $(1, -4)$, then the eccentricity of ellipse is

a. $\frac{3}{13}$
b. $\frac{4}{13}$
c. $\frac{5}{13}$
d. None of these

35 If the tangents are drawn to the ellipse $x^2+2y^2=2$, then the locus of mid-point of the intercept made by the tangents between the coordinate axes is

a. $\frac{1}{2x^2}+\frac{1}{4y^2}=1$
b. $\frac{1}{4x^2}+\frac{1}{2y^2}=1$
c. $\frac{x^2}{2}+\frac{y^2}{4}=1$
d. $\frac{x^2}{4}+\frac{y^2}{2}=1$

36 If the normal at one end of the latusrectum of the ellipse $\frac{x^2}{a^2}+\frac{y^2}{b^2}=1$ passes through the one end of the minor axis, then

a. $e^4-e^2+1=0$
b. $e^2-e+1=0$
c. $e^4+e^2-1=0$
d. None of these

37 If the normal at any point P on the ellipse $\frac{x^2}{a^2}+\frac{y^2}{b^2}=1$, meets the axes in G and g respectively, then $PG : Pg$ is equal to

a. $a : b$
b. $a^2 : b^2$
c. $b^2 : a^2$
d. $b : a$

Working Space

38 The locus of mid-points of the portion of the tangents to the ellipse intercepted between the axes is

a. $a^2y^2 + b^2x^2 = 4x^2y^2$
b. $a^2x^2 + b^2y^2 = 4x^2y^2$
c. $x^2 + y^2 = a^2$
d. $x^2 + y^2 = b^2$

39 The locus of mid-points of a focal chord of the ellipse $\frac{x^2}{a^2} + \frac{y^2}{b^2} = 1$ is

a. $\frac{x^2}{a^2} + \frac{y^2}{b^2} = \frac{ex}{a}$
b. $\frac{x^2}{a^2} - \frac{y^2}{b^2} = \frac{ex}{a}$
c. $x^2 + y^2 = a^2 + b^2$
d. None of these

40 If $y = m_1x$ and $y = m_2x$ are conjugate diameters of an ellipse $\frac{x^2}{a^2} + \frac{y^2}{b^2} = 1$, then

a. $m_1m_2 = \frac{b^2}{a^2}$
b. $m_1m_2 = -\frac{b^2}{a^2}$
c. $\frac{m_1}{m_2} = -\frac{b^2}{a^2}$
d. None of these

Statement Type

Direction (Q. Nos. 41-45) *This section is based on Statement I and Statement II. Select the correct answer from the codes given below.*

a. Both Statement I and Statement II are correct and Statement II is the correct explanation of Statement I
b. Both Statement I and Statement II are correct but Statement II is not the correct explanation of Statement I
c. Statement I is correct but Statement II is incorrect
d. Statement II is correct but Statement I is incorrect

41 **Statement I** Tangent drawn at a point $P\left(\frac{4\sqrt{5}}{3}, 2\right)$ on the ellipse $9x^2 + 16y^2 = 144$ intersects a straight line $x = \frac{16}{\sqrt{7}}$ at M, then PM subtends a right angle at $(-\sqrt{7}, 0)$.

Statement II The portion of tangent to an ellipse between the point of contact and the directrix subtend a right angle at the corresponding focus.

42 **Statement I** Feet of perpendicular drawn from foci of an ellipse $4x^2 + y^2 = 16$ on the line $2\sqrt{3}x + y = 8$ lie on circle $x^2 + y^2 = 16$.

Statement II If perpendicular are drawn from foci of an ellipse to its any tangent, then foot of perpendiculars lie on directors circle of the ellipse.

43 **Statement I** Any chord of the ellipse $x^2 + y^2 + xy = 1$ through (0, 0) is bisected at (0, 0).

Statement II The centre of an ellipse is a point through which every chord is bisected.

44 **Statement I** If $P\left(\frac{3\sqrt{3}}{2}, 1\right)$ is a point on the ellipse $4x^2 + 9y^2 = 36$. Circle drawn AP as diameter touches another circle $x^2 + y^2 = 9$, where $A(-\sqrt{5}, 0)$.

Statement II Circle drawn with focal radius as diameter touches the auxiliary circle.

45 Tangents are drawn from the point $P(-\sqrt{3}, \sqrt{2})$ to an ellipse $4x^2 + y^2 = 4$.

Statement I The tangents are mutually perpendicular.

Statement II The locus of the points from which mutually perpendicular tangents can be drawn to given ellipse is $x^2 + y^2 = 5$.

SET - 2

Revisal Problems for JEE Advanced

Only One Option Correct Type

Working Space

Direction (Q. Nos. 1-15) *This section contains 15 multiple choice questions. Each question has four choices (a), (b), (c) and (d), out of which ONLY ONE option is correct.*

1 From the point P, perpendicular tangents PQ and PR are drawn to ellipse $x^2 + 4y^2 = 4$. Locus of circumcentre of ΔPQR is

a. $x^2 + y^2 = \frac{16}{5}(x^2 + 4y^2)^2$

b. $x^2 + y^2 = \frac{5}{16}(x^2 + 4y^2)^2$

c. $x^2 + 4y^2 = \frac{16}{5}(x^2 + y^2)^2$

d. $x^2 + 4y^2 = \frac{5}{16}(x^2 + y^2)^2$

2 The normal to the ellipse $4x^2 + 5y^2 = 20$ at the point P touches $y^2 = 4x$. The eccentricity angle of point P is

a. $\frac{\pi}{2} + \cos^{-1}\left(\frac{1}{\sqrt{5}}\right)$ b. $\frac{\pi}{2} + \tan^{-1}\left(\frac{1}{\sqrt{5}}\right)$

c. $\pi - \tan^{-1}\left(\frac{1}{\sqrt{5}}\right)$ d. $\pi - \cos^{-1}\left(\frac{1}{\sqrt{5}}\right)$

3 An ellipse with major axis and minor axis of length $10\sqrt{3}$ and 10 respectively, slides along the coordinate axes and always remains confined in the first quadrant. The locus of centre of the ellipse will be the arc of a circle. The length of this arc will be equal to

a. 10π b. 5π

c. $\frac{5\pi}{4}$ d. $\frac{5\pi}{3}$

4 The area of the triangle inscribed in an ellipse bears the ratio $\sqrt{5}:3$ to the area of triangle formed by joining points on the auxilliary circle corresponding to the vertices of the first triangle, then the eccentricity of the ellipse is

a. 2/3 b. 3/4

c. $\sqrt{5}/3$ d. 3/5

5 Tangents are drawn from the points on the line $x - y - 5 = 0$ to $x^2 + 4y^2 = 4$, then all the chords of contact pass through a fixed point, whose coordinates are

a. $\left(\frac{4}{5}, -\frac{1}{5}\right)$ b. $\left(\frac{4}{5}, \frac{1}{5}\right)$

c. $\left(-\frac{4}{5}, -\frac{1}{5}\right)$ d. $\left(-\frac{4}{5}, \frac{1}{5}\right)$

6 For an ellipse $\frac{x^2}{9} + \frac{y^2}{4} = 1$ with vertices A and A', tangent drawn at the point P in the first quadrant meets the Y-axis in Q and the chord $A'P$ meets the Y-axis in M. If O is the origin, then $OQ^2 - MQ^2$ equals

a. 9 b. 13

c. 4 d. 5

7 Point O is the centre of ellipse with major axis AB and minor axis CD. Point F is one focus of the ellipse. If $OF = 6$ and the diameter of inscribed circle of ΔOCF is 2, then the product $(AB)(CD)$ is equal to

a. 65 b. 52

c. 78 d. None of these

Working Space

8 If $\frac{x^2}{f(4a)} + \frac{y^2}{f(4a^2 - 5)} = 1$ represents an ellipse with major axis as Y-axis and f is a decreasing function, then

a. $a \in (-\infty, 1)$
b. $a \in (5, \infty)$
c. $a \in (1, 4)$
d. $a \in (-1, 5)$

9 The equation of largest circle with centre (1, 0) that can be inscribed in the ellipse $x^2 + 4y^2 = 16$, is

a. $2x^2 + 2y^2 - 4x + 7 = 0$
b. $x^2 + y^2 - 2x + 5 = 0$
c. $3x^2 + 3y^2 - 6x - 8 = 0$
d. None of these

10 Let P be a point on the ellipse $\frac{x^2}{a^2} + \frac{y^2}{b^2} = 1 (a > b)$ in the first quadrant whose foci are S_1 and S_2. Then, the least possible value of circumradius of ΔPS_1S_2 is

a. ae
b. be
c. $\frac{ae}{b}$
d. $\frac{ae^2}{b}$

11 Let P be a point on ellipse $9x^2 + 16y^2 = 144$ whose foci are S, S'. If A, B are feet of perpendiculars from S, S' respectively to the tangent at P, and if normal at P passes through S meet the ellipse at Q and O is origin, then $\frac{SP + SQ}{SP \cdot SQ} + \frac{OA \cdot OB}{SA \cdot S'B}$ is

a. 9/24
b. 25/9
c. 24/9
d. 9/16

12 For ellipse $\frac{x^2}{a^2} + \frac{y^2}{b^2} = 1$, four normals cannot be drawn through a point $(h, 0)$ unless

a. $|h| > ae$
b. $|h| \le ae^2$
c. $|h| \le ae$
d. None of these

13 A circle C intersecting the ellipse $\frac{x^2}{16} + \frac{y^2}{9} = 1$. Such that common chord is of maximum length and intersecting the circle $x^2 + y^2 - 6x - 10y + 9 = 0$ orthogonally, then the number of common tangents of circle C and circle $x^2 + y^2 - 10x - 11 = 0$ is

a. 0
b. 2
c. 3
d. 4

14 If semi-major axis of the ellipse $\frac{x^2}{a^2} + \frac{y^2}{b^2} = 1 (a > b)$ is 20 units and $\sum_{i=1}^{20} (S_1P_i)(S_2Q_i) = 5120$, where P_i and Q_i denotes feet of perpendicular drawn from foci S_1 and S_2 on a tangent $T_i = 0$, then the eccentricity of ellipse is

a. $\frac{1}{5}$
b. $\frac{\sqrt{89}}{10}$
c. $\frac{3}{5}$
d. $\frac{\sqrt{7}}{8}$

15 If the ellipse $\frac{x^2}{a^2 - 3} + \frac{y^2}{a + 4} = 1$ is inscribed in a square of side length $a\sqrt{2}$, then a is equal to

a. 4
b. 2
c. 1
d. None of these

One or More than One Options Correct Type

Direction (Q. Nos. 16-20) *This section contains 5 multiple choice questions. Each question has four choices (a), (b), (c) and (d), out of which ONE or MORE THAN ONE are correct.*

16 The tangent at any point P on a standard ellipse with foci S and S' meets the tangents at the vertices A and A' in the points V and V', then

a. $l(AV) \cdot l(A'V') = b$
b. $l(AV) \cdot l(A'V') = b^2$
c. $\angle V'SV = 90°$
d. $V'S'VS$ is cyclic quadrilateral

Working Space

17 Extremities of latusrectum of the ellipse $\frac{x^2}{a^2}+\frac{y^2}{b^2}=1\,(a>b)$ having a major axis $2a$ lies on

a. $x^2 = a(a-y)$
b. $x^2 = a(a+y)$
c. $y^2 = a(a+x)$
d. $y^2 = a(a-x)$

18 A point P moves on XY-plane such that $PS + PS' = 4$, where $S(k, 0)$ and $S'(-k, 0)$, then the locus of P is

a. ellipse, if $k \in (-2, 2)$
b. pair of coincident lines, if $k = \pm 2$
c. empty, if $k \in (-\infty, -2) \cup (2, \infty)$
d. None of the above

19 If P is a point of the ellipse $\frac{x^2}{a^2}+\frac{y^2}{b^2}=1$, whose foci are S and S'. Let $\angle PSS' = \alpha$ and $\angle PS'S = \beta$, then

a. $PS + PS' = 2a$, if $a > b$
b. $PS + PS' = 2b$, if $a < b$
c. $\tan\frac{\alpha}{2}\cdot\tan\frac{\beta}{2}=\frac{1-e}{1+e}$
d. $\tan\frac{\alpha}{2}\cdot\tan\frac{\beta}{2}=\frac{\sqrt{a^2-b^2}}{b^2}\left(a-\sqrt{a^2-b^2}\right)$, when $a > b$

20 For the ellipse $x^2 + 2y^2 = 1$, the set of rational points on it can be of the form

a. $\left(\frac{p^2+q^2}{2p^2+q^2}, \frac{2pq}{p^2+q^2}\right)$
b. $\left(\frac{2p^2-q^2}{2p^2+q^2}, \frac{-2pq}{p^2+q^2}\right)$
c. $\left(\frac{q^2-2p^2}{2p^2-q^2}, \frac{2pq}{2p^2-q^2}\right)$
d. $\left(\frac{2p^2-q^2}{2p^2+q^2}, \frac{2pq}{2p^2+q^2}\right)$

Matching List Type

Direction (Q. Nos. 21 and 22) ***Choices for the correct combination of elements from Column I and Column II are given as options (a), (b), (c) and (d), out of which one is correct.***

21 Consider an ellipse $9x^2 + 16y^2 = 144$. Match the Column I with Column II and mark the correct option from the codes given below.

	Column I		Column II
i.	The product of perpendicular from foci to the tangent at the point whose eccentric angle is 75° on the ellipse.	p.	9
ii.	The sum of squares of the perpendicular distances from two points on the circle with line joining foci as diameter to the tangent of curve at the point whose eccentric angle is $\pi/4$.	q.	16
iii.	The tangent at the point $P(\pi/6)$ meet the tangents at vertices A, A' at L and M, then $(AL)\cdot(A'M)$ is	r.	28
iv.	The tangent at point $Q(\pi/3)$ meet the tangents at vertices A, A' and T, T'. Then, the square of the intercepts made by circle on TT' as diameter on major axis is	s.	32

Codes

	i	ii	iii	iv
a.	p	s	p	r
b.	q	p	q	r
c.	r	q	r	q
d.	s	r	s	p

22 Match the Column I with Column II and mark the correct option from the codes given below.

Working Space

	Column I		Column II
i.	The equation of an ellipse is given by $5x^2 + 5y^2 - 6xy - 8 = 0$. If r_1, r_2 are distances of points on the ellipse which are at maximum and minimum distance from origin, then $r_1 + r_2$ is	p.	15
ii.	If 1 and 8 are the length of perpendicular's chords $y^2 = 4ax$ of the parabola drawn through the vertex, then a^2 is	q.	1/5
iii.	From a point A, common tangents are drawn to the circle $x^2 + y^2 = 8$ and the parabola $y^2 = 8x$. The area of quadrilateral formed by the common tangents and the chords contacts of the circle and parabola is	r.	3
iv	PQ is a variable focal chord of the parabola $y^2 = 36x$ whose vertex is A. Locus of centroid of ΔAPQ is parabola, then its length of latusrectum is	s.	12

Codes

	i	ii	iii	iv		i	ii	iii	iv
a.	r	q	p	s	b.	p	r	s	q
c.	q	s	r	p	d.	s	p	q	r

Comprehension Type

Direction (Q. Nos. 23-28) This section contains 2 paragraphs, each describing theory, experiments, data, etc. Six questions related to the paragraphs have been given. Each question has only one correct answer among the four given options (a), (b), (c) and (d).

Passage I

If foci of an ellipse is $(-1, 2)$ and $(-2, 3)$ and its tangent at a point A is $2x + 3y + 9 = 0$.

23 The length of minor axis of the ellipse will be

a. $\sqrt{27/2}$ units b. $\sqrt{14}$ units c. $2\sqrt{14}$ units d. None of these

24 The area of ellipse will be

a. $3\sqrt{21}\pi$ sq units b. $\sqrt{203}\pi$ sq units c. $21\sqrt{3}\pi$ sq units d. None of these

25 The coordinates of A will be

a. $(-32/9, -17/9)$ b. $(-32/9, -17/27)$ c. $(0, -3)$ d. None of these

Passage II

Let an ellipse having major axis and minor axis parallel to X-axis and Y-axis respectively. Its two foci S and S' are (2, 1), (4, 1) and a line $x + y = 9$ is a tangent to this ellipse at point P.

26 Eccentricity of the ellipse is equal to

a. $1/\sqrt{12}$ b. $1/\sqrt{13}$ c. $1/2$ d. None of these

27 The length of major axis is equal to

a. $\sqrt{13}$ b. $2\sqrt{11}$ c. $\sqrt{52}$ d. $2\sqrt{12}$

28 The latusrectum of ellipse is

a. $12/13$ b. $12/\sqrt{13}$ c. $24/\sqrt{13}$ d. $25/\sqrt{13}$

One Integer Value Correct Type

Direction (Q. Nos. 29 and 30) This section contains 2 questions. When worked out will result in an integer from 0 to 9 (both inclusive).

29 If the curve $x^2 + 3y^2 = 9$ subtends an obtuse angle at the point $(2\alpha, \alpha)$, then a possible integral value of α^2 is

30 From a point perpendicular tangents are drawn to the ellipse $x^2 + 2y^2 = 2$. The chord of contact touches a circle which is concentric with the given ellipse. Then, the ratio of maximum and minimum area of circle is

JEE Main & AIEEE Archive

(Compilation of Last 13 Years' Questions)

1 The area (in sq units) of the quadrilateral formed by the tangents at the end points of the latusrecta to the ellipse $\frac{x^2}{9}+\frac{y^2}{5}=1$ is **(JEE Main 2015)**

a. $\frac{27}{4}$ b. 18

c. $\frac{27}{2}$ d. 27

2 The locus of the foot of perpendicular drawn from the centre of the ellipse $x^2+3y^2=6$ on any tangent to it, is **(JEE Main 2014)**

a. $(x^2-y^2)^2=6x^2+2y^2$ b. $(x^2-y^2)^2=6x^2-2y^2$

c. $(x^2+y^2)^2=6x^2+2y^2$ d. $(x^2+y^2)^2=6x^2-2y^2$

3 The equation of the circle passing through the foci of the ellipse $\frac{x^2}{16}+\frac{y^2}{9}=1$ and having centre at (0, 3) is **(JEE Main 2013)**

a. $x^2+y^2-6y-7=0$ b. $x^2+y^2-6y+7=0$

c. $x^2+y^2-6y-5=0$ d. $x^2+y^2-6y+5=0$

4 An ellipse is drawn by taking a diameter of the circle $(x-1)^2+y^2=1$ as its semi-minor axis and a diameter of the circle $x^2+(y-2)^2=4$ is semi-major axis. If the centre of the ellipse is at the origin and its axes are the coordinate axes, then the equation of the ellipse is **(AIEEE 2012)**

a. $4x^2+y^2=4$ b. $x^2+4y^2=8$

c. $4x^2+y^2=8$ d. $x^2+4y^2=16$

5 Equation of the ellipse whose axes are the axes of coordinates and which passes through the point (–3, 1) and has eccentricity $\sqrt{\frac{2}{5}}$ is **(AIEEE 2011)**

a. $5x^2+3y^2-48=0$ b. $3x^2+5y^2-15=0$

c. $5x^2+3y^2-32=0$ d. $3x^2+5y^2-32=0$

6 The shortest distance between the line $y-x=1$ and curve $x=y^2$ is **(AIEEE 2011, 2009)**

a. $\frac{3\sqrt{2}}{8}$ b. $\frac{8}{3\sqrt{2}}$ c. $\frac{4}{\sqrt{3}}$ d. $\frac{\sqrt{3}}{4}$

7 The ellipse $x^2+4y^2=4$ is inscribed in a rectangle aligned with the coordinate axes, which in turn is inscribed in another ellipse that passes through the point (4, 0). Then, the equation of the ellipse is **(AIEEE 2009)**

a. $x^2+12y^2=16$ b. $4x^2+48y^2=48$

c. $4x^2+64y^2=48$ d. $x^2+16y^2=16$

8 A focus of an ellipse is at the origin. The directrix is the line $x=4$ and the eccentricity is $\frac{1}{2}$, then the length of semi-major axis is **(AIEEE 2008)**

a. $\frac{5}{3}$ b. $\frac{8}{3}$ c. $\frac{2}{3}$ d. $\frac{4}{3}$

9 In an ellipse, the distances between its foci is 6 and minor axis is 8. Then, its eccentricity is

a. $\frac{1}{2}$ b. $\frac{4}{5}$

c. $\frac{1}{\sqrt{5}}$ d. $\frac{3}{5}$

10 Area (in sq units) of the greatest rectangle that can be inscribed in the ellipse $\frac{x^2}{a^2}+\frac{y^2}{b^2}=1$ is **(AIEEE 2005)**

a. $\frac{a}{b}$ b. $\sqrt{ab}$

c. ab 33d. $2ab$

11 An ellipse has OB as semi-minor axis, F and F' its foci and the $\angle FBF$ is a right angle. Then, the eccentricity of the ellipse is **(AIEEE 2005)**

a. $\frac{1}{\sqrt{3}}$ b. $\frac{1}{4}$

c. $\frac{1}{2}$ d. $\frac{1}{\sqrt{2}}$

12 The eccentricity of an ellipse with its centre at the origin, is $\frac{1}{2}$. If one of the directrices is $x=4$, then the equation of the ellipse is **(AIEEE 2004)**

a. $3x^2+4y^2=1$ b. $3x^2+4y^2=12$

c. $4x^2+3y^2=12$ d. $4x^2+3y^2=1$

13 The radius of the circle passing through the foci of the ellipse $\frac{x^2}{16}+\frac{y^2}{9}=1$ and having its centre at (0, 3), is **(AIEEE 2002)**

a. 4 units b. 3 units

c. $\sqrt{12}$ units d. $\frac{7}{2}$ units

14 The equation of the ellipse whose foci are $(\pm 2, 0)$ and eccentricity is 1/2, is **(AIEEE 2002)**

a. $\frac{x^2}{12}+\frac{y^2}{16}=1$

b. $\frac{x^2}{16}+\frac{y^2}{12}=1$

c. $\frac{x^2}{16}+\frac{y^2}{8}=1$

d. None of the above

JEE Advanced & IIT JEE Archive

(Compilation of Last 10 Years' Questions)

1 Suppose that the foci of the ellipse $\frac{x^2}{9}+\frac{y^2}{5}=1$ are $(f_1, 0)$ and $(f_2, 0)$, where $f_1 > 0$ and $f_2 < 0$. Let P_1 and P_2 be two parabolas with a common vertex at (0, 0) with foci at $(f_1, 0)$ and $(2f_2, 0)$, respectively. Let T_1 be a tangent to P_1 which passes through $(2f_2, 0)$ and T_2 be a tangent to P_2 which passes through $(f_1, 0)$. If m_1 is the slope of T_1 and m_2 is the slope of T_2, then the value of $\left(\frac{1}{m_1^2}+m_2^2\right)$ is

(2015, Integer Type)

2 Let E_1 and E_2 be two ellipses whose centres are at the origin. The major axes of E_1 and E_2 lie along the X-axis and Y-axis, respectively. Let S be the circle $x^2+(y-1)^2=2$. The straight line $x+y=3$ touches the curves S, E_1 and E_2 at P, Q and R, respectively.

Suppose that $PQ = PR = \frac{2\sqrt{2}}{3}$. If e_1 and e_2 are the eccentricities of E_1 and E_2 respectively, then the correct expression(s) is/are

(2015, One or More than One Options Correct Type)

a. $e_1^2+e_2^2=\frac{43}{40}$ b. $e_1 e_2=\frac{\sqrt{7}}{2\sqrt{10}}$

c. $|e_1^2-e_2^2|=\frac{5}{8}$ d. $e_1 e_2=\frac{\sqrt{3}}{4}$

3 The equation of the circle passing through the foci of the ellipse $\frac{x^2}{16}+\frac{y^2}{9}=1$ and having centre at (0, 3) is **(2011, Only One Option Correct Type)**

a. $x^2+y^2-6y-7=0$ b. $x^2+y^2-6y+7=0$

c. $x^2+y^2-6y-5=0$ d. $x^2+y^2-6y+5=0$

4 The locus of the foot of perpendicular drawn from the centre of the ellipse $x^2+3y^2=6$ on any tangent to it is **(2011, Only One Option Correct Type)**

a. $(x^2-y^2)^2=6x^2+2y^2$ b. $(x^2-y^2)^2=6x^2-2y^2$

c. $(x^2+y^2)^2=6x^2+2y^2$ d. $(x^2+y^2)^2=6x^2-2y^2$

5 The ellipse $E_1: \frac{x^2}{9}+\frac{y^2}{4}=1$ is inscribed in a rectangle R whose sides are parallel to the coordinate axes. Another ellipse E_2 passing through the point (0, 4) circumscribes the rectangle R. The eccentricity of the ellipse E_2 is

(2010, Only One Option Correct Type)

a. $\frac{\sqrt{2}}{2}$ b. $\frac{\sqrt{3}}{2}$

c. $\frac{1}{2}$ d. $\frac{3}{4}$

6 The line passing through the extremity A of the major axis and extremity B of the minor axis of the ellipse $x^2+9y^2=9$ meets its auxiliary circle at the point M. Then, the area (in sq units) of the triangle with vertices at A, M and the origin O is

(2009, Only One Option Correct Type)

a. $\frac{31}{10}$ b. $\frac{29}{10}$ c. $\frac{21}{10}$ d. $\frac{27}{10}$

7 The normal at a point P on the ellipse $x^2+4y^2=16$ meets the X-axis at Q. If M is the mid-point of the line segment PQ, then the locus of M intersects the latusrectum of the given ellipse at the points **(2009, Only One Option Correct Type)**

a. $\left(\pm\frac{3\sqrt{5}}{2}, \pm\frac{2}{7}\right)$ b. $\left(\pm\frac{3\sqrt{5}}{2}, \pm\frac{\sqrt{19}}{4}\right)$

c. $\left(\pm 2\sqrt{3}, \pm\frac{1}{7}\right)$ d. $\left(\pm 2\sqrt{3}, \pm\frac{4\sqrt{3}}{7}\right)$

Passage for Q. Nos. (8-10)

Tangents are drawn from the point $P(3, 4)$ to the ellipse $\frac{x^2}{9}+\frac{y^2}{4}=1$ touching the ellipse at points A and B. **(2010, Comprehension Type)**

8 The coordinates of A and B are

a. (3, 0) and (0, 2)

b. $\left(-\frac{8}{5}, \frac{2\sqrt{161}}{15}\right)$ and $\left(-\frac{9}{5}, \frac{8}{5}\right)$

c. $\left(-\frac{8}{5}, \frac{2\sqrt{161}}{15}\right)$ and (0, 2)

d. (3, 0) and $\left(-\frac{9}{5}, \frac{8}{5}\right)$

9 The orthocentre of ΔPAB is

a. $\left(5, \frac{8}{7}\right)$ b. $\left(\frac{7}{5}, \frac{25}{8}\right)$

c. $\left(\frac{11}{5}, \frac{8}{5}\right)$ d. $\left(\frac{8}{25}, \frac{7}{5}\right)$

10 The equation of locus of the point whose distance from the point P and the line AB are equal, is

a. $9x^2+y^2-6xy-54x-62y+241=0$

b. $x^2+9y^2+6xy-54x+62y-241=0$

c. $9x^2+9y^2-6xy-54x-62y-241=0$

d. $x^2+y^2-2xy+27x+31y-120=0$

11 Find the equation of the common tangent in 1st quadrant to the circle $x^2+y^2=16$ and the ellipse $\frac{x^2}{25}+\frac{y^2}{4}=1$. Also, find the length of the intercept of the tangent between the coordinate axes.

(2005, Subjective Type)

3. Hyperbola

DPP
DAILY PRACTICE PROBLEMS

DPP-1 Equation of Hyperbola

Only One Option Correct Type

Working Space

Direction (Q. Nos. 1-22) *This section contains 22 multiple choice questions. Each question has four choices (a), (b), (c) and (d), out of which ONLY ONE is correct.*

1 The foci of hyperbola $4x^2 - 9y^2 - 36 = 0$ are

a. $(\pm\sqrt{11}, 0)$
b. $(\pm\sqrt{12}, 0)$
c. $(\pm\sqrt{13}, 0)$
d. $(0, \pm\sqrt{2})$

2 The foci of hyperbola $\frac{x^2}{16} - \frac{(y-2)^2}{9} = 1$ are

a. (5, 2), (−5, 2)
b. (5, 2), (5, − 2)
c. (5, 2), (−5, − 2)
d. None of these

3 If length of the transverse and conjugate axes of a hyperbola are 8 and 6 respectively, then the difference of focal distances of any point of the hyperbola will be

a. 8
b. 6
c. 14
d. 2

4 The equation of hyperbola whose directrix is $2x + y = 1$, focus (1, 1) and $e = \sqrt{3}$, is

a. $7x^2 + 12xy - 2y^2 + 4y - 7 = 0$
b. $11x^2 + 12xy + 2y^2 - 10x - 4y + 1 = 0$
c. $11x^2 + 12xy + 2y^2 - 14x - 14y + 1 = 0$
d. None of the above

5 The equation of hyperbola whose foci are the foci of ellipse $\frac{x^2}{25} + \frac{y^2}{9} = 1$ and eccentricity is 2, is

a. $\frac{x^2}{4} + \frac{y^2}{12} = 1$
b. $\frac{x^2}{4} - \frac{y^2}{12} = 1$
c. $\frac{x^2}{12} + \frac{y^2}{4} = 1$
d. $\frac{x^2}{12} - \frac{y^2}{4} = 1$

6 The eccentricity of an ellipse is $\frac{2}{3}$, latusrectum is 5 and centre (0, 0). The equation of ellipse is

a. $\frac{x^2}{81} + \frac{y^2}{45} = 1$
b. $\frac{4x^2}{81} + \frac{4y^2}{45} = 1$
c. $\frac{x^2}{9} + \frac{y^2}{5} = 1$
d. None of these

Working Space

7 The vertices of a hyperbola are at (0, 0) and (10, 0) and one of its foci is at (18, 0). The equation of hyperbola is

a. $\frac{x^2}{25} - \frac{y^2}{144} = 1$ b. $\frac{(x-5)^2}{25} - \frac{y^2}{144} = 1$

c. $\frac{x^2}{5} - \frac{(y-5)^2}{144} = 1$ d. $\frac{(x-5)^2}{25} - \frac{(y-5)^2}{144} = 1$

8 The equation of hyperbola with eccentricity $\frac{3}{2}$ and foci at $(\pm 2, 0)$ is

a. $\frac{x^2}{4} - \frac{y^2}{5} = \frac{4}{9}$ b. $x^2 - y^2 = 4$

c. $\frac{x^2}{4} - \frac{y^2}{9} = 1$ d. None of these

9 If the centre, vertex and focus of a hyperbola are (0, 0), (4, 0) and (6, 0) respectively, then the equation of hyperbola is

a. $4x^2 - 5y^2 = 8$ b. $4x^2 - 5y^2 = 80$

c. $5x^2 - 4y^2 = 80$ d. $5x^2 - 4y^2 = 8$

10 If the latusrectum of a hyperbola is 8 and eccentricity is $\frac{3}{\sqrt{5}}$, then the equation of a hyperbola is

a. $4x^2 - 5y^2 = 100$ b. $5x^2 - 4y^2 = 100$

c. $4x^2 + 5y^2 = 100$ d. $5x^2 + 4y^2 = 100$

11 The equation $\frac{x^2}{12-\lambda} + \frac{y^2}{8-\lambda} = 1$ represents

a. a hyperbola, if $\lambda < 8$
b. an ellipse, if $\lambda > 0$
c. a hyperbola, if $8 < \lambda < 12$
d. None of the above

12 The equation $\frac{x^2}{1-k} - \frac{y^2}{1+k} = 1, k > 1$ represents a/an

a. circle b. ellipse

c. hyperbola d. None of these

13 The equation of conic with focus (1, – 1) directrix along $x - y + 1 = 0$ and with eccentricity $\sqrt{2}$ is

a. $x^2 - y^2 = 1$ b. $xy = 1$

c. $2xy - 4x + 4y + 1 = 0$ d. $2xy + 4x - 4y - 1 = 0$

14 The eccentricity of $-\frac{x^2}{a^2} + \frac{y^2}{b^2} = 1$ is

a. $e = \sqrt{\frac{a^2 + b^2}{a^2}}$ b. $e = \sqrt{\frac{a^2 - b^2}{a^2}}$

c. $e = \sqrt{\frac{b^2 - a^2}{a^2}}$ d. $e = \sqrt{\frac{a^2 + b^2}{b^2}}$

15 Let e and e' be the eccentricity of two conics S and S' such that $e^2 + e'^2 = 3$, then both S and S' are

a. ellipse b. parabola

c. hyperbola d. None of these

Working Space

16 The equation of hyperbola whose foci are (6, 5), (−4, 5) and eccentricity is $\frac{5}{4}$, is

a. $\frac{(x-1)^2}{16} - \frac{(y-5)^2}{9} = 1$

b. $\frac{x^2}{16} - \frac{y^2}{9} = 1$

c. $\frac{(x-1)^2}{16} - \frac{(y-5)^2}{9} = -1$

d. None of these

17 The eccentricity of hyperbola whose latusrectum is 8 and conjugate axis is equal to half the distance between the foci, is

a. 4/5

b. $4/\sqrt{3}$

c. $2/\sqrt{3}$

d. None of these

18 The locus of point of intersection of the straight lines $\frac{x}{a} + \frac{y}{b} = \lambda$ and $\frac{x}{a} - \frac{y}{b} = \frac{1}{\lambda}$, (where, λ is a variable) is

a. a circle

b. a parabola

c. an ellipse

d. a hyperbola

19 The equation $9x^2 - 16y^2 - 18x + 32y - 151 = 0$ represents a hyperbola

a. the length of the transverse axes is 4

b. the length of latusrectum is 9

c. the equation of directrix is $x = \frac{21}{5}$ and $x = -\frac{11}{5}$

d. None of the above

20 The points of intersection of curves whose parametric equations are $x = t^2 + 1$, $y = 2t$ and $x = 2s$, $y = \frac{2}{s}$ is

a. (1, −3)

b. (2, 2)

c. (−2, 4)

d. (1, 2)

21 In the hyperbola $\frac{x^2}{\cos^2\alpha} - \frac{y^2}{\sin^2\alpha} = 1$, corresponding to any change α, which of the following is not changed?

a. Abscissae of vertices

b. Abscissae of foci

c. Eccentricity

d. Directrices

22 The locus of point of intersection of the lines $\sqrt{3}x - y - 4\sqrt{3}k = 0$ and $\sqrt{3}xk + ky - 4\sqrt{3} = 0$ for different values of k, is a/an

a. ellipse

b. parabola

c. circle

d. hyperbola

Comprehension Type

Direction (Q. Nos. 23-25) *This section contains a paragraph, describing theory, experiments, data, etc. Three questions related to the paragraph have been given. Each question has only one correct answer among the four given options (a), (b), (c) and (d).*

Passage

The graph of conic $x^2 - (y-1)^2 = 1$ has one tangent line with positive slope that passes through the origin, the point of tangency being (a, b).

23 The value of $\sin^{-1}\left(\frac{a}{b}\right)$ is

a. $\frac{5\pi}{12}$

b. $\frac{\pi}{6}$

c. $\frac{\pi}{13}$

d. $\frac{\pi}{4}$

Working Space

24 The length of latusrectum of conic is

a. 1
b. $\sqrt{2}$
c. 2
d. None of these

25 The eccentricity of conic is

a. $\frac{4}{3}$
b. $\sqrt{3}$
c. 2
d. None of these

One Integer Value Correct Type

Direction *(Q. Nos. 26-30) This section contains 5 questions. When worked out will result in an integer from 0 to 9 (both inclusive).*

26 The area of a quadrilateral formed with the foci of hyperbola $\frac{x^2}{a^2} + \frac{y^2}{b^2} = 1$ and $\frac{x^2}{a^2} - \frac{y^2}{b^2} = -1$ is $\lambda(a^2 + b^2)$, then the value of λ is

27 The equation of one of the latusrectum of the hyperbola $(10x - 5)^2 + (10y - 2)^2 = 9(3x + 4y - 7)^2$ is $ax + by + c = 0$, then the value of $|b + 2c|$ is

28 If (5, 12) and (24, 7) are foci of a hyperbola passing through the origin, then the value of $\frac{12e}{\sqrt{386}}$ is (where, e is the eccentricity of hyperbola)

29 PQ is double ordinate of hyperbola $\frac{x^2}{a^2} - \frac{y^2}{b^2} = 1$ such that OPQ is an equilateral triangle, O being the centre of hyperbola, where the eccentricity of hyperbola e satisfy, $\sqrt{3}e > k$, then the value of k is

30 With one focus of the hyperbola $\frac{x^2}{9} - \frac{y^2}{16} = 1$ as the centre, a circle is drawn which is tangent to the hyperbola with no part of the circle being outside the hyperbola. The radius of circle is

DPP-2 Tangent and Its Properties

Only One Option Correct Type

Working Space

Direction *(Q. Nos. 1-18) This section contains 18 multiple choice questions. Each question has four choices (a), (b), (c) and (d), out of which ONLY ONE is correct.*

1 The line $y = x + 2$ touches the hyperbola $5x^2 - 9y^2 = 45$ at the point

a. $(0, 2)$
b. $(3, 1)$
c. $\left(-\frac{9}{2}, -\frac{5}{2}\right)$
d. None of these

2 The line $lx + my + n = 0$ is a tangent to the hyperbola $\frac{x^2}{a^2} - \frac{y^2}{b^2} = 1$, if

a. $a^2l^2 - b^2m^2 = n^2$
b. $a^2l^2 + m^2 = b^2n^2$
c. $a^2 + b^2 = n^2(l^2 + m^2)$
d. $a^2l^2 + b^2m^2 = n^2$

3 The slope of tangent to the hyperbola $2x^2 - 3y^2 = 6$ at $(3, 2)$ is

a. -1
b. 1
c. 0
d. 2

4 A tangent to a hyperbola $\frac{x^2}{a^2} - \frac{y^2}{b^2} = 1$ intercepts a length of unity from each of the coordinate axes, then the point (a, b) lies on

a. $x^2 - y^2 = 2$
b. $x^2 - y^2 = 1$
c. $x^2 - y^2 = -1$
d. None of these

5 If the straight line $x\cos\alpha + y\sin\alpha = p$ is a tangent to the hyperbola $\frac{x^2}{a^2} - \frac{y^2}{b^2} = 1$, then

a. $a^2\cos^2\alpha + b^2\sin^2\alpha = p^2$
b. $a^2\cos^2\alpha - b^2\sin^2\alpha = p^2$
c. $a^2\sin^2\alpha + b^2\cos^2\alpha = p^2$
d. $a^2\sin^2\alpha - b^2\sin^2\alpha = p^2$

6 If $\frac{x^2}{a^2} + \frac{y^2}{b^2} = 1 (a > b)$ and $x^2 - y^2 = c^2$ cut at right angles, then

a. $a^2 + b^2 = 2c^2$
b. $b^2 - a^2 = 2c^2$
c. $a^2 - b^2 = 2c^2$
d. $a^2b^2 = 2c^2$

7 A common tangent to $9x^2 - 16y^2 = 144$ and $x^2 + y^2 = 9$ is

a. $y = \frac{3}{\sqrt{7}}x + \frac{15}{\sqrt{7}}$
b. $y = 3\sqrt{\frac{2}{7}}x + \frac{15}{\sqrt{7}}$
c. $y = \sqrt{\frac{3}{7}}x + 15\sqrt{7}$
d. None of these

Working Space

8 The value of m for which $y = mx + 6$ is a tangent to hyperbola $\frac{x^2}{100} - \frac{y^2}{49} = 1$, is

a. $\sqrt{\frac{17}{20}}$
b. $\sqrt{\frac{20}{17}}$
c. $\sqrt{\frac{3}{20}}$
d. $\sqrt{\frac{13}{20}}$

9 The product of lengths of the perpendicular drawn from foci on any tangent to the hyperbola $\frac{x^2}{a^2} - \frac{y^2}{b^2} = 1$, is

a. a^2
b. b^2
c. a^2b^2
d. a^2/b^2

10 The locus of point of intersection of two lengths of the hyperbola $\frac{x^2}{a^2} - \frac{y^2}{b^2} = 1$, if the product of their slopes is c^2, will be

a. $y^2 - b^2 = c^2(x^2 + a^2)$
b. $y^2 + b^2 = c^2(x^2 - a^2)$
c. $y^2 + a^2 = c^2(x^2 - b^2)$
d. $y^2 - a^2 = c^2(x^2 + b^2)$

11 At which of the following points of hyperbola $x^2 - y^2 = 3$, tangents is parallel to the line $2x + y + 8 = 0$?

a. $(2, 1)$
b. $(2, -1)$
c. $(-2, -1)$
d. None of these

12 The equation of common tangent of the curve $y^2 = 8x$ and $xy = -1$ is

a. $3y = 9x + 2$
b. $y = x + 2$
c. $2y = x + 8$
d. $y = 2x + 1$

13 The locus of a point $P(a, b)$ moving under the condition that the line $y = ax + b$ is a tangent to the hyperbola $\frac{x^2}{a^2} - \frac{y^2}{b^2} = 1$, is

a. a hyperbola
b. a parabola
c. a circle
d. an ellipse

14 The equation of common tangent to the parabola $y^2 = 8x$ and hyperbola $3x^2 - y^2 = 3$ is

a. $2x \pm y + 1 = 0$
b. $x \pm y + 1 = 0$
c. $x \pm 2y + 1 = 0$
d. $x \pm y + 2 = 0$

15 The area (in sq units) of triangle formed by the lines $x - y = 0$, $x + y = 0$ and any tangent to the hyperbola $x^2 - y^2 = a^2$ is

a. a^2
b. $2a^2$
c. $3a^2$
d. $4a^2$

16 The equation of common tangents to the two hyperbola $\frac{x^2}{a^2} - \frac{y^2}{b^2} = 1$ and $\frac{y^2}{a^2} - \frac{x^2}{b^2} = 1$ are

a. $y = \pm x \pm \sqrt{b^2 - a^2}$
b. $y = \pm x \pm \sqrt{a^2 - b^2}$
c. $y = \pm x \pm (a^2 - b^2)$
d. $y = \pm x \pm \sqrt{a^2 + b^2}$

Working Space

17 The locus of point of intersection of two tangents to the hyperbola $\frac{x^2}{a^2} - \frac{y^2}{b^2} = 1$, the sum of whose slopes is c^2, is

a. $(x^2 - a^2)c^2 = y^2 - b^2$
b. $(x^2 + a^2)c^2 = y^2 + b^2$
c. $(x^2 - a^2)c^2 = 2xy$
d. $(x^2 + a^2)c^2 = y^2 - b^2$

18 If $P(a\sec\alpha, b\tan\alpha)$ and $Q(a\sec\beta, b\tan\beta)$ are two points on the hyperbola $\frac{x^2}{a^2} - \frac{y^2}{b^2} = 1$. Such that $\alpha - \beta = 2\theta$, then PQ touches the hyperbola

a. $\frac{x^2}{a^2\sec^2\theta} - \frac{y^2}{b^2} = 1$
b. $\frac{x^2}{a^2} - \frac{y^2}{b^2\sec^2\theta} = 1$
c. $\frac{x^2}{a^2} - \frac{y^2}{b^2} = 1$
d. None of these

Matching List Type

***Direction** (Q. No. 19) Choices for the correct combination of elements from Column I and Column II are given as options (a), (b), (c) and (d), out of which one is correct.*

19 Match the Column I with Column II and mark the correct option from the codes given below.

	Column I		Column II
i.	If the chord of contact of tangents from a point P to the parabola $y^2 = 4ax$ touches the parabola $x^2 = 4by$, the locus of P is	p.	straight line
ii.	A variable circle C has the equation $x^2 + y^2 - 2(t^2 - 3t + 1)x - 2(t^2 + 2t)y + t = 0$, where t is a parameter. The locus of centre of the circle is	q.	circle
iii.	The locus of point of intersection of tangents to an ellipse $\frac{x^2}{a^2} + \frac{y^2}{b^2} = 1$ at two points, the sum of whose eccentric angle is constant, is	r.	parabola
iv.	An ellipse slides between two perpendiculars straight lines, then the locus of its centre is	s.	hyperbola

Codes

	i	ii	iii	iv		i	ii	iii	iv
a.	s	r	p	q	b.	p	s	q	r
c.	q	p	r	s	d.	r	q	s	p

Comprehension Type

***Direction** (Q. Nos. 20-22) This section contains a paragraph, describing theory, experiments, data, etc. Three questions related to the paragraph have been given. Each question has only one correct answer among the four given options (a), (b), (c) and (d).*

Passage

A conic C passes through the point (2, 4) and is such that the segment of any of its tangents at any point contained between the coordinate axes is bisected at the point of tangency. Let S denotes circle described on the foci F_1 and F_2 of the conic C as diameter.

20 Vertex of the conic C is

a. (2, 2), (−2, − 2)
b. $(2\sqrt{2}, 2\sqrt{2}), (-2\sqrt{2}, -2\sqrt{2})$
c. (4, 4), (−4, − 4)
d. $(\sqrt{2}, \sqrt{2}), (-\sqrt{2}, -\sqrt{2})$

Working Space

21 Directors circle of the conic is

a. $x^2 + y^2 = 4$

b. $x^2 + y^2 = 8$

c. $x^2 + y^2 = 2$

d. None of these

22 Equation of the circle S is

a. $x^2 + y^2 = 16$

b. $x^2 + y^2 = 8$

c. $x^2 + y^2 = 32$

d. $x^2 + y^2 = 4$

One Integer Value Correct Type

Direction *(Q. Nos. 23-25) This section contains 3 questions. When worked out will result in an integer from 0 to 9 (both inclusive).*

23 If radii of director circles $\frac{x^2}{a^2} + \frac{y^2}{b^2} = 1$ and $\frac{x^2}{a^2} - \frac{y^2}{b^2} = 1$ are $2r$ and r respectively, and e_e and e_h are the eccentricities of ellipse and hyperbola respectively, then the value of $4e_h^2 - e_e^2$ is

24 If the tangent at point $P(h, k)$ on the hyperbola $\frac{x^2}{a^2} - \frac{y^2}{b^2} = 1$ cuts the circle $x^2 + y^2 = a^2$ at the point $P_1(x_1, y_1)$ and $P_2(x_2, y_2)$, where $\frac{1}{y_1} + \frac{1}{y_2} = \frac{\lambda}{k}$, then the value of λ is

25 If the tangent drawn to the hyperbola $\frac{x^2}{a^2} - \frac{y^2}{b^2} = 1$ at any point P meets the coordinate axes at the points A and B respectively. If the rectangle $OACB$ (O being the origin) is completed, where C lies on $\frac{a^2}{x^2} - \frac{b^2}{y^2} = \lambda$. Then, the value of λ is

DPP-3 Normal, Chord Bisected and Product of Pair of Tangents

Only One Option Correct Type

Working Space

Direction (Q. Nos. 1-18) *This section contains 18 multiple choice questions. Each question has four choices (a), (b), (c) and (d), out of which ONLY ONE is correct.*

1 What will be the equation of chord of hyperbola $25x^2 - 16y^2 = 400$, whose mid-point is (5, 3)?

a. $115x - 117y = 17$
b. $125x - 48y = 481$
c. $127x + 33y = 341$
d. $15x + 121y = 105$

2 The equation of chord of hyperbola $25x^2 - 16y^2 = 400$, which is bisected at (6, 2) is

a. $16x - 75y = 0$
b. $75x - 16y = 400$
c. $25x - 4y = 400$
d. None of these

3 The locus of middle points of the chords of hyperbola $3x^2 - 2y^2 + 4x - 6y = 0$ parallel to $y = 2x$ is

a. $3x - 4y = 4$
b. $3y - 4x + 4 = 0$
c. $4x - 4y = 3$
d. $3x - 4y = 2$

4 The equation of normal to the hyperbola $\frac{x^2}{a^2} - \frac{y^2}{b^2} = 1$ at the point $(a\sec\theta, b\tan\theta)$ is

a. $\frac{ax}{\sec\theta} - \frac{by}{\tan\theta} = a^2 - b^2$
b. $\frac{ax}{\sec\theta} + \frac{by}{\operatorname{cosec}\theta} = a^2 + b^2$
c. $\frac{ax}{\sec\theta} + \frac{by}{\tan\theta} = a^2 + b^2$
d. $\frac{ax}{\sec\theta} - \frac{by}{\operatorname{cosec}\theta} = a^2 + b^2$

5 If $x = 9$ is the chord of chord of contact of the hyperbola $x^2 - y^2 = 9$, then the equation of corresponding pair of tangents is

a. $9x^2 - 8y^2 + 18x - 9 = 0$
b. $9x^2 - 8y^2 - 18x + 9 = 0$
c. $9x^2 - 8y^2 - 18x - 9 = 0$
d. $9x^2 - 8y^2 + 18x + 9 = 0$

Working Space

6 Let $P(a\sec\theta, b\tan\theta)$ and $Q(a\sec\phi, b\tan\phi)$, where $\theta + \phi = \frac{\pi}{2}$, be two points on the hyperbola $\frac{x^2}{a^2} - \frac{y^2}{b^2} = 1$. If (h, k) is the point of intersection of the normals at P and Q, then k is equal to

a. $\frac{a^2 + b^2}{a}$
b. $\frac{a^2 + b^2}{b}$
c. $-\left(\frac{a^2 + b^2}{a}\right)$
d. $-\left(\frac{a^2 + b^2}{b}\right)$

7 The locus of mid-point of the chords of circle $x^2 + y^2 = 16$, which are tangent to the hyperbola $9x^2 - 16y^2 = 144$, is

a. $x^2 + y^2 = 7$
b. $(x^2 + y^2)^2 = 7$
c. $(x^2 + y^2)^2 = 16x^2 - 9y^2$
d. $(x^2 + y^2)^2 = 25$

8 The number of normals which can be drawn from an external point on the hyperbola $\frac{x^2}{a^2} - \frac{y^2}{b^2} = 1$, is

a. 1
b. 4
c. 6
d. 8

9 Line $x\cos\alpha + y\sin\alpha = p$ is a normal to the hyperbola $\frac{x^2}{a^2} - \frac{y^2}{b^2} = 1$, if

a. $a^2\sec^2\alpha - b^2\operatorname{cosec}^2\alpha = \frac{(a^2 + b^2)^2}{p^2}$
b. $a^2\sec^2\alpha + b^2\operatorname{cosec}^2\alpha = \frac{(a^2 + b^2)^2}{p^2}$
c. $a^2\cos^2\alpha - b^2\sin^2\alpha = \frac{(a^2 + b^2)^2}{p^2}$
d. $a^2\cos^2\alpha + b^2\sin^2\alpha = \frac{(a^2 + b^2)^2}{p^2}$

10 If the tangent and normal to $x^2 - y^2 = 4$ at a point cut off intercepts a_1, a_2 on the X-axis respectively and b_1, b_2 on Y-axis respectively, then the value of $a_1a_2 + b_1b_2$ is

a. 1
b. −1
c. 0
d. 4

11 The normal at P to a hyperbola of eccentricity e, intersects its transverse and conjugate axes at L and M respectively. If the locus of mid-point of LM is hyperbola, having eccentricity e_1, then

a. $e_1 = \frac{e}{e^2 - 1}$
b. $e_1 = \sqrt{\frac{e}{e^2 - 1}}$
c. $e_1 = \frac{e}{\sqrt{e^2 - 1}}$
d. None of these

Working Space

12 The line $lx + my + n = 0$ will be a normal to the hyperbola $\frac{x^2}{a^2} - \frac{y^2}{b^2} = 1$, if $\frac{a^2}{l^2} - \frac{b^2}{m^2}$ is equal to

a. $\frac{a^2 + b^2}{n^2}$

b. $\frac{(a^2 + b^2)^2}{n^2}$

c. $\frac{a^2 - b^2}{n^2}$

d. $\frac{(a^2 - b^2)^2}{n^2}$

13 If the normal at θ on the hyperbola $\frac{x^2}{a^2} - \frac{y^2}{b^2} = 1$ meets the transverse axis at G, then $AG \cdot A'G$ is equal to (where, A and A' are the vertices of hyperbola)

a. $a^2(e^4 \sec^2\theta - 1)$

b. $a^2(e^4 \tan^2\theta - 1)$

c. $b^2(e^4 \sec^2\theta - 1)$

d. $b^2(e^4 \sec^2\theta + 1)$

14 The locus of feet of the normals drawn from a point to the hyperbola $\frac{x^2}{a^2} - \frac{y^2}{b^2} = 1$ lie on the curve

a. $a^2y(x + h) + b^2x(y + k) = 0$

b. $b^2y(x + h) + a^2x(y + k) = 0$

c. $a^2y(x - h) + b^2x(y - k) = 0$

d. $b^2y(x - h) + a^2x(y - k) = 0$

15 From points on the circle $x^2 + y^2 = a^2$ tangents are drawn to the hyperbola $x^2 - y^2 = a^2$. Then, the locus of mid-points of the chord of contact of tangents is

a. $(x^2 - y^2)^2 = a^2(x^2 + y^2)$

b. $(x^2 - y^2) = a^2(x^2 + y^2)$

c. $(x^2 - y^2) = a^2(x^2 + y^2)^2$

d. None of the above

16 If the equation $lx^2 + 2mxy + ny^2 = 0$ represents a pair of conjugate diameters of the hyperbola $\frac{x^2}{a^2} - \frac{y^2}{b^2} = 1$, then

a. $a^2l = b^2n$

b. $a^2l = b^2m$

c. $b^2l = a^2n$

d. $al = bn$

17 The pole of line $lx + my + n = 0$ with respect to the hyperbola $\frac{x^2}{a^2} - \frac{y^2}{b^2} = 1$ is

a. $\left(\frac{a^2l}{n}, \frac{b^2m}{n}\right)$

b. $\left(\frac{a^2l}{n}, -\frac{b^2m}{n}\right)$

c. $\left(-\frac{a^2l}{n}, \frac{b^2m}{n}\right)$

d. $\left(-\frac{a^2l}{n}, -\frac{b^2m}{n}\right)$

18 The locus of poles of normal chords of the hyperbola $\frac{x^2}{a^2} - \frac{y^2}{b^2} = 1$ is

a. $a^6y^2 - b^6x^2 = (a^2 + b^2)^2x^2y^2$

b. $a^6y^2 + b^6x^2 = x^2y^2$

c. $a^6y^2 - b^6x^2 = (a^2 + b^2)^2$

d. $a^6y^2 + b^6x^2 = a^2 + b^2$

Working Space

Matching List Type

Direction (Q. No. 19) *Choices for the correct combination of elements from Column I and Column II are given as options (a), (b), (c) and (d), out of which one is correct.*

19 Match the Column I with Column II and mark the correct option from the codes given below.

	Column I		Column II
i.	The curve such that product of the distances of any of its tangent from two given points is constant, can be	p.	circle
ii.	A curve for which the length of the subnormal at any of its point is equal to 2 and the curve passes through (1, 2) can be	q.	parabola
iii.	A curve passes through (1, 4) and is such that the segment joining any point P on the curve and the point of intersection of the normal at P with the X-axis is bisected by the Y-axis. The curve can be	r.	ellipse
iv.	A curve passes through (1, 2) is such that the length of the normal at any of its point is equal to 2. The curve can be	s.	hyperbola

Codes

	i	ii	iii	iv
a.	r,s	q	r	p
b.	p	r	s	q,s
c.	q	s,q	p	r
d.	r	p	q	s,p

Comprehension Type

Direction (Q. Nos. 20-27) *This section contains 3 paragraphs, each describing theory, experiments, data, etc. Eight questions related to the paragraphs have been given. Each question has only one correct answer among the four given options (a), (b), (c) and (d).*

Passage I

If normal to the hyperbola $\frac{x^2}{a^2} - \frac{y^2}{b^2} = 1$ at any point P meets the transverse and conjugate axes in G and g respectively and F is the foot of perpendicular to the normal at P from the centre C.

20 The product of length PF and PG is

a. a^2 b. b^2
c. $2a^2$ d. $2b^2$

21 The product of length PF and Pg is

a. a^2 b. b^2
c. $2a^2$ d. $2b^2$

Passage II

Let $H : x^2 - y^2 = 9$, $P : y^2 = 4(x - 5)$ and $L : x = 9$.

22 If L is the chord of contact of the hyperbola H, then the equation of corresponding pair of tangents is

a. $9x^2 - 8y^2 + 18x - 9 = 0$ b. $9x^2 - 8y^2 - 18x + 9 = 0$
c. $9x^2 - 8y^2 - 18x - 9 = 0$ d. $9x^2 - 8y^2 + 18x + 9 = 0$

Working Space

23 If R is the point of intersection of tangents to H at the extremities of the chord L, then equation of the chord of contact of R with respect to the parabola P is

a. $x = 7$ b. $x = 8$
c. $x = 9$ d. $x = 10$

24 If the chord of contact of R with respect to the parabola P meets the parabola at T and T', S is the focus of parabola, then the area of $\Delta STT'$ is equal to

a. 8 sq units b. 9 sq units
c. 12 sq units d. 16 sq units

Passage III

If normal to the hyperbola $\frac{x^2}{a^2} - \frac{y^2}{b^2} = 1$ at a point $A(a\sec\theta, b\tan\theta)$ meets the transverse and conjugate axes in D and E respectively and F is the foot of perpendicular to the normal at A from the centre C.

25 The value of AD^2 is

a. $\frac{b^2}{a^2}(b^2\operatorname{cosec}^2\theta + a^2\cot^2\theta)$ b. $\frac{a^2}{b^2}(a^2\sec^2\theta + b^2\tan^2\theta)$
c. $\frac{b^2}{a^2}(b^2\sec^2\theta + a^2\tan^2\theta)$ d. $\frac{b^2}{a^2}(b^2\tan^2\theta + a^2\sec^2\theta)$

26 Minimum value of AD is

a. $\frac{b^2}{a}$ b. $\left|\frac{a}{b}(a+b)\right|$
c. $\left|\frac{b}{a}(a-b)\right|$ d. $\left|\frac{a}{b}(a-b)\right|$

27 The value of AE^2 is

a. $\frac{a^2}{b^2}(a^2\tan^2\theta + b^2\sec^2\theta)$
b. $\frac{a^2}{b^2}(a^2\sec^2\theta + b^2\tan^2\theta)$
c. $\frac{b^2}{a^2}(a^2\sec^2\theta + b^2\tan^2\theta)$
d. $\frac{b^2}{a^2}(a^2\tan^2\theta + b^2\sec^2\theta)$

One Integer Value Correct Type

Direction (Q. Nos. 28-30) This section contains 3 questions. When worked out will result in an integer from 0 to 9 (both inclusive).

28 If the locus of mid-points of normal chords of the hyperbola $x^2 - y^2 = a^2$ is $(x^2 - y^2)^2 = \lambda a^2x^2y^2$. Then, the value of λ is

29 If the polars of (x_1, y_1) and (x_2, y_2) with respect to the hyperbola $\frac{x^2}{a^2} - \frac{y^2}{b^2} = 1$ are at the right angles, where $(x_1x_2)b^4 + a^4(y_1y_2) = \lambda$, then λ the value of is

30 If CP and CD represent a pair of conjugate diameters of the hyperbola $x^2 - y^2 = a^2$ has area of $\Delta CPD = \lambda a^2$, then the value of $|4\lambda|$ is

DPP-4 Asymptotes and Rectangular Hyperbola

Only One Option Correct Type

Working Space

Direction (Q. Nos. 1-15) *This section contains 15 multiple choice questions. Each question has four choices (a), (b), (c) and (d), out of which ONLY ONE is correct.*

1 The angle between the asymptotes of $\frac{x^2}{a^2} - \frac{y^2}{b^2} = 1$ is

a. $2\tan^{-1}\left(\frac{b}{a}\right)$
b. $\tan^{-1}\left(\frac{a}{b}\right)$
c. $2\tan^{-1}\left(\frac{a}{b}\right)$
d. $\tan^{-1}\left(\frac{b}{a}\right)$

2 The product of length of perpendiculars drawn from any point on the hyperbola $x^2 - 2y^2 - 2 = 0$ to its asymptotes, is

a. 1/2
b. 2/3
c. 3/2
d. 2

3 The distance between directrices of the hyperbola $x = 8\sec\theta$ and $y = 8\tan\theta$ is

a. $4\sqrt{2}$
b. $6\sqrt{2}$
c. $8\sqrt{2}$
d. $10\sqrt{2}$

4 The equation of asymptote of the hyperbola $2x^2 - 5xy - 3y^2 - 5x - 3y - 21 = 0$ is

a. $2x^2 - 5xy - 3y^2 - 5x - 3y - 18 = 0$
b. $2x^2 - 5xy - 3y^2 - 5x - 3y - 49 = 0$
c. $2x^2 - 5xy - 3y^2 - 5x - 3y - \frac{49}{18} = 0$
d. $2x^2 - 5xy - 3y^2 - 5x - 3y - \frac{18}{49} = 0$

5 The equation of hyperbola whose asymptotes are the straight lines $3x - 4y + 7 = 0$ and $4x + 3y + 1 = 0$ and which passes through the origin, is

a. $(3x - 4y + 7)(4x + 3y + 1) + 7 = 0$
b. $(3x - 4y + 7)(4x + 3y + 1) - 7 = 0$
c. $(3x - 4y + 7)(4x + 3y - 1) + 7 = 0$
d. $(3x + 4y + 7)(4x - 3y + 1) = 0$

6 The asymptotes of a hyperbola having centre at the point (1, 2) are parallel to the lines $2x + 3y = 0$ and $3x + 2y = 0$. If the hyperbola passes through the point (5, 3), its equation is $(2x + 3y - 8)(3x + 2y + 7) = \lambda$, then λ is equal to

a. 148
b. 152
c. 154
d. 160

Working Space

7 The coordinates of the foci of hyperbola $xy = c^2$ is

a. $(\pm c, 0)$
b. $(\pm\sqrt{2}\, c, 0)$
c. $(\pm 2c, 0)$
d. None of these

8 A rectangular hyperbola whose centre is C, is cut by any circle of radius r in four points P, Q, R and S. Then, $CP^2 + CQ^2 + CR^2 + CS^2$ is

a. r^2
b. $2r^2$
c. $3r^2$
d. $4r^2$

9 If the circle $x^2 + y^2 = a^2$ intersects the hyperbola $xy = c^2$ in four points $P(x_1, y_1)$, $Q(x_2, y_2)$, $R(x_3, y_3)$ and $S(x_4, y_4)$, then

a. $x_1 + x_2 + x_3 + x_4 = 0$
b. $y_1 + y_2 + y_3 + y_4 = 0$
c. $x_1 x_2 x_3 x_4 = 2c^4$
d. $y_1 y_2 y_3 y_4 = 2c^4$

10 From any point on the hyperbola $\frac{x^2}{a^2} - \frac{y^2}{b^2} = 1$ tangents are drawn to the hyperbola $\frac{x^2}{a^2} - \frac{y^2}{b^2} = 2$. Then, the area cut off by the chord of contact on the asymptotes, is

a. $2ab$
b. $4ab$
c. $6ab$
d. $8ab$

11 If a triangle is inscribed in a rectangular hyperbola, then the orthocentre of triangle lies on the

a. focus
b. axis
c. directrix
d. curve

12 If e is the eccentricity of $\frac{x^2}{a^2} - \frac{y^2}{b^2} = 1$ and θ be the angle between the asymptotes, then $\sec(\theta/2)$ is equal to

a. $2e$
b. e^2
c. e
d. None of these

13 If normal to the rectangular hyperbola $xy = c^2$ at the point t on it intersects the hyperbola at t_1, then $t^3 t_1$ is equal to

a. 1
b. 2
c. -1
d. -2

14 The locus of point of intersection of tangents at the ends of normal chord of the hyperbola $x^2 - y^2 = a^2$ is

a. $y^4 - x^4 = 4a^2x^2y^2$
b. $y^2 - x^2 = 4a^2x^2y^2$
c. $a^2(y^2 - x^2) = 4x^2y^2$
d. $y^2 + x^2 = 4a^2x^2y^2$

15 If the normal at the point $P(x_i, y_i)$, $i = 1, 2, 3, 4$ on the hyperbola $xy = c^2$ are concurrent at the point $Q(h, k)$, then $\frac{(x_1 + x_2 + x_3 + x_4)(y_1 + y_2 + y_3 + y_4)}{x_1 x_2 x_3 x_4}$ is

a. $\frac{hk}{c^2}$
b. $\frac{hk}{c^4}$
c. $-\frac{hk}{c^2}$
d. $-\frac{hk}{c^4}$

Working Space

Comprehension Type

Direction (Q. Nos. 16-27) This section contains 4 paragraphs, each describing theory, experiments, data, etc. Twelve questions related to the paragraphs have been given. Each question has only one correct answer among the four given options (a), (b), (c) and (d).

Passage I

The normal at any point $P(x_1, y_1)$ of curve is a line perpendicular to tangent at the point $P(x_1, y_1)$. In case of rectangular hyperbola $xy = c^2$, the equation of normal at $(ct, c/t)$ is $xt^3 - yt - ct^4 + c = 0$. The shortest distance between any two curves always exist along the common normal.

16 If normal at (5, 3) of rectangular hyperbola $xy - y - 2x - 2 = 0$ intersect it again at a point

a. (−1, 0)
b. (−1, 1)
c. (0, − 2)
d. (3/4, − 14)

17 The shortest distance between the parabola $2y^2 = 2x - 1$, $2x^2 = 2y - 1$ is

a. $2\sqrt{2}$
b. $\frac{1}{2\sqrt{2}}$
c. 4
d. $\sqrt{\frac{36}{5}}$

18 The number of normals drawn from (7/6, 4) to parabola $y^2 = 2x - 1$ is

a. 1
b. 2
c. 3
d. 4

Passage II

Considering the rectangular hyperbola $xy = (15)!$. The number of points (α, β) lying on it, where

19 $\alpha, \beta \in I$ is

a. 2032
b. 4032
c. 8066
d. 8064

20 $\alpha, \beta \in I^+$ and HCF $(\alpha, \beta) = 1$ is

a. 64
b. 625
c. 8064
d. 49

21 $\alpha, \beta \in I^+$ and α divides β is

a. 96
b. 64
c. 49
d. 2032

Passage III

P, Q are two points on the rectangular hyperbola $(x - 1)(y - 2) = c^2$, 0 is the centre of hyperbola, also tangent at P is perpendicular to OQ and meets OQ at N such that $(OQ)(ON) = 4$.

22 One of the equation of directrix of the hyperbola is

a. $x + y - 2 = 0$
b. $x + y - 2\sqrt{2} = 0$
c. $x + y - 4 = 0$
d. $x + y - 5 = 0$

23 An ellipse confocal with the hyperbola and with eccentricity equal to $\frac{1}{\sqrt{2}}$ is intersected by the line $x + y - 7 = 0$ at A and B, then intersection point of tangents at A and B will lie on

a. $x + y - 8 = 0$
b. $x + y - 9 = 0$
c. $x + y - 11 = 0$
d. $x + y - 13 = 0$

Working Space

24 If S is a focus of ellipse (defined in previous question) and P is a moving point on ellipse, then circle on SP as diameter always touches a circle whose area is

a. 32π sq units
b. 30π sq units
c. 25π sq units
d. 16π sq units

Passage IV

A point P moves such that sum of the slopes of the normals drawn from it to the hyperbola $xy = 16$ is equal to the sum of ordinates of feet of normals, then the locus of $P(h, k)$ is a curve C.

25 The equation of the curve C is

a. $x^2 = 12y$
b. $x^2 = 16y$
c. $y^2 = 16x$
d. $y^2 = 12x$

26 If the tangent to the curve C cut the coordinate axes at A and B. Then, the locus of the middle point of AB is

a. $x^2 + 2y = 0$
b. $x^2 - 2y = 0$
c. $y^2 + 2x = 0$
d. $y^2 - 2x = 0$

27 The area (in sq units) of equilateral inscribed in the curve C having one vertex as the vertex of the curve C, is

a. $76\sqrt{3}$
b. $786\sqrt{3}$
c. $768\sqrt{3}$
d. $687\sqrt{3}$

One Integer Value Correct Type

Direction (Q. Nos. 28-30) *This section contains 3 questions. When worked out will result in an integer from 0 to 9 (both inclusive).*

28 The equation of asymptotes of $xy + 2x + 4y + 6 = 0$ is $xy + 2x + 4y + c = 0$, then the value of c is

29 PQ and RS are two perpendicular chords of the rectangular hyperbola $xy = c^2$. If C is the centre of rectangular hyperbola, then the product of the slopes of CP, CQ, CR and CS is

30 If e is the eccentricity of a hyperbola whose asymptotes are $3x + 4y = 2$ and $4x - 3y + 5 = 0$. Then, the value of e^2 is

SET - 1

Revisal Problems
for JEE Main

Only One Option Correct Type

Working Space

Direction (Q. Nos. 1-30) *This section contains 30 multiple choice questions. Each question has four choices (a), (b), (c) and (d), out of which ONLY ONE is correct.*

1 A point moves in a plane so that its distance PA and PB from two fixed points A and B in the plane satisfy the relation $PA - PB = k$, where $(k \neq 0)$, then the locus of P is

a. a parabola
b. an ellipse
c. a hyperbola
d. None of these

2 Let A and B be two fixed points and P another point in the plane, moves in such a way that $k_1 PA + k_2 PB = k_3$, where k_1, k_2, k_3 are real constants. Then, the locus of P is

a. a circle, if $k_1 = 0$ and $k_2, k_3 > 0$
b. a circle, if $k_1 > 0$, $k_2 < 0$ and $k_3 = 0$
c. an ellipse, if $k_1 = k_2 > 0$ and $k_3 > 0$
d. a hyperbola, if $k_2 = -1$ and $k_1, k_3 > 0$

3 The eccentricity of conjugate hyperbola of the hyperbola $x^2 - 3y^2 = 1$ is

a. 2
b. $2/\sqrt{3}$
c. 4
d. 4/3

4 The eccentricity of conic represented by $x^2 - y^2 - 4x + 4y + 16 = 0$ is

a. 1
b. $\sqrt{2}$
c. 2
d. 1/2

5 The equations of transverse and conjugate axes of a hyperbola are respectively $x + 2y - 3 = 0$, $2x - y + 4 = 0$ and their respectively lengths are $\sqrt{2}$ and $2/\sqrt{3}$. The equation of the hyperbola is

a. $\frac{2}{5}(x + 2y - 3)^2 - \frac{3}{5}(2x - y + 4)^2 = 1$
b. $\frac{2}{5}(2x - y + 4)^2 - \frac{3}{5}(x + 2y - 3)^2 = 1$
c. $2(2x - y + 4)^2 - 3(x + 2y - 3)^2 = 1$
d. $2(2x - y + 4)^2 - (x + 2y - 3)^2 = 1$

6 The locus of centre of a circle which touches two given circles externally is

a. a parabola
b. an ellipse
c. a hyperbola
d. None of these

7 If the latusrectum of a hyperbola forms an equilateral triangle with the vertex at the centre of the hyperbola, then the eccentricity of the hyperbola is

a. $\frac{\sqrt{5}+1}{2}$
b. $\frac{\sqrt{11}+1}{2}$
c. $\frac{\sqrt{13}+1}{2\sqrt{3}}$
d. $\frac{\sqrt{13}+1}{2\sqrt{3}}$

8 If the latusrectum of a hyperbola through one focus subtends 60° angle at the other focus, then its eccentricity e is

a. $\sqrt{2}$
b. $\sqrt{3}$
c. $\sqrt{5}$
d. $\sqrt{6}$

9 If e and e' are the eccentricity of hyperbola and its conjugate, then $\frac{1}{e^2} + \frac{1}{(e')^2}$ is equal to

a. 0
b. 1
c. 2
d. 3

Working Space

10 Let LL' be the latusrectum through the focus of the hyperbola $\frac{x^2}{a^2} - \frac{y^2}{b^2} = 1$ and A' be the farther vertex. If $\Delta A'LL'$ is equilateral, then the eccentricity is

a. $\sqrt{3}$
b. $\sqrt{3} + 1$
c. $\frac{\sqrt{3}+1}{\sqrt{2}}$
d. $\frac{\sqrt{3}+1}{\sqrt{3}}$

11 If the line $2x + \sqrt{6}y = 2$ touches the hyperbola $x^2 - 2y^2 = 4$, then the point of contact is

a. $(-2, \sqrt{6})$
b. $(-\sqrt{5}, 2\sqrt{6})$
c. $(1/2, 1/\sqrt{6})$
d. $(4, -\sqrt{6})$

12 The tangents to the hyperbola $x^2 - y^2 = 3$ are parallel to the straight line $2x + y + 8 = 0$, at which of the following points?

a. (2, 1)
b. (2, – 1)
c. (–2, – 1)
d. (–2, 1)

13 The equation of a tangent to the hyperbola $16x^2 - 25y^2 - 96x + 100y - 356 = 0$, which makes an angle $\pi/4$ with the transverse axis, is

a. $y = x + 2$
b. $y = x - 5$
c. $y = x + 3$
d. $y = x - 2$

14 The equation of chord of the hyperbola $x^2 - y^2 = 9$, which is bisected at (5, – 3), is

a. $5x + 3y = 9$
b. $5x - 3y = 16$
c. $5x + 3y = 16$
d. $5x - 3y = 9$

15 If the line $y = 3x + \lambda$ touches the hyperbola $9x^2 - 5y^2 = 45$, then the value of λ is

a. 36
b. 45
c. 6
d. 15

16 The equation of common tangents to the two hyperbolas $\frac{x^2}{a^2} - \frac{y^2}{b^2} = 1$ and $\frac{y^2}{a^2} - \frac{x^2}{b^2} = 1$ are

a. $y = \pm x \pm \sqrt{b^2 - a^2}$
b. $y = \pm x \pm \sqrt{a^2 - b^2}$
c. $y = \pm x \pm (a^2 - b^2)$
d. $y = \pm x \pm \sqrt{a^2 + b^2}$

17 Tangents drawn from the point (c, d) to the hyperbola $\frac{x^2}{a^2} - \frac{y^2}{b^2} = 1$ makes angles α and β with the X-axis. If $\tan\alpha \cdot \tan\beta = 1$, then $(c^2 - d^2)$ is equal to

a. $a^2 - b^2$
b. $b^2 - a^2$
c. $a^2 + b^2$
d. None of these

18 If the tangent at (α, β) to the hyperbola $\frac{x^2}{a^2} - \frac{y^2}{b^2} = 1$. Cuts the auxiliary circle at points whose ordinates are y_1 and y_2, then $\frac{1}{y_1} + \frac{1}{y_2}$ is equal to

a. $\frac{1}{\alpha}$
b. $\frac{2}{\alpha}$
c. $\frac{1}{\beta}$
d. $\frac{2}{\beta}$

19 If the tangents drawn from a point on the hyperbola $x^2 - y^2 = a^2 - b^2$ to the ellipse $\frac{x^2}{a^2} + \frac{y^2}{b^2} = 1$ make angle α and β with the transverse axis of the hyperbola, then

a. $\tan\alpha \cdot \tan\beta = 1$
b. $\tan\alpha + \tan\beta = 1$
c. $\tan\alpha - \tan\beta = 1$
d. $\tan\alpha \cdot \tan\beta = -1$

20 If $P(a\sec\alpha, b\tan\alpha)$ and $Q(a\sec\beta, b\tan\beta)$ are two points on the hyperbola $\frac{x^2}{a^2} - \frac{y^2}{b^2} = 1$ such that $\alpha - \beta = 2\theta$ (a constant), then PQ touches the hyperbola

a. $\frac{x^2}{a^2\sec^2\theta} - \frac{y^2}{b^2} = 1$
b. $\frac{x^2}{a^2} - \frac{y^2}{b^2\sec^2\theta} = 1$
c. $\frac{x^2}{a^2} - \frac{y^2}{b^2} = \cos^2\theta$
d. None of these

Working Space

21 The equation of the chord joining two points (x_1, y_1) and (x_2, y_2) on the rectangular hyperbola $xy = c^2$ is

a. $\dfrac{x}{x_1 + x_2} + \dfrac{y}{y_1 + y_2} = 1$

b. $\dfrac{x}{x_1 - x_2} + \dfrac{y}{y_1 - y_2} = 1$

c. $\dfrac{x}{y_1 + y_2} + \dfrac{y}{x_1 + x_2} = 1$

d. $\dfrac{x}{y_1 - y_2} + \dfrac{y}{x_1 - x_2} = 1$

22 The product of perpendiculars drawn from any points on a hyperbola $\dfrac{x^2}{a^2} - \dfrac{y^2}{b^2} = 1$ to its asymptotes is

a. $\dfrac{ab}{\sqrt{a} + \sqrt{b}}$

b. $\dfrac{ab}{a^2 + b^2}$

c. $\dfrac{a^2b^2}{a^2 + b^2}$

d. $\dfrac{a^2 + b^2}{a + b}$

23 The locus of point of intersection of tangents at the ends of normal chord of the hyperbola $x^2 - y^2 = a^2$ is

a. $y^4 - x^4 = 4a^2x^2y^2$

b. $y^2 - x^2 = 4a^2x^2y^2$

c. $a^2(y^2 - x^2) = 4x^2y^2$

d. $y^2 + x^2 = 4a^2x^2y^2$

24 If angle subtended by any chord of a rectangular hyperbola at the centre is α and angle between the tangents at ends of chord is β, then

a. $\alpha = 2\beta$

b. $\alpha + \beta = \pi$

c. $\beta = 2\alpha$

d. $\alpha + \beta = \pi/2$

25 If a hyperbola passing through the origin has $3x - 4y - 1 = 0$ and $4x - 3y - 6 = 0$ as its asymptotes, then the equations of its transverse and conjugate axes are

a. $x - y - 5 = 0$ and $x + y + 1 = 0$

b. $x - y = 0$ and $x + y + 5 = 0$

c. $x + y - 5 = 0$ and $x - y - 1 = 0$

d. $x + y - 1 = 0$ and $x - y - 5 = 0$

26 The equations of asymptotes of the hyperbola $2x^2 + 5xy + 2y^2 - 11x - 7y - 4 = 0$ are

a. $2x^2 + 5xy + 2y^2 - 11x - 7y - 5 = 0$

b. $2x^2 + 5xy + 2y^2 - 11x - 7y + 5 = 0$

c. $2x^2 + 5xy + 2y^2 - 11x - 7y = 0$

d. None of the above

27 If $\dfrac{x^2}{a^2} + \dfrac{y^2}{b^2} = 1 (a > b)$ and $x^2 - y^2 = c^2$ cut at right angles, then

a. $a^2 + b^2 = 2c^2$

b. $b^2 - a^2 = 2c^2$

c. $a^2 - b^2 = 2c^2$

d. None of these

28 If a variable straight line $x\cos\alpha + y\sin\alpha = p$ which is a chord of hyperbola $\dfrac{x^2}{a^2} - \dfrac{y^2}{b^2} = 1$ $(b > a)$ subtends a right angle at the centre of the hyperbola, then it always touches a fixed circle whose radius is

a. $\dfrac{ab}{\sqrt{b - 2a}}$

b. $\dfrac{a}{\sqrt{a - b}}$

c. $\dfrac{ab}{\sqrt{b^2 - a^2}}$

d. $\dfrac{ab}{\sqrt{b + a}}$

29 The equation of line passing through the centre of a rectangular hyperbola is $x - y - 1 = 0$. If one of the asymptotes is $3x - 4y - 6 = 0$, then the equation of other asymptotes is

a. $4x + 3y + 17 = 0$

b. $4x - 3y + 8 = 0$

c. $3x - 2y = 15$

d. None of these

30 If $P(x_1, y_1)$, $Q(x_2, y_2)$, $R(x_3, y_3)$ and $S(x_4, y_4)$ are four con-cyclic points on the rectangular hyperbola $xy = c^2$, then the coordinates of orthocentre of the ΔPQR are

a. $(x_4, -y_4)$

b. (x_4, y_4)

c. $(-x_4, -y_4)$

d. $(-x_4, y_4)$

Statement Type

Working Space

Direction *(Q. Nos. 31-35) This section is based on Statement I and Statement II. Select the correct answer from the codes given below.*

a. Both Statement I and Statement II are correct and Statement II is the correct explanation of Statement I
b. Both Statement I and Statement II are correct but Statement II is not the correct explanation of Statement I
c. Statement I is correct but Statement II is incorrect
d. Statement II is correct but Statement I is incorrect

31 **Statement I** Ellipse $\frac{x^2}{25}+\frac{y^2}{16}=1$ and $12x^2-4y^2=27$ intersect each other at right angle.

Statement II Whenever confocal ellipse and hyperbola intersect, they intersect each other orthogonally.

32 **Statement I** $\frac{5}{3}$ and $\frac{5}{4}$ are the eccentricity of two hyperbola which are conjugate to each other.

Statement II If e and e_1 are the eccentricities of two conjugate hyperbola, then $ee_1 > 1$.

33 **Statement I** A bullet is fired and hit a target. An observer in the same plane heard two sounds the crack of the rifle and the Thnd... of the ball striking the tangent the same instant, then locus of the observer is hyperbola, where velocity of sound is smaller than velocity of bullet.

Statement II If difference of distances of a point P from the two fixed points is constant and less than the distance between the fixed points. Then, the locus of P is hyperbola.

34 **Statement I** If a circle $S=0$ intersects a hyperbola $xy=4$ at four points. Three of them are (2, 2), (4, 1) and (6, 2/3). Then, coordinates of the fourth point are $\left(\frac{1}{4}, 16\right)$.

Statement II If a circle $S=0$ intersects a hyperbola $xy=c^2$ at t_1, t_2, t_3, t_4, then $t_1 t_2 t_3 t_4 = 1$.

35 **Statement I** The foci of hyperbola $xy=c^2$ are at $\left(\pm\frac{c}{\sqrt{2}}, \pm\frac{c}{\sqrt{2}}\right)$.

Statement II If axes are rotated through an angle $-45°$, then the equation $x^2-y^2=a^2$ is transformed to $xy=\frac{a^2}{2}$.

SET - 2

Revisal Problems
for JEE Advanced

Only One Option Correct Type

Working Space

Direction (Q. Nos. 1-20) *This section contains 20 multiple choice questions. Each question has four choices (a), (b), (c) and (d), out of which ONLY ONE option is correct.*

1 If $x + iy = \sqrt{\phi + i\psi}$, where $i = \sqrt{-1}$ and ϕ and ψ are non-zero real parameters, then ϕ and ψ are constants, represents two system of rectangular hyperbola which intersect at an angle of

a. $\pi/6$
b. $\pi/3$
c. $\pi/4$
d. $\pi/2$

2 PQ is double ordinate of the ellipse $x^2 + 9y^2 = 9$, the normal at P meets the diameter through Q at R, then the locus of mid-point of PR is

a. a circle
b. a parabola
c. an ellipse
d. a hyperbola

3 Let the major axis of a standard ellipse equals the transverse axis of a standard hyperbola and their director circles have radius equal to $2R$ and R, respectively. If e_1 and e_2 are the eccentricity of the ellipse and hyperbola, then the correct relation is

a. $4e_1^2 - e_2^2 = 6$
b. $e_1^2 - 4e_2^2 = 6$
c. $4e_2^2 - e_1^2 = 6$
d. None of these

4 For each positive integer n, consider the point P with abscissa n on the curve $y^2 - x^2 = 1$. If d_n represents the shortest distance from the point P to the line $y = x$. Then, $\lim_{n \to \infty} (n \cdot d_n)$ has the value equal to

a. $\frac{1}{2\sqrt{2}}$
b. $\frac{1}{2}$
c. $\frac{1}{\sqrt{2}}$
d. 0

5 If a ray of light incident along the line $3x + (5 - 4\sqrt{2})y = 15$ gets reflected from the hyperbola $\frac{x^2}{16} - \frac{y^2}{9} = 1$, then its reflected ray goes along the line

a. $\sqrt{2}x - y + 5 = 0$
b. $\sqrt{2}y - x + 5 = 0$
c. $\sqrt{2}y - x - 5 = 0$
d. $3x - y(4\sqrt{2} + 5) + 15 = 0$

6 If the eccentricity of hyperbola $x^2 - y^2 \sec^2 \alpha = 5$ is $\sqrt{3}$ times the eccentricity of ellipse $x^2 \sec^2 \alpha + y^2 = 25$, then the value of α is

a. $\pi/6$
b. $\pi/4$
c. $\pi/3$
d. $\pi/2$

7 A normal to the hyperbola $x^2 - 4y^2 = 4$ has equal intercepts on positive X and Y-axes. If this normal touches the ellipse $\frac{x^2}{a^2} + \frac{y^2}{b^2} = 1$. Then, $(a^2 + b^2)$ is equal to

a. 5
b. 16
c. 25/3
d. 16/9

8 If a right angled triangle is inscribed in the rectangular hyperbola $xy = c^2$, then the angle between the tangent at the right angled vertex and the hypotenuse of the triangle is

a. 0
b. 45°
c. 60°
d. 90°

9 The equation of a hyperbola, conjugate to the hyperbola $x^2 + 3xy + 2y^2 + 2x + 3y = 0$ is

a. $x^2 + 3xy + 2y^2 + 2x + 3y + 1 = 0$
b. $x^2 + 3xy + 2y^2 + 2x + 3y + 2 = 0$
c. $x^2 + 3xy + 2y^2 + 2x + 3y + 3 = 0$
d. $x^2 + 3xy + 2y^2 + 2x + 3y + 4 = 0$

Working Space

10 The asymptote of hyperbola $\frac{x^2}{a^2}-\frac{y^2}{b^2}=1$ from with any tangent to the hyperbola, a triangle whose area is $a^2 \tan\lambda$ in magnitude, then its eccentricity is

a. $\sec\lambda$ b. $\operatorname{cosec}\lambda$
c. $\sec^2\lambda$ d. $\operatorname{cosec}^2\lambda$

11 If PN is the perpendicular from a point on a rectangular hyperbola $x^2-y^2=a^2$ on any of its asymptotes, then the locus of mid-point of PN is

a. a circle b. a parabola
c. an ellipse d. a hyperbola

12 A conic passes through the point (2, 4) and is such that the segment of any of its tangents at any point contained between the coordinate axes is bisected at the point of tangency. Then, the foci of conic are

a. $(2\sqrt{2}, 0)$ and $(-2\sqrt{2}, 0)$ b. $(2\sqrt{2}, 2\sqrt{2})$ and $(-2\sqrt{2}, -2\sqrt{2})$
c. $(4, 4)$ and $(-4, -4)$ d. $(4\sqrt{2}, 4\sqrt{2})$ and $(-4\sqrt{2}, -4\sqrt{2})$

13 At the point of intersection of the rectangular hyperbola $xy=c^2$ and the parabola $y^2=4ax$ tangents to the rectangular hyperbola and the parabola make an angle θ and ϕ respectively with the X-axis, then

a. $\theta=\tan^{-1}(-2\tan\phi)$ b. $\phi=\tan^{-1}(-2\tan\theta)$
c. $\theta=\frac{1}{2}\tan^{-1}(-\tan\phi)$ d. $\phi=\frac{1}{2}\tan^{-1}(-\tan\theta)$

14 If r is the radius of a circle whose centre is the origin. If the circle does not intersect the hyperbola $xy=1$, then the range of r is

a. $(0, 1)$ b. $(0, \sqrt{2})$
c. $\left(0, \frac{1}{\sqrt{2}}\right)$ d. $(0, 2)$

15 A line is such that its intercepts on the coordinate axes are e and e', where e and e' are the eccentricity of a hyperbola and its conjugate hyperbola. If this line is touching a fixed circle, then the area of circle is

a. $\pi/2$ b. π
c. $3\pi/2$ d. 2π

16 If $f(x)=ax^3+bx^2+cx+d$ (a, b, c, d are rational numbers) and roots of $f(x)=0$ and eccentricities of a parabola and rectangular hyperbola, then $a+b+c+d$ is equal to

a. 0 b. 1
c. 2 d. 3

17 The number of normal(s) of a rectangular hyperbola which can touch its conjugate is equal to

a. 0 b. 2
c. 4 d. 8

18 If from the point (α, α^2) two tangents drawn to any one branch of hyperbola $x^2-4y^2=1$, then

a. $-\frac{1}{2}<\alpha<\frac{1}{2}$ b. $0<\alpha<\frac{1}{2}$
c. $-1<\alpha<1$ d. None of these

19 $P(x, y)$ satisfy $x^2+y^2=1$, let maximum value of $(x+y)^2$ is λ, then the number of tangents drawn from the point $(\lambda, 0)$ to hyperbola $(x-2)^2-y^2=1$ is

a. 0 b. 1
c. 2 d. 4

20 Consider a hyperbola $xy=4$ and a line $y+2x=4$. Let the given line intersect the X-axis at R. If a line through R intersect the hyperbola at S and T. Then, the minimum value of $RS \times RT$ is

a. 8 b. 4
c. 6 d. 2

Working Space

One or More than One Options Correct Type

Direction *(Q. Nos. 21-25) This section contains 5 multiple choice questions. Each question has four choices (a), (b), (c) and (d), out of which ONE or MORE THAN ONE are correct.*

21 Which of the following equations in parametric form can represent a hyperbola, where t is a parameter?

a. $x = \frac{a}{2}\left(t + \frac{1}{t}\right)$ and $y = \frac{b}{2}\left(t - \frac{1}{t}\right)$
b. $\frac{tx}{a} - \frac{y}{b} = 1$ and $\frac{x}{a} + \frac{ty}{b} + t = 0$
c. $x = e^t + e^{-t}$ and $y = e^t - e^{-t}$
d. $x^2 - 6 = 2\cos t$ and $y^2 + 2 = 4\cos^2\frac{t}{2}$

22 Let P and Q be non-zero real numbers. Then, the equation $(px^2 + qy^2 + r)(4x^2 + 4y^2 - 8x - 4) = 0$ represents

a. two straight lines and a circle, when $r = 0$ and p, q are of the opposite sign
b. two circles, when $p = q$ and r is of sign opposite to that of p
c. a hyperbola and an ellipse, when p and q are of opposite sign and $r \neq 0$
d. a circle and an ellipse, when p and q are unequal but of same sign and r of sign opposite to that of p

23 Solution of the differential equation $(1 - x^2)\frac{dy}{dx} + xy = ax$, where $a \in R$, is

a. a conic which is an ellipse or a hyperbola with principal axes parallel to coordinates axes
b. centre of conic is $(0, a)$
c. length of one of the principal axes is 1
d. length of one of the principal axes is equal to 2

24 If θ is eliminated from the equations $a\sec\theta - x\tan\theta = y$ and $b\sec\theta + y\tan\theta = x$ (a and b are constants), then

a. the director circle of the hyperbola $\frac{x^2}{a^2} - \frac{y^2}{b^2} = 1$
b. auxiliary circle of the ellipse $\frac{x^2}{a^2} + \frac{y^2}{b^2} = 1$
c. director circle of the ellipse $\frac{x^2}{a^2} + \frac{y^2}{b^2} = 1$
d. director circle of the circle $x^2 + y^2 = \frac{a^2 + b^2}{2}$

25 The tangent to hyperbola, $x^2 - 3y^2 = 3$ at the point $(\sqrt{3}, 0)$, when associated with two asymptotes constitutes

a. isosceles triangle
b. an equilateral triangle
c. a triangle whose area is $\sqrt{3}$ sq units
d. a right isosceles triangle

Matching List Type

Direction *(Q. Nos. 26 and 27) Choices for the correct combination of elements from Column I and Column II are given as options (a), (b), (c) and (d), out of which one is correct.*

26 Let the circle $(x - 1)^2 + (y - 2)^2 = 25$ cut a rectangular hyperbola with transverse axis along $y = x$ at four points A, B, C and D having coordinates (x_i, y_i), $i = 1, 2, 3, 4$, respectively. O being the centre of hyperbola. Match the Column I with Column II and mark the correct option from the codes given below.

	Column I		Column II
i.	$(x_1 + x_2 + x_3 + x_4)$ is	p.	2
ii.	$(x_1^2 + x_2^2 + x_3^2 + x_4^2)$ is	q.	4
iii.	$(OA^2 + OB^2 + OC^2 + OD^2)$ is	r.	44
iv.	$(y_1^2 + y_2^2 + y_3^2 + y_4^2)$ is	s.	56
		t.	100

Codes

	i	ii	iii	iv
a.	p	r	t	s
b.	q	s	p	t
c.	r	t	q	p
d.	t	p	r	q

27 If the normals at (x_i, y_i) $(i = 1, 2, 3, 4)$ of the hyperbola $xy = c^2$ meet at the point (p, q). Match the Column I with Column II and mark the correct option from the codes given below.

	Column I		Column II
i.	p	p.	$x_1^2 + x_2^2 + x_3^2 + x_4^2$
ii.	q	q.	$y_1^2 + y_2^2 + y_3^2 + y_4^2$
iii.	p^2	r.	$x_1 + x_2 + x_3 + x_4$
iv.	q^2	s.	$y_1 + y_2 + y_3 + y_4$

Codes

	i	ii	iii	iv		i	ii	iii	iv
a.	r	s	p	q	b.	s	p	q	r
c.	p	q	r	s	d.	q	r	s	p

Working Space

Comprehension Type

Direction (Q. Nos. 28-30) *This section contains a paragraph, describing theory, experiments, data, etc. Three questions related to the paragraph have been given. Each question has only one correct answer among the four given options (a), (b), (c) and (d).*

Passage

If from a given point P on a hyperbola $\frac{x^2}{2} - \frac{y^2}{1} = 1$ chords are drawn at right angles to meet the hyperbola at R and T such that all the chords RT are concurrent at the point Q.

28 If the eccentric angle of points P is $\pi/4$, then is the coordinates of point Q is given by

a. $(-6, -3)$ b. $(6, -3)$ c. $(6, 3)$ d. $(-6, 3)$

29 If C is the centre of hyperbola, then one of the bisector of $\angle PCQ$ is

a. $x = 4$ b. $y = 4$ c. $y = 0$ d. $x + 2y = 2$

30 If the point P is a variable, then the locus of Q is given by

a. $\frac{x^2}{18} - \frac{y^2}{9} = 1$ b. $\frac{x^2}{32} - \frac{y^2}{16} = 1$ c. $\frac{x^2}{16} - \frac{y^2}{8} = 1$ d. $\frac{x^2}{16} - \frac{y^2}{4} = 2$

One Integer Value Correct Types

Direction (Q. Nos. 31-35) *This section contains 5 questions. When worked out will result in an integer from 0 to 9 (both inclusive).*

31 If a variable line has its intercepts on the coordinate axes e, e', where $\frac{e}{2}, \frac{e'}{2}$ are the eccentricities of a hyperbola and its conjugate hyperbola, then the line always touches the circle $x^2 + y^2 = r^2$, then the value of r is

32 The graph of $x^2 + y^2 + 6x - 24y + 72 = 0$ and $x^2 - y^2 + 6x + 16y - 46 = 0$ intersect at four points. The sum of distances of there four points from the point $(-3, 12)$ is d, then the number of positive divisors of d is

33 A chord AB which bisected at (1, 1) is drawn to the hyperbola $7x^2 + 8xy - y^2 - 4 = 0$ with centre C, which intersects its asymptotes E and F. If the equation of circumcircle of ΔCEF is $x^2 + y^2 - ax - by + c = 0$. Then, the value of $\frac{23(a - b + c)}{12}$ is

34 The latusrectum subtends a right angle at the centre of hyperbola $\frac{x^2}{a^2} - \frac{y^2}{b^2} = 1$. If the eccentricity is e, then the value of $[e]$, where $[\cdot]$ denotes greatest integer function is

35 A lattice point (x, y) is a point in XY-plane both of whose coordinates are integers. The number of such lattice points on hyperbola $2xy - 5x + y = 55$ is λ. Then, the value of $\frac{\lambda}{4}$ is

JEE Main & AIEEE Archive

(Compilation of Last 13 Years' Questions)

1 The equation of the hyperbola whose foci are $(-2, 0)$ and $(2, 0)$ and eccentricity is 2, is given by

(AIEEE 2011)

a. $-3x^2 + y^2 = 3$
b. $x^2 - 3y^2 = 3$
c. $3x^2 - y^2 = 3$
d. $-x^2 + 3y^2 = 3$

2 For the hyperbola $\frac{x^2}{\cos^2 \alpha} - \frac{y^2}{\sin^2 \alpha} = 1$, which of the following remains constant when α varies?

(AIEEE 2007)

a. Eccentricity
b. Directrix
c. Abscissae of vertices
d. Abscissae of foci

3 The locus of a point $P(\alpha, \beta)$ moving under the condition that the line $y = \alpha x + \beta$ is a tangent to the hyperbola $\frac{x^2}{a^2} - \frac{y^2}{b^2} = 1$, is

(AIEEE 2005)

a. a hyperbola
b. a parabola
c. a circle
d. an ellipse

4 The foci of the ellipse $\frac{x^2}{16} + \frac{y^2}{b^2} = 1$ and the hyperbola $\frac{x^2}{144} - \frac{y^2}{81} = \frac{1}{25}$ coincide. Then, the value of b^2 is

(AIEEE 2003)

a. 1
b. 5
c. 7
d. 9

5 The equation of chord joining two points (x_1, y_1) and (x_2, y_2) on the rectangular hyperbola $xy = c^2$ is

(AIEEE 2002)

a. $\frac{x}{x_1 + x_2} + \frac{y}{y_1 + y_2} = 1$

b. $\frac{x}{x_1 - x_2} + \frac{y}{y_1 - y_2} = 1$

c. $\frac{x}{y_1 + y_2} + \frac{y}{x_1 + x_2} = 1$

d. $\frac{x}{y_1 - y_2} + \frac{y}{x_1 - x_2} = 1$

JEE Advanced & IIT JEE Archive

(Compilation of Last 10 Years' Questions)

1 Consider the hyperbola H: $x^2 - y^2 = 1$ and a circle S with centre $N(x_2, 0)$. Suppose that H and S touch each other at a point $P(x_1, y_1)$ with $x_1 > 1$ and $y_1 > 0$. The common tangent to H and S at P intersects the X-axis at point M. If (l, m) is the centroid of ΔPMN, then the correct expression(s) is/are

(2015, One or More than One Options Correct Type)

a. $\dfrac{dl}{dx_1} = 1 - \dfrac{1}{3x_1^2}$ for $x_1 > 1$

b. $\dfrac{dm}{dx_1} = \dfrac{x_1}{3(\sqrt{x_1^2 - 1})}$ for $x_1 > 1$

c. $\dfrac{dl}{dx_1} = 1 + \dfrac{1}{3x_1^2}$ for $x_1 > 1$

d. $\dfrac{dm}{dy_1} = \dfrac{1}{3}$ for $y_1 > 0$

2 Let $P(6, 3)$ be a point on the hyperbola $\dfrac{x^2}{a^2} - \dfrac{y^2}{b^2} = 1$. If the normal at the point P intersects the X-axis at (9, 0), then the eccentricity of the hyperbola is

(2011, Only One Option Correct Type)

a. $\sqrt{\dfrac{5}{2}}$ b. $\sqrt{\dfrac{3}{2}}$

c. $\sqrt{2}$ d. $\sqrt{3}$

3 Let the eccentricity of the hyperbola $\dfrac{x^2}{a^2} - \dfrac{y^2}{b^2} = 1$ be reciprocal to that of the ellipse $x^2 + 4y^2 = 4$. If the hyperbola passes through a focus of the ellipse, then

(2011, One or More than One Options Correct Type)

a. the equation of the hyperbola is $\dfrac{x^2}{3} - \dfrac{y^2}{2} = 1$

b. a focus of the hyperbola is (2, 0)

c. the eccentricity of the hyperbola is $\sqrt{\dfrac{5}{3}}$

d. the equation of hyperbola is $x^2 - 3y^2 = 3$

Passage for Q. Nos. (4-5)

The circle $x^2 + y^2 - 8x = 0$ and hyperbola $\dfrac{x^2}{9} - \dfrac{y^2}{4} = 1$ intersect at the points A and B.

(2010, Comprehension Type)

4 Equation of a common tangent with positive slope to the circle as well as to the hyperbola is

a. $2x - \sqrt{5}y - 20 = 0$ b. $2x - \sqrt{5}y + 4 = 0$

c. $3x - 4y + 8 = 0$ d. $4x - 3y + 4 = 0$

5 Equation of the circle with AB as its diameter is

a. $x^2 + y^2 - 12x + 24 = 0$

b. $x^2 + y^2 + 12x + 24 = 0$

c. $x^2 + y^2 + 24x - 12 = 0$

d. $x^2 + y^2 - 24x - 12 = 0$

6 The line $2x + y = 1$ is tangent to the hyperbola $\dfrac{x^2}{a^2} - \dfrac{y^2}{b^2} = 1$. If this line passes through the point of intersection of the nearest directrix and the X-axis, then the eccentricity of hyperbola is

(2010, Integer Type)

7 An ellipse intersects the hyperbola $2x^2 - 2y^2 = 1$ orthogonally. The eccentricity of the ellipse is reciprocal to that of the hyperbola. If the axes of the ellipse are along the coordinate axes, then **(2009, One or More than One Options Correct Type)**

a. equation of ellipse is $x^2 + 2y^2 = 2$
b. the foci of ellipse are $(\pm 1, 0)$
c. equation of ellipse is $x^2 + 2y^2 = 4$
d. the foci of ellipse are $(\pm\sqrt{2}, 0)$

8 Consider a branch of the hyperbola, $x^2 - 2y^2 - 2\sqrt{2}x - 4\sqrt{2}y - 6 = 0$ with vertex at the point A. Let B be one of the end points of its latusrectum. If C is the focus of hyperbola nearest to the point A, then the area (in sq unit) of ΔABC is **(2008, Only One Option Correct Type)**

a. $1 - \sqrt{\frac{2}{3}}$ b. $\sqrt{\frac{3}{2}} - 1$
c. $1 + \sqrt{\frac{2}{3}}$ d. $\sqrt{\frac{3}{2}} + 1$

9 A hyperbola, having the transverse axis of length $2\sin\theta$, is confocal with the ellipse $3x^2 + 4y^2 = 12$. Then, its equation is **(2007, Only One Option Correct Type)**

a. $x^2\operatorname{cosec}^2\theta - y^2\sec^2\theta = 1$ b. $x^2\sec^2\theta - y^2\operatorname{cosec}^2\theta = 1$
c. $x^2\sin^2\theta - y^2\cos^2\theta = 1$ d. $x^2\cos^2\theta - y^2\sin^2\theta = 1$

10 If e_1 is the eccentricity of ellipse $\frac{x^2}{16} + \frac{y^2}{25} = 1$ and e_2 is the eccentricity of the hyperbola passing through the foci of ellipse and $e_1 e_2 = 1$, then the equation of hyperbola is **(2006, Only One Option Correct Type)**

a. $\frac{x^2}{9} - \frac{y^2}{16} = 1$ b. $\frac{x^2}{16} - \frac{y^2}{9} = -1$
c. $\frac{x^2}{9} - \frac{y^2}{25} = 1$ d. None of these

11 Tangents are drawn from any point on the hyperbola $\frac{x^2}{9} - \frac{y^2}{4} = 1$ to the circle $x^2 + y^2 = 9$. Find the locus of mid-point of the chord of contact. **(2005, Subjective Type)**

4. Vectors

DPP-1 Introduction to Vectors

Only One Option Correct Type

Working Space

Direction (Q. Nos. 1-10) This section contains 10 multiple choice questions. Each question has four choices (a), (b), (c) and (d), out of which ONLY ONE is correct.

1 If the vector **b** is collinear with the vector $\mathbf{a} = (2\sqrt{2}, -1, 4)$ and $|\mathbf{b}| = 10$, then

a. $\mathbf{a} \pm \mathbf{b} = 0$ b. $\mathbf{a} \pm 2\mathbf{b} = 0$
c. $2\mathbf{a} \pm \mathbf{b} = 0$ d. None of these

2 If **a**, **b**, **c** and **d** are linearly independent set of vectors and $k_1\mathbf{a} + k_2\mathbf{b} + k_3\mathbf{c} + k_4\mathbf{d} = 0$, then

a. $k_1 + k_2 + k_3 + k_4 = 0$
b. $k_1 + k_3 = k_2 + k_4 \neq 0$
c. $k_1 + k_2 = k_3 + k_4$
d. None of the above

3 The position vector of two points A and B are $6\mathbf{a} + 2\mathbf{b}$ and $\mathbf{a} - 3\mathbf{b}$. If a point C divides AB in the ratio 3 : 2, then the position vector of C is

a. $3\mathbf{a} - \mathbf{b}$ b. $3\mathbf{a} + \mathbf{b}$
c. $\mathbf{a} + \mathbf{b}$ d. $\mathbf{a} - \mathbf{b}$

4 If $ABCD$ is a quadrilateral, E and F are the mid-points of AC and BD respectively, then $\mathbf{AB} + \mathbf{AD} + \mathbf{CB} + \mathbf{CD}$ is equal to

a. $\mathbf{EF}$ b. $2\mathbf{EF}$
c. $3\mathbf{EF}$ d. $4\mathbf{EF}$

5 If $ABCDEF$ is a regular hexagon and $\mathbf{AB} + \mathbf{AC} + \mathbf{AD} + \mathbf{AE} + \mathbf{AF} = \lambda\, \mathbf{AD}$, then λ is equal to

a. 0 b. 1
c. 2 d. 3

6 $OABCDE$ is a regular hexagon of side 2 units in the XY-plane. O being the origin and OA taken along the X-axis. A point P is taken on a line parallel to Z-axis through the centre of the hexagon at a distance of 3 units from O. Then, the vector **AP** is

a. $-\hat{\mathbf{i}} + 3\hat{\mathbf{j}} + \sqrt{5}\,\hat{\mathbf{k}}$ b. $\hat{\mathbf{i}} - \sqrt{3}\,\hat{\mathbf{j}} + 5\hat{\mathbf{k}}$
c. $-\hat{\mathbf{i}} + \sqrt{3}\,\hat{\mathbf{j}} + \sqrt{5}\,\hat{\mathbf{k}}$ d. $\hat{\mathbf{i}} + \sqrt{3}\,\hat{\mathbf{j}} + \sqrt{5}\,\hat{\mathbf{k}}$

7 The vertices of triangle are $A(1, 1, 2)$, $B(4, 3, 1)$ and $C(2, 3, 5)$. A vector representing the internal bisector of the $\angle A$ is

a. $\hat{\mathbf{i}} + \hat{\mathbf{j}} + 2\hat{\mathbf{k}}$ b. $2\hat{\mathbf{i}} - 2\hat{\mathbf{j}} + \hat{\mathbf{k}}$
c. $2\hat{\mathbf{i}} + 2\hat{\mathbf{j}} + \hat{\mathbf{k}}$ d. None of these

Working Space

8 The vector **c**, directed along the internal bisector of the angle between the vectors $\mathbf{a} = 7\hat{i} - 4\hat{j} - 4\hat{k}$ and $\mathbf{b} = -2\hat{i} - \hat{j} + 2\hat{k}$ with $|\mathbf{c}| = 5\sqrt{6}$, is

a. $\frac{5}{3}(\hat{i} - 7\hat{j} + 2\hat{k})$

b. $\frac{5}{3}(\hat{i} + 7\hat{j} - 2\hat{k})$

c. $\frac{5}{3}(\hat{i} + 7\hat{j} + 2\hat{k})$

d. $\frac{5}{3}(-\hat{i} - 7\hat{j} + 2\hat{k})$

9 Forces **P, Q** act at *O* and have a resultant **R**. If any transversal cuts their line of action at *A, B, C* respectively, then $\left|\frac{\mathbf{P}}{\mathbf{OA}}\right| + \left|\frac{\mathbf{Q}}{\mathbf{OB}}\right|$ is

a. $\left|\frac{\mathbf{R}}{\mathbf{OC}}\right|$
b. $2\left|\frac{\mathbf{R}}{\mathbf{OC}}\right|$

c. $3\left|\frac{\mathbf{R}}{\mathbf{OC}}\right|$
d. None of these

10 Let $\mathbf{A}(t) + f_1(t)\,\hat{i} + f_2(t)\,\hat{j}$ and $\mathbf{B}(t) = g_1(t)\,\hat{i} + g_2(t)\,\hat{j}$, where f_1, f_2, g_1, g_2 are continuous functions. If $\mathbf{A}(t)$ and $\mathbf{B}(t)$ are non-zero vectors for all t and $\mathbf{A}(0) = 2\hat{i} + 3\hat{j}$, $\mathbf{A}(1) = 6\hat{i} + 2\hat{j}$, $\mathbf{B}(0) = 3\hat{i} + 2\hat{j}$ and $\mathbf{B}(1) = 2\hat{i} + 6\hat{j}$, then **[IIT JEE 2001]**

a. **A** (t) is perpendicular to **B** (t)
b. **A** (t) is parallel to **B** (t) for all t
c. **A** (t) is parallel to **B** (t) for some t
d. None of the above

Comprehension Type

Direction (Q. Nos. 11 and 12) This section contains a paragraph, describing theory, experiments, data, etc. Two questions related to the paragraph have been given. Each question has only one correct answer among the four given options (a), (b), (c) and (d).

Passage

If two vectors **OA** and **OB** are there, then their resultant **OA** + **OB** can be found by completing the parallelogram *OACB* and **OC** = **OA** + **OB**. Also, if $|\mathbf{OA}| = |\mathbf{OB}|$, then the resultant will bisect the angle between them.

11 A vector **C** directed along internal bisector of angle between vectors $\mathbf{A} = 7\hat{i} - 4\hat{j} - 4\hat{k}$ and $\mathbf{B} = -2\hat{i} - \hat{j} + 2\hat{k}$ with $|\mathbf{C}| = 5\sqrt{6}$ is

a. $\frac{5}{3}(\hat{i} - \hat{j} + \hat{k})$
b. $\frac{5}{3}(\hat{i} - 7\hat{j} + 2\hat{k})$

c. $\frac{5}{3}(5\hat{i} + 5\hat{j} + 2\hat{k})$
d. $\frac{5}{3}(-5\hat{i} + 5\hat{j} + 2\hat{k})$

12 If internal and external bisectors of $\angle A$ of ΔABC meet the base *BC* at *D* and *E* respectively, then (*D* and *E* lie on same side of *B*)

a. $BC = \frac{BD + BE}{4}$

b. $BC^2 = BD \times DE$

c. $\frac{2}{BC} = \frac{1}{BD} + \frac{1}{BE}$

d. None of the above

Working Space

One Integer Value Correct Type

Direction (*Q. Nos.* 13-20) *This section contains* 8 *questions. When worked out will result in an integer from* 0 *to* 9 (*both inclusive*).

13 If O and O' are circumcentre and orthocentre of a ΔABC, where $\mathbf{OA} + \mathbf{OB} + \mathbf{OC}$ is $\lambda \mathbf{OO'}$, then the value λ is

14 If A, B, C, P and Q are five coplanar points, where $\mathbf{PA} + \mathbf{PB} + \mathbf{PC} + \mathbf{AQ} + \mathbf{BQ} + \mathbf{CQ}$ is λ $\mathbf{PQ}$, then the value of λ is

15 Let $\mathbf{a}$, $\mathbf{b}$, $\mathbf{c}$ be three non-zero vectors such that no two of these are collinear. If the vector $\mathbf{a} + 2\mathbf{b}$ is collinear with $\mathbf{c}$ and $\mathbf{b} + 3\mathbf{c}$ is collinear with $\mathbf{a}$, where $\mathbf{a} + 2\mathbf{b} + 6\mathbf{c} = \lambda \cdot \mathbf{a}$, then the value of λ is

16 If $\mathbf{a} + \mathbf{b}$ is along the angle bisector of $\mathbf{a}$ and $\mathbf{b}$, where $|\mathbf{a}| = \lambda |\mathbf{b}|$, then the value of λ is

17 If $A(1, -1, -3)$, $B(2, 1, -2)$ and $C(-5, 2, -6)$ are the position vectors of the vertices of ΔABC. The length of the bisector of its internal angle at A is $\frac{\lambda \sqrt{10}}{4}$, then the value of λ is

18 P and Q have position vectors $\mathbf{a}$ and $\mathbf{b}$ relative to the origin O and X, Y divide $\mathbf{PQ}$ internally and externally respectively in the ratio 2 : 1. Vector $\mathbf{XY}$ is $\lambda\, \mathbf{a} + \mu\, \mathbf{b}$, then the value of $|\lambda + \mu|$ is

19 Let $\mathbf{p}$ be the position vector of orthocentre and $\mathbf{g}$ is the position vector of the centroid of ΔABC, where circumcentre is the origin. If $\mathbf{p} = k\mathbf{g}$, then the value of k is

20 A vector $\mathbf{a}$ has component $2p$ and 1 with respect to a rectangular cartesian system. The system is rotated through a certain angle about the origin in the counter clockwise sense. If with to the new system, $\mathbf{a}$ has components $(p + 1)$ and 1, where p takes the values p_1 and p_2. Then, the value of $3|p_1 + p_2|$ is

DPP-2 Scalar or Dot Product

Subjective Questions

Working Space

Direction *(Q. Nos. 1 and 2) These questions are subjective in nature, need to be solved completely on notebook.*

1 i. If $\hat{\mathbf{a}}$ and $\hat{\mathbf{b}}$ are unit vectors inclined at an angle θ, then prove that $\sin\frac{\theta}{2} = \frac{1}{2}|\mathbf{a} - \mathbf{b}|$.

ii. Prove Cauchy-Schwartz inequality $(\mathbf{a}\cdot\mathbf{b})^2 \le |\mathbf{a}|^2 \cdot |\mathbf{b}|^2$.

iii. If the sum of two unit vectors is a unit vector, then prove that the magnitude of their difference is $\sqrt{3}$.

iv. If **a**, **b**, **c** are three mutually perpendicular vectors of equal magnitude, then prove that $\mathbf{a} + \mathbf{b} + \mathbf{c}$ is equally inclined with vectors **a**, **b** and **c**.

v. Prove that the two vectors **a** and **b** are at right angles, iff $|\mathbf{a} + \mathbf{b}|^2 = |\mathbf{a}|^2 + |\mathbf{b}|^2$.

2 i. Prove that the altitudes of a triangle are concurrent.

ii. Using vector method, prove that the angle in a semi-circle is a right angle.

iii. Prove cosine law, i.e. let a, b, c be lengths of the sides opposite respectively to the angles A, B, C of a ΔABC, show that $\cos A = \frac{b^2 + c^2 - a^2}{2bc}$.

iv. Prove projection formula, i.e. let a, b, c be lengths of the sides of a ΔABC, then show that $b = c\cos A + a\cos C$.

v. Using vectors, prove that $\cos(\alpha + \beta) = \cos\alpha\cos\beta - \sin\alpha\sin\beta$.

Only One Option Correct Type

Direction *(Q. Nos. 3-17) This section contains 15 multiple choice questions. Each question has four choices (a), (b), (c) and (d), out of which ONLY ONE is correct.*

3 If $|\mathbf{a}| = 5$, $|\mathbf{a} - \mathbf{b}| = 8$ and $|\mathbf{a} + \mathbf{b}| = 10$, then $|\mathbf{b}|$ is equal to

a. 1
b. $\sqrt{57}$
c. 3
d. None of these

4 Angle between diagonals of a parallelogram whose sides are represented by $\mathbf{a} = 2\hat{\mathbf{i}} + \hat{\mathbf{j}} + \hat{\mathbf{k}}$ and $\mathbf{b} = \hat{\mathbf{i}} - \hat{\mathbf{j}} - \hat{\mathbf{k}}$

a. $\cos^{-1}\left(\frac{1}{3}\right)$
b. $\cos^{-1}\left(\frac{1}{2}\right)$
c. $\cos^{-1}\left(\frac{4}{9}\right)$
d. $\cos^{-1}\left(\frac{5}{9}\right)$

5 A, B, C and D are four points in a plane with position vectors **a**, **b**, **c** and **d**, respectively such that $(\mathbf{a} - \mathbf{d})\cdot(\mathbf{b} - \mathbf{c}) = (\mathbf{b} - \mathbf{d})\cdot(\mathbf{c} - \mathbf{a}) = 0$, then for the ΔABC, D is its

a. incentre
b. circumcentre
c. orthocentre
d. centroid

Working Space

6 Let **a**, **b**, **c** be vectors of lengths 3, 4, 5 respectively and **a** be perpendicular to (**b** + **c**), **b** to (**c** + **a**) and **c** to (**a** + **b**), then the value of (**a** + **b** + **c**) is

a. $2\sqrt{5}$ b. $2\sqrt{2}$
c. $10\sqrt{5}$ d. $5\sqrt{2}$

7 If the unit vectors $\mathbf{e}_1$ and $\mathbf{e}_2$ are inclined at an angle 2θ and $|\mathbf{e}_1 - \mathbf{e}_2| < 1$, then for $\theta \in [0, \pi]$, θ may lie in the interval

a. $\left[0, \frac{\pi}{6}\right)$ b. $\left[\frac{\pi}{6}, \frac{\pi}{2}\right]$
c. $\left(\frac{5\pi}{6}, \pi\right]$ d. $\left[\frac{\pi}{2}, \frac{5\pi}{6}\right]$

8 Let $\mathbf{b} = 4\hat{\mathbf{i}} + 3\hat{\mathbf{j}}$ and **c** be a vector perpendicular to **b** and it lies in the *XY*-plane. A vector in the *XY*-plane having projections 1 and 2 along **b** and **c** is

a. $2\hat{\mathbf{i}} + \hat{\mathbf{j}}$ b. $\hat{\mathbf{i}} - 2\hat{\mathbf{j}}$
c. $\frac{1}{5}(-2\hat{\mathbf{i}} + 11\hat{\mathbf{j}})$ d. None of these

9 In the isosceles ΔABC, $|\mathbf{AB}| = |\mathbf{BC}| = 8$ and a point E divides AB internally in the ratio 1 : 3, then the cosine of angle between **CE** and **CA** is (where, $|\mathbf{CA}| = 12$)

a. $-\frac{3\sqrt{7}}{8}$ b. $\frac{3\sqrt{8}}{17}$
c. $\frac{3\sqrt{7}}{8}$ d. $-\frac{3\sqrt{8}}{17}$

10 If A is 3×3 matrix and **u** is a vector. If $A\mathbf{u}$ and **u** are orthogonal for all real **u**, then matrix A is a

a. singular b. non-singular
c. symmetric d. skew-symmetric

11 Let the cosine of angle between the vectors **p** and **q** be λ such that $2\mathbf{p} + \mathbf{q} = \hat{\mathbf{i}} + \hat{\mathbf{j}}$ and $\mathbf{p} + 2\mathbf{q} = \hat{\mathbf{i}} - \hat{\mathbf{j}}$, then λ is equal to

a. $\frac{5}{9}$ b. $-\frac{4}{5}$
c. $\frac{3}{9}$ d. $\frac{7}{9}$

12 The position vectors of the vertices of triangle are $3\hat{\mathbf{i}} + 4\hat{\mathbf{j}} + 5\hat{\mathbf{k}}$, $\hat{\mathbf{i}} + 7\hat{\mathbf{k}}$ and $5\hat{\mathbf{i}} + 5\hat{\mathbf{j}}$, the distance between the circumcentre and orthocentre is

a. 0 b. $\sqrt{306}$
c. $2\sqrt{306}$ d. $\frac{3}{2}\sqrt{306}$

13 If **a**, **b** and **c** are three vectors, such that $\mathbf{a} + \mathbf{b} + \mathbf{c} = 0$, $|\mathbf{a}| = 1$, $|\mathbf{b}| = 2$, $|\mathbf{c}| = 2$, then $\mathbf{a} \cdot \mathbf{b} + \mathbf{b} \cdot \mathbf{c} + \mathbf{c} \cdot \mathbf{a}$ is

a. $-\frac{9}{2}$ b. 7
c. 1 d. None of these

14 Let **u**, **v** and **w** be such that $|\mathbf{u}| = 1$, $|\mathbf{v}| = 2$, $|\mathbf{w}| = 3$. If the projection **v** along **u** is equal to that of **w** along **u** and **v**, **w** are perpendicular to each other, then $|\mathbf{u} - \mathbf{v} + \mathbf{w}|$ equals

a. $\sqrt{14}$ b. $\sqrt{7}$
c. 2 d. 14

Working Space

15 If **a**, **b** and **c** are unit vectors, then $|\mathbf{a}-\mathbf{b}|^2+|\mathbf{b}-\mathbf{c}|^2+|\mathbf{c}-\mathbf{a}|^2$ does not exceed

a. 4 b. 9
c. 8 d. 6

16 Let two non-collinear unit vectors $\hat{\mathbf{a}}$ and $\hat{\mathbf{b}}$ form an acute angle. A point P moves so that at any time t, the position vector **OP** (where, O is the origin) is given by $\hat{\mathbf{a}}\cos t+\hat{\mathbf{b}}\sin t$. When P is farthest from the origin O, let M be the length of **OP** and **u** be the unit vector along OP. Then,

a. $\hat{\mathbf{u}}=\dfrac{\hat{\mathbf{a}}+\hat{\mathbf{b}}}{|\hat{\mathbf{a}}+\hat{\mathbf{b}}|}$ and $M=(\hat{\mathbf{a}}\cdot\hat{\mathbf{b}}+1)^{1/2}$

b. $\hat{\mathbf{u}}=\dfrac{\hat{\mathbf{a}}+\hat{\mathbf{b}}}{|\hat{\mathbf{a}}+\hat{\mathbf{b}}|}$ and $M=(1+2\hat{\mathbf{a}}\cdot\hat{\mathbf{b}})^{1/2}$

c. $\hat{\mathbf{u}}=\dfrac{\hat{\mathbf{a}}-\hat{\mathbf{b}}}{|\hat{\mathbf{a}}-\hat{\mathbf{b}}|}$ and $M=(1+\hat{\mathbf{a}}\cdot\hat{\mathbf{b}})^{1/2}$

d. $\hat{\mathbf{u}}=\dfrac{\hat{\mathbf{a}}-\hat{\mathbf{b}}}{|\hat{\mathbf{a}}-\hat{\mathbf{b}}|}$ and $M=(1+2\hat{\mathbf{a}}\cdot\hat{\mathbf{b}})^{1/2}$

17 Two adjacent sides of a parallelogram $ABCD$ are given by $\mathbf{AB}=2\hat{\mathbf{i}}+10\hat{\mathbf{j}}+11\hat{\mathbf{k}}$ and $\mathbf{AD}=-\hat{\mathbf{i}}+2\hat{\mathbf{j}}+2\hat{\mathbf{k}}$. The side AD is rotated by an acute angle α in the plane of parallelogram so that AD becomes AD'. If AD' makes a right angle with the side AB, then the cosine of angle α is

a. $\dfrac{8}{9}$ b. $\dfrac{\sqrt{17}}{9}$

c. $\dfrac{1}{9}$ d. $\dfrac{4\sqrt{5}}{9}$

One or More than One Option Correct Type

Direction (Q. Nos. 18-20) *This section contains 3 multiple choice questions. Each question has four choices (a), (b), (c) and (d), out of which ONE or MORE THAN ONE are correct.*

18 If $\mathbf{z}_1=a\hat{\mathbf{i}}+b\hat{\mathbf{j}}$ and $\mathbf{z}_2=c\hat{\mathbf{i}}+d\hat{\mathbf{j}}$ are two vectors in $\hat{\mathbf{i}}$ and $\hat{\mathbf{j}}$ system, where $|\mathbf{z}_1|=|\mathbf{z}_2|=r$ and $\mathbf{z}_1\cdot\mathbf{z}_2=0$, then $\mathbf{w}_1=a\hat{\mathbf{i}}+c\hat{\mathbf{j}}$ and $\mathbf{w}_2=b\hat{\mathbf{i}}+d\hat{\mathbf{j}}$ satisfy

a. $|\mathbf{w}_1|=r$ b. $|\mathbf{w}_2|=r$
c. $\mathbf{w}_1\cdot\mathbf{w}_2=0$ d. None of these

19 **a**, **b** and **c** are mutually perpendicular vectors of equal magnitude, then the angle between $\mathbf{a}+\mathbf{b}+\mathbf{c}$ and **a** is

a. $\cos^{-1}\left(\dfrac{1}{3}\right)$ b. $\cos^{-1}\left(\dfrac{1}{\sqrt{3}}\right)$

c. $\pi-\cos^{-1}\left(\dfrac{1}{\sqrt{3}}\right)$ d. $\tan^{-1}(\sqrt{2})$

20 If unit vectors $\hat{\mathbf{i}}$ and $\hat{\mathbf{j}}$ are at right angles to each other and $\mathbf{p}=3\hat{\mathbf{i}}+4\hat{\mathbf{j}}$, $\mathbf{q}=5\hat{\mathbf{i}}$, $4\mathbf{r}=\mathbf{p}+\mathbf{q}$ and $2\mathbf{s}=\mathbf{p}-\mathbf{q}$, then

a. $|\mathbf{r}+k\mathbf{s}|=|\mathbf{r}-k\mathbf{s}|$ for all real k
b. **r** is perpendicular to **s**
c. $\mathbf{r}+\mathbf{s}$ is perpendicular to $\mathbf{r}-\mathbf{s}$
d. $|\mathbf{r}|=|\mathbf{s}|=|\mathbf{p}|=|\mathbf{q}|$

Comprehension Type

Direction (Q. Nos. 21-23) This section contains a paragraph, describing theory, experiments, data, etc. Three questions related to the paragraph have been given. Each question has only one correct answer among the four given options (a), (b), (c) and (d).

Working Space

Passage

Let **r** be a position vector of a variable point in the cartesian XOY-plane such that $\mathbf{r}\cdot(10\hat{\mathbf{j}} - 8\hat{\mathbf{i}} - \mathbf{r}) = 40$ and $P_1 = \max\{|\mathbf{r} + 2\hat{\mathbf{i}} - 3\hat{\mathbf{j}}|^2\}$, $P_2 = \min\{|\mathbf{r} + 2\hat{\mathbf{i}} - 3\hat{\mathbf{j}}|^2\}$.
A tangent is drawn to curve $y = \frac{8}{x^2}$ at a point D with abscissa 2 intersect X-axis at E.

21 Locus of **r** is

a. a circle with centre (4, 5)
b. a circle with ends of diameter (0, 0), (8, – 10)
c. a circle with radius 1
d. an ellipse whose major axis is 40

22 The value of $p_1 + p_2$ is

a. $8\sqrt{2}$ b. 8
c. $4\sqrt{2}$ d. 18

23 $\mathbf{DE}\cdot\mathbf{OE}$ is

a. 3 b. $3\sqrt{2}$
c. $4\sqrt{2}$ d. 18

One Integer Value Correct Type

Direction (Q. Nos. 24 and 25) This section contains 2 questions. When worked out will result in an integer from 0 to 9 (both inclusive).

24 The position vector of a point P is $\mathbf{r} = x\hat{\mathbf{i}} + y\hat{\mathbf{j}} + z\hat{\mathbf{k}}$, where $x, y, z \in N$ and $\mathbf{a} = \hat{\mathbf{i}} + 2\hat{\mathbf{j}} + \hat{\mathbf{k}}$. If $\mathbf{r}\cdot\mathbf{a} = 20$ and the number of possible position of P is 9λ, then the value of λ is

25 Let $\mathbf{a}\cdot\mathbf{b} > 0$, $\overline{\alpha} = \frac{\hat{\mathbf{i}}}{a} + \frac{4\hat{\mathbf{j}}}{b} + b\hat{\mathbf{k}}$ and $\overline{\beta} = b\hat{\mathbf{i}} + a\hat{\mathbf{j}} + \frac{1}{b}\hat{\mathbf{k}}$, then the maximum value of $\frac{10}{5 + \overline{\alpha}\cdot\overline{\beta}}$ is

DPP-3 Vector or Cross Product

Only One Option Correct Type

Working Space

Direction *(Q. Nos. 1-16) This section contains 16 multiple choice questions. Each question has four choices (a), (b), (c), and (d), out of which ONLY ONE is correct.*

1 For any vectors $\mathbf{a}, \mathbf{b}$; $|\mathbf{a} \times \mathbf{b}|^2 + (\mathbf{a} \cdot \mathbf{b})^2$ is equal to

a. $|\mathbf{a}|^2 |\mathbf{b}|^2$ b. $|\mathbf{a} + \mathbf{b}|$
c. $|\mathbf{a}|^2 - |\mathbf{b}|^2$ d. 0

2 If $\mathbf{a} = \hat{\mathbf{i}} + \hat{\mathbf{j}} + \hat{\mathbf{k}}$, $\mathbf{b} = \hat{\mathbf{i}} + \hat{\mathbf{j}} - \hat{\mathbf{k}}$, then vectors perpendicular to $\mathbf{a}$ and $\mathbf{b}$ is/are

a. $\lambda(\hat{\mathbf{i}} + \hat{\mathbf{j}})$ b. $\lambda(\hat{\mathbf{i}} + \hat{\mathbf{j}} + \hat{\mathbf{k}})$
c. $\lambda(\hat{\mathbf{i}} - \hat{\mathbf{j}})$ d. None of these

3 If $\mathbf{a} \times \mathbf{b} = \mathbf{b} \times \mathbf{c} \neq 0$, then the correct statement is

a. $\mathbf{b} || \mathbf{c}$ b. $\mathbf{a} || \mathbf{b}$
c. $(\mathbf{a} + \mathbf{c}) || \mathbf{b}$ d. None of these

4 If vectors $\mathbf{a}$, $\mathbf{b}$ and $\mathbf{c}$ represent sides BC, CA and AB of ΔABC, then **[IIT JEE 2007]**

a. $\mathbf{a} \cdot \mathbf{b} + \mathbf{b} \cdot \mathbf{c} + \mathbf{c} \cdot \mathbf{a} = 0$ b. $\mathbf{a} \times \mathbf{b} = \mathbf{b} \times \mathbf{c} = \mathbf{c} \times \mathbf{a}$
c. $\mathbf{a} \cdot \mathbf{b} = \mathbf{b} \cdot \mathbf{c} = \mathbf{c} \cdot \mathbf{a}$ d. $\mathbf{a} \times \mathbf{b} + \mathbf{b} \times \mathbf{c} + \mathbf{c} \times \mathbf{a} = 0$

5 If $\mathbf{a} = \hat{\mathbf{i}} + 2\hat{\mathbf{j}} + 3\hat{\mathbf{k}}$, $\mathbf{b} = -\hat{\mathbf{i}} + 2\hat{\mathbf{j}} + \hat{\mathbf{k}}$ and $\mathbf{c} = 3\hat{\mathbf{i}} + \hat{\mathbf{j}}$. If $(\mathbf{a} + t\mathbf{b}) \perp \mathbf{c}$, then t is equal to

a. 5 b. 4
c. 3 d. 2

6 If $\mathbf{a} = 2\hat{\mathbf{i}} - 3\hat{\mathbf{j}} + \hat{\mathbf{k}}$, $\mathbf{b} = -\hat{\mathbf{i}} + \hat{\mathbf{k}}$, $\mathbf{c} = 2\hat{\mathbf{j}} - \hat{\mathbf{k}}$, then the area (in sq units) of parallelogram with diagonals $\mathbf{a} + \mathbf{b}$ and $\mathbf{b} + \mathbf{c}$ will be

a. $\sqrt{21}$ b. $2\sqrt{21}$
c. $\frac{1}{2}\sqrt{21}$ d. None of these

7 The points with position vectors $10\hat{\mathbf{i}} + 3\hat{\mathbf{j}}$, $12\hat{\mathbf{i}} - 15\hat{\mathbf{j}}$ and $a\hat{\mathbf{i}} + 11\hat{\mathbf{j}}$ are collinear, if a is

a. -8 b. 4
c. 2 d. $\frac{82}{9}$

8 Let $\mathbf{OA} = \mathbf{a}$, $\mathbf{OB} = 10\mathbf{a} + 2\mathbf{b}$ and $\mathbf{OC} = \mathbf{b}$, where O, A and C are non-collinear points, p denotes the area of quadrilateral $OABC$ and q denotes the area of the parallelogram with OA and OC as adjacent sides. If $p = kq$, then k is equal to

a. 5 b. 6 c. 4 d. 3

9 In a parallelogram $ABCD$, $\mathbf{AB} = \hat{\mathbf{i}} + \hat{\mathbf{j}} + \hat{\mathbf{k}}$ and diagonal $\mathbf{AC} = \hat{\mathbf{i}} - \hat{\mathbf{j}} + \hat{\mathbf{k}}$ and area of parallelogram is $\sqrt{8}$ sq units, then $\angle BAC$ is equal to

a. $\frac{\pi}{6}$ b. $\frac{\pi}{3}$ c. $\sin^{-1}\left(\frac{\sqrt{8}}{3}\right)$ d. $\cos^{-1}\left(\frac{\sqrt{8}}{3}\right)$

Working Space

10 If $\hat{\mathbf{u}}$ and $\hat{\mathbf{v}}$ are unit vectors and θ is the acute angle between them, then $2\hat{\mathbf{u}} \times 3\hat{\mathbf{v}}$ is a unit vector for **[AIEEE 2007]**

a. no value of θ
b. exactly one value of θ
c. exactly two values of θ
d. more than 2 values of θ

11 Let $\mathbf{u} = \hat{\mathbf{i}} + \hat{\mathbf{j}}$, $\mathbf{v} = \hat{\mathbf{i}} - \hat{\mathbf{j}}$ and $\mathbf{w} = \hat{\mathbf{i}} + 2\hat{\mathbf{j}} + 3\hat{\mathbf{k}}$. If $\hat{\mathbf{n}}$ is a unit vector such that $\mathbf{u} \cdot \hat{\mathbf{n}} = 0$ and $\mathbf{v} \cdot \hat{\mathbf{n}} = 0$, then $|\mathbf{w} \cdot \mathbf{n}|$ is equal to **[AIEEE 2003]**

a. 1 b. 2 c. 3 d. 4

12 For any vector $\mathbf{a}$, the value of $|\mathbf{a} \times \hat{\mathbf{i}}|^2 + |\mathbf{a} \times \hat{\mathbf{j}}|^2 + |\mathbf{a} \times \hat{\mathbf{k}}|^2$ is **[AIEEE 2005]**

a. $4|\mathbf{a}|^2$ b. $2|\mathbf{a}|^2$ c. $|\mathbf{a}|^2$ d. $3|\mathbf{a}|^2$

13 Let $\mathbf{a} = \hat{\mathbf{i}} + \hat{\mathbf{j}}$ and $\mathbf{b} = 2\hat{\mathbf{i}} - \hat{\mathbf{k}}$. The point of intersection of the lines $\mathbf{r} \times \mathbf{a} = \mathbf{b} \times \mathbf{a}$ and $\mathbf{r} \times \mathbf{b} = \mathbf{a} \times \mathbf{b}$ is

a. $-\hat{\mathbf{i}} + \hat{\mathbf{j}} + \hat{\mathbf{k}}$
b. $3\hat{\mathbf{i}} + \hat{\mathbf{j}} - \hat{\mathbf{k}}$
c. $3\hat{\mathbf{i}} - \hat{\mathbf{j}} + \hat{\mathbf{k}}$
d. $\hat{\mathbf{i}} - \hat{\mathbf{j}} + \hat{\mathbf{k}}$

14 If $\mathbf{a}$, $\mathbf{b}$ and $\mathbf{c}$ are pth, qth, rth terms of HP and $\mathbf{u} = (q - r)\hat{\mathbf{i}} + (r - p)\hat{\mathbf{j}} + (p - q)\hat{\mathbf{k}}$, $\mathbf{v} = \frac{\hat{\mathbf{i}}}{a} + \frac{\hat{\mathbf{j}}}{b} + \frac{\hat{\mathbf{k}}}{c}$, then

a. $\mathbf{u}$ and $\mathbf{v}$ are parallel vectors
b. $\mathbf{u}$ and $\mathbf{v}$ are orthogonal vectors
c. $\mathbf{u} \cdot \mathbf{v} = 1$
d. $\mathbf{u} \times \mathbf{v} = \hat{\mathbf{i}} + \hat{\mathbf{j}} + \hat{\mathbf{k}}$

15 If the vector product of a constant vector **OA** with variable vector **OB** in a fixed plane OAB be a constant vector, then the locus of B is

a. a straight line perpendicular to **OA**
b. a circle with centre O radius equal to $|\mathbf{OA}|$
c. a straight line parallel to **OA**
d. None of the above

16 Unit vector perpendicular to the plane of ΔABC with position vectors $\mathbf{a}$, $\mathbf{b}$, $\mathbf{c}$ of the vertices A, B, C is

a. $\dfrac{\mathbf{a} \times \mathbf{b} + \mathbf{b} \times \mathbf{c} + \mathbf{c} \times \mathbf{a}}{\Delta}$
b. $\dfrac{\mathbf{a} \times \mathbf{b} + \mathbf{b} \times \mathbf{c} + \mathbf{c} \times \mathbf{a}}{2\Delta}$
c. $\dfrac{\mathbf{a} \times \mathbf{b} + \mathbf{b} \times \mathbf{c} + \mathbf{c} \times \mathbf{a}}{4\Delta}$
d. None of these

Comprehension Type

Direction *(Q. Nos. 17-21) This section contains 2 paragraphs, each describing theory, experiments, data, etc. Five questions related to the paragraphs have been given. Each question has only one correct answer among the four given options (a), (b), (c) and (d).*

Passage I

Let $\mathbf{a}$, $\mathbf{b}$ be two vectors perpendicular to each other and $|\mathbf{a}| = 2$, $|\mathbf{b}| = 3$ and $\mathbf{c} \times \mathbf{a} = \mathbf{b}$.

17 The least value of $|\mathbf{c} - \mathbf{a}|$ is

a. 1 b. $\frac{1}{2}$ c. $\frac{1}{4}$ d. $\frac{3}{2}$

18 When $|\mathbf{c} - \mathbf{a}|$ is least the value of α (when α is angle between $\mathbf{a}$ and $\mathbf{c}$) equals

a. $\tan^{-1}\left(\frac{1}{2}\right)$ b. $\tan^{-1}\left(\frac{3}{4}\right)$ c. $\cos^{-1}\left(\frac{2}{3}\right)$ d. None of these

Working Space

19 When $|\mathbf{c}-\mathbf{a}|$ attains least value, then the value of $|\mathbf{c}|$ is

a. $\frac{1}{2}$ b. $\frac{7}{2}$

c. $\frac{5}{2}$ d. 4

Passage II

Three vectors **a**, **b** and **c** are forming a right handed system, if $\mathbf{a}\times\mathbf{b}=\mathbf{c}$, $\mathbf{b}\times\mathbf{c}=\mathbf{a}$, $\mathbf{c}\times\mathbf{a}=\mathbf{b}$.

20 If $x=\mathbf{a}+\mathbf{b}-\mathbf{c}$, $y=-\mathbf{a}+\mathbf{b}-2\mathbf{c}$, $z=-\mathbf{a}+2\mathbf{b}-\mathbf{c}$, then a unit vector normal to $\mathbf{x}+\mathbf{y}$ and $\mathbf{y}+\mathbf{z}$ is

a. **a** b. **b**

c. **c** d. None of these

21 If $\mathbf{x}=\mathbf{a}+\mathbf{b}$, $\mathbf{y}=2\mathbf{a}-\mathbf{b}$, then the point of intersection of straight lines $\mathbf{r}\times\mathbf{x}=\mathbf{y}\times\mathbf{x}$, $\mathbf{r}\times\mathbf{y}=\mathbf{x}\times\mathbf{y}$ is

a. $2\mathbf{b}$ b. $3\mathbf{b}$

c. $3\mathbf{a}$ d. None of these

One Integer Value Correct Type

Direction *(Q. Nos. 22-30) This section contains 9 questions. When worked out will result in an integer from 0 to 9 (both inclusive).*

22 If $|\mathbf{a}|=3$, $|\mathbf{b}|=1$ and $|\mathbf{a}+\mathbf{b}|=|\mathbf{a}-\mathbf{b}|$, then the value of $|\mathbf{a}\times\mathbf{b}|$ is

23 If **a**, **b** are vectors perpendicular to each other and $|\mathbf{a}|=2$, $|\mathbf{b}|=3$, $\mathbf{c}\times\mathbf{a}=\mathbf{b}$, then the least value of $2|\mathbf{c}-\mathbf{a}|$ is

24 The arc AC of a circle subtends a right angle at the centre O. The point B divides the arc AC in the ratio 1 : 2. If $\mathbf{OA}=\mathbf{a}$, $\mathbf{OB}=\mathbf{b}$ and $\mathbf{OC}=x\mathbf{a}+y\mathbf{b}$, then the value of $|\sqrt{3}xy|$ is

25 If $\mathbf{a}=\hat{\mathbf{i}}+\hat{\mathbf{j}}+\hat{\mathbf{k}}$, $\mathbf{a}\cdot\mathbf{b}=1$ and $\mathbf{a}\times\mathbf{b}=\hat{\mathbf{j}}-\hat{\mathbf{k}}$, then the value of $|\mathbf{b}|$ is

26 If **a** and **b** are vectors such that $|\mathbf{a}+\mathbf{b}|=\sqrt{29}$ and $\mathbf{a}\times(2\hat{\mathbf{i}}+3\hat{\mathbf{j}}+4\hat{\mathbf{k}})=(2\hat{\mathbf{i}}+3\hat{\mathbf{j}}+4\hat{\mathbf{k}})\times\mathbf{b}$, then a possible value of $|(\mathbf{a}+\mathbf{b})\cdot(-7\hat{\mathbf{i}}+2\hat{\mathbf{j}}+3\hat{\mathbf{k}})|$ is **[IIT JEE 2012]**

27 Let $\mathbf{a}=-\hat{\mathbf{i}}-\hat{\mathbf{k}}$, $\mathbf{b}=-\hat{\mathbf{i}}+\hat{\mathbf{j}}$ and $\mathbf{c}=\hat{\mathbf{i}}+2\hat{\mathbf{j}}+3\hat{\mathbf{k}}$ be three given vectors. If **r** is a vector such that $\mathbf{r}\times\mathbf{b}=\mathbf{c}\times\mathbf{b}$ and $\mathbf{r}\cdot\mathbf{a}=0$, then the value of $\mathbf{r}\cdot\mathbf{b}$ is

28 For any four points P, Q, R, S, if $|\mathbf{PQ}\times\mathbf{RS}-\mathbf{QR}\times\mathbf{PS}+\mathbf{RP}\times\mathbf{QS}|=\lambda$ area of ΔQRS, then the value of λ is

29 From the following system for unknown vectors **x** and **y** (**a** and **b** are given vectors) such that $\mathbf{x}+\mathbf{y}=\mathbf{a}$, $\mathbf{x}\times\mathbf{y}=\mathbf{b}$ and $\mathbf{x}\cdot\mathbf{a}=1$. We can conclude that $\mathbf{x}=\dfrac{\mathbf{a}+\lambda(\mathbf{a}\times\mathbf{b})}{|\mathbf{a}|^2}$, then the value of λ is

30 Three vectors **a** ($|\mathbf{a}|\neq 0$), **b** and **c** are such that $\mathbf{a}\times\mathbf{b}=3\mathbf{a}\times\mathbf{c}$, also $|\mathbf{a}|=|\mathbf{b}|=1$ and $|\mathbf{c}|=\frac{1}{3}$. If the angle between **b** and **c** is 60° and $|\mathbf{b}-3\mathbf{c}|=\lambda|\mathbf{a}|$, then the value of λ is

DPP-4 Scalar Triple Product

Only One Option Correct Type

Working Space

Direction (Q. Nos. 1-21) This section contains 21 multiple choice questions. Each question has four choices (a), (b), (c) and (d), out of which ONLY ONE is correct.

1 The value of $\hat{\mathbf{i}}\cdot(\hat{\mathbf{j}}\times\hat{\mathbf{k}})+\hat{\mathbf{j}}\cdot(\hat{\mathbf{k}}\times\hat{\mathbf{i}})+\hat{\mathbf{k}}\cdot(\hat{\mathbf{i}}\times\hat{\mathbf{j}})$ is

a. 3 b. 2
c. 1 d. 0

2 Which of the following is meaningful?

a. $(\mathbf{u}\cdot\mathbf{v})\cdot\mathbf{w}$ b. $\mathbf{u}\times(\mathbf{v}\cdot\mathbf{w})$
c. $(\mathbf{u}\cdot\mathbf{v})\,\mathbf{w}$ d. None of these

3 If **u**, **v** and **w** are the three non-coplanar vectors, then $(\mathbf{u}+\mathbf{v}-\mathbf{w})\cdot(\mathbf{u}-\mathbf{v})\times(\mathbf{v}-\mathbf{w})$ is equal to

a. $\mathbf{u}\cdot(\mathbf{v}\times\mathbf{w})$ b. $\mathbf{u}\cdot(\mathbf{w}\times\mathbf{v})$
c. $3\mathbf{u}\cdot(\mathbf{v}\times\mathbf{w})$ d. 0

4 If $2\hat{\mathbf{i}}-\hat{\mathbf{j}}+\hat{\mathbf{k}}$, $\hat{\mathbf{i}}+2\hat{\mathbf{j}}-3\hat{\mathbf{k}}$, $3\hat{\mathbf{i}}+p\hat{\mathbf{j}}+5\hat{\mathbf{k}}$ are coplanar, then the value of p is

a. 1 b. −2
c. 4 d. −4

5 If the vectors $a\hat{\mathbf{i}}+\hat{\mathbf{j}}+\hat{\mathbf{k}}$, $\hat{\mathbf{i}}+b\hat{\mathbf{j}}+\hat{\mathbf{k}}$ and $\hat{\mathbf{i}}+\hat{\mathbf{j}}+c\hat{\mathbf{k}}$ $(a\neq b\neq c\neq 1)$ are coplanar, then the value of $\dfrac{1}{1-a}+\dfrac{1}{1-b}+\dfrac{1}{1-c}$ is

a. 1 b. −1
c. 0 d. None of these

6 The volume of parallelopiped whose sides are given by $\mathbf{OA}=2\hat{\mathbf{i}}-2\hat{\mathbf{j}}$, $\mathbf{OB}=\hat{\mathbf{i}}+\hat{\mathbf{j}}-\hat{\mathbf{k}}$, $\mathbf{OC}=3\hat{\mathbf{i}}-\hat{\mathbf{k}}$ is

a. 2 b. 3
c. $\dfrac{2}{7}$ d. $\dfrac{4}{13}$

7 For non-zero vectors **a, b, c**; $|(\mathbf{a}\times\mathbf{b})\cdot\mathbf{c}|=|\mathbf{a}||\mathbf{b}||\mathbf{c}|$ holds if and only if

a. $\mathbf{a}\cdot\mathbf{b}=0,\ \mathbf{b}\cdot\mathbf{c}=0$
b. $\mathbf{b}\cdot\mathbf{c}=0,\ \mathbf{c}\cdot\mathbf{a}=0$
c. $\mathbf{c}\cdot\mathbf{a}=0,\ \mathbf{a}\cdot\mathbf{b}=0$
d. $\mathbf{a}\cdot\mathbf{b}=\mathbf{b}\cdot\mathbf{c}=\mathbf{c}\cdot\mathbf{a}=0$

8 For three vectors **a, b, c**, the value of $[\mathbf{a}+\mathbf{b}\ \ \mathbf{b}+\mathbf{c}\ \ \mathbf{c}+\mathbf{a}]$ is

a. $[\mathbf{a}\ \mathbf{b}\ \mathbf{c}]$ b. 0
c. $[\mathbf{a}\ \mathbf{b}\ \mathbf{c}]^2$ d. $2[\mathbf{a}\ \mathbf{b}\ \mathbf{c}]$

Working Space

9 The position vectors of three vertices A, B, C of a tetrahedron $OABC$ with respect to its vertex O are $6\hat{\mathbf{i}}$, $6\hat{\mathbf{j}}$, $\hat{\mathbf{k}}$, then its volume (in cu units) is

a. 3 b. $\frac{1}{3}$

c. $\frac{1}{6}$ d. 6

10 If $\mathbf{a}, \mathbf{b}, \mathbf{c}$ are unit coplanar vectors, then scalar triple product $[2\mathbf{a} - \mathbf{b}\ 2\mathbf{b} - \mathbf{c}\ 2\mathbf{c} - \mathbf{a}]$ is

a. 0 b. 1

c. $-\sqrt{3}$ d. $[\mathbf{a}\ \mathbf{b}\ \mathbf{c}]$

11 If $|\mathbf{a}| = 1, |\mathbf{b}| = 3$ and $|\mathbf{c}| = 5$, then the value of $[\mathbf{a} - \mathbf{b}\ \mathbf{b} - \mathbf{c}\ \mathbf{c} - \mathbf{a}]$ is

a. 0 b. 1

c. −1 d. None of these

12 If $\mathbf{a}, \mathbf{b}, \mathbf{c}$ are three non-coplanar vectors, then $3\mathbf{a} - 7\mathbf{b} - 4\mathbf{c}$, $3\mathbf{a} - 2\mathbf{b} + \mathbf{c}$ and $\mathbf{a} + \mathbf{b} + \lambda\mathbf{c}$ will be coplanar, if λ is

a. −1 b. 1

c. 3 d. 2

13 If $\mathbf{a}, \mathbf{b}, \mathbf{c}$ are non-coplanar vectors and λ is real number, then the vectors $\mathbf{a} + 2\mathbf{b} + 3\mathbf{c}$, $\lambda\mathbf{b} + 4\mathbf{c}$ and $(2\lambda - 1)\,\mathbf{c}$ are non-coplanar for **[AIEEE 2004]**

a. all except two values of λ

b. all except one value of λ

c. all values of λ

d. no value of λ

14 Let $\mathbf{a} = \hat{\mathbf{i}} - \hat{\mathbf{k}}$, $\mathbf{b} = x\hat{\mathbf{i}} + \hat{\mathbf{j}} + (1 - x)\hat{\mathbf{k}}$ and $\mathbf{c} = y\hat{\mathbf{i}} + x\hat{\mathbf{j}} + (1 + x - y)\hat{\mathbf{k}}$, then $[\mathbf{a}\ \mathbf{b}\ \mathbf{c}]$ depends on **[AIEEE 2005]**

a. neither x nor y b. both x and y

c. only x d. only y

15 Let a, b, c be distinct non-negative numbers. If the vectors $a\hat{\mathbf{i}} + a\hat{\mathbf{j}} + c\hat{\mathbf{k}}$, $\hat{\mathbf{i}} + \hat{\mathbf{k}}$ and $c\hat{\mathbf{i}} + c\hat{\mathbf{j}} + b\hat{\mathbf{k}}$ lie in a plane, then c is **[AIEEE 2005]**

a. HM of a and b

b. 0

c. AM of a and b

d. GM of a and b

16 If $\mathbf{a}, \mathbf{b}$ and $\mathbf{c}$ are non-coplanar vectors and λ is a real number, then $[\lambda(\mathbf{a} + \mathbf{b})\ \lambda^2\mathbf{b}\ \lambda\mathbf{c}] = [a\ \mathbf{a} + \mathbf{c}\ \mathbf{b}]$ for **[AIEEE 2005]**

a. exactly two values of λ

b. exactly one value of λ

c. no value of λ

d. exactly three values of λ

17 Let $\mathbf{a} = \hat{\mathbf{i}} + \hat{\mathbf{j}} + \hat{\mathbf{k}}$, $\mathbf{b} = \hat{\mathbf{i}} - \hat{\mathbf{j}} + 2\hat{\mathbf{k}}$ and $\mathbf{c} = x\hat{\mathbf{i}} + (x - 2)\,\hat{\mathbf{j}} - \hat{\mathbf{k}}$. If the vector $\mathbf{c}$ lies in the plane of $\mathbf{a}$ and $\mathbf{b}$, then x is equal to **[AIEEE 2007]**

a. −4 b. −2

c. 0 d. 1

Working Space

18 If **u**, **v**, **w** are non-coplanar vectors and p, q are real numbers, then the inequality $[3\mathbf{u}\ p\mathbf{v}\ p\mathbf{w}] - [p\mathbf{v}\ \mathbf{w}\ q\mathbf{u}] - [2\mathbf{w}\ q\mathbf{v}\ q\mathbf{u}] = 0$ holds for

a. exactly one value of (p, q)
b. exactly two values of (p, q)
c. more than two but not all values of (p, q)
d. all values of (p, q)

19 If $\mathbf{a} = \hat{\mathbf{i}} + \hat{\mathbf{j}} + \hat{\mathbf{k}}$, $\mathbf{b} = 4\hat{\mathbf{i}} + 3\hat{\mathbf{j}} + 4\hat{\mathbf{k}}$ and $\mathbf{c} = \hat{\mathbf{i}} + \alpha\hat{\mathbf{j}} + \beta\hat{\mathbf{k}}$ are linearly dependent vectors and $|\mathbf{c}| = \sqrt{3}$, then

a. $\alpha = 1, \beta = -1$
b. $\alpha = 1, \beta = \pm 1$
c. $\alpha = -1, \beta = \pm 1$
d. $\alpha = \pm 1, \beta = 1$

20 Let $\mathbf{a} = a_1\hat{\mathbf{i}} + a_2\hat{\mathbf{j}} + a_3\hat{\mathbf{k}}$, $\mathbf{b} = b_1\hat{\mathbf{i}} + b_2\hat{\mathbf{j}} + b_3\hat{\mathbf{k}}$ and $\mathbf{c} = c_1\hat{\mathbf{i}} + c_2\hat{\mathbf{j}} + c_3\hat{\mathbf{k}}$ be three non-zero vectors such that **c** is a unit vector perpendicular to both the vectors **a** and **b**. If the angle between **a** and **b** is $\frac{\pi}{6}$, then $\begin{vmatrix} a_1 & a_2 & a_3 \\ b_1 & b_2 & b_3 \\ c_1 & c_2 & c_3 \end{vmatrix}^2$ is **[IIT JEE 2007]**

a. 0
b. $\frac{1}{4}(a_1^2 + a_2^2 + a_3^2)(b_1^2 + b_2^2 + b_3^2)$
c. 1
d. $\frac{3}{4}(a_1^2 + a_2^2 + a_3^3)(b_1^2 + b_2^2 + b_3^3)(c_1^2 + c_2^2 + c_3^3)$

21 If $\mathbf{a} = \hat{\mathbf{i}} + \hat{\mathbf{j}} + \hat{\mathbf{k}}$, $\mathbf{b} = \hat{\mathbf{i}} - \hat{\mathbf{j}} + \hat{\mathbf{k}}$, $\mathbf{c} = \hat{\mathbf{i}} + 2\hat{\mathbf{j}} - \hat{\mathbf{k}}$, then the value of $\begin{vmatrix} \mathbf{a}\cdot\mathbf{a} & \mathbf{a}\cdot\mathbf{b} & \mathbf{a}\cdot\mathbf{c} \\ \mathbf{b}\cdot\mathbf{a} & \mathbf{b}\cdot\mathbf{b} & \mathbf{b}\cdot\mathbf{c} \\ \mathbf{c}\cdot\mathbf{a} & \mathbf{c}\cdot\mathbf{b} & \mathbf{c}\cdot\mathbf{c} \end{vmatrix}$ is

a. 2
b. 4
c. 16
d. 64

Matching List Type

Direction *(Q. No. 22) Choices for the correct combination of elements from Column I and Column II are given as options (a), (b), (c) and (d), out of which one is correct.*

22 Match the Column I with Column II and mark the correct option from the codes given below.

	Column I		Column II
i.	Given two vectors $\mathbf{a} = 2\hat{\mathbf{i}} - 3\hat{\mathbf{j}} + 6\hat{\mathbf{k}}$, $\mathbf{b} = -2\hat{\mathbf{i}} + 2\hat{\mathbf{j}} - \hat{\mathbf{k}}$ and $\lambda = \frac{\text{Projection of } \mathbf{a} \text{ on } \mathbf{b}}{\text{Projection of } \mathbf{b} \text{ on } \mathbf{a}}$, then the value of 3λ is	p.	0
ii.	If $\mathbf{a} = \hat{\mathbf{i}} + 2\hat{\mathbf{j}} + 3\hat{\mathbf{k}}$, $\mathbf{b} = -\hat{\mathbf{i}} + 2\hat{\mathbf{j}} + \hat{\mathbf{k}}$, $\mathbf{c} = 3\hat{\mathbf{i}} + \hat{\mathbf{j}}$ and $\mathbf{a} + p\mathbf{b}$ is normal to **c**, then p is	q.	7
iii.	Let **a**, **b**, **c** be three non-zero vectors such that $\mathbf{a} + \mathbf{b} + \mathbf{c} = 0$, then $\lambda(\mathbf{a} \times \mathbf{b}) + \mathbf{c} \times \mathbf{b} + \mathbf{a} \times \mathbf{c} = 0$, where λ is equal to	r.	5
iv.	The points whose position vectors are $p\hat{\mathbf{i}} + q\hat{\mathbf{j}} + r\hat{\mathbf{k}}$, $q\hat{\mathbf{i}} + r\hat{\mathbf{j}} + p\hat{\mathbf{k}}$ and $r\hat{\mathbf{i}} + p\hat{\mathbf{j}} + q\hat{\mathbf{k}}$ are collinear, then the value of $(p^2 + q^2 + r^2 - pq - qr - rp)$ is	s.	2

Codes

	i	ii	iii	iv			i	ii	iii	iv
a.	q	r	s	p		b.	r	s	p	q
c.	s	p	q	r		d.	p	q	r	s

Working Space

Comprehension Type

Direction (Q. Nos. 23-25) *This section contains a paragraph, describing theory, experiments, data, etc. Three questions related to the paragraph have been given. Each question has only one correct answer among the four given options (a), (b), (c) and (d).*

Passage

Let **a**, **b**, **c** are non-zero unit vectors inclined pairwise with the same angle θ. p, q, r are non-zero scalars satisfying $\mathbf{a} \times \mathbf{b} + \mathbf{b} \times \mathbf{c} = p\mathbf{a} + q\mathbf{b} + r\mathbf{c}$.

23 Volume of parallelopiped with edges **a**, **b** and **c** is

a. $p + (q + r) \cos \theta$ b. $(p + q + r) \cos \theta$

c. $2p - (q + r) \text{ccs} \, \theta$ d. None of these

24 The value of $\left(\dfrac{q}{p} + 2 \cos \theta\right)$ is

a. 1 b. $2 [\mathbf{a} \; \mathbf{b} \; \mathbf{c}]$

c. 0 d. None of these

25 The value of $|(p + q) \cos \theta + r|$ is

a. $(1 + \cos \theta) \sqrt{1 - 2 \cos \theta}$

b. $2 \sin^2 \dfrac{\theta}{2} | \sqrt{1 + 2 \cos \theta} |$

c. $(1 - \sin \theta) \sqrt{1 + 2 \cos \theta}$

d. None of the above

One Integer Value Correct Type

Direction (Q. Nos. 26-30) *This section contains 5 questions. When worked out will result in an integer from 0 to 9 (both inclusive).*

26 Let **a**, **b**, **c** be three non-coplanar vectors and **p**, **q**, **r** are vectors defined by the relations $\mathbf{p} = \dfrac{\mathbf{b} \times \mathbf{c}}{[\mathbf{a} \; \mathbf{b} \; \mathbf{c}]}$, $\mathbf{q} = \dfrac{\mathbf{c} \times \mathbf{a}}{[\mathbf{a} \; \mathbf{b} \; \mathbf{c}]}$, $\mathbf{r} = \dfrac{\mathbf{a} \times \mathbf{b}}{[\mathbf{a} \; \mathbf{b} \; \mathbf{c}]}$, then the value of expression

$(\mathbf{a} + \mathbf{b}) \cdot \mathbf{p} + (\mathbf{b} + \mathbf{c}) \cdot \mathbf{q} + (\mathbf{c} + \mathbf{a}) \cdot \mathbf{r}$ is

27 Let $\mathbf{v} = 2\hat{\mathbf{i}} + \hat{\mathbf{j}} - \hat{\mathbf{k}}$ and $\mathbf{w} = \hat{\mathbf{i}} + 3\hat{\mathbf{k}}$. If $\hat{\mathbf{u}}$ is unit vector and the maximum value of $[\mathbf{u} \; \mathbf{v} \; \mathbf{w}] = \sqrt{\lambda}$, then the value of $(\lambda - 51)$ is

28 The value of a, so that the volume of parallelopiped the value of formed by $\hat{\mathbf{i}} + a\hat{\mathbf{j}} + \hat{\mathbf{k}}$, $\hat{\mathbf{j}} + a\hat{\mathbf{k}}$ and $a\hat{\mathbf{i}} + \hat{\mathbf{k}}$ becomes minimum is $\dfrac{1}{\sqrt{\lambda}}$, then the value of λ is

29 Let $PR = 3\hat{\mathbf{i}} + \hat{\mathbf{j}} - 2\hat{\mathbf{k}}$ and $SQ = \hat{\mathbf{i}} - 3\hat{\mathbf{j}} + 4\hat{\mathbf{k}}$. Determine the diagonal of a parallelogram $PQRS$ and $PT = \hat{\mathbf{i}} + 2\hat{\mathbf{j}} + 3\hat{\mathbf{k}}$ be another vector. The volume of parallelopiped determined by the vectors PT, PQ and RS is λ, then the value of $(\lambda - 5)$ is **[JEE Advanced 2013]**

30 Consider the set of eight vectors $V = [a\hat{\mathbf{i}} + b\hat{\mathbf{j}} + c\hat{\mathbf{k}} : a, b, c \in \{-1, 1\}]$. Three non-coplanar vectors can be chosen from V in 2^p ways, then the value of p is

DPP-5 Vector Triple Product

Only One Option Correct Type

Working Space

Direction (Q. Nos. 1-16) *This section contains 10 multiple choice questions. Each question has four choices (a), (b), (c). and (d), out of which ONLY ONE is correct.*

1 For any vector $\mathbf{a}$, $\hat{\mathbf{i}} \times (\mathbf{a} \times \hat{\mathbf{i}}) + \hat{\mathbf{j}} \times (\mathbf{a} \times \hat{\mathbf{j}}) + \hat{\mathbf{k}} \times (\mathbf{a} \times \hat{\mathbf{k}})$ is

a. $2\mathbf{a}$
b. $-2\mathbf{a}$
c. $\mathbf{a}$
d. $-\mathbf{a}$

2 Let $\mathbf{a}, \mathbf{b}, \mathbf{c}$ be three vectors of magnitude 1, 1 and 2 respectively. If $\mathbf{a} \times (\mathbf{a} \times \mathbf{c}) + \mathbf{b} = 0$, then the acute angle between $\mathbf{a}$ and $\mathbf{c}$ is

a. $\frac{\pi}{3}$
b. $\frac{\pi}{4}$
c. $\frac{\pi}{6}$
d. None of these

3 Let $\mathbf{a} = 2\hat{\mathbf{i}} + \hat{\mathbf{j}} + \hat{\mathbf{k}}$, $\mathbf{b} = \hat{\mathbf{i}} + 2\hat{\mathbf{j}} - \hat{\mathbf{k}}$ and $\mathbf{c}$ is a unit vector coplanar to them. If $\mathbf{c}$ is perpendicular to $\mathbf{a}$, then $\mathbf{c}$ is equal to

a. $\frac{1}{\sqrt{2}}(-\hat{\mathbf{j}} + \hat{\mathbf{k}})$
b. $-\frac{1}{\sqrt{3}}(\hat{\mathbf{i}} + \hat{\mathbf{j}} + \hat{\mathbf{k}})$
c. $\frac{1}{\sqrt{5}}(\hat{\mathbf{i}} - 2\hat{\mathbf{j}})$
d. $\frac{1}{\sqrt{3}}(\hat{\mathbf{i}} - \hat{\mathbf{j}} - \hat{\mathbf{k}})$

4 Let $\mathbf{a} = 2\hat{\mathbf{i}} + \hat{\mathbf{j}} - 2\hat{\mathbf{k}}$ and $\mathbf{b} = \hat{\mathbf{i}} + \hat{\mathbf{j}}$. If $\mathbf{c}$ is a vector such that $\mathbf{a} \cdot \mathbf{c} = |\mathbf{c}|$, $|\mathbf{c} - \mathbf{a}| = 2\sqrt{2}$ and the angle between $\mathbf{a} \times \mathbf{b}$ and $\mathbf{c}$ is 30°, then $|(\mathbf{a} \times \mathbf{b}) \times \mathbf{c}|$ is equal to

a. $\frac{2}{3}$
b. $\frac{3}{2}$
c. 2
d. 3

5 Let $\mathbf{a}, \mathbf{b}, \mathbf{c}$ be non-zero vectors such that $(\mathbf{a} \times \mathbf{b}) \times \mathbf{c} = \frac{1}{3}|\mathbf{b}||\mathbf{c}|\mathbf{a}$. If θ is the acute angle between the vectors $\mathbf{b}$ and $\mathbf{c}$, then $\sin\theta$ equals

a. $\frac{2}{3}$
b. $\frac{\sqrt{2}}{3}$
c. $\frac{1}{3}$
d. $\frac{2\sqrt{2}}{3}$

6 If $(\mathbf{a} \times \mathbf{b}) \times \mathbf{c} = \mathbf{a} \times (\mathbf{b} \times \mathbf{c})$, where $\mathbf{a} \cdot \mathbf{b}$ and $\mathbf{c}$ are any three vectors such that $\mathbf{a} \cdot \mathbf{b} \neq 0$, $\mathbf{b} \cdot \mathbf{c} \neq 0$, then $\mathbf{a}$ and $\mathbf{c}$ are **[AIEEE 2006]**

a. inclined at angle of $\frac{\pi}{6}$ between them
b. perpendicular
c. parallel
d. inclined at an angle of $\frac{\pi}{3}$ between them

Working Space

7 Let $\mathbf{a} = \hat{\mathbf{j}} - \hat{\mathbf{k}}$ and $\mathbf{c} = \hat{\mathbf{i}} - \hat{\mathbf{j}} - \hat{\mathbf{k}}$. Then, the vector $\mathbf{b}$ satisfying $\mathbf{a} \times \mathbf{b} + \mathbf{c} = 0$ and $\mathbf{a} \cdot \mathbf{b} = 3$, is **[AIEEE 2010]**

a. $-\hat{\mathbf{i}} + \hat{\mathbf{j}} - 2\hat{\mathbf{k}}$
b. $2\hat{\mathbf{i}} - \hat{\mathbf{j}} + 2\hat{\mathbf{k}}$
c. $\hat{\mathbf{i}} - \hat{\mathbf{j}} - 2\hat{\mathbf{k}}$
d. $\hat{\mathbf{i}} + \hat{\mathbf{j}} - 2\hat{\mathbf{k}}$

8 If $\mathbf{a} = \hat{\mathbf{i}} - \hat{\mathbf{j}} - 2\hat{\mathbf{k}}$, $\mathbf{b} = 2\hat{\mathbf{i}} + 4\hat{\mathbf{j}} + \hat{\mathbf{k}}$ and $\mathbf{c} = \lambda\hat{\mathbf{i}} + \hat{\mathbf{j}} + \mu\hat{\mathbf{k}}$ are mutually orthogonal, then (λ, μ) is equal to **[AIEEE 2010]**

a. (−3, 2)
b. (2, − 3)
c. (−2, 3)
d. (3, − 2)

9 If $\mathbf{a} = \frac{1}{\sqrt{10}}(3\hat{\mathbf{i}} + \hat{\mathbf{k}})$ and $\mathbf{b} = \frac{1}{7}(2\hat{\mathbf{i}} + 3\hat{\mathbf{j}} - 6\hat{\mathbf{k}})$, then the value of $(2\mathbf{a} - \mathbf{b}) \cdot \{(\mathbf{a} \times \mathbf{b}) \times (\mathbf{a} + 2\mathbf{b})\}$ is

a. −3
b. 5
c. 3
d. −5

10 The vectors $\mathbf{a}$ and $\mathbf{b}$ are not perpendicular and $\mathbf{c}$ and $\mathbf{d}$ are two vectors satisfying $\mathbf{b} \times \mathbf{c} = \mathbf{b} \times \mathbf{d}$ and $\mathbf{a} \cdot \mathbf{d} = 0$, then the vector $\mathbf{d}$ is equal to **[AIEEE 2011]**

a. $\mathbf{c} + \left(\frac{\mathbf{a} \cdot \mathbf{c}}{\mathbf{a} \cdot \mathbf{b}}\right)\mathbf{b}$
b. $\mathbf{b} + \left(\frac{\mathbf{b} \cdot \mathbf{c}}{\mathbf{b} \cdot \mathbf{a}}\right)\mathbf{c}$
c. $\mathbf{c} - \left(\frac{\mathbf{a} \cdot \mathbf{c}}{\mathbf{a} \cdot \mathbf{b}}\right)\mathbf{b}$
d. $\mathbf{b} - \left(\frac{\mathbf{b} \cdot \mathbf{c}}{\mathbf{b} \cdot \mathbf{a}}\right)\mathbf{c}$

11 Let $\mathbf{a} = 2\hat{\mathbf{i}} - \hat{\mathbf{j}} + \hat{\mathbf{k}}$, $\mathbf{b} = \hat{\mathbf{i}} + 2\hat{\mathbf{j}} - \hat{\mathbf{k}}$ and $\mathbf{c} = \hat{\mathbf{i}} + \hat{\mathbf{j}} - 2\hat{\mathbf{k}}$ be three vectors. A vector in the plane of $\mathbf{b}$ and $\mathbf{c}$ whose projection on $\mathbf{a}$ is of magnitude $\sqrt{\frac{2}{3}}$, is

a. $2\hat{\mathbf{i}} + 3\hat{\mathbf{j}} - 3\hat{\mathbf{k}}$
b. $2\hat{\mathbf{i}} + 3\hat{\mathbf{j}} + 3\hat{\mathbf{k}}$
c. $2\hat{\mathbf{i}} + \hat{\mathbf{j}} + 5\hat{\mathbf{k}}$
d. $2\hat{\mathbf{i}} + \hat{\mathbf{j}} + 3\hat{\mathbf{k}}$

12 Let $\mathbf{a} = \hat{\mathbf{i}} + \hat{\mathbf{j}} + \hat{\mathbf{k}}$, $\mathbf{b} = \hat{\mathbf{i}} - \hat{\mathbf{j}} + \hat{\mathbf{k}}$ and $\mathbf{c} = \hat{\mathbf{i}} - \hat{\mathbf{j}} - \hat{\mathbf{k}}$ be three vectors. A vector $\mathbf{v}$ in the plane of $\mathbf{a}$ and $\mathbf{b}$ whose projection on $\mathbf{c}$ is $\frac{1}{\sqrt{3}}$, is given by **[IIT JEEE 2011]**

a. $\hat{\mathbf{i}} - 3\hat{\mathbf{j}} + 3\hat{\mathbf{k}}$
b. $-3\hat{\mathbf{i}} - 3\hat{\mathbf{j}} - \hat{\mathbf{k}}$
c. $3\hat{\mathbf{i}} - \hat{\mathbf{j}} + 3\hat{\mathbf{k}}$
d. None of these

13 Let $\mathbf{a} = \hat{\mathbf{i}} - \hat{\mathbf{j}}$, $\mathbf{b} = \hat{\mathbf{j}} - \hat{\mathbf{k}}$ and $\mathbf{c} = \hat{\mathbf{k}} - \hat{\mathbf{i}}$. If $\mathbf{d}$ is a unit vector such that $\mathbf{a} \cdot \mathbf{d} = 0 = (\mathbf{b}\ \mathbf{c}\ \mathbf{d})$, then $\mathbf{d}$ is equal to

a. $\pm\frac{(\hat{\mathbf{i}} + \hat{\mathbf{j}} - 2\hat{\mathbf{k}})}{\sqrt{6}}$
b. $\pm\frac{(\hat{\mathbf{i}} + \hat{\mathbf{j}} - \hat{\mathbf{k}})}{\sqrt{3}}$
c. $\pm\frac{(\hat{\mathbf{i}} + \hat{\mathbf{j}} + \hat{\mathbf{k}})}{\sqrt{3}}$
d. $\pm\hat{\mathbf{k}}$

14 If $\mathbf{a}$, $\mathbf{b}$ and $\mathbf{c}$ are non-coplanar unit vectors such that $\mathbf{a} \times (\mathbf{b} \times \mathbf{c}) = \frac{1}{\sqrt{2}}(\mathbf{b} + \mathbf{c})$, then the angle between $\mathbf{a}$ and $\mathbf{b}$ is

a. $\frac{3\pi}{4}$
b. $\frac{\pi}{4}$
c. $\frac{\pi}{2}$
d. π

Working Space

15 Let **p**, **q**, **r** be three mutually perpendicular vector of the same magnitude. If a vector **x** satisfies the equation
$\mathbf{p} \times [(\mathbf{x} - \mathbf{q}) \times \mathbf{p}] + \mathbf{q} \times [(\mathbf{x} - \mathbf{r}) \times \mathbf{q}] + \mathbf{r} \times [(\mathbf{x} - \mathbf{p}) \times \mathbf{r}] = 0$, then x is

a. $\frac{1}{2}(\mathbf{p} + \mathbf{q} - 2\mathbf{r})$
b. $\frac{1}{2}(\mathbf{p} + \mathbf{q} + \mathbf{r})$
c. $\frac{1}{3}(\mathbf{p} + \mathbf{q} + \mathbf{r})$
d. $\frac{1}{3}(2\mathbf{p} + \mathbf{q} - \mathbf{r})$

16 Let $\mathbf{a} = 2\hat{\mathbf{i}} + \hat{\mathbf{j}} - 2\hat{\mathbf{k}}$ and $\mathbf{b} = \hat{\mathbf{i}} + \hat{\mathbf{j}}$. If **c** is a vector such that $\mathbf{a} \cdot \mathbf{c} = |\mathbf{c}|, |\mathbf{c} - \mathbf{a}| = 2\sqrt{2}$ and angle between $(\mathbf{a} \times \mathbf{b})$ and **c** is 30°, then $|(\mathbf{a} \times \mathbf{b}) \times \mathbf{c}|$ is equal to

a. $\frac{2}{3}$
b. 2
c. $\frac{3}{2}$
d. 3

One or More than One Options Correct Type

Direction (Q. Nos. 17-20) This section contains 4 multiple choice questions. Each question has four choices (a), (b), (c) and (d), out of which ONE or MORE THAN ONE are correct.

17 If $\mathbf{a} = x\hat{\mathbf{i}} + y\hat{\mathbf{j}} + z\hat{\mathbf{k}}$, $\mathbf{b} = y\hat{\mathbf{i}} + z\hat{\mathbf{j}} + x\hat{\mathbf{k}}$ and $\mathbf{c} = z\hat{\mathbf{i}} + x\hat{\mathbf{j}} + y\hat{\mathbf{k}}$, then $\mathbf{a} \times (\mathbf{b} \times \mathbf{c})$ is/are

a. parallel to $(y - z)\hat{\mathbf{i}} + (z - x)\hat{\mathbf{j}} + (x - y)\hat{\mathbf{k}}$
b. orthogonal to $\hat{\mathbf{i}} + \hat{\mathbf{j}} + \hat{\mathbf{k}}$
c. orthogonal to $(y + z)\hat{\mathbf{i}} + (z + x)\hat{\mathbf{j}} + (x + y)\hat{\mathbf{k}}$
d. parallel to $\hat{\mathbf{i}} + \hat{\mathbf{j}} + \hat{\mathbf{k}}$

18 If **a**, **b** and **c** are three non-zero vectors, then which of the following statement(s) is/are true?
a. $\mathbf{a} \times (\mathbf{b} \times \mathbf{c})$, $\mathbf{b} \times (\mathbf{c} \times \mathbf{a})$, $\mathbf{c} \times (\mathbf{a} \times \mathbf{b})$ form a right handed system
b. $\mathbf{c}$, $(\mathbf{a} \times \mathbf{b}) \times \mathbf{c}$, $\mathbf{a} \times \mathbf{b}$ form a right handed system
c. $\mathbf{a} \cdot \mathbf{b} + \mathbf{b} \cdot \mathbf{c} + \mathbf{c} \cdot \mathbf{a} < 0$, if $\mathbf{a} + \mathbf{b} + \mathbf{c} = 0$
d. $\frac{(\mathbf{a} \times \mathbf{b}) \cdot (\mathbf{b} \times \mathbf{c})}{(\mathbf{b} \times \mathbf{c}) \cdot (\mathbf{a} \times \mathbf{c})} = -1$, if $\mathbf{a} + \mathbf{b} + \mathbf{c} = 0$

19 Let the unit vectors **a** and **b** be perpendicular and unit vector **c** is inclined at angle α to **a** and **b**. If $\mathbf{c} = l\mathbf{a} + m\mathbf{b} + n(\mathbf{a} \times \mathbf{b})$, then

a. $l = m$
b. $n^2 = 1 - 2l^2$
c. $n^2 = -\cos 2\alpha$
d. $m^2 = \frac{1 + \cos 2\alpha}{2}$

20 Let **a** and **b** be two given perpendicular vectors, which are non-zero. A vector **r** satisfying the equation $\mathbf{r} \times \mathbf{b} = \mathbf{a}$ can be

a. $\mathbf{b} - \frac{(\mathbf{a} \times \mathbf{b})}{|\mathbf{b}|^2}$
b. $2\mathbf{b} - \frac{(\mathbf{a} \times \mathbf{b})}{|\mathbf{b}|^2}$
c. $|\mathbf{a}|\mathbf{b} - \frac{(\mathbf{a} \times \mathbf{b})}{|\mathbf{b}|^2}$
d. $|\mathbf{b}|\mathbf{b} - \frac{(\mathbf{a} \times \mathbf{b})}{|\mathbf{b}|^2}$

Working Space

Comprehension Type

Direction (*Q. Nos.* 21-23) *This section contains a paragraph, describing theory, experiments, data, etc. Three questions related to the paragraph have been given. Each question has only one correct answer among the four given options (a), (b), (c) and (d).*

Passage

Let **x**, **y**, **z** be the vector, such that $|\mathbf{x}| = |\mathbf{y}| = |\mathbf{z}| = \sqrt{2}$ and **x**, **y**, **z** make angles of 60° with each other, also $\mathbf{x} \times (\mathbf{y} \times \mathbf{z}) = \mathbf{a}$, $\mathbf{y} \times (\mathbf{z} \times \mathbf{x}) = \mathbf{b}$ and $\mathbf{x} \times \mathbf{y} = \mathbf{c}$.

21 The value of **x** is

a. $\{(\mathbf{a}+\mathbf{b})\times\mathbf{c}-(\mathbf{a}+\mathbf{b})\}$

b. $\{(\mathbf{a}+\mathbf{b})-(\mathbf{a}+\mathbf{b})\times\mathbf{c}\}$

c. $\frac{1}{2}\{(\mathbf{a}+\mathbf{b})\times\mathbf{c}-(\mathbf{a}+\mathbf{b})\}$

d. None of these

22 The value of **y** is

a. $\frac{1}{2}\{(\mathbf{a}+\mathbf{b})+(\mathbf{a}+\mathbf{b})\times\mathbf{c}\}$

b. $2\{(\mathbf{a}+\mathbf{b})+(\mathbf{a}+\mathbf{b})\times\mathbf{c}\}$

c. $4\{(\mathbf{a}+\mathbf{b})+(\mathbf{a}+\mathbf{b})\times\mathbf{c}\}$

d. None of these

23 The value of **z** is

a. $\frac{1}{2}\{(\mathbf{b}-\mathbf{a})\times\mathbf{c}+(\mathbf{a}+\mathbf{b})\}$

b. $\frac{1}{2}\{(\mathbf{b}-\mathbf{a})+(\mathbf{a}+\mathbf{b})\times\mathbf{c}\}$

c. $\{(\mathbf{b}-\mathbf{a})\times\mathbf{c}+(\mathbf{a}+\mathbf{b})\}$

d. None of these

One Integer Value Correct Type

Direction (*Q. Nos.* 24 *and* 25) *This section contains* 2 *questions. When worked out will result in an integer from* 0 *to* 9 (*both inclusive*).

24 If **a**, **b** and **c** are three unit vectors and every two are inclined to each other at an angle $\cos^{-1}\left(\frac{3}{5}\right)$. If $\mathbf{a}\times\mathbf{b} = p\mathbf{a} + q\mathbf{b} + r\mathbf{c}$, where p, q, r are scalars, then $55q^2$ is equal to

25 If **a** and **b** are vectors in space given by $\mathbf{a} = \frac{\hat{\mathbf{i}} - 2\hat{\mathbf{j}}}{\sqrt{5}}$ and $\mathbf{b} = \frac{2\hat{\mathbf{i}} + \hat{\mathbf{j}} + 3\hat{\mathbf{k}}}{\sqrt{14}}$, then the value of $(2\mathbf{a}+\mathbf{b})\cdot[(\mathbf{a}\times\mathbf{b})\times(\mathbf{a}-2\mathbf{b})]$ is

SET - 1

Revisal Problems for JEE Main

Working Space

Only One Option Correct Type

Direction (Q. Nos. 1-35) This section contains 35 multiple choice questions. Each question has four choices (a), (b), (c) and (d), out of which ONLY ONE option is correct.

1 The vectors **a**, **b** and **c** are equal in length and taken pairwise, they make equal angles. If $\mathbf{a} = \hat{i} + \hat{j}$, $\mathbf{b} = \hat{j} + \hat{k}$ and **c** makes an obtuse angle with X-axis, then **c** is

a. $-\hat{i} + 4\hat{j} - \hat{k}$
b. $\hat{i} + \hat{k}$
c. $\frac{1}{3}(-\hat{i} + 4\hat{j} - \hat{k})$
d. $\frac{1}{3}(\hat{i} - 4\hat{j} + \hat{k})$

2 If the vector $-\hat{i} + \hat{j} - \hat{k}$ bisects the angle between the vector **c** and the vector $3\hat{i} + 4\hat{j}$, then the unit vector in the direction of **c** is

a. $\frac{1}{15}(11\hat{i} + 10\hat{j} + 2\hat{k})$
b. $-\frac{1}{15}(11\hat{i} - 10\hat{j} + 2\hat{k})$
c. $-\frac{1}{15}(11\hat{i} + 10\hat{j} - 2\hat{k})$
d. $-\frac{1}{15}(11\hat{i} + 10\hat{j} + 2\hat{k})$

3 The vector **r** satisfying the conditions that

I. it is perpendicular to $3\hat{i} + 2\hat{j} + 2\hat{k}$ and $18\hat{i} - 22\hat{j} - 5\hat{k}$.
II. it makes an obtuse angle with Y-axis.
III. $|\mathbf{r}| = 14$

a. $2(-2\hat{i} - 3\hat{j} + 6\hat{k})$
b. $2(2\hat{i} - 3\hat{j} + 6\hat{k})$
c. $4\hat{i} + 6\hat{j} - 12\hat{k}$
d. None of these

4 If G is the centroid of ΔABC such that **GB** and **GC** are inclined at an obtuse angle, then

a. $5a^2 > b^2 + c^2$
b. $5c^2 > a^2 + b^2$
c. $5b^2 > a^2 + c^2$
d. None of these

5 Let **a**, **b**, **c** be vectors of equal magnitude such that the angle between **a** and **b** is α, **b** and **c** is β and **c** and **a** is γ, then the minimum value of $\cos\alpha + \cos\beta + \cos\gamma$ is

a. $\frac{1}{2}$
b. $-\frac{1}{2}$
c. $\frac{3}{2}$
d. $-\frac{3}{2}$

6 Let a, b, c denote the lengths of the sides of a triangle such that

$$(a - b)\mathbf{u} + (b - c)\mathbf{v} + (c - a)(\mathbf{u} \times \mathbf{v}) = 0$$

for any two non-collinear vectors **u** and **v**, then the triangle is

a. right angled
b. equilateral
c. isosceles
d. scalene

7 If $\mathbf{r} \times \mathbf{b} = \mathbf{c} \times \mathbf{b}$ and $\mathbf{r} \perp \mathbf{a}$, then **r** is equal to

a. $\frac{\mathbf{a} \times (\mathbf{c} \times \mathbf{b})}{\mathbf{a} \cdot \mathbf{b}}$
b. $\frac{\mathbf{b} \times (\mathbf{a} \times \mathbf{c})}{\mathbf{a} \cdot \mathbf{b}}$
c. $\frac{\mathbf{c} \times (\mathbf{a} \times \mathbf{b})}{\mathbf{a} \cdot \mathbf{b}}$
d. $\frac{\mathbf{c} \times (\mathbf{a} \times \mathbf{b})}{\mathbf{b} \cdot \mathbf{c}}$

Working Space

8 Let $\mathbf{a} = \hat{\mathbf{i}} + \hat{\mathbf{j}} + \hat{\mathbf{k}}$ and $\mathbf{c} = \hat{\mathbf{j}} - \hat{\mathbf{k}}$. If $\mathbf{b}$ is a vector satisfying $\mathbf{a} \times \mathbf{b} = \mathbf{c}$ and $\mathbf{a} \cdot \mathbf{b} = 3$, then $\mathbf{b}$ is equal to

a. $\frac{1}{3}(5\hat{\mathbf{i}} + 2\hat{\mathbf{j}} + 2\hat{\mathbf{k}})$
b. $\frac{1}{3}(5\hat{\mathbf{i}} - 2\hat{\mathbf{j}} - 2\hat{\mathbf{k}})$
c. $3\hat{\mathbf{i}} - \hat{\mathbf{j}} - \hat{\mathbf{k}}$
d. None of these

9 If $\mathbf{a} = \hat{\mathbf{i}} + \hat{\mathbf{j}}$, $\mathbf{b} = 2\hat{\mathbf{j}} - \hat{\mathbf{k}}$ and $\mathbf{r} \times \mathbf{b} = \mathbf{a} \times \mathbf{b}$, $\mathbf{r} \times \mathbf{a} = \mathbf{b} \times \mathbf{a}$, then a unit vector in the direction of $\mathbf{r}$ is

a. $\frac{1}{\sqrt{11}}(\hat{\mathbf{i}} + 3\hat{\mathbf{j}} - \hat{\mathbf{k}})$
b. $\frac{1}{\sqrt{11}}(\hat{\mathbf{i}} - 3\hat{\mathbf{j}} + \hat{\mathbf{k}})$
c. $\frac{1}{\sqrt{3}}(\hat{\mathbf{i}} + \hat{\mathbf{j}} + \hat{\mathbf{k}})$
d. None of these

10 If $\mathbf{a}$, $\mathbf{b}$ and $\mathbf{c}$ are three non-coplanar, non-zero vectors, then $(\mathbf{a} \cdot \mathbf{a})(\mathbf{b} \times \mathbf{c}) + (\mathbf{a} \cdot \mathbf{b})(\mathbf{c} \times \mathbf{a}) + (\mathbf{a} \cdot \mathbf{c})(\mathbf{a} \times \mathbf{b})$ is equal to

a. $[\mathbf{a}\ \mathbf{b}\ \mathbf{c}]\mathbf{c}$
b. $[\mathbf{b}\ \mathbf{c}\ \mathbf{a}]\mathbf{a}$
c. $[\mathbf{c}\ \mathbf{a}\ \mathbf{b}]$
d. None of these

11 If $\mathbf{a}$, $\mathbf{b}$, $\mathbf{c}$ are any three non-coplanar vectors, then $[\mathbf{a} + \mathbf{b} + \mathbf{c}\ \ \mathbf{a} - \mathbf{c}\ \ \mathbf{a} - \mathbf{b}]$ is equal to

a. 0
b. $[\mathbf{a}\ \mathbf{b}\ \mathbf{c}]$
c. $2\,[\mathbf{a}\ \mathbf{b}\ \mathbf{c}]$
d. $-3\,[\mathbf{a}\ \mathbf{b}\ \mathbf{c}]$

12 The vectors

$$\mathbf{u} = (al + a_1l_1)\,\hat{\mathbf{i}} + (am + a_1m_1)\,\hat{\mathbf{j}} + (an + a_1n_1)\,\hat{\mathbf{k}},$$
$$\mathbf{v} = (bl + b_1l_1)\,\hat{\mathbf{i}} + (bm + b_1m_1)\,\hat{\mathbf{j}} + (bn + b_1n_1)\,\hat{\mathbf{k}} \text{ and}$$
$$\mathbf{w} = (cl + c_1l_1)\,\hat{\mathbf{i}} + (cm + c_1m_1)\,\hat{\mathbf{j}} + (cn + c_1n_1)\,\hat{\mathbf{k}}$$

a. form an equilateral triangle
b. are coplanar
c. are collinear
d. are mutually perpendicular

13 The three vectors $\hat{\mathbf{i}} + \hat{\mathbf{j}}$, $\hat{\mathbf{j}} + \hat{\mathbf{k}}$, $\hat{\mathbf{k}} + \hat{\mathbf{i}}$ taken two at a time form three planes. The three unit vectors drawn perpendicular to these planes form a parallelopiped of volume (in cu units)

a. $\frac{1}{3}$
b. 4
c. $\frac{3\sqrt{3}}{4}$
d. $\frac{4}{3\sqrt{3}}$

14 A line passes through the points whose position vectors are $\hat{\mathbf{i}} + \hat{\mathbf{j}} - 2\hat{\mathbf{k}}$ and $\hat{\mathbf{i}} - 3\hat{\mathbf{j}} + \hat{\mathbf{k}}$, the position vector of a point it at a unit distance from the first point is

a. $\frac{1}{5}(6\hat{\mathbf{i}} + \hat{\mathbf{j}} - 3\hat{\mathbf{k}})$
b. $\frac{1}{5}(4\hat{\mathbf{i}} + 9\hat{\mathbf{j}} - 13\hat{\mathbf{k}})$
c. $(\hat{\mathbf{i}} - 4\hat{\mathbf{j}} + 3\hat{\mathbf{k}})$
d. None of these

15 A groove is in the form of a broken line *ABC* and the position vectors of the three points are respectively $2\hat{\mathbf{i}} - 3\hat{\mathbf{j}} + 2\hat{\mathbf{k}}$, $3\hat{\mathbf{i}} + 2\hat{\mathbf{j}} - \hat{\mathbf{k}}$, $\hat{\mathbf{i}} + \hat{\mathbf{j}} + \hat{\mathbf{k}}$. A force of magnitude $24\sqrt{3}$ acts on a particle of unit mass kept at the point *A* and moves it along the groove to the point *C*. If the line of action of the force is parallel to the vector $\hat{\mathbf{i}} + 2\hat{\mathbf{j}} + \hat{\mathbf{k}}$ all along, then the number of units of work done by the force is

a. $144\sqrt{2}$
b. $144\sqrt{3}$
c. $72\sqrt{3}$
d. $72\sqrt{2}$

16 A unit vector perpendicular to the vector $-\hat{\mathbf{i}} + 2\hat{\mathbf{j}} + 2\hat{\mathbf{k}}$ and making equal angles with *X* and *Y*-axes can be

a. $\frac{1}{3}(2\hat{\mathbf{i}} + 2\hat{\mathbf{j}} - \hat{\mathbf{k}})$
b. $\frac{1}{3}(2\hat{\mathbf{i}} - 2\hat{\mathbf{j}} - \hat{\mathbf{k}})$
c. $\frac{1}{3}(2\hat{\mathbf{i}} + 2\hat{\mathbf{j}} + \hat{\mathbf{k}})$
d. $\frac{1}{3}(2\hat{\mathbf{i}} - 2\hat{\mathbf{j}} + \hat{\mathbf{k}})$

17 If **a**, **b** and **c** are three mutually perpendicular vectors, then the projection of the vectors $l\frac{\mathbf{a}}{|\mathbf{a}|}+m\frac{\mathbf{b}}{|\mathbf{b}|}+n\frac{(\mathbf{a}\times\mathbf{b})}{|\mathbf{a}\times\mathbf{b}|}$ along the angle bisector of the vectors **a** and **b** is

a. $\frac{l+m}{\sqrt{2}}$ b. $\sqrt{l^2+m^2+n^2}$

c. $\frac{\sqrt{l^2+m^2}}{\sqrt{l^2+m^2+n^2}}$ d. None of these

18 If **a, b, c** are three vectors of magnitude $\sqrt{3}$, 1, 2 such that $\mathbf{a}\times(\mathbf{a}\times\mathbf{c})+3\mathbf{b}=0$, if θ is the angle between **a** and **c**, then $\cos^2\theta$ is equal to

a. $\frac{3}{4}$ b. $\frac{1}{2}$

c. $\frac{1}{4}$ d. None of these

19 Let us define the length of a vector $a\hat{\mathbf{i}}+b\hat{\mathbf{j}}+c\hat{\mathbf{k}}$ as $|a|+|b|+|c|$. This definition coincides with the usual definition of length of a vector $a\hat{\mathbf{i}}+b\hat{\mathbf{j}}+c\hat{\mathbf{k}}$, iff

a. $a=b=c=0$
b. any two of a, b and c are zero
c. any one of a, b and c is zero
d. $a+b+c=0$

20 If **a, b, c** are vectors such that $|\mathbf{b}|=|\mathbf{c}|$, then $\{(\mathbf{a}+\mathbf{b})\times(\mathbf{a}+\mathbf{c})\}\times(\mathbf{b}\times\mathbf{c})\cdot(\mathbf{b}+\mathbf{c})$ is

a. 0 b. 1

c. −1 d. None of these

21 If the volume of parallelopiped formed by the vectors **a, b, c** as three coterminus edges is 27 cu units, then the volume of the parallelopiped having $\alpha=\mathbf{a}+2\mathbf{b}-\mathbf{c}$, $\beta=\mathbf{a}-\mathbf{b}$ and $\gamma=\mathbf{a}-\mathbf{b}-\mathbf{c}$ as three coterminus edges is

a. 27 b. 9

c. 81 d. None of these

22 If V is the volume of the parallelopiped having three coterminus edges as **a, b** and **c**, then the volume of parallelopiped having three coterminus edges as $\alpha=(\mathbf{a}\cdot\mathbf{a})\mathbf{a}+(\mathbf{a}\cdot\mathbf{b})\mathbf{b}+(\mathbf{a}\cdot\mathbf{c})\mathbf{c}$, $\beta=(\mathbf{a}\cdot\mathbf{b})\mathbf{a}+(\mathbf{b}\cdot\mathbf{b})\mathbf{b}+(\mathbf{b}\cdot\mathbf{c})\mathbf{c}$ and $\gamma=(\mathbf{a}\cdot\mathbf{c})\mathbf{a}+(\mathbf{b}\cdot\mathbf{c})\mathbf{b}+(\mathbf{c}\cdot\mathbf{c})\mathbf{c}$ is

a. V^3 b. $3V$

c. V^2 d. $2V$

23 If $|\mathbf{a}|=5$, $|\mathbf{b}|=3$, $|\mathbf{c}|=4$ and **a** is perpendicular to **b** and **c** such that angle between **b** and **c** is $\frac{5\pi}{6}$, then the volume of the parallelopiped having **a, b** and **c** as three coterminus edges is

a. 30 b. 60

c. 20 d. None of these

24 The vectors **a** and **b** are perpendicular to each other, then a vector **v** in terms of **a** and **b** satisfying the equations $\mathbf{v}\cdot\mathbf{a}=0$, $\mathbf{v}\cdot\mathbf{b}=1$ and $[\mathbf{v}\ \mathbf{a}\ \mathbf{b}]=1$ is

a. $\frac{\mathbf{b}}{|\mathbf{b}|^2}+\frac{\mathbf{a}\times\mathbf{b}}{|\mathbf{a}\times\mathbf{b}|^2}$ b. $\frac{\mathbf{b}}{|\mathbf{b}|}+\frac{\mathbf{a}\times\mathbf{b}}{|\mathbf{a}\times\mathbf{b}|^2}$

c. $\frac{\mathbf{b}}{|\mathbf{b}|^2}+\frac{\mathbf{a}\times\mathbf{b}}{|\mathbf{a}\times\mathbf{b}|}$ d. None of these

25 If the position vector of point P is $\mathbf{r}=x\hat{\mathbf{i}}+y\hat{\mathbf{j}}+z\hat{\mathbf{k}}$, where $x, y, z\in N$ and **a** is a vector given by $\mathbf{a}=\hat{\mathbf{i}}+\hat{\mathbf{j}}+\hat{\mathbf{k}}$, then the total number of possible positions of the point P for which $\mathbf{r}\cdot\mathbf{a}=10$ is

a. 36 b. 72

c. 66 d. None of these

Working Space

Working Space

26 If $\mathbf{a} = \hat{\mathbf{i}} + 2\hat{\mathbf{j}} + 3\hat{\mathbf{k}}$, $\mathbf{b} = 2\hat{\mathbf{i}} + 3\hat{\mathbf{j}} + 2\hat{\mathbf{k}}$, $\mathbf{c}$ is any vector not collinear with $\mathbf{b}$, then magnitude of

$\frac{\mathbf{a} \cdot (\mathbf{b} \times \mathbf{c})}{|\mathbf{b} \times \mathbf{c}|^2} (\mathbf{b} \times \mathbf{c}) - \left(\frac{\mathbf{a} \cdot \mathbf{b}}{|\mathbf{b}|^2}\right) \mathbf{b} + \left(\frac{\mathbf{a} \cdot \mathbf{c}}{|\mathbf{c}|^2}\right) \mathbf{c}$ is

a. $\sqrt{17}$
b. 14
c. 13
d. None of these

27 The unit vector is ZOX plane making angles 45° and 60° respectively, with $\mathbf{a} = 2\hat{\mathbf{i}} + 2\hat{\mathbf{j}} - \hat{\mathbf{k}}$ and $\mathbf{b} = \hat{\mathbf{j}} - \hat{\mathbf{k}}$, is

a. $\frac{1}{\sqrt{2}}(-\hat{\mathbf{i}} + \hat{\mathbf{k}})$
b. $\frac{1}{\sqrt{2}}(\hat{\mathbf{i}} - \hat{\mathbf{k}})$
c. $\frac{\sqrt{3}}{2}(\hat{\mathbf{i}} + \hat{\mathbf{k}})$
d. None of these

28 If $\mathbf{a} = 2\hat{\mathbf{i}} + \hat{\mathbf{j}} + \hat{\mathbf{k}}$, $\mathbf{b} = \hat{\mathbf{i}} + 2\hat{\mathbf{j}} + 2\hat{\mathbf{k}}$, $\mathbf{c} = \hat{\mathbf{i}} + \hat{\mathbf{j}} + 2\hat{\mathbf{k}}$ and $\mathbf{a} \times (\mathbf{b} \times \mathbf{c}) = (1+\alpha)\hat{\mathbf{i}} + \beta(1+\alpha)\hat{\mathbf{j}} + \gamma(1+\alpha)(1+\beta)\hat{\mathbf{k}}$, then

a. $\alpha = -2, \beta = -4, \gamma = -\frac{2}{3}$
b. $\alpha = 2, \beta = -4, \gamma = \frac{2}{3}$
c. $\alpha = -2, \beta = 4, \gamma = \frac{2}{3}$
d. $\alpha = 2, \beta = 4, \gamma = -\frac{2}{3}$

29 If $\mathbf{a}$, $\mathbf{b}$, $\mathbf{c}$ are three non-coplanar, non-zero vectors and $\mathbf{r}$ is any vector in space, then $(\mathbf{a} \times \mathbf{b}) \times (\mathbf{r} \times \mathbf{c}) + (\mathbf{b} \times \mathbf{c}) \times (\mathbf{r} \times \mathbf{a}) + (\mathbf{c} \times \mathbf{a}) \times (\mathbf{r} \times \mathbf{b})$ is

a. $2[\mathbf{a}\ \mathbf{b}\ \mathbf{c}]\mathbf{r}$
b. $3[\mathbf{a}\ \mathbf{b}\ \mathbf{c}]\mathbf{r}$
c. $[\mathbf{a}\ \mathbf{b}\ \mathbf{c}]\mathbf{r}$
d. None of these

30 Let $\mathbf{a}$, $\mathbf{b}$, $\mathbf{c}$ be three vectors such that $[\mathbf{a}\ \mathbf{b}\ \mathbf{c}] = 2$. If $\mathbf{r} = l(\mathbf{b} \times \mathbf{c}) + m(\mathbf{c} \times \mathbf{a}) + n(\mathbf{a} \times \mathbf{b})$ is perpendicular to $\mathbf{a} + \mathbf{b} + \mathbf{c}$, then the value of $(l + m + n)$ is

a. 2
b. 1
c. 0
d. None of these

31 If the vectors $\mathbf{a}$, $\mathbf{b}$, $\mathbf{c}$ are non-coplanar and l, m, n are distinct scalars such that $[l\mathbf{a} + m\mathbf{b} + n\mathbf{c}\ \ l\mathbf{b} + m\mathbf{c} + n\mathbf{a}\ \ l\mathbf{c} + m\mathbf{a} + n\mathbf{b}] = 0$, then

a. $lm + mn + nl = 0$
b. $l + m + n = 0$
c. $l^2 + m^2 + n^2 = 0$
d. $l^3 + m^3 + n^3 = 0$

32 Let $\mathbf{a}$, $\mathbf{b}$, $\mathbf{c}$ be three unit vectors such that $\mathbf{a} \cdot \mathbf{b} = \mathbf{a} \cdot \mathbf{c} = 0$. If the angle between $\mathbf{b}$ and $\mathbf{c}$ is $\frac{\pi}{3}$, then the volume of parallelopiped whose three coterminus edges are $\mathbf{a}$, $\mathbf{b}$, $\mathbf{c}$ is

a. $\frac{\sqrt{3}}{2}$
b. $\frac{1}{2}$
c. 1
d. None of these

33 If $\mathbf{a}$, $\mathbf{b}$, $\mathbf{c}$ are non-coplanar, non-zero vectors, then

$$(\mathbf{a} \times \mathbf{b}) \times (\mathbf{a} \times \mathbf{c}) + (\mathbf{b} \times \mathbf{c}) \times (\mathbf{b} \times \mathbf{a}) + (\mathbf{c} \times \mathbf{a}) \times (\mathbf{c} \times \mathbf{b}) \text{ is}$$

a. $[\mathbf{a}\ \mathbf{b}\ \mathbf{c}]^2 (\mathbf{a} + \mathbf{b} + \mathbf{c})$
b. $[\mathbf{a}\ \mathbf{b}\ \mathbf{c}] (\mathbf{a} + \mathbf{b} + \mathbf{c})$
c. 0
d. None of these

34 If $\mathbf{a}$, $\mathbf{b}$ are non-zero and non-collinear vectors, then $[\mathbf{a}\ \mathbf{b}\ \hat{\mathbf{i}}]\hat{\mathbf{i}} + [\mathbf{a}\ \mathbf{b}\ \hat{\mathbf{j}}]\hat{\mathbf{j}} + [\mathbf{a}\ \mathbf{b}\ \hat{\mathbf{k}}]\hat{\mathbf{k}}$ is

a. $\mathbf{a} + \mathbf{b}$
b. $\mathbf{a} \times \mathbf{b}$
c. $\mathbf{a} - \mathbf{b}$
d. $\mathbf{b} \times \mathbf{a}$

35 If $\mathbf{a}$, $\mathbf{b}$, $\mathbf{c}$, $\mathbf{d}$ are any four vectors, then $(\mathbf{a} \times \mathbf{b}) \times (\mathbf{c} \times \mathbf{d})$ is a vector

a. perpendicular to **a**, **b**, **c**, **d**
b. along the line of intersection of the plane determined by **a**, **b** and the other determined by **c** and **d**
c. equally inclined to **a**, **b**, **c**, **d**
d. None of the above

Statement Type

Working Space

Direction (Q. Nos. 36-40) This section is based on Statement I and Statement II. Select the correct answer from the codes given below.

a. Both Statement I and Statement II are correct and Statement II is the correct explanation of Statement I
b. Both Statement I and Statement II are correct but Statement II is not the correct explanation of Statement I
c. Statement I is correct but Statement II is incorrect
d. Statement II is correct but Statement I is incorrect

36 Let the vectors **PQ**, **QR**, **RS**, **ST**, **TU** and **UP** represent the sides of a regular hexagon.

Statement I $\mathbf{PQ} \times (\mathbf{RS} + \mathbf{ST}) \neq 0$

Statement II $\mathbf{PQ} \times \mathbf{RS} = 0$ and $\mathbf{PQ} \times \mathbf{ST} \neq 0$

37 **Statement I** If $\mathbf{a} = 3\hat{\mathbf{i}} - 3\hat{\mathbf{j}} + \hat{\mathbf{k}}$, $\mathbf{b} = -\hat{\mathbf{i}} + 2\hat{\mathbf{j}} + \hat{\mathbf{k}}$ and $\mathbf{c} = \hat{\mathbf{i}} + \hat{\mathbf{j}} + \hat{\mathbf{k}}$ and $\mathbf{d} = 2\hat{\mathbf{i}} - \hat{\mathbf{j}}$, then there exist real numbers α, β, γ such that $\mathbf{a} = \alpha\mathbf{b} + \beta\mathbf{c} + \gamma\mathbf{d}$.

Statement II **a**, **b**, **c**, **d** are four vectors in a 3-dimensional space. If **b**, **c**, **d** are non-coplanar, then there exist real numbers α, β, γ such that $\mathbf{a} = \alpha\mathbf{b} + \beta\mathbf{c} + \gamma\mathbf{d}$.

38 **Statement I** Let **a**, **b**, **c** and **d** are position vectors of four points *A*, *B*, *C* and *D* and $3\mathbf{a} - 2\mathbf{b} + 5\mathbf{c} - 6\mathbf{d} = 0$, then the points *A, B, C* and *D* are coplanar.

Statement II The non-zero, linearly dependent coinitial vectors (**PQ**, **PR** and **PS**) are coplanar.

39 Let $\hat{\mathbf{a}}$ and $\hat{\mathbf{c}}$ be unit vectors at an angle $\frac{\pi}{3}$ with each other. If $[\mathbf{a} \times (\mathbf{b} \times \hat{\mathbf{i}})] \cdot (\mathbf{a} \times \mathbf{c}) = 5$.

Statement I $[\mathbf{a}\ \mathbf{b}\ \mathbf{c}] = 10$

Statement II $[\mathbf{x}\ \mathbf{y}\ \mathbf{z}] = 0$, if $\mathbf{x} = \mathbf{y}$ or $\mathbf{y} = \mathbf{z}$ or $\mathbf{z} = \mathbf{x}$.

40 **Statement I** For $a = \frac{1}{\sqrt{3}}$, the volume of parallelopiped formed by vectors $\hat{\mathbf{i}} + a\hat{\mathbf{j}}$ and $\hat{\mathbf{j}} + a\hat{\mathbf{k}}$ is maximum (the vectors forms a right handed system).

Statement II The volume of parallelopiped having three coterminus edges **a**, **b** and **c** is $[\mathbf{a}\ \mathbf{b}\ \mathbf{c}]$.

SET - 2

Revisal Problems for JEE Advanced

Only One Option Correct Type

Direction (Q. Nos. 1-14) *This section contains 14 multiple choice questions. Each question has four choices (a), (b), (c) and (d), out of which ONLY ONE option is correct.*

Working Space

1 If **a, b** and **c** are any three vectors forming a linearly independent system, then $\theta \in R$,

$[\mathbf{a} \cos \theta + \mathbf{b} \sin \theta + \mathbf{c} \cos 2\theta, \mathbf{a} \cos\left(\frac{2\pi}{3} + \theta\right) + \mathbf{b} \sin\left(\frac{2\pi}{3} + \theta\right) + \mathbf{c} \cos 2\left(\frac{2\pi}{3} + \theta\right),$

$\mathbf{a} \cos\left(\theta - \frac{2\pi}{3}\right) + b \sin\left(\theta - \frac{2\pi}{3}\right) + \mathbf{c} \cos 2\left(\theta - \frac{2\pi}{3}\right)$ is

a. $[\mathbf{a}\ \mathbf{b}\ \mathbf{c}] \cos \theta$ b. $[\mathbf{a}\ \mathbf{b}\ \mathbf{c}] \cos 2\theta$
c. $[\mathbf{a}\ \mathbf{b}\ \mathbf{c}] \cos 3\theta$ d. None of these

2 If in a ΔABC, $\mathbf{BC} = \frac{\mathbf{e}}{|\mathbf{e}|} - \frac{\mathbf{f}}{|\mathbf{f}|}$ and $\mathbf{AC} = \frac{2\mathbf{e}}{|\mathbf{e}|}$, $|\mathbf{e}| \neq |\mathbf{f}|$, then the value of $\cos 2A + \cos 2B + \cos 2C$ is

a. -1 b. 0
c. 2 d. $-\frac{3}{2}$

3 If **a, b** and **c** are three unit vectors, **a** and **b** are perpendicular to each other and vector **c** is equally inclined to both **a** and **b** at an angle θ. If $\mathbf{c} = \alpha\mathbf{a} + \beta\mathbf{b} + \gamma(\mathbf{a} \times \mathbf{b})$, then ($\alpha, \beta, \gamma$ are constants)

a. $\alpha = \beta = -\cos\theta, \gamma^2 = \cos 2\theta$ b. $\alpha = \beta = \cos\theta, \gamma^2 = \cos 2\theta$
c. $\alpha = \beta = \cos\theta, \gamma^2 = -\cos 2\theta$ d. None of these

4 If $(\mathbf{a} + 3\mathbf{b}) \cdot (7\mathbf{a} - 5\mathbf{b}) = 0$ and $(\mathbf{a} - 4\mathbf{b}) \cdot (7\mathbf{a} - 2\mathbf{b}) = 0$. Then, the angle between **a** and **b** is

a. 60° b. 30°
c. 90° d. None of these

5 The perpendicular distance of a corner of a unit cube from a diagonal not passing through it is

a. $\sqrt{\frac{3}{2}}$ b. $\sqrt{\frac{2}{3}}$
c. $\sqrt{\frac{3}{4}}$ d. $\sqrt{\frac{4}{3}}$

6 If **p** and **q** are two non-collinear vectors such that $(b - c)(\mathbf{p} \times \mathbf{q}) + (c - a)(\mathbf{p}) + (a - b)\mathbf{q} = 0$, where a, b, c are lengths of sides of triangle, then the triangle is

a. right angled b. obtuse angled
c. equilateral d. right angled isosceles

7 A parallelopiped is formed by planes drawn parallel to coordinate axes through the points $A(1, 2, 3)$ and $B(9, 8, 5)$. The volume of that parallelopiped is

a. 192 b. 48
c. 32 d. 96

8 Let **a, b, c** be three non-coplanar vectors and **d** be a non-zero vector, which is perpendicular to $\mathbf{a} + \mathbf{b} + \mathbf{c}$. Now, if $\mathbf{d} = (\sin x)(\mathbf{a} \times \mathbf{b}) + (\cos y)(\mathbf{b} \times \mathbf{c}) + 2(\mathbf{c} \times \mathbf{a})$, then the minimum value of $(x^2 + y^2)$ is

a. π^2 b. $\frac{\pi^2}{2}$
c. $\frac{\pi^2}{4}$ d. $\frac{5\pi^2}{4}$

9 If the four faces of a tetrahedron are represented by the equations; $\mathbf{r}\cdot(\alpha\hat{\mathbf{i}}+\beta\hat{\mathbf{j}})=0$, $\mathbf{r}\cdot(\beta\hat{\mathbf{j}}+\gamma\hat{\mathbf{k}})=0$, $\mathbf{r}\cdot(\gamma\hat{\mathbf{k}}+\alpha\hat{\mathbf{i}})=0$ and $\mathbf{r}\cdot(\alpha\hat{\mathbf{i}}+\beta\hat{\mathbf{j}}+\gamma\hat{\mathbf{k}})=p$. Then, the volume of tetrahedron is

a. $\left|\dfrac{p^3}{6\alpha\beta\gamma}\right|$ b. $\left|\dfrac{4p^3}{6\alpha\beta\gamma}\right|$

c. $\left|\dfrac{3p^3}{6\alpha\beta\gamma}\right|$ d. None of these

10 If two adjacent sides of two rectangles are represented by the vectors $\mathbf{p}=5\mathbf{a}-3\mathbf{b}$, $\mathbf{q}=-\mathbf{a}-2\mathbf{b}$ and $\mathbf{r}=-4\mathbf{a}-\mathbf{b}$, $\mathbf{s}=-\mathbf{a}+\mathbf{b}$ respectively, then the angle between the vectors $\mathbf{x}=\frac{1}{3}(\mathbf{p}+\mathbf{r}+\mathbf{s})$ and $\mathbf{y}=\frac{1}{5}(\mathbf{r}+\mathbf{s})$ is

a. $\pi-\cos^{-1}\left(\dfrac{19}{3\sqrt{5}}\right)$ b. $\cos^{-1}\left(\dfrac{19}{5\sqrt{43}}\right)$

c. $-\cos^{-1}\left(\dfrac{19}{5\sqrt{43}}\right)$ d. None of these

11 If $\mathbf{a}$, $\mathbf{b}$, $\mathbf{c}$ be non-zero vectors such that $\mathbf{a}$ is perpendicular to $\mathbf{b}$ and $\mathbf{c}$ and $|\mathbf{a}|=1$, $|\mathbf{b}|=2$, $|\mathbf{c}|=1$, $\mathbf{b}\cdot\mathbf{c}=1$ and there is a non-zero vector $\mathbf{d}$ coplanar with $\mathbf{a}+\mathbf{b}$ and $2\mathbf{b}-\mathbf{c}$ and $\mathbf{d}\cdot\mathbf{a}=1$, then minimum value of $|\mathbf{d}|$ is

a. $\dfrac{2}{\sqrt{13}}$ b. $\dfrac{3}{\sqrt{13}}$

c. $\dfrac{4}{\sqrt{5}}$ d. $\dfrac{4}{\sqrt{13}}$

12 Let $\mathbf{P}=3ax^2\hat{\mathbf{i}}-2(x-1)\hat{\mathbf{j}}$, $\mathbf{Q}=b(x-1)\hat{\mathbf{i}}+x\hat{\mathbf{j}}$ and $ab<0$, then $\mathbf{P}$ and $\mathbf{Q}$ are parallel for

a. atleast one x in $(-1, 0)$ b. atleast one $x\in(0, 1)$

c. atleast one $x\in(1, 2)$ d. None of these

13 Let $\mathbf{r}$, $\mathbf{a}$, $\mathbf{b}$ and $\mathbf{c}$ be four non-zero vectors such that $\mathbf{r}\cdot\mathbf{a}=0$, $|\mathbf{r}\times\mathbf{b}|=|\mathbf{r}||\mathbf{b}|$, $|\mathbf{r}\times\mathbf{c}|=|\mathbf{r}||\mathbf{c}|$, then $[\mathbf{a}\ \mathbf{b}\ \mathbf{c}]$ is

a. $|\mathbf{a}||\mathbf{b}||\mathbf{c}|$ b. $-|\mathbf{a}||\mathbf{b}||\mathbf{c}|$

c. 0 d. None of these

14 If $[(\mathbf{a}\times\mathbf{b})\times(\mathbf{c}\times\mathbf{d})]\cdot(\mathbf{a}\times\mathbf{d})=0$, then

a. **a, b, c** and **d** are necessarily coplanar

b. either **a** or **d** must lie in the plane of **b** or **c**

c. either **b** or **c** must lie in the plane of **a** and **d**

d. either **a** or **b** must lie in the plane of **c** and **d**

One or More than One Options Correct Type

Direction (Q. Nos. 15-20) *This section contains 6 multiple choice questions. Each question has four choices (a), (b), (c) and (d), out of which ONE or MORE THAN ONE are correct.*

15 If $\mathbf{a}$ and $\mathbf{b}$ are two unit vectors perpendicular to each other and $\mathbf{c}=\lambda_1\mathbf{a}+\lambda_2\mathbf{b}+\lambda_3(\mathbf{a}\times\mathbf{b})$, then which of the following is/are true?

a. $\lambda_1=\mathbf{a}\cdot\mathbf{c}$ b. $\lambda_2=|\mathbf{b}\times\mathbf{a}|$

c. $\lambda_3=|(\mathbf{a}\times\mathbf{b})\times\mathbf{c}|$ d. $\lambda_1+\lambda_2+\lambda_3=(\mathbf{a}+\mathbf{b}+\mathbf{a}\times\mathbf{b})\cdot\mathbf{c}$

16 Given three non-coplanar vectors $\mathbf{OA}=\mathbf{a}$, $\mathbf{OB}=\mathbf{b}$, $\mathbf{OC}=\mathbf{c}$. Let S be the centre of sphere passing through the points O, A, B, C, if $\mathbf{OS}=\mathbf{x}$, then

a. **x** must be linear combination of **a, b, c**

b. **x** must be linear combination of $\mathbf{b}\times\mathbf{c}$, $\mathbf{c}\times\mathbf{a}$ and $\mathbf{a}\times\mathbf{b}$

c. $\mathbf{x}=\dfrac{a^2(\mathbf{b}\times\mathbf{c})+b^2(\mathbf{c}\times\mathbf{a})+c^2(\mathbf{a}\times\mathbf{b})}{2\,[\mathbf{a}\ \mathbf{b}\ \mathbf{c}]}$, $a=|\mathbf{a}|$, $b=|\mathbf{b}|$, $c=|\mathbf{c}|$

d. $\mathbf{x}=\mathbf{a}+\mathbf{b}+\mathbf{c}$

17 If $\mathbf{a}=\hat{\mathbf{i}}+\hat{\mathbf{j}}+\hat{\mathbf{k}}$ and $\mathbf{b}=\hat{\mathbf{i}}-\hat{\mathbf{j}}$, then the vectors $(\mathbf{a}\cdot\hat{\mathbf{i}})\hat{\mathbf{i}}+(\mathbf{a}\cdot\hat{\mathbf{j}})\hat{\mathbf{j}}+(\mathbf{a}\cdot\hat{\mathbf{k}})\hat{\mathbf{k}}$, $(\mathbf{b}\cdot\hat{\mathbf{i}})\hat{\mathbf{i}}+(\mathbf{b}\cdot\hat{\mathbf{j}})\hat{\mathbf{j}}+(\mathbf{b}\cdot\hat{\mathbf{k}})\hat{\mathbf{k}}$ and $\hat{\mathbf{i}}+\hat{\mathbf{j}}-2\hat{\mathbf{k}}$

a. are mutually perpendicular b. are coplanar

c. form a parallelopiped of volume 6 units d. form a parallelopiped of volume 3 units

Working Space

18 If **a** and **b** are any two unit vectors, then the possible integers in the range of $\frac{3|\mathbf{a}+\mathbf{b}|}{2}+2|\mathbf{a}-\mathbf{b}|$, is/are

a. 2 b. 3
c. 4 d. 5

19 If **a** is perpendicular to **b** and p is non-zero scalar such that $p\mathbf{r}+(\mathbf{r}\cdot\mathbf{b})\mathbf{a}=\mathbf{c}$, then **r** satisfy

a. $[\mathbf{r}\ \mathbf{a}\ \mathbf{c}]=0$ b. $p^2\mathbf{r}=p\mathbf{a}-(\mathbf{c}\cdot\mathbf{a})\mathbf{b}$
c. $p^2\mathbf{r}=p\mathbf{b}-(\mathbf{a}\cdot\mathbf{b})\mathbf{c}$ d. $p^2\mathbf{r}=p\mathbf{c}-(\mathbf{b}\cdot\mathbf{c})\mathbf{a}$

20 A vector **d** is equally inclined to three vectors $\mathbf{a}=\hat{\mathbf{i}}-\hat{\mathbf{j}}+\hat{\mathbf{k}}$, $\mathbf{b}=2\hat{\mathbf{i}}+\hat{\mathbf{j}}$ and $\mathbf{c}=3\hat{\mathbf{j}}-2\hat{\mathbf{k}}$. Let **x**, **y**, **z** be three vectors in the plane of **a, b; b, c; c, a** respectively, then

a. $\mathbf{x}\cdot\mathbf{d}=14$ b. $\mathbf{y}\cdot\mathbf{d}=3$
c. $\mathbf{z}\cdot\mathbf{d}=0$ d. $\mathbf{r}\cdot\mathbf{d}=0$ and $\mathbf{r}=\lambda\mathbf{x}+\mu\mathbf{y}+\delta\mathbf{z}$

Matching List Type

***Direction** (Q. Nos. 21-24) Choices for the correct combination of elements from Column I and Column II are given as options (a), (b), (c) and (d), out of which one is correct.*

21 Match the Column I with Column II and mark the correct option from the codes given below.

	Column I		Column II
i.	Let **a**, **b**, **c** be the three vectors such that $\mathbf{a}\cdot(\mathbf{b}+\mathbf{c})=\mathbf{b}\cdot(\mathbf{c}+\mathbf{a})=0$ and $|\mathbf{a}|=1$, $|\mathbf{b}|=4$, $|\mathbf{c}|=8$, then $|\mathbf{a}+\mathbf{b}+\mathbf{c}|$ is	p.	$\frac{5}{2}$
ii.	If $\mathbf{a}_1,\mathbf{a}_2,\mathbf{a}_3$ are vectors reciprocal to the non-coplanar vectors $\mathbf{b}_1,\mathbf{b}_2,\mathbf{b}_3$, then $[\mathbf{a}_1\ \mathbf{a}_2\ \mathbf{a}_3][\mathbf{b}_1\ \mathbf{b}_2\ \mathbf{b}_3]$ is	q.	9
iii.	*ABCD* is a quadrilateral with $\mathbf{AB}=\mathbf{a}$, $\mathbf{AD}=\mathbf{b}$ and $\mathbf{AC}=2\mathbf{a}+3\mathbf{b}$. If its area is α times the area of parallelogram with *AB*, *AD* as its adjacent sides, then α is	r.	8
iv.	If $\mathbf{d}=x(\mathbf{a}\times\mathbf{b})+y(\mathbf{b}\times\mathbf{c})+z(\mathbf{c}\times\mathbf{a})$ and $[\mathbf{a}\ \mathbf{b}\ \mathbf{c}]=\frac{1}{8}$, then $x+y+z=R(\mathbf{a}+\mathbf{b}+\mathbf{c})\cdot\mathbf{d}$, where R is	s.	1

Codes

	i	ii	iii	iv
a.	q	s	p	r
b.	r	p	q	s
c.	s	q	r	p
d.	t	r	s	q

22 Given two vectors $\mathbf{a}=\hat{\mathbf{i}}+2\hat{\mathbf{j}}+2\hat{\mathbf{k}}$ and $\mathbf{b}=\hat{\mathbf{i}}+\hat{\mathbf{j}}+2\hat{\mathbf{k}}$. Match the Column I with Column II and mark the correct option from the codes given below.

	Column I		Column II
i.	A vector coplanar with **a** and **b**	p.	$-3\hat{\mathbf{i}}+3\hat{\mathbf{j}}+4\hat{\mathbf{k}}$
ii.	A vector which is perpendicular to both **a** and **b**	q.	$2\hat{\mathbf{i}}-2\hat{\mathbf{j}}+3\hat{\mathbf{k}}$
iii.	A vector which is equally inclined to **a** and **b**	r.	$\hat{\mathbf{i}}+\hat{\mathbf{j}}$
iv.	A vector which forms a triangle with **a** and **b**	s.	$\hat{\mathbf{i}}-\hat{\mathbf{j}}+5\hat{\mathbf{k}}$

Codes

	i	ii	iii	iv
a.	r	p	q	s,p
b.	p,r	q	s	p
c.	s	r	r	p,q
d.	q	s	p	s,p

23 Match the Column I with Column II and mark the correct option from the codes given below.

Working Space

	Column I		Column II
i.	If a_1, a_2 are two values of a for which $a\hat{\mathbf{i}} + b\hat{\mathbf{j}} + \hat{\mathbf{k}}$ whose magnitude is $\sqrt{2}$, is linearly dependent with $\hat{\mathbf{i}} + 2\hat{\mathbf{j}}$ and $\hat{\mathbf{j}} - 2\hat{\mathbf{k}}$, then $\left[\left\|\frac{1}{a_1} + \frac{1}{a_2}\right\|\right]$ is equal to (where, [·] denotes greatest integer function)	p.	0
ii.	Let $A(\alpha\hat{\mathbf{i}} - 2\hat{\mathbf{j}} + \hat{\mathbf{k}})$, $B(\hat{\mathbf{i}} - 3\hat{\mathbf{j}})$ and minimum length of projection of **AB** on $\hat{\mathbf{i}} + \hat{\mathbf{j}} + \hat{\mathbf{k}}$ is $\frac{8}{\sqrt{3}}$, then least value of $\|\beta\|$ for which $y = x^2 - (\alpha + 1)x + \beta^2$ does not intersect X-axis at two distinct points, is	q.	2
iii.	If a, b and c are the magnitudes of **a, b, c** respectively, **a** is perpendicular to **b** and $\mathbf{a} + 2\mathbf{b} - 3\mathbf{c} = 0$, then $\frac{(3c + a)(3c - a)}{b^2}$ is equal to	r.	4
iv.	If $\mathbf{a} \cdot (\mathbf{b} \times \mathbf{c}) = 1$, then the value of $(\mathbf{a} + 2\mathbf{b} + 3\mathbf{c}) \cdot (\mathbf{a} \times \mathbf{b} + \mathbf{b} \times \mathbf{c} + \mathbf{c} \times \mathbf{a})$ is	s.	6

Codes

	i	ii	iii	iv
a.	q	r	r	s
b.	p	s	s	q
c.	r	p	q	s
d.	s	q	p	r

24 Match the Column I with Column II and mark the correct option from the codes given below.

	Column I		Column II
i.	Let $\|\mathbf{a}\| = \|\mathbf{b}\| = 2$, $\mathbf{x} = \mathbf{a} + \mathbf{b}$, $\mathbf{y} = \mathbf{a} - \mathbf{b}$. If $\|\mathbf{x} \times \mathbf{y}\| = 2\{\lambda - (\mathbf{a} \cdot \mathbf{b})^2\}^{1/2}$, then the value of λ is	p.	4
ii.	The non-zero value of λ for which angle between $\lambda\hat{\mathbf{i}} + \hat{\mathbf{j}} + \hat{\mathbf{k}}$ and $\hat{\mathbf{i}} + \lambda\hat{\mathbf{j}} + \hat{\mathbf{k}}$ is $\frac{\pi}{3}$, is	q.	42
iii.	If $\|\mathbf{a}\| = \|\mathbf{b}\| = 1$ and $\|\mathbf{c}\| = 2$, then the maximum value of $\|\mathbf{a} - 2\mathbf{b}\|^2 + \|\mathbf{b} - 2\mathbf{c}\|^2 + \|\mathbf{c} - 2\mathbf{a}\|^2$ is	r.	16
		s.	7

Codes

	i	ii	iii
a.	r	p	q
b.	s	q	p
c.	p	r	s
d.	q	s	r

Working Space

Comprehension Type

Direction (*Q. Nos.* 25-35) *This section contains 3 paragraphs, each describing theory, experiments, data, etc. Eleven questions related to the paragraphs have been given. Each question has only one correct answer among the four given options (a), (b), (c) and (d).*

Passage I

If AP, BQ and CR are the altitudes of acute ΔABC and $9\mathbf{AP} + 4\mathbf{BQ} + 7\mathbf{CR} = 0$.

25 $\angle ACB$ is equal to

a. $\frac{\pi}{4}$ b. $\frac{\pi}{3}$

c. $\cos^{-1}\left(\frac{1}{3\sqrt{7}}\right)$ d. $\cos^{-1}\left(\frac{1}{\sqrt{7}}\right)$

26 $\angle ABC$ is equal to

a. $\cos^{-1}\left(\frac{2}{\sqrt{7}}\right)$ b. $\frac{\pi}{2}$

c. $\cos^{-1}\left(\frac{\sqrt{7}}{3}\right)$ d. $\frac{\pi}{3}$

27 $\angle BAC$ is equal to

a. $\cos^{-1}\left(\frac{2}{\sqrt{7}}\right)$ b. $\cos^{-1}\left(\frac{1}{2\sqrt{7}}\right)$

c. $\cos^{-1}\left(\frac{1}{\sqrt{7}}\right)$ d. $\frac{\pi}{3}$

Passage II

Let r be the variable point satisfying $\mathbf{r}\cdot\mathbf{n}_1 = d_1$, $\mathbf{r}\cdot\mathbf{n}_2 = d_2$, $\mathbf{r}\cdot\mathbf{n}_3 = d_3$, where $\mathbf{n}_1$, $\mathbf{n}_2$ and $\mathbf{n}_3$ are non-coplanar vectors.

28 The position vector of the point of intersection of three planes is

a. $\frac{1}{[\mathbf{n}_1\ \mathbf{n}_2\ \mathbf{n}_3]}[d_3(\mathbf{n}_1\times\mathbf{n}_2)+d_1(\mathbf{n}_2\times\mathbf{n}_3)+d_2(\mathbf{n}_1\times\mathbf{n}_3)]$

b. $\frac{4}{[\mathbf{n}_1\ \mathbf{n}_2\ \mathbf{n}_3]}[d_3(\mathbf{n}_1\times\mathbf{n}_2)+d_1(\mathbf{n}_2\times\mathbf{n}_3)+d_2(\mathbf{n}_1\times\mathbf{n}_3)]$

c. $\frac{-4}{[\mathbf{n}_1\ \mathbf{n}_2\ \mathbf{n}_3]}[d_3(\mathbf{n}_1\times\mathbf{n}_2)+d_1(\mathbf{n}_2\times\mathbf{n}_3)+d_2(\mathbf{n}_1\times\mathbf{n}_3)]$

d. None of the above

29 If the planes $\mathbf{r}\cdot\mathbf{n}_1 = d_1$, $\mathbf{r}\cdot\mathbf{n}_2 = d_2$ and $\mathbf{r}\cdot\mathbf{n}_3 = d_3$, have a common line of intersection, then $d_1(\mathbf{n}_1\times\mathbf{n}_2)+d_2(\mathbf{n}_2\times\mathbf{n}_3)+d_3(\mathbf{n}_1\times\mathbf{n}_2)$ is

a. $[\mathbf{n}_1\ \mathbf{n}_2\ \mathbf{n}_3]$ b. $4\,[\mathbf{n}_1\ \mathbf{n}_2\ \mathbf{n}_3]$

c. $2\,[\mathbf{n}_1\ \mathbf{n}_2\ \mathbf{n}_3]$ d. None of these

Passage III

If a, b, c are three given non-coplanar vectors and any arbitrary vector r is in space, where

$$\Delta_1 = \begin{vmatrix} \mathbf{r}\cdot\mathbf{a} & \mathbf{b}\cdot\mathbf{a} & \mathbf{c}\cdot\mathbf{a} \\ \mathbf{r}\cdot\mathbf{b} & \mathbf{b}\cdot\mathbf{b} & \mathbf{c}\cdot\mathbf{b} \\ \mathbf{r}\cdot\mathbf{c} & \mathbf{b}\cdot\mathbf{c} & \mathbf{c}\cdot\mathbf{c} \end{vmatrix}, \quad \Delta_2 = \begin{vmatrix} \mathbf{a}\cdot\mathbf{a} & \mathbf{r}\cdot\mathbf{a} & \mathbf{c}\cdot\mathbf{a} \\ \mathbf{a}\cdot\mathbf{b} & \mathbf{r}\cdot\mathbf{b} & \mathbf{c}\cdot\mathbf{b} \\ \mathbf{a}\cdot\mathbf{c} & \mathbf{r}\cdot\mathbf{c} & \mathbf{c}\cdot\mathbf{c} \end{vmatrix},$$

$$\Delta_3 = \begin{vmatrix} \mathbf{a}\cdot\mathbf{a} & \mathbf{b}\cdot\mathbf{a} & \mathbf{r}\cdot\mathbf{a} \\ \mathbf{a}\cdot\mathbf{b} & \mathbf{b}\cdot\mathbf{b} & \mathbf{r}\cdot\mathbf{b} \\ \mathbf{a}\cdot\mathbf{c} & \mathbf{b}\cdot\mathbf{c} & \mathbf{r}\cdot\mathbf{c} \end{vmatrix}, \quad \Delta = \begin{vmatrix} \mathbf{a}\cdot\mathbf{a} & \mathbf{b}\cdot\mathbf{a} & \mathbf{c}\cdot\mathbf{a} \\ \mathbf{a}\cdot\mathbf{b} & \mathbf{b}\cdot\mathbf{b} & \mathbf{c}\cdot\mathbf{b} \\ \mathbf{a}\cdot\mathbf{c} & \mathbf{b}\cdot\mathbf{c} & \mathbf{c}\cdot\mathbf{c} \end{vmatrix}.$$

30 The vector **r** is expressible in the form

a. $\mathbf{r} = \frac{\Delta_1}{2\Delta}\mathbf{a} + \frac{\Delta_2}{2\Delta}\mathbf{b} + \frac{\Delta_3}{2\Delta}\mathbf{c}$ b. $\mathbf{r} = \frac{2\Delta_1}{\Delta}\mathbf{a} + \frac{2\Delta_2}{\Delta}\mathbf{b} + \frac{2\Delta_3}{\Delta}\mathbf{c}$

c. $\mathbf{r} = \frac{\Delta}{\Delta_1}\mathbf{a} + \frac{\Delta}{\Delta_2}\mathbf{b} + \frac{\Delta}{\Delta_3}\mathbf{c}$ d. $\mathbf{r} = \frac{\Delta_1}{\Delta}\mathbf{a} + \frac{\Delta_2}{\Delta}\mathbf{b} + \frac{\Delta_3}{\Delta}\mathbf{c}$

31 The vector **r** is expressible as

a. $\mathbf{r} = \dfrac{[\mathbf{r\ b\ c}]}{2\,[\mathbf{a\ b\ c}]}\mathbf{a} + \dfrac{[\mathbf{r\ c\ a}]}{2\,[\mathbf{a\ b\ c}]}\mathbf{b} + \dfrac{[\mathbf{r\ a\ b}]}{2\,[\mathbf{a\ b\ c}]}\mathbf{c}$

b. $\mathbf{r} = \dfrac{2\,[\mathbf{r\ b\ c}]}{[\mathbf{a\ b\ c}]}\mathbf{a} + \dfrac{2\,[\mathbf{r\ c\ a}]}{[\mathbf{a\ b\ c}]}\mathbf{b} + \dfrac{2\,[\mathbf{r\ a\ b}]}{[\mathbf{a\ b\ c}]}\mathbf{c}$

c. $\mathbf{r} = \dfrac{1}{[\mathbf{a\ b\ c}]}([\mathbf{r\ b\ c}]\mathbf{a} + [\mathbf{r\ c\ a}]\mathbf{b} + [\mathbf{r\ a\ b}]\mathbf{c})$

d. None of the above

32 If vector is expressible as $\mathbf{r} = x\mathbf{a} + y\mathbf{b} + g\mathbf{c}$, then

a. $\mathbf{a} = \dfrac{1}{[\mathbf{a\ b\ c}]}[(\mathbf{a}\cdot\mathbf{a})(\mathbf{b}\times\mathbf{c}) + (\mathbf{b}\cdot\mathbf{b})(\mathbf{c}\times\mathbf{a}) + (\mathbf{c}\cdot\mathbf{c})(\mathbf{a}\times\mathbf{b})]$

b. $\mathbf{a} = \dfrac{1}{[\mathbf{a\ b\ c}]}[(\mathbf{a}\cdot\mathbf{a})(\mathbf{b}\times\mathbf{c}) + (\mathbf{b}\cdot\mathbf{a})(\mathbf{c}\times\mathbf{a}) + (\mathbf{c}\cdot\mathbf{a})(\mathbf{a}\times\mathbf{b})]$

c. $\mathbf{a} = (\mathbf{a}\cdot\mathbf{a})(\mathbf{b}\times\mathbf{c}) + (\mathbf{a}\cdot\mathbf{b})(\mathbf{c}\times\mathbf{a}) + (\mathbf{c}\cdot\mathbf{a})(\mathbf{a}\times\mathbf{b})$

d. None of the above

33 The value of $\begin{vmatrix} \mathbf{a} & \mathbf{b} & \mathbf{c} \\ \mathbf{a}\cdot\mathbf{p} & \mathbf{b}\cdot\mathbf{p} & \mathbf{c}\cdot\mathbf{p} \\ \mathbf{a}\cdot\mathbf{q} & \mathbf{b}\cdot\mathbf{q} & \mathbf{c}\cdot\mathbf{q} \end{vmatrix}$ is

a. $(\mathbf{p}\times\mathbf{q})\,[\mathbf{a}\times\mathbf{b}\ \ \mathbf{b}\times\mathbf{c}\ \ \mathbf{c}\times\mathbf{a}]$
b. $2\,(\mathbf{p}\times\mathbf{q})\,[\mathbf{a}\times\mathbf{b}\ \ \mathbf{b}\times\mathbf{c}\ \ \mathbf{c}\times\mathbf{a}]$
c. $4\,(\mathbf{p}\times\mathbf{q})\,[\mathbf{a}\times\mathbf{b}\ \ \mathbf{b}\times\mathbf{c}\ \ \mathbf{c}\times\mathbf{a}]$
d. $(\mathbf{p}\times\mathbf{q})\sqrt{|[\mathbf{a}\times\mathbf{b}\ \mathbf{b}\times\mathbf{c}\ \mathbf{c}\times\mathbf{a}]|}$

Passage IV

For any three vectors a, b, c; $\mathbf{a}\cdot(\mathbf{b}\times\mathbf{c})$ is called as scalar triple product. It is also denoted by [a b c]. If a, b and c are coplanar, then [a b c] = 0. If a, b, c are any three non-coplanar and non-zero vectors such that $\mathbf{b}\times\mathbf{c}=\mathbf{a}$, $\mathbf{a}\times\mathbf{b}=\mathbf{c}$ and $\mathbf{c}\times\mathbf{a}=\mathbf{b}$.

34 $|\mathbf{a}|$ is equal to

a. 1
b. 2
c. 3
d. 4

35 $|\mathbf{a}| + |\mathbf{b}| + |\mathbf{c}|$ is

a. 0
b. 2
c. 3
d. 4

One Integer Value Correct Type

Direction (*Q. Nos.* 36-40) *This section contains* 5 *questions. When worked out will result in an integer from* 0 *to* 9 (*both inclusive*).

36 Let $A(2\hat{i} + 3\hat{j} + 5\hat{k})$, $B(-\hat{i} + 3\hat{j} + 2\hat{k})$ and $C(\lambda\hat{i} + 5\hat{j} + \mu\hat{k})$ are vertices of a triangle and its median through A is equally inclined to the positive directions of the axes, the value of $2\lambda - \mu$ is

37 If **V** is the volume of parallelopiped having three coterminous edges as **a**, **b** and **c** and the volume of the parallelopiped having three coterminous edges as $\alpha = (\mathbf{a}\cdot\mathbf{a})\mathbf{a} + (\mathbf{a}\cdot\mathbf{b})\mathbf{b} + (\mathbf{a}\cdot\mathbf{c})\mathbf{c}$, $\beta = (\mathbf{b}\cdot\mathbf{a})\mathbf{a} + (\mathbf{b}\cdot\mathbf{b})\mathbf{b} + (\mathbf{b}\cdot\mathbf{c})\mathbf{c}$ and $\gamma = (\mathbf{c}\cdot\mathbf{a})\mathbf{a} + (\mathbf{c}\cdot\mathbf{b})\mathbf{b} + (\mathbf{c}\cdot\mathbf{c})\mathbf{c}$ is V^{λ}, then the value of λ is

38 If $\mathbf{a}\times\mathbf{b}=\mathbf{c}$, $\mathbf{b}\times\mathbf{c}=\mathbf{a}$, $\mathbf{c}\times\mathbf{a}=\mathbf{b}$. If vectors **a**, **b** and **c** are forming a right handed system, then the volume of tetrahedron formed by vectors $3\mathbf{a} - 2\mathbf{b} + 2\mathbf{c}$, $-\mathbf{a} - 2\mathbf{c}$ and $2\mathbf{a} - 3\mathbf{b} + 4\mathbf{c}$ is

39 If **a** and **b** are vectors perpendicular to each other and $|\mathbf{a}| = 2$, $|\mathbf{b}| = 3$, $\mathbf{c}\times\mathbf{a} = \mathbf{b}$, then the least value of $2|\mathbf{c} - \mathbf{a}|$ is

40 If **a**, **b** and **c** are three units vectors and every two are inclined to each other at an angle $\cos^{-1}\left(\dfrac{3}{5}\right)$ and $\mathbf{a}\times\mathbf{b} = p\mathbf{a} + q\mathbf{b} + r\mathbf{c}$, where p, q, r are scalars, then the value of $55q^2$ is

Working Space

JEE Main & AIEEE Archive

(Compilation of Last 13 Years' Questions)

1. Let **a**, **b** and **c** be three non-zero vectors such that no two of them are collinear and $(\mathbf{a} \times \mathbf{b}) \times \mathbf{c} = \frac{1}{3}|\mathbf{b}||\mathbf{c}|\mathbf{a}$. If θ is the angle between vectors **b** and **c**, then the value of $\sin\theta$ is **(JEE Main 2015)**

a. $\frac{2\sqrt{2}}{3}$ b. $-\frac{\sqrt{2}}{3}$

c. $\frac{2}{3}$ d. $-\frac{2\sqrt{3}}{3}$

2 If $[\mathbf{a} \times \mathbf{b}\ \ \mathbf{b} \times \mathbf{c}\ \ \mathbf{c} \times \mathbf{a}] = \lambda\,[\mathbf{a}\ \mathbf{b}\ \mathbf{c}]^2$, then λ is equal to **(JEE Main 2014)**

a. 0 b. 1

c. 2 d. 3

3 If the vectors $\mathbf{AB} = 3\hat{i} + 4\hat{k}$ and $\mathbf{AC} = 5\hat{i} - 2\hat{j} + 4\hat{k}$ are the sides of a ΔABC, then the length of the median through A is **(JEE Main 2013, AIEEE 2003)**

a. $\sqrt{18}$ b. $\sqrt{72}$

c. $\sqrt{33}$ d. $\sqrt{45}$

4 Let **a** and **b** be two unit vectors. If the vectors $\mathbf{c} = \mathbf{a} + 2\mathbf{b}$ and $\mathbf{d} = 5\mathbf{a} - 4\mathbf{b}$ are perpendicular to each other, then the angle between **a** and **b** is **(AIEEE 2012)**

a. $\frac{\pi}{6}$ b. $\frac{\pi}{2}$

c. $\frac{\pi}{3}$ d. $\frac{\pi}{4}$

5 Let $ABCD$ be a parallelogram such that $\mathbf{AB} = \mathbf{q}$, $\mathbf{AD} = \mathbf{p}$ and $\angle BAD$ be an acute angle. If **r** is the vector that coincides with the altitude directed from the vertex B to the side AD, then **r** is given by **(AIEEE 2012)**

a. $\mathbf{r} = 3\mathbf{q}\ \frac{3(\mathbf{p}\cdot\mathbf{q})}{(\mathbf{p}\cdot\mathbf{p})}\mathbf{p}$ b. $\mathbf{r} = -\mathbf{q} + \left(\frac{\mathbf{p}\cdot\mathbf{q}}{\mathbf{p}\cdot\mathbf{p}}\right)\mathbf{p}$

c. $\mathbf{r} = \mathbf{q} - \left(\frac{\mathbf{p}\cdot\mathbf{q}}{\mathbf{p}\cdot\mathbf{p}}\right)\mathbf{p}$ d. $\mathbf{r} = -3\mathbf{q} + \frac{3(\mathbf{p}\cdot\mathbf{q})}{(\mathbf{p}\cdot\mathbf{p})}\mathbf{p}$

6 If $\mathbf{a} = \frac{1}{\sqrt{10}}(3\hat{i} + \hat{k})$ and $\mathbf{b} = \frac{1}{7}(2\hat{i} + 3\hat{j} - 6\hat{k})$, then the value of $(2\mathbf{a} - \mathbf{b})\cdot[(\mathbf{a} \times \mathbf{b}) \times (\mathbf{a} + 2\,\mathbf{b})]$ is **(AIEEE 2011)**

a. –3 b. 5

c. 3 d. –5

7 The vectors **a** and **b** are not perpendicular and **c** and **d** are two vectors satisfying $\mathbf{b} \times \mathbf{c} = \mathbf{b} \times \mathbf{d}$ and $\mathbf{a}\cdot\mathbf{b} = 0$ Then, the vector **d** is equal to **(AIEEE 2011)**

a. $\mathbf{c} + \left(\frac{\mathbf{a}\cdot\mathbf{c}}{\mathbf{a}\cdot\mathbf{b}}\right)\mathbf{b}$ b. $\mathbf{b} + \left(\frac{\mathbf{b}\cdot\mathbf{c}}{\mathbf{a}\cdot\mathbf{b}}\right)\mathbf{c}$

c. $\mathbf{c} - \left(\frac{\mathbf{a}\cdot\mathbf{c}}{\mathbf{a}\cdot\mathbf{b}}\right)\mathbf{b}$ d. $\mathbf{b} - \left(\frac{\mathbf{b}\cdot\mathbf{c}}{\mathbf{a}\cdot\mathbf{b}}\right)\mathbf{c}$

8 If the vectors $p\hat{i} + \hat{j} + \hat{k}$, $\hat{i} + q\hat{j} + \hat{k}$ and $\hat{i} + \hat{j} + r\hat{k}$ (where, $p \neq q \neq r \neq 1$) are coplanar, then the value of $pqr - (p + q + r)$ is **(AIEEE 2011)**

a. – 2 b. 2 c. 0 d. –1

9 Let **a**, **b** and **c** be three non-zero vectors which are pairwise non-collinear. If $\mathbf{a} + 3\mathbf{b}$ is collinear with **c** and $\mathbf{b} + 2\mathbf{c}$ is collinear with **a**, then $\mathbf{a} + 3\mathbf{b} + 6\mathbf{c}$ is equal to **(AIEEE 2011)**

a. $\mathbf{a} + \mathbf{c}$ b. $\mathbf{a}$

c. $\mathbf{c}$ d. 0

10 Let $\mathbf{a} = \hat{j} - \hat{k}$ and $\mathbf{a} = \hat{i} - \mathbf{p} - \hat{k}$. Then, the vector **b** satisfying $\mathbf{a} \times \mathbf{b} + \mathbf{c} = 0$ and $\mathbf{a}\cdot\mathbf{b} = 3$ is **(AIEEE 2010)**

a. $-\hat{i} + \hat{j} - 2\hat{k}$ b. $2\hat{i} - \hat{j} + 2\hat{k}$

c. $\hat{i} - \hat{j} - 2\hat{k}$ d. $\hat{i} + \hat{j} - 2\hat{k}$

11 If the vectors $\mathbf{a} = \hat{i} - \hat{j} + 2\hat{k}$, $\mathbf{b} = 2\hat{i} + 4\hat{j} + \hat{k}$ and $\mathbf{c} = \lambda\hat{i} + \hat{j} + \mu\hat{k}$ are mutually orthogonal, then (λ, μ) is equal to **(AIEEE 2010)**

a. (– 3, 2) b. (2, – 3)

c. (– 2, 3) d. (3, – 2)

12 If **u**, **v** and **w** are non-coplanar vectors and p, q are real numbers, then the equality $[3\mathbf{u}\ p\mathbf{v}\ p\mathbf{w}] - [p\mathbf{v}\ \mathbf{w}\ q\mathbf{u}] - [2\mathbf{w}\ q\mathbf{v}\ q\mathbf{u}] = 0$ holds for **(AIEEE 2009)**

a. exactly two values of (p, q)

b. more than two but not all values of (p, q)

c. all values of (p, q)

d. exactly one value of (p, q)

13 The vector $\mathbf{a} = \alpha\hat{i} + 2\hat{j} + \beta\hat{k}$ lies in the plane of the vectors $\mathbf{b} = \hat{i} + \hat{j}$ and $\mathbf{c} = \hat{j} + \hat{k}$ and bisects the angle between **b** and **c**. Then, which one of the following gives possible values of α and β? **(AIEEE 2008)**

a. $\alpha = 1, \beta = 1$ b. $\alpha = 2, \beta = 2$

c. $\alpha = 1, \beta = 2$ d. $\alpha = 2, \beta = 1$

14 The non-zero vectors **a**, **b** and **c** are related by $\mathbf{a} = 8\mathbf{b}$ and $\mathbf{c} = -7\mathbf{b}$. Then, the angle between **a** and **c** is **(AIEEE 2008)**

a. π b. 0

c. $\frac{\pi}{4}$ d. $\frac{\pi}{2}$

15 If **u** and **v** are unit vectors and θ is the acute angle between them, then $2\mathbf{u} \times 3\mathbf{v}$ is a unit vector for **(AIEEE 2007)**

a. exactly two values of θ

b. more than two values of θ

c. no value of θ

d. exactly one value of θ

16 Let $\mathbf{a} = \hat{\mathbf{i}} + \hat{\mathbf{j}} + \hat{\mathbf{k}}, \mathbf{b} = \hat{\mathbf{i}} - \hat{\mathbf{j}} + 2\hat{\mathbf{k}}$ and $\mathbf{c} = x\hat{\mathbf{i}} + (x-2)\hat{\mathbf{j}} - \hat{\mathbf{k}}$. If the vector **c** lies in the plane of **a** and **b**, then x is equal to **(AIEEE 2007)**

a. 0 b. 1
c. – 4 d. –2

17 If $(\mathbf{a} \times \mathbf{b}) \times \mathbf{c} = \mathbf{a} \times (\mathbf{b} \times \mathbf{c})$, where **a**, **b** and **c** are any three vectors such that $\mathbf{a} \cdot \mathbf{b} \neq 0, \mathbf{b} \cdot \mathbf{c} \neq 0$, then **a** and **c** are **(AIEEE 2006)**

a. inclined at an angle of $\frac{\pi}{6}$ between them
b. perpendicular
c. parallel
d. inclined at an angle of $\frac{\pi}{3}$ between them

18 The values of a, for which the points, A, B, C with position vectors $2\hat{\mathbf{i}} - \hat{\mathbf{j}} + \hat{\mathbf{k}}, \hat{\mathbf{i}} - 3\hat{\mathbf{j}} - 5\hat{\mathbf{k}}$ and $a\mathbf{i} - 3\mathbf{j} + \mathbf{k}$ respectively are the vertices of a right angled triangle with $C = \frac{\pi}{2}$ are **(AIEEE 2006)**

a. –2 and –1 b. –2 and 1
c. 2 and –1 d. 2 and 1

19 If C is the mid-point of AB and P is any point outside AB, then **(AIEEE 2005)**

a. $\mathbf{PA} + \mathbf{PB} + \mathbf{PC} = 0$
b. $\mathbf{PA} + \mathbf{PB} + 2\mathbf{PC} = 0$
c. $\mathbf{PA} + \mathbf{PB} = \mathbf{PC}$
d. $\mathbf{PA} + \mathbf{PB} = 2\mathbf{PC}$

20 The distance between the line $\mathbf{r} = 2\hat{\mathbf{i}} - 2\hat{\mathbf{j}} + 3\hat{\mathbf{k}} + \lambda(\hat{\mathbf{i}} - \hat{\mathbf{i}} + 4\hat{\mathbf{k}})$ and the plane $\mathbf{r} \cdot (\hat{\mathbf{i}} + 5\hat{\mathbf{j}} + \hat{\mathbf{k}}) = 5$ is **(AIEEE 2005)**

a. $\frac{10}{3}$ b. $\frac{3}{10}$
c. $\frac{10}{3\sqrt{3}}$ d. $\frac{10}{9}$

21 For any vector **a**, the value of $(\mathbf{a} \times \hat{\mathbf{i}})^2 + (\mathbf{a} \times \hat{\mathbf{j}})^2 + (\mathbf{a} \times \hat{\mathbf{k}})^2$ is equal to **(AIEEE 2005)**

a. $4\mathbf{a}^2$ b. $2\mathbf{a}^2$
c. $\mathbf{a}^2$ d. $3\mathbf{a}^2$

22 Let $\mathbf{a} = \hat{\mathbf{i}} - \hat{\mathbf{k}}, \mathbf{b} = x\hat{\mathbf{i}} + \hat{\mathbf{j}} + (1-x)\hat{\mathbf{k}}$ and $\mathbf{c} = y\hat{\mathbf{i}} + x\hat{\mathbf{j}} + (1+x-y)\hat{\mathbf{k}}$. Then, $[\mathbf{a}\ \mathbf{b}\ \mathbf{c}]$ depends on **(AIEEE 2005)**

a. neither x nor y b. both x and y
c. only x d. only y

23 Let a, b and c be distinct non-negative numbers. If the vectors $a\hat{\mathbf{i}} + a\hat{\mathbf{j}} + c\hat{\mathbf{k}}, \hat{\mathbf{i}} + \hat{\mathbf{k}}$ and $c\hat{\mathbf{i}} + c\hat{\mathbf{j}} + b\hat{\mathbf{k}}$ lie in a plane, then c is **(AIEEE 2005)**

a. the harmonic mean of a and b
b. equal to zero
c. the arithmetic mean of a and b
d. the geometric mean of a and b

24 If **a**, **b**, **c** are non-coplanar vectors and λ is a real number, then $[\lambda(\mathbf{a}+\mathbf{b})\quad \lambda^2\mathbf{b}\quad \lambda\mathbf{c}] = [\mathbf{a}\quad \mathbf{b}+\mathbf{c}\quad \mathbf{b}]$ for **(AIEEE 2005)**

a. exactly two values of λ
b. exactly three values of λ
c. no value of λ
d. exactly one value of λ

25 If **a**, **b** and **c** are three non-zero vectors such that no two of these are collinear. If the vector $\mathbf{a} + 2\mathbf{b}$ is collinear with **c** and $\mathbf{b} + 3\mathbf{c}$ is collinear with **a** (λ being some non-zero scalar), then $\mathbf{a} + 2\mathbf{b} + 6\mathbf{c}$ is equal to **(AIEEE 2004)**

a. $\lambda\mathbf{a}$ b. $\lambda\mathbf{b}$
c. $\lambda\mathbf{c}$ d. 0

26 A particle is acted upon by constant forces $4\hat{\mathbf{i}} + \hat{\mathbf{j}} - 3\hat{\mathbf{k}}$ and $3\hat{\mathbf{i}} + \hat{\mathbf{j}} - \hat{\mathbf{k}}$ which displace it from a point $\hat{\mathbf{i}} + 2\hat{\mathbf{j}} + 3\hat{\mathbf{k}}$ to the point $5\hat{\mathbf{i}} + 4\hat{\mathbf{j}} + \hat{\mathbf{k}}$. The work done in standard units by the forces is given by **(AIEEE 2004)**

a. 40 units b. 30 units
c. 25 units d. 15 units

27 If **a**, **b**, **c** are non-coplanar vectors and λ is a real number, then the vectors $\mathbf{a} + 2\mathbf{b} + 3\mathbf{c}, \lambda\mathbf{b} + 4\mathbf{c}$ and $(2\lambda - 1)\mathbf{c}$ are non-coplanar for **(AIEEE 2004)**

a. all values of λ
b. all except one value of λ
c. all except two values of λ
d. no value of λ

28 Let **u**, **v**, **w** be such that $|\mathbf{u}| = 1, |\mathbf{v}| = 2, |\mathbf{w}| = 3$. If the projection **v** along **u** is equal to that of **w** along **u** and **v**, **w** are perpendicular to each other, then $|\mathbf{u} - \mathbf{v} + \mathbf{w}|$ is equal to **(AIEEE 2004)**

a. 2 b. $\sqrt{7}$
c. $\sqrt{14}$ d. 14

29 Let **a**, **b** and **c** be non-zero vectors such that $(\mathbf{a} \times \mathbf{b}) \times \mathbf{c} = \frac{1}{3}|\mathbf{b}||\mathbf{c}|\mathbf{a}$. If θ is an acute angle between the vectors **b** and **c**, then $\sin\theta$ is equal to **(AIEEE 2004)**

a. $\frac{1}{3}$ b. $\frac{\sqrt{2}}{3}$
c. $\frac{2}{3}$ d. $\frac{2\sqrt{2}}{3}$

30 A tetrahedron has vertices at $O(0, 0, 0)$, $A(1, 2, 1)$, $B(2, 1, 3)$ and $C(-1, 1, 2)$. Then, the angle between the faces OAB and ABC will be **(AIEEE 2003)**

a. $\cos^{-1}\left(\frac{19}{35}\right)$
b. $\cos^{-1}\left(\frac{17}{31}\right)$
c. $30°$
d. $90°$

31 $\mathbf{a}, \mathbf{b}, \mathbf{c}$ are three vectors, such that $\mathbf{a} + \mathbf{b} + \mathbf{c} = 0$, $|\mathbf{a}| = 1, |\mathbf{b}| = 2, |\mathbf{c}| = 3$, then $\mathbf{a} \cdot \mathbf{b} + \mathbf{b} \cdot \mathbf{c} + \mathbf{c} \cdot \mathbf{a}$ is equal to **(AIEEE 2003)**

a. 0 b. −7
c. 7 d. 1

32 If $\mathbf{u}, \mathbf{v}$ and $\mathbf{w}$ are three non-coplanar vectors, then $(\mathbf{u} + \mathbf{v} - \mathbf{w}) \cdot [(\mathbf{u} - \mathbf{v}) \times (\mathbf{v} - \mathbf{w})]$ is equal to

a. 0 b. $\mathbf{u} \cdot \mathbf{v} \times \mathbf{w}$ **(AIEEE 2003)**
c. $\mathbf{u} \cdot \mathbf{w} \times \mathbf{v}$ d. $3\mathbf{u} \cdot \mathbf{v} \times \mathbf{w}$

33 Consider points A, B, C and D with position vectors $7\hat{\mathbf{i}} - 4\hat{\mathbf{j}} + 7\hat{\mathbf{k}}, \hat{\mathbf{i}} - 6\hat{\mathbf{j}} + 10\hat{\mathbf{k}}, -\hat{\mathbf{i}} - 3\hat{\mathbf{j}} + 4\hat{\mathbf{k}}$ and $5\hat{\mathbf{i}} - \hat{\mathbf{j}} + 5\hat{\mathbf{k}}$, respectively. Then, $ABCD$ is a **(AIEEE 2003)**

a. square b. rhombus
c. rectangle d. None of these

34 If $\begin{vmatrix} a & a^2 & 1+a^3 \\ b & b^2 & 1+b^3 \\ c & c^2 & 1+c^3 \end{vmatrix} = 0$ and vectors $(1, a, a^2), (1, b, b^2)$ and $(1, c, c^2)$ are non-coplanar, then the product abc is equal to **(AIEEE 2003)**

a. 2 b. −1
c. 1 d. 0

35 Let $\mathbf{u} = \hat{\mathbf{i}} + \hat{\mathbf{j}}, \mathbf{v} = \hat{\mathbf{i}} - \hat{\mathbf{j}}$ and $\mathbf{w} = \hat{\mathbf{i}} + 2\hat{\mathbf{j}} + 3\hat{\mathbf{k}}$. If $\mathbf{n}$ is a unit vector such that $\mathbf{u} \cdot \mathbf{n} = 0$ and $\mathbf{v} \cdot \mathbf{n} = 0$, then $|\mathbf{w} \cdot \mathbf{n}|$ is equal to **(AIEEE 2003)**

a. 0 b. 1
c. 2 d. 3

36 Given, two vectors are $\hat{\mathbf{i}} - \hat{\mathbf{j}}$ and $\hat{\mathbf{i}} + 2\hat{\mathbf{j}}$, the unit vector coplanar with the two vectors and perpendicular to first is **(AIEEE 2002)**

a. $\frac{1}{\sqrt{2}}(\hat{\mathbf{i}} + \hat{\mathbf{j}})$

b. $\frac{1}{\sqrt{5}}(2\hat{\mathbf{i}} + \hat{\mathbf{j}})$

c. $\pm\frac{1}{\sqrt{2}}(\hat{\mathbf{i}} + \hat{\mathbf{j}})$

d. None of the above

37 The vector $\hat{\mathbf{i}} + x\hat{\mathbf{j}} + 3\hat{\mathbf{k}}$ is rotated through an angle θ and doubled in magnitude, then it becomes $4\hat{\mathbf{i}} + (4x - 2)\hat{\mathbf{j}} + 2\hat{\mathbf{k}}$. The values of x are **(AIEEE 2002)**

a. $-\frac{2}{3}, 2$ b. $\frac{1}{3}, 2$
c. $\frac{2}{3}, 0$ d. 2, 7

38 If the vectors $\mathbf{a}, \mathbf{b}$ and $\mathbf{c}$ form the sides BC, CA and AB respectively of a ΔABC, then

a. $\mathbf{a} \cdot \mathbf{b} = \mathbf{b} \cdot \mathbf{c} = \mathbf{c} \cdot \mathbf{b} = 0$ **(AIEEE 2002)**
b. $\mathbf{a} \times \mathbf{b} = \mathbf{b} \times \mathbf{c} = \mathbf{c} \times \mathbf{a}$
c. $\mathbf{a} \cdot \mathbf{b} = \mathbf{b} \cdot \mathbf{c} = \mathbf{c} \cdot \mathbf{a} = 0$
d. $\mathbf{a} \times \mathbf{a} + \mathbf{a} \times \mathbf{c} + \mathbf{a} \times \mathbf{a} = 0$

39 If the vectors $\mathbf{c}, \mathbf{a} = x\hat{\mathbf{i}} + y\hat{\mathbf{j}} + z\hat{\mathbf{k}}$ and $\mathbf{b} = \mathbf{j}$ are such that $\mathbf{a}, \mathbf{c}$ and $\mathbf{b}$ form a right handed system, then $\mathbf{c}$ is equal to **(AIEEE 2002)**

a. $z\hat{\mathbf{i}} - x\hat{\mathbf{k}}$ b. 0
c. $y\hat{\mathbf{j}}$ d. $-z\hat{\mathbf{i}} + x\hat{\mathbf{k}}$

JEE Advanced & IIT JEE Archive

(Compilation of Last 10 Years' Questions)

1 Let ΔPQR be a triangle. Let $\mathbf{a} = \mathbf{QR}$, $\mathbf{b} = \mathbf{RP}$ and $\mathbf{c} = \mathbf{PQ}$. If $|\mathbf{a}| = 12$, $|\mathbf{b}| = 4\sqrt{3}$ and $\mathbf{b} \cdot \mathbf{c} = 24$, then which of the following is/are true? **(2015, One or More than One Options Correct Type)**

a. $\frac{|\mathbf{c}|^2}{2} - |\mathbf{a}| = 12$

b. $\frac{|\mathbf{c}|^2}{2} + |\mathbf{a}| = 30$

c. $|\mathbf{a} \times \mathbf{b} + \mathbf{c} \times \mathbf{a}| = 48\sqrt{3}$

d. $\mathbf{a} \cdot \mathbf{b} = -72$

2 Suppose that $\mathbf{p}, \mathbf{q}$ and $\mathbf{r}$ are three non-coplanar vectors in R^3. Let the components of a vector $\mathbf{s}$ along $\mathbf{p}, \mathbf{q}$ and $\mathbf{r}$ be 4, 3 and 5, respectively. If the components of this vector $\mathbf{s}$ along $(-\mathbf{p} + \mathbf{q} + \mathbf{r})$, $(\mathbf{p} - \mathbf{q} + \mathbf{r})$ and $(-\mathbf{p} - \mathbf{q} + \mathbf{r})$ are x, y and z respectively, then the value of $2x + y + z$ is **(2015, Integer Type)**

3 Let $\mathbf{a} = \hat{i} + \hat{j} + \hat{k}$, $\mathbf{b} = \hat{i} - \hat{j} + \hat{k}$ and $\mathbf{c} = \hat{i} - \hat{j} - \hat{k}$ be three vectors. A vector $\mathbf{v}$ in the plane of $\mathbf{a}$ and $\mathbf{b}$, whose projection on $\mathbf{c}$ is $\frac{1}{\sqrt{3}}$, is given by **(2011, Only One Option Correct Type)**

a. $\hat{i} - 3\hat{j} + 3\hat{k}$

b. $-3\hat{i} - 3\hat{j} - \hat{k}$

c. $3\hat{i} - \hat{j} + 3\hat{k}$

d. $\hat{i} + 3\hat{j} - 3\hat{k}$

4 The vector(s) which is/are coplanar with vectors $\hat{i} + \hat{j} + 2\hat{k}$ and $\hat{i} + 2\hat{j} + \hat{k}$, are perpendicular to the vector $\hat{i} + \hat{j} + \hat{k}$ is/are **(2011, One or More than One Options Correct Type)**

a. $\hat{j} - \hat{k}$ b. $-\hat{i} + \hat{j}$

c. $\hat{i} - \hat{j}$ d. $-\hat{j} + \hat{k}$

5 Let $\mathbf{a} = -\hat{i} - \hat{k}$, $\mathbf{b} = -\hat{i} + \hat{j}$ and $\mathbf{a} = \hat{i} + 2\hat{j} + 3\hat{k}$ be three given vectors. If $\mathbf{r}$ is a vector such that $\mathbf{r} \times \mathbf{b} = \mathbf{c} \times \mathbf{b}$ and $\mathbf{r} \cdot \mathbf{a} = 0$, then the value of $\mathbf{r} \cdot \mathbf{b}$ is **(2011, Integer Type)**

6 Let P, Q, R and S be the points on the plane with position vectors $-2\hat{i} - \hat{j}$, $4\hat{i}$, $3\hat{i} + 3\hat{j}$ and $-3\hat{i} + 2\hat{j}$ respectively. The quadrilateral $PQRS$ must be a **(2010, Only One Option Correct Type)**

a. parallelogram, which is neither a rhombus nor a rectangle

b. square

c. rectangle but not a square

d. rhombus but not a square

7 Two adjacent sides of a parallelogram $ABCD$ are given by $\mathbf{AD} = 2\hat{i} + 10\hat{j} + 11\hat{k}$ and $\mathbf{AD} = -\hat{i} + 2\hat{j} + 2\hat{k}$. The side AD is rotated by an acute angle α in the plane of the parallelogram so that AD becomes AD'. If AD' makes a right angle with the side AB, then the cosine of the angle α is given by **(2010, Only One Option Correct Type)**

a. $\frac{8}{9}$ b. $\frac{\sqrt{17}}{9}$

c. $\frac{1}{9}$ d. $\frac{4\sqrt{5}}{9}$

8 If $\mathbf{a}$ and $\mathbf{b}$ are vectors in space given by $\mathbf{a} = \frac{\hat{i} - 2\hat{j}}{\sqrt{5}}$ and $\mathbf{b} = \frac{2\hat{i} + \hat{j} + 3\hat{k}}{\sqrt{14}}$, then the value of $(2\mathbf{a} + \mathbf{b}) \cdot [(\mathbf{a} \times \mathbf{b}) \times (\mathbf{a} - 2\mathbf{b})]$ is **(2010, Integer Type)**

9 If $\mathbf{a}$, $\mathbf{b}$, $\mathbf{c}$ and $\mathbf{d}$ are the unit vectors such that $(\mathbf{a} \times \mathbf{b}) \cdot (\mathbf{c} \times \mathbf{d}) = 1$ and $\mathbf{a} \cdot \mathbf{c} = \frac{1}{2}$, then **(2009, Only One Option Correct Type)**

a. $\mathbf{a}, \mathbf{b}, \mathbf{c}$ are non-coplanar

b. $\mathbf{a}, \mathbf{b}, \mathbf{d}$ are non-coplanar

c. $\mathbf{b}, \mathbf{d}$ are non-parallel

d. $\mathbf{a}, \mathbf{d}$ are parallel and $\mathbf{b}, \mathbf{c}$ are parallel

10 The edges of a parallelopiped are of unit length and are parallel to non-coplanar unit vector $\hat{\mathbf{a}}, \hat{\mathbf{b}}, \hat{\mathbf{c}}$ such that $\hat{\mathbf{a}} \cdot \hat{\mathbf{b}} = \hat{\mathbf{b}} \cdot \hat{\mathbf{c}} = \hat{\mathbf{c}} \cdot \hat{\mathbf{a}} = \frac{1}{2}$. Then, the volume of the parallelopiped is **(2008, Only One Option Correct Type)**

a. $\frac{1}{\sqrt{2}}$ cu unit b. $\frac{1}{2\sqrt{2}}$ cu unit

c. $\frac{\sqrt{3}}{2}$ cu unit d. $\frac{1}{\sqrt{3}}$ cu unit

11 Let two non-collinear unit vectors $\hat{\mathbf{a}}$ and $\hat{\mathbf{b}}$ form an acute angle. A point P moves so that at any time t the position vector $\mathbf{OP}$ (where, O is the origin) is given by $\hat{\mathbf{a}} \cos t + \hat{\mathbf{b}} \sin t$. When P is farthest from origin O, let M be the length of $\mathbf{OP}$ and $\hat{\mathbf{u}}$ be the unit vector along $\mathbf{OP}$. Then, **(2008, Only One Option Correct Type)**

a. $\hat{\mathbf{u}} = \frac{\hat{\mathbf{a}} + \hat{\mathbf{b}}}{|\hat{\mathbf{a}} + \hat{\mathbf{b}}|}$ and $M = (1 + \hat{\mathbf{a}} \cdot \hat{\mathbf{b}})^{1/2}$

b. $\hat{\mathbf{u}} = \frac{\hat{\mathbf{a}} - \hat{\mathbf{b}}}{|\hat{\mathbf{a}} - \hat{\mathbf{b}}|}$ and $M = (1 + \hat{\mathbf{a}} \cdot \hat{\mathbf{b}})^{1/2}$

c. $\hat{\mathbf{u}} = \frac{\hat{\mathbf{a}} + \hat{\mathbf{b}}}{|\hat{\mathbf{a}} + \hat{\mathbf{b}}|}$ and $M = (1 + 2\hat{\mathbf{a}} \cdot \hat{\mathbf{b}})^{1/2}$

d. $\hat{\mathbf{u}} = \frac{\hat{\mathbf{a}} - \hat{\mathbf{b}}}{|\hat{\mathbf{a}} - \hat{\mathbf{b}}|}$ and $M = (1 + 2\hat{\mathbf{a}} \cdot \hat{\mathbf{b}})^{1/2}$

12 The number of distinct real values of λ, for which the vectors $-\lambda^2\hat{i}+\hat{j}+\hat{k}$, $\hat{i}-\lambda^2\hat{j}+\hat{k}$ and $\hat{i}+\hat{j}-\lambda^2\hat{k}$ are coplanar, is **(2007, Only One Option Correct Type)**

a. 0 b. 1 c. 2 d. 3

13 Let **a**, **b**, **c** be unit vectors such that $\mathbf{a}+\mathbf{b}+\mathbf{c}=0$. Which one of the following is correct? **(2007, Only One Option Correct Type)**

a. $\mathbf{a}\times\mathbf{b}=\mathbf{b}\times\mathbf{c}=\mathbf{c}\times\mathbf{a}=0$
b. $\mathbf{a}\times\mathbf{b}=\mathbf{b}\times\mathbf{c}=\mathbf{c}\times\mathbf{a}\neq 0$
c. $\mathbf{b}\times\mathbf{b}=\mathbf{b}\times\mathbf{c}=\mathbf{a}\times\mathbf{c}=0$
d. $\mathbf{a}\times\mathbf{b}$, $\mathbf{b}\times\mathbf{c}$, $\mathbf{c}\times\mathbf{a}$ are mutually perpendicular

14 Let the vectors **PQ**, **QR**, **RS**, **ST**, **TU** and **UP** represent the sides of a regular hexagon.

Statement I $\mathbf{PQ}\times(\mathbf{RS}+\mathbf{ST})\neq 0$.

Statement II $\mathbf{PQ}\times\mathbf{RS}=0$ and $\mathbf{PQ}\times\mathbf{ST}\neq 0$.

(2007, Statement Type)

a. Both Statement I and Statement II are correct and Statement II is the correct explanation of Statement I
b. Both Statement I and Statement II are correct but Statement II is not the correct explanation of Statement I
c. Statement I is correct but Statement II is incorrect
d. Statement II is correct but Statement I is incorrect

15 Let $\mathbf{a}=\hat{i}+2\hat{j}+\hat{k}$, $\mathbf{b}=\hat{i}-\hat{j}+\hat{k}$, $\mathbf{c}=\hat{i}+\hat{j}-\hat{k}$. A vector coplanar to **a** and **b** has a projection along **c** of magnitude $\frac{1}{\sqrt{3}}$, then the vector is **(2006, Only One Option Correct Type)**

a. $4\hat{i}-\hat{j}+4\hat{k}$ b. $4\hat{i}+\hat{j}-4\hat{k}$
c. $2\hat{i}+\hat{j}+\hat{k}$ d. None of these

16 Let **A** be vector parallel to line of intersection of planes P_1 and P_2 through origin. P_1 is parallel to the vectors $2\hat{j}+3\hat{k}$ and $4\hat{j}-3\hat{k}$ and P_2 is parallel to $\hat{j}-\hat{k}$ and $3\hat{i}+3\hat{j}$, then the angle between vectors **A** and $2\hat{i}+\hat{j}-2\hat{k}$ is **(2006, One or More than One Options Correct Type)**

a. $\frac{\pi}{2}$ b. $\frac{\pi}{4}$
c. $\frac{\pi}{6}$ d. $\frac{3\pi}{4}$

17 If **a**, **b**, **c** are three non-zero, non-coplanar vectors and

$\mathbf{b}_1=\mathbf{b}-\frac{\mathbf{b}\cdot\mathbf{a}}{|\mathbf{a}|^2}\mathbf{a}$, $\mathbf{b}_2=\mathbf{b}+\frac{\mathbf{b}\cdot\mathbf{a}}{|\mathbf{a}|^2}\mathbf{a}$,

$\mathbf{c}_1=\mathbf{c}-\frac{\mathbf{c}\cdot\mathbf{a}}{|\mathbf{a}|^2}\mathbf{a}-\frac{\mathbf{c}\cdot\mathbf{b}}{|\mathbf{b}|^2}\mathbf{b}$,

$\mathbf{c}_2=\mathbf{c}-\frac{\mathbf{c}\cdot\mathbf{a}}{|\mathbf{a}|^2}\mathbf{a}-\frac{\mathbf{c}\cdot\mathbf{b}}{|\mathbf{b}|^2}\mathbf{b}_1$,

$\mathbf{c}_3=\mathbf{c}-\frac{\mathbf{c}\cdot\mathbf{a}}{|\mathbf{a}|^2}\mathbf{a}-\frac{\mathbf{c}\cdot\mathbf{b}}{|\mathbf{b}|^2}\mathbf{b}_2$,

$\mathbf{c}_4=\mathbf{a}-\frac{\mathbf{c}\cdot\mathbf{a}}{|\mathbf{a}|^2}\mathbf{a}$.

Then, which of the following is a set of mutually orthogonal vectors? **(2005, Only One Option Correct Type)**

a. $\{\mathbf{a},\mathbf{b}_1,\mathbf{c}_1\}$
b. $\{\mathbf{a},\mathbf{b}_1,\mathbf{c}_2\}$
c. $\{\mathbf{c},\mathbf{b}_2,\mathbf{c}_3\}$
d. $\{\mathbf{a},\mathbf{b}_2,\mathbf{c}_2\}$

JEE Archive

5. 3D Geometry

DPP-1 Direction Ratios, Direction Cosines and Equation of Straight Line

Only One Option Correct Type

Working Space

Direction (Q. Nos. 1-22) This section contains 22 multiple choice questions. Each question has four choices (a), (b), (c) and (d), out of which ONLY ONE is correct.

1 If direction ratios of two straight lines are (5, – 12, 13) and (– 3, 4, 5), then the angle between them is

a. $\cos^{-1}(2/65)$ b. $\cos^{-1}(3/65)$

c. $\cos^{-1}(1/65)$ d. $\pi/3$

2 Angle between the diagonals of a cube is

a. $\pi/3$ b. $\pi/2$

c. $\cos^{-1}(1/3)$ d. $\cos^{-1}(1/\sqrt{3})$

3 If a line makes angle $\alpha, \beta, \gamma, \delta$ with four diagonals of a cube, then $\cos^2\alpha + \cos^2\beta + \cos^2\gamma + \cos^2\delta$ is equal to

a. 3 b. 4

c. 4/3 d. 3/4

4 A straight line makes an angle of 60° with each of *Y* and *Z*-axes, inclined with *X*-axis at an angle

a. 45° b. 60°

c. 75° d. 30°

5 In what ratio *YZ*-plane divides the line segment joining the points (2, 4, 5) and (3, 5, – 4) is mark?

a. 2 : 3 b. 3 : 2

c. – 2 : 3 d. 1 : 2

6 Line passes through (3, 4, 5) and (4, 6, 3) has a projection on line passing through (– 1, 2, 4) and (1, 0, 5) will be

a. – 4/3 b. 2/3

c. 1/3 d. 1/2

7 The angle between the pair of lines with direction ratios (1, 1, 2) and $(\sqrt{3}-1, -\sqrt{3}-1, 4)$ is

a. $\pi/6$ b. $\pi/4$

c. $\pi/3$ d. $\pi/2$

Working Space

8 If $\cos\alpha, \cos\beta, \cos\gamma$ are the direction cosines, then $\cos 2\alpha + \cos 2\beta + \cos 2\gamma$ is equal to

a. -1 b. 0
c. 4 d. 3

9 A line is making $\alpha, \beta. \gamma$ angles from X, Y, Z-axes. If value of α, β, γ are respectively θ, 60°, 30°, then $\sin\theta$ is equal to

a. 1 b. -2
c. 0 d. 1/2

10 The plane YZ divides the joining of $A(-2, 3, 5)$ and $B(a, 2, 5)$ in the ratio of 2 : 3, then a is equal to

a. 3 b. -1
c. 2 d. None of these

11 The angle between the lines whose direction cosines are given by $l + m + n = 0$ and $l^2 + m^2 + n^2$ is

a. $\pi/4$ b. $\pi/3$
c. $\pi/6$ d. $\pi/2$

12 If a line makes an angle of $\pi/4$ with the positive directions of each of X-axis and Y-axis, then the angle that the line makes with the positive direction of the Z-axis, is **[AIEEE 2007]**

a. $\pi/4$ b. $\pi/2$
c. $\pi/6$ d. $\pi/3$

13 The angle between lines $\frac{x+1}{2} = \frac{y-1}{5} = \frac{z+3}{4}$ and $\frac{x-1}{1} = \frac{y+2}{2} = \frac{z-3}{-3}$ is

a. 45° b. 30°
c. 60° d. 90°

14 If the lines $\frac{x-1}{-3} = \frac{y-2}{2k} = \frac{z-3}{2}$ and $\frac{x-1}{3k} = \frac{y-5}{1} = \frac{Z-6}{-5}$ perpendicular to each other, then k is equal to

a. 5/7 b. 7/5
c. $-7/10$ d. $-10/7$

15 The angle between the lines $2x = 3y = -z$ and $6x = -y = -4z$ is

a. $\pi/6$ b. $\pi/3$
c. $\pi/4$ d. $\pi/2$

16 If the straight lines $\frac{x-1}{k} = \frac{y-2}{2} = \frac{z-3}{3}$ and $\frac{x-2}{3} = \frac{y-3}{k} = \frac{z-1}{2}$ intersect $\frac{x-2}{3} = \frac{y-3}{k} = \frac{z-1}{2}$ intersect at a point, then the integer k is **[AIEEE 2008]**

a. -5 b. 5
c. 2 d. -2

17 If a plane cuts off intercepts $OA = a$, $OB = b$, $OC = c$ from the coordinate axes, then the area (in sq units) of a ΔABC is

a. $\frac{1}{2}\sqrt{b^2c^2 + c^2a^2 + a^2b^2}$ b. $\frac{1}{2}(bc + ca + ab)$
c. $\frac{1}{2}abc$ d. $\frac{1}{2}\sqrt{(b-c)^2 + (c-a)^2 + (a-b)^2}$

Working Space

18 The cartesian form of the equation of a line whose vector form is given by

$$\mathbf{r} = (2\hat{\mathbf{i}} - \hat{\mathbf{j}} + 4\hat{\mathbf{k}}) + \lambda(\hat{\mathbf{i}} + \hat{\mathbf{j}} - 2\hat{\mathbf{k}}) \text{ is}$$

a. $\frac{x-2}{1} = \frac{y+1}{1} = \frac{z-4}{-2}$ b. $\frac{x+2}{1} = \frac{y-1}{1} = \frac{z+4}{-2}$

c. $\frac{x-2}{-1} = \frac{y+1}{1} = \frac{z-4}{-2}$ d. $\frac{x+2}{1} = \frac{y-1}{-1} = \frac{z+4}{-2}$

19 Let the points $A(a, b, c)$ and $B(a', b', c')$ be at distances r and r' from the origin. The line AB passes through the origin, when

a. $\frac{a'}{a} = \frac{b'}{b} = \frac{c'}{c}$ b. $aa' + bb' + cc' = rr'$

c. $aa' + bb' + cc' = r^2 + (r')^2$ d. None of these

20 The acute angle between the lines $\frac{x-1}{l} = \frac{y+1}{m} = \frac{z}{n}$ and $\frac{x+1}{m} = \frac{y-3}{n} = \frac{z-1}{l}$ where $l > m > n$ and l, m, n are the roots of the cubic equation $x^3 + x^2 - 4x = 4$, is

a. $\frac{\pi}{2}$ b. $\cos^{-1}\left(\frac{1}{4}\right)$

c. $\cos^{-1}\left(\frac{4}{9}\right)$ d. $\cos^{-1}\left(\frac{5}{9}\right)$

21 The line $\frac{x-x_1}{0} = \frac{y-y_1}{1} = \frac{z-z_1}{2}$ is

a. parallel to X-axis
b. perpendicular to X-axis
c. perpendicular to YOZ plane
d. parallel to Y-axis

22 The vector equations of the two lines L_1 and L_2 are given

$$L_1 : \mathbf{r} = (2\hat{\mathbf{i}} + 9\hat{\mathbf{j}} + 13\hat{\mathbf{k}}) + \lambda(\hat{\mathbf{i}} + 2\hat{\mathbf{j}} + 3\hat{\mathbf{k}})$$

$$L_2 : \mathbf{r} = (-3\hat{\mathbf{i}} + 7\hat{\mathbf{j}} + p\hat{\mathbf{k}}) + \mu(-\hat{\mathbf{i}} + 2\hat{\mathbf{j}} - 3\hat{\mathbf{k}}),$$

then the lines L_1 and L_2 are

a. skew lines for all $p \in R$
b. intersecting for all $p \in R$ and the point of intersection is $(-1, 3, 4)$
c. intersecting lines for $p = -2$
d. intersecting for all real $p \in R$

Comprehension Type

Direction (Q. Nos. 23-28) This section contains 2 paragraphs, each describing theory, experiments, data, etc. Six questions related to the paragraphs have been given. Each question has only one correct answer among the four given options (a), (b), (c) and (d).

Passage I

Consider the lines

$$L_1 : \frac{x+1}{3} = \frac{y+2}{1} = \frac{z+1}{2}$$

and

$$L_2 : \frac{x-2}{1} = \frac{y+2}{2} = \frac{z-3}{3}.$$

[IIT JEE 2008]

Working Space

23 The unit vector perpendicular to both the lines L_1 and L_2 is

a. $\dfrac{-\hat{\mathbf{i}}+7\hat{\mathbf{j}}+7\hat{\mathbf{k}}}{\sqrt{99}}$ b. $\dfrac{-\hat{\mathbf{i}}-7\hat{\mathbf{j}}+5\hat{\mathbf{k}}}{5\sqrt{3}}$

c. $\dfrac{-\hat{\mathbf{i}}+7\hat{\mathbf{j}}+5\hat{\mathbf{k}}}{5\sqrt{3}}$ d. $\dfrac{7\hat{\mathbf{i}}-7\hat{\mathbf{j}}-\hat{\mathbf{k}}}{\sqrt{99}}$

24 The shortest distance between L_1 and L_2 is

a. 0 b. $\dfrac{17}{\sqrt{13}}$

c. $\dfrac{41}{5\sqrt{3}}$ d. $\dfrac{17}{5\sqrt{3}}$

25 The distance of the point (1, 1, 1) from the plane passing through the point $(-1, -2, -1)$ and whose normal is perpendicular to both the lines L_1 and L_2 is

a. $\dfrac{2}{\sqrt{75}}$ b. $\dfrac{7}{\sqrt{75}}$

c. $\dfrac{13}{\sqrt{75}}$ d. $\dfrac{23}{\sqrt{75}}$

Passage II

Consider a triangle pyramid *ABCD* the position vector of whose angular points are $A(3, 0, 1)$; $B(-1, 4, 1)$; $C(5, 2, 3)$ and $D(0, -5, 4)$. Let G be the point of intersection of the medians of the ΔBCD.

26 The length of vector **AG** is

a. $\sqrt{17}$ b. $\dfrac{\sqrt{51}}{3}$ c. $\dfrac{\sqrt{51}}{9}$ d. $\dfrac{\sqrt{59}}{4}$

27 Area of ΔABC (in sq units) is

a. 24 b. $8\sqrt{6}$

c. $4\sqrt{6}$ d. None of these

28 The length of perpendicular form the vertex D on the opposite face is

a. $\dfrac{14}{\sqrt{6}}$ b. $\dfrac{2}{\sqrt{6}}$

c. $\dfrac{3}{\sqrt{6}}$ d. None of these

One Integer Value Correct Type

Direction (Q. Nos. 29-35) This section contains 7 questions. When worked out will result in an integer from 0 to 9 (both inclusive).

29 A line from the origin meets the lines $\dfrac{x-2}{1}=\dfrac{y-1}{2}=\dfrac{z+1}{1}$ and $\dfrac{x-\frac{8}{3}}{2}=\dfrac{y+3}{-1}=\dfrac{z-1}{1}$ at P and Q respectively. If length $PQ = d$, then the value of d^2 is

30 If the line $\dfrac{x-2}{-1}=\dfrac{y+2}{1}=\dfrac{z+k}{4}$ is one of the angle bisectors of the lines $\dfrac{x}{1}=\dfrac{y}{-2}=\dfrac{z}{3}$ and $\dfrac{x}{-2}=\dfrac{y}{3}=\dfrac{z}{1}$, then the value of k is

Working Space

31 Foot of perpendicular drawn from (1, 2, 3) to the line $\frac{x-2}{2}=\frac{y-1}{3}=\frac{z-2}{4}$ is (x, y, z), then $\frac{2x+3y+4z}{4}$ is

32 If the lines $\frac{x-2}{1}=\frac{y-3}{1}=\frac{z-4}{k}$ and $\frac{x-1}{k}=\frac{y-4}{-2}=\frac{z-5}{-1}$ intersect each other then the value of k (where, $k>0$) is

33 The number of real values of k for which the lines $\frac{x}{1}=\frac{y-1}{k}=\frac{z}{-1}$ and $\frac{x-k}{2k}=\frac{y-k}{3k-1}=\frac{z-2}{k}$ are coplanar, is

34 If the shortest distance between the lines $\frac{x-1}{1}=\frac{y-1}{1}=\frac{z-1}{1}$ and $\frac{x-2}{1}=\frac{y-3}{1}=\frac{z-4}{1}$ is $\sqrt{k}$, then the value of k is

35 The straight lines whose direction cosines are given by $al+bm+cn=0$ and $fmn+gln+hlm=0$ are perpendicular, if $\frac{f}{a}+\frac{g}{b}+\frac{h}{c}$ is

DPP-2 Equation of Plane and Sphere

Only One Option Correct Type

Working Space

Direction *(Q. Nos. 1-34) This section contains 34 multiple choice questions. Each question has four choices (a), (b), (c) and (d), out of which ONLY ONE is correct.*

1 The locus represented by $xy + yz = 0$ is

a. a pair of perpendicular planes
b. a pair of parallel planes
c. a pair of perpendicular lines
d. a pair of parallel lines

2 The equation of plane which passes through $(2, -3, 1)$ and is normal to the line joining the points $(3, 4, -1)$ and $(2, -1, 5)$, then

a. $x + 5y - 6z + 19 = 0$
b. $x - 5y + 6z - 19 = 0$
c. $x + 5y + 6z + 19 = 0$
d. $x - 5y - 6z - 19 = 0$

3 A variable plane passes through a fixed point (1, 2, 3). The locus of the foot of the perpendicular drawn from the origin to this plane is

a. $x^2 + y^2 + z^2 - x - 2y - 3z = 0$
b. $x^2 + 2y^2 + 3z^2 - x - 2y - 3z = 0$
c. $x^2 + 4y^2 + 9z^2 + x + 2y + 3 = 0$
d. $x^2 + y^2 + z^2 + x + 2y + 3z = 0$

4 The equation of plane passing through the point $(1, -3, -2)$ and perpendicular to planes $x + 2y + 3z = 5$ and $3x + 3y + 2z = 8$, is

a. $5x - 7y + 3z + 20 = 0$
b. $2x - 4y - 3z + 8 = 0$
c. $2x + 4y + 3z + 8 = 0$
d. None of the above

5 The reflection of the point $(2, -1, 3)$ in the plane $3x - 2y - z = 9$ is

a. $\left(\frac{26}{7}, \frac{15}{7}, \frac{17}{7}\right)$
b. $\left(\frac{26}{7}, -\frac{15}{7}, \frac{17}{7}\right)$
c. $\left(\frac{15}{7}, \frac{26}{7}, -\frac{17}{7}\right)$
d. $\left(\frac{26}{7}, \frac{17}{7}, -\frac{15}{7}\right)$

6 The distance of the point $(-1, -5, -10)$ from the point of intersection of the line, $\frac{x-2}{3} = \frac{y+1}{4} = \frac{z-2}{12}$ and the plane $x - y + z = 5$, is

a. 10
b. 11
c. 12
d. 13

Working Space

7 The distance of the point (1, − 2, 3) from the plane $x - y + z = 5$ measured parallel to the line $\frac{x}{2} = \frac{y}{3} = \frac{z}{-6}$, is

a. 1

b. $\frac{6}{7}$

c. $\frac{7}{6}$

d. None of these

8 The radius of the circular section of the sphere $|\mathbf{r}| = 5$ by the plane $\mathbf{r} \cdot (\hat{\mathbf{i}} + \hat{\mathbf{j}} + \hat{\mathbf{k}}) = 3\sqrt{3}$ is

a. 3 units

b. 4 units

c. 5 units

d. None of these

9 Equation of the plane passing through $A(x_1, y_1, z_1)$ and containing the line $\frac{x - x_2}{d_1} = \frac{y - y_2}{d_2} = \frac{z - z_2}{d_3}$ is

a. $\begin{vmatrix} x - x_1 & y - y_1 & z - z_1 \\ x_2 - x_1 & y_2 - y_1 & z_2 - z_1 \\ d_1 & d_2 & d_3 \end{vmatrix} = 0$

b. $\begin{vmatrix} x & y & z \\ x_2 - x_1 & y_2 - y_1 & z_2 - z_1 \\ d_1 & d_2 & d_3 \end{vmatrix} = 0$

c. $\begin{vmatrix} x - d_1 & y - d_2 & z - d_3 \\ x_1 & y_1 & z_1 \\ x_2 & y_2 & z_2 \end{vmatrix} = 0$

d. $\begin{vmatrix} x - d_2 & y - d_2 & z - d_3 \\ x_1 - x_2 & y_2 - y_1 & z_2 - z_1 \\ d_1 & d_2 & d_3 \end{vmatrix} = 0$

10 The equation of line $x + y + z - 1 = 0$, $4x + y - 2z + 2 = 0$ written in the symmetrical form is

a. $\frac{x+1}{1} = \frac{y-2}{-2} = \frac{z}{1}$

b. $\frac{x}{1} = \frac{y}{-2} = \frac{z}{1}$

c. $\frac{x}{1} = \frac{y}{-2} = \frac{z+1}{2}$

d. $\frac{x - \frac{1}{2}}{1} = \frac{y-1}{-2} = \frac{z - \frac{1}{2}}{1}$

11 The actue angle that the vector $2\hat{\mathbf{i}} - 2\hat{\mathbf{j}} + \hat{\mathbf{k}}$ makes with the plane contained by the two vectors $2\hat{\mathbf{i}} + 3\hat{\mathbf{j}} - \hat{\mathbf{k}}$ and $\hat{\mathbf{i}} - \hat{\mathbf{j}} + 2\hat{\mathbf{k}}$ is given by

a. $\cos^{-1}\left(\frac{1}{\sqrt{3}}\right)$

b. $\sin^{-1}\left(\frac{1}{\sqrt{2}}\right)$

c. $\tan^{-1}(\sqrt{2})$

d. $\cot^{-1}(\sqrt{2})$

12 A plane meets the coordinate axes in A, B, C and (α, β, γ) is the centroid of the ΔABC. Then, the equation of plane is

a. $\frac{x}{\alpha} + \frac{y}{\beta} + \frac{z}{\gamma} = 3$

b. $\frac{x}{\alpha} + \frac{y}{\beta} = \frac{z}{\gamma} = 1$

c. $\frac{3x}{\alpha} + \frac{3y}{\beta} + \frac{3z}{\gamma} = 1$

d. $\alpha x + \beta y + \gamma z = 1$

13 A point moves so that the sum of the squares of its distance from the six faces of a cube given by $x = \pm 1$, $y = \pm 1$, $z = \pm 1$ is 10 units. The locus of the point is

a. $x^2 + y^2 + z^2 = 1$

b. $x^2 + y^2 + z^2 = 2$

c. $x + y + z = 1$

d. $x + y + z = 2$

Working Space

14 Two system of rectangular axes have the same origin. If a plane cuts them at distances a, b, c and a_1, b_1, c_1 from the origin, then

a. $\frac{1}{a^2}+\frac{1}{b^2}+\frac{1}{c^2}=\frac{1}{(a_1)^2}+\frac{1}{(b_1)^2}+\frac{1}{(c_1)^2}$

b. $\frac{1}{a^2}-\frac{1}{b^2}+\frac{1}{c^2}=\frac{1}{(a_1)^2}-\frac{1}{(b_1)^2}+\frac{1}{(c_1)^2}$

c. $a^2+b^2+c^2=(a_1)^2+(b_1)^2+(c_1)^2$

d. $a^2-b^2+c^2=(a_1)^2-(b_1)^2+(c_1)^2$

15 Equation of plane which passes through the point of intersection of lines $\frac{x-1}{3}=\frac{y-2}{1}=\frac{z-3}{2}$ and $\frac{x-3}{1}=\frac{y-1}{2}=\frac{z-2}{3}$, and at greatest distance from the point (0, 0, 0) is

a. $4x+3y+5z=25$ b. $4x+3y+5z=50$
c. $3x+4y+5z=49$ d. $x+7y-5z=2$

16 The angle between a plane $2x-y+z=6$ and a plane perpendicular to the planes $x+y+2z=7$ and $x-y=3$ is

a. $\frac{\pi}{4}$ b. $\frac{\pi}{3}$
c. $\frac{\pi}{6}$ d. $\frac{\pi}{2}$

17 The non-zero value of a for which the lines $2x-y+3z+4=0=ax+y-z+2$ and $x-3y+z=0=x+2y+z+1$ are coplanar, is

a. −2 b. 4
c. 6 d. 0

18 If the lines $\frac{x}{1}=\frac{y}{2}=\frac{z}{3}$, $\frac{x-1}{3}=\frac{y-2}{-1}=\frac{z-3}{4}$ and $\frac{x+k}{3}=\frac{y-1}{2}=\frac{z-2}{h}$ are concurrent, then

a. $h=-2$ and $k=-6$ b. $h=\frac{1}{2}$ and $k=2$
c. $h=6$ and $k=2$ d. $h=2$ and $k=\frac{1}{2}$

19 Which of the following planes intersects the planes $x-y+2z=3$ and $4x+3y-z=1$ along the same line?

a. $11x+10y-5z=0$ b. $7x+7y+4z=0$
c. $5x+2y-z=2$ d. None of these

20 The coplanar points A, B, C, D are $(2-x, 2, 2)(2, 2-y, 2)$, $(2, 2, 2-z)$ and (1, 1, 1) respectively, then

a. $\frac{1}{x}+\frac{1}{y}+\frac{1}{z}=1$ b. $x+y+z=1$
c. $\frac{1}{1-x}+\frac{1}{1-y}+\frac{1}{1-z}=1$ d. None of these

21 The direction ractios of a normal to the plane through (1, 0, 0), (0, 1, 0) which makes an angle of $\pi/4$ with the plane $x+y=3$ are

a. $(1, \sqrt{2}, 1)$ b. $(1, 1, \sqrt{2})$
c. (1, 1, 2) d. $(\sqrt{2}, 1, 1)$

Working Space

22 The vector equation of plane passing through the origin and the line of intersection of the plane $\mathbf{r}\cdot\mathbf{a}=\lambda$ and $\mathbf{r}\cdot\mathbf{b}=\mu$, is

a. $\mathbf{r}\cdot(\lambda\mathbf{a}-\mu\mathbf{b})=0$
b. $\mathbf{r}\cdot(\lambda\mathbf{b}-\mu\mathbf{a})=0$
c. $\mathbf{r}\cdot(\lambda\mathbf{a}+\mu\mathbf{b})=0$
d. $\mathbf{r}\cdot(\lambda\mathbf{b}+\mu\mathbf{a})=0$

23 The equation of the plane containing the lines $\mathbf{r}=\mathbf{a}_1+\lambda\mathbf{b}$ and $\mathbf{r}=\mathbf{a}_2=\mu\mathbf{b}$ is

a. $\mathbf{r}\cdot(\mathbf{a}_1-\mathbf{a}_2)\times\mathbf{b}=[\mathbf{a}_1\,\mathbf{a}_2\,\mathbf{a}_3]$
b. $\mathbf{r}\cdot(\mathbf{a}_2-\mathbf{a}_1)\times\mathbf{b}=[\mathbf{a}_1\,\mathbf{a}_2\,\mathbf{b}]$
c. $\mathbf{r}\cdot(\mathbf{a}_1+\mathbf{a}_2)\times\mathbf{b}=[\mathbf{a}_2\,\mathbf{a}_1\,\mathbf{b}]$
d. None of the above

24 The equation of plane containing the lines $\mathbf{r}=\mathbf{a}_1+\lambda\mathbf{a}_2$ and $\mathbf{r}=\mathbf{a}_2+\lambda\mathbf{a}_1$, is

a. $[\mathbf{r}\,\mathbf{a}_1\,\mathbf{a}_2]=0$
b. $[\mathbf{r}\,\mathbf{a}_1\,\mathbf{a}_2]=\mathbf{a}_1\cdot\mathbf{a}_2$
c. $[\mathbf{r}\,\mathbf{a}_2\,\mathbf{a}_1]=\mathbf{a}_1\cdot\mathbf{a}_2$
d. None of these

25 A straight line $\mathbf{r}=\mathbf{a}+\lambda\mathbf{b}$ meets the plane $\mathbf{r}\cdot\mathbf{n}=0$ in P. The position vector of P is

a. $\mathbf{a}+\left(\dfrac{\mathbf{a}\cdot\mathbf{n}}{\mathbf{b}\cdot\mathbf{n}}\right)\mathbf{b}$
b. $\mathbf{a}-\left(\dfrac{\mathbf{a}\cdot\mathbf{n}}{\mathbf{b}\cdot\mathbf{n}}\right)\mathbf{b}$
c. $\mathbf{a}-\left(\dfrac{\mathbf{b}\cdot\mathbf{n}}{\mathbf{a}\cdot\mathbf{n}}\right)\mathbf{b}$
d. None of these

26 The length of perpendicular from the origin to the plane passing through the point **a** and containing the line $\mathbf{r}=\mathbf{b}+\lambda\mathbf{c}$ is

a. $\dfrac{[\mathbf{a}\,\mathbf{b}\,\mathbf{c}]}{|\mathbf{a}\times\mathbf{b}+\mathbf{b}\times\mathbf{c}+\mathbf{c}\times\mathbf{a}|}$
b. $\dfrac{[\mathbf{a}\,\mathbf{b}\,\mathbf{c}]}{|\mathbf{a}\times\mathbf{b}+\mathbf{b}\times\mathbf{c}|}$
c. $\dfrac{[\mathbf{a}\,\mathbf{b}\,\mathbf{c}]}{|\mathbf{b}\times\mathbf{c}+\mathbf{c}\times\mathbf{a}|}$
d. $\dfrac{[\mathbf{a}\,\mathbf{b}\,\mathbf{c}]}{|\mathbf{c}\times\mathbf{a}+\mathbf{a}\times\mathbf{b}|}$

27 The equation of plane passing through the line of intersection of planes $ax+by+cz+d=0$, $a'x+b'y+c'z+d'=0$ and parallel to the line $y=0$, $z=0$ is

a. $(ab'-a'b)x+(bc'-b'c)y+(ad'-a'd)z=0$
b. $(ab'-a'b)x+(bc'-b'c)y+(ad'-a'd)=0$
c. $(ab'-a'b)x+(ac'-a'c)y+(ad'-a'd)z=0$
d. None of the above

28 The lines $\mathbf{r}=\mathbf{a}+\lambda(\mathbf{b}\times\mathbf{c})$ and $\mathbf{r}=\mathbf{b}+u(\mathbf{c}\times\mathbf{a})$ will intersect, if

a. $\mathbf{a}\times\mathbf{c}=\mathbf{b}\times\mathbf{c}$
b. $\mathbf{a}\cdot\mathbf{c}=\mathbf{b}\cdot\mathbf{c}$
c. $\mathbf{b}\times\mathbf{a}=\mathbf{c}\times\mathbf{a}$
d. None of these

29 The lines $\mathbf{r}=\mathbf{a}_1+\lambda\mathbf{b}_1$ and $r=\mathbf{a}_2+\mu\mathbf{b}_2$ are coplanar, if

a. $\mathbf{a}_1\times\mathbf{a}_2=0$
b. $\mathbf{b}_1\times\mathbf{b}_2=0$
c. $(\mathbf{a}_2-\mathbf{a}_1)\times(\mathbf{b}_1\times\mathbf{b}_2)=0$
d. $[\mathbf{a}_1\,\mathbf{b}_1\,\mathbf{b}_2]=[\mathbf{a}_2\,\mathbf{b}_1\,\mathbf{b}_2]$

30 The radius of circle in which the sphere $x^2+y^2+z^2+2x-2y-4z-19=0$ is cut by the plane $x+2y+2z+7=0$, is

a. 4
b. 1
c. 2
d. 3

31 The shortest distance from the plane $12x+4y+3z=327$ to the sphere $x^2+y^2+z^2+4x-2y-6z=155$ is

a. 39
b. 26
c. $11\frac{4}{13}$
d. 13

Working Space

32 The equation of sphere passing through the point (1, 3, – 2) and the circle $x^2 + y^2 + z^2 = 25$, $x = 0$ is

a. $x^2 + y^2 + z^2 + 11x + 25 = 0$
b. $x^2 + y^2 + z^2 - 11x + 25 = 0$
c. $x^2 + y^2 + z^2 + 11x - 25\ 0$
d. $x^2 + y^2 + z^2 - 11x - 25 = 0$

33 Radius of the circle $\mathbf{r}^2 + \mathbf{r} \cdot (2\hat{\mathbf{i}} - 2\hat{\mathbf{j}} - 4\hat{\mathbf{k}}) - 19 = 0$, $\mathbf{r} \cdot (\hat{\mathbf{i}} - 2\hat{\mathbf{j}} + 2\hat{\mathbf{k}}) + 8 = 0$ is

a. 5 b. 4
c. 3 d. 2

34 The equation of the line passing through (1, 1, 1) and parallel to the plane $2x + 3y + z + 5 = 0$ is

a. $\frac{x-1}{1} = \frac{y-1}{2} = \frac{z-1}{1}$
b. $\frac{x-1}{-1} = \frac{y-1}{1} = \frac{z-1}{1}$
c. $\frac{x-1}{3} = \frac{y-1}{2} = \frac{z-1}{1}$
d. $\frac{x-1}{2} = \frac{y-1}{3} = \frac{z-1}{1}$

Comprehension Type

***Direction** (Q. Nos. 35-39) This section contains 2 paragraphs, each describing theory, experiments, data, etc. Five questions related to the paragraphs have been given. Each question has only one correct answer among the four given options (a), (b), (c) and (d).*

Passage I

Consider a plane $x + y - z = 1$ and the point $A(1, 2, 3)$.
A line L has the equation $x = 1 + 3r$, $y = 2 - r$ and $z = 3 + 4r$.

35 The coordinates of a point B of line L, such that AB is parallel to the plane, is

a. (10, – 1, 15) b. (– 5, 4, – 5)
c. (4, 1, 7) d. (– 8, 5, – 9)

36 Equation of the plane containing the line L and the point A has the equation

a. $x - 3y + 5 = 0$ b. $x + 3y - 7 = 0$
c. $3x - y - 1 = 0$ d. $3x + y - 5 = 0$

Passage II

Given four points $A(2, 1, 0)$, $B(1, 0, 1)$, $C(3, 0, 1)$ and $D(0, 0, 2)$. The point D lies on a line L orthogonal to the plane determined by the point A, B and C.

37 Equation of the plane BC is

a. $x + y + z = 3$ b. $y + z - 1 = 0$
c. $x + z - 1 = 0$ d. $2y + z - 1 = 0$

38 Equation of the line L is

a. $\mathbf{r} = 2\hat{\mathbf{k}} + \lambda(\hat{\mathbf{i}} + \hat{\mathbf{k}})$ b. $\mathbf{r} = 2\hat{\mathbf{k}} + \lambda(2\hat{\mathbf{i}} + \hat{\mathbf{k}})$
c. $\mathbf{r} = 2\hat{\mathbf{k}} + \lambda(\hat{\mathbf{j}} + \hat{\mathbf{k}})$ d. None of these

39 Perpendicular distance of D from the plane ABC is

a. $\sqrt{2}$ b. $1/2$
c. 2 d. $1/\sqrt{2}$

One Integer Value Correct Type

Direction (Q. Nos. 40-45) *This section contains 6 questions. When worked out will result in an integer from 0 to 9 (both inclusive).*

Working Space

40 If the plane $x + y + z + d = 0$ bisects the line segment joining the points (1, 2, 3) and (3, 2, 1), then $|d|$ is equal to

41 If the intercept made by the plane $\mathbf{r} \cdot \mathbf{n} = q$ on X-axis is $\lambda\left[\frac{a}{\hat{\mathbf{i}} \cdot \mathbf{n}}\right]$, then the value of λ is

42 The distance between the parallel planes given by the equations, $\mathbf{r} \cdot (2\hat{\mathbf{i}} - 2\hat{\mathbf{j}} + \hat{\mathbf{k}}) + 3 = 0$ and $\mathbf{r} \cdot (4\hat{\mathbf{i}} - 4\hat{\mathbf{j}} + 2\hat{\mathbf{k}}) + 5 = 0$ is λ, then the value of 6λ is

43 If the plane $2x - 3y + 6z - 11 = 0$ makes an angle $\sin^{-1}(k)$ with X-axis, then the value of $7k$ is

44 A variable plane forms a tetrahedron of constant volume $64k^3$ with the coordinate planes and the origin. If the locus of centroid of tetrahedron is $xyz = \lambda k^2$, then the value of λ is

45 Let $ABCD$ be a tetrahedron such that the edges AB, AC and AD are mutually perpendicular. If the area of ΔABC, ΔACD and ΔADB are 3, 4 and 5 sq units respectively, where the area of ΔBCD is $\lambda\sqrt{2}$, then the value of λ is

SET - 1

Revisal Problems for JEE Main

Only One Option Correct Type

Working Space

Direction *(Q. Nos. 1-30) This section contains 30 multiple choice questions. Each question has four choices (a), (b), (c) and (d), out of which ONLY ONE option is correct.*

1 A line makes the same angle θ, with each of the X and Z-axes. If the angle β, which it makes with Y-axis, is such that $\sin^2 \beta = 3\sin^2 \theta$, then $\cos^2 \theta$ is equal to

a. $\frac{2}{5}$ b. $\frac{1}{5}$

c. $\frac{3}{5}$ d. $\frac{2}{3}$

2 If l_1, m_1, n_1 and l_2, m_2, n_2 are direction cosiness of the two lines inclined to each other at an angle θ, then the direction cosines of the internal bisector of the angle between these lines are

a. $\frac{l_1 + l_2}{2\cos\theta/2}, \frac{m_1 + m_2}{2\cos\theta/2}, \frac{n_1 + n_2}{2\cos\theta/2}$

b. $\frac{l_1 + l_2}{2\sin\theta/2}, \frac{m_1 + m_2}{2\sin\theta/2}, \frac{n_1 + n_2}{2\sin\theta/2}$

c. $\frac{l_1 - l_2}{2\cos\theta/2}, \frac{m_1 - m_2}{2\cos\theta/2}, \frac{n_1 - n_2}{2\cos\theta/2}$

d. $\frac{l_1 - l_2}{2\sin\theta/2}, \frac{m_1 - m_2}{2\sin\theta/2}, \frac{n_1 - n_2}{2\sin\theta/2}$

3 The length of perpendicular from (1, 2, 3) to the line $\frac{x-6}{3} = \frac{y-7}{2} = \frac{z-7}{-2}$ is

a. 4 b. 5

c. 6 d. 7

4 A line with direction cosines proportional to 2, 1, 2 meets each of the lines $x = y + a = z$ and $x + a = 2y = 2z$. The coordinates of each of the points of intersection are given by

a. $(2a, 3a, a), (2a, a, a)$ b. $(3a, 2a, 3a), (a, a, a)$

c. $(3a, 2a, 3a), (a, a, 2a)$ d. $(3a, 3a, 3a), (a, a, a)$

5 The two lines $x = ay + b, z = cy + d$ and $x = a'y + b', z = c'y + d'$ will be perpendicular if and only if

a. $aa' + cc' + 1 = 0$

b. $aa' + bb' + cc' = 1$

c. $aa' + bb' + cc' = 0$

d. $(\bar{a} + b')(b + b')(\bar{c} + c') = 0$

6 If the lines $\frac{x-1}{2} = \frac{y+2}{3} = \frac{z-1}{4}$ and $\frac{x-3}{1} = \frac{y-k}{2} = \frac{z}{1}$ intersect, then the value of k is

a. $\frac{3}{2}$ b. $\frac{9}{2}$

c. $-\frac{2}{9}$ d. $-\frac{3}{2}$

7 A plane which passes through the point (3, 2, 0) and the line $\frac{x-3}{1} = \frac{y-6}{5} = \frac{z-4}{4}$ is

a. $x - y + z = 1$ b. $x + y + z = 5$

c. $x + 2y - z = 1$ d. None of these

Working Space

8 The shortest distance from the plane $12x + 4y + 3z = 327$ to the sphere $x^2 + y^2 + z^2 + 4x - 2y - 6z = 155$ is

a. 39 b. 26 c. $11\frac{4}{13}$ d. 13

9 Two systems of rectangular axes have the same origin. If a plane cuts them at distances a, b, c and a', b', c' from the origin, then

a. $\frac{1}{a^2} + \frac{1}{b^2} + \frac{1}{c^2} - \frac{1}{(a')^2} - \frac{1}{(b')^2} - \frac{1}{(c')^2} = 0$

b. $\frac{1}{a^2} + \frac{1}{b^2} + \frac{1}{c^2} + \frac{1}{(a')^2} + \frac{1}{(b')^2} + \frac{1}{(c')^2} = 0$

c. $\frac{1}{a^2} - \frac{1}{b^2} + \frac{1}{c^2} + \frac{1}{(a')^2} - \frac{1}{(b')^2} + \frac{1}{(c')^2} = 0$

d. None of the above

10 If the equation of a plane is $lx + my + nz = p$ is in the normal form, then which one of the following is not true?

a. l, m and n are the direction cosines of the normal of the plane
b. p is the length of perpendicular from the origin to the plane
c. The plane passes through origin for all values of p
d. $l^2 + m^2 + n^2 = 1$

11 The equation $ax + by + c = 0$ represents a plane perpendicular to the

a. XY-plane b. YZ-plane
c. ZX-plane d. None of these

12 If a plane meets the coordinate axes at A, B and C in such a way that the centroid of ΔABC is at the point (1, 2, 3), the equation of the plane is

a. $x + \frac{y}{2} + \frac{z}{3} = 1$ b. $\frac{x}{3} + \frac{y}{6} + \frac{z}{9} = 1$
c. $\frac{x}{1} + \frac{y}{2} + \frac{z}{3} = \frac{1}{3}$ d. None of these

13 A non-zero vector **a** is parallel to the line of intersection of the plane determined by the vectors, $\hat{\mathbf{i}}, \hat{\mathbf{i}} + \hat{\mathbf{j}}$ and the plane determined by the vectors $\hat{\mathbf{i}} - \hat{\mathbf{j}}, \hat{\mathbf{i}} + \hat{\mathbf{k}}$. The angle between **a** and $\hat{\mathbf{i}} - 2\hat{\mathbf{j}} + 2\hat{\mathbf{k}}$ is

a. $\frac{\pi}{3}$ b. $\frac{\pi}{4}$
c. $\frac{\pi}{5}$ d. None of these

14 The distance between the line $\mathbf{r} = 2\hat{\mathbf{i}} - 2\hat{\mathbf{j}} + 3\hat{\mathbf{k}} + \lambda(\hat{\mathbf{i}} - \hat{\mathbf{j}} + 4\hat{\mathbf{k}})$ and the plane $\mathbf{r} \cdot (\hat{\mathbf{i}} + 5\hat{\mathbf{j}} + \hat{\mathbf{k}}) = 5$ is

a. $\frac{10}{9}$ b. $\frac{10}{3\sqrt{3}}$
c. $\frac{10}{3}$ d. None of these

15 The position vector of the centre of the circle $|\mathbf{r}| = 5, \mathbf{r} \cdot (\hat{\mathbf{i}} + \hat{\mathbf{j}} + \hat{\mathbf{k}}) = 3\sqrt{3}$ is

a. $\sqrt{3}(\hat{\mathbf{i}} + \hat{\mathbf{j}} + \hat{\mathbf{k}})$ b. $\hat{\mathbf{i}} + \hat{\mathbf{j}} + \hat{\mathbf{k}}$
c. $3(\hat{\mathbf{i}} + \hat{\mathbf{j}} + \hat{\mathbf{k}})$ d. None of these

16 Given a line $L: \frac{x-1}{3} = \frac{y+1}{2} = \frac{z-3}{-1}$ and a plane $P: x - 2y - z = 0$ on the following assertions, the only one that is always true, is

a. L is perpendicular to P b. L lies in P
c. L is parallel to P d. None of these

17 Equation of the plane that contains the lines $\mathbf{r} = (\hat{\mathbf{i}} + \hat{\mathbf{j}}) + \lambda(\hat{\mathbf{i}} + 2\hat{\mathbf{j}} + \hat{\mathbf{k}})$ and $\mathbf{r} = (\hat{\mathbf{i}} + \hat{\mathbf{j}}) + \mu(-\hat{\mathbf{i}} + \hat{\mathbf{j}} - 2\hat{\mathbf{k}})$ is

a. $\mathbf{r} \cdot (-\hat{\mathbf{i}} + \hat{\mathbf{j}} + \hat{\mathbf{k}}) = 0$ b. $\mathbf{r} \cdot (\hat{\mathbf{i}} + \hat{\mathbf{j}} + \hat{\mathbf{k}}) = 0$
c. $\mathbf{r} \cdot (2\hat{\mathbf{i}} + \hat{\mathbf{j}} - 3\hat{\mathbf{k}}) = -4$ d. $\mathbf{r} \times (-\hat{\mathbf{i}} + \hat{\mathbf{j}} + \hat{\mathbf{k}}) = 0$

Working Space

18 The equation of the sphere inscribed in a tetrahedron whose faces are $x = 0, y = 0, z = 0$ and $x + 2y + 2z = 4$ is

a. $x^2 + y^2 + z^2 + 4x + 4y + 4z + 8 + 0$

b. $x^2 + y^2 + z^2 - 4x - 4y - 4z + 8 = 0$

c. $x^2 + y^2 + z^2 + 4x + 4y + 4z - 8 = 0$

d. $x^2 + y^2 + z^2 - 4x - 4y - 4z - 8 = 0$

19 The point of contact of the plane $\mathbf{r} \cdot (2\hat{\mathbf{i}} - 2\hat{\mathbf{j}} + \hat{\mathbf{k}}) = 6$ and the sphere $|\mathbf{r} - (\hat{\mathbf{i}} + 2\hat{\mathbf{j}} + \hat{\mathbf{k}})| = 6$ is

a. $3\hat{\mathbf{i}}$ b. $3\hat{\mathbf{j}}$

c. $3\hat{\mathbf{k}}$ d. $3\hat{\mathbf{i}} - 2\hat{\mathbf{j}} + \hat{\mathbf{k}}$

20 If **a**, **b**, **c** are three non-coplanar vectors, then the length of projection of vector **a** in the plane of the vectors **b** and **c** may be given by

a. $\dfrac{|\mathbf{a} \cdot (\mathbf{b} \times \mathbf{c})|}{|\mathbf{b} \times \mathbf{c}|}$ b. $\dfrac{|\mathbf{a} \times (\mathbf{b} \times \mathbf{c})|}{|\mathbf{b} \times \mathbf{c}|}$

c. $\dfrac{[\mathbf{a}\ \mathbf{b}\ \mathbf{c}]}{\mathbf{b} \cdot \mathbf{c}}$ d. None of these

21 A plane p passes through the point (1, 1, 1). If b, c, a are the direction ratios of a normal to the plane, where a, b, c $(a < b < c)$ are the prime factors of 2001, then the equation of plane p is

a. $29x + 31y + 3z = 63$ b. $23x + 29y - 29z = 23$

c. $23x + 29y + 3z = 55$ d. None of these

22 The direction ratio of normal to the plane through (1, 0, 0), (0, 1, 0) which makes an angle $\pi/4$ with the plane $x + y = 3$ are proportional to

a. $1, \sqrt{2}, 1$ b. $1, 1, \sqrt{2}$

c. 1, 1,2 d. $\sqrt{2}, 1, 1$

23 The lines $\dfrac{x-2}{1} = \dfrac{y-3}{1} = \dfrac{z-4}{-k}$ and $\dfrac{x-1}{k} = \dfrac{y-4}{2} = \dfrac{z-5}{1}$ are coplanar, if

a. $k = 3, -3$ b. $k = 0, -1$

c. $k = 1, -1$ d. $k = 0, -3$

24 A tetrahedron has vertices $O(0, 0, 0)$, $A(1, 2, 1)$, $B(2, 1, 3)$, $C(-1, 1, 2)$. Then, the angle between the faces OAB and ABC will be

a. 90° b. $\cos^{-1}\left(\dfrac{19}{35}\right)$

c. $\cos^{-1}\left(\dfrac{17}{31}\right)$ d. 30°

25 The value of k such that $\dfrac{x-4}{1} = \dfrac{y-2}{1} = \dfrac{z-k}{2}$ lies in the plane $2x - 4y + 7 = 7$ is

a. 7 b. -7

c. 4 d. no real value

26 The intersection of the spheres $x^2 + y^2 + z^2 + 7x - 2y - z = 13$ and $x^2 + y^2 + z^2 - 3x + 3y + 4z = 8$ is the same as the intersection of one of the sphere and the plane, is

a. $2x - y - z = 1$ b. $x - 2y - z = 1$

c. $x - y - 2z = 1$ d. $x - y - z = 1$

27 The directions ratios of the noraml to the plane passing through the points $(1, -2, 3)$, $(-1, 2, -1)$ and parallel to the line $\dfrac{x-2}{2} = \dfrac{y+1}{3} = \dfrac{z}{4}$ are proportional to

a. 2, 3, 4 b. 4, 0, 7

c. $-2, 0, -1$ d. $2, 0, -1$

28 If the stright lines $= 1 + s$, $y = -3 - \lambda s$, $z = 1 + \lambda s$ and $x = t/2$, $y = 1 + t$, $z = 2 - t$ with parameters s and t respectively, are coplanar, then λ equals

a. 0 b. -1

c. $-1/2$ d. -2

Working Space

29 A plane P makes intercepts 3 and 4 respectively on X and Z-axes. If P is parallel to Y-axis, then its equation is

a. $3x + 3z = 12$
b. $3x + 4z = 12$
c. $4x + 3z = 12$
d. $4x - 3z = 12$

30 The equation $12x^2 - 2y^2 - 6z^2 - 2xy - 8yz + 6yx = 0$ represents

a. a pair of straight lines
b. a pair of planes
c. a sphere
d. a pair of planes not passing through the origin

Statement Type

Direction (Q. Nos. 31-35) ***This section is based on Statement I and Statement II. Select the correct answer from the codes given below.***

a. Both Statement I and Statement II are correct and Statement II is the correct explanation of Statement I
b. Both Statement I and Statement II are correct but Statement II is not the correct explanation of Statement I
c. Statement I is correct but Statement II is incorrect
d. Statement II is correct but Statement I is incorrect

31 **Statement I** If $\alpha, \beta\, \gamma$ are the angles which a half ray makes with the positive direction of the axes, then $\sin^2 \alpha + \sin^2 \beta + \sin^2 \gamma = 2$.

Statement II If l, m, n are the direction cosines of a line, then $l^2 + m^2 + n^2 = 1$.

32 **Statement I** The locus represented by $xy + yz = 0$ is a pair of perpendicular planes.

Statement II If $a_1x + b_1y + c_1z + d_1 = 0$, $a_2x + b_2y + c_2z + d_2 = 0$ are perpendicular, then $a_1a_2 + b_1b_2 + c_1c_2 = 1$.

33 **Statement I** The equation $2x^2 - 6y^2 + 4z^2 + 18yz + 2zx + xy = 0$ represents a pair of perpendicular planes.

Statement II A pair of planes given by $ax^2 + by^2 + cz^2 + 2fyz + 2gzx + 2hxy = 0$ are perpendicular, if $a + b + c = 0$.

34 **Statement I** The shortest distance between the skew lines $\frac{x+3}{-4} = \frac{y-6}{3} = \frac{z}{2}$ and $\frac{x+2}{-4} = \frac{y}{1} = \frac{z-7}{1}$ is 9.

Statement II Two lines are skew lines, if there exists no plane passing through them.

35 Consider the planes $3x - 6y - 2z = 15$ and $2x + y - 2z = 5$.

Statement I The parametric equations of the lines of intersection of the given planes are $x = 3 + 14t$, $y = 1 + 2t$, $z = 15t$.

Statement II A vector $14\hat{\mathbf{i}} + 2\hat{\mathbf{j}} + 15\hat{\mathbf{k}}$ is parallel to the line of intersection of given planes.

SET - 2

Revisal Problems
for JEE Advanced

Only One Option Correct Type

Direction (Q. Nos. 1-14) *This section contains 14 multiple choice questions. Each question has four choices (a), (b), (c) and (d), out of which ONLY ONE option is correct.*

Working Space

1 In a three dimensional coordinate system P, Q and R are the images of a point $A(a, b, c)$ in the XY, YZ and ZX-plane respectively. If G is the centroid of ΔPQR, then the area of ΔAOG is (O where, the origin) is

a. 0
b. $a^2 + b^2 + c^2$
c. $\frac{2}{3}(a^2 + b^2 + c^2)$
d. None of these

2 If $a + b + c = 0$ and

$$(b + c)(y + z) - ax = b - c$$
$$(c + a)(z + x) - by = c - a$$
$$(a + b)(x + y) - cz = a - b$$

then $x : y : z$ is equal to

a. $a : b : c$
b. $bc : ca : ab$
c. $b + c : a + b : c + a$
d. $b - c : c - a : a - b$

3 The value of λ such that the system $x - 2y + z = -4$, $2x - y + 2z = 2$, $x + y + \lambda z = 4$ has no solution, is

a. 0
b. 1
c. -1
d. 3

4 A plane $2x + 3y + 5z = 1$ has a point P which is at minimum distance from the line joining $A(1, 0, -3)$ and $B(1, -5, 7)$, then distance AP is equal to

a. $3\sqrt{5}$
b. $2\sqrt{5}$
c. $4\sqrt{5}$
d. None of these

5 The locus of a point which moves in such a way that its distance from the line $\frac{x}{1} = \frac{y}{1} = \frac{z}{-1}$ is twice the distance from the plane $x + y + z = 0$, is

a. $x^2 + y^2 + z^2 - 5x - 3y - 3z = 0$
b. $x^2 + y^2 + z^2 + 5x + 3y + 3z = 0$
c. $x^2 + y^2 + z^2 - 5xy - 3yz - 3zx = 0$
d. None of the above

6 A variable plane intercepts on the axes, the sum of whose squares is 9. If locus of the foot of perpendicular from origin to plane is $(x^2 + y^2 + z^2)^2 \cdot (x^{-2} + y^{-2} + z^{-2}) = k$, then k is equal to

a. 3
b. 9
c. 27
d. 81

7 A ΔABC has vertices $A(1, -2, 2)$, $B(1, 4, 0)$ and $C(-4, 1, 1)$. The vector BM, where M is the foot of the altitude drawn from B to AC, is

a. $-\frac{20}{3}\hat{\mathbf{i}} - 10\hat{\mathbf{j}} + \frac{10}{3}\hat{\mathbf{k}}$
b. $-\frac{10}{3}\hat{\mathbf{i}} - \frac{30}{7}\hat{\mathbf{j}} + \frac{10}{7}\hat{\mathbf{k}}$
c. $-\frac{20}{7}\hat{\mathbf{i}} - \frac{30}{7}\hat{\mathbf{j}} + \frac{10}{7}\hat{\mathbf{k}}$
d. None of the above

Working Space

8 Angle between the lines $3x + 2y + z - 5 = 0 = x + y - 2z - 3$ and $x - y + 4z = 0 = x + y - 4z$ is equal to

a. $\cos^{-1}\left(\frac{29}{5\sqrt{51}}\right)$

b. $\cos^{-1}\left(\frac{51}{5\sqrt{29}}\right)$

c. $\cos^{-1}\left(\frac{1}{3}\right)$

d. $\frac{\pi}{2}$

9 A cube $C = \{(x, y, z) | 0 \le x, y, z \le 1\}$ is cut by a sharp knife along the plane $x = y, y = z, z = x$, if no piece is moved untill all three cuts are made, the number of pieces is

a. 6

b. 7

c. 8

d. 27

10 Let $A(0, 0, 1)$, $B(0, 1, 0)$ and $C(1, 1, 1)$ are the points in a plane. Then, the equation of plane perpendicular to the plane ABC and passing through A and B is

a. $-x + y + z + 1 = 0$

b. $x - y - z - 1 = 0$

c. $-x + y + z - 1 = 0$

d. $2x + y + z - 1 = 0$

11 A ray of light is sent through the point $P(1, 2, 3)$ and is reflected on the XY-plane. If the reflected ray passes through the point $Q(3, 2, 5)$, then the equation of the reflected ray is

a. $\frac{x-3}{1} = \frac{y-2}{0} = \frac{z-5}{1}$

b. $\frac{x-3}{1} = \frac{y-2}{0} = \frac{z-5}{-4}$

c. $\frac{x-3}{1} = \frac{y-2}{0} = \frac{z-5}{4}$

d. $\frac{x-1}{1} = \frac{y-2}{0} = \frac{z-3}{4}$

12 The shortest distance between any two opposite edges of the tetrahedron formed by planes $x + y = 0, y + z = 0$ $z + x = 0, x + y + z = a$ is constant, equal to

a. $2a$

b. $2a/\sqrt{6}$

c. $a/\sqrt{6}$

d. $2a/\sqrt{3}$

13 The shortest distance from (1, 1, 1) to the line of intersection of the pair of planes $xy + yz + zx + y^2 = 0$ is

a. $\sqrt{\frac{8}{3}}$

b. $\frac{2}{\sqrt{3}}$

c. $\frac{1}{\sqrt{3}}$

d. $\frac{2}{3}$

14 $A = \begin{bmatrix} l_1 & m_1 & n_1 \\ l_2 & m_2 & n_2 \\ l_3 & m_3 & n_3 \end{bmatrix}$ and $B = \begin{bmatrix} p_1 & q_1 & r_1 \\ p_2 & q_2 & r_2 \\ p_3 & q_3 & r_3 \end{bmatrix}$, where p_i, q_i, r_i are the cofactors of the elements l_i, m_i, n_i for $i = 1, 2, 3$. If $(l_1, m_1, n_1), (l_2, m_2, n_2)$ and (l_3, m_3, n_3) are direction cosines of three mutually perpendicular lines, then $(p_1, q_1, r_1), (p_2, q_2, r_2)$ and (p_3, q_3, r_3) are

a. the direction cosines of three mutually perpendicular lines

b. the direction ratios of three mutualy perpendicular lines which are not direction cosines

c. the direction cosines of three lines which need not be perpendicular

d. the direction ratios but not the direction cosines of three lines which need to be perpendicular

One or More than One Options Correct Type

Direction (Q. Nos. 15-19) This section contains 5 multiple choice questions. Each question has four choices (a), (b), (c) and (d), out of which ONE or MORE THAN ONE are correct.

15 Consider the planes $P_1 : 2x + y + z + 4 = 0$, $P_2 : y - z + 4 = 0$, $P_3 : 3x + 2y + z + 8 = 0$. Let L_1, L_2, L_3 be the lines of intersection of the planes P_2 and P_3; P_3 and P_1; P_1 and P_2 respectively. Then,

a. atleast two of the lines L_1, L_2 and L_3 are non-parallel

b. atleast two of the lines L_1, L_2 and L_3 are parallel

c. the three planes intersect in a line

d. the three planes form a triangular prism

Working Space

16 A and B are two points (x_1, y_1, z_1) and (x_2, y_2, z_2) in space. Let P and Q be feet of the perpendiculars drawn from A and B to a line L whose direction ratios are l, m, n and Q be the angle between AB and L, then

a. $PQ = AB\cos\theta$
b. $PQ = |(x_2 - x_1)|l + (y_2 - y_1)m + (z_2 - z_1)n|$
c. $PQ = \dfrac{|(x_2 - x_1)|l + (y_2 - y_1)m + (z_2 - z_1)n|}{\sqrt{l^2 + m^2 + n^2}}$
d. AB and PQ are always coplanar

17 If the lines $\dfrac{x}{1} = \dfrac{y}{2} = \dfrac{z}{3}$ intersects the lines
$3\beta^2 x + 3(1-2\alpha)y + z + 3 = -\dfrac{1}{2}\{6\alpha^2 x + 3(1-2\beta)y + 2z\}$, then point $(\alpha, \beta, 1)$ lie on the plane

a. $2x - y + z = 4$
b. $x + y - z = 2$
c. $x - 2y = 0$
d. $2x - y = 0$

18 Let PM be the perpendicular from the point $P(1, 2, 3)$ to XY-plane. If **OP** makes an angle θ with the positive direction of Z-axis on **OM** makes an angle ϕ with the positive direction of X-axis, where O is the origin and θ and ϕ acute angles, then

a. $\tan\theta = \dfrac{\sqrt{5}}{3}$
b. $\sin\theta \cdot \sin\phi = \dfrac{2}{\sqrt{14}}$
c. $\tan\phi = 2$
d. $\cos\theta \cdot \cos\phi = \dfrac{1}{\sqrt{14}}$

19 A variable plane which remains at a constant distance p from the origin (O) cuts the coordinate axes in A, B, C.

a. Locus of centroid of tetrahedron $OABC$ is $x^2y^2 + y^2z^2 + z^2x^2 = \dfrac{16}{p^2}x^2y^2z^2$
b. Locus of centroid of tetrahedron $OABC$ is $x^2y^2 + y^2z^2 + z^2x^2 = \dfrac{4}{p^2}x^2y^2z^2$
c. Parametric equation of the centroid of tetrahedron is of the form
$\left(\dfrac{p}{4}\cdot\sec\alpha\sec\beta, \dfrac{p}{4}\cdot\sec\alpha\,\text{cosec}\beta, \dfrac{p}{4}\text{cosec}\,\alpha\right); \alpha,\beta \in (0, 2\pi) - \left\{\dfrac{\pi}{2}, \pi, 3\dfrac{\pi}{2}\right\}$
d. None of the above

Matching List Type

Direction **(Q. Nos. 20-24)** ***Choices for the correct combination of elements from Column I and Column II are given as options (a), (b), (c) and (d), out of which one is correct.***

20 The directional cosines of two lines are connected by relations $l + m + n = 0$ and $4l$ is the harmonic mean between m and n, then match the Column I with Column II and mark the correct option from the codes given below.

	Column I		Column II
i.	$\dfrac{l_1}{l_2} + \dfrac{m_1}{m_2} + \dfrac{n_1}{n_2}$ is	p.	$-1/2$
ii.	$l_1l_2 + m_1m_2 + n_1n_2$ is	q.	$-\sqrt{6}/9$
iii.	$l_1m_1n_1 + l_2m_2n_2$ is	r.	$\sqrt{6}/18$
iv.	$(l_1 + l_2)(m_1 + m_2)(n_1 + n_2)$ is	s.	$-3/2$

Codes

	i	ii	iii	iv
a.	r	s	p	q
b.	s	p	q	r
c.	p	q	r	s
d.	q	r	s	p

21 Consider the following linear equations: $ax + by + cz = 0$, $bx + cy + az = 0$ and $cx = ay + bz = 0$, then match the Column I with Column II and mark the correct option from the codes given below.

Working Space

	Column I		Column II
i.	$a + b + c \neq 0$ and $a^2 + b^2 + c^2 = ab + bc + ca$	p.	The equations represent planes meeting only a single point.
ii.	$a + b + c = 0$ and $a^2 + b^2 + c^2 \neq ab + bc + ca$	q.	The equations represent the line $x = y = z$.
iii.	$a + b + c \neq 0$ and $a^2 + b^2 + c^2 \neq ab + bc + ca$	r.	The equations represent identical planes.
iv.	$a + b + c = 0$ and $a^2 + b^2 + c^2 = ab + bc + ca$	s.	The equations represent the whole the three dimensional space.

Codes

	i	ii	iii	iv		i	ii	iii	iv
a.	p	s	r	q	b.	q	p	s	r
c.	r	q	p	s	d.	s	r	q	p

22 Match the Column I with Column II and mark the correct option from the codes given below.

	Column I		Column II
i.	P is a point on the segment of the line segment (3, 3, 5) and (4, 6, 7) dividing it in the ratio k :1 and also projection OP (O is the origin) on coordinate axes are $\frac{32}{9}, \frac{42}{9}, \frac{55}{9}$, then the value of k is	p.	19
ii.	$A(3, 4, 6)$, $B(0, 4, 3)$, $C(a, 6, b)$ are the vertices of a ΔABC such that the median through A is equally inclined to the coordinate axes, then $(a + b)$ is	q.	35
iii.	The line $\frac{x+1}{-1} = \frac{y-12}{5} = \frac{z-7}{2}$ intersects the curve $yz = k$, $x = 0$, then the value of k is	r.	$\sqrt{593}$
iv.	Given $\lvert \mathbf{p} \rvert = 2\sqrt{2}$, $\lvert \mathbf{q} \rvert = 3$ angle between $\mathbf{p}$ and $\mathbf{q}$ is $\pi/4$. If $\mathbf{a} = \mathbf{p} - 3\mathbf{q}$, $\mathbf{b} = 5\mathbf{p} + 2\mathbf{q}$ form the adjacent sides of a parallelogram, then the length of a diagonal is	s.	12
		t.	5/4

Codes

	i	ii	iii	iv		i	ii	iii	iv
a.	r	s	t	p	b.	q	r	p	t
c.	p	q	s	t	d.	t	p	q	r

23 Match the Column I with Column II and mark the correct option from the codes given below.

	Column I		Column II
i.	The angle between the lines whose direction cosines are given by the equation $l^2 + m^2 + n^2 = 0$, $l + m + n = 0$ is θ, then the value of $4\sqrt{3}\sin\theta$, is	p.	1
ii.	The plane passing through the point $(-2, -2, -2)$ and containing the line joining the points (1, 1, 1) and (1, – 1, 2) makes intercepts on the coordinate axes, the sum of whose length is	q.	6
iii.	A line segment has length 63 and direction ratios are (3, –2, 6). If the line makes an obtuse angle with X-axis and the components of the vector are a, b,c, then the value of $(2a + b - c)$ is	r.	12
iv.	If the planes $x = cy + bz$, $y = az + cx$ and $g = bx + ay$ pass through a line, then the value of $a^2 + b^2 + c^2 + 2abc$ is	s.	18

Codes

	i	ii	iii	iv		i	ii	iii	iv
a.	p	r	s	p	b.	q	r	s	p
c.	r	p	q	r	d.	s	p	r	q

24 Match the Column I with Column II and mark the correct option from the codes given below.

Working Space

	Column I		Column II
i.	If the plane $ax - by + cz = 0$ contains the line $\frac{x-a}{a} = \frac{y-2d}{b} = \frac{z-c}{c}$, then $\frac{b}{d}$ is	p.	0
ii.	The distance of the point (1, – 2, 3) from the plane $x - y + z - 5 = 0$ measured parallel to $\frac{x}{2} = \frac{y}{3} = \frac{z-1}{-6}$ is equal to	q.	1
iii.	If the straight lines $\frac{x-2}{1} = \frac{y-3}{1} = \frac{4-z}{K}$ and $\frac{x-1}{k} = \frac{y-4}{2} = \frac{z-5}{1}$ intersect, then k is	r.	2
iv.	Value (s) of k for which the planes $kx + 4y + z = 0$, $4x + ky + 2z = 0$ and $2x + 2y = z = 0$ intersect in a straight line.	s.	1/3

Codes

	i	ii	iii	iv		i	ii	iii	iv
a.	r	q	p	r	b.	s	r	q	s
c.	p	s	r	p	d.	q	p	s	q

Comprehension Type

***Direction** (Q. Nos. 25-32) This section contains 3 paragraphs, each describing theory, experiments, data, etc. Eight questions related to the paragraphs have been given. Each question has only one correct answer among the four given options (a), (b), (c) and (d).*

Passage I

Let P_1 denotes the plane consisting of all points that are equidistant from the points $A(-4, 2, 1)$ and $B(2, -4, 3)$ and P_2 be the plane $x - y + \lambda z = 1$, where $\lambda \in R$.

25 The plane P_1 is parallel to plane P_2,

a. if $\lambda = 1$
b. if $\lambda = 1/3$
c. if $\lambda = 1/3$
d. for no value of λ

26 If the angle between the planes P_1, P_2 is $\frac{\pi}{4}$, then the product of all possible values of λ is

a. –17
b. 17
c. $\frac{24}{17}$
d. –2

27 If the line L with equation $\frac{x-1}{1} = \frac{y+2}{3} = \frac{z-7}{-1}$ intersects the plane P_1 at the point $C(\alpha, \beta, \gamma)$, then $(\alpha + \beta + \gamma)$ is equal to

a. –15
b. – 11
c. 12
d. 13

Passage II

Three vertices of parallelogram taken in order are $A(2, -1, 4)$, $B(1, 0, -1)$, $C(1, 2, 3)$ and D is fourth vertex.

28 The distance between the parallel lines AB and CD is

a. $\sqrt{6}$
b. $3\sqrt{\frac{6}{5}}$
c. $2\sqrt{2}$
d. 3

29 The orthogonal projections of the parallelogram on the three coordinate planes XY, YZ and ZX respectively are

a. 14, 4, 2
b. 2, 4, 12
c. 4, 2, 14
d. 2, 14, 4

Working Space

Passage III

Consider a plane $x + y - z = 1$ and the point $A(1, 2, -3)$. A line L has the equation $x = 1 + 3r$, $y = 2 - r$ and $z = 3 + 4r$.

30 The coordinates of a point B on line L such that AB is parallel to the plane are

a. $(10, -1, 15)$ b. $(-5, 4, -5)$
c. $(4, 1, 7)$ d. $(-8, 5, -9)$

31 The equation of the plane containing the line L and the point A has the equation

a. $x - 3y + 5 = 0$ b. $x + 3y - 7 = 0$
c. $3x - y - 1 = 0$ d. $3x + y - 5 = 0$

32 The distance between the points on the line which are at a distance of $4/\sqrt{3}$ from the plane is

a. $4\sqrt{26}$ b. 20
c. $10\sqrt{13}$ d. 30

One Integer Value Correct Type

Direction (Q. Nos. 33-40) *This section contains 8 questions. When worked out will result in an integer from 0 to 9 (both inclusive).*

33 If the distance between the plane $Ax - 2y + z = d$ and the plane containing the lines $\frac{x-1}{2} = \frac{y-2}{3} = \frac{y-3}{4}$ and $\frac{x-2}{3} = \frac{y-3}{4} = \frac{z-4}{5}$ is $\sqrt{6}$, then $|d|$ is

34 Consider the lines $L_1 : \frac{x-7}{3} = \frac{y-7}{2} = \frac{z-3}{1}$ and $L_2 : \frac{x-1}{2} = \frac{y+1}{4} = \frac{z+1}{3}$. If a line L whose direction ratios are (2, 2, 1) intersect the lines L_1 and L_2 at A and B, then find the distance $\frac{AB}{2}$.

35 Through a point $P(\alpha, \beta, \gamma)$, a plane is drawn at right angles to OP to meet axes in A, B, C. If area of ΔABC is Δ, then the value of $\left[\frac{\alpha\beta\gamma\,\Delta}{p^5}\right]$ is

(where $P = |\mathbf{OP}|$ and $[x]$ is the greatest integer function)

36 The value of λ do the planes $x - y + z + 1 = 0$, $\lambda x + 3y + 2z - 3 = 0$, $3x + \lambda y + z - 2 = 0$ form a triangular prism must be

37 If the lattice point $P(x, y, z)$; $x, y, z > 0$ and $x, y, z \in I$ with least value of z such that the P lies on the planes $7x + 6y + 2z = 272$ and $x - y + z = 16$, then the value of $(x + y + z - 42)$ is

38 The shortest distance between the lines given by $4x + 4y - 5z - 12 = 0$, $8x + 12y - 13z - 32 = 0$ and $\frac{x-2}{1} = \frac{y+2}{2} = \frac{z-3}{3}$ is d, then the value of $\frac{d^2}{4}$ is

39 If the lines $\frac{x-2}{3} = \frac{y+1}{2} = \frac{z-1}{-1}$ intersect the curve $xy = c^2$ and $z = 0$, then the value of $[|c|]$ is (where, $[\cdot]$ denotes the greatest integer function)

40 If the perpendicular distance of a corner of a unit cube from a diagonal not passing through it is d, then the value of $3d^2$ is

JEE Main & AIEEE Archive

(Compilation of Last 13 Years' Questions)

1 The distance of the point (1, 0, 2) from the point of intersection of the line $\frac{x-2}{3}=\frac{y+1}{4}=\frac{z-2}{12}$ and the plane $x-y+z=16$ is **(JEE Main 2015)**

a. $2\sqrt{14}$ b. 8
c. $3\sqrt{21}$ d. 13

2 The equation of the plane containing the line $2x-5y+z=3$, $x+y+4z=5$ and parallel to the plane $x+3y+6z=1$ is **(JEE Main 2015)**

a. $2x+6y+12z=13$
b. $x+3y+6z=-7$
c. $x+3y+6z=7$
d. $2x+6y+12z=-13$

3 The angle between the lines, whose direction cosines satisfy the equation $l+m+n=0$ and $l^2=m^2+n^2$ is **(JEE Main 2014)**

a. $\frac{\pi}{3}$ b. $\frac{\pi}{4}$
c. $\frac{\pi}{6}$ d. $\frac{\pi}{2}$

4 The image of the line $\frac{x-1}{3}=\frac{y-3}{1}=\frac{z-4}{-5}$ in the plane $2x-y+z+3=0$ is the line **(JEE Main 2014)**

a. $\frac{x+3}{3}=\frac{y-5}{1}=\frac{z-2}{-5}$
b. $\frac{x+3}{-3}=\frac{y-5}{-1}=\frac{z+2}{5}$
c. $\frac{x-3}{3}+\frac{y+5}{1}=\frac{z-2}{-5}$
d. $\frac{x-3}{-3}=\frac{y+5}{-1}=\frac{z-2}{5}$

5 Distance between two parallel planes $2x+y+2z=8$ and $4x+2y+4z+5=0$ is **(JEE Main 2013)**

a. $\frac{3}{2}$ b. $\frac{5}{2}$
c. $\frac{7}{2}$ d. $\frac{9}{2}$

6 If the lines $\frac{x-2}{1}=\frac{y-3}{1}=\frac{z-4}{-k}$ and $\frac{x-1}{k}=\frac{y-4}{2}=\frac{z-5}{1}$ are coplanar, then k can have **(JEE Main 2013)**

a. any value
b. exactly one value
c. exactly two values
d. exactly three values

7 An equation of a plane parallel to the plane $x-2y+2z-5=0$ and at a unit distance from the origin is **(AIEEE 2012)**

a. $x-2y+2z-3=0$
b. $x-2y+2z+1=0$
c. $x-2y+2z-1=0$
d. $x-2y+2z+5=0$

8 If the line $\frac{x-1}{2}=\frac{y+1}{3}=\frac{z-1}{4}$ and $\frac{x-3}{1}=\frac{y-k}{2}=\frac{z}{1}$ intersect, then k is equal to **(AIEEE 2012)**

a. -1 b. $\frac{2}{9}$
c. $\frac{9}{2}$ d. 0

9 If the angle between the line $x=\frac{y-1}{2}=\frac{z-3}{\lambda}$ and the plane $x+2y+3z=4$ is $\cos^{-1}\left(\sqrt{\frac{5}{14}}\right)$, then λ is equal to **(AIEEE 2011)**

a. $\frac{3}{2}$ b. $\frac{2}{5}$
c. $\frac{5}{3}$ d. $\frac{2}{3}$

10 **Statement I** The point $A(1, 0, 7)$ is the mirror image of the point $B(1, 6, 3)$ in the line $\frac{x}{1}=\frac{y-1}{2}=\frac{z-2}{3}$.

Statement II The line $\frac{x}{1}=\frac{y-1}{2}=\frac{z-2}{3}$ bisects the line segment joining $A(1, 0, 7)$ and $B(1, 6, 3)$. **(AIEEE 2011)**

a. Both Statement I and Statement II are correct and Statement II is the correct explanation of Statement I
b. Both Statement I and Statement II are correct but Statement II is not the correct explanation of Statement I
c. Statement I is correct but Statement II is incorrect
d. Statement II is correct but Statement I is incorrect

11 The length of the perpendicular drawn from the point $(3, -1, 11)$ to the line $\frac{x}{2}=\frac{y-2}{3}=\frac{z-3}{4}$ is **(AIEEE 2011)**

a. $\sqrt{66}$ b. $\sqrt{29}$
c. $\sqrt{33}$ d. $\sqrt{53}$

12 The distance of the point $(1, -5, 9)$ from the plane $x-y+z=5$ measured along a straight line $x=y=z$, is **(AIEEE 2011)**

a. $3\sqrt{5}$ b. $10\sqrt{3}$
c. $5\sqrt{3}$ d. $3\sqrt{10}$

13 A line AB in three-dimensional space makes angles of 45° and 120° with the positive X-axis and the positive Y-axis, respectively. If AB makes an acute angle θ with the positive Z-axis, then θ is equal to **(AIEEE 2010)**

a. 30° b. 45°
c. 60° d. 75°

14 **Statement I** The point $A(3, 1, 6)$ is the mirror image of the point $B(1, 3, 4)$ in the plane $x - y + z = 5$.
Statement II The plane $x - y + z = 5$ bisects the line segment joining $A(3, 1, 6)$ and $B(1, 3, 4)$ **(AIEEE 2010)**

a. Both Statement I and Statement II are correct and Statement II is the correct explanation of Statement I
b. Both Statement I and Statement II are correct but Statement II is not the correct explanation of Statement I
c. Statement I is correct but Statement II is incorrect
d. Statement II is correct but Statement I is incorrect

15 Let the line $\frac{x-2}{3} = \frac{y-1}{-5} = \frac{z+2}{2}$ lies in the plane $x + 3y - \alpha z + \beta = 0$. Then, (α, β) is equal to **(AIEEE 2009)**

a. $(6, -17)$ b. $(-6, 7)$
c. $(5, -15)$ d. $(-5, 15)$

16 The projections of a vector on the three coordinate axes are $6, -3$ and 2, respectively. The direction cosines of the vector are **(AIEEE 2009)**

a. $6, -3, 2$ b. $\frac{6}{5}, -\frac{3}{5}, \frac{2}{5}$
c. $\frac{6}{7}, -\frac{3}{7}, \frac{2}{7}$ d. $-\frac{6}{7}, -\frac{3}{7}, \frac{2}{7}$

17 The line passing through the points $(5, 1, a)$ and $(3, b, 1)$ crosses the YZ-plane at the point $\left(0, \frac{17}{2}, -\frac{13}{2}\right)$. Then, **(AIEEE 2008)**

a. $a = 8, b = 2$ b. $a = 2, b = 8$
c. $a = 4, b = 6$ d. $a = 6, b = 4$

18 If the straight lines $\frac{x-1}{k} = \frac{y-2}{2} = \frac{z-3}{3}$ and $\frac{x-2}{3} = \frac{y-3}{k} = \frac{z-1}{2}$ intersect at a point, then the integer k is equal to **(AIEEE 2008)**

a. -2 b. -5
c. 5 d. 2

19 Let L be the line of intersection of the planes $2x + 3y + z = 1$ and $x + 3y + 2z = 2$. If L makes an angle α with the positive X-axis, then $\cos\alpha$ is equal to **(AIEEE 2007)**

a. $\frac{1}{\sqrt{3}}$ b. $\frac{1}{2}$
c. 1 d. $\frac{1}{\sqrt{2}}$

20 If a line makes an angle of $\frac{\pi}{4}$ with the positive directions of each of X-axis and Y-axis, then the angle that the line makes with the positive direction of the Z-axis is **(AIEEE 2007)**

a. $\frac{\pi}{6}$ b. $\frac{\pi}{3}$
c. $\frac{\pi}{4}$ d. $\frac{\pi}{2}$

21 If $(2, 3, 5)$ is one end of a diameter of the sphere $x^2 + y^2 + z^2 - 6x - 12y - 2z + 20 = 0$, then the coordinates of the other end of the diameter are **(AIEEE 2007)**

a. $(4, 9, -3)$ b. $(4, -3, 3)$
c. $(4, 3, 5)$ d. $(4, 3, -3)$

22 The two lines $x = ay + b, z = cy + d$ and $x = a'y + b', z = c'y + d'$ are perpendicular to each other, if **(AIEEE 2006, 2003)**

a. $aa' + cc' = 1$ b. $\frac{a}{a'} + \frac{c}{c'} = -1$
c. $\frac{a}{a'} + \frac{c}{c'} = 1$ d. $aa' + cc' = -1$

23 The image of the point $(-1, 3, 4)$ in the plane $x - 2y = 0$ is **(AIEEE 2006)**

a. $(15, 11, 4)$ b. $\left(-\frac{17}{3}, -\frac{19}{3}, 1\right)$
c. $(8, 4, 4)$ d. $\left(\frac{9}{5}, -\frac{13}{5}, 4\right)$

24 If the plane $2ax - 3ay + 4az + 6 = 0$ passes through the mid-point of the line joining the centres of the spheres $x^2 + y^2 + z^2 + 6x - 8y - 2z = 13$ and $x^2 + y^2 + z^2 - 10x + 4y - 2z = 8$, then a is equal to **(AIEEE 2005)**

a. 2 b. -2
c. 1 d. -1

25 If the angle θ between the line $\frac{x+1}{1} = \frac{y-1}{2} = \frac{z-2}{2}$ and the plane $2x - y + \sqrt{\lambda}z + 4 = 0$ is such that $\sin\theta = \frac{1}{3}$. The value of λ is **(AIEEE 2005)**

a. $-\frac{4}{3}$ b. $\frac{3}{4}$
c. $-\frac{3}{5}$ d. $\frac{5}{3}$

26 The angle between the lines $2x = 3y = -z$ and $6x = -y = -4z$ is **(AIEEE 2005)**

a. 30° b. 45°
c. 90° d. 0°

27 The plane $x + 2y - z = 4$ cuts the sphere $x^2 + y^2 + z^2 - x + z - 2 = 0$ in a circle of radius **(AIEEE 2005)**

a. $\sqrt{2}$ b. 2
c. 1 d. 3

28 A line makes the same angle θ with each of the X and Z-axes. If the angle β, which it makes with Y-axis, is such that $\sin^2 \beta = 3\sin^2 \theta$, then $\cos^2 \theta$ is equal to **(AIEEE 2004)**

a. 2/3 b. 1/5
c. 3/5 d. 2/5

29 Distance between two parallel planes, $2x + y + 2z = 8$ and $4x + 2y + 4z + 5 = 0$ is **(AIEEE 2004)**

a. $\frac{3}{2}$ b. $\frac{5}{2}$
c. $\frac{7}{2}$ d. $\frac{9}{2}$

30 A line with direction cosines proportional to 2, 1, 2 meets each of the lines $x = y + a = z$ and $x + a = 2y = 2z$. The coordinates of each of the points of intersection are given by **(AIEEE 2004)**

a. $(3a, 3a, 3a), (a, a, a)$
b. $(3a, 2a, 3a), (a, a, a)$
c. $(3a, 2a, 3a), (a, a, 2a)$
d. $(2a, 3a, 3a), (2a, a, a)$

31 If the straight lines $x = 1 + s, y = -3 - \lambda s, z = 1 + \lambda s$ and $x = \frac{t}{2}, y = 1 + t, z = 2 - t$, with parameters s and t respectively are coplanar, then λ is equal to **(AIEEE 2004)**

a. -2 b. -1
c. $-\frac{1}{2}$ d. 0

32 The intersection of the spheres $x^2 + y^2 + z^2 + 7x - 2y - z = 13$ and $x^2 + y^2 + z^2 - 3x + 3y + 4z = 8$ is the same as the intersection of one of the sphere and the plane. **(AIEEE 2004)**

a. $x - y - z = 1$
b. $x - 2y - z = 1$
c. $x - y - 2z = 1$
d. $2x - y - z = 1$

33 The radius of the circle in which the sphere $x^2 + y^2 + z^2 + 2x - 2y - 4z - 19 = 0$ is cut by the plane $x + 2y + 2z + 7 = 0$, is **(AIEEE 2003)**

a. 1 b. 2
c. 3 d. 4

34 The lines $\frac{x-2}{1} = \frac{y-3}{1} = \frac{z-4}{-k}$ and $\frac{x-1}{k} = \frac{y-4}{2} = \frac{z-5}{1}$ are coplanar, if **(AIEEE 2003)**

a. $k = 0$ or -1 b. $k = 1$ or -1
c. $k = 0$ or -3 d. $k = 3$ or -3

35 The shortest distance from the plane $12x + 4y + 3z = 327$ to the sphere $x^2 + y^2 + z^2 + 4x - 2y - 6z = 155$ is **(AIEEE 2003)**

a. 26 b. $11\frac{4}{13}$
c. 13 d. 39

36 Two systems of rectangular axes have the same origin. If a plane cuts them at distances a, b, c and a', b', c' from the origin, then **(AIEEE 2003)**

a. $\frac{1}{a^2} + \frac{1}{b^2} + \frac{1}{c^2} + \frac{1}{a'^2} + \frac{1}{b'^2} + \frac{1}{c'^2} = 0$
b. $\frac{1}{a^2} + \frac{1}{b^2} - \frac{1}{c^2} + \frac{1}{a'^2} + \frac{1}{b'^2} - \frac{1}{c'^2} = 0$
c. $\frac{1}{a^2} - \frac{1}{b^2} - \frac{1}{c^2} + \frac{1}{a'^2} - \frac{1}{b'^2} - \frac{1}{c'^2} = 0$
d. $\frac{1}{a^2} + \frac{1}{b^2} + \frac{1}{c^2} - \frac{1}{a'^2} - \frac{1}{b'^2} - \frac{1}{c'^2} = 0$

37 A parallelopiped is formed by planes drawn through the points (2, 3, 5) and (5, 9, 7), parallel to the coordinate planes. The length of a diagonal of the parallelopiped is **(AIEEE 2002)**

a. 7 units b. $\sqrt{38}$ units
c. $\sqrt{155}$ units d. None of these

38 The equation of the plane containing the line $\frac{x - x_1}{l} = \frac{y - y_1}{m} = \frac{z - z_1}{n}$ is $a(x - x_1) + b(y - y_1) + c(z - z_1) = 0$, where **(AIEEE 2002)**

a. $ax_1 + by_1 + cz_1 = 0$ b. $al + bm + cn = 0$
c. $\frac{a}{l} = \frac{b}{m} = \frac{c}{n}$ d. $lx_1 + my_1 + nz_1 = 0$

39 The centre of the circle given by $\mathbf{r} \cdot (\hat{\mathbf{i}} + 2\hat{\mathbf{j}} + 2\hat{\mathbf{k}}) = 15$ and $|\mathbf{r} - (\hat{\mathbf{j}} + 2\hat{\mathbf{k}})| = 4$ is **(AIEEE 2002)**

a. (0, 1, 2) b. (1, 3, 4)
c. (−1, 3, 4) d. None of these

JEE Advanced & IIT JEE Archive

(Compilation of Last 10 Years' Questions)

1 In R^3, consider the planes $P_1 : y = 0$ and $P_2 : x + z = 1$. Let P_3 be a plane, different from P_1 and P_2, which passes through the intersection of P_1 and P_2. If the distance of the point (0, 1, 0) from P_3 is 1 and the distance of a point (α, β, γ) from P_3 is 2, then which of the following relation(s) is/are correct?

(2015 Adv., One or More than One Options Correct Type)

a. $2\alpha + \beta + 2\gamma + 2 = 0$
b. $2\alpha - \beta + 2\gamma + 4 = 0$
c. $2\alpha + \beta - 2\gamma - 10 = 0$
d. $2\alpha - \beta + 2\gamma - 8 = 0$

2 In R^3, let L be a straight line passing through the origin. Suppose that all the points on L are at a constant distance from the two planes, $P_1 : x + 2y - z + 1 = 0$ and $P_2 : 2x - y + z - 1 = 0$. Let M be the locus of the foot of the perpendiculars drawn from the points on L to the plane P_1. Which of the following point(s) lie(s) on M?

(2015 Adv., One or More than One Options Correct Type)

a. $\left(0, -\frac{5}{6}, -\frac{2}{3}\right)$ b. $\left(-\frac{1}{6}, -\frac{1}{3}, \frac{1}{6}\right)$
c. $\left(-\frac{5}{6}, 0, \frac{1}{6}\right)$ d. $\left(-\frac{1}{3}, 0, \frac{2}{3}\right)$

3 Let $P(3, 2, 6)$ be a point in space and Q be a point on the line $\mathbf{r} = (\hat{\mathbf{i}} - \hat{\mathbf{j}} + 2\hat{\mathbf{k}}) + \mu(-3\hat{\mathbf{i}} + \hat{\mathbf{j}} + 5\hat{\mathbf{k}})$. Then, the value of μ for which the vector **PQ** is parallel to the plane $x - 4y + 3z = 1$ is

(2009, Only One Option Correct Type)

a. $\frac{1}{4}$ b. $-\frac{1}{4}$
c. $\frac{1}{8}$ d. $-\frac{1}{8}$

4 A line with positive direction cosines passes through the point $P(2, -1, 2)$ and makes equal angles with the coordinate axes. The line meets the plane $2x + y + z = 9$ at point Q. The length of the line segment PQ equals

(2009, Only One Option Correct Type)

a. 1 b. $\sqrt{2}$
c. $\sqrt{3}$ d. 2

5 Consider three planes.

$P_1 : x - y + z = 1$
$P_2 : x + y - z = -1$
and $P_3 : x - 3y + 3z = 2$

Let L_1, L_2, L_3 be the lines of intersection of the planes P_2 and P_3, P_3 and P_1, P_1 and P_2, respectively.

Statement I At least two of the lines L_1, L_2 and L_3 are non-parallel.

Statement II The three planes do not have a common point.

(2008, , Statement Type)

a. Both Statement I and Statement II are correct and Statement II is the correct explanation of Statement I
b. Both Statement I and Statement II are correct but Statement II is not the correct explanation of Statement I
c. Statement I is correct but Statement II is incorrect
d. Statement II is correct but Statement I is incorrect

Passage for Q. Nos. (6-8)

Read the following passage and answer the questions.

Consider the lines

$$L_1 : \frac{x+1}{3} = \frac{y+2}{1} = \frac{z+1}{2},$$

$$L_2 : \frac{x-2}{\text{[illegible]}} = \frac{y+2}{2} = \frac{z-3}{3}$$

(2008, Comprehension Type)

6 The unit vector perpendicular to both L_1 and L_2 is

a. $\frac{-\hat{\mathbf{i}} + 7\hat{\mathbf{j}} + 7\hat{\mathbf{k}}}{\sqrt{99}}$
b. $\frac{-\hat{\mathbf{i}} - 7\hat{\mathbf{j}} + 5\hat{\mathbf{k}}}{5\sqrt{3}}$
c. $\frac{-\hat{\mathbf{i}} + 7\hat{\mathbf{j}} + 5\hat{\mathbf{k}}}{5\sqrt{3}}$
d. $\frac{7\hat{\mathbf{i}} - 7\hat{\mathbf{j}} - \hat{\mathbf{k}}}{\sqrt{99}}$

7 The shortest distance between L_1 and L_2 is

a. 0 unit b. $17/\sqrt{3}$ unit
c. $41/5\sqrt{3}$ unit d. $17/5\sqrt{3}$ unit

8 The distance of the point (1, 1, 1) from the plane passing through the point $(-1, -2, -1)$ and whose normal is perpendicular to both the lines L_1 and L_2 is

a. $\frac{2}{\sqrt{75}}$ unit b. $\frac{7}{\sqrt{75}}$ unit
c. $\frac{13}{\sqrt{75}}$ unit d. $\frac{23}{\sqrt{75}}$ unit

9 Consider the following linear equations are $ax + by + cz = 0, bx + cy + az = 0, cx + ay + bz = 0.$

(2007, Matching Type)

	Column I		Column II
i.	$a+b+c \neq 0$ and $a^2+b^2+c^2 = ab+bc+ca$	p.	The equations represent planes meeting only at a single point
ii.	$a+b+c = 0$ and $a^2+b^2+c^2 \neq ab+bc+ca$	q.	The equations represent the line $x = y = z$
iii.	$a+b+c \neq 0$ and $a^2+b^2+c^2 \neq ab+bc+ca$	r.	The equations represent identical planes
iv.	$a+b+c = 0$ and $a^2+b^2+c^2 = ab+bc+ca$	s.	The equations represent the whole of the three dimensional space

Codes

	i	ii	iii	iv
a.	p,s	p,q	r	r,s
b.	r	q	p	s
c.	p,q	p	q	r
d.	s	s	q	p,r

10 Consider the planes,

$$3x - 6y - 2z = 15$$

and $$2x + y - 2z = 5.$$

Statement I The parametric equations of the line of intersection of the given planes are $x = 3 + 14t,\ y = 1 + 2t,\ z = 15t.$

Statement II The vectors $14\hat{i} + 2\hat{j} + 15\hat{k}$ is parallel to the line of intersection of the given planes. **(2007, Statement Type)**

a. Both Statement I and Statement II are correct and Statement II is the correct explanation of Statement I

b. Both Statement I and Statement II are correct but Statement II is not the correct explanation of Statement I

c. Statement I is correct but Statement II is incorrect

d. Statement II is correct but Statement I is incorrect

11 A plane passes through $(1, -2, 1)$ and is perpendicular to two planes $2x - 2y + z = 0$ and $x - y + 2z = 4$, then the distance of the plane from the point $(1, 2, 2)$ is

(2006, Only One Option Correct Type)

a. 0 b. 1 c. $\sqrt{2}$ d. $2\sqrt{2}$

12 A variable plane $\frac{x}{a} + \frac{y}{b} + \frac{z}{c} = 1$ at a unit distance from origin cuts the coordinate axes at A, B and C. Centroid (x, y, z) satisfies the equation $\frac{1}{x^2} + \frac{1}{y^2} + \frac{1}{z^2} = K$. The value of K is

(2005, Only One Option Correct Type)

a. 9 b. 3

c. 1/9 d. 1/3

13 Find the equation of the plane containing the line $2x - y + z - 3 = 0, 3x + y + z = 5$ and at a distance of $\frac{1}{\sqrt{6}}$ from the point $(2, 1, -1)$.

(2005, Subjective Type)

Answers *with* Explanations

Answers

1. Parabola

DPP-1 Standard Equation of Conics

1. (i) Parabola (ii) Hyperbola (iii) Ellipse (iv) Parabola
2. (i) Focus (2,0), Directrix $\Rightarrow x=-2$, Latusrectum $=8$

   (ii) Focus $\left(0,\frac{3}{2}\right)$, Directrix $\Rightarrow y=-\frac{3}{2}$, Latusrectum $=6$

   (iii) Focus $(-3,0)$, Directrix $\Rightarrow x=3$, Latusrectum $=12$

   (iv) Focus $(0,-4)$, Directrix $\Rightarrow y=4$, Latusrectum $=16$
3.

	Vertex	Axis	Directrix	Latusrectum
(i)	(3,4)	$y=4$	$x=\frac{11}{4}$	1
(ii)	$\left(-\frac{7}{2},-\frac{5}{2}\right)$	$y=\frac{5}{2}$	$x=-\frac{11}{4}$	3
(iii)	(1,2)	$x=1$	$y=\frac{7}{4}$	1
(iv)	$\left(\frac{3}{2},-\frac{11}{8}\right)$	$x=\frac{3}{2}$	$y=-\frac{7}{8}$	2

4. $4x^2+y^2+4xy+4x+32y+16=0$
5. $16x^2+y^2+8xy+96x-554y-1879=0$
6. $4x^2+y^2-4xy+8x+46y-71=0$
7. $(3x+4y-4)^2=20(4x-3y+7)$
8. $y^2=2ax$

9. *a*	10. *d*	11. *d*	12. *d*	13. *b*	14. *a*	15. *c*	16. *d*	17. *a,b,c,d*	18. *a,d*
19. *b,c,d*	20. *c,d*	21. *a*	22. *b*	23. *a*	24. *c*	25. *a*			

DPP-2 Equation of Chord, Focal Chord and Its Properties

1. *a*	2. *c*	3. *d*	4. *a*	5. *c*	6. *a*	7. *c*	8. *d*	9. *d*	10. *b*
11. *a*	12. *c*	13. *d*	14. *b*	15. *b*	16. *a*	17. *c*	18. *d*	19. *a*	20. *b*
21. *a*	22. *a*	23. *c*	24. *c*	25. *d*	26. 2	27. 4	28. 2	29. 3	30. 4
31. 7									

DPP-3 Equation of Tangent and Its Properties

1. *d*	2. *d*	3. *a*	4. *b*	5. *c*	6. *c*	7. *b*	8. *b*	9. *d*	10. *c*
11. *b*	12. *c*	13. *a*	14. *a*	15. *d*	16. *c*	17. *b*	18. *d*	19. *b*	20. *d*
21. *a,c*	22. *b,c*	23. *b,c,d*	24. *a,d*	25. *a,b*	26. *a*	27. *a*	28. *b*	29. *c*	30. *b*
31. *d*	32. 4	33. 1	34. 2	35. 3					

DPP-4 Equation of Normal and Its Properties

1. *b*	2. *a*	3. *b*	4. *b*	5. *d*	6. *c*	7. *a*	8. *b*	9. *b*	10. *c*
11. *d*	12. *a*	13. *b*	14. *a*	15. *a*	16. *b*	17. *a*	18. *b*	19. *b*	20. *a*
21. *c*	22. *c*	23. *c*	24. *c*	25. *a*	26. *c*	27. *d*	28. *c*	29. *a*	30. *a*
31. *c*	32. 3	33. 4	34. 2	35. 4					

DPP-5 Equation of $S'=T$, $T=0$ and $SS'=T^2$

1. *c*	2. *a*	3. *d*	4. *a*	5. *a*	6. *d*	7. *b*	8. *a*	9. *a*	10. *c*
11. *c*	12. *a*	13. *c*	14. *a*	15. *d*	16. 5	17. 4	18. 8	19. 2	20. 4

Revisal Problems (JEE Main)

1. *b* 2. *c* 3. *d* 4. *b* 5. *c* 6. *d* 7. *b* 8. *c* 9. *c* 10. *d*
11. *b* 12. *a* 13. *b* 14. *a* 15. *b* 16. *a* 17. *c* 18. *b* 19. *a* 20. *b*
21. *d* 22. *c* 23. *d* 24. *a* 25. *c* 26. *c* 27. *b* 28. *a* 29. *d* 30. *b*
31. *b* 32. *c* 33. *c* 34. *a* 35. *c* 36. *d* 37. *d* 38. *b* 39. *b* 40. *d*
41. *a* 42. *d* 43. *a* 44. *a* 45. *a* 46. *d* 47. *d* 48. *a* 49. *a* 50. *a*

Revisal Problems (JEE Advanced)

1. *b* 2. *b* 3. *c* 4. *a* 5. *b* 6. *b* 7. *a* 8. *b* 9. *a* 10. *b*
11. *a* 12. *a* 13. *b* 14. *c* 15. *c* 16. *a* 17. *a* 18. *c* 19. *a* 20. *d*
21. *a* 22. *d* 23. *a* 24. *c* 25. *a* 26. *b* 27. *b* 28. *b* 29. *d* 30. *a*
31. *c* 32. *a* 33. *a* 34. *b* 35. *c* 36. *c* 37. *b* 38. *a* 39. *b* 40. *a*
41. *a* 42. *d* 43. *b* 44. *b* 45. *a* 46. *a* 47. *b* 48. *c* 49. *d* 50. *a*
51. *a* 52. *b* 53. *b* 54. *b* 55. *a* 56. *c* 57. 6 58. 3 59. 8 60. 2
61. 4 66. 2 63. 8 64. 5 65. 1

JEE Main & AIEEE Archive

1. *d* 2. *a* 3. *b* 4. *b* 5. *c* 6. *c* 7. *d* 8. *d* 9. *a* 10. *a*
11. *d*

JEE Advanced & IIT JEE Archive

1. 2 2. *a,d* 3. *d* 4. *a* 5. *c* 6. *a,b,d* 7. 4 8. 2 9. *b* 10. *d*
11. *c,d* 12. *a,d* 13. *b,c* 14. *a,b* 15. *a* 16. *a*

2. Ellipse

DPP-1 Equation of Ellipse

1. (i) Length of major axis $=10$, Length of minor axis $=8$, Foci $=(\pm 3,0)$, Vertices $=(\pm 5,0)$, and Eccentricity $=\dfrac{3}{5}$

(ii) Length of major axis $=2\sqrt{3}$, Length of minor axis $=2\sqrt{2}$, Foci $=(0,\pm 1)$, Vertices $=(0,\pm\sqrt{3})$ and Eccentricity $=\dfrac{1}{\sqrt{3}}$

(iii) Length of major axis $=2$, Length of minor axis $=1$, Foci $=\left(1\pm\dfrac{\sqrt{3}}{2},0\right)$, Vertices $=(2,0)$ and $(0,0)$ and Eccentricity $=\dfrac{\sqrt{3}}{2}$

2. Eccentricity $=\dfrac{\sqrt{3}}{2}$, Vertices $=(1,-2)$ and $(-3,2)$, Foci $=(\pm\sqrt{3}+1,-2)$, Directrix $=\pm\dfrac{4}{\sqrt{3}}-1$, Length of latusrectum $=1$, Equation of latusrectum $=(x=\pm\sqrt{3}-1)$.

3. (i) $\dfrac{x^2}{25}+\dfrac{y^2}{9}=1$ (ii) $\dfrac{x^2}{36}+\dfrac{y^2}{100}=1$, (iii) $\dfrac{x^2}{9}+\dfrac{y^2}{25}=1$ (iv) $\dfrac{x^2}{16}+\dfrac{y^2}{12}=1$

4. *c* 5. *a* 6. *d* 7. *b* 8. *c* 9. *d* 10. *c* 11. *a* 12. *b* 13. *c*
14. *c* 15. *a* 16. *b* 17. *a* 18. *a* 19. *c* 20. 5 21. 6 22. 0 23. 5
24. 1 25. 3

DPP-2 Equation of Chord, Focal Chord and Its Properties

1. *a* 2. *d* 3. *a* 4. *c* 5. *a* 6. *b* 7. *c* 8. *b* 9. *c* 10. *a*
11. *a* 12. *b* 13. *a* 14. *c* 15. 2

DPP-3 Tangent and Its Properties

1. *a* 2. *b* 3. *d* 4. *a* 5. *b* 6. *c* 7. *a* 8. *c* 9. *d* 10. *c*
11. *a* 12. *b* 13. *c* 14. *c* 15. *c* 16. *b* 17. *d* 18. *d* 19. *b* 20. *d*
21. *c* 22. *b* 23. *d* 24. *a* 25. *b* 26. *c* 27. *d* 28. *a* 29. 3 30. 5

DPP-4 Normal and Its Properties

1. *b*	2. *b*	3. *c*	4. *a*	5. *c*	6. *c*	7. *c*	8. *b*	9. *d*	10. *b*
11. *c*	12. *c*	13. *a*	14. *a*	15. *c*	16. *a*	17. *c*	18. 3	19. 0	20. 4

DPP-5 Equation of $S' = T$, $T = 0$ and $SS' = T^2$, Diameter and Conjugate Diameter

1. *c*	2. *a*	3. *b*	4. *b*	5. *a*	6. *c*	7. *a*	8. *c*	9. *a*	10. *c*
11. *a*	12. *a*	13. *a*	14. *b*	15. *c*					

Revisal Problems (JEE Main)

1. *b*	2. *c*	3. *b*	4. *b*	5. *a*	6. *c*	7. *a*	8. *a*	9. *b*	10. *d*
11. *c*	12. *a*	13. *b*	14. *a*	15. *c*	16. *c*	17. *d*	18. *b*	19. *a*	20. *b*
21. *a*	22. *b*	23. *d*	24. *b*	25. *a*	26. *c*	27. *d*	28. *c*	29. *b*	30. *b*
31. *d*	32. *c*	33. *c*	34. *c*	35. *a*	36. *c*	37. *c*	38. *a*	39. *a*	40. *b*
41. *d*	42. *c*	43. *a*	44. *a*	45. *a*					

Revisal Problems (JEE Advanced)

1. *b*	2. *d*	3. *d*	4. *a*	5. *a*	6. *c*	7. *a*	8. *d*	9. *c*	10. *a*
11. *c*	12. *b*	13. *b*	14. *c*	15. *d*	16. *b,c*	17. *a,b*	18. *a,b,c*	19. *a,b,c*	20. *b,d*
21. *a*	22. *a*	23. *c*	24. *d*	25. *b*	26. *b*	27. *c*	28. *c*	29. 2	30. 4

JEE Main & AIEEE Archive

1. *d*	2. *c*	3. *a*	4. *d*	5. *d*	6. *a*	7. *a*	8. *b*	9. *d*	10. *d*
11. *d*	12. *b*	13. *a*	14. *b*						

JEE Advanced & IIT JEE Archive

1. 4	2. *a,b*	3. *a*	4. *c*	5. *c*	6. *d*	7. *c*	8. *d*	9. *c*	10. *a*
11. $\frac{14\sqrt{3}}{3}$ units									

3. Hyperbola

DPP-1 Equation of Hyperbola

1. *c*	2. *a*	3. *a*	4. *d*	5. *b*	6. *b*	7. *b*	8. *a*	9. *c*	10. *a*
11. *c*	12. *d*	13. *c*	14. *d*	15. *c*	16. *a*	17. *c*	18. *d*	19. *c*	20. *b*
21. *b*	22. *d*	23. *d*	24. *c*	25. *d*	26. 2	27. 6	28. 1	29. 2	30. 2

DPP-2 Tangent and Its Properties

1. *c*	2. *a*	3. *b*	4. *b*	5. *b*	6. *c*	7. *b*	8. *a*	9. *b*	10. *b*
11. *d*	12. *b*	13. *a*	14. *a*	15. *a*	16. *b*	17. *c*	18. *a*	19. *a*	20. *b*
21. *d*	22. *c*	23. 6	24. 2	25. 1					

DPP-3 Normal, Chord Bisected and Product of Pair of Tangents

1. *b*	2. *d*	3. *a*	4. *c*	5. *b*	6. *d*	7. *c*	8. *b*	9. *a*	10. *c*
11. *c*	12. *b*	13. *a*	14. *c*	15. *a*	16. *a*	17. *c*	18. *a*	19. *a*	20. *b*
21. *a*	22. *b*	23. *c*	24. *c*	25. *c*	26. *a*	27. *a*	28. 4	29. 0	30. 2

DPP-4 Asymptotes and Rectangular Hyperbola

1. *a*	2. *b*	3. *c*	4. *d*	5. *b*	6. *c*	7. *b*	8. *d*	9. *a*	10. *b*
11. *d*	12. *c*	13. *c*	14. *c*	15. *d*	16. *d*	17. *b*	18. *a*	19. *d*	20. *a*
21. *a*	22. *d*	23. *c*	24. *d*	25. *b*	26. *a*	27. *c*	28. 8	29. 1	30. 2

Revisal Problems (JEE Main)

1. *c* **2.** *d* **3.** *a* **4.** *b* **5.** *b* **6.** *c* **7.** *d* **8.** *b* **9.** *b* **10.** *d*
11. *d* **12.** *b* **13.** *a* **14.** *c* **15.** *c* **16.** *b* **17.** *c* **18.** *d* **19.** *a* **20.** *a*
21. *a* **22.** *c* **23.** *c* **24.** *b* **25.** *c* **26.** *d* **27.** *c* **28.** *c* **29.** *a* **30.** *c*
31. *a* **32.** *b* **33.** *a* **34.** *d* **35.** *d*

Revisal Problems (JEE Advanced)

1. *d* **2.** *c* **3.** *c* **4.** *a* **5.** *d* **6.** *b* **7.** *c* **8.** *d* **9.** *b* **10.** *a*
11. *d* **12.** *c* **13.** *a* **14.** *b* **15.** *b* **16.** *a* **17.** *c* **18.** *a* **19.** *a* **20.** *a*
21. *a,c,d* **22.** *a,b,d* **23.** *a,b,d* **24.** *c,d* **25.** *b,c* **26.** *a* **27.** *a* **28.** *b* **29.** *c* **30.** *a*
31. 2 **32.** 9 **33.** 4 **34.** 1 **35.** 4

JEE Main & AIEEE Archive

1. *c* **2.** *d* **3.** *a* **4.** *c* **5.** *a*

JEE Advanced & IIT JEE Archive

1. *a,b,d* **2.** *b* **3.** *b,d* **4.** *b* **5.** *a* **6.** 2 **7.** *a,b* **8.** *b* **9.** *a* **10.** *b*
11. $\frac{x^2}{9}-\frac{y^2}{4}=\frac{(x^2+y^2)^2}{81}$

4. Vectors

DPP-1 Introduction to Vectors

1. *c* **2.** *a* **3.** *a* **4.** *d* **5.** *d* **6.** *c* **7.** *c* **8.** *a* **9.** *a* **10.** *c*
11. *b* **12.** *c* **13.** 1 **14.** 3 **15.** 0 **16.** 1 **17.** 3 **18.** 0 **19.** 3 **20.** 2

DPP-2 Scalar or Dot Product

3. *b* **4.** *a* **5.** *c* **6.** *d* **7.** *a* **8.** *c* **9.** *c* **10.** *a* **11.** *b* **12.** *b*
13. *a* **14.** *a* **15.** *b* **16.** *a* **17.** *b* **18.** *a,b,c* **19.** *b,d* **20.** *a,b,d* **21.** *c* **22.** *d*
23. *a* **24.** 9 **25.** 1

DPP-3 Vector or Cross Product

1. *a* **2.** *c* **3.** *c* **4.** *b* **5.** *a* **6.** *c* **7.** *d* **8.** *b* **9.** *c* **10.** *b*
11. *c* **12.** *b* **13.** *b* **14.** *b* **15.** *c* **16.** *b* **17.** *d* **18.** *b* **19.** *c* **20.** *d*
21. *c* **22.** 3 **23.** 3 **24.** 6 **25.** 1 **26.** 4 **27.** 9 **28.** 4 **29.** 1 **30.** 1

DPP-4 Scalar Triple Product

1. *a* **2.** *c* **3.** *a* **4.** *d* **5.** *a* **6.** *a* **7.** *d* **8.** *d* **9.** *d* **10.** *a*
11. *a* **12.** *d* **13.** *a* **14.** *a* **15.** *d* **16.** *c* **17.** *b* **18.** *a* **19.** *d* **20.** *b*
21. *c* **22.** *a* **23.** *a* **24.** *c* **25.** *b* **26.** 3 **27.** 8 **28.** 3 **29.** 5 **30.** 5

DPP-5 Vector Triple Product

1. *a* **2.** *c* **3.** *a* **4.** *b* **5.** *d* **6.** *c* **7.** *a* **8.** *a* **9.** *d* **10.** *c*
11. *a* **12.** *c* **13.** *a* **14.** *a* **15.** *b* **16.** *c* **17.** *a,b,c* **18.** *b,c,d* **19.** *a,b,c,d*
20. *a,b,c,d* **21.** *c* **22.** *a* **23.** *b* **24.** 9 **25.** 5

Revisal Problems (JEE Main)

1. *c* **2.** *d* **3.** *a* **4.** *a* **5.** *d* **6.** *b* **7.** *a* **8.** *b* **9.** *a* **10.** *b*
11. *d* **12.** *b* **13.** *b* **14.** *d* **15.** *d* **16.** *a* **17.** *a* **18.** *a* **19.** *b* **20.** *a*
21. *c* **22.** *a* **23.** *a* **24.** *a* **25.** *a* **26.** *d* **27.** *b* **28.** *a* **29.** *a* **30.** *c*
31. *b* **32.** *a* **33.** *b* **34.** *b* **35.** *b* **36.** *c* **37.** *b* **38.** *a* **39.** *d* **40.** *a*

Revisal Problems (JEE Advanced)

1. *d* 2. *a* 3. *c* 4. *a* 5. *b* 6. *c* 7. *d* 8. *d* 9. *a* 10. *b*
11. *d* 12. *b* 13. *c* 14. *c* 15. *a,d* 16. *a,b,c* 17. *a,c* 18. *b,c,d* 19. *a,d* 20. *c,d*
21. *a* 22. *b* 23. *a* 24. *a* 25. *b* 26. *a* 27. *b* 28. *a* 29. *d* 30. *d*
31. *c* 32. *b* 33. *d* 34. *a* 35. *c* 36. 4 37. 3 38. 2 39. 3 40. 9

JEE Main & AIEEE Archive

1. *a* 2. *b* 3. *c* 4. *c* 5. *b* 6. *d* 7. *c* 8. *a* 9. *d* 10. *a*
11. *a* 12. *d* 13. *a* 14. *a* 15. *d* 16. *d* 17. *c* 18. *d* 19. *d* 20. *c*
21. *b* 22. *a* 23. *d* 24. *c* 25. *d* 26. *a* 27. *c* 28. *c* 29. *d* 30. *a*
31. *b* 32. *b* 33. *d* 34. *b* 35. *d* 36. *a* 37. *a* 38. *b* 39. *a*

JEE Advanced & IIT JEE Archive

1. *a,c,d* 2. 9 3. *c* 4. *a,d* 5. 9 6. *a* 7. *b* 8. 5 9. *c* 10. *a*
11. *a* 12. *c* 13. *b* 14. *c* 15. *a* 16. *b,d* 17. *b*

5. 3D Geometry

DPP-1 Direction Ratios, Direction Cosines and Equation of Straight Line

1. *c* 2. *c* 3. *c* 4. *a* 5. *c* 6. *a* 7. *c* 8. *a* 9. *a* 10. *a*
11. *b* 12. *b* 13. *d* 14. *d* 15. *d* 16. *a* 17. *a* 18. *a* 19. *a* 20. *c*
21. *b* 22. *c* 23. *b* 24. *d* 25. *c* 26. *b* 27. *c* 28. *a* 29. 6 30. 8
31. 5 32. 3 33. 1 34. 2 35. 0

DPP-2 Equation of Plane and Sphere

1. *a* 2. *a* 3. *a* 4. *a* 5. *b* 6. *d* 7. *a* 8. *b* 9. *a* 10. *a*
11. *d* 12. *a* 13. *b* 14. *a* 15. *b* 16. *d* 17. *a* 18. *d* 19. *a* 20. *a*
21. *b* 22. *b* 23. *b* 24. *a* 25. *b* 26. *c* 27. *d* 28. *b* 29. *d* 30. *d*
31. *d* 32. *c* 33. *b* 34. *b* 35. *d* 36. *b* 37. *b* 38. *c* 39. *d* 40. 6
41. 1 42. 1 43. 2 44. 6 45. 5

Revisal Problems (JEE Main)

1. *c* 2. *a* 3. *d* 4. *b* 5. *a* 6. *b* 7. *a* 8. *d* 9. *c* 10. *c*
11. *a* 12. *b* 13. *b* 14. *b* 15. *a* 16. *b* 17. *a* 18. *b* 19. *a* 20. *b*
21. *c* 22. *b* 23. *d* 24. *b* 25. *a* 26. *a* 27. *d* 28. *d* 29. *c* 30. *b*
31. *a* 32. *c* 33. *d* 34. *b* 35. *d*

Revisal Problems (JEE Advanced)

1. *a* 2. *d* 3. *b* 4. *b* 5. *b* 6. *b* 7. *c* 8. *a* 9. *a* 10. *d*
11. *c* 12. *b* 13. *a* 14. *a* 15. *b,c* 16. *a,c* 17. *a,b,c* 18. *a,b,c* 19. *a* 20. *b*
21. *c* 22. *d* 23. *b* 24. *a* 25. *b* 26. *d* 27. *c* 28. *c* 29. *d* 30. *d*
31. *b* 32. *a* 33. 6 34. 9 35. 0 36. 4 37. 4 38. 6 39. 2 40. 2

JEE Main & AIEEE Archive

1. *d* 2. *c* 3. *a* 4. *a* 5. *c* 6. *c* 7. *a* 8. *c* 9. *d* 10. *a*
11. *d* 12. *b* 13. *c* 14. *a* 15. *b* 16. *c* 17. *d* 18. *b* 19. *a* 20. *d*
21. *a* 22. *d* 23. *d* 24. *b* 25. *d* 26. *c* 27. *c* 28. *c* 29. *c* 30. *b*
31. *a* 32. *d* 33. *c* 34. *c* 35. *c* 36. *d* 37. *a* 38. *b* 39. *b*

JEE Advanced & IIT JEE Archive

1. *b,d* 2. *a,b* 3. *a* 4. *c* 5. *d* 6. *b* 7. *d* 8. *c* 9. *b* 10. *d*
11. *d* 12. *a* 13. 0

1. Parabola

DPP-1 Standard Equation of Conics

1. (i) We have, $\sqrt{px} + \sqrt{qy} = 1$

$\Rightarrow \quad (\sqrt{px} + \sqrt{qy})^2 = (1)^2$

$\Rightarrow \quad px + qy + 2\sqrt{pqxy} = 1$

$\Rightarrow \quad (px + qy - 1)^2 = 4pqxy$

$\Rightarrow p^2x^2 + q^2y^2 + 1 + 2pqxy - 2qy - 2px = 4pqxy$

$\Rightarrow p^2x^2 + q^2y^2 - 2pqxy - 2px - 2qy + 1 = 0$

On comparing with $ax^2 + 2hxy + by^2 + 2gx + 2fy + c = 0$, we get

$a = p^2, b = q^2, c = 1, g = -p, f = -q$ and $h = -pq$

$\therefore \quad \Delta = abc + 2fgh - af^2 - bg^2 - ch^2$

$= p^2q^2 - 2p^2q^2 - p^2q^2 - p^2q^2 - p^2q^2$

$= -4p^2q^2$

$\neq 0$

and $h^2 - ab = p^2q^2 - p^2q^2 = 0$

Thus, we have $\Delta \neq 0$ and $h^2 = ab$.
Hence, the given equation represents a parabola.

(ii) We have,

$$2xy + 4x - 6y + 17 = 0$$

On comparing with $ax^2 + 2hxy + by^2 + 2gx + 2fy + c = 0$, we get

$a = 0, h = 1, b = 0, g = 2, f = -3, c = 17$

$\therefore \ \Delta = abc + 2fgh - af^2 - bg^2 - ch^2$

$= 0 \times 0 \times 17 + 2 \times (-3)(2)(1) - 0(-3)^2 - 0(2)^2 - 17(1)^2$

$= -12 - 17 = -29 \neq 0$

and $\quad h^2 - ab = (1)^2 - (0)(0) = 1$

$\Rightarrow \quad h^2 - ab > 0$

Thus, we have $\Delta \neq 0$ and $h^2 > ab$.

Hence, the given equation represents a hyperbola.

(iii) We have,

$$13x^2 - 18xy + 37y^2 + 2x + 14y - 2 = 0$$

On comparing with $ax^2 + 2hxy + by^2 + 2gx + 2fy + c = 0$, we get

$a = 13, b = 37, h = -9, g = 1, f = 7, c = -2$

$\therefore \Delta = abc + 2fgh - af^2 - bg^2 - ch^2$

$= 13 \times 37 \times (-2) + 2 \times 7 \times 1 \times (-9) - 37(1)^2 - (-2)(-9)^2$

$= -962 - 126 - 37 + 162 \neq 0$

and $\quad h^2 - ab = (-9)^2 - (13)(37)$

$= 81 - 481 = -400$

$\Rightarrow \quad h^2 - ab < 0$

Thus, we have $\Delta \neq 0$ and $h^2 < ab$.
Hence, the given equation represents an ellipse.

(iv) We have, $\quad x = t^2 + t + 1$

and $\quad y = t^2 - t + 1$

$\Rightarrow \quad x + y = 2(t^2 + 1)$

and $\quad x - y = 2t$

Now, $\quad (x + y) = 2\left\{\left(\frac{x - y}{2}\right)^2 + 1\right\}$

$\Rightarrow \quad 2(x + y) = (x - y)^2 + 4$

$\Rightarrow \quad x^2 + y^2 - 2xy - 2x - 2y + 4 = 0$

On comparing with

$ax^2 + 2hxy + by^2 + 2gx + 2fy + c = 0$, we get

$a = 1, b = 1, c = 4, h = -1, g = -1$ and $f = -1$

$\therefore \quad \Delta = abc + 2fgh - af^2 - bg^2 - ch^2$

$= 4 - 2 - 1 - 1 - 4$

$= -4$

$\neq 0$

and $h^2 - ab = 1 - 1 = 0$

Thus, we have

$$\Delta \neq 0 \quad \text{and} \quad h^2 = ab$$

So, the given equations represent a parabola.

2. (i) Given equation of parabola is $y^2 = 8x$.

On comparing with $y^2 = 4ax$, we get $a = 2$

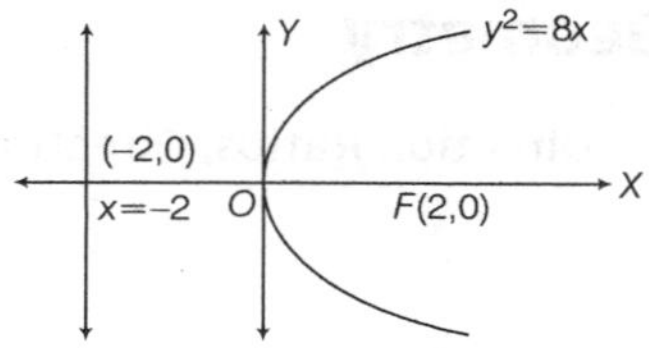

$\therefore$ Focus $= (a, 0) = (2, 0)$

Equation of directrix $\Rightarrow x = -a \quad \Rightarrow \quad x = -2$

Length of the latusrectum $= 4a = 4 \times 2 = 8$

(ii) Given equation of parabola is $x^2 = 6y$.

On comparing with $x^2 = 4ay$, we have $a = \frac{3}{2}$.

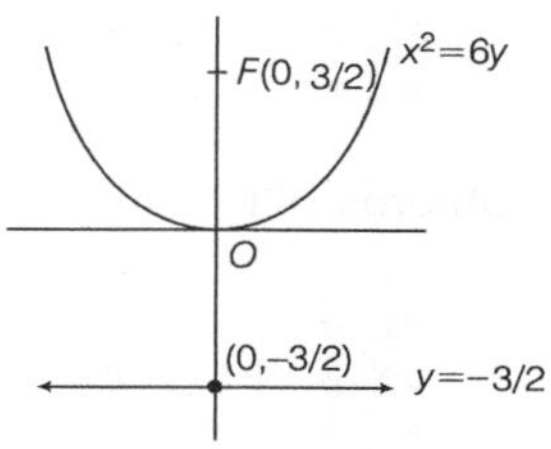

$\therefore \quad$ Focus $= (0, a) = \left(0, \frac{3}{2}\right)$

Equation of directrix $\Rightarrow y = -a \Rightarrow y = \frac{-3}{2}$

Length of latusrectum $= 4a = 4 \times \frac{3}{2} = 6$

(iii) Given equation of parabola $y^2 = -12x$

On comparing with $y^2 = -4ax$, we get $a = -3$.

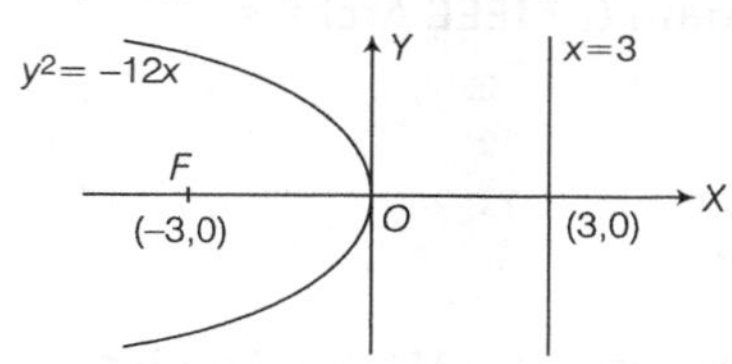

$\therefore \quad$ Focus $= (-a, 0) = (-3, 0)$

Equation of directrix $\Rightarrow x = a \quad \Rightarrow \quad x = 3$

Length of latusrectum $= 4a = 4 \times 3 = 12$

(iv) Given equation of parabola is $x^2 = -16y$.

On comparing with $x^2 = -4ay$, we get $a = 4$

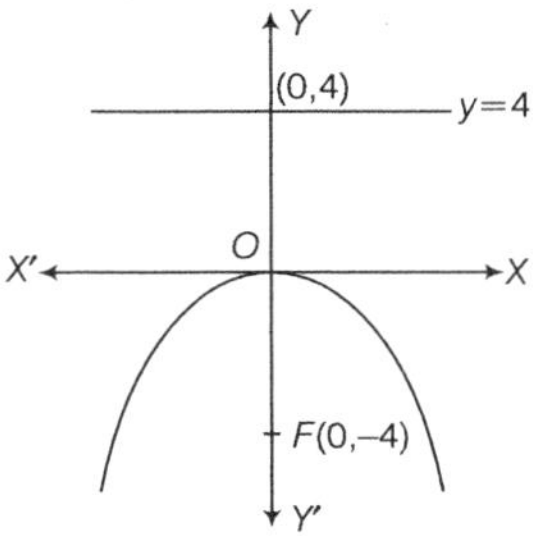

$\therefore$ Focus $= (0, -a) = (0, -4)$

Equation of directrix $\Rightarrow y = a \Rightarrow y = 4$

Length of latusrectum $= 4a = 4 \times 4 = 16$

3. (i) The given equation is

$$y^2 - 8y - x + 19 = 0$$

$$\Rightarrow \quad y^2 - 8y = x - 19$$

$$\Rightarrow \quad (y-4)^2 - 16 = x - 19$$

$$\Rightarrow \quad (y-4)^2 = (x - 3) \quad \ldots(i)$$

Shifting the origin to the point (3, 4) without rotating the axes and assuming new coordinates X and Y, we have

$$x = X + 3 \quad \text{and} \quad y = Y + 4 \quad \ldots(ii)$$

Using these relations, Eq. (i) reduces to

$$Y^2 = X \quad \ldots(iii)$$

On comparing with $Y^2 = 4aX$, we get

$$a = \frac{1}{4}$$

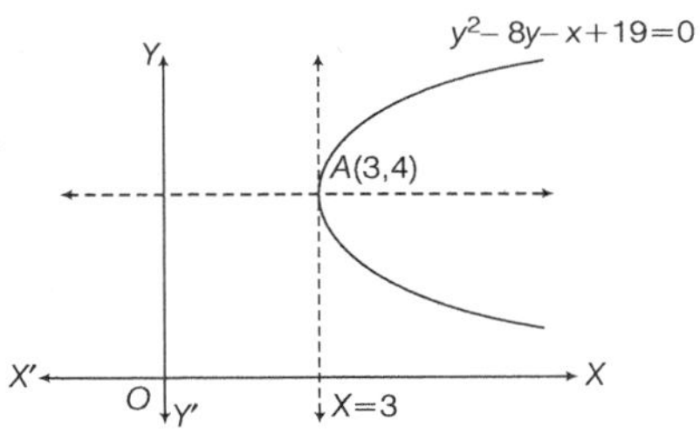

Vertex The coordinates of the vertex with respect to the new axes are $X = 0$ and $Y = 0$.

So, the coordinates of the vertex with respect to the old axes are (3, 4).

Axis The equation of parabola (iii) with respect to the new axes is $Y = 0$. So, the equation of axis with respect to the old axes is $y = 4$.

Focus The coordinates of the focus with respect to the new axes are $(X = a, Y = 0)$, i.e. $(X = 1/4, Y = 0)$.

So, the coordinates of the focus with respect to the old axes are (13/4, 4).

Directrix The equation of the directrix with respect to the new axes is $X = -a$, i.e. $x = -1/4$.

So, the equation of the directrix with respect to the old axes is

$$x = -\frac{1}{4} + 3 \Rightarrow x = \frac{11}{4}$$

Latusrectum The length of the latusrectum of the parabola (iii) is equal to $4a = 1$.

(ii) The given equation is

$$4y^2 + 12x - 20y + 67 = 0$$

$$\Rightarrow \quad y^2 + 3x - 5y - \frac{67}{4} = 0$$

$$\Rightarrow \quad y^2 - 5y = -3x - \frac{67}{4}$$

$$\Rightarrow \quad y^2 - 5y + \left(\frac{5}{2}\right)^2 = -3x - \frac{67}{4} + \left(\frac{5}{2}\right)^2$$

$$\Rightarrow \quad \left(y - \frac{5}{2}\right)^2 = -3x - \frac{42}{4}$$

$$\Rightarrow \quad \left(y - \frac{5}{2}\right)^2 = -3\left(x + \frac{7}{2}\right) \quad \ldots(i)$$

Shifting the origin to the point $(-7/2, 5/2)$ without rotating the axes and denoting the new coordinates with respect to these axes by X and Y, we have

$$x = X - \frac{7}{2}, \; y = Y + \frac{5}{2} \quad \ldots(ii)$$

Using these relations, Eq. (i) reduces to

$$Y^2 = -3X \quad \ldots(iii)$$

This is of the form $Y^2 = -4aX$. On comparing, we get

$$4a = 3 \Rightarrow a = 3/4$$

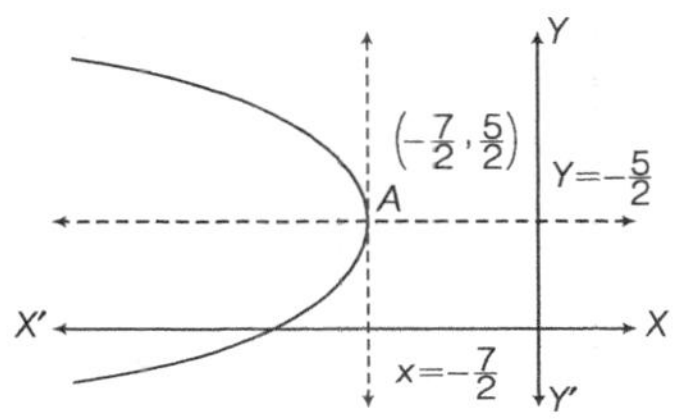

Vertex The coordinates of the vertex with respect to the new axes are $[X = 0, Y = 0]$. So, the coordinates of the vertex with respect to the old axes are

$$\left(\frac{-7}{2}, \frac{-5}{2}\right) \quad \text{[putting } X = 0, Y = 0 \text{ in Eq. (ii)]}$$

Axis The equation of the axis of the parabola with respect to the new axis is $Y = 0$. So, the equation of the axis with respect to the old axes is

$$y = 5/2 \quad \text{[putting } Y = 0 \text{ in Eq. (ii)]}$$

Focus The coordinates of the focus with respect to the new axes are $(X = -a, Y = 0)$, i.e. $(X = -3/4, Y = 0)$.

So, the coordinates of the focus with respect to the old axes are

$$\left(\frac{-17}{4}, \frac{5}{2}\right)$$

[putting $X = -3/4$ and $Y = 0$ in Eq. (ii)]

Directrix The equation of the directrix with respect to the new axes is $x = a$, i.e. $X = 3/4$, So, the equation of the directrix with respect to the old axes is

$$x = -11/4 \quad \text{[putting } x = 3/4 \text{ in Eq. (ii)]}$$

Latusrectum The length of the latusrectum of the given parabola is $4a = 3$.

(iii) The given equation is

$$y = x^2 - 2x + 3$$

$$\Rightarrow \quad x^2 - 2x = y - 3$$

$$\Rightarrow \quad x^2 - 2x + 1 = y - 3 + 1$$

$$\Rightarrow \quad (x-1)^2 = y - 2 \quad \ldots(i)$$

Shifting the origin to the point (1, 2) without rotating the axes and denoting the new coordinates with respect to these axes by X and Y, we have

$$x = X + 1, \; y = Y + 2 \quad \ldots(ii)$$

Using these relations, Eq. (i) reduces to

$$X^2 = Y \quad \ldots(iii)$$

This is of the form $X^2 = 4aY$. On comparing, we get $4a = 1$, i.e. $a = 1/4$.

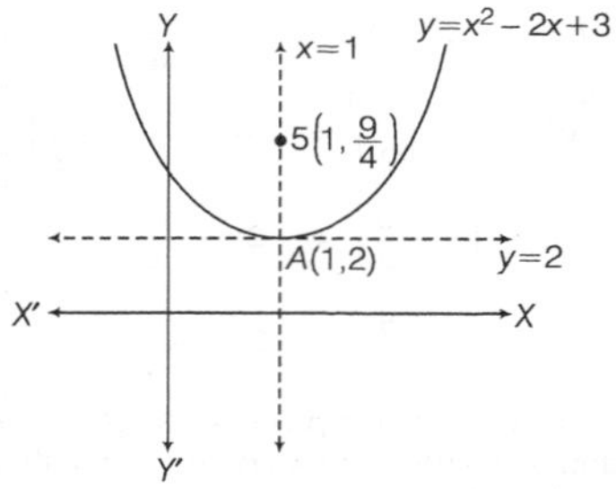

Vertex The coordinates of the vertex with respect to the new axes are $(X = 0, Y = 0)$. So, the coordinates of the vertex with respect to the old axes are $(1, 2)$. [putting $X = 0, Y = 0$ in Eq. (ii)]

Axis The equation of the axis of the parabola with respect to the new axes is $X = 0$. So, the equation of the axis with respect to the old axes is $x = 1$. [putting $X = 0$ in Eq. (ii)]

Focus The coordinates of the focus with respect to the new axes are $(X = 0, Y = a)$, i.e. $(X = 0, Y = 1/4)$.

So, the coordinates of the focus with respect to the old axes are $(1, 9/4)$. [putting $X = 0, Y = 1/4$ in Eq. (ii)]

Directrix The equation of the directrix with respect to the new axes is $Y = -a$, i.e. $Y = -1/4$. So, the equation of the directrix with respect to the old axes is

$$y = \frac{-1}{4} + 2 \quad \text{or} \quad y = \frac{7}{4}$$

[putting $Y = \frac{-1}{4}$ in Eq. (iii)]

Latusrectum The length of the latusrectum of the given parabola is equal to $4a = 1$.

(iv) The given equation is

$$x^2 + 2y - 3x + 5 = 0$$

$$\Rightarrow \quad x^2 - 3x = -2y - 5$$

$$\Rightarrow \quad x^2 - 3x + \frac{9}{4} = -2y - 5 + \frac{9}{4}$$

$$\Rightarrow \quad \left(x - \frac{3}{2}\right)^2 = -2\left(y + \frac{11}{8}\right) \qquad \text{...(i)}$$

Shifting the origin to the point $(3/2, -11/8)$ without rotating the coordinate axes and denoting the new coordinates with respect to the new axes by X and Y, we have

$$x = X + \frac{3}{2}$$

$$y = Y - \frac{11}{8} \qquad \text{...(ii)}$$

Using these relations, Eq. (i) reduces to

$$X^2 = -2Y \qquad \text{...(iii)}$$

This is the form of $X^2 = -4aY$. On comparing, we get $4a = 2$, i.e. $a = \frac{1}{2}$.

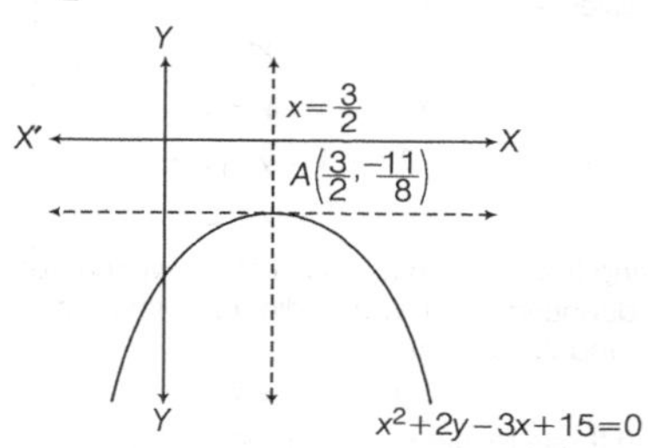

Vertex The coordinates of the vertex with respect to the new axes are $(X = 0, Y = 0)$. So, the coordinates of the vertex with respect to the old axes are $(3/2, -11/8)$. [putting $X = 0, Y = 0$ in Eq. (ii)]

Axis The equation of the axis of the parabola with respect to the new axes in $X = 0$. So, the equation of the axis with respect to the old axes is

$$x = 3/2 \qquad \text{[putting } X = 0 \text{ in Eq. (ii)]}$$

Focus The coordinates of the focus with respect to the new axes are $(X = 0, Y = -a)$, i.e. $(X = 0, Y = -1/2)$.

So, the coorinates of the focus with respect to the old axes are $(3/2, -15/8)$. [putting $X = 0, Y = -1/2$ in Eq. (ii)]

Directrix The equation of the directrix with respect to the new axes is $Y = a$, i.e. $Y = 1/2$. So, the equation of the directrix with respect to the old axes is $y = -7/8$. [putting $Y = 1/2$ in Eq. (ii)].

Latusrectum The length of the latusrectum of the given parabola is equal to $4a = 2$.

4. Let $P(x, y)$ be any point on the parabola whose focus is $(-1, -2)$ and the directrix $x - 2y + 3 = 0$. Draw PM perpendicular to $x - 2y + 3 = 0$.

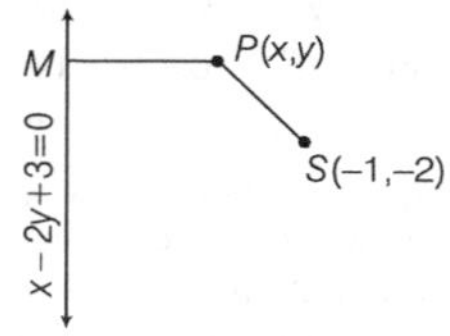

Then, $SP = SM$

$$\Rightarrow \quad SP^2 = SM^2$$

$$\Rightarrow \quad (x + 1)^2 + (y + 2)^2 = \left|\frac{x - 2y + 3}{\sqrt{1 + 4}}\right|^2$$

$$\Rightarrow \quad 5[x^2 + 2x + 1 + y^2 + 4y + 4] = x^2 + 4y^2 + 9 - 4xy - 12y + 6x$$

$$\therefore \quad 4x^2 + y^2 + 4xy + 4x + 32y + 16 = 0$$

5. In a parabola, vertex is the mid-point of the focus and the point of intersection of the axis and directrix. So, let (x_1, y_1) be the coordinates of the point of intersection of the axis and directrix. Then, $(2, -3)$ is the mid-point of the line segment joining $(0, 5)$ and (x_1, y_1).

$$\therefore \quad \frac{x_1 + 0}{2} = 2 \quad \text{and} \quad \frac{y_1 + 5}{2} = -3$$

$$\Rightarrow \quad x_1 = 4 \quad \text{and} \quad y_1 = -11$$

Let A be the vertex and S be the focus the required parabola. Then,

$$m_1 = \text{Slope of } AS = \frac{5 + 3}{0 - 2} = -4$$

Let m_2 be the slope of the directrix.

Then, $m_1 m_2 = -1 \Rightarrow m_2 = \frac{-1}{m_1} = \frac{1}{4}$

Thus, the directrix passes through $(4, -11)$ and has slope $\frac{1}{4}$.

Therefore, the equation of directrix is

$$y + 11 = \frac{1}{4}(x - 4) \Rightarrow x - 4y - 48 = 0$$

Let $P(x, y)$ be any point on the required parabola and let PM be the length of the perpendicular from P on the directrix. Then,

$$SP = PM$$

$$\Rightarrow \quad (SP)^2 = (PM)^2$$

$$\Rightarrow \quad (x - 0)^2 + (y - 5)^2 = \left(\frac{x - 4y - 48}{\sqrt{1^2 + (-4)^2}}\right)^2$$

$$\Rightarrow \quad 17[x^2 + y^2 + 25 - 10y] = x^2 + 16y^2 + 2304 - 8xy - 96x + 384y$$

$$\therefore \quad 16x^2 + y^2 + 8xy + 96x - 554y - 1879 = 0$$

6. Let (x_1, y_1) be the coordinates of the point of intersection of the axis and the directrix. Then, the vertex is the mid-point of the line segment joining (x_1, y_1) and the focus $(1, -1)$.

$$\therefore \quad \frac{x_1 + 1}{2} = 2 \quad \text{and} \quad \frac{y_1 + (-1)}{2} = 1$$

$$\Rightarrow \quad x_1 = 3, \quad y_1 = 3$$

Thus, the directrix meets the axis at (3, 3).

Let m_1 be the slope of the axis.

Then, m_1 = Slope of the line joining the focus and the vertex.

$$\Rightarrow \quad m_1 = \frac{1+1}{2-1} = 2 \quad \ldots\text{(ii)}$$

$\therefore$ Slope of the directrix = $-1/2$

[$\because$ directrix is perpendicular to the axis]

Thus, the directrix passes through (3, 3) and has slope $-1/2$.

So, its equation is

$$y - 3 = -\frac{1}{2}(x - 3)$$

$$\Rightarrow \quad x + 2y - 9 = 0$$

Let $P(x, y)$ be a point on the parabola. Then,

Distance of P from the focus = Distance of P from the directrix

$$\Rightarrow \quad \sqrt{(x-1)^2 + (y+1)^2} = \left|\frac{x + 2y - 9}{\sqrt{1^2 + 2^2}}\right|$$

$$\Rightarrow \quad (x-1)^2 + (y-1)^2 = \frac{(x + 2y - 9)^2}{5}$$

$$\Rightarrow \quad 5x^2 + 5y^2 - 10x + 10y + 10 = x^2 + 4y^2 + 81 + 4xy - 18x - 36y$$

$$\therefore \quad 4x^2 + y^2 - 4xy + 8x + 46y - 71 = 0$$

7. Let $P(x, y)$ be any point on the parabola, draw PN perpendicular to its axis and PM perpendicular to the tangent at the vertex.

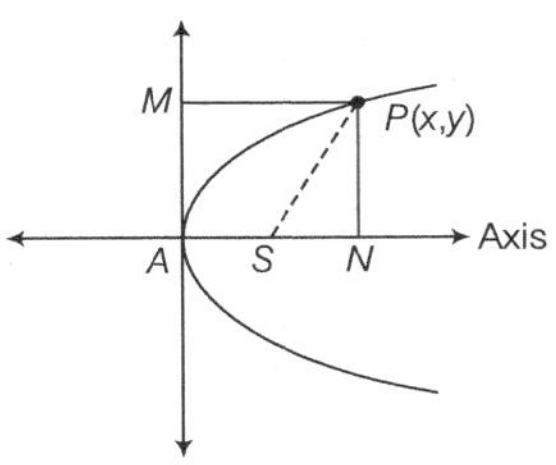

$$\text{Then,} \quad Y^2 = 4ax$$

$$\Rightarrow \quad PN^2 = \text{Latusrectum} \times PM \quad \ldots\text{(i)}$$

Now, PN = Length of perpendicular from $P(x, y)$ on the axis $3x + 4y - 4 = 0$

$$= \frac{3x + 4y - 4}{\sqrt{9 - 16}} = \frac{3x + 4y - 4}{5}$$

PM = Length of perpendicular from $P(x, y)$ on the tangent at the vertex A

$$= \frac{4x - 3y + 7}{\sqrt{16 + 9}} = \frac{4x - 3y + 7}{5}$$

Latusrectum = 4 [given]

From Eq. (i), $\left(\frac{3x + 4y - 4}{5}\right)^2 = 4\left(\frac{4x - 3y + 7}{5}\right)$

$$\therefore \quad (3x + 4y - 4)^2 = 20(4x - 3y + 7)$$

which is the required equation of parabola.

8. Let $P(x, y)$ be the coordinates of the other end of the chord OP..

Let the coordinates of mid-point of chord is (h, k). Then,

$$\frac{0 + x}{2} = h$$

$$\text{and} \quad \frac{0 + y}{2} = k$$

$$\Rightarrow \quad x = 2h$$

$$\text{and} \quad y = 2k$$

But (x, y) lies on the parabola.

$$\therefore \quad y^2 = 4ax$$

$$\Rightarrow \quad (2k)^2 = 4a(2h)$$

$$\Rightarrow \quad 4k^2 = 8ah$$

$$\Rightarrow \quad k^2 = 2ah$$

Hence, the required locus is $y^2 = 2ax$.

9. (a) Here, $h = at^2, \; k = \dfrac{4at - 2at}{3}$

$$h = at^2, \; k = \frac{2at}{3}$$

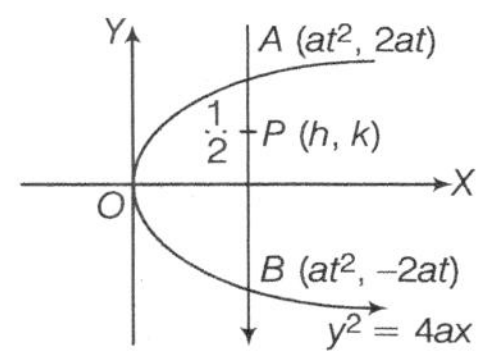

$$\therefore \quad t = \frac{3k}{2a} \text{ and } h = at^2$$

$$\Rightarrow \quad h = a \cdot \frac{9k^2}{4a^2}$$

$$\text{or} \quad 9k^2 = 4ah$$

$$\therefore \text{ locus, is} \quad 9y^2 = 4ax \qquad \text{where } 4a = l$$

$$\therefore \quad y^2 = \frac{l}{9}x$$

Hence, length of latusrectum = $l/9$

10. (d) $OC = r + 1 \;\Rightarrow\; OC = CP + 1$

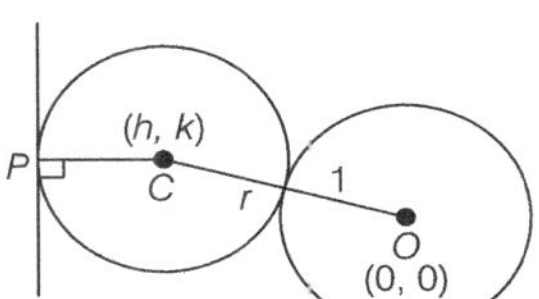

$$\Rightarrow \quad \sqrt{h^2 + k^2} = \frac{|3h - 4k - 10|}{5} + 1$$

$$\Rightarrow \quad \sqrt{h^2 + k^2} = \frac{3h - 4k - 5}{5}$$

$$\text{or} \quad \sqrt{h^2 + k^2} = \frac{-3h - 4k - 15}{5}$$

On squaring both sides, we get

$$h^2 + k^2 = \frac{9h^2 + 16k^2 + 25 - 24hk + 40k - 30h}{25}$$

$$\Rightarrow \quad 25h^2 + 25k^2 = 9h^2 + 16k^2 - 24hk - 30h + 40k + 25$$

$$\Rightarrow \quad 16h^2 + 9k^2 + 24hk + 30h - 40k - 25 = 0$$

$$\text{or} \quad 16x^2 + 24xy + 9y^2 + 30x - 40y - 25 = 0$$

Here, $a = 16, b = 9, h = 12$

and $h^2 = ab \;\Rightarrow\;$ locus of centre is parabola

11. (d) Since, $y = m(x - a)$ passes through the focus (0, 0) of the parabola. Thus, for this to be focal chord $m \in R - \{0\}$.

12. (d) Here, $(x-1)^2 = 8y$ has vertex $V(1, 0)$ and focus $F(1, 2)$

$\therefore$ Equation of circle,

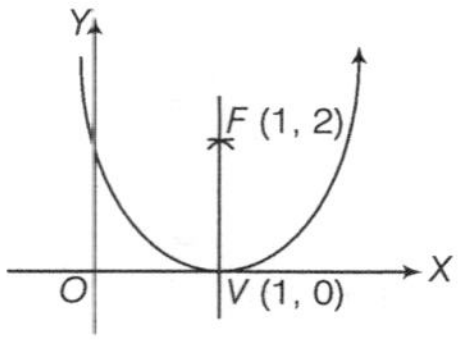

$$(x-1)^2 + (y-2)^2 = 2^2$$

$$\Rightarrow \quad x^2 + y^2 - 2x - 4y + 1 = 0$$

13. (b) We have, $x^2 - 2 = -2\cos t$ and $y = 4\cos^2\frac{t}{2}$

$$\Rightarrow \quad x^2 - 2 = -2\cos t \text{ and } y = 4\left[\frac{1+\cos t}{2}\right]$$

$$\Rightarrow \quad x^2 - 2 = -2\cos t \text{ and } y = 2(1+\cos t)$$

$$\Rightarrow \quad y = 2\left[1 + \frac{x^2-2}{-2}\right] \Rightarrow y = 2 - x^2 + 2$$

$$\Rightarrow \quad x^2 = 4 - y$$

$$\therefore \quad x^2 = -(y-4)$$

which is an equation of parabola having vertex at (0, 4).

14. (a) We have,

$$y^2 - 5y = 3x - 6 \quad \ldots \text{(i)}$$

For y-intercept, $x = 0$. By putting $x = 0$ in Eq. (i), we have

$$y^2 - 5y = -6$$

$$\Rightarrow \quad y^2 - 5y + 6 = 0$$

$$\Rightarrow \quad (y-2)(y-3) = 0$$

$$\Rightarrow \quad y = 2, 3$$

Hence, parabola intersect Y-axis at (0, 2) and (0, 3).

$$\therefore \quad Y\text{-intercept} = \sqrt{(0-0)^2 + (3-2)^2} = 1$$

15. (c) The equation of parabola is

$$y^2 - kx + 8 = 0$$

$$\Rightarrow \quad y^2 = k\left(x - \frac{8}{k}\right)$$

$$\Rightarrow \quad (y-0)^2 = k\left(x - \frac{8}{k}\right)$$

The equation of the directrix of this parabola is

$$x - \frac{8}{k} = \frac{-k}{4} \quad [\text{using, } x = -a]$$

$$\Rightarrow \quad x = \frac{8}{k} - \frac{k}{4}$$

But the equation of the directrix is given as $x - 1 = 0$.

$$\therefore \quad \frac{8}{k} - \frac{k}{4} = 1$$

$$\Rightarrow \quad k^2 + 4k - 32 = 0$$

$$\Rightarrow \quad k = -8, 4$$

16. (d) The equation of parabola is

$$\sqrt{(x-2)^2 + (y-3)^2} = \left|\frac{3x - 4y + 7}{\sqrt{(3)^2 + (-4)^2}}\right|$$

Clearly, its focus is (2, 3) and the equation of directrix is $3x - 4y + 7$.

$\therefore$ Length of the Latusrectum $= 2 \times$ Length of perpendicular from the focus (2, 3) on the directrix

$$\therefore \text{ Length of the Latusrectum} = 2\left|\frac{3 \times 2 - 4 \times 3 + 7}{\sqrt{(3)^2 + (-4)^2}}\right| = \frac{2}{5}$$

17. (a,b,c,d) Any point on the parabola is $P(2t^2, 4t)$.

Therefore, the mid-point of $S(2, 0)$ and $P(2t^2, 4t)$ is

$$R\left(\frac{2+2t^2}{2}, 2t\right) \equiv (h, k)$$

$$\therefore \quad h = \frac{2+2t^2}{2}, k = 2t$$

Eliminate t i.e. $\quad 2x = 2\left(1 + \frac{y^2}{4}\right) = 2 + \frac{y^2}{2}$

i.e. $\quad 4x = 4 + y^2$

i.e. $\quad y^2 = 4(x-1)$

It is a parabola with vertex at (1, 0) and latusrectum 2.

The directrix is $\quad x - 1 = -1, \Rightarrow x = 0$

The focus is $\quad x - 1 = 1, \Rightarrow x = 2$

i.e. its coordinates $\quad$ (2, 0).

18. (a,d) Here, Equation of VP is

$$\frac{y-0}{x-0} = \frac{2at}{at^2}$$

$$\Rightarrow \quad y = \frac{2}{t}x$$

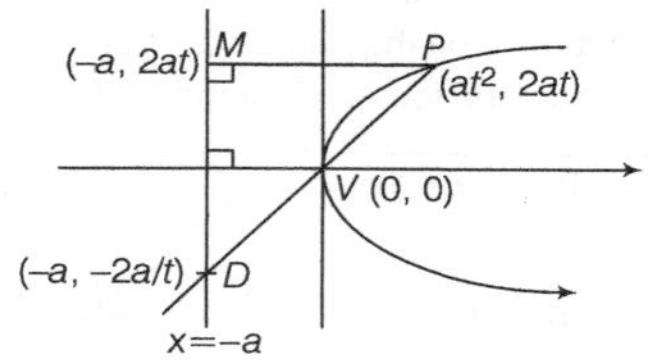

which meets directrix at $x = -a$

$$\therefore \quad y = -\frac{2a}{t}$$

$$D\left(-a, \frac{-2a}{t}\right)$$

Equation of circle having DM as diameter is

$$(x+a)(x+a) + \left(y + \frac{2a}{t}\right)(y - 2at) = 0$$

For point of intersection used with X-axis, put $y = 0$

$$\therefore \quad (x+a)^2 - 4a^2 = 0$$

$$x + a = \pm 2a, -2a$$

$$\therefore \quad x = a, -3a$$

19. (b,c,d) Here, $y = \tan(\tan^{-1} x) = x, x \in R$

Since, $PF = PM$

$\therefore$ locus of P is parabola.

Having, focus as (1, 0) and directrix as $x - y = 0$

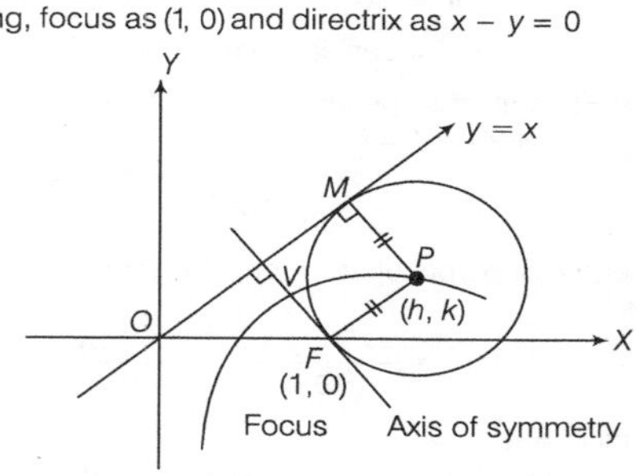

Distance between focus and directrix $= 2a$

$$\Rightarrow \quad 2a = \frac{|1-0|}{\sqrt{2}}$$

$$\text{Length of latusrectum} = 4a = \frac{2}{\sqrt{2}} = \sqrt{2}$$

Axis of symmetry is,

$$\frac{y-0}{x-1} = -1$$

$\Rightarrow \quad y = -x + 1$

or $\quad x + y = 1$

Vertex $\left(1 + \frac{1}{2\sqrt{2}}\cos 135°, 0 + \frac{1}{2\sqrt{2}}\sin 135°\right) = \left(\frac{3}{4}, \frac{1}{4}\right)$

20. (c,d) Let $A(t_1^2, 2t_1)$ and $B(t_2^2, 2t_2^2)$ be two points on $y^2 = 4x$. The coordinates of the centre of the circle are $\left(\frac{t_1^2 + t_2^2}{2}, t_1 + t_2\right)$.

Since, the circle with diameter AB touches the axis of the parabola and is of radius r.

$\therefore \quad t_1 + t_2 = \pm r$

$\therefore \quad$ Slope of $AB = \frac{2}{t_2 + t_1} = \pm\frac{2}{r}$

21. (a)

(i) $xy + a^2 - ax - ay = 0$

$\Rightarrow \quad x(y - a) - a(y - a) = 0$

$\Rightarrow \quad (x - a)(y - a) = 0$

$\therefore$ pair of straight line.

(ii) $2x^2 - 72xy + 23y^2 - 4x - 28y - 48 = 0$

Here, $a = 2, h = -36, b = 23$

$\Rightarrow \quad \Delta \neq 0$ and $h^2 > ab$

$\Rightarrow$ Hyperbola

(iii) $6x^2 - 5xy - 6y^2 + 14x + 5y + 4 = 0$

Here, $\Delta = abc + 2fgh - af^2 - bg^2 - ch^2$

$\Delta = 0$

$\Rightarrow$ pair of straight lines

(iv) $14x^2 - 4xy + 11y^2 - 44x - 58y + 71 = 0$

Here, $\Delta \neq 0$

and $h^2 < ab \Rightarrow$ Ellipse

(v) $4x^2 - 4xy + y^2 - 12x + 16y + 9 = 0$

Here, $\quad \Delta = 0$

$\Rightarrow$ pair of straight lines

Explanations for Q. Nos. (22-25)

Given equation of curve is

$$S : 9x^2 - 24xy + 16y^2 - 20x - 15y - 60 = 0$$

On comparing with $ax^2 + 2hxy + by^2 + 2gx + 2fy + c = 0$, we have

$$a = 9, h = -12, b = 16, g = -10, f = -\frac{15}{2}, c = -60$$

$\therefore \quad \Delta = abc + 2fgh - af^2 - bg^2 - ch^2$

$$= 9 \times 16 \times (-60) + 2\left(\frac{-15}{2}\right) \times (-10) \times (-12)$$

$$- 9\left(\frac{-15}{2}\right)^2 - 16(-10)^2 - (-60)(-12)^2$$

$$= -8640 - 1800 - \frac{2025}{4} - 1600 + 8640$$

$\Rightarrow \quad \Delta \neq 0$

$\therefore \quad h^2 - ab = (-12)^2 - (9)(16) = 0$

Hence, the given equation is a parabola.

Now, $9x^2 - 24xy + 16y^2 - 20x - 15y - 60 = 0$

$\Rightarrow \quad (3x - 4y)^2 = 20x + 15y + 60$

$\Rightarrow \quad (3x - 4y)^2 = 5(4x + 3y + 12)$

$\Rightarrow \quad \left(\frac{3x - 4y}{\sqrt{3^2 + 4^2}}\right)^2 = \frac{4x + 3y + 12}{5} \quad \ldots(i)$

Let $\frac{3x - 4y}{5} = y$ and $\frac{4x + 3y + 12}{5} = x$

$\therefore$ Eq. (i) reduces to $\quad y^2 = x \quad \ldots(ii)$

Comparing this equation with $y^2 = 4ax$, we obtain that $4a = 1$

$\Rightarrow \quad a = \frac{1}{4}$

$\therefore$ latusrectum = 1

Axis is $y = 0 \Rightarrow 3x - 4y = 0$

Vertex is $x = 0, y = 0$

$\Rightarrow \quad 4x + 3y + 12 = 0$ and $3x - 4y = 0$

On solving, $\quad x = -\frac{48}{25}, y = -\frac{36}{25}$

i.e. vertex $\left(\frac{-48}{25}, \frac{-36}{25}\right)$

Directrix is $\quad x = -\frac{1}{4}$

$\Rightarrow \quad \frac{4x + 3y + 12}{5} = -\frac{1}{4}$

or $\quad 4x + 3y + \frac{53}{4} = 0$

22. (b) **23.** (a) **24.** (c) **25.** (a)

DPP-2 Equation of Chord, Focal Chord and Its Properties

1. (a) Let $P(at_1^2, 2at_1)$ and $Q(at_2^2, 2at_2)$ be the end points of a focal chord PQ which makes an angle α with the axis of parabola.

Then, $\quad PQ = a(t_2 - t_1)^2$

Also, $\quad \tan\alpha =$ Slope of PQ

$\Rightarrow \quad \tan\alpha = \frac{2}{t_2 + t_1}$

$\Rightarrow \quad t_1 + t_2 = 2\cot\alpha$

$\therefore$ Facal chord, $PQ = a(t_2 - t_1)^2$

$$= a\{(t_2 + t_1)^2 - 4t_1 t_2\}$$

$$= a\{4\cot^2\alpha + 4\}$$

$[\because t_2 + t_1 = 2\cot\alpha$ and $t_1 t_2 = -1]$

$$= 4a\,\text{cosec}^2\alpha$$

2. (c) If the point $(-2a, a + 1)$ lies in the shaded region as shown below.

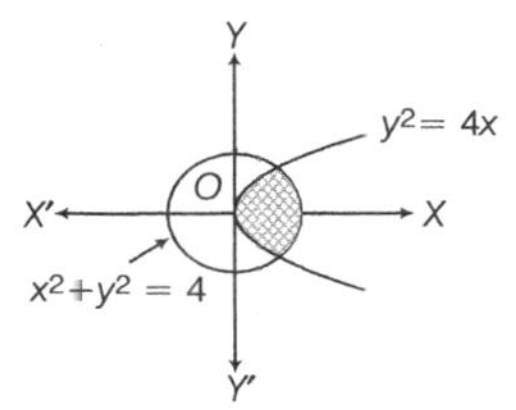

Then, $\quad 4a^2 + (a + 1)^2 - 4 < 0$

and $\quad (a + 1)^2 - 4(-2a) < 0$

$\Rightarrow \quad 5a^2 + 2a - 3 < 0$

and $\quad a^2 + 10a + 1 < 0$

$\Rightarrow \quad (a + 1)(5a - 3) < 0$

and $\quad (a + 5 - 2\sqrt{6})(a + 5 + 2\sqrt{6}) < 0$

$$\Rightarrow \quad -1 < a < \frac{3}{5}$$

and $$-5 - 2\sqrt{6} < a < -5 + 2\sqrt{6}$$

$$\Rightarrow \quad -1 < a < -5 + 2\sqrt{6}$$

$$\Rightarrow \quad a \in (-1, -5 + 2\sqrt{6})$$

3. (d) The equation of circle described on PQ as diameter is

$$(x - at^2)\left(x - \frac{a}{t^2}\right) + (y - 2at)\left(y + \frac{2a}{t}\right) = 0$$

Solving it with $x = -a$, we have

$$(-a - at^2)\left(-a - \frac{a}{t^2}\right) + (y - 2at)\left(y + \frac{2a}{t}\right) = 0$$

$$\Rightarrow \quad y^2 - 2a\left(t - \frac{1}{t}\right)y + a^2\left(t - \frac{1}{t}\right) = 0$$

which is a perfect square.
Hence, $x = -a$ touches the circle.

4. (a) If $A(at_1^2, at_1)$, $B(at_2^2, 2at_2)$ is the extremities of a focal chord for the parabola $y^2 = 4ax$, then $t_1 t_2 = -1$

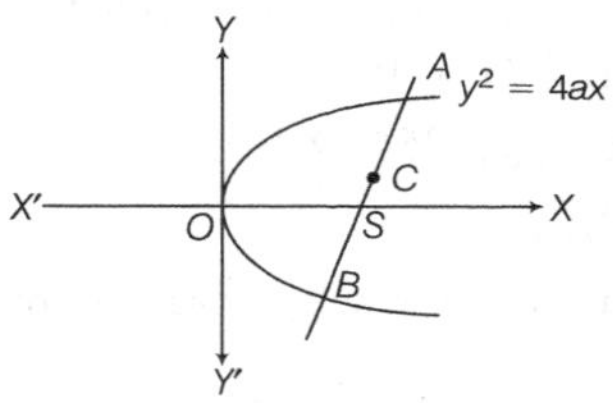

We want to find locus of point $C(\alpha, \beta)$, where C is the centre of the circle having AB as the diameter.

$$\Rightarrow \quad \alpha = \frac{a}{2}(t_1^2 + t_2^2) \quad \text{and} \quad \beta = a(t_1 + t_2)$$

By eliminating t_1 and t_2, we get

$$\beta^2 = a^2(t_1^2 + t_2^2 + 2t_1 t_2)$$

$$\Rightarrow \quad \beta^2 = a^2\left(\frac{2\alpha}{a} - 2\right)$$

$$\Rightarrow \quad \beta^2 = 2a(\alpha - a)$$

$$\therefore \quad y^2 = 2a(x - a)$$

5. (c) The length of focal chord making an angle α with the X-axis is $4a\operatorname{cosec}^2\alpha$. For $\alpha \in [0, \pi/4]$,
its maximum length is $4a \times 2 = 8a$ units.

6. (a) For $y^2 = 32x$, $a = 8$.

Then, any point on the parabola is $(8t^2, 16t)$ comparing this with the point $P(2, -8)$, we get $t = -\frac{1}{2}$.

Now, if PQ is the focal chord and point Q is $(8t_1^2, 16t_1)$, then

$t_1 = -\frac{1}{t} = = 2.$

Hence, point Q has coordinates (32, 32).

7. (c) A parabola and a circle can intersect in at most 4 distinct points and from these 4 points, we get

$${}^4C_2 = 6 \text{ common chords.}$$

8. (d) $\tan\theta = \dfrac{t_1/2}{t_1^2/4} = \dfrac{2}{t_1}$

$$\tan(90° - \theta) = \frac{2}{t_2}$$

$$\therefore \quad t_1 = 2\cot\theta$$

and $$t_2 = 2\tan\theta \qquad \text{(i)}$$

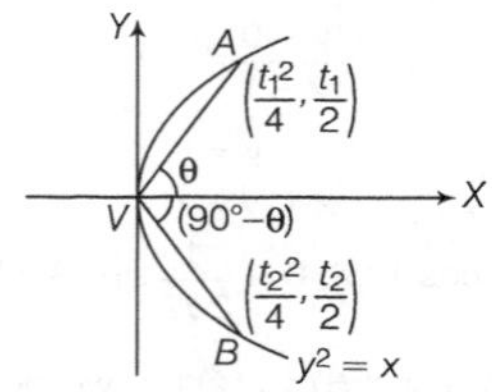

Now, $$\left|\frac{VA}{VB}\right| = \frac{\sqrt{\dfrac{t_1^4}{16} + \dfrac{t_1^2}{4}}}{\sqrt{\dfrac{t_2^4}{16} + \dfrac{t_2^2}{4}}} = \frac{\sqrt{\cot^4\theta + \cot^2\theta}}{\sqrt{\tan^4\theta + \tan^2\theta}}$$

$$= \frac{\cot\theta \cdot \operatorname{cosec}\theta}{\tan\theta \cdot \sec\theta} = \cot^3\theta$$

9. (d) Let $P(at_1^2, 2at_1)$ and $Q(at_2^2, 2at_2)$ be the end points of a focal chord of the parabola $y^2 = 4ax$. Then,

$$PQ = a(t_2 - t_1)^2$$

The equation of PQ is

$$(t_1 + t_2)y = 2x - 2a. \qquad [\because t_1 t_2 = -1]$$

It is given that $PQ = c$ and it is at a distance b from the vertex.

$$\therefore \quad a(t_2 - t_1)^2 = c$$

and $$\left|\frac{-2a}{\sqrt{(t_1 + t_2)^2 + 4}}\right| = b$$

$$\Rightarrow \quad (t_2 - t_1)^2 = \frac{c}{a}$$

and $$\frac{2a}{t_2 - t_1} = b \qquad [\because t_1 t_2 = 1]$$

$$\Rightarrow \quad \left(\frac{2a}{b}\right)^2 = \frac{c}{a}$$

$$\therefore \quad 4a^3 = b^2 c$$

10. (b) Tangent at point P is

$$ty = x + at^2 \qquad \text{...(i)}$$

The line perpendicular to Eq. (i) passes through $(a, 0)$. Therefore,

$$y - 0 = -t(x - a)$$

or $$tx + y = ta \qquad \text{...(ii)}$$

or $$y = t(a - x)$$

The equation of OP is

$$y - \frac{2}{t}x = 0$$

or $$y = \frac{2}{t}x \qquad \text{...(iii)}$$

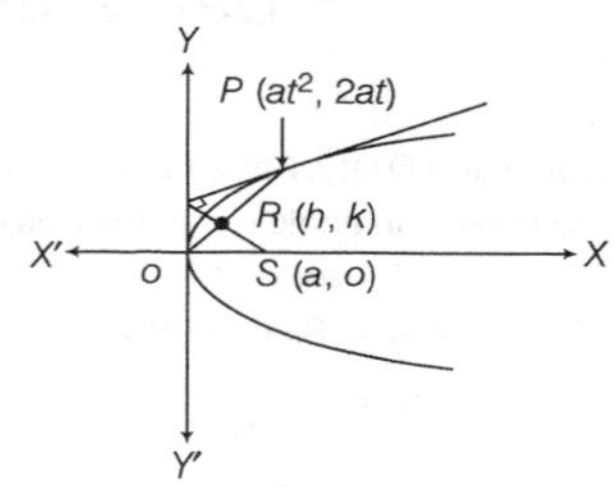

From Eqs. (ii) and (iii) eliminating t, we get

$$y^2 = 2x(a - x)$$

$$\therefore \quad 2x^2 + y^2 - 2ax = 0$$

11. (a) Here, $M_{OA} \cdot M_{OB} = -1$

$$\Rightarrow \quad \frac{2}{t_1} \cdot \frac{2}{t_2} = -1$$

$$\Rightarrow \quad t_1 t_2 = -4 \quad \ldots \text{(i)}$$

$$M_{OP} \cdot M_{AB} = -1$$

$$\Rightarrow \quad \frac{k}{h} \cdot \frac{2}{t_1 + t_2} = -1$$

$$\Rightarrow \quad t_1 + t_2 = \frac{-2k}{h} \quad \ldots \text{(ii)}$$

Equation of chord AB, is

$y(t_1 + t_2) = 2x + 2at_1t_2$, passes through (h, k)

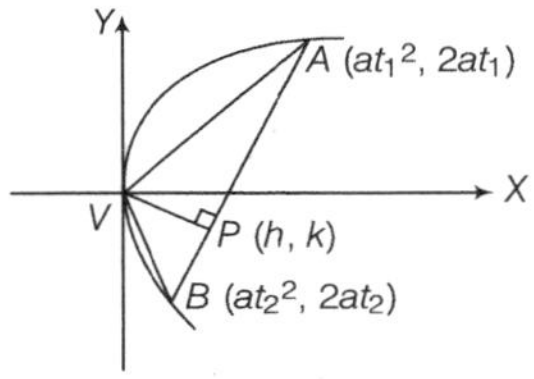

$$\Rightarrow \quad k(t_1 + t_2) = 2h + 2at_1t_2 \quad \ldots \text{(iii)}$$

From Eqs. (i), (ii) and (iii)

$$k \cdot \left(\frac{-2k}{h}\right) = 2h - 8a$$

$$\Rightarrow \quad -y^2 = x^2 - 4ax$$

or $\quad x^2 + y^2 - 4ax = 0$ is required locus.

12. (c) Let PQ be a double ordinate of $x^2 = 4by$ and let $R(h, k)$ be a point of trisection. Let the coordinates of P be (x, y). Then,

$$x = 3k \text{ and } y = h$$

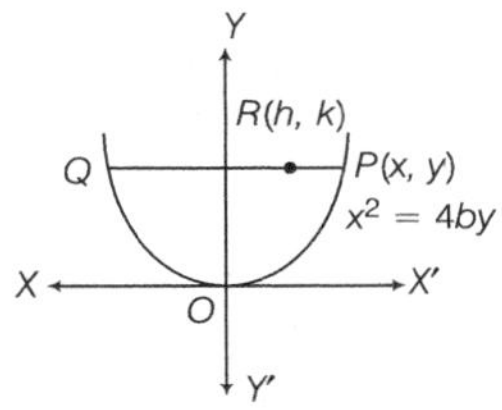

But (x, y) lies on $x^2 = 4by$

$$\therefore \quad 9k^2 = 4bh$$

Hence, the locus of (h, k) is $9x^2 = 4by$.

13. (d) Let $P(at^2, 2at)$ be a point on the parabola $y^2 = 4ax$ and S be the focus of the parabola.

Then, $\quad SP = a + at^2$

[$\because$ focal distance $= x + a$]

Clearly, SP is least for $t = 0$.

Hence, the abscissae of P is

$$at^2 = a \times 0 = 0$$

14. (b) Let A be the vertex and AP be a chord of $x^2 = 4ay$ such that slope of AP is $\tan\alpha$. Let coordinates of P be $(2at, at^2)$. Then,

$$\text{Slope of } AP = \frac{at^2}{2at} = \frac{t}{2}$$

$$\tan\alpha = \frac{t}{2}$$

$$\Rightarrow \quad t = 2\tan\alpha$$

Now, $\quad AP = \sqrt{(2at - 0)^2 + (at^2 - 0)^2} = at\sqrt{4 + t^2}$

$$= 2a\tan\alpha\sqrt{4 + 4\tan^2\alpha}$$

$$= 4a\tan\alpha\sec\alpha$$

15. (b) Since, $\quad (FT)^2 = (FP)(FQ)$

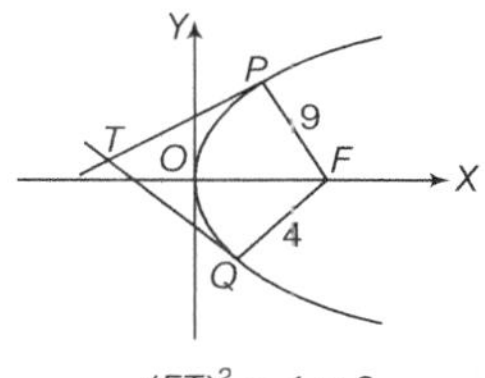

$$\Rightarrow \quad (FT)^2 = 4 \times 9$$

$$FT = 6$$

16. (a) We know that the semi-latusrectum of a parabola is the harmonic mean of the segments of a focal chord.

$$\therefore \quad 2a = \frac{2bk}{b + k}$$

$$\Rightarrow \quad k = \frac{ab}{b - a}$$

17. (c) Parabolas $y^2 = 4ax$ and $x^2 = 4ay$ intersect at points $O\ (0,0)$ and $P\ (4a, 4a)$.

It is given that the line $2bx + 3cy + 4d = 0$ passes through O and P.

$$\therefore \quad 4d = 0$$

and $\quad 8ab + 12ac + 4d = 0$

$$\Leftrightarrow \quad d = 0 \text{ and } 2b + 3c = 0$$

$$\Leftrightarrow \quad d^2 + (2b + 3c)^2 = 0$$

18. (d) We know that the vertex of the parabola given by $y = ax^2 + bx + c$ is at $\left(\frac{-b}{2a}, \frac{-D}{4a}\right)$ So, the coordinates of the vertex of the parabola $y = \frac{a^3x^2}{3} + \frac{a^2x}{3} - 2a$ are $\left(\frac{-3}{4a}, \frac{-35a}{16}\right)$.

Thus, if (h, k) are the coordinates of the vertex .

Then, $\quad h = -\frac{3}{4a}$ and $k = -\frac{35a}{16}$

$$\Rightarrow \quad hk = \frac{105}{64}$$

Hence, the locus of (h, k) is $xy = \frac{105}{64}$.

19. (a) The focus of the parabola $y^2 = 2px$ is at $(p/2, 0)$ and its directrix is $x = -p/2$.

It is given that the circle touches the directrix.

$\therefore$ Radius = Distance between the point $(p/2, 0)$ and the line $x = -p/2$.

$$\Rightarrow \quad \text{Radius} = p$$

So, the equation of circle is

$$\left(x - \frac{p}{2}\right)^2 + (y - 0)^2 = p^2$$

$$\Rightarrow \quad x^2 - y^2 - px - \frac{3p^2}{4} = 0$$

On solving this equation with $y^2 = 2px$, we get

$$x^2 + px - \frac{3p^2}{4} = 0$$

$$\Rightarrow \quad (2x - p)(2x + 3p) = 0$$

$$\Rightarrow \quad x = \frac{p}{2}, x = -\frac{3p}{2}$$

Putting $x = p/2$ is $y^2 = 2px$, we obtain $y = \pm p$.

For $x = -3p/2$, y is imaginary.

Hence, the circle and the parabola intersect at $(p/2, p)$ and $(p/2, -p)$.

Explanations for Q. Nos. (20-22)

Let $y = m_1x$ and $y = m_2x$ be the equations of chords OP and OQ respectively, such that $OP \perp OQ$

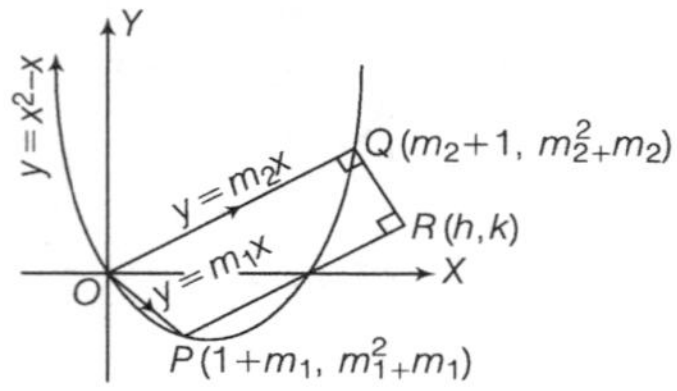

$\therefore \quad m_1m_2 = -1$

Solving $y = m_1x$ with the equation of the parabola $y = x^2 - x$, we obtain that the coordinates of P are $(1 + m_1, m_1 + m_1^2)$

Similarly, $Q\,(1 + m_2, m_2 + m_2^2)$

Let the coordinates of R be (h, k).

Since, diagonals of a rectangle bisects each other.

$\therefore \quad \frac{h}{2} = \frac{m_1 + m_2 + 2}{2}$ and $\frac{k}{2} = \frac{m_1^2 + m_2^2 + m_1 + m_2}{2}$

$\Rightarrow \quad m_1 + m_2 = h - 2$ and $m_1^2 + m_2^2 + m_1 + m_2 = k$

$\Rightarrow m_1 + m_2 = h - 2$ and $(m_1 + m_2)^2 - 2m_1m_1 + (m_1 + m_2) = k$

$\therefore \quad (h-2)^2 + 2 + (h-2) = k, \{m_1m_2 = -1\}$

$\Rightarrow \quad h^2 - 4h + 4 + h = k$

or $\quad h^2 - 3h + 4 = k$

Hence, the locus of (h, k) is $y = x^2 - 3x + 4$.

Hence, $y = x^2 - 3x + 4$

$\Rightarrow \quad y = x^2 - 3x + \frac{9}{4} - \frac{9}{4} + 4$

or $\quad y = \left(x - \frac{3}{2}\right)^2 + \frac{7}{4}$

or $\quad \left(y - \frac{7}{4}\right) = \left(x - \frac{3}{2}\right)^2$

$\therefore$ Length of latusrectum = 1 and axis is

$$x - \frac{3}{2} = 0$$

or $\quad x = \frac{3}{2}$

20. (b) **21.** (a) **22.** (a)

23. (c) Given, OP and OQ are at right angles.

$\therefore \quad t_1t_2 = -4$

Equation of PQ is

$$y(t_1 + t_2) - 2x - 2at_1t_2 = 0$$

It cuts the X-axis $y = 0$, where $2x = -2at_1t_2 = 8a$

$\therefore x = 4$, as $a = 1$.

Hence, the fixed point is (4, 0).

24. (c) If (x, y) is the mid-point of PQ, then

$$2x = a(t_1^2 + t_2^2)$$

and $\quad 2y = 2a(t_1 + t_2)$

$\Rightarrow \quad 2x = a[(t_1 + t_2)^2 - 2t_1t_2]$

$$= a\left[\frac{y^2}{a^2} + 8\right]$$

$\Rightarrow \quad 2ax = y^2 + 8a^2$

$\therefore \quad y^2 = 2(x - 4)$

25. (d) **Directrix** $x - 4 = -\frac{1}{2} \Rightarrow x = \frac{7}{2}$.

26. (2) Equation of chord passing through $P\,(4a, 0)$ is

$$y = m(x - 4a) \quad \text{...(i)}$$

where, m is the shape of the chord.

Converting $y^2 = 4ax$ into homogeneous from chord (i)

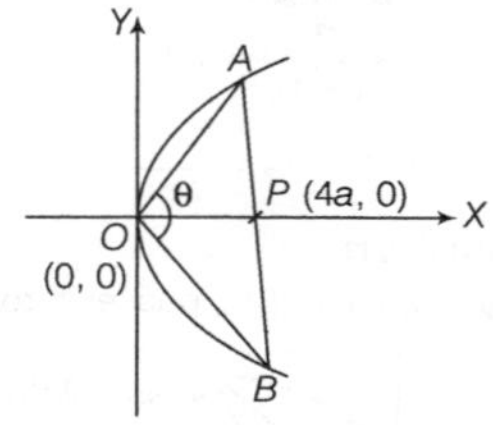

$\Rightarrow \quad y^2 = 4ax\left(\frac{mx - y}{4a}\right)$

$\Rightarrow \quad my^2 - mx^2 + xy = 0$

Clearly, coefficient of x^2 + coefficient of $y^2 = 0 \Rightarrow \theta = 90^\circ$

$\therefore \quad 2\sin^2\theta + 4\cos^2\theta = 2$

27. (4) Length of the perpendiculars, $l_1 = \frac{4a\cos\theta}{\sin^2\theta}$

Replacing θ by $90^\circ - \theta$ for perpendicular chord, $l_2 = \frac{4a\sin\theta}{\cos^2\theta}$

$\therefore \quad l_1l_2 = \frac{16a^2}{\sin\theta\cos\theta}$

$\therefore \quad (l_1l_2)^{4/3} = \frac{(4a)^{8/3}}{(\sin\theta\cos\theta)^{4/3}} \quad \text{... (i)}$

Now, $\quad l_1^{2/3} = \left(\frac{4a\cos\theta}{\sin^2\theta}\right)^{2/3} = (4a)^{2/3}\frac{\cos^{2/3}\theta}{\sin^{4/3}\theta} \quad \text{... (ii)}$

$\Rightarrow \quad l_2^{2/3} = \left(\frac{4a\sin\theta}{\cos^2\theta}\right)^{2/3} = (4a)^{2/3}\frac{\sin^{2/3}\theta}{\cos^{4/3}\theta} \quad \text{... (iii)}$

Now, given that

$$(l_1l_2)^{4/3} = 4a^2\lambda \cdot (l_1^{2/3} + l_2^{2/3})$$

$\Rightarrow \quad \frac{(4a)^{8/3}}{(\sin\theta\cos\theta)^{4/3}} = 4a^2\lambda\left[(4a)^{2/3}\frac{\cos^{2/3}\theta}{\sin^{4/3}\theta} + (4a)^{2/3}\frac{\sin^{2/3}\theta}{\cos^{4/3}\theta}\right]$

[from Eqs. (i), (ii) and (iii)]

$\Rightarrow \quad \frac{(4a)^{8/3}}{(\sin\theta\cos\theta)^{4/3}} = 4a^2\lambda \cdot (4a)^{2/3}\left[\frac{\cos^{2/3}\theta}{\sin^{4/3}\theta} + \frac{\sin^{2/3}\theta}{\cos^{4/3}\theta}\right]$

$\Rightarrow \quad \frac{(4a)^{8/3}}{(\sin\theta\cos\theta)^{4/3}} = 4a^2\lambda \cdot (4a)^{2/3}\left[\frac{1}{(\sin\theta\cos\theta)^{4/3}}\right]$

$\Rightarrow \quad (4a)^2 = 4a^2\lambda$

$\therefore \quad \lambda = 4$

28. (2) Difference of the ordinates

$$d = \left|2at + \frac{2a}{t}\right| = 2a\left|t + \frac{1}{t}\right|$$

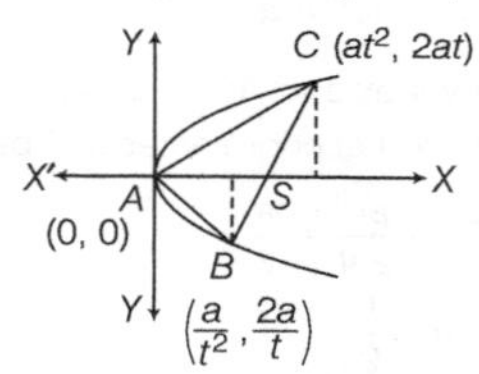

Now, $\quad \text{area}(\Delta) = \frac{1}{2}\begin{vmatrix} at^2 & 2at & 1 \\ a/t^2 & -2a/t & 1 \\ 0 & 0 & 1 \end{vmatrix} = a^2\left[t + \frac{1}{t}\right]$

$\Rightarrow \quad 2a\left(t + \frac{1}{t}\right) = \frac{2\Delta}{a}$

But according to the question, the distance of B and C from the axis of the parabola is $\lambda \frac{\Delta}{a}$.

Now, $$\frac{\lambda\Delta}{a} = \frac{2\Delta}{a}$$

$\therefore$ $$\lambda = 2$$

29. (3) PQ being focal chord, $t_1 t_2 = -1$ or $t_2 = \frac{-1}{t_1}$. Let (x, y) be the centroid of ΔAPQ, then

$$3x = a\left(t_1^2 + \frac{1}{t_1^2}\right)$$

and $$3y = 2a\left(t_1 - \frac{1}{t_1}\right)$$

On eliminating t_1, we get

$$9y^2 = 4a\,(3x - 2a)$$

$\Rightarrow$ $$y^2 = \frac{4a \cdot 3}{9}\left(x - \frac{2a}{3}\right)$$

Hence, it is a parabola whose latusrectum is $\frac{12}{9}a$.

But according to the question,

$$\text{latusrectum} = \frac{4a}{\lambda}$$

Then, $$\frac{4a}{\lambda} = \frac{12a}{9}$$

$\therefore$ $$\lambda = 3$$

30. (4) Equation of chord PQ, is

$$y\,(t_1 + t_2) = 2x + 2at_1t_2 \qquad \ldots\text{(i)}$$

As it passes through focus $(a, 0)$

$\Rightarrow$ $$t_1t_2 = -1$$

Now, $$P = \left|\frac{2at_1t_2}{\sqrt{4 + (t_1 + t_2)^2}}\right|, a = 1$$

$$= \left|\frac{2\,t_1t_2}{\sqrt{4 + (t_1 + t_2)^2}}\right|$$

$$= \left|\frac{-2}{\sqrt{(t_1 + t_2)^2 - 4t_1\,t_2}}\right|$$

$$= \frac{2}{|t_1 - t_2|} \qquad \ldots\text{(ii)}$$

We have, $$l = PQ = a\,(t_2 - t_1)^2 = (t_1 - t_2)^2$$

From Eq. (ii),

$$l = \left(\frac{2}{p}\right)^2 = \frac{4}{p^2} = \frac{\lambda}{p^2}$$

$\Rightarrow$ $$\lambda = 4$$

31. (7) The coordinate of Q are $\left(\frac{t^2}{2}, 2t\right)$.

$$R\left(\frac{1}{t^2}, \frac{-2}{t}\right)$$

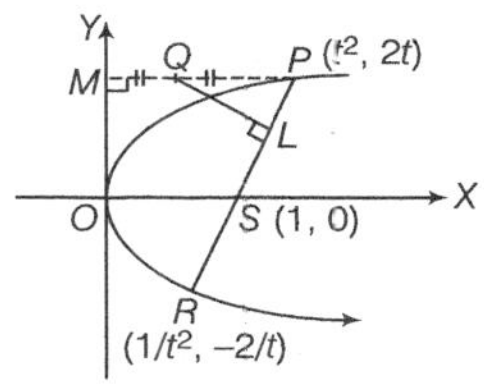

Equation of PR, is

$$y\left(t - \frac{1}{t}\right) = 2x - 2$$

or $$2x - y\left(t - \frac{1}{t}\right) - 2 = 0$$

Now, $$QL = \left|\frac{t^2 - 2t\left(t - \frac{1}{t}\right) - 2}{\sqrt{4 + \left(t - \frac{1}{t}\right)^2}}\right|$$

$$= \frac{t^2}{t + \frac{1}{t}} = \frac{t^3}{t^2 + 1}$$

$\therefore$ $$\frac{d}{dt}(QL) = \frac{t^4 + 3t^2}{(t^2 + 1)^2} > 0, \text{ for all } t \in [1, 3]$$

$\therefore$ QL is maximum when $t = 3$.

The maximum value of QL is given by

$$QL = \frac{3^3}{3^2 + 1} = \frac{27}{10} = \frac{\lambda}{\mu}$$

$\therefore$ $$\lambda - 2\mu = 7$$

DPP-3 Equation of Tangent and Its Properties

1. (d) If $y = 2x - 3$ touches the parabola $y^2 = 4a\left(x - \frac{1}{3}\right)$, then the equation

$$(2x - 3)^2 = 4a\left(x - \frac{1}{3}\right)$$

$\Rightarrow$ $$4x^2 - 12x + 9 = 4ax - \frac{4a}{3}$$

$\Rightarrow$ $$12x^2 - 36x + 27 = 12ax - 4a$$

$\Rightarrow$ $$12x^2 - (36 + 12a)x + 27 + 4a = 0$$

The above equation must have equal roots.

$\therefore$ $$(36 + 12a)^2 - 4 \times 12(27 + 4a) = 0$$

$$1296 + 144a^2 + 864a - 1296 - 192a = 0$$

$\Rightarrow$ $$144a^2 + 672a = 0$$

$\Rightarrow$ $$a(144a + 672) = 0$$

$\therefore$ $$a = 0, \frac{-14}{3}$$

2. (d) Given equation of parabola is

$$y = (x - 3)^2 \qquad \ldots\text{(i)}$$

On differentiating Eq. (i), we get

$\therefore$ $$y' = 2(x - 3)$$

Since, tangent is parallel to the chord joining the points (3,0) and (4,1).

$\therefore$ $$2(x - 3) = \frac{1 - 0}{4 - 3}$$

$\Rightarrow$ $$x - 3 = \frac{1}{2} \quad \Rightarrow \quad x = \frac{7}{2}$$

On putting $x = \frac{7}{2}$ in Eq. (i), we get

$$y = \frac{1}{4}$$

Hence, equation of tangent is $y - \frac{1}{4} = 1\left(x - \frac{7}{2}\right)$

$\Rightarrow$ $$4x - 4y = 13$$

3. (a) Combined equation of the pair of tangents drawn from an external point (x_1, y_1) to the parabola $y^2 = 4ax$ is

$$(y^2 - 4ax)(y_1^2 - 4ax_1) = [yy_1 - 2a(x + x_1)]^2$$

Hence, equation of the combined tangent is

$$(y^2 - 8x)[0 - 8(-2)] = [y \times 0 - 4(x - 2)]^2$$

$\Rightarrow \quad 16(y^2 - 8x) = 16(x - 2)^2$

$\Rightarrow \quad y^2 - 8x = (x - 2)^2$

$\Rightarrow \quad y^2 - 8x = x^2 - 4x + 4$

$\Rightarrow \quad y^2 = x^2 + 4x + 4$

$\Rightarrow \quad y^2 = (x + 2)^2$

$\therefore \quad y = \pm (x + 2)$

4. (b) The equation of any tangent of slope m to the parabola $y^2 = 4ax$ is $y = mx + \frac{a}{m}$,

if it passes through $P(-a, 2a)$, then

$$2a = -ma + \frac{a}{m}$$

$\Rightarrow \quad 2am = -m^2a + a$

$\Rightarrow \quad m^2a + 2am - a$

Let m_1 and m_2 be the roots of equation.

Then, $\quad m_1 + m_2 = \frac{-2a}{a} = -2$

and $\quad m_1m_2 = \frac{-a}{a} = -1$

Let θ be the angle between the tangents drawn from P. then

$$\tan\theta = \frac{m_1 - m_2}{1 + m_1m_2} = \frac{\sqrt{(m_1 + m_2)^2 - 4m_1m_2}}{1 + m_1m_2}$$

$$= \frac{\sqrt{(-2)^2 - 4(-1)}}{1 + (-1)} = \frac{\sqrt{4 + 4}}{0} = \infty$$

$\Rightarrow \quad \tan\theta = \infty$

$\therefore \quad \theta = \frac{\pi}{2}$

5. (c) Equation of chord of contact of tangents from $T(h, k)$ is

$$yk = 2(x + h) \quad \text{...(i)}$$

Also, equation of chord is

$$4x - 7y + 10 = 0 \quad \text{...(ii)}$$

Since, Eqs. (i) and (ii) are same.

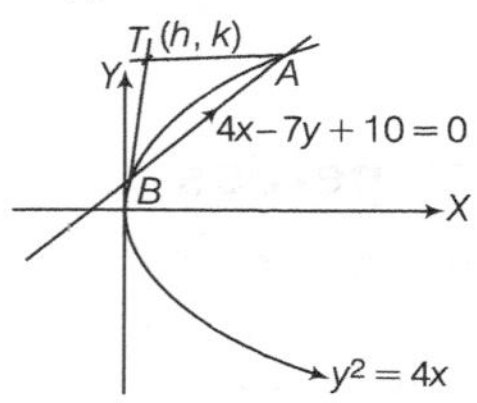

$\therefore \quad \frac{2}{4} = \frac{-k}{-7} = \frac{2h}{10}$

$\therefore \quad h = \frac{5}{2}, k = \frac{7}{2}$

$\Rightarrow$ Point of intersection point of tangents is $\left(\frac{5}{2}, \frac{7}{2}\right)$.

6. (c) The equation of tangent to $y^2 = 4x$ having slope m is

$$y = mx + \frac{1}{m}$$

It will touch $\quad x^2 = 4y,$

if $x^2 = 4\left(mx + \frac{1}{m}\right)$ has equal roots.

$\Rightarrow \quad x^2 = 4mx + \frac{4}{m}$ has equal roots.

$\Rightarrow \quad x^2 - 4mx - \frac{4}{m} = 0$ has equal roots.

Now, $\quad (-4m)^2 - 4\left(\frac{-4}{m}\right) = 0 \quad [\because D = 0]$

$\Rightarrow \quad 16m^2 + \frac{16}{m} = 0$

$\Rightarrow \quad 16m^2 = \frac{-16}{m}$

$\Rightarrow \quad m^3 = -1$

$\Rightarrow \quad m = -1$

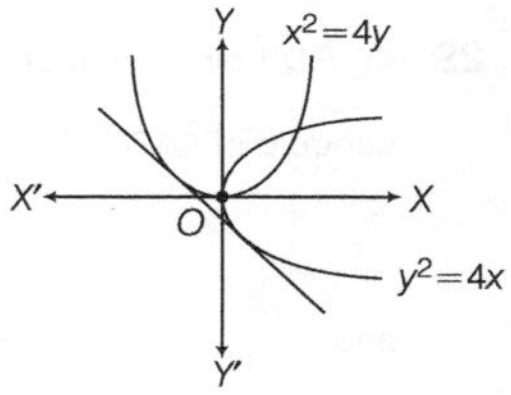

Hence, the equation of common tangent is

$$y = (-1)x + \frac{1}{(-1)}$$

$\Rightarrow \quad y = -x - 1$

$\Rightarrow \quad x + y + 1 = 0$

7. (b) Let the given points B and C be $(0, 0)$ and $(a, 0)$. According to the given condition,

$$\tan\alpha + \tan\beta = m \quad \text{[where, } m \text{ is a constant]}$$

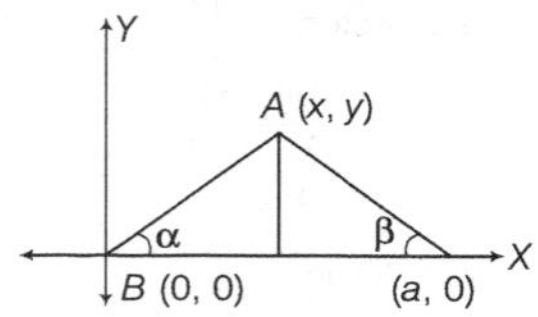

$\Rightarrow \quad \frac{y}{x} + \frac{y}{a - x} = m$

$\Rightarrow \quad y\left[\frac{a}{x(a - x)}\right] = m$

$\Rightarrow \quad ay = mx(a - x)$

which is the equation of a parabola.

8. (b) Equation of PN : $x = at^2$

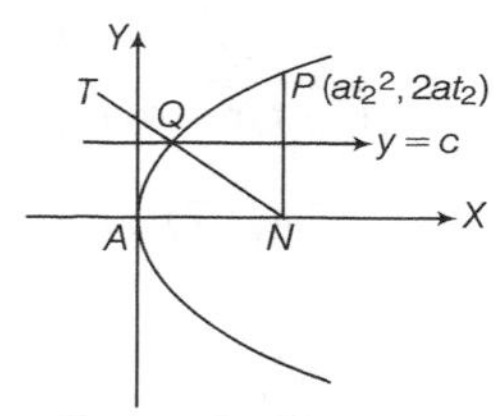

$y = c$ bisects PN

$\therefore \quad c = at$

Which cuts the parabola at Q.

$\Rightarrow \quad c^2 = 4ax$

$\Rightarrow \quad x = \frac{c^2}{4a}$

$\therefore \quad Q\left(\frac{c^2}{4a}, c\right) = Q\left(\frac{at^2}{4}, at\right)$

$\therefore$ Equation of NQ : $y = \frac{-4}{3t}(x - at^2)$

Which cuts $x = 0$, at $\left(0, \frac{4t}{3}\right)$.

$\therefore \quad AT = \frac{4at}{3}$ and $NP = 2at$

$\Rightarrow \quad \frac{AT}{NP} = \frac{4at/3}{2at} = \frac{2}{3}$

9. (d) The equation of a tangent to $y^2 = 8x$ is $y = mx + \frac{2}{m}$ ...(i)

if it touches $xy = -1$, then

$$x\left(mx + \frac{2}{m}\right) = -1 \text{ must have equal roots.}$$

$\Rightarrow$ $$mx^2 + \frac{2}{m}x + 1 = 0 \text{ must have equal roots}$$

$\Rightarrow$ $$\frac{4}{m^2} - 4m = 0$$

$\Rightarrow$ $$m^3 - 1 = 0$$

$\Rightarrow$ $$m = 1$$

On substituting the value of m in Eq. (i), we get

$y = x + 2$ as the equation of common tangent.

10. (c) Let $y = mx + \frac{1}{m}$ be a tangent to the parabola. If it touches the circle $(x-3)^2 + (y-0)^2 = 3^2$, then

$$\left|\frac{3m + \frac{1}{m}}{\sqrt{m^2+1}}\right| = 3$$

$\Rightarrow$ $$(3m^2 + 1) = 9m^2 + 9m^4$$

$\Rightarrow$ $$m = \pm\frac{1}{\sqrt{3}}$$

Hence, the equation of common tangents are $y = \pm\frac{x}{\sqrt{3}} \pm \sqrt{3}$.

11. (b) Let the coordinates of B and C be $(at_1^2, 2at_1)$ and $(at_2^2, 2at_2)$, respectively.

Then, the coordinates of A are $[at_1t_2, a(t_1 + t_2)]$

The equation of any tangent of $y^2 = 4ax$ is $ty = x + at^2$

$\therefore$ $$p_1 = \frac{at_1t_2 - a(t_1+t_2)t + at^2}{\sqrt{1+t^2}},$$

$$p_2 = \frac{at_1^2 - 2att_1 + at^2}{\sqrt{1+t^2}}$$

and $$p_3 = \frac{at_2^2 - 2at_1t_2 + at^2}{\sqrt{1+t^2}}$$

Clearly, $p_2p_3 = p_1^2$

Hence, p_2, p_1, p_3 are in GP.

12. (c) By definition, $SL = TN$

13. (a) Equation of tangent at P

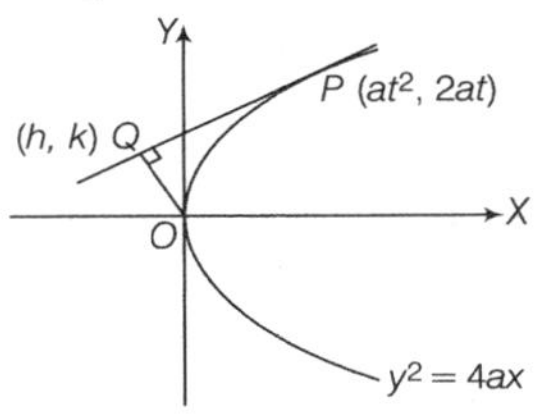

$\Rightarrow$ $$yt = x + at^2$$

or $$x - yt + at^2 = 0$$

as (h, k) is foot of Perpendicular from (0, 0)

$\Rightarrow$ $$\frac{h-0}{1} = \frac{k-0}{-t} = \frac{-(at^2)}{1+t^2}$$

$\therefore$ $$t = -\frac{k}{h} \text{ and } h = -\frac{at^2}{1+t^2}$$

$\Rightarrow$ $$h = -\frac{a\cdot\frac{k^2}{h^2}}{1+\frac{k^2}{h^2}} \text{ or } x(x^2 + y^2) = -ay^2$$

14. (a) Since, circle and parabola touch externally.

$\Rightarrow a$ and $-b$ are of opposite sign

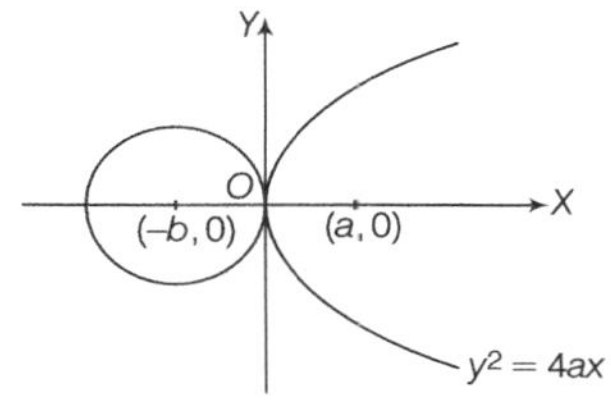

$\Rightarrow$ $$a(-b) < 0$$

$\Rightarrow$ $$ab > 0$$

15. (d) $(x-a)^2 = -4a(y+a)$

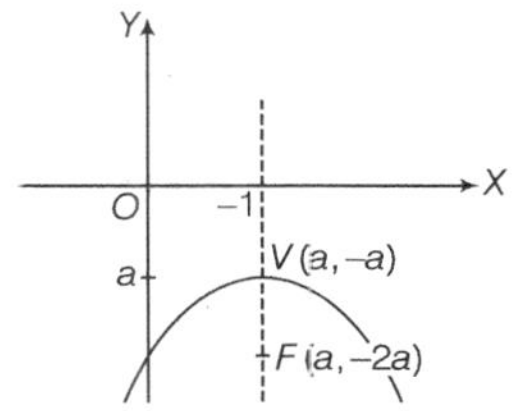

As, the directrix of parabola is X-axis.

$\therefore$ Tangents drawn from the origin is 90°.

16. (c) Equation of chord AB, is $y(t_1 + t_2) = 2x + 2at_1t_2$ passing through $(-1, -1)$

$\Rightarrow$ $$-(t_1 + t_2) = -2 + 2t_1t_2$$

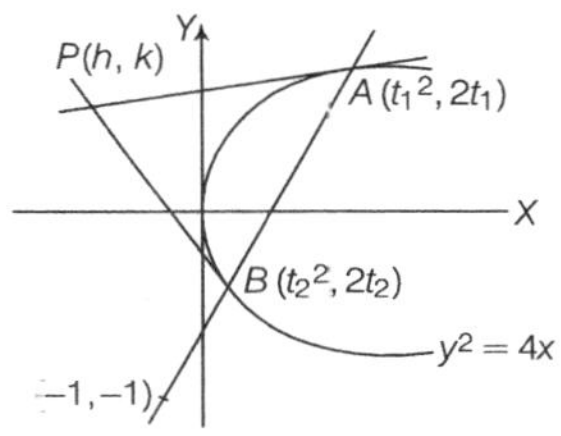

$\Rightarrow$ $(t_1 + t_1) + 2t_1t_2 - 2 = 0$ ...(i)

as, $$P(h, k) = P(-t_1t_2, t_1 + t_2)$$

$\therefore$ $h = -t_1t_2$ and $k = t_1 + t_2$ ...(ii)

From Eqs. (i) and (ii), we get

$\Rightarrow$ $$k - 2h - 2 = 0$$

or $$y = 2(x+1)$$

17. (b) The equation of any tangent of slope m to the parabola $y^2 = 4x$ is

$$y = mx + \frac{1}{m}$$

If it passes through $P(1, 4)$, then

$$4 = m + \frac{1}{m}$$

$\Rightarrow$ $$4m = m^2 + 1$$

$\Rightarrow$ $$m^2 - 4m + 1 = 0$$

Let m_1 and m_2 be the roots of equation.

Then, $m_1 + m_2 = -4$

and $m_1m_2 = 1$

$\therefore$ $$m_1 - m_2 = \sqrt{(m_1+m_2)^2 - 4m_1m_2}$$

$$= \sqrt{(-4)^2 - 4(1)}$$

$$= \sqrt{12} = 2\sqrt{3}$$

Let θ be the angle between the tangents drawn from P, then

$$\tan\theta = \frac{m_1 - m_2}{1 + m_1 m_2} = \frac{2\sqrt{3}}{1+1} = \sqrt{3}$$

$\Rightarrow \quad \theta = \tan^{-1}(\sqrt{3})$

$\therefore \quad \theta = \frac{\pi}{3}$

18. (d) The given curve is $y = x^2 + 6$

The equation of tangent at (1, 7) is

$$\frac{1}{2}(y + 7) = x(1) + 6$$

$\Rightarrow \quad 2x - y + 5 = 0 \quad \ldots(\text{i})$

According to the question Eq. (i) touches the circle $x^2 + y^2 + 16x + 12y + C = 0$ at Q [centre of circle of $(-8, -6)$].

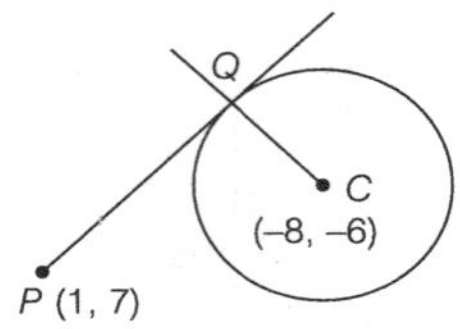

Then, the equation of CQ which is perpendicular to Eq. (i) and passes through $(-8, -6)$ is

$$y + 6 = \frac{-1}{2}(x + 8)$$

$\Rightarrow \quad x + 2y + 20 = 0 \quad \ldots(\text{ii})$

Now, Q is the point of intersection of Eqs. (i) and (ii).

i.e. $\quad x = -6$ and $y = -7$

$\therefore$ The required is point $(-6, -7)$.

19. (b) The parabola is $(y - 1)^2 = 4(x - 1)$ whose directrix is the Y-axis or $x = 0$.

Any point on this parabola is $P(t^2 + 1, 2t + 1), t \in R$.

$\therefore$ The equation of tangent at $P(t^2 + 1, 2t + 1)$ is

$$t(y - 1) = (x - 1) + t^2$$

or $\quad x - ty + (t^2 + t - 1) = 0 \quad \ldots(\text{i})$

Tangent meets the directrix at $Q\left(0, \frac{t^2 + t - 1}{t}\right)$.

Now, Q is the mid-point of PR. Therefore,

$$\frac{h + t^2 + 1}{2} = 0$$

and $\quad \frac{k + 2t + 1}{2} = \frac{t^2 + t - 1}{t}$

or $\quad t^2 = -1 - h$

and $\quad kt + 2r^2 + t = 2t^2 + 2t - 2$

or $\quad t^2 = -1 - h$

and $\quad 2 = t(1 - k)$

On eliminating t, we have

$$4 = (-1 - h)(1 - k)^2$$

$\Rightarrow \quad (x + 1)(y - 1)^2 + 4 = 0$

20. (d) The point of intersection of tangents at $P(at_1^2, 2at_1)$ and $Q(at_2^2, 2at_2)$ is

$$R(h, k) \equiv \{at_1t_2, a(t_1 + t_2)\}$$

$\therefore \quad t_1 + t_2 = \frac{k}{a}$ and $t_1t_2 = \frac{h}{a}$

Now, $\quad \tan 45^\circ = \left|\frac{\left(\frac{1}{t_1}\right) - \left(\frac{1}{t_2}\right)}{1 + \left(\frac{1}{t_1t_2}\right)}\right|$

or $\quad 1 = \left|\frac{t_2 - t_1}{t_1t_2 + 1}\right| = \left|\frac{\sqrt{(t_1 + t_2)^2 - 4t_1t_2}}{t_1t_2 + 1}\right|$

$\Rightarrow \quad 1 = \left|\frac{\sqrt{\left(\frac{k^2}{a^2}\right) - 4\left(\frac{h}{a}\right)}}{\left(\frac{h}{a}\right) + 1}\right|$

$\Rightarrow \quad k^2 - 4ah = (h + a)^2$

$\Rightarrow \quad x^2 - y^2 + 6ax + a^2 = 0$

which is a hyperbola.

21. (a,c) Let $P(at_1^2, 2at_1)$ and $Q(at_2^2, 2at_2)$ be the extremities of a focal chord of the parabola $y^2 = 4ax$.

$R[at_1t_2, a(t_1 + t_2)]$ is the point of intersection of tangents at P and Q.

Since, PQ is a focal chord of the parabola.

$\therefore \quad t_1t_2 = -1$

Thus, the coordinates of R are $(-a, a(t_1 + t_2))$.Clearly, it lies on the directrix.

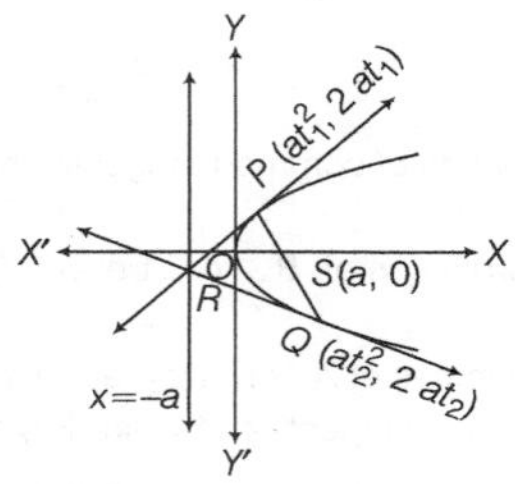

The equation of tangents PR and PQ are

$$t_1y = x + at_1^2$$

and $\quad t_2y = x + at_2^2$

Now, $\quad m_1 = \text{Slope of } PR = \frac{1}{t_1}$

and $\quad m_2 = \text{Slope of } QR = \frac{1}{t_2}$

$\therefore \quad m_1m_2 = \frac{1}{t_1t_2} = -1 \quad [\because t_1t_2 = -1]$

Hence, the tangents at P and Q intersect at right angles on the directrix.

22. (b,c) The equation of tangent to the parabola $y^2 = 4ax$ at $(at^2, 2at)$ is

$$ty = x + at^2$$

Here, $a = 3$. So, the equation of the tangent at $(3t^2, 6t)$ to the parabola $y^2 = 12x$ is

$$ty = x + 3t^2 \quad \ldots(\text{i})$$

The slope of Eq, (i) is $\frac{1}{t}$ and that of the given line is 3. Therefore,

$$\frac{\frac{1}{t} - 3}{1 + \frac{1}{t} \times 3} = \pm \tan 45^\circ = \pm 1$$

$\Rightarrow \quad \frac{1 - 3t}{t + 3} = \pm 1$

$\Rightarrow \quad t = -\frac{1}{2}, 2$

Now, equation of the tangents are

$$x - 2y + 12 = 0$$

and $\quad 4x + 2y + 3 = 0$

23. (b,c,d) Here, axis of parabola is $y = x$

Here, $$VF = \frac{|1+1-1|}{\sqrt{2}} = \frac{1}{\sqrt{2}}$$

$\therefore$ length of latusrectum $= \frac{4}{\sqrt{2}}$

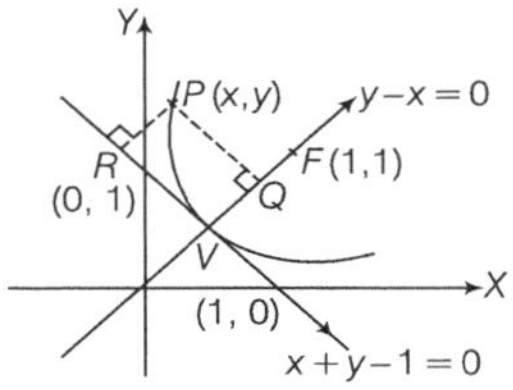

$\Rightarrow$ Equation of parabola, is

$$(PQ)^2 = 4a\,(PR)$$

$$\Rightarrow \quad \left(\frac{x-y}{\sqrt{2}}\right)^2 = \frac{4}{\sqrt{2}}\left(\frac{x+y-1}{\sqrt{2}}\right)$$

or $$(x-y)^2 = 4\,(x+y-1)$$

$\Rightarrow$ Vertex $\left(\frac{1}{2}, \frac{1}{2}\right)$, and length of latusrectum $= 2\sqrt{2}$.

24. (a,d) As, $P\,(t^2, 2t)$

and $$t^2 = 2t$$

$$\Rightarrow \quad t = 2$$

Equation of Normal at P is

$$y + xt = 2at + at^3$$

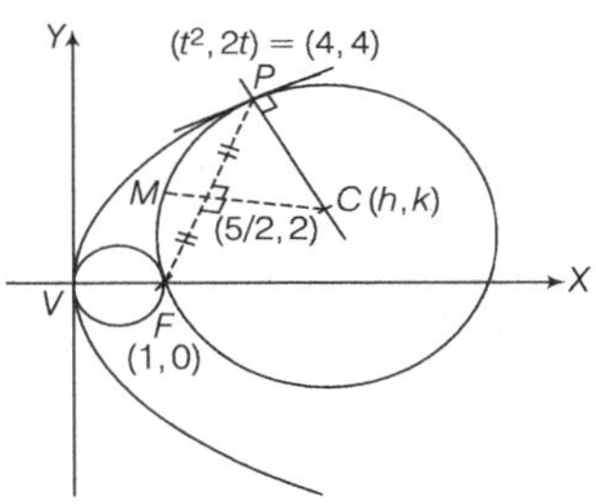

$$\Rightarrow \quad y + 2x = 4 + 8$$

$$\Rightarrow \quad 2x + y = 12 \text{ passes through } (h, k)$$

$$\Rightarrow \quad 2h + k = 12 \quad \ldots\text{(i)}$$

and $$m_{CM} \cdot m_{PF} = -1$$

$$\Rightarrow \quad \frac{k-2}{h-\frac{5}{2}} \cdot \frac{4-0}{4-1} = -1$$

$$\Rightarrow \quad 4k - 8 = -3\left(h - \frac{5}{2}\right)$$

$$\Rightarrow \quad 3h + 4k = \frac{31}{2} \quad \ldots\text{(ii)}$$

On solving Eqs. (i) and (ii), we get $h = \frac{13}{2}, k = -1$

$\therefore$ Equation of circle, is

$$x^2 + y^2 - 13x + 2y + 12 = 0$$

and $$x^2 + y^2 - x = 0$$

when VF is diameter.

25. (a,b) The equations of tangent to $y = x^2$ is

$$x = my + \frac{1}{4m} \quad \ldots\text{(i)}$$

The equation of tangent to $y = -(x-2)^2$ is

$$x - 2 = my - \frac{1}{4m} \quad \ldots\text{(ii)}$$

Lines given in Eqs. (i) and (ii) represent the same line.

$$\therefore \quad \frac{1}{4m} = 2 - \frac{1}{4m}$$

$$\Rightarrow \quad m = \frac{1}{4}$$

On substituting $m = \frac{1}{4}$ in Eq. (i), we get $y = 4\,(x-1)$ as a common tangent.

Clearly, $y = x^2$ and $y = -(x-2)^2$ represent parabolas having vertex at (0, 0) and (2, 0) respectively and touching X - axis. So, X - axis, i.e. $y = 0$ is also a common tangent.

26. (a) Given equation of curve is $y = ax^2 + c$

$$\therefore \quad \frac{dy}{dx} = 2ax = 1$$

Therefore, the point of contact of the tangent is

$$\left(\frac{1}{2a}, \frac{1}{4a} + c\right)$$

Since, it lies on $y = x$, we have

$$c = \frac{1}{4a}$$

Thus, $c = \frac{1}{8}$ when $a = 2$.

27. (a) If (1, 1) is the point of contact, then $a = \frac{1}{2}$.

28. (b) If $c = 2$, then the point of contact is

$$\left(\frac{1}{2}a, \frac{1}{4}a + 2\right)$$

Since, it lies on the line $y = x$, we have

$$\frac{1}{2a} = \frac{1}{4a} + 2 \text{ i.e. } a = \frac{1}{8}$$

$\therefore$ The point of contact is (4, 4).

Explanations for Q. Nos. (29-31)

Solving the circle $x^2 + y^2 = 9$ and the parabola $y^2 = 8x$, we get

$$x^2 + 8x - 9 = 0$$

$$\Rightarrow \quad x = 1, -9$$

$$\Rightarrow \quad x = 1 \quad [x = -9 \text{ is not possible}]$$

$$\Rightarrow \quad y^2 = 8$$

$$\Rightarrow \quad y = \pm 2\sqrt{2}$$

Hence, the points of intersection are $P\,(1, 2\sqrt{2})$ and $Q\,(1, -2\sqrt{2})$.

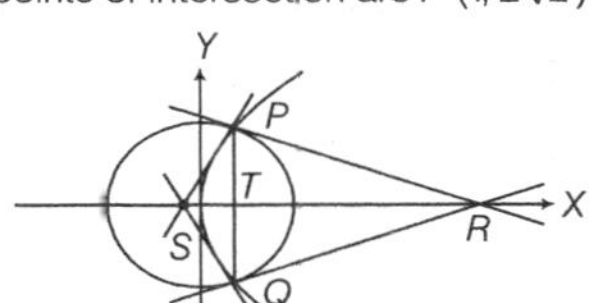

Tangent to the parabola at point P is $2\sqrt{2}\;y = 4\,(x+1)$.

It meets the X- axis at S (−1, 0).

Tangent to the circle at point P is $(1)x + 2\sqrt{2}\;y = 9$.

It meets the X- axis at R (9, 0).

Now, $$\frac{\text{Ar}\,(\Delta PQS)}{\text{Ar}\,(\Delta PQR)} = \frac{\frac{1}{2}PQ \times ST}{\frac{1}{2}PQ \times TR} = \frac{ST}{TR} = \frac{2}{8} = \frac{1}{4}$$

For ΔPRS, Ar $(\Delta PRS) = \Delta = \frac{1}{2} \times SR \times PT$

$$= \frac{1}{2} \times 10 \times 2\sqrt{2}$$

$$\therefore \quad \Delta = 10\sqrt{2},\; a = PS = 2\sqrt{3},$$

$$b = PR = 6\sqrt{2},\; c = SR = 10$$

$\therefore$ Radius of circumcircle $= R = \dfrac{abc}{2\Delta} = \dfrac{2\sqrt{3} \times 6\sqrt{2} \times 10}{4 \times 10\sqrt{2}} = 3\sqrt{3}$

Thus, radius of incircle of $\Delta\, PQR$ is

$$\frac{\text{Area of } \Delta PQR}{\text{Semi - perimeter of } \Delta PQR} = \frac{\Delta}{s}$$

We have, $\quad a = PR = 6\sqrt{2}$,

$\qquad b = QP = PR = 6\sqrt{2}$

and $\quad c = PQ = 4\sqrt{2}$

Also, $\quad \Delta = \dfrac{1}{2} \times PQ \times TR = 16\sqrt{2}$

$\Rightarrow \quad s = \dfrac{6\sqrt{2} + 6\sqrt{2} + 4\sqrt{2}}{2} = 8\sqrt{2}$

$\therefore \quad r = \dfrac{16\sqrt{2}}{8\sqrt{2}} = 2$

29. (c) **30.** (b) **31.** (d)

32. (4) The given parabola is $y^2 = 4ax$. ...(i)

Its vertex is (0, 0).

Let $P\,(at_1^2, 2at_1)$ and $Q\,(at_2^2, 2at_2)$ be the ends of a chord of Eq. (i) such that $\angle PAQ$ = Right angle

$$\text{Slope of } AP = \frac{2at_1 - 0}{at_1^2 - 0} = \frac{2}{t_1}$$

$$\text{Slope of } AQ = \frac{2at_2 - 0}{at_2^2 - 0} = \frac{2}{t_2}$$

$\because \quad AP \perp AQ$

$\therefore \quad \dfrac{2}{t_1} \cdot \dfrac{2}{t_2} = -1$

$\Rightarrow \quad t_1 t_2 = -4$ ...(ii)

The tangents at P and Q intersect at $R\,[at_1t_2,\, a(t_1 + t_2)]$, i.e. at $[-4a,\, a(t_1 + t_2)]$ which for all values of t_1 and t_2 lies on $x = -4a$.

So, $|\lambda| = 4$

33. (1) Here, $A\,(at_1t_2,\, a(t_1 + t_2))$

$\qquad B\,(at_1t_3,\, a\,(t_1 + t_3))$

$\qquad C\,(at_2t_3,\, a\,(t_2 + t_3))$

The equation of altitudes through the vertex A, is

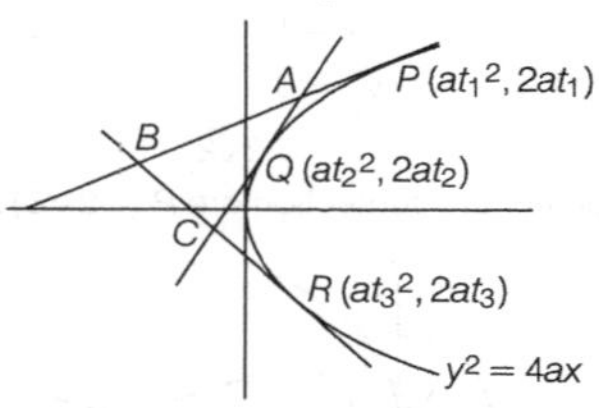

$$y - a\,(t_1 + t_2) = -t_3\,(x - at_1t_2)$$

or $\quad y + xt_3 = a\,(t_1 + t_2 + t_1t_2t_3)$ ...(i)

Similarly, the equation of altitudes from B and C, are

$\qquad y + xt_2 = a\,(t_3 + t_1 + t_1t_2t_3)$ ...(ii)

and $\quad y + xt_1 = a\,(t_2 + t_3 + t_1t_2t_3)$ ...(iii)

Solving Eqs. (i) and (ii), we get

$$x = -a \text{ and } y = a\,(t_1 + t_2 + t_3 + t_1t_2t_3)$$

$\therefore \quad x = \lambda a \Rightarrow \lambda = -1$ (as, $a = 1$)

or $\quad |\lambda| = 1$

34. (2) The equation of the tangent of slope m_1 to $y^2 = 4a\,(x + a)$ is

$$y = m_1(x + a) + \frac{a}{m_1} \quad \text{...(i)}$$

The equation of tangent of slope m_2 to

$y^2 = 4b\,(x + b)$ is

$$y = m_2\,(x + b) + \frac{b}{m_2} \quad \text{...(ii)}$$

Since, Eqs. (i) and (ii) are perpendicular

$\Rightarrow \quad m_2 = -\dfrac{1}{m_1}$

Putting $\quad m_2 = -\dfrac{1}{m_1}$ in Eq. (i), we get

$$y = -\frac{1}{m_1}(x + b) - bm_1 \quad \text{...(iii)}$$

Subtracting Eqs. (iii) and (i), we get

$$0 = \left(m_1 + \frac{1}{m_1}\right)x + a\left(m_1 + \frac{1}{m_1}\right) + b\left(m_1 + \frac{1}{m_1}\right)$$

$\Rightarrow \quad 0 = (x + a + b)\left(m_1 + \dfrac{1}{m_1}\right)$

$\Rightarrow \quad x + a + b = 0$

$\therefore \quad \lambda = \mu = 1$

$\Rightarrow \quad \lambda + \mu = 2$

35. (3) Let the two equal parabolas be $y^2 = 4ax$ and $x^2 = 4ay$.

The equation of any tangent to $y^2 = 4ax$ is $y = mx + \dfrac{a}{m}$ at the point $\left(\dfrac{a}{m^2}, \dfrac{2a}{m}\right)$.

Now, $\quad y = mx + \dfrac{a}{m}$

$\Rightarrow \quad x = \dfrac{y}{m} - \dfrac{a}{m^2}$

If it touches the parabola $x^2 = 4ay$, then

$$\frac{-a}{m^2} = \frac{a}{1/m} \quad \left[\because C = \frac{a}{m}\right]$$

$\Rightarrow \quad m^3 = -1$

$\Rightarrow \quad m = -1$

The common tangent touches $y^2 = 4ax$ at $P\left(\dfrac{a}{m^2}, \dfrac{2a}{m}\right)$. So, the coordinates of the point of contact are $P(a, -2a)$.

The common tangent (i) touches $x^2 = 4ay$ at point Q whose coordinates are $\left(\dfrac{2a}{1/m}, \dfrac{a}{1/m^2}\right)$.

$$\left[\because x = my + \frac{a}{m} \text{ touches } x^2 = 4ay \text{ at } \left(\frac{2a}{m}, \frac{a}{m^2}\right)\right]$$

$$= Q\,(2am, am^2) \equiv Q\,(-2a, a)$$

$\therefore$ Length of the common tangent

$$= PQ = \sqrt{(a + 2a)^2 + (-2a - a)^2}$$

$$= 3\sqrt{2}\,a$$

On comparing with $\lambda\sqrt{2}\,a$, we get

$$\lambda = 3$$

DPP-4 Equation of Normal and Its Properties

1. (b) Normal to the parabola $y^2 = 4x$ having slope m is

$$y = mx - 2m - m^3$$

The given normal is

$$y = 2x + (c - 4)$$

Therefore, $m = 2$ and $c - 4 = -2m - m^3$

$$\therefore \quad c = 4 - 2(2) - (2)^3$$
$$= 4 - 4 - 8 = -8$$

2. (a) The equation of normal to the parabola at $(at_1^2, -2at_1)$ is

$$y + 2at_1 = \frac{2at_1}{2a}(x - at_1^2)$$

$$\Rightarrow \quad y + 2at_1 = t_1x - at_1^3$$

$$\Rightarrow \quad y - t_1x = -2at_1 - at_1^3 \quad \ldots(i)$$

It passes through $(at_2^2, 2at_2)$, then

$$2at_2 - t_1(at_2^2) = -2at_1 - at_1^3$$

$$\Rightarrow \quad 2a(t_2 + t_1) = at_1(t_2^2 - t_1^2)$$

$$\Rightarrow \quad \frac{2}{t_1} = \frac{(t_2 + t_1)(t_2 - t_1)}{(t_2 + t_1)}$$

$$\Rightarrow \quad \frac{2}{t_1} = t_2 - t_1$$

$$\therefore \quad t_2 = \frac{2}{t_1} + t_1$$

3. (b) The equation of any normal to $y^2 = 12x$ is

$$y = mx - 6m - 3m^3 \quad \ldots(i)$$

[using, $y = mx - 2am - am^3$]

The line is $x + y = k$

$$\Rightarrow \quad y = k - x \quad \ldots(ii)$$

On comparing Eqs. (i) and (ii), we get

$$m = -1$$

and $k = -6m - 3m^2 = (-1)(-6) - 3(-1)^3$

$$= 6 + 3 = 9$$

4. (b) The equation of normal to the parabola $y^2 = 4ax$ at point $(am^2, 2am)$ is given by

$$y - 2am = -\frac{2am}{2a}(x - am^2)$$

$$\Rightarrow \quad y - 2am = -xm + am^3$$

$$\therefore \quad y = -mx + 2am + am^3$$

5. (d) We have, $a = 1$

Normal at $(m^2, -2m)$ is $y = mx - 2m - m^3$.

Given that normal makes equal angle with axes, its slope is $m = \pm 1$.

Therefore, point P is $(m^2, 2m) \equiv (1, \pm 2)$

Hence, from the given points (1, 2) is correct option.

6. (c) The equation of normal to $y^2 = -8x$ is

$$y = mx + 4m + 2m^3 \quad \ldots(i)$$

[using, $y = mx - 2am - am^3$]

The given line is $2x + y + \lambda = 0$

$$\Rightarrow \quad y = -2x - \lambda \quad \ldots(ii)$$

On comparing Eqs. (i) and (ii), we get

$$m = -2 \text{ and } \lambda = -4m - 2m^3$$

$$\therefore \quad \lambda = 8 + 16 = 24$$

7. (a) Let $P(at^2, 2at)$ be a point on the parabola $y^2 = 4ax$.

Then, $at^2 = 2at$

$$\Rightarrow \quad t = 2 \quad [\because x_1 = y_1]$$

So, the coordinates of P are $(4a, 4a)$. Let PQ be a normal chord normal at P to the parabola.

Let the coordinates of Q be $(at_1^2, 2at_1)$.

Then, $t_1 = t - \frac{2}{t}$

$$= -2 - \frac{2}{2} = -3$$

So, the coordinates of Q are $(9a, -6a)$.

Now, slope of SP × slope of $SQ = \frac{4a}{3a} \times \frac{-6a}{8a}$

$$= -1$$

Thus, the normal chord makes a right angle at the focus.

8. (b) Tangent at P, is

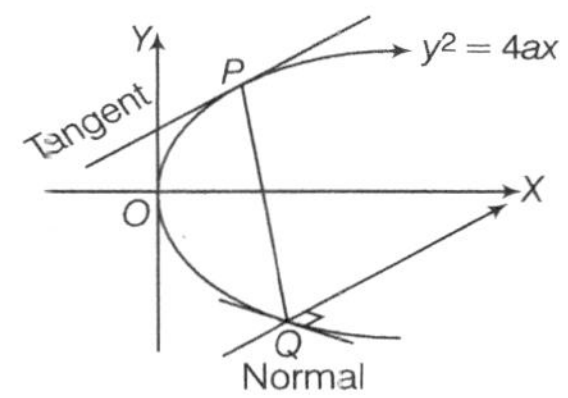

$$y = x + a \quad \ldots(i)$$

and equation of normal at Q, is

$$y = x - 2a - a$$

or $y = x - 3a \quad \ldots(ii)$

From Eqs. (i) and (ii), distance between parallel lines is $\left|\frac{4a}{\sqrt{2}}\right| = 2\sqrt{2}a$

9. (b) Since, the normal at $(ap^2, 2ap)$ to $y^2 = 4ax$ meets the parabola at $(aq^2, 2aq)$.

Therefore, $q = -p - \frac{2}{p} \quad \ldots(i)$

Since, $OP \perp OQ$, therefore,

$$\frac{2ap - 0}{ap^2 - 0} \times \frac{2aq - 0}{aq^2 - 0} = 1$$

$$\Rightarrow \quad pq = -4$$

$$\Rightarrow \quad p\left(-p - \frac{2}{p}\right) = -4 \quad \text{[by Eq. (i)]}$$

$$\therefore \quad p^2 = 2$$

10. (c) Given, a line $y = x - 1$ and a parabola $y^2 = 4x$.

On solving these, we have

$$(x - 1)^2 = 4x$$

$$\Rightarrow \quad x^2 - 6x + 1 = 0$$

$$\Rightarrow \quad x = 3 \pm \sqrt{8}$$

and $y = 2 \pm \sqrt{8}$

Suppose point D is (x_3, y_3). Then,

$$y_1 + y_2 + y_3 = 0$$

$$2 + \sqrt{8} + 2 - \sqrt{8} + y_3 = 0$$

$$\Rightarrow \quad y_3 = -4$$

Then, $x_3 = 4$

11. (d) Let $P(at_1^2, 2at_1)$ and $Q(at_2^2, at_2)$ be the extremities of a focal chord of the parabola $y^2 = 4ax$. The tangents at P and Q intersect at $[at_1t_2, a(t_1 + t_2)]$. Therefore,

$$x_1 = at_1t_2$$

and $$y_1 = a(t_1 + t_2)$$

$\Rightarrow$ $$x_1 = -a$$

and $$y_1 = a(t_1 + t_2)$$

$[\because PQ$ is a focal chord $t_1t_2 = -1]$

The normals at P and Q intersect at

$$[2a + a(t_1^2 + t_2^2 + t_1t_2), -at_1t_2(t_1 + t_2)]$$

$\therefore$ $$x_2 = 2a + a(t_1^2 + t_2^2 + t_1t_2)$$

and $$y_2 = -at_1t_2(t_1 + t_2)$$

Clearly, $$y_1 = y_2$$

12. (a) Here, $$m = \frac{2at_2 - 2at_1}{at_2^2 - at_1^2} = \frac{2}{t_2 + t_2} \quad \ldots(i)$$

and $$h = 2a + a(t_1^2 + t_2^2 + t_1t_2) \quad \ldots(ii)$$

$$k = -at_1t_2(t_1 + t_2) \quad \ldots(iii)$$

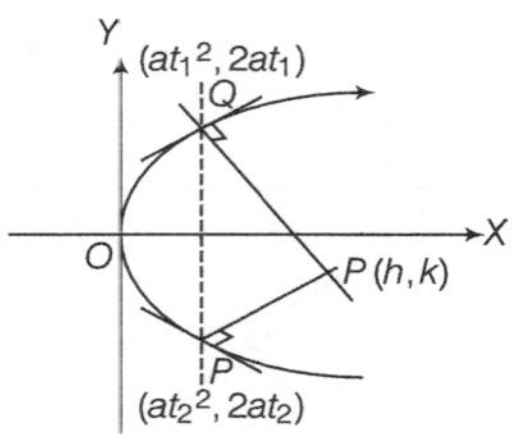

From Eqs. (i), (ii) and (iii), we get

$$h = 2a + a\left[\frac{4}{m^2} + \frac{km}{2a}\right]$$

or $$2m^2x = 4am^2 + 8a + ym^3$$

13. (b) We have,

$$y^2 = 4ax$$

$\Rightarrow$ $$2y\frac{dy}{dx} = 4a \Rightarrow \frac{dy}{dx} = \frac{2a}{y}$$

At (x_1, y_1), we have

$$\text{subtangent} = \frac{y_1}{dy/dx} = \frac{y_1}{2a/y_1} = \frac{y_1^2}{2a}$$

$$\text{subnormal} = y_1\frac{dy}{dx} = 2a$$

Clearly, $$y_1^2 = \left(\frac{y_1^2}{2a}\right) \times 2a$$

i.e. (Ordinate)^2 = Subtangent × Subnormal

Hence, subtangent, ordinate and subnormal are in GP.

14. (a) The equation to any chord which is normal, is $y = mx - 2am - am^3$.

$\Rightarrow$ $$mx - y = 2am + am^3$$

Let the equation of parabola is $y^2 = 4ax$.

The straight lines joining the origin to the intersection of these two, given by the equation

$$y^2(2am + am^3) - 4ax(mx - y) = 0$$

If these be at right angles, then

$$2am + am^3 - 4am = 0$$

$[\because$ coefficient of y^2 + coefficient of $x^2 = 0]$

$\Rightarrow$ $$m = \pm\sqrt{2}$$

$\Rightarrow$ $$\tan\theta = \sqrt{2}$$

$\therefore$ $$\theta = \tan^{-1}\sqrt{2}$$

15. (a) Equation of normal, is $y + xt = 2at + at^3$, it passes through the focus $(a, 0)$ is

$\Rightarrow$ $$0 = -at + 2at + at^3$$

$\Rightarrow$ $$at^3 + at = 0$$

$\Rightarrow$ $$at(t^2 + 1) = 0$$

$\Rightarrow$ $$t = 0$$

$\therefore$ There is only 1 real normal.

16. (b) Let $P(h, k)$, $(k \neq 0)$ through which three normals are drawn to the parabola $y^2 = 4ax (a = 1)$ such that product of slopes of two is α_1, i.e. $m_1m_2 = \alpha$.

Any normal to the parabola is

$$y = mx - 2am - am^3$$

$\Rightarrow$ $$am^3 + 0m^2 + m(2a - h) + k = 0$$

$\Rightarrow$ $$m^3 + 0m^2 + m(2a - h) + k = 0, \text{ as } a = 1 \quad \ldots(i)$$

$$m_1m_2m_3 = -k$$

$\therefore$ $$m_3 = \frac{-k}{m_1m_2} = \frac{-k}{\alpha}, \quad \text{[by Eq. (i)]}$$

But m_3 is a root of Eq. (i).

$\therefore$ $$k^3 + (2 - h)k\alpha^2 - k\alpha^3 = 0$$

Hence, locus of P is $y^2 + (2 - x)\alpha^2 - \alpha^3 = 0$ $[k \neq 0$, cancelled$]$

Above is same as $y^2 - 4x = 0$, as given.

Hence, on comparing $\alpha^2 = 4$

and $$2\alpha^2 - \alpha^3 = 0$$

or $$\alpha = 0$$

$\therefore$ $\alpha = 2$, which satisfies both.

17. (a) Given, $$\frac{2at_1}{2at_2} = \frac{1}{2}$$

$\Rightarrow$ $$2t_1 = t_2, \; y^2 = 12x$$

$\therefore$ $$a = 3$$

Now, $$h = 2a + a(t_1^2 + t_2^2 + t_1t_2)$$

$$= 6 + 3 \cdot 7t_1^2 \quad [\because t_2 = 2t_1]$$

and $$k = -at_1t_2(t_1 + t_2) = -6t_1^2 \cdot 3t_1 = 18t_1^3$$

$\Rightarrow$ $$\left(\frac{h - 6}{21}\right)^3 = t_1^6 = \left(\frac{k}{18}\right)^2$$

$\therefore$ Locus of (h, k) is

$$y^2 = \frac{12}{343}(x - 6)^2$$

i.e. $$x = 3\left[7\left(\frac{y}{18}\right)^{2/3} + 2\right].$$

18. (b) Let $P(at^2, 2at)$ be a point on the parabola $y^2 = 4ax$ such that it cuts the parabola again at $Q(at_1^2, at_1)$. Then,

$$t_1 = -t - \frac{2}{t}$$

$\Rightarrow$ $$|t| = \left|t + \frac{2}{t}\right|$$

$\Rightarrow$ $$|t_1| \geq \left|2\sqrt{t \times \frac{2}{t}}\right| \quad [\because \text{AM} \geq \text{GM}]$$

$\Rightarrow$ $$|t| = 2\sqrt{2}$$

Now, $$OQ = \sqrt{(at_1^2 - 0)^2 + (2at_1 - 0)^2}$$

$\Rightarrow$ $$OQ = at_1\sqrt{t_1^2 + 4}$$

$\Rightarrow$ $$OQ \geq 2a\sqrt{2}\sqrt{8 + 4}$$

$\therefore$ $$OQ \geq 4\sqrt{6}a$$

19. (b) Let $(2t, t^2)$ is a point on the parabola $x^2 = 4y$.

$\therefore$ Normal at $(2t, t^2)$ is

$$t^3 + t(2 - y) - x = 0 \quad \ldots(i)$$

Let three normals passes through (h, k), then

$$t^3 + t(2 - k) - h = 0$$

$$t_1 + t_2 + t_3 = 0$$

$$t_1t_2 + t_2t_3 + t_3t_1 = 2 - k$$

and $$t_1t_2t_3 = h$$

Given that normals cuts the lines $y = 2$,

then from Eq. (i)

$$x = t^3$$

Then, abscissae of three points are t_1^3, t_2^3, t_3^3 are in AP. [given]

$\therefore$ $$2t_2^3 = t_1^3 + t_3^3$$

$$= (t + t_3)^3 - 3t_1t_3(t_1 + t_3)$$

$\Rightarrow$ $$2t_2^3 = -t_2^3 + 3t_1t_2t_3$$

$\therefore$ $$t_2^2 = t_1t_3$$

$\therefore$ Slope of tangents at three conormal points, i.e. t_1, t_2, t_3 are in GP.

20. (a) The two parabolas are given by

$$y^2 = 4ax \quad \ldots(i)$$

and $$y^2 = 4c(x - b) \quad \ldots(ii)$$

Equation to any normal of Eq. (i) is

$$y = mx - 2am - am^3 \quad \ldots(iii)$$

$$y = m(x - b) - 2am - am^3 \quad \ldots(iv)$$

If there is any common normal, then Eqs. (iii) and (iv) must be identical. As the coefficients of x and y are equal, so the constant terms will also be equal, hence

$$-2am - am^3 = -bm - 2mc - cm^3$$

$\Rightarrow$ $$m[m^2(c - a) - 2a + b + 2c] = 0$$

So, either $m = 0$ or $m^2 = \dfrac{2a - b - c}{c - a}$

If $m = 0$, the common normal is the X-axis.

If $$m^2 = \frac{2a - b - 2c}{c - a}$$

$$m = \sqrt{\frac{2(a - c) - b}{c - a}} = \sqrt{\left(-2 - \frac{b}{c - a}\right)}$$

then, $$-2 - \frac{b}{c - a} > 0, \quad \frac{-b}{c - a} > 2$$

$\Rightarrow$ $$\frac{b}{a - c} > 2$$

21. (c) The equation of normal at $P(at^2, 2at)$ to the parabola $y^2 = 4ax$ is

$$y = -tx + 2at = at^3 \quad \ldots(i)$$

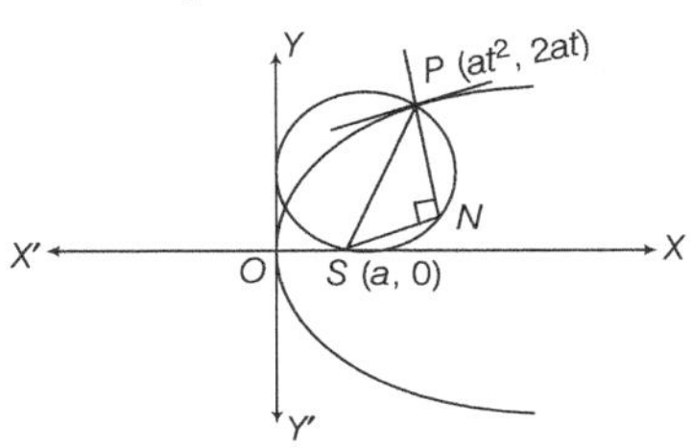

Let this normal meets the circle on SP as diameter at N. Then,

$$\angle SNP = \frac{\pi}{2} \quad \text{[an angle in semicircle]}$$

$\therefore$ $$PN^2 = SP^2 - SN^2$$

SN is perpendicular to normal. Now,

$$SP = a + at^2$$

and $$SN = \frac{|at - 2at - at^3|}{\sqrt{1 + t^2}} = at\sqrt{1 + t^2}$$

$\therefore$ $$PN^2 = a^2(1 + t^2)^2 - a^2t^2(1 + t^2) = a^2(1 + t^2)$$

$\therefore$ $$PN = a\sqrt{1 + t^2}$$

22. (c) Tangent and normal at $P(at^2, 2at)$ to the parabola $y^2 = 4ax$ are respectively,

$$ty = x + at^2 \quad \ldots(i)$$

and $$y = -tx + 2at + at^3 \quad \ldots(ii)$$

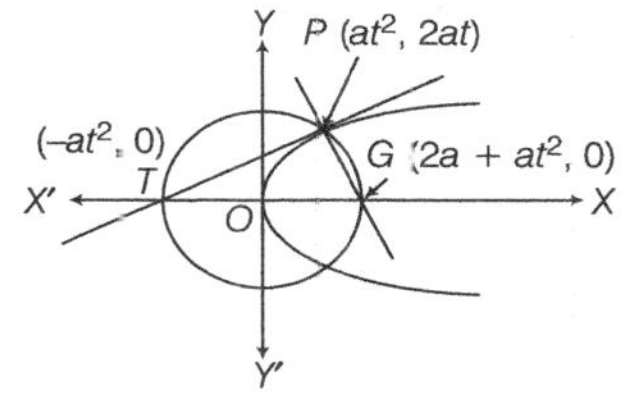

Eqs. (i) and (ii) meet the X-axis, where $y = 0$.

From Eq. (i), $$x = -at^2$$

Therefore, T is $(-at^2, 0)$.

From Eq. (ii), $$tx = 2at + at^3$$

Therefore, G is $(2a + at^2, 0)$.

Mid-point of $TG \equiv \left(\dfrac{2a + at^2 - at^2}{2}, 0\right) \equiv O(a, 0)$

Since, $\angle TPG = 90°$, the centre of circle through PTG is $(a, 0)$.

If θ is the angle between tangents at P to the parabola and circle through P, T and G, then $(90° - \theta)$ is the angle between PT and OP.

$$\text{Slope of } PT = \frac{2at}{2at^2} = \frac{1}{t}$$

$$\text{Slope of } OP = \frac{2at}{a(t^2 - 1)} = \frac{2t}{t^2 - 1}$$

Now, $$\tan(90° - \theta) = \frac{\left|\dfrac{1}{t} - \dfrac{2t}{t^2 - 1}\right|}{\left|1 + \dfrac{1}{t}\left(\dfrac{2t}{t^2 - 1}\right)\right|} = \frac{1}{t}$$

$\Rightarrow$ $$\cot\theta = \frac{1}{t} \Rightarrow \tan\theta = t$$

$\therefore$ $$\theta = \tan^{-1}t$$

23. (c) $t_2 = -t_1 - \dfrac{2}{t_1} \Rightarrow t_1t_2 = -t_1^2 - 2$

The equation of a line through P parallel to AQ is

$$y - 2at_1 = \frac{2}{t_2}(x - at_1^2)$$

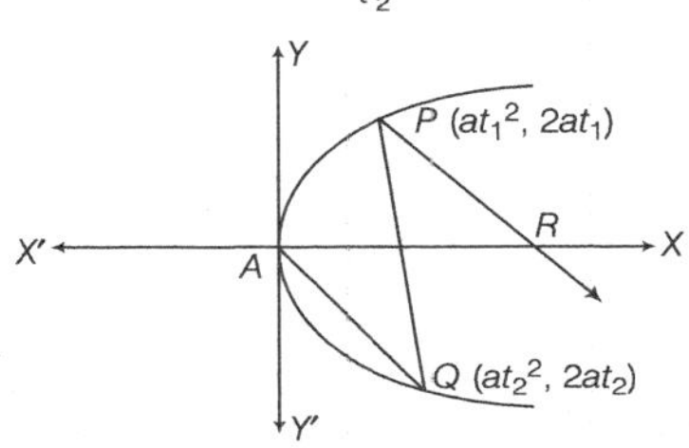

Put $y = 0$, then $x = at_1^2 - at_1t_2 = at_1^2 - a(-2 - t_1^2)$

$$= 2a + 2at_1^2 = 2(a + at_1^2)$$

which is twice the focal distance of P.

24. (c) For, $y^2 = 4ax$

$$\text{Area} = \frac{1}{2}(4a)(4a) = 8a^2$$

Here, $a = 2$

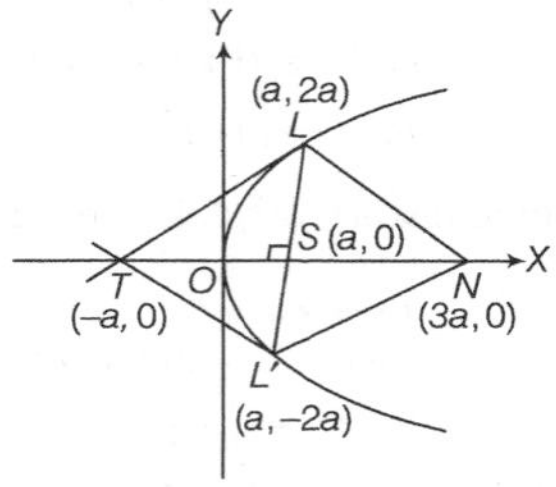

$\Rightarrow$ $A = 32$ sq units

25. (a) $y = mx + \dfrac{a}{m}$

$$y = mx - 2am - am^3$$

$$d = \left|\frac{\frac{a}{m} + 2am + am^3}{\sqrt{1+m^2}}\right|$$

$$= a\left|\frac{(m^2+1)^2}{m\sqrt{1+m^2}}\right|$$

$$= a\left|\frac{(1+m^2)\sqrt{1+m^2}}{m}\right|$$

Put $m = \tan\theta$

$$\Rightarrow \quad d = a\cdot\left|\frac{\sec^2\theta\cdot\sec\theta}{\tan\theta}\right|$$

$$= a\cdot\sec^2\theta\cdot\operatorname{cosec}\theta$$

Here, $a = 3$, $\theta = 30°$

$$\Rightarrow \quad d = 2\cdot\frac{4}{3}\cdot 2 = \frac{16}{3} \text{ sq units}$$

26. (c) Suppose that the two points on the parabola are $P(at_1^2, 2at_1)$ and $Q(at_2^2, 2at_2)$ and the equations of normals at these points are

$$y = -t_1x + 2at_1 + at_1^3 \quad \ldots\text{(i)}$$

and $$y = -t_2x + 2at_2 + at_2^3 \quad \ldots\text{(ii)}$$

Since, the normals are perpendicular.

$$\therefore \quad (-t_1)(-t_2) = -1$$

or $$t_1t_2 = -1 \quad \ldots\text{(iii)}$$

If (h, k) be their point of intersection, then on solving Eqs. (i) and (ii), we get

$$h = 2a + a(t_1^2 + t_2^2 + t_1t_2),$$

$$k = -at_1t_2(t_1 + t_2)$$

We have to eliminate t_2 and t_2 by the help of Eq. (iii) to get the locus

$$k = a(t_1 + t_2)$$

$$\because \quad t_1t_2 = -1$$

$$\Rightarrow \quad h - 2a = a[(t_1+t_2)^2 - t_1t_2] = a\left[\frac{k^2}{a^2} + 1\right]$$

or $$a(h - 3a) = k^2$$

Hence, the locus is $y^2 = a(x - 3a)$

Now, for real values of k, h will always be greater than or equal to $3a$, i.e. abscissa h can never be smaller than $3a$.

27. (d) Minimum value of abscissa can be $3a$ and in that case the ordinate k will be zero.

Explanations for (Q. Nos. 28 and 29)

Let 3 normals pass through $(h, k) \Rightarrow k = mh - 2am - am^3$

or $$am^3 + m(2a - h) + k = 0$$

$$\therefore \quad m_1 + m_2 + m_3 = 0$$

$$m_1m_2 + m_2m_3 + m_3m_1 = \frac{2a - h}{a}$$

$$m_1m_2m_3 = \frac{-k}{a}$$

Intersection point of normals on X-axis, at P, Q and R.

$$(2a + am_i^2, 0), \ i = 1, 2, 3$$

as, OP, OQ, OR are in AP

$$\Rightarrow \quad 2(OQ) = OP + OR$$

$$\Rightarrow \quad 2(2a + m_2^2) = 2a + m_1^2 + 2a + m_3^2$$

$$\Rightarrow \quad 2m_2^2 = m_1^2 + m_3^2$$

$$\Rightarrow \quad 2m_2^2 = (m_1 + m_3)^2 - 2m_1m_3$$

$$\Rightarrow \quad 2m_2^2 = m_2^2 - 2m_1m_3$$

or $$m_2^2 = -2m_1m_3 \Rightarrow m_2^3 = -2m_1m_2m_3$$

$$\Rightarrow \quad m_2^3 = -2\left(\frac{-k}{a}\right)$$

$$m_2 = \left(\frac{2k}{a}\right)^{1/3}$$

But m_2 is a root of the equation.

$$\Rightarrow \quad am^3 + m(2a - h) + k = 0$$

$$\therefore \quad am_2^3 + m_2(2a - h) + k = 0$$

$$\Rightarrow \quad a\cdot\left(\frac{2k}{a}\right) + \left(\frac{2k}{a}\right)^{1/3}(2a - h) + k = 0$$

$$\Rightarrow \quad \left(\frac{2k}{a}\right)^{1/3}(2a - h) = -3k$$

or $$27ak^2 = 2(h - 2a)^3$$

Hence, locus of (h, k) is the curve $\Rightarrow 27ay^2 = 2(x - 2a)^3$

28. (c) **29.** (a)

Explanations for (Q. Nos. 30 and 31)

$$\tan^{-1}m_1^2 + \tan^{-1}m_2^2 + \tan^{-1}m_3^2 = \alpha$$

$$\Rightarrow \quad \tan^{-1}\left\{\frac{m_1^2 + m_2^2 + m_3^2 - m_1^2m_2^2m_3^2}{1 - m_1^2m_2^2 - m_2^2m_3^2 - m_3^2m_1^2}\right\} = \alpha$$

$$\Rightarrow (\Sigma m_1)^2 - 2(\Sigma m_1m_2) - (m_1m_2m_3)^2$$

$$= \tan\alpha\,[1 - (\Sigma m_1m_2)^2 + 2m_1m_2m_3\Sigma m_1]$$

$$\Rightarrow 0 - 2\left(\frac{2a - h}{a}\right) - \left(-\frac{k}{a}\right)^2 = \tan\alpha\left[1 - \left(\frac{2a - h}{a}\right)^2 + 0\right]$$

or $$2ah - 4a^2 - k^2 = \tan\alpha(-3a^2 + 4ah - h^2)$$

or $x^2\tan\alpha + 2ax\cdot(1 - 2\tan\alpha) - y^2 + a^2(3\tan\alpha - 4) = 0$

$$\tan^1 m_1^2 + \tan^{-1}m_2^2 + \tan^{-1}m_3^2$$

$$= \tan^{-1}m_1 + \tan^{-1}m_2 + \tan^{-1}m_3$$

$$\Rightarrow \quad \frac{0 - 2\left(\frac{2a-h}{a}\right) - \frac{k^2}{a^2}}{1 - \left\{\left(\frac{2a-h}{a}\right)^2 - 0\right\}} = \frac{0 + \frac{k}{a}}{1 - \left(\frac{2a-h}{a}\right)}$$

$$\Rightarrow \quad \frac{-2\left(\frac{2a-h}{a}\right) - \frac{k^2}{a^2}}{1 + \frac{2a-h}{a}} = \frac{k}{a}$$

$$\Rightarrow \quad -2a(2a - h) - k^2 = k(3a - h)$$

or $$k^2 - 2ah + 3ak - hk + 4a^2 = 0$$

$\therefore$ locus of point (h, k) is $y^2 - 2ax + 3ay - xy + 4a^2 = 0$

30. (a) **31.** (c)

32. (3) The equation of any normal to $y^2 = 4ax$ is $y = mx - 2m - m^3$.

If it passes through $(2k, 0)$, then

$$m^3 + 2m(1-k) = 0$$

$\Rightarrow$ $m = 0$ or $m^2 + 2(1-k) = 0$

Clearly, $m = 0$ gives $y = 0$, i.e. X-axis as a normal. Other two normals will be perpendicular, if product of the roots of

$m^2 + 2(1-k) = 0$ is -1

i.e. $2(1-k) = -1 \Rightarrow k = \frac{3}{2}$

Hence, the value of $2k$ is $2 \times \frac{3}{2} = 3$.

33. (4) Any normal to $y^2 = 4ax$ is $y^2 = mx - 2am - am^3$.

Any normal to $x^2 = 4ay$ is obtained from above by interchanging x and y.

$\therefore$ Normal to $x^2 = 8y$, $(a = 2)$ is $x = my - 4m - 2m^3$

If it passes through (h, k), then

$$h = mk - 4m - 2m^3$$

or $2m^3 + 0m^2 + m(4-k) + h = 0$ ...(i)

which shows that three normals will pass through (h, k). If two of them are perpendicular, then

$$m_2 m_3 = -1.$$

But $m_1 m_2 m_3 = -h/2$

Since, m_1 is a root of Eq. (i), it will satisfy it

$$2\left(\frac{h^3}{8}\right) + \frac{h}{2}(4-k) + h = 0$$

$\Rightarrow$ $h^2 + 2(4-k) + 4 = 0.$

$\therefore$ Locus of (h, k) is $x^2 - 2y + 12 = 0$

But according to the question,

locus of (h, k) is $x^2 + 2(4-y) + \lambda = 0$

$\therefore$ $x^2 - 2y + 12 = x^2 + 2(4-y) + \lambda \Rightarrow \lambda = 4$

34. (2) Let $y = mx - 2am - am^3$ passes (h, k)

$\Rightarrow$ $k = mh - 2am - am^3$

$\Rightarrow am^3 + m(2a - h) + k = 0$, having roots m_1, m_2, m_3

$\Rightarrow$ $m_1 + m_2 + m_3 = 0$

$$m_1 m_2 + m_2 m_3 + m_3 m_1 = \frac{2a-h}{a}$$

$$m_1 \cdot m_2 \cdot m_3 = -\frac{k}{a}$$

We have,

$$\tan^{-1} m_1 + \tan^{-1} m_2 + \tan^{-1} m_3 = \text{constant } (= \lambda \text{ say})$$

$\Rightarrow$ $\tan^{-1}\left[\frac{\Sigma m_1 - m_1 m_2 m_3}{1 - \Sigma m_1 m_2}\right] = \lambda$

$\Rightarrow$ $\frac{k}{2a-h} = \tan \lambda$

or $2x + (\cot \lambda) y - 2a = 0$

$\Rightarrow$ $\mu = 2$

35. (4) Let the normal at $P(at^2, 2at)$ meets the parabola again at $Q(at_1^2, 2at_1)$.

Then, $t_1 = -t - \frac{2}{t}$.

Now, distance of Q from the axis of the parabola

$$= |2at_1| = 2a|t_1|$$

$$= 2a\left|-t - \frac{2}{t}\right|$$

$$= 2a\left|t + \frac{2}{t}\right| \geq 2a\left|2\sqrt{t \times \frac{2}{t}}\right| \geq 4a\sqrt{2} \quad [\because \text{AM} \geq \text{GM}]$$

But according to the question,

Distance of Q from the axis of the parabola $= \lambda\sqrt{2a}$

But $4a\sqrt{2} = \lambda\sqrt{2a}$

$\therefore$ $\lambda = 4$

DPP-5 Equation of $S' = T$, $T = 0$ and $SS' = T^2$

1. (c) We have, $m_1 m_2 = \frac{a}{h}$

If the tangents are perpendicular, then

$$m_1 m_2 = -1 \Rightarrow \frac{a}{h} = -1$$

$\Rightarrow$ $h = -a$

Hence, the locus of (h, k) is $x = -a$, which is the directrix of parabola.

2. (a) We know that the locus of the point of intersection of tangents to $y^2 = 4ax$ which meet at an angle α is given by

$$y^2 - 4ax = (a + x)^2 \tan^2 \alpha \quad \text{...(i)}$$

Here, $\alpha = 45°$

Putting $\alpha = 45°$ in Eq. (i), we get

$$y^2 - 4ax = (a + x)^2$$

Hence, the required locus is

$$y^2 - 4ax = (a + x)^2.$$

3. (d) Equation of the chord of contact of tangents drawn from a point (3, 4) to the parabola $y^2 = 2x$ is

$$yy_1 = 2(x + x_1)$$

$\Rightarrow$ $y(4) = 2 \times \frac{1}{2}(x + 3)$

$\Rightarrow$ $4y = x + 3$

$\therefore$ $x - 4y + 3 = 0$

4. (a) Let the tangents intersect at $P(h, k)$. Then, $lx + my + n = 0$ will be chord of contact of P. That means $lx + my + n = 0$ and $yk - 2ax - 2ah = 0$ will represent the same line.

Thus, $\frac{k}{m} = \frac{-2a}{l} = \frac{-2ah}{n}$

$\Rightarrow$ $h = \frac{n}{l}, k = \frac{-2am}{l}$

5. (a) If (h, k) be the point of intersection of tangents, then

$$ky = 2a(x + h) \quad \text{...(i)}$$

is same as normal

$$y = mx - 2am - am^3 \quad \text{...(ii)}$$

On comparing Eqs. (i) and (ii), we get

$$m = \frac{2a}{k}, -2am - am^3 = 2a \cdot \frac{h}{k}$$

On eliminating the variable m between the above two relations, we get

$$-2a\left(\frac{2a}{k}\right) - a \cdot \frac{8a^3}{k^3} = 2a \cdot \frac{h}{k}$$

$\Rightarrow$ $h + 2a + \frac{4a^3}{k^2} = 0$

$\Rightarrow$ $k^2(h + 2a) + 4a^3 = 0$

Hence, the required locus in

$$y^2(x + 2a) + 4a^3 = 0$$

6. (d) Equation of the chord bisected at (x_1, y_1) is

$$yy_1 - 2a(x + x_1) = y_1^2 - 4ax_1$$

Given, $(x_1, y_1) = (2, -3)$ and $a = 2$, then

$$-3y - 4x - 8 = 9 - 16$$

$$\Rightarrow \quad 3y + 4x + 1 = 0$$

or $$4x + 3y + 1 = 0$$

7. (b) Let (h, k) be the coordinates of mid-point of one of the chords which pass through the focus $(a, 0)$. Then, the equation of the chord whose mid-point is (h, k), is

$$ky - 2a(x + h) = k^2 - 4ah$$

$$\Rightarrow \quad ky - 2ax + 2ah - k^2 = 0$$

If the line passes through the focus $(a, 0)$, then

$$-2a^2 + 2ah - k^2 = 0$$

$$\Rightarrow \quad k^2 - 2ah + 2a^2 = 0$$

Hence, the locus of (h, k) is $y^2 - 2ax + 2a^2 = 0$.

8. (a) We know that, the polar of the point (x_1, y_1) with respect to the parabola $y^2 = 4ax$ is $yy_1 = 2a(x + x_1)$.

Given, $y^2 = 6x$ and $(2, 1)$.

$$\therefore \quad y = 6\left(\frac{x + 2}{2}\right)$$

$$\Rightarrow \quad y = 3x + 6$$

9. (a) Let the pole be (x_1, y_1). Then, the equation of polar with respect to $y^2 = 4ax$ is

$$yy_1 = 2a(x + x_1)$$

$$\Rightarrow \quad 2ax - yy_1 + 2ax_1 = 0 \quad \ldots(i)$$

Clearly, Eq. (i) and $lx + my + n = 0$ represent the same line.

$$\therefore \quad \frac{2a}{l} = \frac{-y_1}{m} = \frac{2ax_1}{n} \quad \Rightarrow \quad x_1 = \frac{n}{l}$$

and $$y_1 = \frac{-2am}{l}$$

Hence, the pole of given line is $\left(\frac{n}{l}, \frac{-2am}{l}\right)$.

10. (c) Let $P(h, k)$ be the pole of a focal chord of the parabola $y^2 = 4ax$. Then, the equation of chord is $ky - 2a(x + h) = 0$.

It passes through $(a, 0)$.

$$\therefore \quad a + h = 0$$

Hence, the locus of (h, k) is $x + a = 0$, i.e. $x = -a$.

Clearly, it is the directrix of the parabola.

Explanations for Q. Nos. (11 and 12)

Let $y = mx - 2am - am^3 \ldots$(i) be a normal chord of the parabola $y^2 = 4ax$. Such that it is normal at point $P(am^2, -2am)$ to the parabola and cuts the parabola again at Q.

Let $R(h, k)$ be the mid-point of PQ.

Then, its equation is $ky - 2a(x + h) = k^2 - 4ah \quad \{s' = T\}$

or $$ky - 2ax = k^2 - 2ah \quad \ldots(ii)$$

Clearly, Eqs. (i) and (ii) represents the same line.

$$\therefore \quad \frac{l}{k} = \frac{m}{2a} = \frac{-2am - am^3}{k^2 - 2ah}$$

$$\Rightarrow m = \frac{2a}{k} \text{ and } k^2 - 2ah = -2a^2(2 + m^2)$$

$$\Rightarrow \quad k^2 - 2ah = -2a^2\left(2 + \frac{4a^2}{k^2}\right)$$

$$\Rightarrow \quad k^2(k^2 - 2ah) + 4a^2(k^2 + 2a^2) = 0$$

Hence, locus of (h, k) is

$$y^2(y^2 - 2ax) + 4a^2(y^2 + 2a^2) = 0$$

Let (h, k) be a point. Then, the polar of this point with respect to $y^2 = 4x$, is

$$ky = 2(x + h) \text{ or } 2x - ky + 2h = 0 \quad \ldots(i)$$

This touches the parabola $x^2 = 4y$. So, it should be of the form

$$x = my + \frac{b}{m} \quad \ldots(ii)$$

Comparing Eqs. (i) and (ii), we get

$$m = \frac{k}{2a} \text{ and } \frac{b}{m} = -\frac{2ah}{2a}$$

$$\Rightarrow \quad -\frac{b}{h} = \frac{k}{2a}$$

$$\Rightarrow \quad hk = -2ab$$

or $$xy = -2ab$$

$$\Rightarrow \quad \lambda = -2 \text{ or } |\lambda| = 2$$

11. (c) **12.** (a)

Explanations for Q. Nos. (13-15)

The equation of any tangent to the parabola $y^2 = 4ax$ is $y = mx + \frac{a}{m}$.

This line will touch the circle $x^2 + y^2 = \frac{a^2}{2}$.

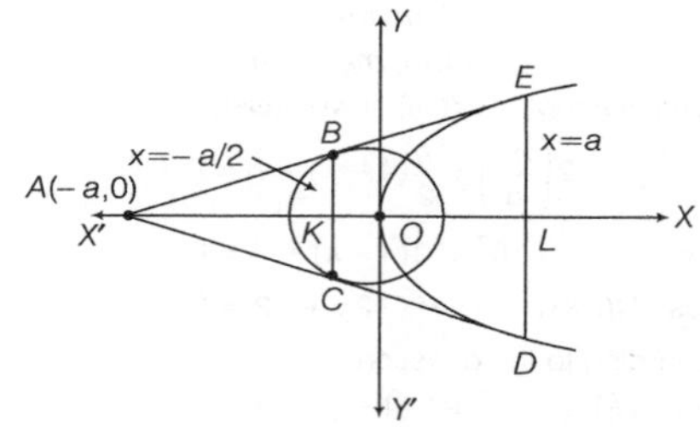

If $\frac{a}{m} = \pm \frac{a}{\sqrt{2}} \sqrt{m^2 + 1} \qquad \left[c = \pm r\sqrt{1 + m^2}\right]$

$$\Rightarrow \quad \frac{a^2}{m^2} = \frac{a^2}{2}(m^2 + 1)$$

$$\Rightarrow \quad 2 = m^4 + m^2$$

$$\Rightarrow \quad m^4 + m^2 - 2 = 0$$

$$\Rightarrow \quad (m^2 + 2)(m^2 - 1) = 0$$

$$\therefore \quad m = 1, -1$$

Thus, the two tangents (common one) are $\mathbf{y = x + a}$ and $\mathbf{y = -x - a}$. These two tangents intersect each other at $(-a, 0)$.

The chord of contact of the circle with respect to $A(-a, 0)$ is

$$(-a)x + (0)y = \frac{a^2}{2}$$

$$\Rightarrow \quad x = \frac{-a}{2}$$

and the chord of contact of the parabola with respect to $A(-a, 0)$ is

$$(0)y = 2a(x - a)$$

or $$x = a$$

Note that DE is the latusrectum of the parabola $y^2 = 4ax$.

$\therefore$ Its length is $4a$.

The chords of contact are clearly parallel to each other. So, the required quadrilateral is a trapezium.

$$\therefore \quad \text{Area (trap } BCDE) = \frac{1}{2}(BC + DE) \times KL$$

$$= \frac{1}{2} \times (a + 4a)\left(\frac{3a}{2}\right)$$

$$A = \frac{15a^2}{4} \text{ sq units}$$

13. (c) **14.** (a) **15.** (d)

16. (5) Let QR be the chord of contact of tangents drawn from a point $P(-1, -1)$ to the parabola $y^2 = 4x$. Then, the length of

$$QR = \frac{1}{a}\sqrt{\{(y_1)^2 - 4ax_1\}\{y_1^2 + 4a^2)}$$

$$= \frac{1}{1}\sqrt{\{(-1)^2 - 4 \times 1 \times (-1)\}[(-1)^2 + 4(-1)^2]}$$

$$= \sqrt{5 \times 5} = 5 \text{ units}$$

17. (4) Area of triangle $= \frac{1}{2a}(y_1^2 - 4ax_1)^{3/2} = \frac{1}{2}(1 + 3)^{3/2}$

$$= 4 \text{ sq units}$$

18. (8) Given, equation of parabola is $y^2 = 8x$.

Put $y = \frac{4x + 4}{3}$ in the equation of parabola.

$$\therefore \quad \left(\frac{4x + 4}{3}\right)^2 = 8x$$

$$\Rightarrow \quad \frac{16(x + 1)^2}{9} = 8x$$

$$\Rightarrow \quad 2x^2 - 5x + 2 = 0$$

$$\Rightarrow \quad x = 2, \frac{1}{2}$$

when $\quad x = 2, y = 4$

and $\quad x = \frac{1}{2}, y = 2$

Mid-point of (2, 4) and $\left(\frac{1}{2}, 2\right)$ is $\left(\frac{5}{4}, 3\right)$.

On comparing with (λ, μ), we get

$$\lambda = \frac{5}{4}, \mu = 3$$

$$\therefore \quad 4\lambda + \mu = 4 \times \frac{5}{4} + 3$$

$$= 8$$

19. (2) Let (h,k) be a point then the polar of this point with respect to $y^2 = 4ax$ is

$$ky = 2(x + h) \text{ or } 2ax - ky + 2ah = 0 \quad \ldots\text{(i)}$$

This touches the parabola $x^2 = 4y$

So, it should be of the form

$$x = my + b/m \quad \ldots\text{(ii)}$$

Comparing (i) and (ii), we get

$$m = k/2a \text{ and } b/m = -2ah/2a$$

$$\Rightarrow \quad -b/h = k/2a$$

$$\Rightarrow \quad hk = -2ab$$

or $\quad xy = -2ab$, where $ab = 1$

$$\Rightarrow \quad \lambda = -2$$

or $\quad |\lambda| = 2$

20. (4) The equation of any tangent to $y^2 = 4ax$ is

$$y = mx + \frac{a}{m} \quad \ldots\text{(i)}$$

Let (h, k) be the pole of Eq. (i) with respect to the parabola $y^2 = 4bx$. Then, the equation of polar is

$$ky = 2b(x + h)$$

$$\Rightarrow \quad 2bx - ky + 2bh = 0 \quad \ldots\text{(ii)}$$

Clearly, Eqs. (i) and (iii) represent the same straight line.

$$\therefore \quad \frac{2b}{m} = \frac{-k}{-l} = \frac{2bh}{a/m}$$

$$\Rightarrow \quad k = \frac{2b}{m}$$

and $\quad h = \frac{a}{m^2}$

$$\Rightarrow \quad m = \frac{2b}{k}$$

and $\quad m^2 = \frac{a}{h}$

$$\Rightarrow \quad \left(\frac{2b}{k}\right)^2 = \frac{a}{h}$$

$$\Rightarrow \quad k^2 = \frac{4b^2h}{a}$$

Hence, the locus of (h, k) is $y^2 = \frac{4b^2x}{a}$.

But according to the question,

$$y^2 = \frac{\lambda b^2 x}{a}$$

$$\therefore \quad \frac{4b^2x}{a} = \frac{\lambda b^2 x}{a}$$

$$\Rightarrow \quad \lambda = 4$$

Revisal Problems (JEE Main)

1. (b) Comparing the given equation with the general equation of second degree, i.e.

$$ax^2 + 2hxy + by^2 + 2gx + 2fy + c = 0$$

we have,

$$a = \lambda, b = 1, c = 2, f = \frac{3}{2}, g = \frac{\lambda}{2} \text{ and } h = 2.$$

$$\therefore \quad \Delta = abc + 2fgh - af^2 - bg^2 - ch^2$$

$$= 2\lambda + 3\lambda - \frac{9\lambda}{4} - \frac{\lambda^2}{4} - 8$$

$$= -\frac{1}{4}(\lambda^2 - 11\lambda + 32)$$

and $\quad h^2 - ab = 4 - \lambda$

If the given equation represents a parabola, we must have $\Delta \neq 0$ and $h^2 = ab$.

Clearly, $\lambda = 4$ satisfies these two conditions.

2. (c) The equation of parabola is

$$y^2 - kx + 8 = 0$$

$$\Rightarrow \quad y^2 = k\left(x - \frac{8}{k}\right)$$

$$\Rightarrow \quad (y - 0)^2 = k\left(x - \frac{8}{k}\right)$$

The equation of directrix of this parabola is

$$x - \frac{8}{k} = -\frac{k}{4} \quad [\because x = -a]$$

$$\Rightarrow \quad x = \frac{8}{k} - \frac{k}{4}$$

But the equation of directrix is given as $x - 1 = 0$.

$$\therefore \quad \frac{8}{k} - \frac{k}{4} = 1$$

$$\Rightarrow \quad k^2 + 4k - 32 = 0$$

$$\Rightarrow \quad k = -8, 4$$

3. (d) We have, $y^2 + 4y + 4x + 2 = 0$

$\Rightarrow \quad (y+2)^2 + 4x + 2 - 4 = 0$

$\Rightarrow \quad (y+2)^2 = -4\left(x - \frac{1}{2}\right) \quad \ldots(i)$

Clearly, it represents a parabola with vertex at $\left(\frac{1}{2}, -2\right)$ and latusrectum = 4. Shifting the origin at $\left(\frac{1}{2}, -1\right)$, we have

$$x = X + \frac{1}{2} \text{ and } y = Y - 2 \quad \ldots(ii)$$

On substituting these in Eq. (i), we get

$$Y^2 = -4X$$

It represents a parabola whose directrix has the equation

$X = a$

$\Rightarrow \quad X = 1$

or $\quad x - \frac{1}{2} = 1$

$\Rightarrow \quad x = \frac{3}{2}$

4. (b) The equation of parabola is

$$2\{(x-a)^2 + (y-a)^2\} = (x+y)^2$$

$\Rightarrow \quad \sqrt{(x-a)^2 + (y-a)^2} = \left|\frac{x+y}{\sqrt{2}}\right|$

$\Rightarrow \quad \sqrt{(x-a)^2 + (y-a)^2} = \left|\frac{x+y}{\sqrt{(1)^2 + (1)^2}}\right|$

Clearly, its focus is (a, a) and the equation of directrix is $x + y = 0$,

$\therefore$ Length of the latusrectum = 2 × Length of perpendicular from the focus (a, a) on the directrix

$\Rightarrow$ Length of the latusrectum $= 2\left|\frac{a+a}{\sqrt{1^2+1^2}}\right| = 2\sqrt{2}a$

5. (c)

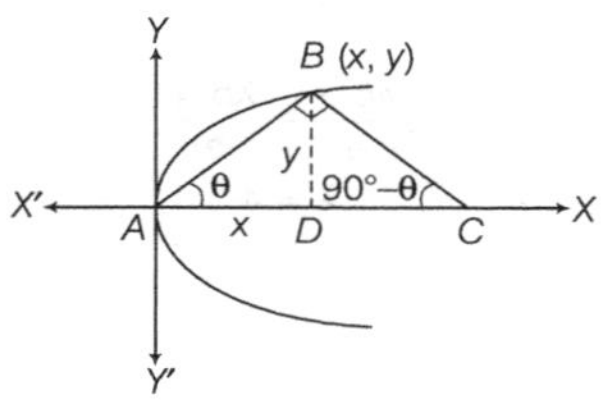

In ΔABD, we have $\quad \tan\theta = \frac{y}{x}$

In ΔBCD, we have

$$\tan(90° - \theta) = \frac{y}{DC}$$

$\therefore \quad CD = y\tan\theta = \frac{y^2}{x}$

$= \frac{4ax}{x} = 4a \quad [\because y^2 = 4ax]$

6. (d) The equation of parabola is given by

$$\sqrt{(x-0)^2 + (y-1)^2} = \left|\frac{x+2}{\sqrt{1+6}}\right|$$

$\Rightarrow \quad x^2 = (y-1)^2 = (x+2)^2$

$\Rightarrow \quad (y-1)^2 = 4(x+1)$

The parametric coordinates of any point on this parabola are given by

$$x + 1 = t^2 \text{ and } y - 1 = 2t.$$

i.e. $(t^2 - 1, 2t + 1)$

7. (b) The general equation of a parabola having its axis parallel to Y-axis is

$$y = ax^2 + bx + c \quad \ldots(i)$$

This touched by the line $y = x$ at $x = 1$. Therefore, slope of the tangent at $(1, 1) = 1$

and $x = ax^2 + bx + c$ must have equal roots.

$\Rightarrow \quad \left(\frac{dy}{dx}\right)_{(1,1)} = -1$ and $(b-1)^2 = 4ac$

$\Rightarrow \quad 2a + b = 1$ and $(b-1)^2 = 4ac$

Also (1, 1) lies O on Eq. (i).

$\therefore \quad a + b + c = 1$

Now, $\quad 2a + b = 1$

and $\quad a + b + c = 1$

$\Rightarrow \quad a - c = 0 \Rightarrow a = c$

and $\quad a + b + c = 1$

$\Rightarrow \quad 2c + b = 1$

$\therefore \quad 2f(0) + f'(0) = 1 \quad [\because f(0) = c \text{ and } f'(0) = b]$

8. (c) Since, AA', BB' and CC' are focal chords, therefore

$$t_i' = \frac{-1}{t_i}, i = 1, 2, 3$$

$\Rightarrow \quad t_1', t_2', t_3'$ are in HP, as t_1, t_2, t_3 are in AP.

9. (c) We have, $x^2 + y^2 - 2xy - 4x - 4y + 4 = 0$

$\Rightarrow \quad (x-y)^2 = 4(x+y-1)$

$\Rightarrow \quad \left\{\frac{x-y}{\sqrt{1^2-1^2}}\right\}^2 = 2\sqrt{2}\left\{\frac{x+y-1}{\sqrt{1^2-1^2}}\right\} \quad \ldots(i)$

Let $\quad \frac{x-y}{\sqrt{1^2+1^2}} = Y$

and $\quad \frac{x+y-1}{\sqrt{1^2+1^2}} = X$

Then, Eq. (i) reduces to $Y^2 = 2\sqrt{2}X$. The coordinates of vertex of this parabola are $(X = 0, Y = 0)$.

So, the coordinates of vertex of the given parabola are given by

$$\frac{x+y-1}{\sqrt{1^2+1^2}} = 0$$

and $\quad \frac{x-y}{\sqrt{1^2+1^2}} = 0$

$\Rightarrow \quad x + y = 1$

and $\quad x - y = 0$

$\Rightarrow \quad x = y = \frac{1}{2}$

Hence, the coordinates of the vertex are (1/2, 1/2).

10. (d) Two parabolas are equal, if the length of their latusrectum are equal.

Length of the latusrectum of $y^2 = \lambda x$ is λ.

The equation of second parabola is

$$25\{(x-3)^2 + (y+2)^2\} = (3x - 4y - 2)^2$$

$\Rightarrow \quad \sqrt{(x-3)^2 + (y+2)^2} = \left|\frac{3x - 4y + 2}{\sqrt{3^2 + 4^2}}\right|$

Clearly, it represents a parabola having focus at (3, – 2) and equation of the directrix as $3x - 4y - 2 = 0$.

$\therefore$ Length of the latusrectum

= 2 (Distance between focus and directrix)

$= 2\left|\frac{3 \times 3 - 4 \times -2 - 2}{\sqrt{3^2 + (-4)^2}}\right| = 6$

Thus, the two parabolas are equal to $\lambda = 6$.

11. (b) Let $S(a, 0)$ be the focus of parabola $y^2 = 4ax$ and $P(at^2, 2at)$ be a point on it.

Then, the equation of a circle on SP as diameter is

$$(x-a)(x-at^2) + (y-0)(y-2at) = 0$$

It meets Y-axis at $x = 0$.

$\therefore \quad y^2 - 2at\,y + a^2t^2 = 0$

$\Rightarrow \quad (y - at)^2 = 0$

which shows that Y-axis meets the circle in two coincident points. Hence, the circle touches the tangent at the vertex.

12. (a) Let a be the radius of circle. Then, its equation is

$$(x+4)^2 + y^2 = a^2$$

This cuts the parabola $y^2 = 8x$ at points $A(x_1, y_1)$ and $B(x_2, y_2)$.

The abscissae of A and B are the roots of the equation

$$(x+4)^2 + 8x = a^2$$

or $\quad x^2 = 16x + 16 - r^2 = 0$

$\therefore \quad x_1x_2 = 16 - r^2$

The ordinates of A and B are given by $y_1^2 = 8x_1$ and $y_2^2 = 8x_2$, respectively.

$\therefore \quad y_1 = \pm 2\sqrt{2}x_1$

and $\quad y_2 \pm = 2\sqrt{2}x_2$

Since, AB subtends a right angle at the vertex of parabola.

Then, $\quad \dfrac{y_1}{x_1} \times \dfrac{y_2}{x_2} = -1$

$\Rightarrow \quad x_1x_2 + y_1y_2 = 0$

$\Rightarrow \quad x_1x_2 + 8x_1x_2 = 0$

$\Rightarrow \quad x_1x_2 = 0$

$\Rightarrow \quad 16 - a^2 = 0$

Then, $\quad a = 4$

13. (b) The equation of any tangent to $y^2 = 8ax$ is

$$y = mx + \frac{2a}{m} \quad \ldots \text{(i)}$$

If it touches $x^2 + y^2 = 2a^2$, then

$$\left(\frac{2a}{m}\right)^2 = 2a^2(1+m^2) \quad [\text{using, } c^2 = a^2(1+m^2)]$$

$\Rightarrow \quad 2 = m^2(m^2+1)$

$\Rightarrow \quad m^4 + m^2 - 2 = 0$

$\Rightarrow \quad (m^2+2)(m^2-1) = 0$

$\Rightarrow \quad m^2 - 1 = 0$

$\therefore \quad m = \pm 1$

Putting the values of m in Eq. (i), we get

$y = \pm (x + 2a)$ as the equations of common tangents.

14. (a) The equation of tangent to $y^2 = 8x$ at $P(2, 4)$ is

$$4y = 4(x+2)$$

$\Rightarrow \quad x - y + 2 = 0 \quad \ldots \text{(i)}$

Let (x_1, y_1) be the mid-point of chord QR.

Then, the equation of QR is

$$yy_1 - 4(x + x_1) - 5 = y_1^2 - 8x_1 = -5$$

$\Rightarrow \quad 4x - yy_1 - 4x_1 + y_1^2 = 0 \quad \ldots \text{(ii)}$

Clearly, Eqs. (i) and (ii) represent the same line.

$\therefore \quad \dfrac{4}{1} = \dfrac{-y_1}{-1} = -\dfrac{4x_1 + y_1^2}{2}$

$\Rightarrow \quad y_1 = 4$ and $8 = -4x_1 + y_1^2$

$\Rightarrow \quad y_1 = 4$ and $x_1 = 2$

15. (b) Let $y = mx + c$ be the common tangent to the two parabolas.

Then, the equations

$$mx + c = x^2 - 3x + 10$$

and $\quad mx + c = x^2 - x + 1$

must have equal roots.

i.e. $\quad x^2 - x(3+m) + 10 - c = 0$

and $\quad x^2 - x(1+m) + 1 - c = 0$

must have equal roots.

Now, $\quad (3+m)^2 = 4(10-c)$

and $\quad (1+m)^2 = 4(1-c)$

Then, $\quad (3+m)^2 - (1+m)^2 = 36$

$\therefore \quad m = 7$

16. (a) The combined equation of the pair of tangents drawn from (3, 4) to the parabola $y^2 - 2y + 4x = 0$ is

$$(y^2 - 2y + 4x)(16 - 8 + 2) = \{4y - (y+4) + 2(x+3)\}^2$$

$\Rightarrow \quad 4x^2 - 12xy - 11y^2 - 72x + 12y + 4 = 0$

Let θ be the angle between the lines given by this equation.

Then, $\quad \tan\theta = \left|\dfrac{2\sqrt{36+44}}{4-11}\right| = \dfrac{8\sqrt{5}}{7}$

$$\left[\because \tan\theta = \frac{2\sqrt{h^2 - ab}}{a+b}\right]$$

$\therefore \quad \theta = \tan^{-1}\left(\dfrac{8\sqrt{5}}{7}\right)$

17. (c) Clearly, two given tangents pass through $(-a, -b)$, which is a point on the directrix of the parabola $y^2 = 4ax$. So, the angle between the tangents is $\pi/2$.

Hence, $m_1m_2 = -1$

18. (b) Let r be the radius of the largest circle passing through the focus (1, 0) of $y^2 = 4x$.

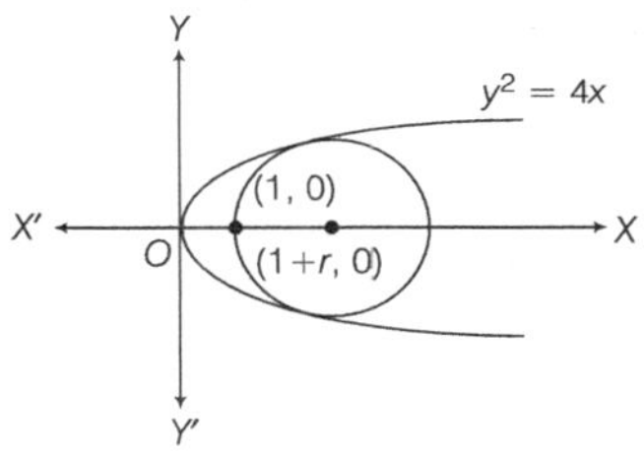

Clearly, centre of circle will be on X-axis and its coordinates are $(1 + r, 0)$.

The equation of circle is

$$(x - 1 - r)^2 + y^2 = r^2$$

It touches $y^2 = 4x$. Therefore, the equation $(x - r - 1)^2 + 4x = r^2$ must have equal roots.

$\Rightarrow \quad 4(1-r)^2 - 4(2r+1) = 0$

$\Rightarrow \quad r = 4$

19. (a) The equation of tangent to $y^2 = 4ax$ at (x_1, y_1) is

$$yy_1 = 2a(x + x_1)$$

$\Rightarrow \quad 2ax - yy_1 + 2ax_1 = 0 \quad \ldots \text{(i)}$

Let (h, k) be the mid-point of QR.

Then, the equation of QR is

$$ky - 2a(x+h) - 4ab = k^2 - 4a(h+b) \quad [\text{using } S' = T]$$

$\Rightarrow \quad 2ax - ky + k^2 - 2ah = 0 \quad \ldots \text{(ii)}$

Clearly, Eqs. (i) and (ii) represent the same line.

$\therefore \quad \frac{1}{1} = \frac{y_1}{k} = \frac{2ax_1}{k^2 - 2ah}$

$\Rightarrow \quad k = y_1$ and $k^2 - 2ah = 2ax_1$

$\Rightarrow \quad k = y_1$ and $y_1^2 - 2ah = 2ax_1$

$\Rightarrow \quad k = y_1$ and $4ax_1 - 2ah = 2ax_1$

$[\because (x_1, y_1)$ lies on $y^2 = 4ax \Rightarrow y_1^2 = 4ax_1]$

$\Rightarrow \quad k = y_1$ and $h = x_1$

Hence, the mid-point of QR is (x_1, y_1).

20. (b) The equation of double ordinate through the focus is $x = 4$. It is given that the point $(a, 2a)$ is an interior point of $y^2 - 16x = 0$.

$\therefore \quad 4a^2 - 16a < 0$

$\Rightarrow \quad a^2 - 4a < 0$

$\Rightarrow \quad a \in (0, 4) \quad \dots(i)$

The point $(a, 2a)$ will lie in the region bounded by the double ordinate and the parabola, if the origin and $(a, 2a)$ lie on the same side of $x = 4$.

$\therefore \quad (0 - 4)(a - 4) > 0$

$\Rightarrow \quad a < 4$

$\Rightarrow \quad a \in (-\infty, 4) \quad \dots(ii)$

From Eqs. (i) and (ii), we get $a \in (0, 4)$.

21. (d) The equation of the parabolas are

$$y^2 - 4x - 8y + 40 = 0 \quad \dots(i)$$

and

$$x^2 - 8x - 4y + 40 = 0 \quad \dots(ii)$$

We observe that, if we interchange x and y in Eq. (i), we obtain Eq. (ii). So, the two parabolas are symmetric about $y = x$.

If the two parabolas intersect on $y = x$, then the minimum distance between them is zero.

On solving $y = x$ and $y^2 - 4x - 8y + 4 = 0$,

we get

$$x^2 - 12x + 40 = 0$$

which has imaginary roots.

So, the parabolas do not intersect. Consequently, distance between them is not zero.

Minimum distance between the two parabolas is the distance between tangents to the two parabolas which are parallel to $y = x$.

On differentiating Eq. (ii) w.r.t. x, we get

$$2x - 8 - 4\frac{dy}{dx} = 0$$

$\Rightarrow \quad \frac{dy}{dx} = \frac{x - 4}{2}$

If the tangent is parallel to $y = x$, then

$$\frac{x - 4}{2} = 1$$

$\Rightarrow \quad x = 6$

On putting $x = 6$ in Eq. (ii), we get

$$y = 7$$

Thus, the coordinates of a point on parabola (ii) are $P(6, 7)$. The corresponding point on parabola (i) is $Q(7, 6)$.

Hence, required distance $= PQ = \sqrt{1 + 1} = \sqrt{2}$

22. (c) Tangent to $y^2 = 4(x + 1)$, is

$$y = m(x + 1) + \frac{1}{m} \quad \dots(i)$$

and tangent perpendicular to Eq. (i) for $y^2 = 8(x + 2)$, is

$$y = -\frac{1}{m}(x + 2) - 2m \quad \dots(ii)$$

Subtracting Eqs. (i) and (ii) we get

$\Rightarrow \quad 0 = x\left(m + \frac{1}{m}\right) + 3\left(m + \frac{1}{m}\right)$

$\Rightarrow \quad \left(m + \frac{1}{m}\right)(x + 3) = 0$

$\Rightarrow x + 3 = 0$, locus of the point of intersection of tangents

23. (d) Let $P(x_1, y_1)$ be the point of contact of the two parabolas. Tangents at P to the two parabolas are

$$yy_1 = 2a(x + x_1) - 4a\lambda_1$$

$\Rightarrow \quad 2ax - yy_1 = 2a(2\lambda_1 - x_1) \quad \dots(i)$

and $\quad xx_1 = 2a(y + y_1) - 4a\lambda_2$

$\Rightarrow \quad xx_1 - 2ay = 2a(y_1 - 2\lambda_2) \quad \dots(ii)$

Clearly, Eqs. (i) and (ii) represent the same line.

$\therefore \quad \frac{2a}{x_1} = \frac{y_1}{2a}$

$\Rightarrow \quad x_1 y_1 = 4a^2$

Hence, the locus of (x_1, y_1) is $xy = 4a^2$, which is a hyperbola.

24. (a) We have,

$$AS = \text{Focal distance of } A = 1 + t^2$$

Now, $\quad CS = AC - AS = \frac{25}{4} - (1 + t^2)$

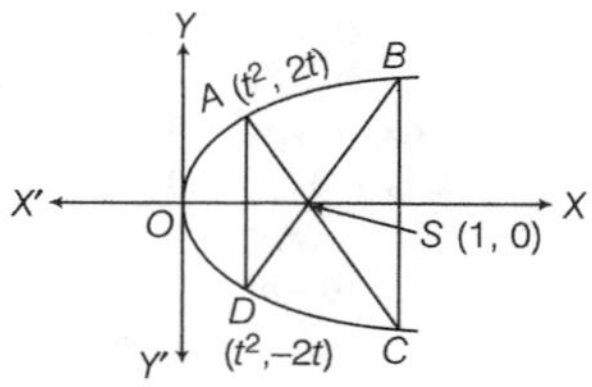

If PQ is a focal chord of parabola $y^2 = 4ax$, then

$$\frac{1}{SP} + \frac{1}{SQ} = \frac{1}{a}$$

$\therefore \quad \frac{1}{AS} + \frac{1}{CS} = \frac{1}{1}$

$\Rightarrow \quad \frac{1}{1 + t^2} + \frac{1}{\frac{25}{4} - (1 + t^2)} = 1$

$\Rightarrow \quad \frac{1}{1 + t^2} + \frac{4}{25 - 4(1 + t^2)} = 1$

$\Rightarrow \quad 25(1 + t^2) - 4(1 + t^2)^2 = 25$

$\Rightarrow \quad 4(1 + t^2)^2 - 25(1 + t^2) + 25 = 0$

$\Rightarrow \quad [(1 + t^2) - 5][4(1 + t^2) - 5] = 0$

$\Rightarrow \quad 1 + t^2 = 5, \frac{5}{4}$

$\Rightarrow \quad t^2 = 4, \frac{1}{4}$

$\Rightarrow \quad t = \pm 2, \pm \frac{1}{2}$

Thus, the coordinates of A, B, C, D are

$$A \equiv \left(\frac{1}{4}, 1\right), B = (4, 4), C \equiv (4, -4) \text{ and } D \equiv \left(\frac{1}{4}, -1\right).$$

Then, $AD = 2$, $BC = 8$ and distance between AD and $BC = \frac{15}{4}$

$\therefore$ Area of trapezium $ABCD = \frac{1}{2}(2 + 8) \times \frac{15}{4} = \frac{75}{4}$ sq units

25. (c) The equation of any tangent of slope m to the parabola $y^2 = 4x$ is

$$y = mx + \frac{1}{m} \qquad \ldots(i)$$

If it passes through $P(1, 4)$, then

$$4 = m + \frac{1}{m}$$

$$\Rightarrow \quad m^2 - 4m + 1 = 0$$

Let m_1, m_2 be the roots of this equation.

Then, $\quad m_1 + m_2 = 4 \quad$ and $\quad m_1 m_2 = 1$

Let θ be the angle between the tangents drawn from P, then

$$\tan\theta = \frac{m_1 - m_2}{1 + m_1 m_2} = \frac{\sqrt{(m_1 + m_2)^2 - 4m_1 m_2}}{1 + m_1 m_2}$$

$$\Rightarrow \quad \tan\theta = \frac{\sqrt{16 - 4}}{2} = \sqrt{3}$$

$$\therefore \quad \theta = \frac{\pi}{3}$$

26. (c) We know that, the tangents at the ends of a focal chord are perpendicular. Thus, the chord of contact are perpendicular tangents is a focal chord. Also, latusrectum is a focal chord of minimum length. Hence, the minimum length of AB is $4a$.

27. (b) Since, the tangents at the end points of a focal chord intersect at right angles on the directrix. So, the triangle is right-angled. The equations of tangents at $L(a, 2a)$ and $L'(a, -2a)$ are $x - y + a = 0$ and $x + y + a = 0$ respectively. These two intersect at $P(-a, 0)$.

We have, $PL = 2\sqrt{2}a = PL'$

Hence, the triangle is right-angled isosceles.

28. (a) The equation of tangent of slope m to the parabola $y^2 = 4x$ is $y = mx + \frac{1}{m}$. This will be a chord of the circle $x^2 + y^2 = 4$, if length of the perpendicular from the centre (0, 0) is less than the radius.

$$\therefore \quad \left|\frac{1}{m\sqrt{m^2 + 1}}\right| < 2$$

$$\Rightarrow \quad 4m^4 + 4m^2 - 1 > 0$$

$$\Rightarrow \quad \left(m^2 - \frac{\sqrt{2} - 1}{2}\right)\left(m^2 + \frac{\sqrt{2} + 1}{2}\right) > 0$$

$$\Rightarrow \quad m^2 - \frac{\sqrt{2} - 1}{2} > 0$$

$$\Rightarrow \quad \left(m - \sqrt{\frac{\sqrt{2} - 1}{2}}\right)\left(m + \sqrt{\frac{\sqrt{2} - 1}{2}}\right) > 0$$

$$\Rightarrow \quad m \in \left(-\infty, -\sqrt{\frac{\sqrt{2} - 1}{2}}\right) \cup \left(\sqrt{\frac{\sqrt{2} - 1}{2}}, \infty\right)$$

29. (d) Let $P(t, 4 - 2t)$ be any point on $2x + y = 4$. The chord of contact of tangents drawn from P to $y^2 = -4x$ is

$$(4 - 2t)y = -2(x + t)$$

or $$(2x + 4y) - 2t(y - 1) = 0$$

Clearly, it represents a line passing through the point of intersection of the lines $2x + 4y = 0$ and $y - 1 = 0$, i.e. $(-2, 1)$.

30. (b)

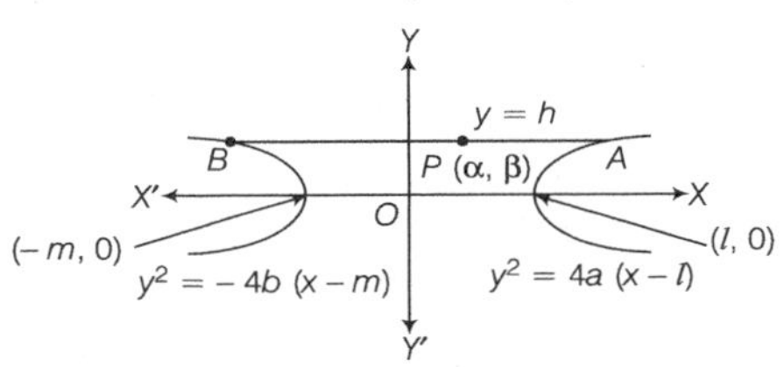

Let the two parabolas be $y^2 = 4a(x - l)$ and $y^2 = -4b(x + m)$.

Further, let $y = h$ be a line parallel to the common axis. This intersects the two parabolas at

$$A\left(\frac{h^2}{4a} + l, h\right) \text{ and } B\left(-\frac{h^2}{4b} - m, h\right), \text{ respectively.}$$

Let $P(\alpha, \beta)$ be the mid-point of AB. Then,

$$2\alpha = \frac{h^2}{4}\left(\frac{1}{a} - \frac{1}{b}\right) + l - m \text{ and } \beta = h$$

$$\Rightarrow \quad 2\alpha = \frac{\beta^2}{4}\left(\frac{1}{a} - \frac{1}{b}\right) + l - m$$

Thus, the locus of (α, β) is

$$2x = \frac{y^2}{4}\left(\frac{1}{a} - \frac{1}{b}\right) + l - m$$

Clearly, it represents a parabola, if $a \neq b$.

If $a = b$, then we have

$2x = l - m$, which represents a line.

31. (b) Since, PQ subtends right angle at vertex.

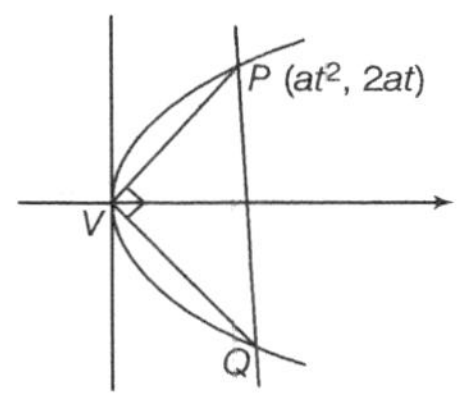

$$\Rightarrow \quad t^2 = 2$$

or $$t = \pm\sqrt{2}$$

Equation of normal is $\quad y + xt = 2at + at^3$

$$\therefore \quad \text{slope} = \pm\sqrt{2}$$

32. (c) Let $y + t_1 x = 2t_1 + t_1^3$ be a normal at $P(t_1^2, 2t_1)$ to the parabola $y^2 = 4x$. Suppose it cuts the parabola again at $Q(t_2^2, 2t_2)$. Then, $t_2 = -t_1 - \frac{2}{t_1}$.

The equation of normal at Q is $y + t_2 x = 2t_2 + t_2^3$

Suppose it makes an angle θ with the positive direction of X-axis.

$$\therefore \quad \tan\theta = -t_2$$

$$\Rightarrow \quad \tan\theta = t_1 + \frac{2}{t_1}$$

$$\Rightarrow \quad |\tan\theta| = \left|t_1 + \frac{2}{t_1}\right| \geq 2\sqrt{t_1 \times \frac{2}{t_1}} \qquad [\because \text{AM} \geq \text{GM}]$$

$$\Rightarrow \quad |\tan\theta| \geq 2\sqrt{2}$$

$$\Rightarrow \quad \theta > 60°$$

33. (c) The equation of a tangent to $y^2 = 4ax$ is

$$y = mx + \frac{a}{m}$$

$$\Rightarrow \quad x = \frac{y}{m} - \frac{a}{m^2} \qquad \ldots(i)$$

this will be a normal to the parabola $x^2 = 4by$, if it is of the form

$$x = \frac{y}{m} - \frac{2b}{m} - \frac{b}{m^3}$$

$$\therefore \quad \frac{-a}{m^2} = -\frac{2b}{m} - \frac{b}{m^3}$$

$$\Rightarrow \quad 2bm^2 - am + b = 0$$

Since, m is real.

$$\therefore \quad a^2 - 8b^2 \geq 0$$

$$\Rightarrow \quad a^2 \geq 8b^2$$

34. (a) The equation of parabola is $(y-1)^2 = 4(x+1)$.

The equation of any normal to this parabola is

$$y - 1 = m(x+1) - 2m - m^3$$

If it passes through $(-2, 1)$. Then,

$$0 = -m - 2m - m^3$$

$\Rightarrow$ $m^3 + 3m = 0$ $m \Rightarrow 0$ $[\because m^2 + 3 \neq 0]$

So, there is only one normal passing through $(-2, 1)$.

35. (c) Since, normal P and R meet at Q

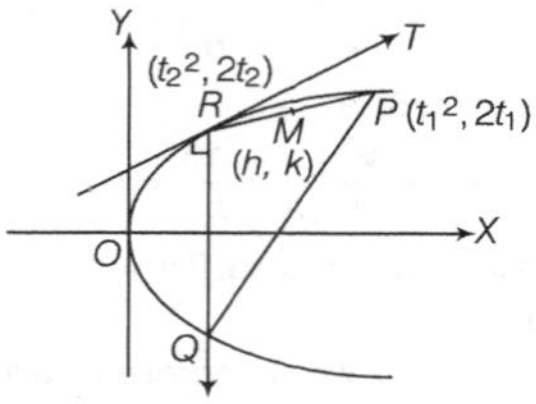

$\Rightarrow$ $t_1 t_2 = 2$ ...(ii)

Mid-point of PR is (h, k)

$$h = \frac{t_1^2 + t_2^2}{2}, k = t_1 + t_2 \quad \text{...(ii)}$$

$\therefore$ $2h = (t_1 + t_2)^2 - 2t_1 t_2$

$\Rightarrow$ $2h = k^2 - 4$ or $y^2 = 2x + 4$

$$y^2 = 2(x+2)$$

$\therefore$ Vertex is $(-2, 0)$

36. (d) Let $y = mx + \frac{2}{m}$ be a tangent to $y^2 = 8x$. Since, the circle

$$x^2 + y^2 - 2x - 6y + 2 = 0$$

intersects the parabola $y^2 = 8x$ orthogonally. Therefore, the tangent to parabola must pass through the centre of circle, i.e. $(1, 3)$.

$\therefore$ $3 = m + \frac{2}{m}$

$\Rightarrow$ $m^2 - 3m + 2 = 0$

$\Rightarrow$ $m = 1, 2$

Thus, the equation of tangents are

$$y = x + 2 \text{ and } 2x - y + 1 = 0.$$

37. (d) We know that the sum of ordinates of the feet of normals drawn to the parabola $y^2 = 4ax$ is zero.

$\therefore$ $y_1 + y_2 + y_3 = 0$

Also, $P(x_1, y_1)$, $Q(x_2, y_2)$ and $R(x_3, y_3)$ lie on the parabola.

$\therefore$ $y_1^2 = 4ax_1$, $y_2^2 = 4ax_2$ and $y_3^2 = 4ax_3$

$\Rightarrow$ $x_1 = \frac{y_1^2}{4a}$, $x_2 = \frac{y_2^2}{4a}$ and $x_3 = \frac{y_3^2}{4a}$

$$\therefore \frac{x_1 - x_2}{y_3} + \frac{x_2 - x_3}{y_1} + \frac{x_3 - x_1}{y_2}$$

$$= \frac{\frac{y_1^2}{4a} - \frac{y_2^2}{4a}}{y_1 + y_2} - \frac{\frac{y_2^2}{4a} - \frac{y_3^2}{4a}}{y_2 + y_3} - \frac{\frac{y_3^2}{4a} - \frac{y_1^2}{4a}}{y_3 + y_1}$$

$$= -\frac{1}{4a}[(y_1 - y_2) + (y_2 - y_3) + (y_3 - y_1)] = 0$$

38. (b) End points of double ordinate $x = 4$ of parabola $y^2 = 4x$ are $A(4, 4)$ and $B(4, -4)$.

Let t_1 and t_2 be the parameters for A and B respectively. Then, $t_1 = 2$ and $t_2 = -2$. Let t_1' and t_2' be the parameters for P and P' respectively. Then,

$$t_1' = -t_1 - \frac{2}{t_1} = -3 \text{ and } t_2' = -t_2 - \frac{2}{t_2} = 3$$

So, the coordinates of P and P' are $(9, -6)$ and $(9, 6)$ respectively.

$\therefore$ $PP' = 12$ units

39. (b) Let $P(h, 0)$ be a point on the axis of the parabola. The equation of any normal to the parabola $y^2 = 4ax$ is $y = mx - 2am - am^3$.

If it passes through $P(h, 0)$, then

$$0 = mh - 2am - am^3$$

$\Rightarrow$ $am^3 + m(2a - h) = 0$

$\Rightarrow$ $m\{m^2 - (h - 2a)\} = 0$

this equation must give three distinct values of m.

$\therefore$ $h > 2a$

Hence, the required set is $\{(x, 0) : x > 2a\}$.

40. (d) The equation of any normal to $(x-2)^2 = 4(y-3)$ is

$$(x - 2) = m(y - 3) - 2m - m^3$$

If it passes through $(1, h/2)$, then

$$1 - 2 = m\left(\frac{h}{2} - 3\right) - 2m - m^3$$

$\Rightarrow$ $2m^3 + m(10 - h) - 2 = 0$

This equation will give three distinct values of m_1, if $f'(m) = 0$ has two distinct roots, where

$$f(m) = 2m^3 + m(10 - h) - 2$$

$\Rightarrow$ $f'(m) = 6m^2 + (10 - h)$

$\therefore$ $f'(m) = 0$

$\Rightarrow$ $m = \pm\sqrt{\frac{h - 10}{6}}$

The values of m are real and distinct, if $h > 0$, i.e. $h \in (10, \infty)$.

41. (a) We have, $2(x-1)^2 + 2(y-1)^2 = (x + y + 2)^2$

$\Rightarrow$ $\sqrt{(x-1)^2 + (y-1)^2} = \left|\frac{x + y + 2}{\sqrt{1+1}}\right|^2$

Clearly, it represents a parabola having its focus at $(1, 1)$ and directrix $x + y + 2 = 0$.

The equation of axis is $y - 1 = 1(x - 1)$, i.e. $y = x$.

$\therefore$ Semi-latusrectum = Length of Perpendicular from $(1, 1)$ on the directrix

$\Rightarrow$ Semi-latusrectum $= \left|\frac{1 + 1 + 2}{\sqrt{1+1}}\right| = 2\sqrt{2}$

The coordinates of vertex are $(0, 0)$.

So, the equation of axis in parametric form is

$$\frac{x - 0}{\cos \pi/4} = \frac{y - 0}{\sin \pi/4} \quad \text{...(i)}$$

We know that three distinct normals can be drawn from a point $(h, 0)$ on the axis of the parabola $y^2 = 4ax$, if

$$h > 2a (= \text{semi-latusrectum})$$

The coordinates of a point on the axis Eq. (i) at a distance $2\sqrt{2}$ units from the vertex are given by

$$\frac{x}{\cos \pi/4} = \frac{y}{\sin \pi/4} = 2\sqrt{2}$$

$\Rightarrow$ $x = 2, y = 2$

Hence, three normals can be drawn from (h, k), if $h > 2$.

42. (d) We have, $y^2 = 4(x + 2)$

this represents a parabola with vertex $(-2, 0)$ and axis is X-axis. We know that three distinct normals can be drawn from a point on the axis of a parabola, if its distance from the vertex is more than the semi-latusrectum.

So, three distinct normals can be drawn to the parabola $y^2 = 4(x + 2)$, if its distance from the vertex is more than 2 units.

Let $(h, 0)$ be such a point. Then,

$$h + 2 > 2 \Rightarrow h > 0$$

Hence, the required set is $\{(x, 0) : x > 0\}$.

43. (a) Let $P(h, k)$ be any point in the plane of the parabola. The equation of a normal to the parabola $y^2 = 4ax$ is

$$y^2 = mx - 2am - am^3$$

If it passes through (h, k), then

$$k = mh - 2am - am^3$$

$$\Rightarrow \quad am^3 + m(2a - h) + k = 0 \qquad \ldots(i)$$

This is a cubic equation is m. So, it gives three values of m.
We have to find the condition, so that all the three values are real and distinct. We know that a cubic equation $f(x) = 0$ has three distinct real roots, if $f'(x) = 0$ has two distinct real roots, say α and β such that $f(\alpha) f(\beta) < 0$.

In this case, we have

$$f(m) = am^3 + m(2a - h) + k$$

$$\therefore \quad f'(m) = 3am^2 + (2a - h)$$

Clearly, $f'(m) = 0$ will have distinct real roots, if

$$\frac{h - 2a}{3a} > 0 \Rightarrow h - 2a > 0 \Rightarrow h > 2a$$

Two distinct roots of $f'(m) = 0$ are

$$\alpha = \sqrt{\frac{h - 2a}{3a}} \text{ and } \beta = -\sqrt{\frac{h - 2a}{3a}}$$

Now, $f(\alpha) f(\beta) < 0$

$$\Rightarrow \quad f(\alpha) f(-\alpha) < 0 \qquad [\because \beta = -\alpha]$$

$$\Rightarrow \quad \{a\alpha^3 + (2a - h)\alpha + k\} \{-a\alpha^3 - (2a - h)\alpha + k\} < 0$$

$$\Rightarrow \quad k^2 - \{a\alpha^2 + (2a - h)\}^2 \alpha^2 < 0$$

$$\Rightarrow \quad k^2 - \left\{\frac{h - 2a}{3} + (2a - h)\right\}^2 \left\{\frac{h - 2a}{3a}\right\} < 0$$

$$\Rightarrow \quad k^2 - \left(\frac{4a - 2h}{3}\right)^2 \left(\frac{h - 2a}{3a}\right) < 0$$

$$\Rightarrow \quad k^2 - \frac{4}{27a}(h - 2a)^3 < 0$$

$$\Rightarrow \quad 27ak^2 < 4(h - 2a)^3$$

Hence, the required conditions are

$$h > 2a \text{ and } 27ak^2 < 4(h - 2a)^3.$$

44. (a) $P(at^2, 2at)$ be any point on $y^2 = 4ax$

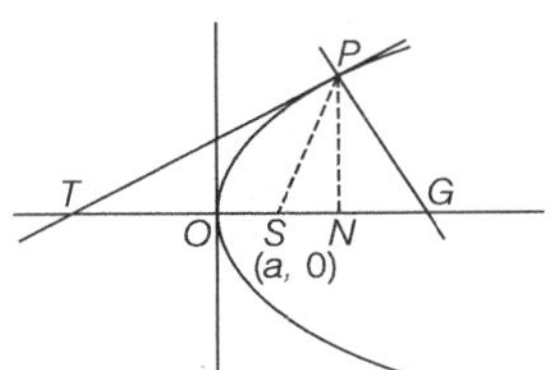

Normal at P meets the axis at $G = (2a + at^2, 0)$

But $SP = a + at^2$

$$\Rightarrow \quad SG = SP$$

$\therefore$ in ΔSPG, $\angle SPG = \angle SGP$

$\therefore$ Statement I is correct

Draw a line parallel to axis through P.

Then $\angle SGP = \angle LPG$

$$\Rightarrow \quad \angle LPG = \angle SPG$$

Hence, normal at P bisects, the external angle between SP and PM.

45. (a) Given, parabola $y^2 = 8(x - 3)$ vertex $(3, 0)$ focus $(5, 0)$

$\therefore$ N foot from focus $(5, 0)$ to tangent PQ lies $x = 3$.

46. (d) $\left(\frac{5}{4}, \frac{1}{2}\right)$ lies on the line $2x + y - 3 = 0$, hence locus will be a straight line. However, Statement II is correct.

47. (d) Area of triangle formed by t_1, t_2, t_3 is

$$\frac{1}{2}|t_1 - t_2||t_2 - t_3||t_3 - t_1| \neq 0$$

$\therefore$ Statement I is false.

48. (a) The equations, $yt = x + at^2$ and $x + sy = 2bs + bs^3$, where s and t are parameter, s represents the same lies.

$$\Rightarrow \quad \frac{1}{1} = \frac{-t}{s} = \frac{-at^2}{2bs + bs^3}$$

or $\quad t = -s$ and $2b + bs^2 = at$

$\Rightarrow \quad bs^2 + as + 2b = 0$ has real and distinct roots.

i.e. $\quad D > 0$

$$\Rightarrow \quad a^2 - 8b^2 > 0$$

or $\quad a^2 > 8b^2$

49. (a) As tangents and normals are perpendicular at end of a focal chord.

$\Rightarrow$ Statement I and Statement II are correct and Statement II is correct explanation of Statement I.

50. (a) The slope of ray $= \frac{y}{x}$

Slope of normal at $P(x, y) = -\frac{dx}{dy}$

$$\Rightarrow \quad \frac{\frac{y}{x} + \frac{dx}{dy}}{1 - \frac{y}{x} \cdot \frac{dx}{dy}} = -\frac{dx}{dy}$$

$\Rightarrow y\left(\frac{dy}{dx}\right)^2 + 2x \cdot \frac{dy}{dx} = y$ is a differential equation of the curve which is satisfied by $y^2 = 2x + 1$

Revisal Problems (JEE Advanced)

1. (b) Equidistant points from focus are symmetric about axis of parabola.

$\therefore$ tangent at vertex is parallel to the line joining the two points

$\therefore$ its slope $= \frac{3 - 2}{2 - 3} = -1$.

2. (b) Given parabolas are symmetric about the line $y = x$. So, they have a common normal with slope -1, it meets the parabolas at $\left(-\frac{1}{2}, \frac{13}{4}\right), \left(\frac{13}{4}, -\frac{1}{2}\right)$. Hence, the required circle, is

$$x^2 + y^2 - \frac{11}{4}x - \frac{11}{4}y - \frac{13}{4} = 0.$$

3. (c) Here, $y = ax^2 + bx + c$ has vertex $(4, 2)$

$$\Rightarrow \quad b = -8a$$

and $\quad \frac{4ac - b^2}{4a} = 2$

$$\Rightarrow \quad c = 2 + 16a$$

$\therefore$ product of $abc = \Delta = -16(a^2 + 8a^3)$

$$\frac{d\Delta}{da} = -16(2a + 24a^2) < 0 \ \forall \ a \in [1, 3]$$

$$\Rightarrow \quad \Delta_{max} = -144, \ \Delta_{min} = -3600$$

$\therefore$ difference $= 3456$.

4. (a) Equation of parabola whose directrix

X-axis is $\quad (x-3)^2 = -56(y+14) \quad \ldots$(i)

$\because \quad 2a = 28 \Rightarrow a = 14$

where vertex is $(3, -14)$

Directrix is parallel to Y-axis

$\Rightarrow \quad 2a = 3 \Rightarrow a = \frac{3}{2}$ or vertex $\left(\frac{3}{2}, -28\right)$

$\Rightarrow$ Equation of parabola $(y+28)^2 = 6\left(x - \frac{3}{2}\right) \quad \ldots$(ii)

Solving Eqs. (i) & (ii), we get $y = \pm x$

$\therefore \quad |t| = 1.$

5. (b) Let (p, p^2) be a point on $y = x^2 \quad \ldots$(i)

Equation of normal, $\quad y - p^2 = -\frac{1}{2p}(x-p) \quad \ldots$(ii)

Solving Eqs. (i) & (ii), we get

$$q = -p - \frac{1}{2p}$$

$$A = \int_q^p \left(p^2 - \frac{1}{2p}(x-p) - x^2\right) dx$$

$$\frac{dA}{dP} = 0 \Rightarrow \phi = \frac{1}{2} \Rightarrow P\left(\frac{1}{2}, \frac{1}{4}\right)$$

6. (b) Required area is half of the area of triangle formed by the points (4, 6), (10, 8), (2, 4)

$\therefore$ required area = 2.

7. (a) Line PQ is $y(t_1 + t_2) = 2x + 2at_1t_2 \quad \ldots$(i)

and $\quad \lambda y = -x + b \quad \ldots$(ii)

Since Eqs. (i) and (ii) are same

$$\Rightarrow \quad \frac{t_1 + t_2}{\lambda} = \frac{2}{-1} = \frac{2at_1t_2}{b}$$

$\therefore b = -at_1t_2$ and $2a \le -at_1t_2 \le 4a$

$\therefore \quad -4 \le t_1t_2 \le -2.$

8. (b) Let, $AB: y = 2x - 17$

$CD \Rightarrow y = 2x + c$

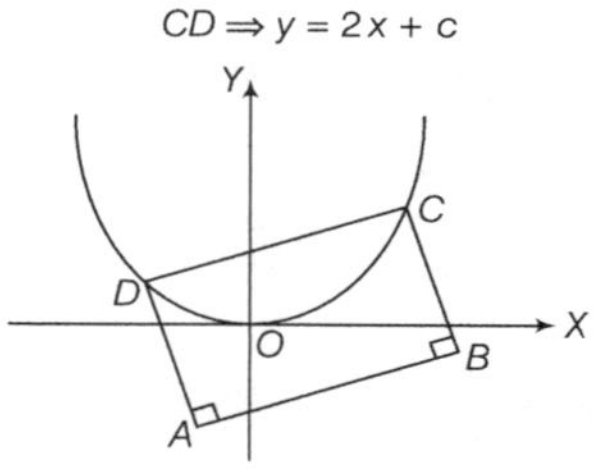

where $C(x_1, y_1)$ and $D(x_2, y_2)$

$m_{CD} = 2$

$\Rightarrow \quad \frac{y_2 - y_1}{x_2 - x_1} = 2 \quad \ldots$(i)

Let, $\quad CD = a = \sqrt{(x_1 - x_2)^2 + (y_1 - y_2)^2}$

$\Rightarrow \quad a^2 = 5(x_1 - x_2)^2 \quad$ from (i)

$= 5[(x_1 + x_2)^2 - 4x_1x_2] \quad \ldots$(ii)

$\therefore \quad y = x^2, y = 2x + c$

$\Rightarrow \quad x^2 - 2x - c = 0$

$\Rightarrow \quad x_1 + x_2 = 2$

and $\quad x_1x_2 = -c \quad \ldots$(iii)

from Eqs. (ii) & (iii)

$a^2 = 5[4 + 4c] \quad \ldots$(iv)

Let, $A = (k, 2k - 17)$ by any point on $y = 2x - 17$

$\therefore \quad AD = a = \frac{|2k - 2k + 17 + c|}{\sqrt{5}}$

$\Rightarrow \quad a^2 = \frac{(17+6)^2}{5} \quad \ldots$(v)

from (iv) & (v)

$\Rightarrow \quad a^2 = 1280.$

9. (a) Normal is $y = mx - 2am - am^3$

$\Rightarrow \quad 0 = 2km - 2m - m^3$

Let m_1, m_2, m_3 be the roots

Then $\quad m_3 = 0$

and $\quad m_1 m_2 = -1$

$\Rightarrow \quad k = 3/2$

10. (b) $\tan A = \frac{k}{a-h}$, $\tan B = \frac{k}{a+h}$

$\tan A + \tan B = \lambda \quad$ [constant]

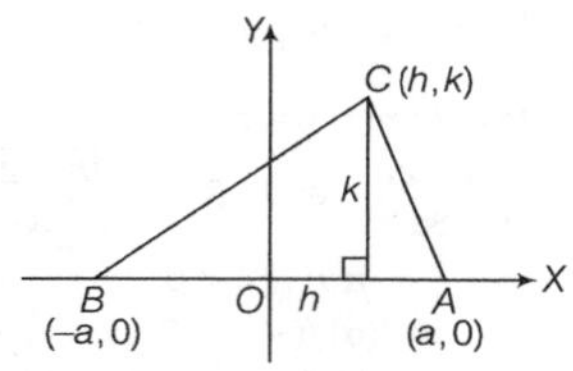

$\Rightarrow \quad \frac{k}{a-h} + \frac{k}{a+h} = \lambda \quad$ [hours]

$\Rightarrow \quad x^2 = a^2 - \frac{2a}{c}y \quad$ [parabola]

11. (a) *Case I*

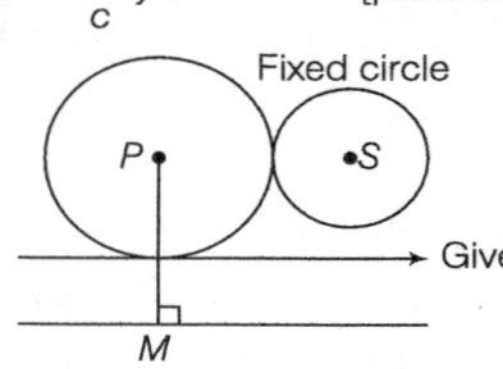

$PS = R + r_1 = PM$

$\therefore$ locus of P is a parabola

Case II

$R - r_1 = PM$

$\therefore$ locus of P is a parabola

12. (a) $P(at_1^2, 2at_1)$, $P'(-bt_2^2, -2bt_2)$

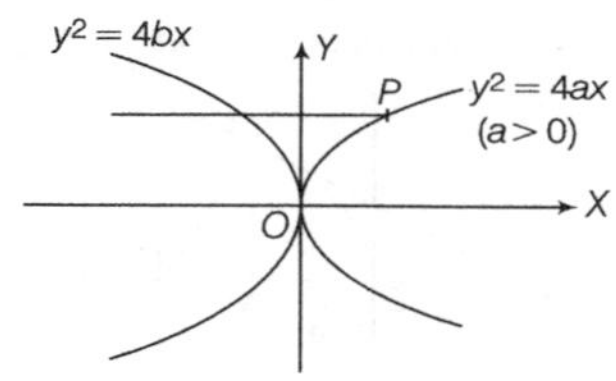

Slope of $PP' = 0$

$\Rightarrow \quad at_1 + bt_2 = 0 \quad \ldots$(i)

Mid-point of $PP' = \left(\frac{at_1^2 - bt_1^2}{2}, at_1 - bt_2\right) = (h, k)$

put $\quad t_2 = -\frac{at_1^2}{b}$

$$\Rightarrow \quad \frac{k}{a} = \frac{2\left(at_1 + \frac{at_1}{b}b\right)}{at_1^2 - b\left(-\frac{at_1}{b}\right)^2} = \frac{4at_1}{a(b-a)t_1^2}$$

$\therefore \quad t_1 = \frac{4h}{k(b-a)}$

$\therefore \quad k = at_1 - bt_2 = at_1 + rt_1 = 2a\frac{4h}{k(b-a)}$

$\Rightarrow \quad k^2 = \frac{8a}{b-a}h$

$\therefore$ locus of (h, k) is $y^2 = \frac{8a}{b-a} \cdot x$, a parabola.

13. (b) $y^2 = 4a\left(x + a\sin\frac{x}{a}\right)$

$$2y\frac{dy}{dx} = 4a\left(1 + a\cdot\frac{1}{a}\cdot\cos\frac{x}{a}\right)$$

$\because$ tangent is parallel to X-axis.

$\therefore \quad \frac{dy}{dx} = 0$

$\Rightarrow \quad \cos\frac{x}{a} = -1, \ \sin\frac{x}{a} = 0$

$\therefore \quad y^2 = 4a\left(x + a\sin\frac{x}{a}\right)$

$\Rightarrow \quad y^2 = 4ax$ (parabola).

14. (c) $CF = CN$

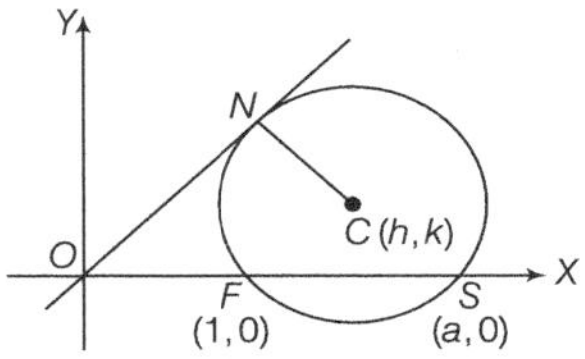

$\Rightarrow$ locus of C is a parabola with focus (1, 0) and directrix $y = x$.

$\Rightarrow$ length of latus rectum = 2 (distance from focus to directrix)

$$= 2\left(\frac{1}{\sqrt{2}}\right) = \sqrt{2}.$$

15. (c) $y^2 = 4b\,[x - (2a - b)]$

or $\quad y^2 = 4bX \quad$ where $X = x - (2a - b)$

and $x^2 + 4a\,(y - 2b - a)$

or $\quad x^2 = -4aY,$

where $Y = y - 2b - a$ for $y^2 = 4bx,$

extremities of latus-rectum $(b, 2b)$ and $(b, -2b)$ w.r.t. XY-axis i.e. $(2a, 2b)$ and $(2a, -2b)$ w.r.t. XY-axis for $x^2 = -4ay$, extremities of lactus rectum $(2a, -a)$ and $(-2a, -a)$ w.r.t. XY-axis i.e. $(2a, 2b)$ and $(-2a, 2b)$ w.r.t. XY-axis.

Hence, the common end of the lactus rectum $(2a, 2b)$ now for I parabola,

$$2y\frac{dy}{dx} = 4b$$

$\Rightarrow \quad \frac{dy}{dx} = \frac{2b}{y_1} = 1$ at $(2a, 2b)$

also for 2nd parabola, $2x = -4a\cdot\frac{dy}{dx}$

or $\quad \frac{dy}{dx} = -\frac{x}{2a} = -1$ at $(2a, 2b)$

Hence, parabolas intersect orthogonally at $(2a, 2b)$

16. (a) $PM = 1 + t^2$

$$PS = 1 + t^2$$

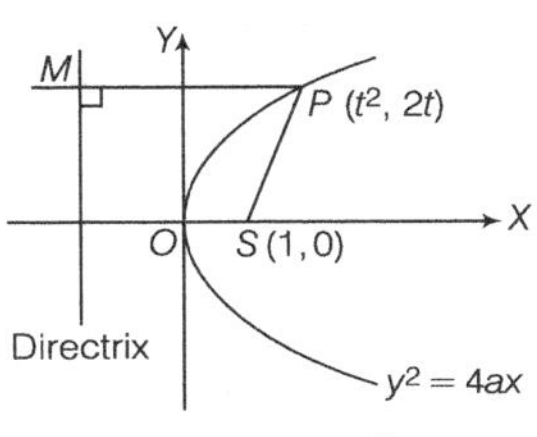

$$MS = 1 + t^2$$

$\Rightarrow \quad 2^2 + 4t^2 = (1 + t^2)^2$

$\therefore \quad PM = 1 + t^2 = 4$

$\therefore$ Area of $\Delta PMS = \sqrt{\frac{3}{4}}\,(4)^2 = 4\sqrt{3}.$

17. (a) $(x - x_1)^2 + (y - y_1)^2 = r^2$

X-axis intercept $= 2a$

$\Rightarrow \quad 2\sqrt{x_1^2 - (x_1^2 + y_1^2 - r^2)} = 2a$

$\Rightarrow \quad r^2 - y_1^2 = a^2$

$\therefore \quad y_1^2 = r^2 - a^2$

Passes, through $(0, b)$

$\therefore \quad x_1^2 + (b - y_1)^2 = r^2$

$\Rightarrow \quad x_1^2 + b^2 - 2by_1 + y_1^2 - r^2 = 0$

$\Rightarrow \quad x_1^2 + b^2 - 2by_1 - a^2 = 0$

$\therefore$ locus of (x_1, y_1)

$\Rightarrow \quad x^2 - 2by + b^2 = x^2.$

18. (c) Here,

$$(h - c)^2 + k^2 = (c\cos\theta - a)^2 \quad \ldots\text{(i)}$$

$$(h + c)^2 + k^2 = c\,(\cos\theta + a)^2 \quad \ldots\text{(ii)}$$

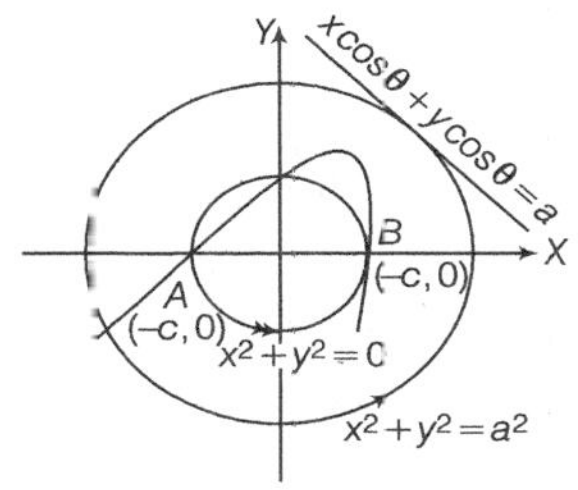

Subtracting

$$4ch = 4ac\cos\theta$$

$\Rightarrow \quad h = a\cos\theta \quad \ldots\text{(iii)}$

adding, we get

$$2\,(c^2 + h^2 + k^2) = 2\,(c^2\cos^2\theta + a^2) \quad \ldots\text{(iv)}$$

put, $\cos\theta = \frac{h}{a}$, we get

$$c^2 + h^2 + k^2 = c^2\cdot\frac{h^2}{a^2} + a^2$$

$\Rightarrow \quad c^2a^2 + h^2a^2 + a^2k^2 = c^2h^2 + a^4$

or $\quad (a^2 - c^2)\,h^2 + k^2a^2 = a^2(a^2 - c^2)$

or $\quad \frac{h^2}{a^2} + \frac{k^2}{a^2 - c^2} = 1$

$\therefore$ locus is

$$\frac{x^2}{a^2} + \frac{y^2}{a^2 - c^2} = 1$$

19. (a) $y^2 = k^2x$

$\Rightarrow \quad y^2 = 4ax, \ k^2 = 4a$

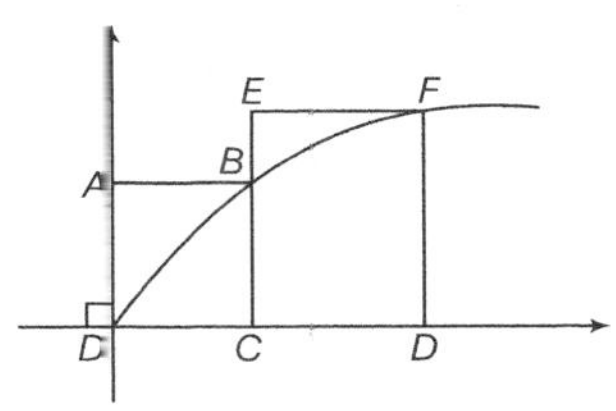

$$B\,(at_1^2, 2at_1),\ F\,(a_2^2, 2t_2)$$

$$(t_1, t_2 > 0)$$

to find $\quad \frac{FG}{BC} = \frac{2at_2}{2nt_1} = \frac{t_2}{t_1}$

now, $\quad DC = BC$

$\Rightarrow \quad at_1^2 = 2at \Rightarrow t_1 = 2$

also, $\quad at_2^2 - at_1^2 = 2at_2$

$\Rightarrow \quad t_2^2 - 4 = 2t_2$

or $\quad t_2^2 - 2t_2 - 4 = 0$

$$t_2 = \frac{2 \pm \sqrt{4+16}}{2}, \text{ neglecting } t_2 = \frac{2-\sqrt{20}}{2} \text{ as } t_2 > 0$$

$\therefore \quad t_2 = \dfrac{2+\sqrt{20}}{2}$

$\Rightarrow \quad \dfrac{t_2}{t_1} = \dfrac{\sqrt{5}+1}{2}.$

20. (d) $\tan\theta = \dfrac{2}{t_1}$ ...(i)

also, $\quad \dfrac{2}{t_1} \cdot \dfrac{2}{t_2} = -1$

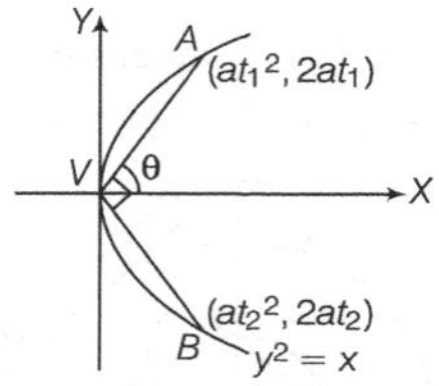

$\therefore \quad t_1 t_2 = -4$

$|VA| = at_1\sqrt{4+t_1^2}$

using, $\quad t_1 t_2 = -4$

$$|VB| = \frac{4a}{t_1}\sqrt{\frac{16}{t_1^2} + 4} = \frac{8a}{t_1^2}\sqrt{4+t_1^2}$$

$\therefore \quad \dfrac{|VA|}{|VB|} = \dfrac{at_1^3\sqrt{4+t_1^2}}{8a\sqrt{4+t_1^2}} = \dfrac{t_1^3}{8}$

also, $\quad \tan\theta = \dfrac{2}{t_1}$

$\Rightarrow \quad t_1^3 = 8\cot^3\theta$

$\therefore \quad \dfrac{|VA|}{|VB|} = \cot^3\theta$

21. (a) Chord of feet of perpendicular as (h, k) is

$$hx + ky = h^2 + k^2 \qquad \text{...(i)}$$

Homogenise $y^2 = 4ax$, with the help of (i) and use coefficient of x^2 + coefficient of $y^2 = 0$

$\Rightarrow \quad x^2 + y^2 - 4ax = 0.$

22. (d) Putting $x^2 = -\dfrac{y}{a}$ in the circle $x^2 + (y-1)^2 = 1$, we get

$$\frac{y}{a} + y^2 - 2y = 0$$

(Note that for $a < 0$, they cannot intersect other than the origin)
Hence, we get

$$y = 0 \quad \text{or} \quad y = 2 - \frac{1}{a}$$

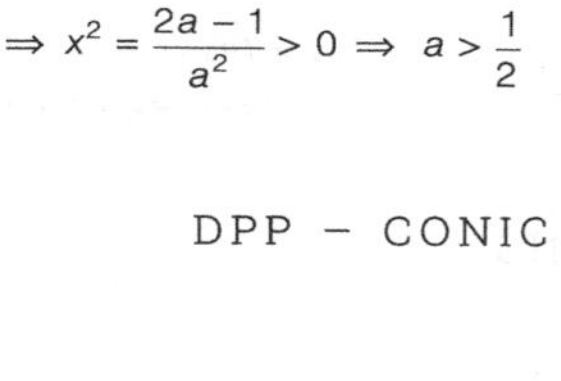

On substituting $y = 2 - \dfrac{1}{a}$ in $y = ax^2$, we get

$$ax^2 = 2 - \frac{1}{a} \Rightarrow x^2 = \frac{2a-1}{a^2} > 0 \Rightarrow a > \frac{1}{2}$$

23. (a) $t_2 = -t_1 - \dfrac{2}{t_1}$ or $t_1 t_2 = -t_1^2 - 2$

The equation of line through P parallel to AQ is

$$y - 2at_1 = \frac{2}{t_2}(x - at_1^2)$$

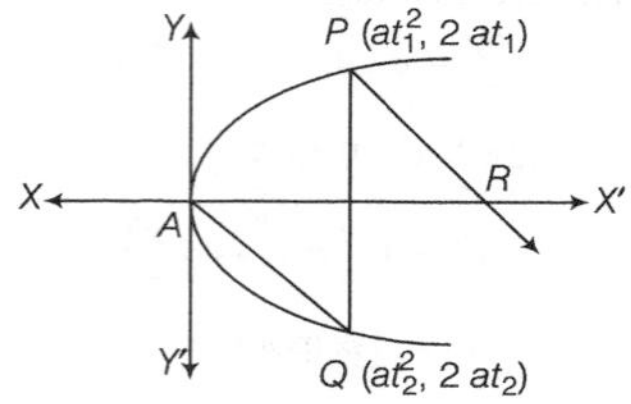

Put $y = 0$, then, $x = at_1^2 - at_1 t_2 = at_1^2 - a(-2 - t_1^2)$

$$= 2a + 2at_1^2 = 2(a + at_1^2)$$

which is twice the focal distance of P.

24. (c) The chord of contact of (h, k) is $ky = 2a(x + h)$.

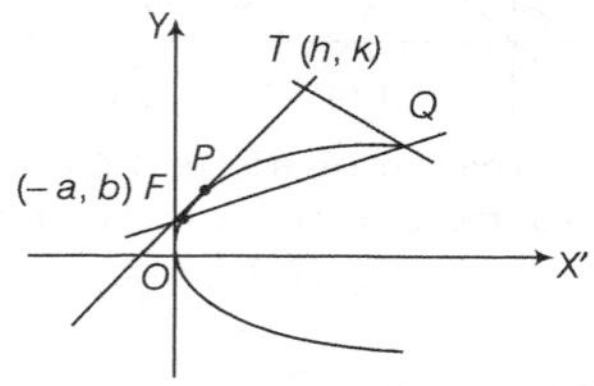

It passes through $(-a, b)$. Therefore, $bk = 2a(-a + h)$
Hence, the locus is $by = 2a(x - a)$.

25. (a) The slope of tangent at P is $1/t_1$ and that at Q is $1/t_2$. Therefore, $\cot\theta_1 = t_1$ and $\cot\theta_2 = t_2$

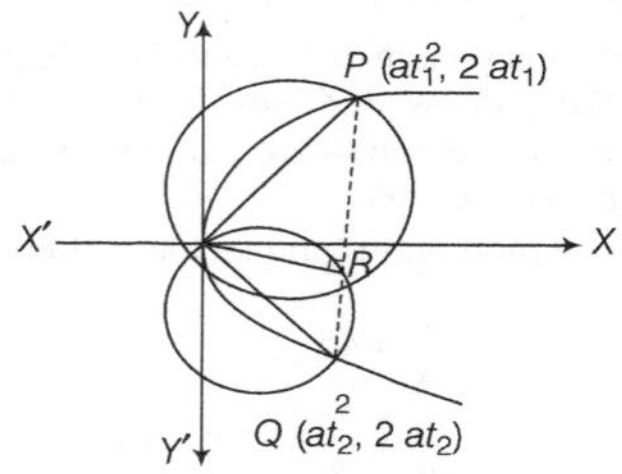

Slope of $PQ = \dfrac{2}{t_1 + t_2}$

$\therefore$ Slope of $OR = -\left(\dfrac{t_1 - t_2}{2}\right) = \tan\phi$ [angle in semicircle is 90°]

or $\quad \tan\phi = -\dfrac{1}{2}[\cot\theta_1 + \cot\theta_2]$

$\therefore \quad \cot\theta_1 + \cot\theta_2 = -2\tan\phi$

26. (b) By hypothesis, the tangents through (h, k) are equally inclined with the line whose inclination with X-axis is α.

Then, $\quad \theta_2 - \alpha = \alpha - \theta_1$

or $\quad \theta_1 + \theta_2 = 2\alpha$

$\Rightarrow \quad \tan(\theta_1 + \theta_2) = \tan 2\alpha$

$\Rightarrow \quad \dfrac{\tan\theta_1 + \tan\theta_2}{1 - \tan\theta_1\tan\theta_2} = \tan 2\alpha = \dfrac{m_1 + m_2}{1 - m_1 m_2}$

where, m_1 and m_2 are the slopes of tangents.
Putting the values of m_1 and m_2, we have

$$\frac{\left(\frac{k}{h}\right)}{\left(1 - \frac{a}{h}\right)} = \tan 2\alpha \quad \text{or} \quad \frac{k}{h-a} = \tan 2\alpha$$

Generalising and simplifying, the required locus of (h, k) is

$$y = (x - \alpha)\tan 2\alpha$$

i.e. a straight line.

27. (b) Equation of normal is

$$y + xt = 2at + at^3 \quad \ldots(i)$$

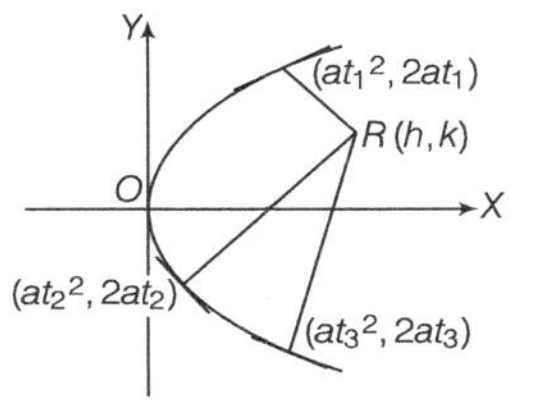

as (i) passes through (h, k), $(at_2^2, 2at_2)$

So, $at^3 + t(2a - h) - k = 0$ has roots t_1, t_2, t_3 ...(ii)

Here, $a = 1$

$\Rightarrow \quad t_1 + t_2 + t_3 = 0$ ...(iii)

also $\quad \dfrac{2}{t_1 + t_2} = 2 \Rightarrow t_1 + t_2 = 1$ ...(iv)

from Eqs. (iii) & (iv)

$$t_3 = -1$$

Put, $t_3 = -1$ in Eq. (ii), we get

$$-1 - 1(2 - h) - k = 0$$

$\Rightarrow$ locus of $p(h, k)$ is $x - y = 3$.

28. (b) $SS_1 = T^2$

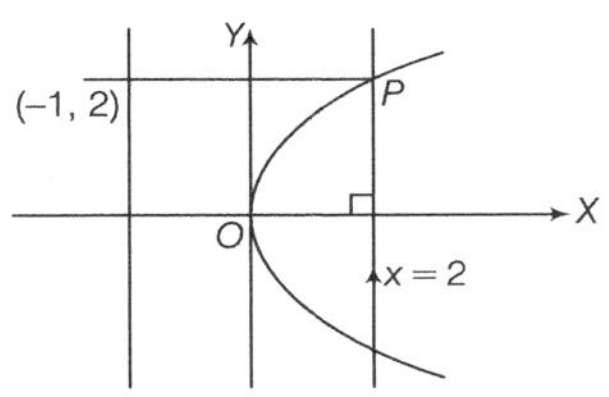

$\Rightarrow \quad (y^2 - 4x)(y_1^2 - 4x_1) = (yy_1 - 2(x_1 + x_1))^2$

$\Rightarrow \quad (y^2 - 4x)(4 + 4) = (2y - 2(x - 1))^2$

$\quad = 4(y - x + 1)^2 \cdot x + 2 = 0$

$\Rightarrow \quad 2(y^2 - 4x) = (y - x + 1)^2$

Solving with the line $x = 2$, we get

$$2(y^2 - 8) = (y - 1)^2$$

or $\quad 2(y^2 - 8) = y^2 - 2y + 1$

or $\quad y^2 + 2y - 17 = 0$

where, $y_1 + y_2 = -2$, $y_1 y_2 = -17$

Now, $\quad |y_1 - y_2|^2 = (y_1 + y_2)^2 - 4y_1y_2$

$\quad = 4 - 4(-17) = 72$

$\therefore \quad |y_1 - y_2| = 6\sqrt{2}$

29. (d) $4y = x^2 - 8$

$\Rightarrow \quad 4\dfrac{dy}{dx} = 2x$

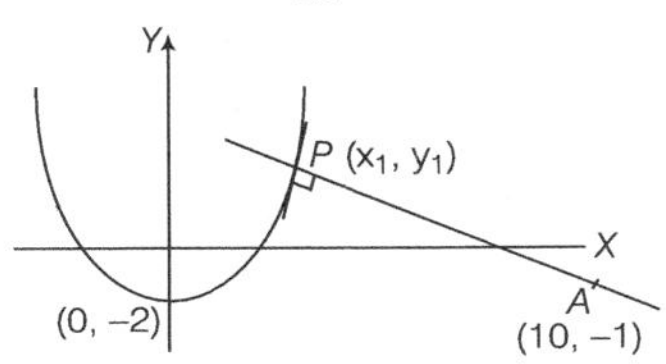

$$\left.\frac{dy}{dx}\right|_{(x_1 y_1)} = \frac{x_1}{2}$$

$\therefore$ Slope of normal $= \dfrac{2}{x_1}$ but slope of normal $= \dfrac{y_1 + 1}{x_1 - 10}$

$\therefore \quad \dfrac{y_1 + 1}{x_1 - 10} = -\dfrac{2}{x_1}$

$\Rightarrow \quad x_1y_1 + x_1 = -2x_1 + 20$

$\Rightarrow \quad x_1y_1 + 3x_1 = 20$

$\quad y_1 = \dfrac{x_1^2 - 8}{4}$

$\Rightarrow \quad x_1\left(\dfrac{x_1^2 - 8}{4} + 3\right) = 0$

$\Rightarrow \quad x_1(x_1^2 + 4) = 80$

$\Rightarrow \quad x_1(x_1^2 + 4) = 80$

$\quad x_1^3 + 4x_1 - 80 = 0$

$\Rightarrow \quad x_1^3(x_1 - 4) + 4x_1(x_1 - 4) + 20(x_1 - 4) = 0$

$\quad (x_1 - 4)[x_1^3 + 4x_1 + 20] = 0$

$\therefore \quad x_1 = 4, y_1 = 2$

Equation of PA, is

$$y + 1 = -\frac{1}{2}(x - 10)$$

$\Rightarrow \quad 2y + 2 = -x + 10$

or $\quad x + 2y - 8 = 0.$

30. (a)

(i) If the parameter t_1, t_2, t_3 corresponding to the points A, B, C respectively, then

$$t_3 = 3, t_1 + t_2 = -3, t_1t_2 = 2 \quad \ldots(i)$$

$$AB^2 = (t_1 - t_2)^2[(t_1 + t_2)^2 + 4] = 13$$

$$AB = \sqrt{13}$$

$\Rightarrow \quad$ (i) $\to$ (r)

(ii) Equation of AB, is $2x - (t_1 + t_2)y + 2t_1t_2 = 0$. Then distance of $C(9, 6)$ from this line is $\dfrac{40}{\sqrt{13}}$.

$\therefore \quad \Delta ABC = \dfrac{\sqrt{13}}{2} \cdot \dfrac{40}{\sqrt{13}} = 20$

(ii) $\to$ (P)

(iii) Distance $(0, 0)$ from $2x + 3y + 4 = 0 \Rightarrow \dfrac{4}{\sqrt{13}}$

(iii) $\to$ (q)

(iv) The line $2x + 3y + 4 = 0$ makes intercepts of $2, \dfrac{4}{3}$ on the axis.

i.e. $\quad$ area is $\left(\dfrac{4}{3}\right)$, (iv) $\to$ (s).

31. (c)

(i) $(\pm 4, \pm 4)(0, 0)$ are the '5' points of intersection.

(ii) Centre is $(2, 2)$ move at right angles to $y = x$ by 'r' units to reach $y^2 = 4|x|$, $x^2 = 4|y|$.

$\therefore \quad \dfrac{x - 2}{-\frac{1}{\sqrt{2}}} = \dfrac{y - 2}{\frac{1}{\sqrt{2}}} = r$

$\Rightarrow (x, y) = \left(2 - \dfrac{r}{\sqrt{2}}, 2 + \dfrac{r}{\sqrt{2}}\right)$ lies on $y^2 = 4x$

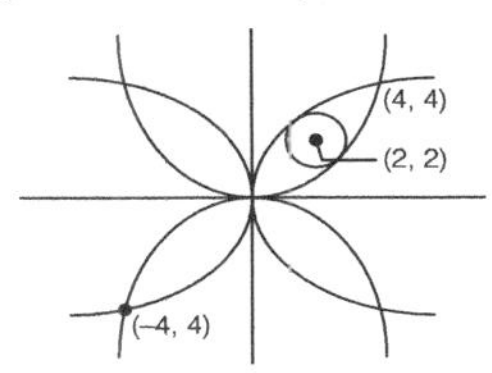

$\therefore \quad \left(2 + \frac{r}{2}\right)^2 = 4\left(2 - \frac{r}{\sqrt{2}}\right)$

$\frac{r^2}{2} + \frac{4r}{\sqrt{2}} + 4 = 8 - \frac{4r}{\sqrt{2}}$

$\Rightarrow \quad r^2 + \frac{16}{\sqrt{2}} r - 8 = 0$

$\Rightarrow \quad r = \frac{-\frac{16}{\sqrt{2}} + \sqrt{128 + 32}}{2}$

$= -4\sqrt{2} + 2\sqrt{10}$

(i) → (r), (ii) → (s), (iii) → (q), (iv) → (q)

32. (a)

(i) $y = |1 - e^x| - 2|$ to $x^2 + 4y = 0$ mutually perpendicular tangents to $x^2 + 4y = 0$ always intersect on $y = 1$.

$\therefore \quad ||1 - e^x| - 2| = 1$

$\Rightarrow \quad |1 - e^x| = 3, 1$

$\Rightarrow \quad e^x = 2, 4$

∴ two points exists (i) → (*p*)

(ii) Distance between focus and vertex $= \frac{1}{4}(12) = 3$

∴ (ii) → (*r*)

(iii) $y^2 = 8x \Rightarrow \left(\frac{dy}{dx}\right) = \sqrt{2}$

∴ focal chord is $y - 0 = \sqrt{2}(x - 2)$ meets at $Q(4, 2\sqrt{2})$, $y = 2\sqrt{2}$.

∴ $PQ = 3$ (iii) → (*r*)

(iv) Fours is $\left(\frac{1}{2}, \frac{1}{2}\right)$ and is a fixed point.

∴ a circle of radius of 0

(iv) → (*s*)

33. (a) Equation of tangent of slope m to $y^2 = 4x$, is

$$y = mx + \frac{1}{m} \quad \ldots(i)$$

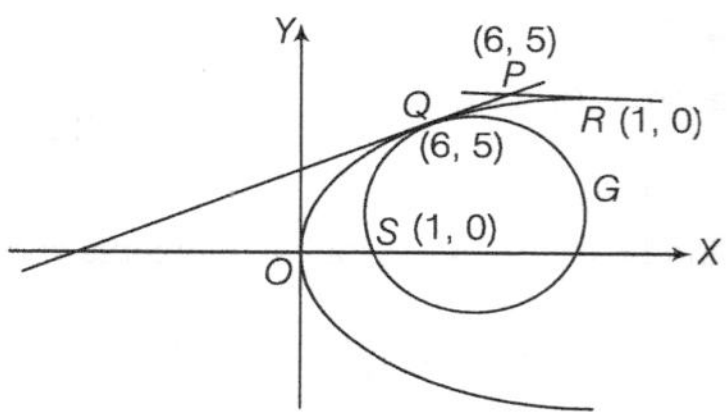

as (i) passes through P (6,5), so

$5 = 6m + \frac{1}{m}$

$\Rightarrow \quad 6m^2 - 5m + 1 = 0$

$\Rightarrow \quad m = \frac{1}{2}$ or $\frac{1}{3}$

Point of contact are $\left(\frac{1}{m_1^2}, \frac{2}{m_1}\right)$ and $\left(\frac{1}{m_2^2}, \frac{2}{m_2}\right)$.

Hence, P (4, 4) and Q (9, 6)

Area of $\Delta PQR = \frac{1}{2}\begin{vmatrix} 6 & 5 & 1 \\ 4 & 4 & 1 \\ 9 & 6 & 1 \end{vmatrix} = \frac{1}{2}$

34. (b) $y = \frac{1}{2}x + 2$

$\Rightarrow \quad x - 2y + 4 = 0 \quad \ldots(i)$

and $\quad y = \frac{1}{3}x + 3$

$\Rightarrow \quad x - 3y + 9 = 0$

Now, equation of circle C_2 touching $x - 3y + 9 = 0$ at (9, 6) is as above circle passes through (1, 0), so

$64 + 36 + 10\lambda = 0$

$\Rightarrow \quad \lambda = -10$

C_2 is $x^2 + y^2 - 28x + 18y + 27 = 0 \quad \ldots(ii)$

$r_2 = 5\sqrt{10}$

35. (c) Equation of C_1

$(x - 4)^2 + (y - 4)^2 + \lambda(x - 2y + 4) = 0$

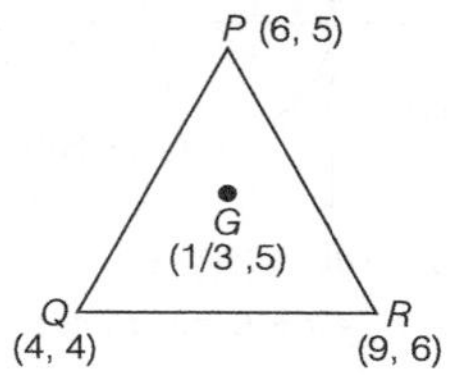

as above circle passes through (1, 0)

$9 + 16 + 5\lambda = 0$

$\Rightarrow \quad \lambda = -5$

Now, C_1 is

$x^2 + y^2 - 13x + 2y + 12 = 0 \quad \ldots(iv)$

Common chord of (iii) and (iv), is

$15x - 16y - 15 = 0 \quad \ldots(v)$

also centroid of ΔPQR is $\left(\frac{1}{3}, 5\right)$

Clearly, $\left(\frac{1}{3}, 5\right)$ satisfies (v)

36. (c) Any point on the given parabola is $(t^2, 2t)$. The equation of tangent at (1, 2) [point at the upper end of the latusrectum] is $x - y + 1 = 0$.

The image (h, k) of the point $(t^2, 2t)$ on $x - y + 1 = 0$ is given by

$$\frac{h - t^2}{1} = \frac{k - 2t}{-1} = \frac{-2(t^2 - 2t + 1)}{1 + t}$$

$\therefore \quad h = t^2 - t^2 + 2t - 1 = 2t - 1$

and $\quad k = 2t^2 + t^2 - 2t + 1 = t^2 + 1$

Eliminating t from, $h = 2t - 1$ and $k = t^2 + 1$,

we get

$(h + 1)^2 = 4(k - 1)$.

The required equation of reflection is

$(x + 1)^2 = 4(y - 1)$.

37. (b) The image of $(t^2, 2t)$ in the tangent line and then eliminate t to get the image now family of circle touching the parabola $y^2 = 4x$ at (1, 2)

$(x - 1)^2 + (y - 2)^2 + \lambda(x - y + 1) = 0$

it passes through (1, 0)

$\Rightarrow \quad 4 + 2\lambda = 0$

$\Rightarrow \quad \lambda = -2$

Hence circle is

$x^2 + y^2 - 4x - 2y + 3 = 0$

$\therefore \quad r = \sqrt{2}$

38. (a) Area $\int_0^2 \left(\frac{y^2}{4} - (y - 1)\right) dy = \left(\frac{y^3}{12} - \frac{y^2}{2} + y\right)_0^2$

$= \frac{8}{12} - 2 + 2$

$= \frac{2}{3}$

39. (b) Since, no point of the parabola is below the X-axis,

$$D = a^2 - 4 \le 0$$

Therefore, the maximum value of a is 2. The equation of parabola, when $a = 2$ is $y = x^2 + 2x + 1$.

It intersects the Y-axis at (0, 1). The equation of tangent at (0, 1) is

$$y = 2x + 1$$

Since, $y = 2x + 1$ touches the circle $x^2 + y^2 = r^2$, we get

$$r = \frac{1}{\sqrt{5}}$$

40. (a) The equation of tangent at (0,1) to the parabola $y = x^2 + ax + 1$ is

$$y - 1 = a(x - 0)$$

$$\Rightarrow \quad ax - y + 1 = 0$$

As it touches the circle, we get

$$r = \frac{1}{\sqrt{a^2 + 1}}$$

The radius is maximum, when $a = 0$. Therefore, the equation of the tangent is $y = 1$. Therefore, the slope of tangent is 0.

41. (a) The equation of tangent is $y = ax + 1$.

The intercepts are $-1/a$ and 1.

Therefore, the area of triangle bounded by the tangent and the axes is $\frac{1}{2}\left|\frac{-1}{a}\cdot 1\right| = \frac{1}{2|a|}$.

It is minimum, when $a = 2$.

Therefore, minimum area $= \frac{1}{4}$.

42. (d) $A(a, a^2 = 3)$ Equation of tangent L is $S_1 = 0$, is

$$2(a^2 - 3)x - ay - a^3 + 3a = 0$$

43. (b) The line L meets the parabola C $y = x^2 - 3$ at the points for which $x^2 - 3 = \frac{2(a^2 - 3)}{a}x - a^2 + 3$

$$\Rightarrow \quad (x - a)(ax + 6 - a^2) = 0$$

But $x = 1$ and $x \ne a$

$$\therefore \quad x = \frac{a^2 - 6}{a} = 1$$

$$\Rightarrow \quad a = -2, 3$$

44. (b) If $a > 0$, then $a = 3$

$A(3, 6)$, $B = (1, -2)$ equation of C is

$$y = x^2 - 3$$

or

$$x^2 = y + 3$$

45. (a) Let (x, x_0^2) be any point on the curve

Slope of tangent $= m = \frac{x_0^2 + 1}{x_0 - t}$

Slope of tangent at 'x_0' $= 2x_0 = m$

$$\therefore \quad x_0 = t \pm \sqrt{Ht^2}$$

Let, $a = t - \sqrt{1 + t^2}$, $b = t + \sqrt{1 + t^2}$

Area enclosed by tangent lines and parabola is equivalent to the area below the parabola form $[a, b]$ and above the line $y = -1$ minus the area of two Δ's first with vertices $(t, -1)$, (a, a^2), $(a, -1)$ and the second with vertices $(t, -1)$, (b, b^2) and $(b, -1)$

Sum of area of two triangles

$$= \frac{(a^2 + 1)(t - a)}{2} + \frac{(b^2 + 1)(b - t)}{2}$$

$$= 2(1 + t^2)^{3/2}$$

$$S(t) = \int_a^b (x^2 + 1)\,dx - 2(1 + t^2)^{3/2} = \frac{b^3 - a^3}{3} + (b - a) - 2(1 + t^2)^{3/2}$$

$$= \frac{8}{3}(1 + t^2)^{3/2} - 2(1 + t^2)^{3/2} = \frac{3}{2}(1 + t^2)^{3/2}$$

46. (a) $\frac{s(t)}{\sqrt{t}}$ is minimum $\Rightarrow \frac{d}{dt}\left(\frac{s(t)}{\sqrt{t}}\right) = 0$

$$\Rightarrow \quad t = \frac{1}{\sqrt{5}}$$

47. (b) $\min\left(\frac{s(t)}{\sqrt{t}}\right) = 4 \cdot (5)^{-5/4} \cdot \sqrt{6}$

48. (c) $C_1 = y = (x - ^-)^2 + |k^2 - 3k + 2| + \frac{1}{4}$

$$C_2 : (x - 1)^2 = -4(y - 2)$$

for no point of intersection $|k^2 - 3k + 2| + \frac{1}{4} > 2$

$$\Rightarrow \quad |k^2 - 3k + 2| > \frac{7}{4}$$

49. (d) $f(k)$ is least when $k = \frac{3}{2}$

Least value of $f(k) = \frac{1}{4}$

foci $s_1\left(1, \frac{1}{2}\right)$ and $s_2(1, 1) \Rightarrow s_1s_2 = \frac{1}{2}$

50. (a) $g(k) = f^3(k) + \frac{1}{f^2(k)}$

A.M. $\ge$ G.M.

$$\frac{g(k)}{2} \ge \sqrt{f(k)} \ge \frac{1}{2}$$

$$\Rightarrow \quad g|k| \ge 1$$

Explanations for Q. Nos. (51-53)

O is the point of intersection of perpendicular tangent, so O lies on directrix. Also AB is focal chord. Let equation of directrix is and S is the focus lying on AB.

Now $\angle ASO = \angle BSO = 90°$

$$\Rightarrow \quad OS \perp AB$$

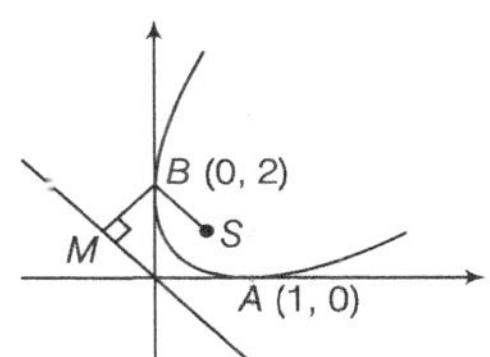

Since, AB is $\quad y + 2x = 2$

Hence, $\quad S = \left(\frac{4}{5}, \frac{2}{5}\right)$

Now, $BS = BM$

$$\Rightarrow \quad \sqrt{\frac{16}{25} + \frac{64}{25}} = \left|\frac{m \cdot (0) - 2}{\sqrt{1 + m^2}}\right|$$

$$\Rightarrow \quad m = -\frac{1}{2}, m \ne \frac{1}{2}$$

So directrix is $y = -\frac{1}{2}x$

Similarly we get axis, $y = 2x - \frac{6}{5}$ and vertex $\left(\frac{16}{25}, \frac{2}{25}\right)$ and latus rectum is $\left(16, \frac{\sqrt{5}}{25}\right)$.

51. (a) **52.** (b), **53.** (b)

Explanations for Q. Nos. (54-56)

$$y^2 = 4|x|$$

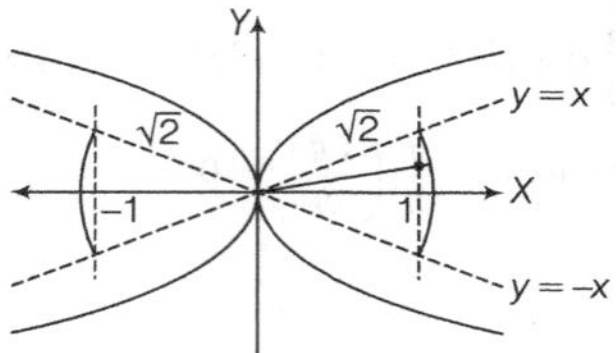

$y^2 = 4x$

Here $y^2 = 4x$, directrix, $x = -1$

$\therefore$ point should lie for $y^2 = -4x$

$\Rightarrow \quad 1 < \pi < \sqrt{2}$

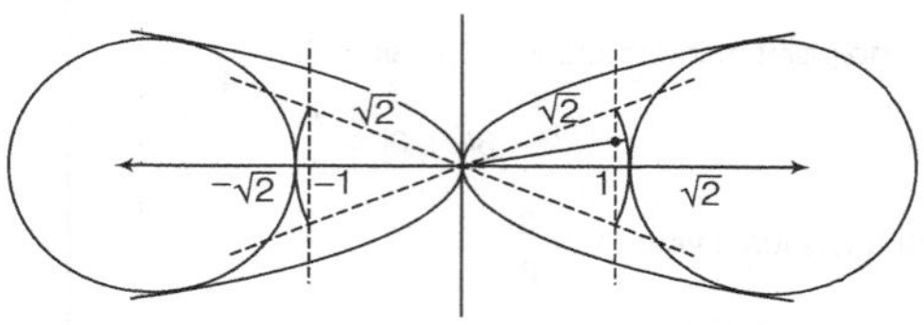

$(x-a)^2 + y^2 = R^2$ touches $x^2 + y^2 = 1$ externally.

$a = 1 + y \Rightarrow y^2 = 4x$ and $(x-a)^2 + 4x - R^2 = 0$ has unique root.

$\Rightarrow \quad R^2 = 4(a-1)$

or $\quad R^2 = 4R \Rightarrow R = 4$

54. (b) **55.** (a), **56.** (c)

57. (6) Parabolas are equal, if their latusrectum are equal.

Leatusrectum of $y^2 = \lambda x$ is $\lambda\ (x-3)^2 + (y-(-2))^2$

$$= \left(\frac{3x - 4y - 2}{5}\right)^2$$

$\therefore$ focus $(3, -2)$

directrix, $3x - 4y - 2 = 0$

$$3z = 2a = 3 \Rightarrow \frac{\lambda}{2} = 3 \Rightarrow \lambda = 6$$

58. (3) Let $P(t) \Rightarrow Q\left(-\frac{1}{4}\right)$

$$P' = -t - \frac{2}{t},\ Q' = \frac{1}{t} + 2t$$

$$\Rightarrow \quad P'Q' = a\left|3\left(t + \frac{1}{t}\right)\right|\sqrt{\left(t - \frac{1}{t}\right)^2 + 4} = 3a\left(t + \frac{1}{t}\right)^2$$

and $\quad PQ = a\left(t + \frac{1}{t}\right)^2$

$$\Rightarrow \quad \frac{P'Q'}{PQ} = 3.$$

59. (8) The curve is $(y+1)^2 = 4(x-1)$

Equation of the normal to the given curve is

$$y + 1 = m(x-1) - 2m - m^3$$

which passes through (h, k)

$\Rightarrow \quad m^3 + m(3-h) + (1+k) = 0$

$m_1m_2m_3 = -(1+k) \Rightarrow m_3 = 1 + k \quad [\because m_1m_2 = -1]$

$\Rightarrow \quad (k+1)^3 + (1+k)(3-h) + (1+k) = 0$

$\Rightarrow \quad (k+1)^3 + 3 - h + 1 = 0$

or $\quad (y+1)^2 = x - 4$

$\therefore \quad \frac{\mu}{2} = -1, \mu = -2, \lambda = 4$

$\therefore \quad \lambda - 2\mu = 8$

60. (2) Since PT bisects $\angle NPS$.

$\therefore \quad \angle NPS = \angle PSX$

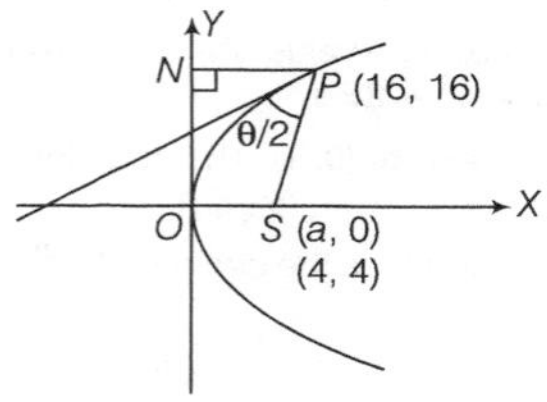

$\Rightarrow \quad \tan^{-1}\left(\frac{16}{12}\right) = \tan^{-1}\left(\frac{4}{3}\right) = \theta$

$\Rightarrow \quad \tan\theta = \frac{4}{3} = \frac{2\tan\theta/2}{1 - \tan^2\theta/2}$

$\Rightarrow \quad \tan\frac{\theta}{2} = \frac{1}{2}$

$\therefore \quad \angle TPS = \cot^{-1}(2)$

$\therefore \quad k = 2$

61. (4) Let r be the radius of the largest circle passing through the focus $(1, 0)$ of $y^2 = 4x$.

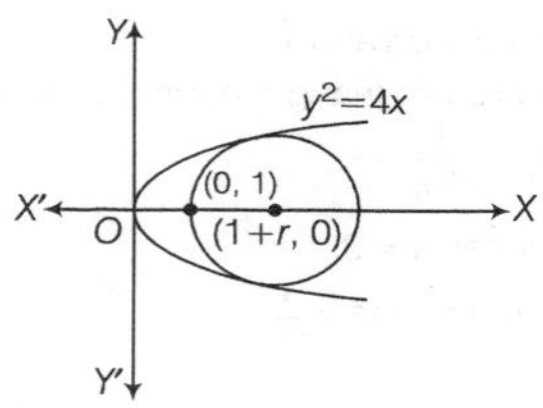

Clearly, centre of the circle will be on X-axis and its coordinates are $(1 + r, 0)$.

The equation of circle is $(x - 1 - r)^2 + y^2 = r^2$.

It touches $y^2 = 4x$. Therefore, the equation

$(x - r - 1)^2 + 4x = r^2$ must have equal roots.

$\Rightarrow 4(1-r)^2 - 4(2r+1) = 0$

$\therefore \quad r = 4$

62. (2) Let $P(h, k)$, $(k \neq 0)$ through which normals are drawn to the parabola

$$y^2 = 4ax(a = 1)$$

such that product of slopes of two is α,

$$m_1 m_2 = \alpha$$

Now, any normal to the parabola is

$$y = mx - 2am - am^3$$

If it passes through the point (h, k), then

$$k = mh - 2am - am^3$$

$\Rightarrow \quad am^3 + 0m^2 + m(2a - h) + k = 0$

$\Rightarrow \quad m^3 + 0m^2 + m(2a - h) + k = 0$, as $a = 1$ $\quad$ ...(i)

$m_1 m_2 m_2 = -k$

$\therefore \quad m_3 = \frac{-k}{m_1 m_2} = \frac{-k}{\alpha}$ $\quad$ [by Eq. (i)]

But m_3 is a root of Eq. (i).

$\therefore \quad k^3 + (2-h)k\alpha^2 - k\alpha^3 = 0$

Hence, the locus of P is

$y^2 + (2-x)\alpha^2 - \alpha^3 = 0$ $\quad$ $[k \neq 0$, cancelled$]$

Above is same as $y^2 - 4x = 0$, as given.

Hence, on comparing $\alpha^2 = 4$ and $2\alpha^2 - \alpha^3 = 0$

and $\quad \alpha = 2$

$\therefore \alpha = 2$, which satisfies both.

63. (8) Length of latus rectum is $4\sqrt{2} = 4a$

$\Rightarrow \quad a = \sqrt{2}$

Equation of directrix is $x + y = 2$
whose distance from (4, 4) is $2\sqrt{2}$.

$\therefore \quad \left|\frac{4 + 4 - \lambda}{\sqrt{2}}\right| = 2\sqrt{2}$

$\Rightarrow \quad 8 - \lambda = \pm 4$

$\therefore \quad \lambda = 4, 12$

$\Rightarrow \quad \frac{\lambda_1 + \lambda_2}{2} = 8$

64. (5) $m = \left(\frac{dy}{dx}\right)_{(1,1)} = b + 2$

Equation of the tangent, is

$(y - 1) = (b + 2)(x - 1)$

$\Rightarrow \quad (b + 2)x - y - (b + 1) = 0$

$\Rightarrow \quad \frac{(b + 1)^2}{2|b + 2|} = 2$

$\Rightarrow \quad b^2 + 2b + 1 = \pm 4(b + 2)$

$\Rightarrow \quad b^2 - 2b - 7 = 0$

and $\quad b^2 - 6b + 9 = 0$

$\Rightarrow \quad b_1 = -3$

$b_2 = 1 - \sqrt{2}$

$b_3 = 1 + \sqrt{2}$

$\Rightarrow \quad |b_1| + |b_2 + b_3| = 5$

65. (1) Slope of $OA = t$

$\Rightarrow$ Slope of $AB = -\frac{1}{t}$

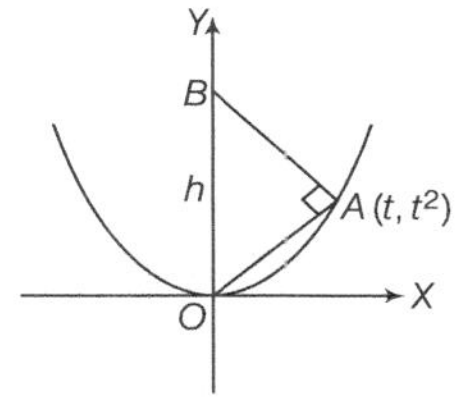

Equation of AB,

$$y - t^2 = -\frac{1}{t}(x - t^2)$$

put, $x = 0$

$\Rightarrow \quad h = t^2 - 1$

as $x \to 0, t \to 0$

$\Rightarrow \quad h \to 1$

JEE Main & AIEEE Archive

1. (d) Any point on the parabola $x^2 = 8y$ is $(4t, 2t^2)$. Point P divides the line segment joining of $O(0, 0)$ and $Q(4t, 2t^2)$ in the ratio 1 : 3. Apply the section formula for internal division.

Equation of parabola is

$$x^2 = 8y \quad \ldots(i)$$

Let any point Q on the parabola (i) is $(4t, 2t^2)$. Let $P(h, k)$ be the point which divides the line segment joining (0,0) and $(4t, 2t^2)$ in the ratio 1 : 3.

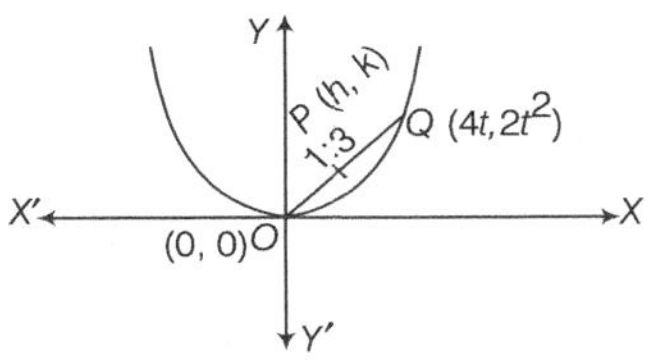

$\therefore \quad h = \frac{1 \times 4t + 3 \times 0}{4}$

$\Rightarrow \quad h = t$

and $k = \frac{1 \times 2t^2 + 3 \times 0}{4} \Rightarrow k = \frac{t^2}{2}$

$\Rightarrow \quad k = \frac{1}{2}h^2 \quad [\because t = h]$

$\Rightarrow \quad 2k = h^2 \Rightarrow 2y = x^2$

which is required locus.

2. (a) For parabola, $y^2 = 4x$

Let $y = mx + \frac{1}{m}$ be the tangent line and it touches the parabola

$x^2 = -32y$.

$\therefore \quad x^2 = -32\left(mx + \frac{1}{m}\right)$

$\Rightarrow \quad x^2 + 32mx + \frac{32}{m} = 0$

$\because \quad D = 0$

$\therefore \quad (32m)^2 - 4 \cdot \left(\frac{32}{m}\right) = 0$

$\Rightarrow \quad m^3 = \frac{1}{8} \Rightarrow m = \frac{1}{2}$

3. (b) Equation of circle can be rewritten as $x^2 + y^2 = \frac{5}{2}$

Here, centre is (0, 0) and radius is $\sqrt{\frac{5}{2}}$.

Let common tangent be $\quad y = mx + \frac{\sqrt{5}}{m}$

The perpendicular from centre to the tangent is equal to radius.

$\therefore \quad \frac{\frac{\sqrt{5}}{m}}{\sqrt{1 + m^2}} = \sqrt{\frac{5}{2}}$

$\Rightarrow \quad m\sqrt{1 + m^2} = \sqrt{2}$

$\Rightarrow \quad m^2(1 + m^2) = 2$

$\Rightarrow \quad m^4 + m^2 - 2 = 0$

$\Rightarrow \quad (m^2 + 2)(m^2 - 1) = 0$

$\Rightarrow \quad m = \pm 1 \quad [\because m^2 + 2 \neq 0, \text{ as } m \in R]$

$\therefore \quad y = \pm (x + \sqrt{5})$

Both statements are correct as $m = \pm 1$ satisfies the given equation of Statement II but Statement II is not a correct explanation of Statement I.

4. (b) **Statement I**

Analysis given, a parabola $y^2 = 16\sqrt{3}x$ and an ellipse $2x^2 + y^2 = 4$.

To find The equation of common tangent to the given parabola and the ellipse. This can be very easily done by comparing the standard equation of tangents. Standard equation of tangent with slope 'm' to the parabola $y^2 = 16\sqrt{3}x$ is

$$y = mx + \frac{4\sqrt{3}}{m} \quad \ldots(i)$$

Standard equation of tangent with slope m to the ellipse $\frac{x^2}{2} + \frac{y^2}{4} = 1$ is

$$y = mx \pm \sqrt{2m^2 + 4} \quad \ldots\text{(ii)}$$

If a line L is a common tangent to both parabola and ellipse, then L should be tangent to parabola, i.e. its equation should be like Eq. (i).

L should be tangent to ellipse i.e. its equation should be like Eq. (ii).

i.e. L must be like both of the Eqs. (i) and (ii).

Hence, comparing Eqs. (i) and (ii), we get

$$\frac{4\sqrt{3}}{m} = \pm \sqrt{2m^2 + 4}$$

On squaring both sides, we get

$$\Rightarrow \quad m^2(2m^2 + 4) = 48$$

$$\Rightarrow \quad m^4 + 2m^2 - 24 = 0$$

$$\Rightarrow \quad (m^2 + 6)(m^2 - 4) = 0$$

$$\Rightarrow \quad m^2 = 4 \quad [\because m^2 \neq -6]$$

$$\therefore \quad m = \pm 2$$

Substituting $m = \pm 2$ in the Eq. (i), we get the required equation of the common tangents as

$$y = 2x + 2\sqrt{3} \quad \text{and} \quad y = -2x - 2\sqrt{3}$$

Hence, Statement I is correct.

Statement II

In Statement II, we have already seen that, if the line $y = mx + \frac{4\sqrt{3}}{m}$ is a common tangent to the parabola $y^2 = 16\sqrt{3}x$ and the ellipse $\frac{x^2}{2} + \frac{y^2}{4} = 1$, then it satisfies the equation $m^4 + 2m^2 - 24 = 0$.

Hence, Statement II is also correct but is not able to explain the Statement I. It is an intermediate step in the final answer.

5. (c) We know that, the locus of point P from which two perpendicular tangents are drawn to the parabola, is the directrix of the parabola.

Hence, the required locus is $x = -1$.

6. (c) Since, the vertex is the mid-point of the focus and foot of the directrix vertex of the parabola at (1, 0).

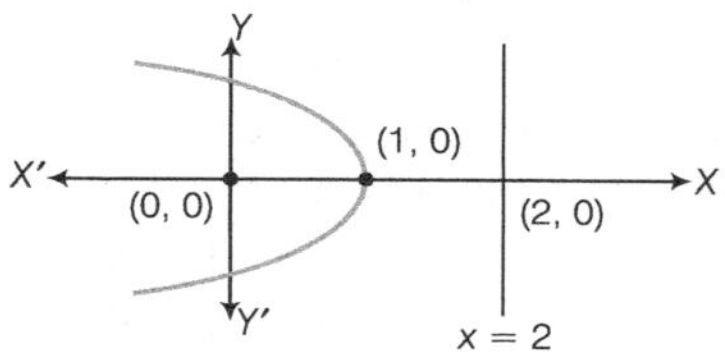

7. (d) Since, perpendicular tangents intersect on the directrix, then point must lie on the directrix $x = -2$.

Hence, the required point is $(-2, 0)$.

8. (d) The given equation of parabola is

$$y = \frac{a^3x^2}{3} + \frac{a^2x}{2} - 2a$$

$$\Rightarrow y + 2a = \frac{a^3}{3}\left(x^2 + \frac{3}{2a}x\right)$$

$$\Rightarrow y + 2a = \frac{a^3}{3}\left(x^2 + \frac{3}{2a}x + \frac{9}{16a^2} - \frac{9}{16a^2}\right)$$

$$\Rightarrow y + 2a = \frac{a^3}{3}\left(x + \frac{3}{4a}\right) - \frac{9}{16a^2} \times \frac{a^3}{3}$$

$$\Rightarrow \quad y + 2a + \frac{3a}{16} = \frac{a^3}{3}\left(x + \frac{3}{4a}\right)^2$$

$$\Rightarrow \quad \left(y + \frac{35a}{16}\right) = \frac{a^3}{3}\left(x + \frac{3}{4a}\right)^2$$

Thus, the vertices of parabola is $\left(-\frac{3}{4a}, -\frac{35a}{16}\right)$.

Let $h = -\frac{3}{4a}$ and $k = -\frac{35a}{16}$

Now, $\quad hk = \frac{105}{64}$

Thus, the locus of vertices of a parabola is

$$xy = \frac{105}{64}$$

9. (a) Given equation of parabolas are

$$y^2 = 4ax$$

and $\quad x^2 = 4ay$

The point of intersection of parabolas are $A(0, 0)$ and $B(4a, 4a)$. Also, given line $2bx + 3cy + 4d = 0$ passes through the points A and B, respectively.

$$\therefore \quad d = 0 \quad \ldots\text{(i)}$$

and $\quad 2b \cdot 4a + 3c \cdot 4a + 4d = 0$

$$\Rightarrow \quad 2ab + 3ac + d = 0$$

$$\Rightarrow \quad a(2b + 3c) = 0 \quad [\because d = 0]$$

$$\Rightarrow \quad 2b + 3c = 0 \quad \ldots\text{(ii)}$$

On squaring and adding Eqs. (i) and (ii), we get

$$d^2 + (2b + 3c)^2 = 0$$

10. (a) Equation of the normal at point $(bt_1^2, 2bt_1)$ on parabola is

$$y = -t_1x + 2bt_1 + bt_1^3$$

It also passes through $(bt_2^2, 2bt_2)$, then

$$2bt_2 = -t_1 \cdot bt_2^2 + 2bt_1 + bt_1^3$$

$$\Rightarrow \quad 2t_2 - 2t_1 = -t_1(t_2^2 - t_1^2)$$

$$= -t_1(t_2 + t_1)(t_2 - t_1)$$

$$\Rightarrow \quad 2 = -t_1(t_2 + t_1)$$

$$\therefore \quad t_2 = -t_1 - \frac{2}{t_1}$$

11. (d) Given equation of parabola can be rewritten as

$$(y + 2)^2 = -4\left(x - \frac{1}{2}\right)$$

Let $\quad y + 2 = Y$

and $\quad x - \frac{1}{2} = X$

$$\therefore \quad Y^2 = -4X$$

Here, $\quad a = 1$

$\therefore$ Equation of directrix is $X = a$.

$$\therefore \quad x - \frac{1}{2} = 1$$

$$\Rightarrow \quad x = \frac{3}{2}$$

JEE Advanced & IIT JEE Archive

1. (2) End points of latusrectum are $(a, \pm 2a)$ i.e. $(1, \pm 2)$.

Equation of normal at (x_1, y_1) is

$$\frac{y - y_1}{x - x_1} = -\frac{y_1}{2a}$$

i.e. $$\frac{y-2}{x-1} = -\frac{2}{2}$$

and $$\frac{y+2}{x-1} = \frac{2}{2}$$

$\Rightarrow$ $x + y = 3$ and $x - y = 3$

which is tangent to $(x-3)^2 + (y+2)^2 = r^2$.

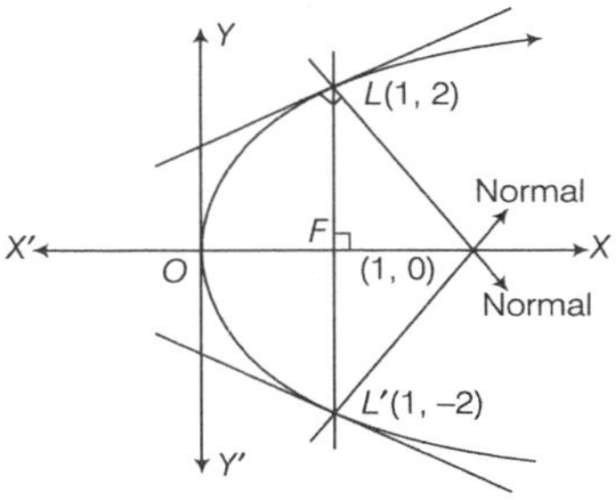

$\therefore$ Length of perpendicular from centre = Radius

$$\Rightarrow \quad \frac{|3-2-3|}{\sqrt{1^2+1^2}} = r$$

$$\therefore \quad r^2 = 2$$

2. (a,d) Since, $\angle POQ = 90°$

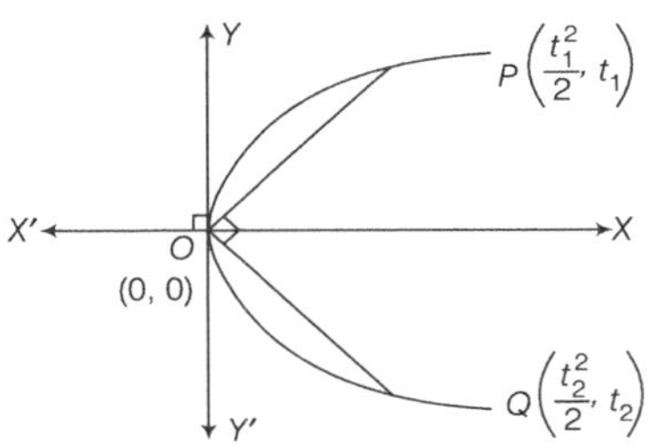

$$\Rightarrow \quad \frac{t_1 - 0}{\frac{t_1^2}{2} - 0} \cdot \frac{t_2 - 0}{\frac{t_2^2}{2} - 0} = -1$$

$$\Rightarrow \quad t_1 t_2 = -4 \quad \ldots \text{(i)}$$

$$\because \quad \text{ar}(\Delta OPQ) = 3\sqrt{2}$$

$$\therefore \quad \frac{1}{2}\begin{vmatrix} 0 & 0 & 1 \\ t_1^2/2 & t_1 & 1 \\ t_2^2/2 & t_2 & 1 \end{vmatrix} = \pm 3\sqrt{2}$$

$$\Rightarrow \quad \frac{1}{2}\left(\frac{t_1^2 t_2}{2} - \frac{t_1 t_2^2}{2}\right) = \pm 3\sqrt{2}$$

$$\Rightarrow \quad \frac{1}{4}(-4t_1 + 4t_2) = \pm 3\sqrt{2}$$

$$\Rightarrow \quad t_1 + \frac{4}{t_1} = 3\sqrt{2} \quad [\because t_1 > 0 \text{ for } P$$

$$\Rightarrow \quad t_1^2 - 3\sqrt{2}t_1 + 4 = 0$$

$$\Rightarrow \quad (t_1 - 2\sqrt{2})(t_1 - \sqrt{2}) = 0$$

$$\Rightarrow \quad t_1 = \sqrt{2} \text{ or } 2\sqrt{2}$$

$$\therefore \quad P(1, \sqrt{2}) \text{ or } P(4, 2\sqrt{2})$$

3. (d) Since, equation of normal to the parabola $y^2 = 4ax$ is $y + xt = 2at + at^3$ passes through $(3, 0)$.

$$\Rightarrow \quad 3t = 2t + t^3 \quad [\because a = 1]$$

$$\Rightarrow \quad t = 0, 1, -1$$

$\therefore$ Coordinates of the normals are $P(1,2), Q(0,0), R(1,-2)$.

Area of $\Delta PQR = \frac{1}{2} \times 1 \times 4 = 2$

Centroid of $\Delta PQR = \left(\frac{2}{3}, 0\right)$

Equation of circle passing through P, Q, R is

$$(x-1)(x-1) + (y-2)(y+2) + \lambda(x-1) = 0$$

$$\Rightarrow \quad 1 - 4 - \lambda = 0 \quad \Rightarrow \quad \lambda = -3$$

$\therefore$ Required equation of circle is

$$x^2 + y^2 - 5x = 0$$

$\therefore$ Centre $\left(\frac{5}{2}, 0\right)$ and radius $\frac{5}{2}$.

Hence, (i) → (p), (ii) → (q), (iii) → (s), (iv) → (r)

4. (a) Equation of circle can be rewritten as

$$x^2 + y^2 = \frac{5}{2},$$

centre is (0, 0) and radius is $\sqrt{\frac{5}{2}}$.

Let common tangent be

$$y = mx + \frac{\sqrt{5}}{m} \Rightarrow m^2x - my + \sqrt{5} = 0$$

The perpendicular from centre to the tangent is equal to radius of the circle.

$$\therefore \quad \frac{\frac{\sqrt{5}}{m}}{\sqrt{1+m^2}} = \sqrt{\frac{5}{2}}$$

$$\Rightarrow \quad m\sqrt{1+m^2} = \sqrt{2}$$

$$\Rightarrow \quad m^2(1+m^2) = 2$$

5. (c) By section formula,

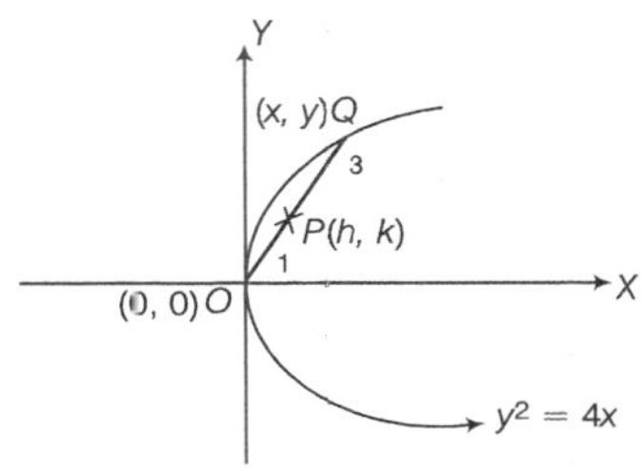

$$h = \frac{x+0}{4}, k = \frac{y+0}{4}$$

$$\therefore \quad x = 4h, y = 4k$$

Substituting in $y^2 = 4x$, $(4k)^2 = 4(4h) \Rightarrow k^2 = h$

or $y^2 = x$ is required locus.

$$\Rightarrow \quad m^4 + m^2 - 2 = 0$$

$$\Rightarrow \quad (m^2 + 2)(m^2 - 1) = 0$$

$$\Rightarrow \quad m = \pm 1 \quad [\because m^2 + 2 \neq 0, \text{ as } m \in R]$$

$\therefore$ $y = \pm(x + \sqrt{5})$, both statements are correct as $m = \pm 1$ satisfies the given equation of Statement II.

6. (a,b,d) Normal to $y^2 = 4x$, is

$$y = mx - 2m - m^3$$

which passes through (9, 6)

$$\Rightarrow \quad 6 = 9m - 2m - m^3$$

$$\Rightarrow \quad m^3 - 7m + 6 = 0$$

$$\Rightarrow \quad m = 1, 2, -3$$

$\therefore$ Equation of normals are

$$y - x + 3 = 0$$
$$y + 3x - 33 = 0$$
and
$$y - 2x + 12 = 0$$

7. (4) **Concept Involved** Parametric coordinates for $y^2 = 4ax$ are $(at^2, 2at)$.

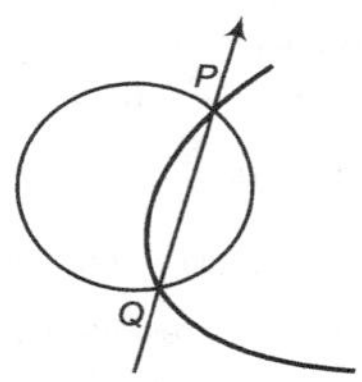

Description of Situation As the circle intersects the parabola at P and Q. Thus, points P and Q should satisfy circle.

$P(2t^2, 4t)$ should lie on $x^2 + y^2 - 2x - 4y = 0$

$$\Rightarrow \quad 4t^4 + 16t^2 - 4t^2 - 16t = 0$$

$$\Rightarrow \quad 4t^4 + 12t^2 - 16t = 0$$

$$\Rightarrow \quad 4t(t^3 + 3t - 4) = 0$$

$$\Rightarrow \quad 4t(t-1)(t^2 + t + 4) = 0$$

$$\therefore \quad t = 0, 1$$

$$\Rightarrow \quad P(2, 4)$$

Thus, area of $\Delta OPS = \frac{1}{2} \cdot OS \times PQ$

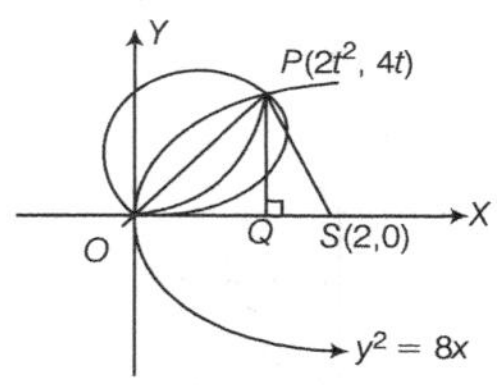

$$= \frac{1}{2} \cdot (2) \cdot (4) = 4 \text{ sq units}$$

8. (2) As, we know area of triangle formed by three points on parabola is twice the area of triangle formed by corresponding tangents

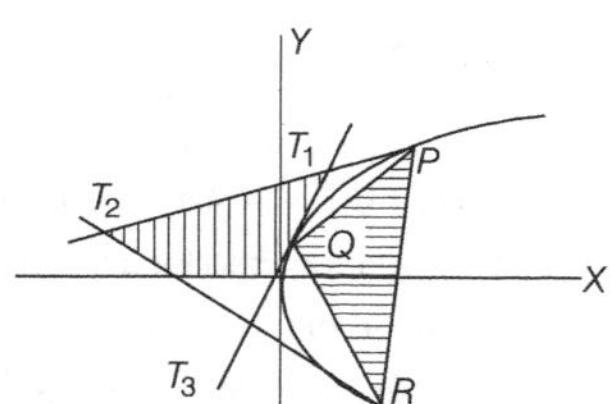

i.e. area of ΔPQR = 2 area of $\Delta T_1T_2T_3$.

$$\therefore \quad \Delta_1 = 2\Delta_2$$

or

$$\frac{\Delta_1}{\Delta_2} = 2$$

9. (b) Since, $R\left(-a, a\left(t - \frac{1}{t}\right)\right)$ lies on $y = 2x + a$

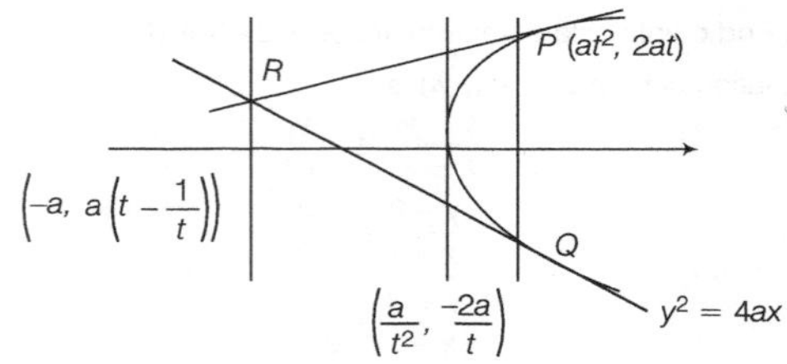

$$\Rightarrow \quad a \cdot \left(t - \frac{1}{t}\right) = -2a + a$$

$$\Rightarrow \quad t - \frac{1}{t} = -1$$

Thus, length of focal chord

$$= a\left(t + \frac{1}{t}\right)^2$$

$$= a\left\{\left(t - \frac{1}{t}\right)^2 + 4\right\}$$

$$= 5a$$

10. (d) Since, $m_{OP} = \frac{2at - 0}{at^2 - 0} = \frac{2}{t}$

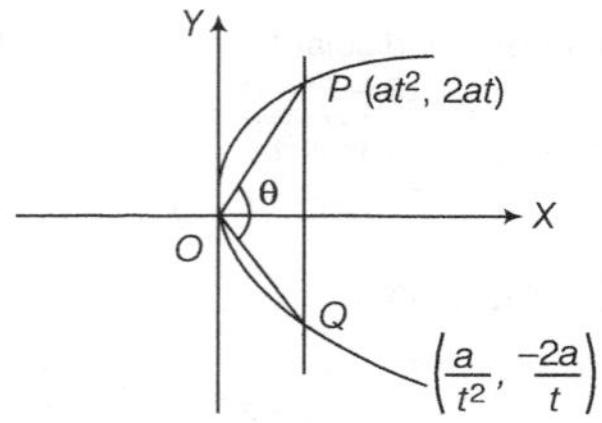

$$m_{OQ} = \frac{-2a/t}{a/t^2} = -2t$$

$$\therefore \quad \tan\theta = \frac{\frac{2}{t} + 2t}{1 - \frac{2}{t} \cdot 2t} = \frac{2\left(t + \frac{1}{t}\right)}{1 - 4}, \text{ where}$$

$$t + \frac{1}{t} = \sqrt{5} = -\frac{2\sqrt{5}}{3}$$

11. (c,d) Here, coordinate $M = \left(\frac{t_1^2 + t_2^2}{2}, t_1 + t_2\right)$ i.e. mid-point of chord AB.

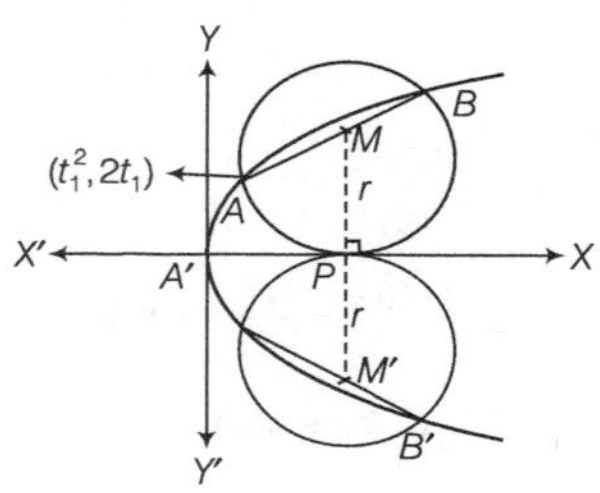

$$MP = t_1 + t_2 = r \qquad \text{...(i)}$$

Also, $$m_{AB} = \frac{2t_2 - 2t_1}{t_2^2 - t_1^2} = \frac{2}{t_2 + t_1} \qquad \text{[when } AB \text{ is chord]}$$

$$\Rightarrow \quad m_{AB} = \frac{2}{r} \qquad \text{[from Eq. (i)]}$$

Also, $$m_{A'B'} = -\frac{2}{r} \qquad \text{[when } A'B' \text{ is chord]}$$

12. (a,d) Equation of tangent and normal at point $P(at^2, 2at)$ is

$$ty = x = at^2$$

and $$y = -tx + 2at + at^2$$

Let centroid of ΔPTN is $R(h, k)$.

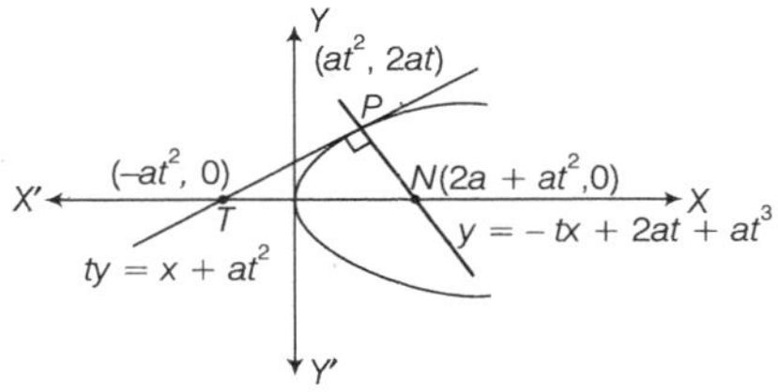

$\therefore$ $$h = \frac{at^2 + (-at^2) + 2a + at^2}{3}$$

and $$k = \frac{2at}{3}$$

$\Rightarrow$ $$3h = 2a + a \cdot \left(\frac{3k}{2a}\right)^2$$

$\Rightarrow$ $$3h = 2a + \frac{9k^2}{4a}$$

$\Rightarrow$ $$9k^2 = 4a(3h - 2a)$$

$\therefore$ Locus of centroid is

$$y^2 = \frac{4a}{3}\left(x - \frac{2a}{3}\right)$$

$\therefore$ Vertex $\left(\frac{2a}{3}, 0\right)$ directrix $x - \frac{2a}{3} = -\frac{a}{3}$

$\Rightarrow$ $$x = \frac{a}{3}$$

and latusrectum $= \frac{4a}{3}$

$\therefore$ Focus $\left(\frac{a}{3} + \frac{2a}{3}, 0\right)$ i.e. $(a, 0)$.

13. (b,c) The equation $x^2 + 4y^2 = 4$ represents an ellipse with 2 and 1 as semi-major and semi-minor axes and eccentricity $\frac{\sqrt{3}}{2}$. Thus, the ends of latusrectum are $\left(\sqrt{3}, \frac{1}{2}\right)$ and $\left(\sqrt{3}, -\frac{1}{2}\right)$, $\left(-\sqrt{3}, \frac{1}{2}\right)$ and $\left(-\sqrt{3}, -\frac{1}{2}\right)$.

According to the question, we consider only $P\left(-\sqrt{3}, -\frac{1}{2}\right)$ and $Q\left(\sqrt{3}, -\frac{1}{2}\right)$.

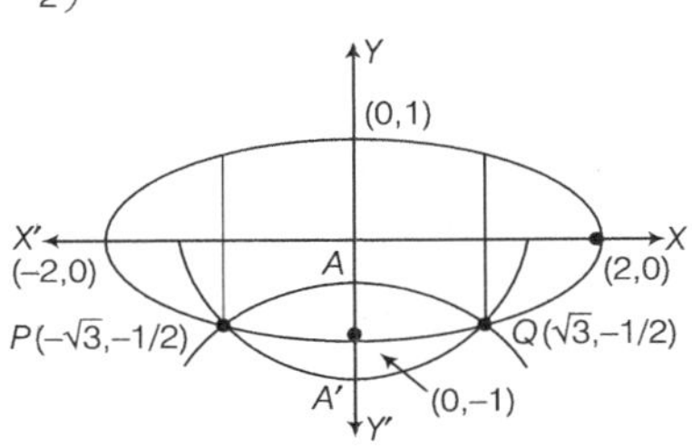

Now, $$PQ = 2\sqrt{3}$$

Thus, the coordinates of the vertex of the parabolas are $A\left(0, \frac{-1+\sqrt{3}}{2}\right)$ and $A'\left(0, \frac{-1-\sqrt{3}}{2}\right)$ and corresponding equations are

$$(x - 0)^2 = -4 \cdot \frac{\sqrt{3}}{2}\left(y + \frac{1-\sqrt{3}}{2}\right)$$

and $$(x - 0)^2 = 4 \cdot \frac{\sqrt{3}}{2}\left(y - \frac{-1-\sqrt{3}}{2}\right)$$

i.e. $$x^2 + 2\sqrt{3}y = 3 - \sqrt{3}$$

and $$x^2 - 2\sqrt{3}y = 3 + \sqrt{3}$$

14. (a,b) The equation of tangent to $y = x^2$, be $y = mx - \frac{m^2}{4}$. Putting in $y = -x^2 + 4x - 4$, we should only get one value of x *ie*, Discriminant must be zero.

$\therefore$ $$mx - \frac{m^2}{4} = -x^2 + 4x - 4$$

$\Rightarrow$ $$x^2 + x(m - 4) + 4 - \frac{m^2}{4} = 0$$

$$D = 0$$

Now, $(m - 4)^2 - (16 - m^2) = 0$

$\Rightarrow$ $$2m(m - 4) = 0 \Rightarrow m = 0, 4$$

$\therefore$ $y = 0$ and $y = 4(x - 1)$ are the required tangents.

15. (a) Since, distance of vertex from origin is $\sqrt{2}$ and focus is $2\sqrt{2}$.

$\therefore$ $V(1, 1)$ and $F(2, 2)$ (i.e. lying on $y = x$)

where, length of latusrectum

$$= 4a$$
$$= 4\sqrt{2} \qquad [\because a = \sqrt{2}]$$

$\therefore$ By definition of parabola

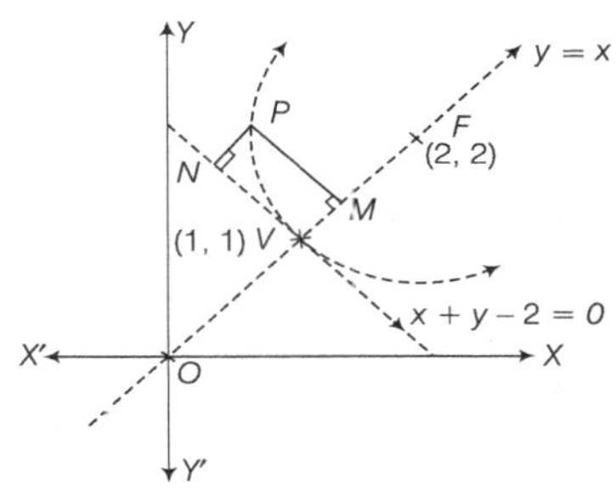

$$PM^2 = (4a)(PN)$$

where, PN is length of perpendicular upon $x + y - 2 = 0$ (i.e. tangent at vertex).

$\Rightarrow$ $$\frac{(x-y)^2}{2} = 4\sqrt{2}\left(\frac{x+y-2}{\sqrt{2}}\right)$$

$\Rightarrow$ $$(x - y)^2 = 8(x + y - 2)$$

16. (a) Let the tangent to parabola be $y = mx + a/m$, if it touches the other curve, then $D = 0$, to get the value of m.

For parabola, $y^2 = 4x$

Let $y = mx + \frac{1}{m}$ be tangent line and it touches the parabola $x^2 = -32y$.

$\therefore$ $$x^2 = -32\left(mx + \frac{1}{m}\right)$$

$\Rightarrow$ $$x^2 + 32mx + \frac{32}{m} = 0$$

$\because$ $$D = 0$$

$\therefore$ $$(32m)^2 - 4 \cdot \left(\frac{32}{m}\right) = 0$$

$\Rightarrow$ $$m^3 = \frac{1}{8}$$

$\therefore$ $$m = 1/2$$

2. Ellipse

DPP-1 Equation of Ellipse

1. (i) Given equation of ellipse is

$$16x^2 + 25y^2 = 400$$

$$\Rightarrow \quad \frac{16x^2}{400} + \frac{25y^2}{400} = 1$$

$$\Rightarrow \quad \frac{x^2}{25} + \frac{y^2}{16} = 1$$

On comparing it with $\frac{x^2}{a^2} + \frac{y^2}{b^2} = 1$, we get

$$a = 5$$

and $b = 4 \, (a > b)$

$\therefore$ Length of major axis $= 2a = 10$

Length of minor axis $= 2b = 8$

Vertices $= (\pm a, 0) = (\pm 5, 0)$

$$\text{Eccentricity} = e\sqrt{1 - \frac{b^2}{a^2}} = \sqrt{1 - \frac{16}{25}} = \frac{3}{5}$$

Foci $= (\pm ae, 0) = (\pm 3, 0)$

(ii) Given equation of ellipse is

$$3x^2 + 2y^2 = 6$$

$$\Rightarrow \quad \frac{3x^2}{6} + \frac{2y^2}{6} = 1$$

$$\Rightarrow \quad \frac{x^2}{2} + \frac{y^2}{3} = 1$$

On comparing it with $\frac{x^2}{a^2} + \frac{y^2}{b^2} = 1$, we get

$$a = \sqrt{2} \text{ and } b = \sqrt{3} \, (a < b)$$

$\therefore$ Length of major axis $= 2b = 2\sqrt{3}$

Length of minor axis $= 2a = 2\sqrt{2}$

Vertices $= (0, \pm b) = (0, \pm \sqrt{3})$

$$\text{Eccentricity} = e = \sqrt{1 - \frac{a^2}{b^2}}$$

$$= \sqrt{1 - \frac{2}{3}} = \frac{1}{\sqrt{3}}$$

Foci $(0, \pm be) = (0, \pm 1)$

(iii) Given equation of ellipse is

$$x^2 + 4y^2 - 2x = 0$$

$$\Rightarrow \quad (x^2 - 2x) + 4y^2 = 0$$

$$\Rightarrow \quad (x-1)^2 + 4y^2 = 1$$

$$\Rightarrow \quad \frac{(x-1)^2}{1} + \frac{y^2}{\frac{1}{4}} = 1$$

On comparing it with $\frac{x^2}{a^2} + \frac{y^2}{b^2} = 1$, we get

$$a = 1 \text{ and } b = \frac{1}{2} \qquad [a > b]$$

$\therefore$ Length of major axis $= 2a = 2$

Length minor axis $= 2b = 1$

Vertices $= (1 + 1, 0)$ and $(1 - 1, 0) = (2, 0)$ and $(0, 0)$

$$\text{Eccentricity} = e = \sqrt{1 - \frac{b^2}{a^2}}$$

$$= \sqrt{1 - \frac{1}{4}} = \frac{\sqrt{3}}{2}$$

$$\text{Foci} = (1 \pm ae, 0) = \left(1 \pm \frac{\sqrt{3}}{2}, 0\right)$$

2. Given equation of ellipse is

$$x^2 + 4y^2 + 2x + 16y + 13 = 0$$

$$\Rightarrow \quad (x^2 + 2x) + 4(y^2 + 4y) + 13 = 0$$

$$\Rightarrow \quad (x+1)^2 - 1 + 4(y+2)^2 - 16 + 13 = 0$$

$$\Rightarrow \quad (x+1)^2 + 4(y+2)^2 = 4$$

$$\Rightarrow \quad \frac{(x+1)^2}{4} + \frac{(y+2)^2}{1} = 1$$

Shifting the origin at $(-1, -2)$, we have

$$x = X - 1 \text{ and } y = Y - 2 \qquad \ldots \text{(i)}$$

The equation of ellipse with reference to new axes is

$$\frac{X^2}{2^2} + \frac{Y^2}{1^2} = 1$$

$$\Rightarrow \quad a = 2 \text{ and } b = 1 \qquad [a > b]$$

$$\therefore \quad \text{Eccentricity} = e = \sqrt{1 - \frac{b^2}{a^2}} = \sqrt{1 - \frac{1}{4}} = \frac{\sqrt{3}}{2}$$

The coordinates of vertices with reference to new axes are $(X = \pm a, Y = 0) = (X = \pm 2, Y = 0)$

Substituting these in Eq. (i), we obtain the coordinates the vertices as $(\pm 2 - 1, -2) = (1, -2)$ and $(-3, 2)$.

The coordinates of the foci with reference to new axes are $(X = \pm ae, Y = 0) = (X = \pm \sqrt{3}, Y = 0)$.

Substituting these in Eq. (i), we obtain the coordinates of foci as $(\pm \sqrt{3} + 1, -2)$.

Equation of directrices with reference to new origin are $X = \pm \frac{a}{e}$

$\Rightarrow X = \pm \frac{4}{\sqrt{3}}$. Putting these values in Eq. (i), we get equation of directrices as $x = \pm \frac{4}{\sqrt{3}} - 1$

$$\text{Length of latusrectum} = \frac{2b^2}{a} = \frac{2(1)^2}{2} = 1$$

Equation of latusrectum with reference to new origin are $X = \pm ae$, i.e. $x = \pm \sqrt{3}$. Hence, equation of latusrectum will be $x = \pm \sqrt{3} - 1$.

3. (i) Let the equation of required ellipse be $\frac{x^2}{a^2} + \frac{y^2}{b^2} = 1$ ...(i)

The coordinates of its vertices and foci are $(\pm a, 0)$ and $(\pm ae, 0)$ respectively.

$$\therefore \quad a = 5 \text{ and } ae = 4 \Rightarrow e = \frac{4}{5}$$

Now, $\quad b^2 = a^2(1 - e^2) \Rightarrow b^2 = 25\left(1 - \frac{16}{25}\right) = 9$

Substituting the values of a^2 and b^2 in Eq. (i), we get

$$\frac{x^2}{25} + \frac{y^2}{9} = 1$$

which is the equation of required ellipse.

(ii) Let the equation of required ellipse be $\frac{x^2}{a^2} + \frac{y^2}{b^2} = 1$.

Since, the vertices of ellipse are on Y-axis. So, the coordinates of the vertices are $(0, \pm b)$.

$$\therefore \quad b = 10$$

The eccentricity e is given by $a^2 = b^2(1 - e^2)$

$$\Rightarrow \quad a^2 = 100\left(1 - \frac{16}{25}\right) = 36$$

Hence, required equation of ellipse is $\frac{x^2}{36} + \frac{y^2}{100} = 1$.

(iii) Let the equation of required ellipse is $\frac{x^2}{a^2} + \frac{y^2}{b^2} = 1$.

Since, the foci of the ellipse are on Y-axis. So, the coordinates of the foci are $(0, \pm be)$.

$\therefore \quad be = 4 \quad \ldots(i)$

Eccentricity $e = \frac{4}{5} \Rightarrow b = 5$

Also, $a^2 = b^2(1 - e^2)$

$\Rightarrow a^2 = 25\left(1 - \frac{16}{25}\right) \Rightarrow a^2 = 9$

Hence, required equation of the ellipse is $\frac{x^2}{9} + \frac{y^2}{25} = 1$.

(iv) Let the equation of the ellipse is $\frac{x^2}{a^2} + \frac{y^2}{b^2} = 1$.

Since, the foci of the ellipse are on X-axis. So, the coordinates are. So, the coordinates of the foci are $(\pm ae, 0)$.

$\therefore \quad ae = 2$

Eccentricity, $e = \frac{1}{2}$

$\Rightarrow a = 4$

Also, $b^2 = a^2(1 - e^2)$

$\Rightarrow b^2 = 16\left(1 - \frac{1}{4}\right) \Rightarrow b^2 = 12$

Hence, required equation of ellipse is $\frac{x^2}{16} + \frac{y^2}{12} = 1$.

4. (c) Let $2a$ and $2b$ be the major and minor axes of the ellipse. Then, its equation will be

$$\frac{(x-2)^2}{a^2} + \frac{(y+3)^2}{b^2} = 1$$

we have, $a = \sqrt{(4-2)^2 + (-3+3)^2} = 2$

Since, the distance between the focus and the centre of ellipse is equal to ae, therefore

$$ae = \sqrt{(2-3)^2 + (-3+3)^2}$$

$\Rightarrow ae = 1$

$\Rightarrow e = \frac{1}{a} \Rightarrow e = \frac{1}{2}$

5. (a) We have, $4(x - 2y + 1)^2 + 9(2x + y + 2)^2 = 180$

$$\Rightarrow \frac{\left(\frac{x - 2y + 1}{\sqrt{1+4}}\right)^2}{3^2} + \frac{\left(\frac{2x + y + 2}{\sqrt{4+1}}\right)^2}{2^2} = 1$$

$$\Rightarrow \frac{X^2}{3^2} + \frac{Y^2}{2^2} = 1$$

where, $X = \frac{x - 2y + 1}{\sqrt{5}}$ and $Y = \frac{2x + y + 2}{\sqrt{5}}$

This is of the form

$$\frac{X^2}{a^2} + \frac{Y^2}{b^2} = 1$$

On comparing, we get

$$a = 3 \text{ and } b = 2$$

Let e be the eccentricity of ellipse. Then,

$$e = \sqrt{1 - \frac{b^2}{a^2}} \Rightarrow e = \sqrt{1 - \frac{4}{9}} = \frac{\sqrt{5}}{3}$$

6. (d) Since, $1^2 + 2^2 = 5 < 9$ and $2^2 + 1^2 = 5 < 9$, both P and Q lie inside C. Also,

$$\frac{1^2}{9} + \frac{2^2}{4} = \frac{1}{9} + 1 > 1 \text{ and } \frac{2^2}{9} + \frac{1}{4} = \frac{25}{36} < 9$$

Hence, P lies outisde E and Q lies inside E. Thus, P lies inside C but outside E.

7. (b) The given equation of ellipse is

$$(x-3)^2 + (y-4)^2 = \frac{y^2}{9}$$

$$\Rightarrow (x-3)^2 + y^2 - 8y + 16 - \frac{y^2}{9} = 0$$

$$\Rightarrow (x-3)^2 + \frac{8}{9}y^2 - 8y + 16 = 0$$

$$\Rightarrow (x-3)^2 + \frac{8}{9}(y^2 - 9y) + 16 = 0$$

$$\Rightarrow (x-3)^2 + \frac{8}{9}\left[\left(y - \frac{9}{2}\right)^2 - \frac{81}{4}\right] + 16 = 0$$

$$\Rightarrow (x-3)^2 + \frac{8}{9}\left(y - \frac{9}{2}\right)^2 - 18 + 16 = 0$$

$$\Rightarrow (x-3)^2 + \frac{8}{9}\left(y - \frac{9}{2}\right)^2 = 2$$

$$\Rightarrow \frac{(x-3)^2}{2} + \frac{\left(y - \frac{9}{2}\right)^2}{\frac{9}{4}} = 1$$

$$\therefore e = \sqrt{1 - \frac{a^2}{b^2}} = \sqrt{1 - \frac{2}{9} \times 4} = \frac{1}{3}$$

8. (c) Given equation is

$$2x^2 + 3y^2 - 8x - 18y + 35 = k$$

$$(\sqrt{2}x - 2\sqrt{2})^2 + (\sqrt{3}y - 3\sqrt{3})^2 = k$$

So, no locus if $k < 0$, an ellipse for $k > 0$ and a point for $k = 0$.

9. (d) $\frac{(x-h)^2}{M} + \frac{(y-k)^2}{N} = 1$ has,

Major axis on the line $y = 2 \Rightarrow k = 2$

Minor axis on the line

$x = -1$

$\Rightarrow h = -1$

Major axis has length

$2b = 10$

$\Rightarrow b = 5$

$\therefore N = b^2 = 25$

Minor axis has length

$2a = 4$

$\Rightarrow a = 2$

$\therefore M = a^2 = 4$

10. (c) Let the ellipse be $\frac{x^2}{a^2} + \frac{y^2}{b^2} = 1$ and the circle be $x^2 + y^2 = a^2e^2$

Radius of circle $= ae$

The point of intersection of the circle and the ellipse is $\left(\frac{a}{e}\sqrt{2e^2 - 1}, \frac{a}{e}(1 - e^2)\right)$.

$$\text{Now, area of } \Delta PF_1F_2 = \frac{1}{2}\begin{vmatrix} \frac{a}{e}\sqrt{2e^2-1} & \frac{a}{e}(1-e^2) & 1 \\ ae & 0 & 1 \\ -ae & 0 & 1 \end{vmatrix} = \frac{1}{2}$$

$$\Rightarrow \left|\frac{a}{e}(1 - e^2)(2ae)\right| = 30$$

$$\Rightarrow a^2(1 - e^2) = 30 \quad \text{[given]}$$

$$\Rightarrow a^2e^2 = a^2 - 30$$

$$= \left(\frac{17}{2}\right)^2 - 30 - \frac{169}{4}$$

$$\therefore 2ae = 13$$

11. (a) Given equation $3y = x^2 + 4x - 9$ can be written as

$$3y = (x+2)^2 - 13$$

$$\Rightarrow \quad (x+2)^2 = 3\left(y + \frac{13}{3}\right)$$

which is an equaticn of parabola, $X^2 = 4ay$

$$\therefore \quad 4a = 3 \Rightarrow a = \frac{3}{4}$$

Hence, the length of latusrectum $= 4a = 4 \times \frac{3}{4} = 3$

Also, the equation $x^2 + 4y^2 - 6x + 16y = 24$ can be written as

$$x^2 + 4y^2 - 6x + 16y = 24$$

$$\Rightarrow \quad (x-3)^2 - 9 + 4[(y+2)^2 - 4] = 24$$

$$\Rightarrow \quad (x-3)^2 + 4(y+2)^2 = 49$$

$$\Rightarrow \quad \frac{(x-3)^2}{7^2} + \frac{(y+2)^2}{(7/2)^2} = 1$$

which is an equation of ellipse.

$$\frac{X^2}{a^2} + \frac{Y^2}{b^2} = 1$$

$$\therefore \quad a = 7 \text{ and } b = \frac{7}{2}$$

Hence, the length of latusrectum $= \frac{2b^2}{a} = \frac{2 \times \frac{49}{4}}{7} = \frac{7}{2}$

$\therefore$ Positive difference between the length of latusrectum of both conics $= \frac{7}{2} - 3 = \frac{1}{2}$

12. (b) The area of the ellipse, A will be given by $A = \pi ab$

where, a is constant and b varies.

$$\therefore \quad A^2 = \pi^2 a^2 b^2 = \pi^2 a^2 (a^2 - c^2)$$

For A to be maximum C must be minimum; A and B must be same.

$\Rightarrow$ Ellipse become circle.

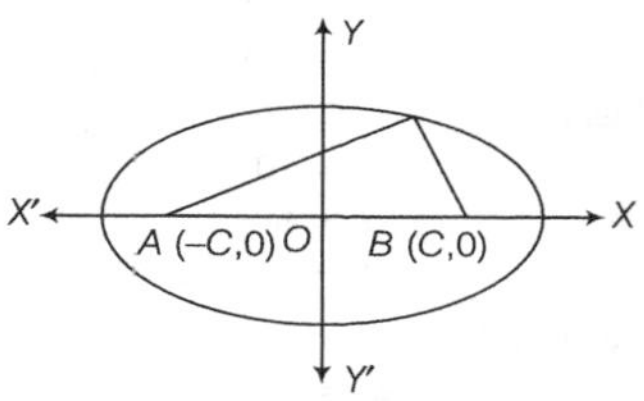

13. (c) If two foci are $S(5, 12)$ and $S'(-12, 5)$ and the ellipse passes through the origin O, then

$$SO = \sqrt{25 + 144} = 13$$

$$S'O = \sqrt{144 + 25} = 13$$

and $\quad SS' = \sqrt{338}$

If the conic is an ellipse, then

$$SO + S'O = 2a$$

and $\quad SS' = 2ae$

$$\therefore \quad e = \frac{SS'}{SO + S'O} = \frac{\sqrt{338}}{26} = \frac{1}{\sqrt{2}}$$

14. (c) Given equation of ellipse is $\frac{x^2}{\tan^2 \alpha} + \frac{y^2}{\sec^2 \alpha} = 1.$

Since, $\alpha \in \left(0, \frac{\pi}{2}\right) \Rightarrow \sec \alpha > \tan \alpha$

$$\therefore \quad a = \tan \alpha \text{ and } b = \sec \alpha$$

$$\Rightarrow \quad e = \sqrt{1 - \frac{a^2}{b^2}} = \sqrt{1 - \frac{\tan^2 \alpha}{\sec^2 \alpha}} = \cos \alpha$$

and $\quad$ foci $= (0, \pm be)$

$$= (0, \pm \sec \alpha \cdot \cos \alpha) = (0, \pm 1)$$

Hence, foci remians constant as α varies.

15. (a)

(i) Since ellipse intersect the lines $2x - 3y = 6$ and $4x + 5y = 20$ on the X-axis and Y-axis respectively. So, their intersecting points will be (3, 0) and (0, 4).

$$\therefore \quad a = 3 \text{ and } b = 4$$

So, $\quad$ eccentricity, $e = \sqrt{1 - \frac{a^2}{b^2}}$

$$= \sqrt{1 - \frac{9}{16}} = \frac{\sqrt{7}}{4}$$

(ii) Let AB be the rod making an angle θ with OX and $P(x, y)$ be the point on it such that $AP = 8$ units. Then

$$PB = AB - AP = (20 - 8) = 12 \text{ units.}$$

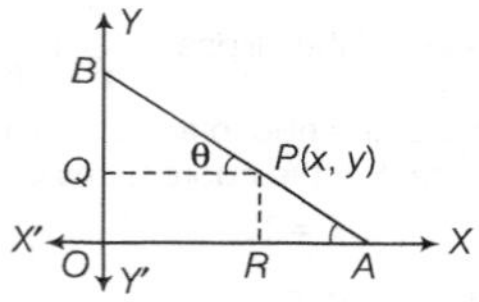

In ΔPBQ,

$$\cos \theta = \frac{PQ}{PB} = \frac{x}{12}$$

In ΔPRA,

$$\sin \theta = \frac{PR}{PA} = \frac{y}{8}$$

Since, $\sin^2 \theta + \cos^2 \theta = 1$, we get

$$\left(\frac{y}{8}\right)^2 + \left(\frac{x}{12}\right)^2 = 1$$

$$\Rightarrow \quad \frac{x^2}{144} + \frac{y^2}{64} = 1$$

$$\therefore \quad e = \sqrt{1 - \frac{64}{144}}$$

$$= \sqrt{\frac{80}{144}} = \frac{\sqrt{5}}{3}$$

(iii) We have,

$$PQ = BP$$

$$\Rightarrow \quad 2ae = \sqrt{a^2e^2 + b^2}$$

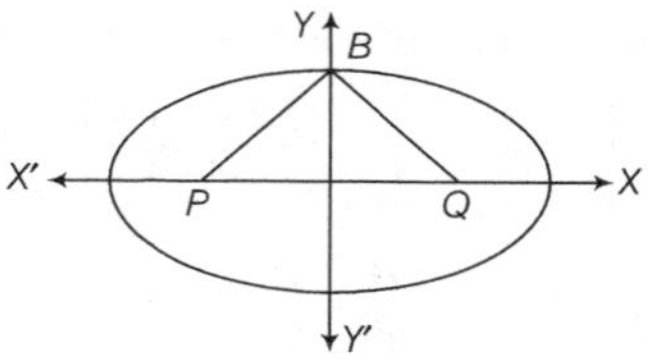

$$\Rightarrow \quad 2ae = a$$

$$\therefore \quad e = \frac{1}{2}$$

(iv) Since, there are exactly two points on the ellipse $\frac{x^2}{a^2} + \frac{y^2}{b^2} = 1,$ whose distance the centre is the same, the points would either be the end points of the major axis or of the minor axis.

But $\sqrt{(a^2 + 2b^2)/2} > b$. So, the points are the vertices of the major axis.

So, $\quad a = \sqrt{\frac{a^2 + 2b^2}{2}}$

$$\Rightarrow \quad a^2 = 2b^2$$

or $\quad e = \sqrt{1 - \frac{b^2}{a^2}} = \frac{1}{\sqrt{2}}$

Hence, (i) $\to$ (s), (ii) $\to$ (r), (iii) $\to$ (p), (iv) $\to$ (q)

Explanations for Q. Nos. (16 and 17)

Equation of the ellipse is

$$\frac{\{(x-3y+3)/\sqrt{1+9}\}^2}{6}+\frac{\{(3x+y-1)/\sqrt{9+1}\}^2}{9}=1$$

or $$\frac{(x-3y+3)^2}{60}+\frac{(3x+y-1)^2}{90}=1$$

or $$3(x-3y+3)^2+2(3x+y-1)^2=180$$

On comparing it with given equation, we get

$$a=3 \text{ and } b=2$$

16. (b) **17.** (a)

18. (a) Let the fixed straight rod be along the coordinate axes and AB be the rod of length a such that its extremities $A(a', 0)$ and $B(0, b')$ are on the coordinate axes. Then,

$$a'^2+b'^2=a^2 \quad \ldots\text{(i)}$$

Let $P(h, k)$ be a point marked on the rod such that it divides the rod AB in the ratio $\lambda : 1$. Then,

$$h=\frac{a'}{\lambda+1}$$

and $$k=\frac{p'\lambda}{\lambda+1}$$

$\Rightarrow$ $$a'=(\lambda+1)h$$

and $$b'=\frac{(\lambda+1)k}{\lambda}$$

On substituting these values in Eq. (i), we get

$$(\lambda+1)^2h^2+\frac{(\lambda+1)^2k^2}{\lambda^2}=a^2$$

Hence, the locus of $P(h, k)$ is

$$(\lambda+1)^2x^2+\frac{(\lambda+1)^2}{\lambda^2}y^2=a^2$$

$\Rightarrow$ $$\frac{a^2}{(\lambda+1)^2}=x^2+\frac{y^2}{\lambda^2}$$

19. (c) We know that

$$\frac{x^2}{\left[\frac{a}{(\lambda+1)}\right]^2}+\frac{y^2}{\left[\frac{\lambda a}{\lambda+1}\right]^2}=1,$$

which represents an ellipse.

Now, eccentricity $$=\sqrt{1-\frac{b^2}{a^2}}=\sqrt{1-\frac{\left(\frac{\lambda a}{(\lambda+1)}\right)^2}{\frac{a}{(\lambda+1)^2}}}$$

$$=\sqrt{1-(\lambda)^2}=\sqrt{1-\frac{1}{4}} \quad [\because \lambda=1/2]$$

$$=\frac{\sqrt{3}}{2}$$

20. (5) We know that,

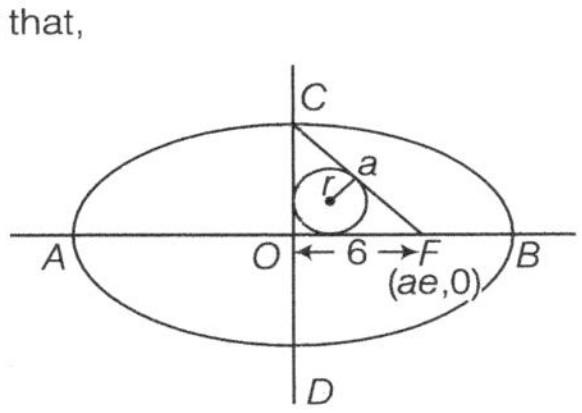

$$OF=ae$$

$\Rightarrow$ $$ae=6$$

$\Rightarrow$ $$a^2e^2=36$$

$\Rightarrow$ $$a^2-b^2=36 \quad \ldots\text{(i)}$$

Using, $r=(s-a)\tan\frac{A}{2}$ in ΔOCF, we get

$$1=(s-a)\tan 45°, \text{ when } a=CF$$

$\Rightarrow$ $$2=2(s-a)$$

$\Rightarrow$ $$2=2s-2a$$

$\Rightarrow$ $$2=2s-AB$$

$\Rightarrow$ $$2=OF+FC+CO-AB$$

$\Rightarrow$ $$2=6+\frac{AB}{2}+\frac{CD}{2}-AB$$

$\Rightarrow$ $$\frac{AB-CD}{2}=4 \Rightarrow 2(a-b)=8$$

$\Rightarrow$ $$a-b=4 \quad \ldots\text{(ii)}$$

From Eqs. (i) and (ii), we get

$$a=\frac{13}{2} \text{ and } b=\frac{5}{2}$$

$\Rightarrow$ $$AB=13 \text{ and } CD=5$$

Now, $$\frac{(AB)(CD)}{13}=\frac{13\times 5}{13}=5$$

21. (6) Given equation

$$\frac{x^2}{8-t}+\frac{y^2}{t-4}=1 \text{ will represent an ellipse, if}$$

$$8-t>0, t-4>0 \text{ and } 8-t\neq t-4$$

$\Rightarrow$ $$t<8, t>4 \text{ and } t\neq 6$$

$\therefore$ $$t\in(4, 8)-\{6\}$$

On comparing, we get $\lambda=6$

22. (0) Let $S(-1, 1)$ be the focus and ZZ' be the directrix, and $P(x, y)$ be any point on the ellipse and PM be perpendicular from P on the directrix. Then, by definition, we have

$$SP=ePM$$

$\Rightarrow$ $$SP^2=e^2PM^2$$

$\Rightarrow$ $$(x+1)^2+(y-1)^2=\left(\frac{1}{2}\right)^2\left|\frac{x-y+3}{\sqrt{1+1}}\right| \quad \left[\because e=\frac{1}{2}\right]$$

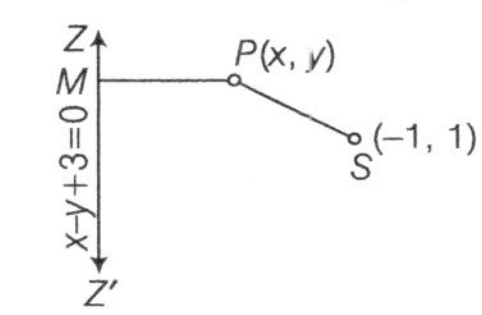

$\Rightarrow$ $$7x^2+7y^2+2xy+10x-10y+7=0$$

On comparing it with given equation, we get

$$\lambda=7, \mu=7$$

$\therefore$ $$|\lambda-u|=|7-7|=0$$

23. (5) The distance between two flag-posts $=10$

$\Rightarrow$ $$2a=10$$

$\Rightarrow$ $$a=5$$

and distance between the flag-posts $=8$

$\Rightarrow$ $$2ae=8$$

$\Rightarrow$ $$ae=4$$

$\Rightarrow$ $$e=\frac{4}{5}$$

Also, $$b^2=a^2(1-e^2)$$

$\Rightarrow$ $$b^2=25\left(1-\frac{16}{25}\right)$$

$\Rightarrow$ $$b^2=9$$

$\Rightarrow$ $$b=3$$

Now, area of ellipse $=\pi ab$

$\Rightarrow$ $$\pi\times 5\times 3=3\lambda\pi$$

$\therefore$ $$\lambda=5$$

24. (1) Given equation is

$$(10x-5)^2+(10y-5)^2=(3x+4y-1)^2$$

$$\Rightarrow \left(x-\frac{1}{2}\right)^2+\left(y-\frac{1}{2}\right)^2=\left(\frac{3x+4y-1}{10}\right)^2$$

$$\Rightarrow \sqrt{\left(x-\frac{1}{2}\right)^2+\left(y-\frac{1}{2}\right)^2}=\left|\frac{3x+4y-1}{10}\right|$$

$$\Rightarrow \sqrt{\left(x-\frac{1}{2}\right)^2+\left(y-\frac{1}{2}\right)^2}=\frac{1}{2}\left|\frac{3x+4y-1}{5}\right|$$

$$\therefore \quad e=\frac{1}{2}$$

Length of latusrectum

$$=2(e)\times \text{Distance of}\left(\frac{1}{2},\frac{1}{2}\right)\text{from the line } 3x+4y-1=0$$

$$=2\times\frac{1}{2}\times\left|\frac{3\times\frac{1}{2}+4\times\frac{1}{2}-1}{5}\right|=\frac{1}{2}$$

Hence, $\lambda=\frac{1}{2}$

$$\Rightarrow \quad 2\lambda=2\times\frac{1}{2}=1$$

25. (3) Let $P(2\cos\theta,\sin\theta)$ be the variable point on the ellipse $\frac{x^2}{4}+\frac{y^2}{1}=1$.

Then, Area of $\Delta\, PS_1S_2=\frac{1}{2}\begin{vmatrix}2\cos\theta & \sin\theta & 1\\ \sqrt{3} & 0 & 1\\ -3 & 0 & 1\end{vmatrix}$

$$A=\frac{1}{2}\sin\theta\,(2\sqrt{3})$$

$$=\sqrt{3}\sin\theta$$

Area is maximum, when $\sin\theta=1$, $\left(\theta=\frac{\pi}{2}\right)$

$$\therefore \quad A^2_{max}=3$$

DPP-2 Equation of Chord, Focal Chord and Its Properties

1. (a) Given equation of directrix, $x=0, e=\frac{1}{2}$ and focus $(3, 0)$.

Equation of ellipse will be

$$\sqrt{(x-3)^2+y^2}=\frac{1}{2}|x|$$

$$\Rightarrow (x-3)^2+y^2=\frac{x^2}{4}$$

$$\Rightarrow 4(x-3)^2+4y^2=x^2$$

$$\Rightarrow 3x^2-24x+4y^2+36=0$$

$$\Rightarrow 3(x-4)^2+4y^2=12$$

$$\Rightarrow \frac{(x-4)^2}{4}+\frac{y^2}{3}=1$$

$\therefore \quad a=2, b=\sqrt{3}$, centre $(4, 0)$

$\therefore$ Equation of auxiliary circle is $(x-4)^2+y^2=4$.

i.e. $\quad x^2+y^2-8x+12=0$

2. (d) The equation of a chord joining points having eccentric angles α and β is given by

$$\frac{x}{a}\cos\left(\frac{\alpha+\beta}{2}\right)+\frac{y}{b}\sin\left(\frac{\alpha+\beta}{2}\right)-\cos\left(\frac{\alpha-\beta}{2}\right)$$

If it pasess through $(ae, 0)$, then

$$e\cos\left(\frac{\alpha+\beta}{2}\right)=\cos\left(\frac{\alpha-\beta}{2}\right)$$

$$=\frac{\cos\left(\frac{\alpha-\beta}{2}\right)}{\cos\left(\frac{\alpha+\beta}{2}\right)}$$

$$=\frac{2\sin\left(\frac{\alpha+\beta}{2}\right)\cos\left(\frac{\alpha-\beta}{2}\right)}{2\sin\left(\frac{\alpha+\beta}{2}\right)\cos\left(\frac{\alpha+\beta}{2}\right)}$$

$$=\frac{\sin\alpha+\sin\beta}{\sin(\alpha+\beta)}$$

3. (a) Area of circle $=\pi a^2$ and area of ellipse $=\pi ab$

$$\therefore \quad \frac{\pi a^2-\pi ab}{\pi a^2}=\frac{2}{3}$$

$$\Rightarrow \quad 1-\frac{b}{a}=\frac{2}{3}$$

$$\Rightarrow \quad \frac{b}{a}=\frac{1}{3}$$

$$\therefore \text{Eccentricity } e=\sqrt{1-\frac{b^2}{a^2}}=\sqrt{1-\frac{1}{9}}=\frac{2\sqrt{2}}{3}$$

4. (c) We have, $4\cos\theta=2h$ and $5(1+\sin\theta)=2k$

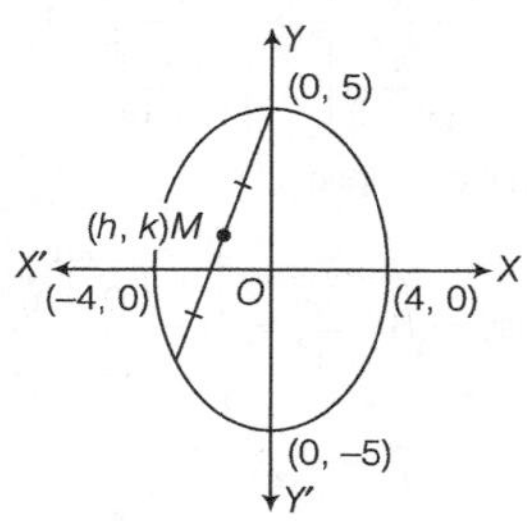

We know that,

$$\cos^2\theta+\sin^2\theta=1$$

$$\Rightarrow \frac{h^2}{4}+\left(\frac{2k}{5}-1\right)^2=1$$

$$\Rightarrow \frac{h^2}{4}+\frac{\left(k-\frac{5}{2}\right)^2}{\frac{25}{4}}=1$$

$$\Rightarrow \frac{x^2}{4}+\frac{\left(y-\frac{5}{2}\right)^2}{\frac{25}{4}}=1 \quad \ldots\text{(i)}$$

Put $X = x$ and $Y = y - \frac{5}{2}$ in Eq. (i), we get

$$\frac{X^2}{4} + \frac{Y^2}{\frac{25}{4}} = 1$$

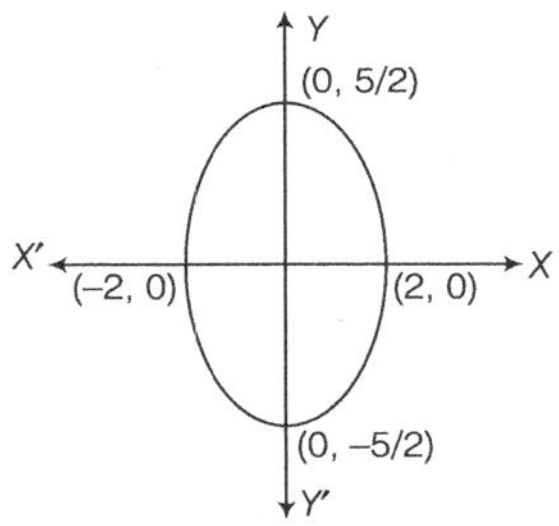

$$\therefore \quad e = \sqrt{1 - \frac{4}{25} \times 4} = \frac{3}{5}$$

$$\therefore \quad F_1 = \left(0, \frac{3}{2}\right), F_2 = \left(0, -\frac{3}{2}\right) \text{ in } XY.$$

Hence, in xy foci are (0, 4) and (0, 1).

5. (a) The coordinates of any point on the ellipse $\frac{x^2}{a^2} + \frac{y^2}{b^2} = 1$, whose eccentric angle is θ are $(a \cos \theta, b \sin \theta)$.

The coordinates of end points of latusrectum are $\left(ae, \pm \frac{b^2}{a}\right)$.

$$\therefore \quad a \cos \theta = ae$$

$$\text{and} \quad b \sin \theta = \pm \frac{b^2}{a}$$

$$\Rightarrow \quad \tan \theta = \pm \frac{b}{ae}$$

$$\therefore \quad \theta = \tan^{-1}\left(\pm \frac{b}{ae}\right)$$

6. (b) Let the points of intersection of the line and the ellipse be $(a \cos \theta, b \sin \theta)$ and $\left[a \cos\left(\frac{\pi}{2} + \theta\right), b \sin\left(\frac{\pi}{2} + \theta\right)\right]$. Since, they lie on the given line $lx + my + n = 0$, we have

$$la \cos \theta + mb \sin \theta + n = 0$$

$$\Rightarrow \quad la \cos \theta + mb \sin \theta = -n$$

$$\text{and} \quad -la \sin \theta + mb \cos \theta + n = 0$$

$$\Rightarrow \quad la \sin \theta - mb \cos \theta = n$$

On squaring and adding, we get

$$a^2l^2 + b^2m^2 = 2n^2$$

7. (c) The coordiantes of end points of the focal chord are $(a \cos \alpha, b \sin \alpha)$ and $(a \cos \beta, b \sin \beta)$. Therefore, the equation by the focal chord is

$$\frac{x}{a} \cos\left(\frac{\alpha + \beta}{2}\right) + \frac{y}{b} \sin\left(\frac{\alpha + \beta}{2}\right) = \cos\left(\frac{\alpha - \beta}{2}\right)$$

This passes through $(ae, 0)$.

$$\therefore \quad \frac{ae}{a} \cos \frac{\alpha + \beta}{2} = \cos\left(\frac{\alpha - \beta}{2}\right)$$

$$\Rightarrow \quad \frac{\cos\left(\frac{\alpha + \beta}{2}\right)}{\cos\left(\frac{\alpha - \beta}{2}\right)} = \frac{1}{e}$$

$$\Rightarrow \quad \frac{\cos\left(\frac{\alpha + \beta}{2}\right) - \cos\left(\frac{\alpha - \beta}{2}\right)}{\cos\left(\frac{\alpha + \beta}{2}\right) + \cos\left(\frac{\alpha - \beta}{2}\right)} = \frac{1 - e}{1 + e}$$

$$\Rightarrow \quad \frac{2 \sin \frac{\alpha}{2} \sin \frac{\beta}{2}}{2 \cos \frac{\alpha}{2} \cos \frac{\beta}{2}} = \frac{1 - e}{1 + e}$$

$$\Rightarrow \quad -\tan \frac{\alpha}{2} \tan \frac{\beta}{2} = \frac{1 - e}{1 + e}$$

$$\Rightarrow \quad \tan \frac{\alpha}{2} \tan \frac{\beta}{2} = \frac{e - 1}{e + 1}$$

8. (b) Let PQ and RS be two chords of the ellipse such that the eccentric angles of P, Q, R and S are α, β, γ and δ respectively. Suppose, PQ and RS meet major axis at $(C, 0)$ and $(-C, 0)$ respectively. Then,

$$\tan \frac{\alpha}{2} + \tan \frac{\beta}{2} = \frac{c - a}{c + a}$$

$$\text{and} \quad \tan \frac{\gamma}{2} \tan \frac{\delta}{2} = \frac{-c - a}{-c + a} = \frac{c + a}{c - a}$$

$$\therefore \quad \tan \frac{\alpha}{2} \tan \frac{\beta}{2} \tan \frac{\gamma}{2} \tan \frac{\delta}{2} = \frac{c - a}{c + a} \times \frac{c + a}{c - a} = 1$$

9. (c) The given ellipse is

$$\frac{x^2}{25} + \frac{y^2}{16} = 1$$

We know that, $\frac{1}{SP} + \frac{1}{SQ} = \frac{2(a)}{b^2}$

$$\Rightarrow \quad \frac{1}{2} + \frac{1}{SQ} = \frac{2 \times (5)}{16} \quad [\because PS = 2]$$

$$\Rightarrow \quad \frac{1}{SQ} = \frac{1}{8}$$

$$\Rightarrow \quad SQ = 8$$

Now, $PQ = PS + SQ$
$= 2 + 8 = 10$

10. (a) (i) The locus is

$$\frac{x^2}{16} + \frac{y^2}{36} = 1$$

$$\Rightarrow \quad e = \sqrt{1 - \frac{16}{36}} = \sqrt{\frac{20}{36}} = \frac{\sqrt{5}}{3}$$

$$\Rightarrow \quad 9e = 9 \times \frac{\sqrt{5}}{3} = 3\sqrt{5}$$

(ii) $3(x^2 + 2x + 1) + 2(y^2 - 2y + 1) = 3 + 2 + 1$

$$\text{or} \quad \frac{(x + 1)^2}{2} + \frac{(y - 1)^2}{3} = 1$$

$$\text{or} \quad e = \sqrt{1 - \frac{2}{3}} = \frac{1}{\sqrt{3}}$$

$$\therefore \quad a = \sqrt{3}, b = \sqrt{2}$$

$$\therefore \quad \text{Area} = \frac{1}{2} \times 2\sqrt{3} \times \sqrt{2} = \sqrt{6}$$

(iii) Eliminating θ from $x = 1 + 4 \cos \theta$ and $y = 2 + 3 \sin \theta$, we get

$$\frac{(x - 1)^2}{16} + \frac{(y - 2)^2}{9} = 1$$

where, $a = 4$ and e is $\frac{\sqrt{7}}{4}$. Therefore, $ae = \sqrt{7}$

So, the distance between the foci is $2\sqrt{7}$.

Hence, (i) → (r), (i) → (p), (iii) → (q)

11. (a) (i) → (q), (ii) → (s), (iii) → (p), (iv) → (r)

$$C_1 : 9x^2 + 8y^2 - 36x - 16y - 28 = 0$$

$$C_2 : 4x^2 + 9y^2 + 8x - 36y + 4 = 0$$

$$C_1 = 9(x^2 - 4x) + 8(y^2 - 2y) = 28$$

$$\Rightarrow \quad 9(x-2)^2 + 8(y-1)^2 = 72$$

or $\quad C_1 : \dfrac{(x-2)^2}{8} + \dfrac{(y-1)^2}{9} = 1$

and $\quad C_2 : \dfrac{(x+1)^2}{9} + \dfrac{(y-2)^2}{4} = 1$

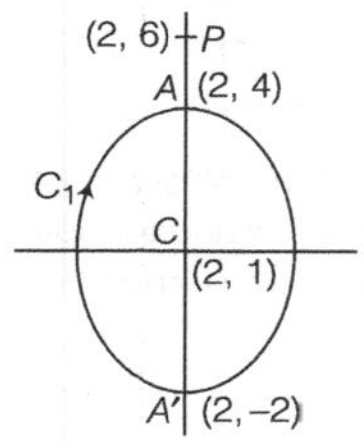

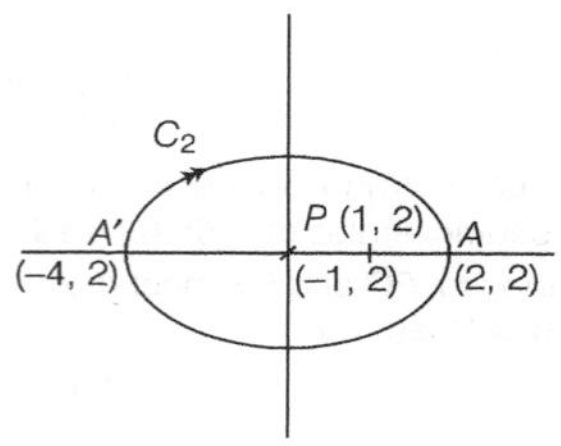

Minimum distance from (2, 6) to $C_1 = 2$

Maximum distance from (2, 6) to $C_1 = 8$

Minimum distance from (1, 2) to $C_2 = 1$

Maximum distance from (1, 2) to $C_2 = 5$

Explanations for Q. Nos. (12-14)

Let PQ be the diameter of length d and let $OP = r$

$\therefore \quad P(r\cos\theta, r\sin\theta)$

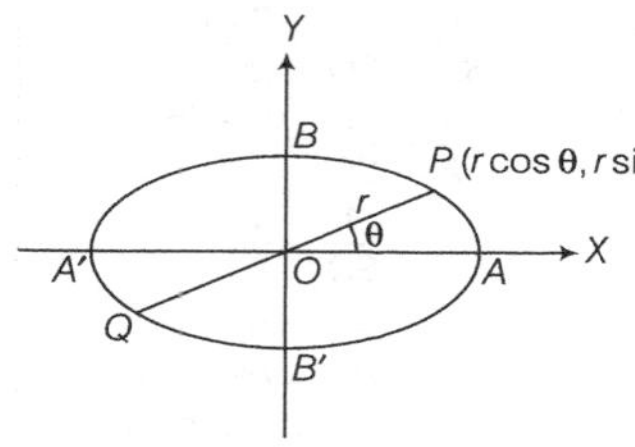

Substitute the coordinate of P in the equation of the ellipse to get

$$r^2 = \frac{a^2b^2}{a^2\sin^2\theta + b^2\cos^2\theta} \qquad \ldots(i)$$

12. (b) $\therefore d^2 = (2r)^2 = \dfrac{(2a)^2 + (2b)^2}{2} \Rightarrow \dfrac{a^2+b^2}{2} = r^2$

using (i), $\quad \tan\theta = \pm\dfrac{b}{a}$

13. (a) Also, $(2r)^2 = d^2 = \sqrt{(2a)^2(2b)^2} \Rightarrow r^2 = ab$,

using (i)

$\Rightarrow$ to get $\tan\theta = \pm\sqrt{\dfrac{b}{a}}$

14. (c) $(2r)^2 = d^2 = \dfrac{2(2a)^2\cdot(2b)^2}{(2a)^2 + (2b)^2} \Rightarrow r^2 = \dfrac{2a^2b^2}{a^2+b^2}$

using (i), $\tan\theta = \pm 1 \Rightarrow \theta = \pm\dfrac{\pi}{4}$

15. (2) We have $\dfrac{(x-4)^2}{16} + \dfrac{(y-3)^2}{9} = 1$ any point P on the curve is $P[4(1+\cos\theta), 3(1+\sin\theta)]$

The reflection of the point P about the line $x - y - 2 = 0$ can be found using

$$\frac{h - 4(1+\cos\theta)}{1} = \frac{k - 3(1+\sin\theta)}{-1}$$

$$= \frac{-2[4(1+\cos\theta) - 3(1+\sin\theta) - 2]}{2}$$

$\therefore \quad h = 4(1+\cos\theta) + 4\cos\theta - 3\sin\theta - 1$

$\Rightarrow \quad h = 2 + 4\cos\theta$

and $\quad k = 3(1+\sin\theta) + 4\cos\theta - 3\sin\theta - 1$

$\Rightarrow \quad k = 2 + 4\cos\theta$

$\therefore \quad \left(\dfrac{h-5}{3}\right)^2 + \left(\dfrac{k-2}{4}\right)^2 = 1$

$\therefore$ Equation of image is $16(x-5)^2 + 9(y-2)^2 = 144$

$\Rightarrow \quad 16x^2 + 9y^2 - 160x - 36y + 292 = 0$

On comparing with given equation, we get

$k_1 = -160, k_2 = 292$

$\therefore \quad k_1 + k_2 = 132 \Rightarrow$ units digit = 2.

DPP-3 Tangent and Its Properties

1. (a) Equation of the line is

$$lx + my + n = 0$$

$$\Rightarrow \quad y = -\frac{l}{m}x + \left(-\frac{n}{m}\right) \qquad \ldots(i)$$

We know that the line $y = mx + c$ touches the ellipse $\dfrac{x^2}{a^2} + \dfrac{y^2}{b^2} = 1$, if

$$c^2 = a^2m^2 + b^2$$

So, line (i) will touch the ellipse $\dfrac{x^2}{a^2} + \dfrac{y^2}{b^2} = 1$, if

$$\left(-\frac{n}{m}\right)^2 = a^2\left(-\frac{l}{m}\right) + b^2$$

$$\therefore \quad n^2 = a^2l^2 + b^2m^2$$

2. (b) We know that the line $y = mx + c$ is a tangent to the ellipse $\dfrac{x^2}{a^2} + \dfrac{y^2}{b^2} = 1$, if

$$c^2 = a^2m^2 + b^2$$

In this case, we have

$$c = -\frac{p}{\sin\alpha}, m = -\frac{\cos\alpha}{\sin\alpha}$$

So, the given line will be a tangent, if

$$\frac{p^2}{\sin^2\alpha} = a^2\frac{\cos^2\alpha}{\sin^2\alpha} + b^2$$

$$\Rightarrow \quad p^2 = a^2\cos^2\alpha + b^2\sin^2\alpha$$

3. (d) Equation of the line is

$$y = x + \lambda$$

Comparing it with $y = mx + c$, we get

$$m = 1 \text{ and } c = \lambda$$

The equation of ellipse is

$$9x^2 + 16y^2 = 144$$

or $\quad \dfrac{x^2}{16} + \dfrac{y^2}{9} = 1$

On comparing this with $\dfrac{x^2}{a^2} + \dfrac{y^2}{b^2} = 1$, we get

$$a^2 = 16 \text{ and } b^2 = 9$$

If the line touches the ellipse, then

$$c^2 = a^2m^2 + b^2$$

$\Rightarrow \quad \lambda^2 = 16 + 9$

$\Rightarrow \quad \lambda = \pm 5$

4. (a) The equation of tangents to $\frac{x^2}{a^2} + \frac{y^2}{b^2} = 1$ are given by $y = mx \pm \sqrt{a^2m^2 + b^2}$, where m is the slope of the tangent.

For the ellipse $4x^2 + 3y^2 = 5$, we have

$$a = \frac{\sqrt{5}}{2}, b = \frac{\sqrt{5}}{\sqrt{3}}, m = \tan 60° = \sqrt{3} \quad \text{[given]}$$

So, the equations of tangents are

$$y = \sqrt{3}x \pm \sqrt{\frac{5}{4} \times 3 + \frac{5}{3}}$$

$$\Rightarrow \quad y = \sqrt{3}x \pm \sqrt{\frac{65}{12}}$$

5. (b) Equation of the given ellipse may be written in the form

$$\frac{x^2}{5/4} + \frac{y^2}{5/3} = 1$$

Hence, $a^2 = \frac{5}{4}, b^2 = \frac{5}{3}$

The slope of given line is 3, therefore the slope of the parallel line is also 3. The equations of tangents to the ellipse

$\frac{x^2}{a^2} + \frac{y^2}{b^2} = 1$ are

$$y = mx \pm \sqrt{a^2m^2 + b^2}$$

In this case, we have

$$m = 3, a^2 = \frac{5}{4}, b^2 = \frac{5}{3}$$

So, the required equations of the tangents are

$$y = 3x \pm \sqrt{\frac{155}{12}}$$

6. (c) The ellipse is $\frac{x^2}{4} + \frac{y^2}{3} = 1$

$$\therefore \quad a^2 = 4 \text{ and } b^2 = 3$$

Slope of the line $y + 2x = 4$ is -2.

Any tangent to the ellipse is

$$y = mx \pm \sqrt{a^2m^2 + b^2}$$

Given, it is perpendicular to given line, then $m = 1/2$.

$$\therefore \quad y = \frac{1}{2}x \pm 2$$

$$\therefore \quad 2y = x \pm 4$$

7. (a) Ellipse $\frac{x^2}{9} + \frac{y^2}{4} = 1$

Tangent at $(3, -2)$ is

$$\frac{x(3)}{9} + \frac{y(-2)}{4} = 1$$

$$\Rightarrow \quad \frac{x}{3} - \frac{y}{2} = 1$$

$$\therefore \quad 2x - 3y = 6$$

8. (c) Let the point be (x_1, y_1). Then $y_1 = 2$.

Since (x_1, y_1) lies on $x^2 + 4y^2 = 25$

$$\Rightarrow \quad x_1^2 + 4y_1^2 = 25$$

$$\therefore \quad x_1 = \pm 3$$

So the points are $(3, 2)$ and $(-3, 2)$

$\therefore$ Equation of tangent at $(3, 2)$ and $(-3, 2)$ is $3x + 8y = 25$ and $-3x + 8y = 25$

9. (d) Equation of line $x + 2y - 4 = 0$ ...(i)

Equation of tangent at any point is

$$\frac{x}{a}\cos\theta + \frac{y}{b}\sin\theta = 1$$

$$\Rightarrow \quad \frac{x}{2}\cos\theta + \frac{y}{\sqrt{3}}\sin\theta = 1 \quad \text{...(ii)}$$

On comparing Eqs. (i) and (ii), we get

$$\cos\theta = \frac{1}{2} \text{ and } \sin\theta = \frac{\sqrt{3}}{2}$$

The point of contact $(a\cos\theta, b\sin\theta)$ is $\left(1, \frac{3}{2}\right)$.

10. (c) The equation of tangent having slope m is

$mx - y + \sqrt{a^2m^2 + b^2} = 0$.

$$\therefore \quad p_1p_2 = \left[\frac{mae + \sqrt{a^2m^2 + b^2}}{\sqrt{1 + m^2}} \cdot \frac{-mea + \sqrt{a^2m^2 + b^2}}{\sqrt{1 + m^2}}\right]$$

$$= \left[\frac{a^2m^2 + b^2 - m^2a^2e^2}{1 + m^2}\right]$$

$$= \left[\frac{a^2m^2 + b^2 - m(a^2 - b^2)}{1 + m^2}\right] = \left[\frac{b^2(1 + m^2)}{1 + m^2}\right]$$

$$= b^2$$

11. (a) Let the equation of tangent be

$$y = mx \pm \sqrt{16m^2 + 9}$$

It passes through the point $(2, 3)$. Therefore,

$$3 = 2m + \sqrt{16m^2 + 9}$$

$$\Rightarrow \quad (3 - 2m)^2 = 16m^2 + 9$$

$$\Rightarrow \quad 12m^2 + 12m = 0$$

$$\Rightarrow \quad m = 0, -1$$

The tangents are $y = 3$ or $y = -x + 5$.

12. (b) Let us consider the extremities of latusrectum in the 1st and 4th quadrants as $L(ae, b^2/a)$, $L'(ae, -b^2/a)$ and the corresponding directrix is

$$x = ae \quad \text{...(i)}$$

Tangent at L is $\frac{x \cdot ae}{a^2} + \frac{yb^2/a}{b^2}$

or $\quad ex + y = a \quad$ ...(ii)

On solving Eqs (i) and (ii), we get the point $(a/e, 0)$ which lies on $x = a/e$.

Similarly, we may consider the tangents at $(-ae, b^2/a)$ and $(-ae, -b^2/a)$ and the corresponding directrix as $x = -a/e$.

13. (c) Any tangent to the ellipse is

$$\frac{x\cos\theta}{a} + \frac{y\sin\theta}{b} = 1$$

On putting $y = 0$ then $x = 0$, we get the points A and B, where it meets the axes as

$$A\left(\frac{a}{\cos\theta}, 0\right), B\left(0, \frac{b}{\sin\theta}\right)$$

If (h, k) be mid-point of AB, then

$$2h = \frac{a}{\cos\theta}$$

$$2k = \frac{b}{\sin\theta}$$

$$\therefore \quad \cos\theta = \frac{a}{2h}$$

$$\sin\theta = \frac{b}{2k}$$

Since, $\quad \cos^2\theta + \sin^2\theta = 1$

$$\therefore \quad \frac{a^2}{4h^2} + \frac{b^2}{4k^2} = 1$$

$$\Rightarrow \quad a^2y^2 + b^2x^2 = 4x^2y^2$$

14. (c) Tangent at $P(a \cos \theta, b \sin \theta)$ is

$$\frac{x}{a} \cos \theta + \frac{y}{b} \sin \theta = 1$$

Now, $\quad PF_1 = a + ex, PF_2 = a - ex$

$\Rightarrow \quad PF_1 - PF_2 = 2ex = 2e \cdot a \cos \theta$

$\therefore \quad (PF_1 - PF_2)^2 = 4a^2 e^2 \cos^2 \theta \quad \ldots(i)$

We know tangent at $P(\theta)$ is $\frac{x}{a} \cos \theta + \frac{y}{b} \sin \theta = 1$

$$\therefore \quad d = \frac{1}{\sqrt{\frac{\cos^2 \theta}{a^2} + \frac{\sin^2 \theta}{2b^2}}}$$

$$\therefore \quad \frac{1}{d^2} = \frac{\cos^2 \theta}{a^2} + \frac{\sin^2 \theta}{b^2}$$

or $$\frac{b^2}{d^2} = \frac{b^2}{a^2} \cos^2 \theta + \sin^2 \theta$$

$$\therefore \quad 1 - \frac{b^2}{d^2} = 1 - \frac{b^2}{a^2} \cos^2 \theta - \sin^2 \theta = \cos^2 \theta \left(1 - \frac{b^2}{a^2}\right)$$

$$= \cos^2 \theta \cdot e^2$$

$$\Rightarrow \quad 4a^2 \left(1 - \frac{b^2}{d^2}\right) = 4a^2 \cos^2 \theta \cdot e^2 \quad \ldots(ii)$$

On comparing Eqs. (i) and (ii), we get

$$(PF_1 - PF_2)^2 = 4a^2(1 - b^2/d^2)$$

Hence, $\quad \lambda = 4$

15. (c) The coordinates of the point of intersection of tangents at points having eccentric angle α and β, are

$$\left[\frac{a \cos\left(\frac{\alpha + \beta}{2}\right)}{\cos\left(\frac{\alpha - \beta}{2}\right)}, \frac{b \sin\left(\frac{\alpha + \beta}{2}\right)}{\cos\left(\frac{\alpha - \beta}{2}\right)}\right]$$

Thus, if (h, k) is the point of intersection of tangents at points whose eccentric angles differ by $\frac{\pi}{2}$, i.e. $\alpha - \beta = \frac{\pi}{2}$, then

$$h = \sqrt{2} a \cos\left(\frac{\alpha + \beta}{2}\right) \text{ and } k = \sqrt{2} b \sin\left(\frac{\alpha + \beta}{2}\right)$$

$$\Rightarrow \quad \frac{h^2}{2a^2} + \frac{k^2}{2b^2} = 1$$

Hence, the locus of (h, k) is

$$\frac{x^2}{2a^2} + \frac{y^2}{2b^2} = 1$$

i.e. $$\frac{x^2}{a^2} + \frac{y^2}{b^2} = 2$$

16. (b) Let (h, k) be the foot of perpendicular drawn from the focus $S(ae, 0)$ on tangent

$$y = mx + \sqrt{a^2m^2 + b^2} \quad \ldots(i)$$

at point P on the ellipse $\frac{x^2}{a^2} + \frac{y^2}{b^2} = 1$.

Then, $\quad k = mh + \sqrt{a^2m^2 + b^2}$

$\Rightarrow \quad (k - mh)^2 = a^2m^2 + b^2 \quad \ldots(ii)$

Also, SP is perpendicular to Eq. (i).

$$\therefore \quad m \times \frac{k - 0}{h - ae} = -1$$

$\Rightarrow \quad h + mk = ae \quad \ldots(iii)$

From Eqs. (ii) and (iii), we get

$$(k - mh)^2 + (h + mk)^2 = a^2m^2 + b^2 + a^2e^2$$

$\Rightarrow \quad (h^2 + k^2)(1 + m^2) = a^2(1 + m^2)$

$\Rightarrow \quad h^2 + k^2 = a^2$

Hence, the locus of (h, k) is $x^2 + y^2 = a$.

17. (d) Ellipse is $\frac{x^2}{16} + \frac{y^2}{4} = 1$; $a^2 = 16, b^2 = 4$

Let the equation of circle with centre $C(1, 0)$ be $(x - 1)^2 + y^2 = r^2$.

Since, the circle is to be largest, so it will touch the ellipse at some point $P(a \cos \theta, b \sin \theta)$,

where tangent is

$$\frac{x \cos \theta}{a} + \frac{y \sin \theta}{b} = 1$$

whose slope of $\quad m_1 = -\frac{b}{a} \frac{\cos \theta}{\sin \theta} = -\frac{1}{2} \frac{\cos \theta}{2 \sin \theta} \quad \ldots(i)$

Also, $\quad m_2 = \text{Slope of } CP$

$$= \frac{b \sin \theta}{(a \cos \theta - 1)} = \frac{2 \sin \theta}{4 \cos \theta - 1} \quad \ldots(ii)$$

But $\quad m_1 m_2 = -1$

$$\therefore \quad -\frac{1}{2} \frac{\sin \theta}{\cos \theta} \times \frac{2 \sin \theta}{4 \cos \theta - 1} = -1$$

$\Rightarrow \quad \cos \theta = 1/2$

$\therefore$ Radius of circle is $\{(a \cos \theta - 1)^2 + b^2 \sin^2 \theta\}^{1/2}$.

or $$\left\{\left(\frac{4}{3} - 1\right)^2 + 4\left(1 - \frac{1}{a}\right)^2\right\}^{1/2} = \sqrt{\frac{33}{9}} = \sqrt{\frac{11}{3}}$$

18. (d) The equation of a tangent to $x^2 + 4y^2 = 4$ at $(2 \cos \theta, \sin \theta)$ is

$$2x \cos \theta + 4y \sin \theta = 4$$

or $$x \cos \theta + 2y \sin \theta = 2 \quad \ldots(i)$$

This cuts the ellipse $x^2 + 2y^2 = 6$ at P and Q.

Let $R(h, k)$ be the point of intersection of tangents at P and Q. Then, PQ is the chord of contact of tangents drawn from $R(h, k)$ to the ellipse $x^2 + 2y^2 = 6$. Therefore, the equation of PQ is

$$hx + 2ky = 6 \quad \ldots(ii)$$

Clearly, Eqs. (i) and (ii) represent the same line.

$$\therefore \quad \frac{h}{\cos \theta} = \frac{2k}{2 \sin \theta} = \frac{6}{2}$$

$\Rightarrow \quad h = 3 \cos \theta, k = 3 \sin \theta$

$\Rightarrow \quad h^2 + k^2 = 9$

$\Rightarrow (h, k)$ lies on $x^2 + y^2 = 9$

which is the director of circle of the ellipse $x^2 + 2y^2 = 6$.

Hence, the angle between the tangents is a right angle, i.e. 90°.

19. (b) The equation of two ellipses are

$$C_1 : \frac{x^2}{4} + \frac{y^2}{1} = 1 \quad \ldots(i)$$

and $$C_2 : \frac{x^2}{6} + \frac{y^2}{3} = 1 \quad \ldots(ii)$$

Suppose tangent at P and Q to the ellipse (ii) intersect at $R(h, k)$. Then, PQ is the chord of contact of tangents drawn from $R(h, k)$

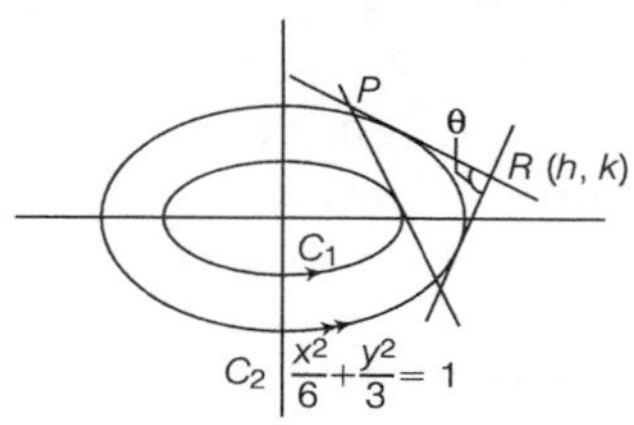

$\therefore$ Equation of PQ is

$$\frac{xh}{6} + \frac{yk}{3} = 1 \quad \ldots(iii)$$

$$\Rightarrow \quad \frac{ky}{3} = -\frac{hx}{6} + 1$$

or $$y = -\frac{h}{2k} x + \frac{3}{k}$$

This touches the ellipse (i).

$\therefore \quad \frac{9}{k^2} = 4\left(-\frac{h}{2k}\right)^2 + 1, \quad \{\text{using}: c^2 = a^2m^2 + b^2\}$

$\Rightarrow \quad h^2 + k^2 = 9$

$\Rightarrow$ (h, k) lies on the directors circle of ellipse (ii).

Hence, the angle between the tangents at P and Q to a ellipse is $90°$.

20. (d) Let PQ be any tangent to the ellipse $\frac{x^2}{a^2} + \frac{y^2}{b^2} = 1$ which cuts the ellipse $\frac{x^2}{a(a+b)} + \frac{y^2}{b(a+b)} = 1$ ...(i)

at P and Q.

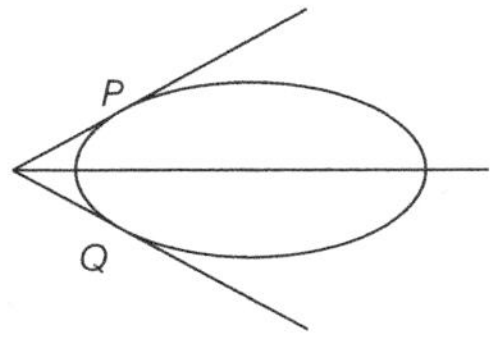

Assume that tangents P and Q are at right angles, then the point $A(h, k)$ will lie on the director circle by 2nd ellipse.

$\therefore$ (h, k) will lie on $x^2 + y^2 = A^2 + B^2$

$$x^2 + y^2 = a(a+b) + b(a+b) = (a+b)^2$$

Since (h, k) lies on it.

$\therefore \quad h^2 + k^2 = (a+b)^2$ ...(ii)

Also, PQ is chord of contact of (h, k) w.r.t. Eq. (i)

$\therefore \quad \frac{hx}{a(a+b)} + \frac{ky}{b(a+b)} = 1$

It will be tangent to $\frac{x^2}{a^2} + \frac{y^2}{b^2} = 1$, if

$$a^2l^2 + b^2m^2 = n^2$$

$\Rightarrow a^2 \cdot \frac{h^2}{a^2(a+b)^2} + b^2 \cdot \frac{k^2}{b^2(a+b)^2} = 1$

or $\quad h^2 + k^2 = (a+b)^2$

which is correct by Eq. (ii).

Therefore, assumption that tangents subtends an angle of $90°$ is correct.

21. (c) Solving the curves $y^2 = 2x$ and $\frac{x^2}{9} + \frac{y^2}{4} = 1$ for the points of intersection, we have

$$4x^2 + 18x - 36 = 0$$

$\Rightarrow \quad x = 3/2, -6$

But from $y^2 = 2x$, we have $x > 0$

$\therefore \quad x = 3/2$

at which $\quad y^2 = 2 \cdot \frac{3}{2} \Rightarrow y = \pm\sqrt{3}$

$\therefore \quad P\left(\frac{3}{2}, \sqrt{3}\right)$ and $Q\left(\frac{3}{2}, -\sqrt{3}\right)$

Now, equation of tangents at P and Q to ellipse $\frac{x^2}{9} + \frac{y^2}{4} = 1$ is

$$\frac{x}{9}\left(\frac{3}{2}\right) + \frac{y}{4}(\pm\sqrt{3}) = 1$$

which intersect at (6, 0).

Equation of tangents at P and Q to parabola $y^2 = 2x$ will be $y(\pm\sqrt{3}) = x + \frac{3}{2}$ which cut X-axis at $\left(-\frac{3}{2}, 0\right)$.

$\therefore \quad \frac{\text{Area } \Delta PQS}{\text{Area } \Delta PQR} = \frac{\frac{1}{2}PQ \cdot MS}{\frac{1}{2}PQ \cdot MR} = \frac{MS}{MR} = \frac{2}{3}$

22. (b) Area of quadrilateral $PQRS$

$$= \frac{1}{2}PQ(MS + MR)$$

$$= \frac{1}{2}2\sqrt{3}\,[6 - (-3/2)]$$

$$= \frac{15\sqrt{3}}{2} \text{ sq units}$$

23. (d) Upper end of latusrectrum of parabola is $\left(\frac{1}{2}, 1\right)$

and equation of tangent at $\left(\frac{1}{2}, 1\right)$ to $y^2 = 2x$ is $y = x + \frac{1}{2}$.

$\therefore$ The equation of circle is

$$\left(x - \frac{1}{2}\right)^2 + (y-1)^2 + \lambda\left(y - x - \frac{1}{2}\right) = 0$$

As above circle passes through (0, 0), so

$$\frac{1}{4} + 1 - \frac{\lambda}{2} = 0$$

$\Rightarrow \quad \lambda = 5/2$

$\therefore$ The equation of required circle is

$$\left(x - \frac{1}{2}\right)^2 + (y-1)^2 + \frac{5}{2}\left(y - x - \frac{1}{2}\right) = 0$$

$\Rightarrow \quad 2x^2 + 2y^2 - 7x + y = 0$

24. (a) Equation of tangent is

$$(y - 6) = -\left(\frac{4-3}{6-4}\right)x - 4$$

i.e. $\quad x + 2y - 16 = 0$

So, $\quad \frac{\alpha - (-1)}{1} = \frac{\beta - 0}{2} = -\left[\frac{1(-1) + 2(0) - 16}{(1)^2 + (2)^2}\right]$

$\Rightarrow \quad (\alpha, \beta) \equiv \left(\frac{12}{5}, \frac{34}{5}\right)$

25. (b) Centre $\equiv \left(\frac{3-1}{2}, \frac{4+0}{2}\right) \equiv (1, 2)$

and radius $= \sqrt{(4-1)^2 + (6-2)^2} = 5$

So, circle is $x^2 + y^2 - 2x - 4y - 20 = 0$.

26. (c) Radius = 5

and $\quad 2ae = \sqrt{(3+1)^2 + (4-0)^2} = 4\sqrt{2}$

So, $\quad b^2 = a^2 - a^2e^2$

$\Rightarrow \quad b^2 = 25 - (2\sqrt{2})^2 = 17$

$\therefore \quad b^2 = 17$

$\Rightarrow \quad b = \sqrt{17}$

27. (d) The directrices are at distances

i.e. $\quad \frac{a}{e} = \frac{5}{2\sqrt{2}/5} = \frac{25}{2\sqrt{2}}$ from centre (1, 2)

and perpendicular to the line joining foci.

Let its equation be

$$x + y + k = 0$$

So, $\quad \frac{|1 + 2 + k|}{\sqrt{2}} = \frac{25}{2\sqrt{2}}$

$\Rightarrow \quad k = \frac{19}{2}, -\frac{31}{2}$

Hence, the required equations are

$$x + y - \frac{31}{2} = 0$$

and $\quad x + y + \frac{19}{2} = 0$

28. (a) Let the point of contact of tangent at P be $(16-2\beta, \beta)$.

Now, we know

$$SP = ePM$$

$$\Rightarrow \quad (16-2\beta-3)^2 + (\beta-4)^2 = \left(\frac{2\sqrt{2}}{5}\right)^2 \frac{(16-2\beta+\beta-31/2)^2}{2}$$

$$\Rightarrow \quad 25(5\beta^2 - 60\beta + 185) = 4\beta^2 - 4\beta + 1$$

$$\Rightarrow \quad (11\beta - 68)^2 = 0$$

$$\Rightarrow \quad \beta = \frac{68}{11}$$

So, $\quad (16-2\beta) = \dfrac{40}{11}$

$\therefore$ Required point of contact is $\left(\dfrac{40}{11}, \dfrac{68}{11}\right)$.

29. (3) Here, $y = \dfrac{x}{2} + 2$ is tangent on the ellipse

Then, $\quad 4 = 4\left(\dfrac{1}{4}\right) + b^2$

$$\Rightarrow \quad b^2 = 3$$

30. (5) Equation of tangent to $\dfrac{x^2}{6} + \dfrac{y^2}{1} = 1$

$$\Rightarrow \quad y = mx \pm \sqrt{6m^2+1} \qquad \ldots\text{(i)}$$

If (i) is also a tangent to the ellipse

$$\frac{x^2}{5} + \frac{y^2}{6} = 1, \text{ then}$$

$$(6m^2+1) = 5m^2 + 6$$

$$\Rightarrow \quad m^2 = 5$$

DPP-4 Normal and Its Properties

1. (b) The equation of the normal at (x_1, y_1) to the given ellipse is $\dfrac{a^2x}{x_1} - \dfrac{b^2y}{y_1} = a^2 - b^2$.

Here, $x_1 = ae$ and $y_1 = \dfrac{b^2}{a}$.

So, the equation of the normal at the positive end of the latusrectum is

$$\frac{a^2x}{ae} - \frac{b^2y}{\frac{b^2}{a}} = a^2e^2 \qquad [\because b^2 = a^2(1-e^2)]$$

$$\Rightarrow \quad \frac{ax}{e} - ay = a^2e^2$$

$$\therefore \quad x - ey - e^3a = 0$$

2. (b) The normal at $P(5\cos\theta, 3\sin\theta)$ is

$$\frac{5x}{\cos\theta} - \frac{3y}{\sin\theta} = 16.$$

Now, this normal has the equation $5x - 3y = 8\sqrt{2}$. Then,

$$\frac{5\sin\theta}{3\cos\theta} = \frac{5}{3}$$

$$\Rightarrow \quad \tan\theta = 1$$

$$\therefore \quad \theta = \frac{\pi}{4}$$

3. (c) Here, equation of normal at P, is

$$3x\sec\theta - y\,\mathrm{cosec}\,\theta = 8 \qquad \ldots\text{(i)}$$

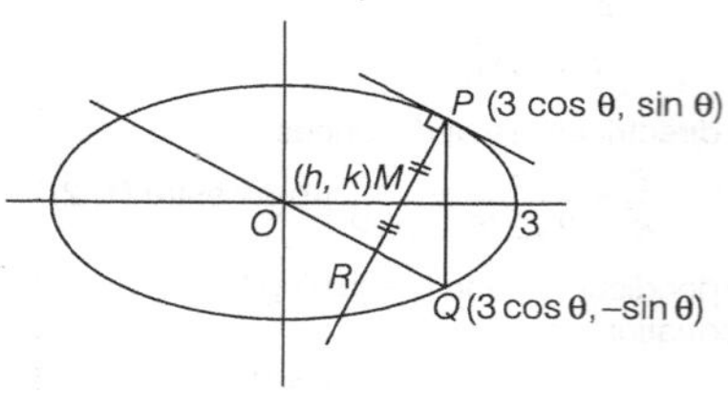

Equation of diameter OQ, is

$$x\sin\theta + 3y\cos\theta = 0 \qquad \ldots\text{(ii)}$$

R is the point of intersection of (i) & (ii),

$$\therefore \quad R\left(\frac{12}{5}\cos\theta, -\frac{4}{5}\sin\theta\right)$$

as $M(h, k)$ is the mid-point

$$\Rightarrow \quad h = \frac{3\cos\theta + \frac{12}{5}\cos\theta}{2}, \; k = \frac{\sin\theta - \frac{4}{5}\sin\theta}{2}$$

$$\therefore \quad \frac{10}{27}h = \cos\theta \;\&\; 10k = \sin\theta$$

Squarring and adding, we get

$$\left(\frac{10h}{27}\right)^2 + \frac{(10k)^2}{1} = 1$$

or $\quad \dfrac{100x^2}{(27)^2} + 100y^2 = 1$

$\therefore$ locus is ellipse

4. (a) Equation of tangent to the ellipse will be

$$\frac{x\cos\theta}{a} + \frac{y\sin\theta}{b} = 1$$

$$\therefore \quad d_1 = \left|\frac{ab}{\sqrt{b^2\cos^2\theta + a^2\sin^2\theta}}\right| \qquad \ldots\text{(i)}$$

Equation of normal to the ellipse will be

$$\frac{ax}{\cos\theta} - \frac{by}{\sin\theta} = a^2 - b^2$$

$$\therefore \quad d_2 = \left|\frac{(a^2-b^2)\sin\theta\cos\theta}{\sqrt{a^2\sin^2\theta + b^2\cos^2\theta}}\right| \qquad \ldots\text{(ii)}$$

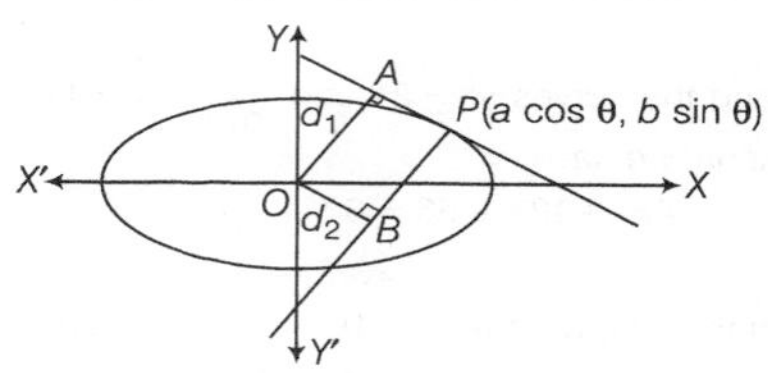

$\therefore$ Area of rectangle $OAPB$ will be given as

$$A = d_1d_2 = \frac{ab(a^2-b^2)}{2\left(\frac{a^2}{2} + \frac{b^2}{2}\right)}$$

where $\theta = \dfrac{\pi}{4}$

$$= \frac{ab(a^2-b^2)}{a^2+b^2}$$

5. (c) Normal at point $P(a\cos\theta, b\sin\theta)$ is $\dfrac{ax}{\cos\theta} - \dfrac{by}{\sin\theta} = a^2 - b^2$.

It meets the axes at

$$Q = \left(\frac{(a^2-b^2)\cos\theta}{a}, 0\right) \text{ and } R = \left(0, -\frac{(a^2-b^2)\sin\theta}{b}\right)$$

Let $T(h, k)$ be a mid-point of QR. Then,

$$2h = \frac{(a^2 - b^2)\cos\theta}{a}$$

and $$2k = \frac{-(a^2 - b^2)\sin\theta}{b}$$

or $$\cos^2\theta + \sin^2\theta = \frac{4h^2a^2}{(a^2 - b^2)^2} + \frac{4k^2b^2}{(a^2 - b^2)^2} = 1$$

Therefore, the locus is

$$\frac{x^2}{\frac{(a^2 - b^2)^2}{4a^2}} + \frac{y^2}{\frac{(a^2 - b^2)^2}{4b^2}} = 1$$

which is an ellipse, having eccentricity e', given by

$$e'^2 = 1 - \frac{(a^2 - b^2)^2/4a^2}{(a^2 - b^2)^2/4b^2} = 1 - \frac{b^2}{a^2} = e^2$$

$\Rightarrow$ $e' = e$

6. (c) We know that,

$$PG = \frac{b}{a}\sqrt{b^2\cos^2\theta + a^2\sin^2\theta} \quad \ldots(i)$$

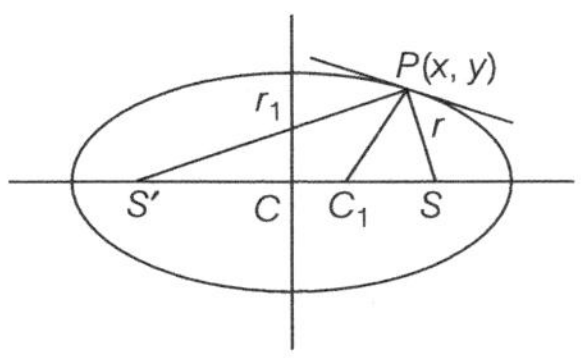

Let S and S' be the foci and P be any point (x, y), then

$rr_1 = SP \cdot S'P$

$= (a - ex)(a + ex)$

$= a^2 - e^2x^2$

$= a^2 - e^2(a^2\cos^2\theta)$ $\quad [\because x = a\cos\theta$ for point $P]$

$= a^2 + (b^2 - a^2)\cos^2\theta$ $\quad [\because b^2 = a^2 - a^2e^2]$

$= a^2(1 - \cos^2\theta) + b^2\cos^2\theta$

$= a^2\sin^2\theta + b^2\cos^2\theta$

$\therefore$ $\sqrt{rr_1} = \sqrt{a^2\sin^2\theta + b^2\cos^2\theta}$

Hence, from Eq. (i), we get

$$PG = \frac{b}{a}\sqrt{rr_1}$$

7. (c) Normal is given by $4x\sec\phi - 2y\operatorname{cosec}\phi = 12$

Coordinates of $Q = (3\cos\phi, 0)$

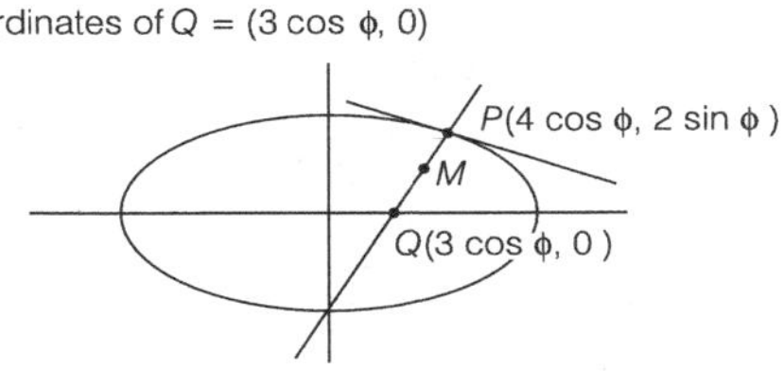

Let $M \equiv (\alpha, \beta)$

$\therefore$ $$\alpha = \frac{3\cos\phi + 4\cos d\ \phi}{2} = \frac{7\cos\phi}{2}$$

or $$\cos\phi = \frac{2}{7}\alpha$$

and $$\beta = \frac{2\sin\phi}{2} \Rightarrow \sin\phi = \beta$$

Now, $$\cos^2\phi + \sin^2\phi = 1$$

$\Rightarrow$ $$\frac{4}{49}\alpha^2 + \beta^2 = 1$$

$\Rightarrow$ $$\frac{4}{49}x^2 + y^2 = 1$$

Now, the latusrectum of given ellipse $x^2 + 4y^2 = 16$ is $x \pm 2\sqrt{3}$.

Solving $x = \pm 2\sqrt{3}$ and $\frac{4}{49}x^2 + y^2 = 1$.

We have, $$\frac{48}{49} + y^2 = 1$$

$\Rightarrow$ $$y^2 = \frac{1}{49} \Rightarrow y = \pm\frac{1}{7}$$

$\therefore$ The point of intersection are $\left(= 2\sqrt{3}, \pm\frac{1}{7}\right)$.

8. (b) The coordinates of extremity of the latusrectum which lies in the first quadrant are $\left(ae, \frac{b^2}{a}\right)$.

The equation of normal at (x_1, y_1) is

$$\frac{a^2x}{x_1} - \frac{b^2y}{y_1} = (a^2 - b^2)$$

Therefore, the equation of normal at $\left(ae, \frac{b^2}{a}\right)$ is

$$\frac{a^2x}{ae} - \frac{b^2y}{b^2/a} = a^2 - b^2$$

$\Rightarrow$ $ax - aey = e(a^2 - b^2)$

$\Rightarrow$ $ax - aey = ea^2e^2$

$\Rightarrow$ $x - ey = ae^3$

This passes through the extremity of the minor axis, i.e. $(0, -b)$.

$\therefore$ $0 + eb - ae^3 = 0 \Rightarrow b = ae^2$

$\Rightarrow$ $b^2 = a^2e^4$

$\Rightarrow$ $a^2(1 - e^2) = a^2e^4$

$\Rightarrow$ $e^4 + e^2 - 1 = 0$

9. (d) Let the line $lx + my + n = 0$ be normal to the ellipse at the point $(a\cos\theta, b\sin\theta)$. Then, the equation of the normal at $(a\cos\theta, b\sin\theta)$ is

$$ax\sec\theta - by\operatorname{cosec}\theta = a^2 - b^2 \quad \ldots(i)$$

If this is to be same as the line $lx + my + n = 0$, then

$$\frac{a\sec\theta}{l} = -\frac{b\operatorname{cosec}\theta}{m} = \frac{a^2 - b^2}{-n}$$

$\Rightarrow$ $$\cos\theta = \frac{-an}{l(a^2 - b^2)} \text{ and } \sin\theta = \frac{bn}{m(a^2 - b^2)}$$

On squaring and adding, we get

$$\frac{a^2n^2}{l^2(a^2 - b^2)^2} + \frac{b^2n^2}{m^2(a^2 - b^2)^2} = 1$$

$\Rightarrow$ $$\frac{a^2}{l^2} + \frac{b^2}{m^2} = \frac{(a^2 - b^2)^2}{n^2}$$

$\Rightarrow$ $$n^2 = \left[\frac{a^2}{l^2} + \frac{b^2}{m^2}\right] = (a^2 - b^2)^2$$

10. (b) Equation of normal to the ellipse at P is

$$ax\sec\theta - by\operatorname{cosec}\theta = a^2 - b^2 \quad \ldots(i)$$

Equation of normal to the cricle $x^2 + y^2 = a^2$ at point θ is

$$y = x\tan\theta \quad \ldots(ii)$$

Substituting the value of x in Eq. (i), we get

$$ax\sec\theta - bx\tan\theta\operatorname{cosec}\theta = a^2 - b^2$$

$$(a - b)x\sec\theta = a^2 - b^2$$

$$\frac{(a - b)}{a^2 - b^2}x = \cos\theta$$

$\Rightarrow$ $$x = \frac{(a^2 - b^2)}{a - b}\cos\theta = (a + b)\cos\theta$$

$\Rightarrow$ $$y = (a + b)\sin\theta$$

Hence, required locus is $x^2 + y^2 = (a + b)^2$.

11. (c) The equation of normals to the ellipse $\frac{x^2}{a^2}+\frac{y^2}{b^2}=1$ at the points whose eccentric angles are θ and $\frac{\pi}{2}+\theta$ are respectively,

$$ax\sec\theta - by\,\text{cosec}\,\theta = a^2-b^2$$

and $$-ax\,\text{cosec}\,\theta - by\sec\theta = a^2-b^2$$

Since, ω is the angle between these two normals, we have

$$\tan\omega = \left|\frac{\frac{a}{b}\tan\theta+\frac{a}{b}\cot\theta}{1-\frac{a^2}{b^2}}\right| = \left|\frac{ab(\tan\theta+\cot\theta)}{b^2+a^2}\right|$$

$$= \left|\frac{2ab}{\sin 2\theta\,(b^2-a^2)}\right| = \frac{2ab}{(a^2-b^2)\sin 2\theta}$$

$$\Rightarrow \frac{2\cot\omega}{\sin 2\theta} = \frac{a^2-b^2}{ab} = \frac{e^2}{\sqrt{1-e^2}}$$

12. (c) Tangent to parabola $yy_1 = 2a(x+x_1)$, $a=1$.

Normal to ellipse $\frac{ax}{\cos\phi}-\frac{by}{\sin\phi}=a^2-b^2$, $a^2=5$, $b^2=4$.

or $$x - ty = -t^2 \qquad \text{(tangent)} \ldots\text{(i)}$$

$$\frac{\sqrt{5}x}{\cos\phi}-\frac{2y}{\sin\phi}=1 \qquad \text{(normal)}$$

or $$\sqrt{5}\sin\phi\, x - 2\cos\phi\, y = \sin\phi\cos\phi \qquad \ldots\text{(ii)}$$

On comparing Eqs. (i) and (ii), we get

$$\frac{\sqrt{5}\sin\phi}{1}=\frac{2\cos\phi}{t}=\frac{\sin\phi\cos\phi}{-t^2} \qquad \ldots\text{(iii)}$$

$\therefore$ $$(2\cos\phi)^2 = -(\sqrt{5}\sin\phi)(\sin\phi\cos\phi)$$

or $$4\cos^2\phi = -(\sqrt{5}\sin\phi)(1-\cos^2\phi)$$

or $$\cos\phi(\sqrt{5}\cos^2\phi - 4\cos\phi - \sqrt{5}) = 0$$

Above gives

$$\cos\phi = 0, -\frac{1}{\sqrt{5}}, \sqrt{5} \text{ (rejected)}$$

or $$\cos\phi = 0, -\cos\alpha, \text{ where } \cos\alpha = \frac{1}{\sqrt{5}}$$

or $$\tan\alpha = 2 \text{ or } \alpha = \tan^{-1}2$$

$\therefore$ $$\phi = \pm\frac{\pi}{2}, \pi-\alpha, \pi+\alpha \qquad \ldots\text{(iv)}$$

$\Rightarrow$ $$\tan\phi = \pm\infty, -2, 2$$

Again from Eq. (iii), $t = \frac{2}{\sqrt{5}}\cdot\frac{1}{\tan\phi}$

$\therefore$ $$t = 0, -\frac{1}{\sqrt{5}}, \frac{1}{\sqrt{5}}$$

13. (a) $SP\cdot S'P = a^2\sin^2\theta + b^2\cos^2\theta = a^2+\cos^2\theta(b^2-a^2)$

$$(SP\cdot S'P)_{max} = b^2$$

$\Rightarrow$ $$\theta = \frac{\pi}{2} \text{ or } \frac{3\pi}{2}$$

$\therefore$ $$P(0, \pm 1)$$

14. (a) $X: -\frac{x^2}{a^2}+\frac{y^2}{b^2}=1$

$$be = 1$$

$\Rightarrow$ $$a^2+b^2=1$$

and $$\frac{a^2}{b}=2$$

$\therefore$ $$b^2+2b-1=0$$

$\Rightarrow$ $$b=\sqrt{2}-1,\ a^2 = 2(\sqrt{2}-1)$$

$\therefore$ $$X: \frac{-x^2}{2(\sqrt{2}-1)}+\frac{y^2}{(\sqrt{2}-1)^2}=1$$

15. (c) Let, $k=\sqrt{2}-1$

$\Rightarrow$ $$-\frac{x^2}{k^2}+\frac{y^2}{2k}=1$$

$$\text{Area of } \Delta BDE = \frac{1}{2}(2)(2+2\sqrt{2})$$
$$= 2(1+\sqrt{2})$$

16. (a) Let P be $(a\cos\theta, b\sin\theta)$. Normal at P intersect the major and minor axes at G and G'. Then, the coordinates of G and G' are $\left(\frac{a^2-b^2}{a}\cos\theta, 0\right)$ and $\left(0, -\frac{a^2-b^2}{a}\sin\theta\right)$ respectively.

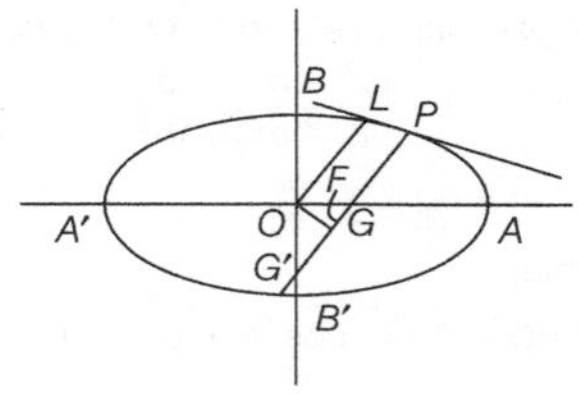

$\therefore$ $$PG^2 = \left(a\cos\theta - \frac{a^2-b^2}{a}\cos\theta\right)^2 + b^2\sin^2\theta$$

$$= \frac{b^4\cos^2\theta}{a^2}+b^2\sin^2\theta$$

$$= \frac{b^2}{a^2}(b^2\cos^2\theta + a^2\sin^2\theta)$$

Similarly, $PG'^2 = \frac{a^2}{b^2}(b^2\cos^2\theta+a^2\sin^2\theta)$

Again, OF is perpendicular from O on the normal and OL is perpendicular from C on the tangent,

$$\frac{x}{a}\cos\theta+\frac{y}{b}\sin\theta = 1$$

Clearly,

$$PF = OL = \frac{1}{\sqrt{\frac{\cos^2\theta}{a^2}+\frac{\sin^2\theta}{b^2}}}$$

$$= \frac{ab}{\sqrt{b^2\cos^2\theta+a^2\sin^2\theta}}$$

Then, $$PF\cdot PG = \frac{ab}{\sqrt{b^2\cos^2\theta+a^2\sin^2\theta}}\cdot\frac{b}{a}\sqrt{b^2\cos^2\theta+a^2\sin^2\theta}$$

$$= b^2$$

17. (c) Let P be the point (x_1, y_1) the normal at which is

$$\frac{x-x_1}{x_1/a^2} = \frac{y-y_1}{y_1/b^2}$$

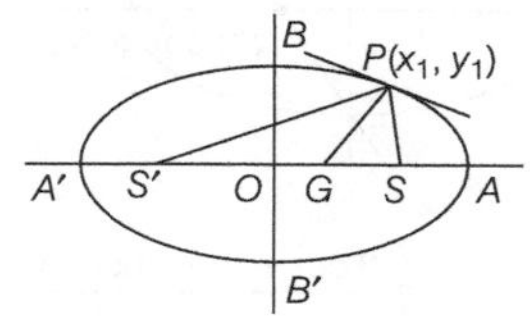

It meets the major axis $Y=0$ at point

$$G\left[x_1\left(1-\frac{b^2}{a^2}\right), 0\right] \text{ or } (e^2x_1, 0) \qquad \left[\because e^2 = 1-\frac{b^2}{a^2}\right]$$

$\therefore$ $$OG = e^2x_1$$

$\therefore$ $S'G = S'O + OG = ae + e^2x_1 = e(a+ex_1) = e\cdot S'P$

and $SG = OS - OG = ae - e^2x_1 = e(a-ex_1) = e\cdot SP$

$\therefore$ $$\frac{S'G}{SG} = \frac{e\cdot S'P}{e\cdot SP} = \frac{S'P}{SP}$$

18. (3) The normal at $(\sqrt{2}\cos\theta, \sin\theta)$ on the ellipse is

$$\sqrt{2}\, x\sec\theta - y\,\text{cosec}\,\theta = 1$$

It passes through the centre $\left(\frac{3}{2}, \sqrt{2}\right)$ of the circle

$\therefore \quad \frac{3}{\sqrt{2}}\sec\theta - \sqrt{2}\,\text{cosec}\,\theta = 1$

Solving, $\theta = \frac{\pi}{4}$

The nearest point on the ellipse is $\left(1, \frac{1}{\sqrt{2}}\right)$.

Its distance from the centre of circle is $\left(\frac{\sqrt{3}}{2}\right)$,

and Radius of circle $= \sqrt{\left(\frac{3}{2}\right)^2 + 2 - 4} = \frac{1}{2}$

$\therefore$ Shortest distance $= \frac{\sqrt{3}-1}{2}$

$\Rightarrow \quad k = 3$

19. (0) We have,

$$\frac{x^2}{14} + \frac{y^2}{5} = 1 \quad \ldots\text{(i)}$$

The equation of normal to this ellipse at $P(\sqrt{14}\cos\theta, \sqrt{5}\sin\theta)$ is $\sqrt{14}\sec\theta\, x - \sqrt{5}\,\text{cosec}\,\theta\, y = 9$

This meets the ellipse (i) at $Q(\sqrt{14}\cos 2\theta, \sqrt{5}\sin 2\theta)$.

$\therefore \quad 14\sec\cos 2\theta - 5\,\text{cosec}\sin 2\theta = 9$

$\Rightarrow \quad 18\cos^2\theta - 9\cos\theta - 14 = 0$

$\Rightarrow \quad (6\cos\theta - 7)(3\cos\theta + 2) = 0$

$\Rightarrow \quad 3\cos\theta + 2 = 0$

20. (4) For the ellipse

$$\frac{x^2}{16} + \frac{y^2}{9} = 1$$

the equation of director circle is $x^2 + y^2 = 25$. The director circle will cut the ellipse $\frac{x^2}{50} + \frac{y^2}{20} = 1$ at four points.

Hence, the number of points is 4.

DPP-5 Equation of $S' = T$, $T = 0$, $SS' = T^2$, Diameter and Conjugate Diameter

1. (c) Equation of chord of contact from at a point (3, 3) to the ellipse $\frac{x^2}{9} + \frac{y^2}{\frac{9}{4}} = 1$ is given by $\frac{3x}{9} + \frac{2y}{\frac{9}{4}} = 1$

$\Rightarrow \quad 3x + 8y = 9$

2. (a) Let (x_1, y_1) be the point of intersection of tangents to the ellipse where it is intersected by the line $lx + my + n = 0$.

Then, the equation of the chord of contact of tangents is

$$\frac{xx_1}{a^2} + \frac{yy_1}{b^2} = 1 \quad \ldots\text{(i)}$$

Clearly, $lx + my + n = 0$ and (i) represents the same line

$\therefore \quad \frac{x_1}{a^2 l} = \frac{y_1}{b^2 m} = -\frac{1}{n}$

$\Rightarrow \quad (x_1, y_1) = \left(-\frac{a^2 l}{n}, -\frac{b^2 m}{n}\right)$

3. (b) Let $P(h, k)$ be the given point. Then, the chord of contact of tangents drawn from P to the ellipse $\frac{x^2}{a^2} + \frac{y^2}{b^2} = 1$ is

$$\frac{hx}{a^2} + \frac{ky}{b^2} = 1 \quad \ldots\text{(i)}$$

This subtends a right angle at the centre $C(0, 0)$ of the ellipse. The combined equation of the pair of straight lines joining C to the points of intersection of Eq. (i) and the ellipse is

$$\frac{x^2}{a^2} + \frac{y^2}{b^2} = \left(\frac{hx}{a^2} + \frac{ky}{b^2}\right)^2$$

this equation represents a pair of perpendicular straight lines

$\therefore \quad \frac{1}{a^2} - \frac{h^2}{a^4} + \frac{1}{b^2} - \frac{k^2}{b^4} = 0$

$\Rightarrow \quad \frac{h^2}{a^4} + \frac{k^2}{b^4} = \frac{1}{a^2} + \frac{1}{b^2}$

4. (b) The equation of ellipse is

$$\frac{x^2}{10} + \frac{y^2}{6} = 1$$

or $\quad 3x^2 + 5y^2 = 30 \quad \ldots\text{(i)}$

Let (h, k) be the mid-point of chord intercepted on the line $2x - y + 3 = 0$ by the ellipse (i). Then, the equation of chord is

$$T = S'$$

i.e. $\quad 3hx + 5ky - 30 = 3h^2 + 5k^2 - 30$

or $\quad 3hx + 5ky = 3h^2 + 5k^2 \quad \ldots\text{(ii)}$

The equation of line is $2x - y + 3 = 0 \quad \ldots\text{(iii)}$

Clearly, Eqs. (ii) and (iii) represent the same lines.

$\therefore \quad \frac{3h}{2} = -5k = \frac{3h^2 + 5k^2}{-3} = \lambda \text{ (say)}$

$\Rightarrow \quad h = \frac{2\lambda}{3}, k = \frac{-\lambda}{5}$ and $3h^2 + 5k^2 = -3\lambda$

$\Rightarrow \quad 3\left(\frac{4\lambda^2}{9}\right) + 5\left(\frac{\lambda^2}{25}\right) = -3\lambda$

$\Rightarrow \quad \frac{4\lambda^2}{3} + \frac{\lambda^2}{5} = -3\lambda$

$\Rightarrow \quad 23\lambda^2 = -45\lambda$

$\Rightarrow \quad \lambda = 0, \lambda = -\frac{45}{23}$

But $\lambda = 0$ is not possible. So, $\lambda = -\frac{45}{23}$.

Now, $h = \frac{2\lambda}{3}, k = -\frac{\lambda}{5} \Rightarrow h = -\frac{30}{23}, k = \frac{9}{23}$

Hence, the required coordinates of the mid-point are $\left(-\frac{30}{23}, \frac{9}{23}\right)$.

5. (a) Let (h, k) be the mid-point of a focal chord of the ellipse $\frac{x^2}{a^2} + \frac{y^2}{b^2} = 1$.

Then, its equation is $\frac{hx}{a^2} + \frac{ky}{b^2} = \frac{h^2}{a^2} + \frac{k^2}{b^2}$

It passes through the focus $S(ae, 0)$.

$\therefore \quad \frac{he}{a} = \frac{h^2}{a^2} + \frac{k^2}{b^2}$

Hence, the locus of (h, k) is $\frac{xe}{a} = \frac{x^2}{a^2} + \frac{y^2}{b^2}$.

6. (c) Let CP and CD be two conjugate semi-diameters of the ellipse $\frac{x^2}{a^2}+\frac{y^2}{b^2}=1$. Then, the eccentric angles of P and D and θ and $\frac{\pi}{2}+\theta$, respectively. So, the coordinates of P and D are $(a\cos\theta, b\sin\theta)$ and $(-a\sin\theta, b\cos\theta)$ respectively.

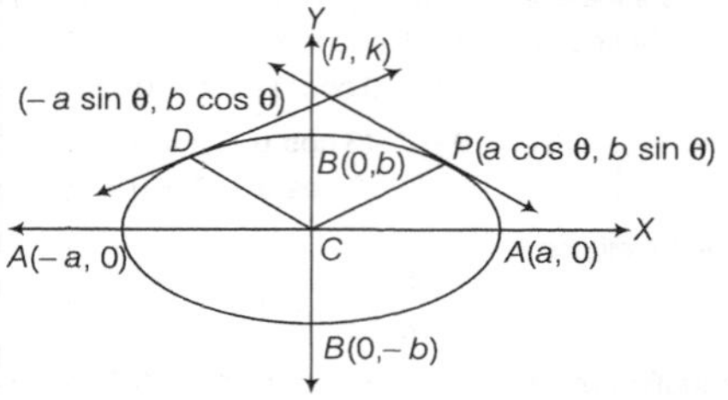

Equations of the tangents at P and D are

$$\frac{x}{a}\cos\theta+\frac{y}{b}\sin\theta=1 \quad \ldots(i)$$

and
$$-\frac{x}{a}\sin\theta+\frac{y}{b}\cos\theta=1 \quad \ldots(ii)$$

Let (h, k) be the point of intersection of Eqs. (i) and (ii). Then,

$$\frac{h}{a}\cos\theta+\frac{k}{b}\sin\theta=1$$

and
$$\frac{-h}{a}\sin\theta+\frac{k}{b}\cos\theta=1$$

$$\Rightarrow \left(\frac{h}{a}\cos\theta+\frac{k}{b}\sin\theta\right)^2+\left(-\frac{h}{a}\sin\theta+\frac{k}{b}\cos\theta\right)^2=1+1$$

$$\Rightarrow \frac{h^2}{a^2}+\frac{k^2}{b^2}=2$$

Hence, the locus of (h, k) is $\frac{x^2}{a^2}+\frac{y^2}{b^2}=2$, which represents an ellipse.

7. (a) Let (h, k) be the mid-point of PD. Then,

$$2h=a\cos\theta-a\sin\theta$$

and
$$2k=b\sin\theta+b\cos\theta$$

$$\Rightarrow \frac{2h}{a}=\cos\theta-\sin\theta$$

and
$$\frac{2k}{b}=\sin\theta+\cos\theta$$

$$\Rightarrow \frac{4h^2}{a^2}+\frac{4k^2}{b^2}=(\cos\theta-\sin\theta)^2+(\sin\theta+\cos\theta)^2$$

$$\Rightarrow \frac{h^2}{a^2}+\frac{k^2}{b^2}=\frac{1}{2}$$

Hence, the locus of (h, k) is $\frac{x^2}{a^2}+\frac{y^2}{b^2}=\frac{1}{2}$.

8. (c) Let $P(a\cos\theta, b\sin\theta)$ and $D(-a\sin\theta, b\cos\theta)$ be the extremities of the conjugate semi-diameters CP and CD of ellipse $\frac{x^2}{a^2}+\frac{y^2}{b^2}=1$. The equations of the circles with CP and CQ as diameters are

$$x(x-a\cos\theta)+y(y-b\sin\theta)=0$$

and
$$x(x+a\sin\theta)+y(y-b\cos\theta)=0$$

$$\Rightarrow x^2+y^2=ax\cos\theta+by\sin\theta \quad \ldots(i)$$

and
$$x^2+y^2=-ax\sin\theta+by\cos\theta \quad \ldots(ii)$$

Let $R(h, k)$ be the intersection of Eqs. (i) and (ii), then

$$h^2+k^2=ah\cos\theta+bk\sin\theta$$

and
$$h^2+k^2=-ah\sin\theta+bk\cos\theta$$

$$\Rightarrow (h^2+k^2)^2+(h^2+k^2)^2=(ah\cos\theta+bk\sin\theta)^2+(-ah\sin\theta+bk\cos\theta)^2$$

$$\Rightarrow 2(h^2+k^2)^2=a^2h^2+b^2k^2$$

Hence, locus of R is

$$2(x^2+y^2)^2=a^2x^2+b^2y^2$$

9. (a) Let (h, k) be the pole of a tangent to the ellipse with respect to the parabola $y^2=4ax$. Then, the equation of polar is

$$ky=2a(x+h) \text{ or } y=\left(\frac{2a}{k}\right)x+\frac{2ah}{k}$$

This touches the ellipse $\frac{x^2}{a^2}+\frac{y^2}{b^2}=1$.

$$\therefore \left(\frac{2ah}{k}\right)^2=a^2\left(\frac{2a}{k}\right)^2+b^2$$

$$\Rightarrow b^2k^2=4a^2h^2-4a^4$$

Hence, the locus of (h, k) is $b^2y^2=4a^2(x^2-a^2)$.

10. (c) Any normal to the ellipse is $\frac{ax}{\cos\theta}-\frac{by}{\sin\theta}=a^2-b^2 \quad \ldots(i)$

If its pole is (h, k), then Eq. (i) is same as polar of (h, k), i.e.

$$\frac{hx}{a^2}+\frac{ky}{b^2}=1 \quad \ldots(ii)$$

On comparing Eqs. (i) and (ii), we get

$$\frac{a}{\cos\theta}\cdot\frac{a^2}{h}=\frac{-b}{\sin\theta}\cdot\frac{b^2}{k}=\frac{a^2-b^2}{1}$$

$$\therefore \frac{a^3}{h}=(a^2-b^2)\cos\theta,\ \frac{b^3}{k}=-(a^2-b^2)\sin\theta$$

On squaring and adding, we get

$$\frac{a^6}{h^2}+\frac{b^6}{k^2}=(a^2-b^2)^2(\cos^2\theta+\sin^2\theta)=(a^2-b^2)^2$$

$\therefore$ Locus is $\frac{a^6}{x^2}+\frac{b^6}{y^2}=(a^2-b^2)^2$

11. (a) The equation of required chord is given by $T=S'$

or
$$\frac{2x}{36}+\frac{y}{9}-1=\frac{4}{36}+\frac{1}{9}-1$$

$$\Rightarrow x+2y-18=-14$$

$$\therefore x+2y-4=0$$

12. (a) Let $P(h, k)$ be a point such that the chord of contact of tangents drawn from it to the ellipse $\frac{x^2}{a^2}+\frac{y^2}{b^2}=1$ touches the circle $x^2+y^2=c^2$.

The equation of the chord of contact of tangents drawn from $P(h,x)$ to the ellipse $\frac{x^2}{a^2}+\frac{y^2}{b^2}=1$, is

$$\frac{hx}{a^2}+\frac{ky}{b^2}=1$$

$$\Rightarrow y=-\left(\frac{b^2h}{a^2k}\right)x+\frac{b^2}{k}$$

It touches the circle $x^2+y^2=c^2$

$$\therefore \left(\frac{b}{k}\right)^2=c^2\left(1+\frac{b^4h^2}{a^4k^2}\right)$$

$$\Rightarrow \frac{b^2}{k^2}=c^2+\frac{b^4h^2c^2}{a^4k^2}$$

or
$$a^4b^4=a^4c^2k^2+b^2h^2c^2$$

$\therefore$ locus of (h, k) is

$$b^4c^2x^2+a^4c^2y^2=a^4b^4$$

13. (a)

(i) We know that,

$$S'P + SP = 2a = 2\sqrt{20} = 4\sqrt{5}$$

(ii) We have,

$$2x^2 + 3y^2 - 4x - 12y + 13 = 0$$

$\Rightarrow$ $2(x^2 - 2x) + 3(y^2 - 4y) + 13 = 0$

$\Rightarrow$ $2(x-1)^2 - 2 + 3(y-2)^2 - 12 + 13 = 0$

$\Rightarrow$ $2(x-1)^2 + 3(y-2)^2 = 1$

$\Rightarrow$ $\dfrac{(x-1)^2}{\frac{1}{2}} + \dfrac{(y-2)^2}{\frac{1}{3}} = 1$

$\therefore$ Eccentricity, $e = \sqrt{1 - \dfrac{\left(\frac{1}{3}\right)}{\left(\frac{1}{2}\right)}} = \dfrac{1}{\sqrt{3}}$

(iii) Any point on the line $x - y - 5 = 0$ will be of the form $(t, t-5)$, where $t \in R$. The chord of contact of this point with respect to the curve $x^2 + 4y^2 = 4$ is $tx + 4(t-5)y - 4 = 0$

or $\qquad (-20y - 4) + t(x + 4y) = 0$

which is a family of straight lines. Each member of this family passes through the point of intersection of straight lines $-20y - 4 = 0$ and $x + 4y = 0$, which is $\left(\dfrac{4}{5}, -\dfrac{1}{5}\right)$.

Hence, the abscissa of the point is $\dfrac{4}{5}$.

(iv) Given equation of ellipse is

$$3x^2 + 4y^2 = 12$$

$\Rightarrow$ $\dfrac{x^2}{4} + \dfrac{y^2}{3} = 1$

Sum of distance of any point from its directrix

$$= \frac{2a}{e} = \frac{2 \times 2}{\sqrt{1 - \frac{3}{4}}} = 8$$

Hence, (i) $\to$ (q), (ii) $\to$ (s), (iii) $\to$ (p), (iv) $\to$ (r)

14. (b) as $P(t^2, 2t)$ lies inside ellipse $S' = T$ and $T < 0$

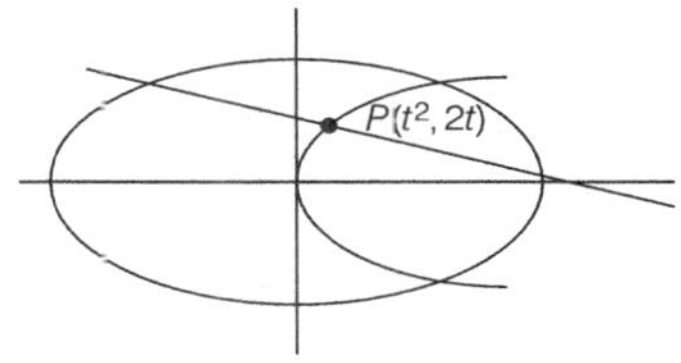

$\Rightarrow$ $\dfrac{x(t^2)}{16} + \dfrac{y(2t)}{4} = \dfrac{t^4}{16} + \dfrac{4t^2}{4}$ $\qquad$ ...(i)

$\Rightarrow$ $\dfrac{t^4}{16} + \dfrac{4t^2}{4} - 1 < 0$

$\Rightarrow$ $t^4 + 16t^2 - 16 < 0$ $\qquad$ ...(ii)

as (i), passes through $(a, 0)$ and $\dfrac{at^2}{16} = \dfrac{t^4}{16} + t^2$

$\Rightarrow$ $t^2[t^2 + (16 - 9)] = 0,$

where, $-8 - 4\sqrt{5} < t^2 < -8 + 4\sqrt{5}$

$\therefore$ $0 < t^2 < -8 + 4\sqrt{5}$

$\Rightarrow$ $t^2 = 0$ or $t^2 = a - 16$

$\therefore$ $0 < a - 16 < -8 + 4\sqrt{5}$

Clearly, $\quad a - 16 > 0 \Rightarrow a > 16$

and $\quad a - 16 < -8 + 4\sqrt{5} \Rightarrow a < 8 + 4\sqrt{5}$

$\therefore$ $16 < a < 8 + 4\sqrt{5}$.

15. (c) Solving, $\dfrac{x^2}{16} + \dfrac{4x}{4} = 1$

$\Rightarrow$ $x^2 + 16x - 16 = 0$

$\Rightarrow$ $x = -8 \pm 4\sqrt{5}$

$\Rightarrow$ $x = -8 + 4\sqrt{5}$ is possible

$\therefore$ $y^2 = 4(-8 + 4\sqrt{5})$

$\Rightarrow$ $y = 2\sqrt{4\sqrt{5} - 8}$

$\Rightarrow$ $y = 4\sqrt{\sqrt{5} - 2}$

$\therefore$ Length of common chord $= 8\sqrt{\sqrt{5} - 2}$.

Revisal Problems (JEE Main)

1. (b) We have, $\qquad PA + PB = 4$

$\therefore$ P lies on the ellipse having its foci at A and B and length of the major axis $= 4$

2. (c) Let the fixed straight lines be along the coordiantes axes and AB be the bar of length l such that its extremities $A(a, 0)$ and $B(0, b)$ are on the coordinate axes. Then,

$$a^2 + b^2 = l^2 \qquad \text{...(i)}$$

Let $P(h, k)$ be a point marked on the bar such that it divides the bar AB in the ratio $\lambda : 1$.

Then, $h = \dfrac{a}{\lambda + 1}$ and $k = \dfrac{b\lambda}{\lambda + 1}$

$\Rightarrow$ $a = (\lambda + 1)h$ and $b = \dfrac{(\lambda + 1)k}{\lambda}$

Substituting these values in Eq. (i), we get

$$(\lambda + 1)^2 h^2 + \frac{(\lambda + 1)^2 k^2}{\lambda^2} = l^2$$

Hence, the locus of $P(h, k)$ is $(\lambda + 1)^2 x^2 + \dfrac{(\lambda + 1)^2}{\lambda^2} y^2 = l^2$

or $\dfrac{x^2}{\left(\frac{l}{\lambda + 1}\right)^2} + \dfrac{y^2}{\left(\frac{\lambda l}{\lambda + 1}\right)^2} = 1$, which represents an ellipse.

3. (b) We have, $e = \dfrac{1}{2}$ and equation of directrix is $x = 4$.

We know that, equation of directrix is $x = a/e$.

$\therefore$ $a/e = 4$

$\Rightarrow$ $a = 4e$

$\Rightarrow$ $a = 4 \times \dfrac{1}{2} = 2$

Also, $\quad e = \sqrt{1 - \dfrac{b^2}{a^2}}$

$\Rightarrow$ $\dfrac{1}{4} = 1 - \dfrac{b^2}{4}$

$\Rightarrow$ $b^2 = 3$

$\therefore$ Equation of ellipse is

$$\frac{x^2}{4} + \frac{y^2}{3} = 1$$

or $\qquad 3x^2 + 4y^2 = 12$

4. (b) Let the equation of ellipse be $\dfrac{x^2}{a^2} + \dfrac{y^2}{b^2} = 1$ and e be the eccentricity of ellipse.

It is given that distance between foci $= 2h$

$\therefore$ $2ae = 2h \Rightarrow ae = h$ $\qquad$ ...(i)

Focal distance of the one end of minor axis, say $(0, b)$ is k.

$\therefore \quad a + e(0) = k \Rightarrow a = k \quad \ldots$(ii)

From Eqs. (i) and (ii), we have

$$b^2 = a^2(1 - e^2)$$
$$= a^2 - (ae)^2 = k^2 - h^2$$

Hence, the equation of ellipse is $\frac{x^2}{k^2} + \frac{y^2}{k^2 - h^2} = 1$.

5. (a) Given equation of ellipse is

$$\frac{x^2}{10 - a} + \frac{y^2}{4 - a} = 1$$

$\therefore \quad 10 - a > 0$ and $4 - a > 0$

$\Rightarrow \quad a < 10$ and $a < 4$

$\therefore \quad a < 4$

6. (c) We have,

Distance of the point from Y-axis = 3 [Distance of P from (4, 0)]

$$\Rightarrow \frac{\text{Distance of } P \text{ from } (4, 0)}{\text{Distance of } P \text{ from } Y\text{-axis}} = \frac{1}{3}$$

$\therefore$ Locus of P is an ellipse with eccentricity $e = \frac{1}{3}$.

7. (a) Clearly, the race course is an ellipse with flag-posts as its foci. If a and b are semi-major and minor axes of the ellipse, then

$$2a = 10 \text{ and } 2ae = 8$$

$$\Rightarrow a = 5, e = \frac{4}{5}$$

Now, $\quad b^2 = a^2(1 - e^2)$

$$= 25\left(1 - \frac{16}{25}\right) = 9$$

$$\Rightarrow b = 3$$

Hence, the area of the ellipse $= \pi ab = 15\pi$

8. (a) Let $P(a \cos \alpha, b \sin \alpha)$ and $Q(a \cos \beta, b \sin \beta)$ be the extremities of the focal chord PQ. Then,

$$e \cos\left(\frac{\alpha + \beta}{2}\right) = \cos\left(\frac{\alpha - \beta}{2}\right)$$

$$\Rightarrow e^2 \cos^2\left(\frac{\alpha + \beta}{2}\right) = \cos^2\left(\frac{\alpha - \beta}{2}\right)$$

$$\Rightarrow e^2\{1 + \cos(\alpha + \beta)\} = \{1 + \cos(\alpha - \beta)\}$$

$$\Rightarrow e^2 \cos(\alpha + \beta) = 1 - e^2 + \cos(\alpha - \beta)$$

Now, $\quad SP = a - ae \cos \alpha$

and $\quad SQ = a - ae \cos \beta$

$\therefore$ HM of SP and SQ

$$= 2\, SP \cdot SQ = \frac{2a^2(1 - e\cos\alpha)(1 - e\cos\beta)}{2a - ae(\cos\alpha + \cos\beta)}$$

$$= \frac{2a(1 - e\cos\alpha - e\cos\beta + e^2\cos\alpha\cos\beta)}{2 - e(\cos\alpha + \cos\beta)}$$

$$= \frac{\left[2a\left\{1 - 2e\cos\left(\frac{\alpha+\beta}{2}\right)\cdot\cos\left(\frac{\alpha-\beta}{2}\right) + \frac{e^2}{2}\cos(\alpha+\beta) + \frac{e^2}{2}\cos(\alpha-\beta)\right\}\right]}{2 - 2e\cos\frac{\alpha+\beta}{2}\cos\frac{\alpha-\beta}{2}}$$

$$= \frac{\left[2a\left\{1 - 2\cos^2\frac{\alpha-\beta}{2} + \frac{1}{2} - \frac{e^2}{2} + \frac{1}{2}\cos(\alpha-\beta) + \frac{e^2}{2}\cos(\alpha-\beta)\right\}\right]}{2 - 2\cos^2\frac{\alpha-\beta}{2}}$$

$$= \frac{\left[a\{2 - 2 - 2\cos(\alpha-\beta) + 1 - e^2 + \cos(\alpha-\beta) + e^2\cos(\alpha-\beta)\}\right]}{1 - \cos(\alpha-\beta)}$$

$$= \frac{a[1 - \cos(\alpha-\beta) - e^2\{1 - \cos(\alpha-\beta)\}]}{1 - \cos(\alpha-\beta)}$$

$$= a(1 - e^2) = a \times \frac{b^2}{a^2} = \frac{b^2}{a}$$

9. (b) We have, $\quad 16x^2 + 25y^2 = 400$

$$\Rightarrow \frac{x^2}{25} + \frac{y^2}{16} = 1$$

$\therefore$ Length of latusrectum $= 2 \times \frac{16}{5} = \frac{32}{5}$

We know that, the semi-latusrectum of an ellipse is the HM of the segments of a focal chord.

Now, HM of SP and SQ = Semi-latusrectum

$$\Rightarrow \frac{2SP \cdot SQ}{SP + SQ} = \frac{16}{5}$$

$$\Rightarrow \frac{2 \times 8 \times SQ}{8 + SQ} = \frac{16}{5}$$

$$\Rightarrow 5SQ = 8 + SQ$$

$$\therefore SQ = 2$$

10. (d) We have, $\quad 16x^2 + 25y^2 = 400$

$$\Rightarrow \frac{x^2}{25} + \frac{y^2}{16} = 1$$

$\therefore$ Semi-letusrectum $= \frac{16}{5}$

Now, HM of SP and SQ = Semi-latusrectum

$$\Rightarrow \frac{2\,SP \cdot SQ}{SP + SQ} = \frac{16}{5}$$

$$\Rightarrow \frac{2 \times 16 \times SQ}{16 + SQ} = \frac{16}{5}$$

$$\Rightarrow 10\,SQ = 16 + SQ$$

$$\Rightarrow SQ = \frac{16}{9}$$

But $\quad SQ + S'Q$ = Major axis

$$\Rightarrow \frac{16}{9} + S'Q = 10$$

$$\Rightarrow S'Q = \frac{74}{9}$$

11. (c) The major axis passes through $S(-1, -1)$ and is perpendicular to the directrix $x - y + 3 = 0$.

So, the equation of major axis is $x + y + 2 = 0$.

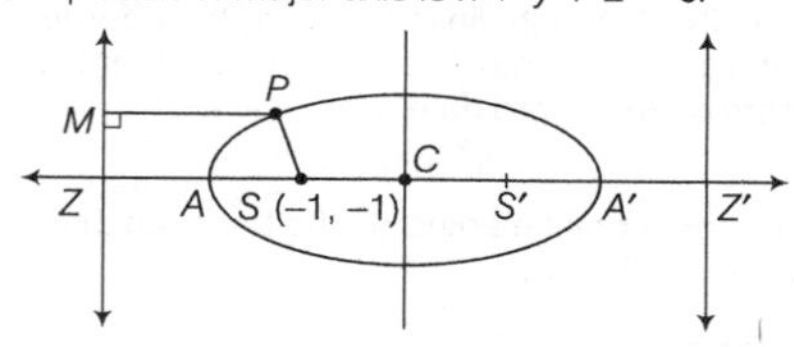

The axis meets the directrix at Z. Then, Z is $\left(-\frac{5}{2}, \frac{11}{2}\right)$. A and A' divide ZS in the ratio $1 : e$, i.e. $1 : \frac{1}{2}$ or $2 : 1$ internally and externally, respectively. Therefore, we have $A\left(-\frac{3}{2}, -\frac{1}{2}\right)$ and $A'\left(\frac{1}{2}, -\frac{5}{2}\right)$.

Thus, centre C (mid-point of AA') $= \left(-\frac{1}{2}, -\frac{3}{2}\right)$.

12. (a) We have, $4x^2 + 9y^2 - 8x - 36y + 4 = 0$

$\Rightarrow \quad 4(x-1)^2 + 9(y-2)^2 = 36$

$\Rightarrow \quad \frac{(x-1)^2}{9} + \frac{(y-2)^2}{4} = 1 \quad \ldots(i)$

Clearly, the coordinates of C are (1, 2). Suppose CA makes an angle θ with the major axis.

Then, the coordinates of A and B are

$$A \equiv (1 + CA\cos\theta, 2 + CA\sin\theta)$$

$$B \equiv \left[1 + CB\cos\left(\frac{\pi}{2}+\theta\right), 2 + CB\sin\left(\frac{\pi}{2}+\theta\right)\right]$$

$$= (1 - CB\sin\theta, 2 + CB\cos\theta)$$

Since, A and B lie on Eq. (i), therefore

$$\frac{CA^2}{9}\cos^2\theta + \frac{CA^2}{4}\sin^2\theta = 1$$

and $\quad \frac{CB^2}{9}\sin^2\theta + \frac{CB^2}{4}\cos^2\theta = 1$

$\Rightarrow \quad \frac{1}{36}(4\cos^2\theta + 9\sin^2\theta) = CA^{-2}$

and $\quad \frac{1}{36}(4\sin^2\theta + 9\cos^2\theta) = CB^{-2}$

$\Rightarrow \quad CA^{-2} + CB^{-2} = \frac{13}{36}$

or $\quad \frac{1}{CA^2} + \frac{1}{CB^2} = \frac{13}{36}$

13. (b) Let $P(a\cos\theta, b\sin\theta)$ be the variable point on the ellipse $\frac{x^2}{a^2} + \frac{y^2}{b^2} = 1$. Then, A = Area of ΔPF_1F_2

$$= \frac{1}{2}\begin{vmatrix} a\cos\theta & b\sin\theta & 1 \\ ae & 0 & 1 \\ -ae & 0 & 1 \end{vmatrix}$$

$$= \frac{1}{2} b\sin\theta \times 2ae = abe\sin\theta$$

Clearly, it is maximum when $\theta = \frac{\pi}{2}$ and the maximum value of A is abe.

14. (a) Let the ellipse be $\frac{x^2}{a^2} + \frac{y^2}{b^2} = 1$ and

$$\frac{x\cos\theta}{a} + \frac{y\sin\theta}{b} = 1 \quad \ldots(i)$$

be a tangent to it at point $(a\cos\theta, b\sin\theta)$. Then,

p = Length of the perpendicular from S $(ae, 0)$ on Eq. (i)

$$= \left|\frac{e\cos\theta - 1}{\sqrt{\frac{\cos^2\theta}{a^2} + \frac{\sin^2\theta}{b^2}}}\right|$$

p' = Length of the perpendicular from O (0, 0) on Eq. (i)

$$= \left|\frac{1}{\sqrt{\frac{\cos^2\theta}{a^2} + \frac{\sin^2\theta}{b^2}}}\right|$$

and $\quad r = ae\cos\theta - a$

Clearly, $\quad rp' = ap$

15. (c) We have, $16x^2 + 25y^2 = 400$

$\Rightarrow \quad \frac{x^2}{25} + \frac{y^2}{16} = 1$

$\Rightarrow \quad a^2 = 25$ and $b^2 = 16$

This equation represents an ellipse with eccentricity e given by

$$e^2 = 1 - \frac{b^2}{a^2} = 1 - \frac{16}{25} \Rightarrow e = \frac{3}{5}$$

So, the coordinates of foci are $(\pm ae, 0)$, i.e. (3, 0) and (– 3, 0). Thus, F_1 and F_2 are the foci of the ellipse.

Since, the sum of the focal distance of a point on an ellipse is equal to its major axis.

$\therefore \quad PF_1 + PF_2 = 2a = 10$

16. (c) We know that the line $y = mx + c$ touches the curve $\frac{x^2}{a^2} + \frac{y^2}{b^2} = 1$ iff $c^2 = a^2m^2 + b^2$.

Here, $\quad a^2 = 4, b^2 = 1, m = 4$

$\therefore \quad c^2 = 64 + 1 \Rightarrow c = \pm\sqrt{65}$

17. (d) The equations of tangents at two points having eccentric angles θ_1 and θ_2 are

$$\frac{x}{a}\cos\theta_1 + \frac{y}{b}\sin\theta_1 = 1 \quad \ldots(i)$$

and $\quad \frac{x}{a}\cos\theta_2 + \frac{y}{b}\sin\theta_2 = 1 \quad \ldots(ii)$

The point of intersection of Eqs. (i) and (ii) is

$$\left(\frac{a\cos\left(\frac{\theta_1+\theta_2}{2}\right)}{\cos\left(\frac{\theta_1-\theta_2}{2}\right)}, \frac{b\sin\left(\frac{\theta_1+\theta_2}{2}\right)}{\cos\left(\frac{\theta_1-\theta_2}{2}\right)}\right)$$

It is given that $\theta_1 + \theta_2 = k$ = constant. Therefore, if (x_1, y_1) is the point of intersection Eqs. (i) and (ii), then

$$x_1 = \frac{a\cos k}{\cos\left(\frac{\theta_1-\theta_2}{2}\right)} \text{ and } y_1 = \frac{b\sin k}{\cos\left(\frac{\theta_1-\theta_2}{2}\right)}$$

$\Rightarrow \quad \frac{x_1}{y_1} = \frac{a}{b}\cot k \Rightarrow y_1 = \left(\frac{b}{a}\tan k\right)x_1$

$\therefore (x_1, y_1)$ lies on the straight line $y = \left(\frac{b}{a}\tan k\right)x$.

18. (b) Let $P(a\cos\theta_1, b\sin\theta_1)$ and $Q(a\cos\theta_2, b\sin\theta_2)$ be two points on the ellipse.

Then, $\quad m_1$ = Slope of $OP = \frac{b}{a}\tan\theta_1$

and $\quad m_2$ = Slope of $OQ = \frac{b}{a}\tan\theta_2$

$\therefore \quad m_1m_2 = \frac{b^2}{a^2}\tan\theta_1\tan\theta_2$

$\Rightarrow \quad m_1m_2 = \frac{b^2}{a^2} \times \frac{-a^2}{b^2} \quad \left[\because \tan\theta_1\tan\theta_2 = -\frac{a^2}{b^2}\right]$

$\Rightarrow \quad m_1m_2 = -1$

$\therefore \quad \angle POQ = \frac{\pi}{2}$

Hence, PQ makes a right angle at the centre the ellipse.

19. (a) The equation of the tangent at $P(a\cos\theta, b\sin\theta)$ to the ellipse $\frac{x^2}{a^2} + \frac{y^2}{b^2} = 1$ is

$$\frac{x}{a}\cos\theta + \frac{y}{b}\sin\theta = 1 \quad \ldots(i)$$

The combined equation of the chords joining the points of intersection of Eq (i) and the auxiliary circle $x^2 + y^2 = a^2$ is

$$x^2 + y^2 = a^2\left(\frac{x}{a}\cos\theta + \frac{y}{b}\sin\theta\right)^2$$

The lines given by this equation are at right angles.

$\therefore \quad$ Coefficient of x^2 + Coefficient of $y^2 = 0$

$\Rightarrow \quad 1 - a^2\left(\frac{\cos^2\theta}{a^2}\right) + 1 - a^2\left(\frac{\sin^2\theta}{b^2}\right) = 0$

$\Rightarrow \quad \sin^2\theta + 1 - \frac{a^2}{b^2}\sin^2\theta = 0$

$\Rightarrow \quad \sin^2\theta\left(1 - \frac{a^2}{b^2}\right) + 1 = 0$

$$\Rightarrow \quad -\sin^2\theta\,(a^2-b^2)+b^2=0$$
$$\Rightarrow \quad -a^2e^2\sin^2\theta+a^2(1-e^2)=0$$
$$\Rightarrow \quad 1=e^2(1+\sin^2\theta)$$
$$\Rightarrow \quad e=(1+\sin^2\theta)^{-1/2}$$

20. (b) We know that the product of perpendiculars from two foci of an ellipse upon any tangent is equal to the square of the semi-minor axis.

$$\therefore \quad (S_1F_1)(S_2F_2)=b^2 \Rightarrow (S_1F_1)(S_2F_2)=3 \qquad [\because b=\sqrt{3}]$$

21. (a) The given point is $\left(a\cos\frac{\pi}{4}, b\sin\frac{\pi}{4}\right)$ i.e. $\left(\frac{a}{\sqrt2},\frac{b}{\sqrt2}\right)$.

So, the equation of the tangent at this point is

$$\frac{x}{a}+\frac{y}{b}=\sqrt2 \qquad \ldots(i)$$

$\therefore P_1$ = Length of the perpendicular from (0, 0) on line (i)

$$\Rightarrow \quad P_1=\left|\frac{\frac{0}{a}+\frac{0}{b}-\sqrt2}{\sqrt{\frac{1}{a^2}+\frac{1}{b^2}}}\right|=\frac{\sqrt2\,ab}{\sqrt{a^2+b^2}}$$

Equation of the normal at $\left(\frac{a}{\sqrt2},\frac{b}{\sqrt2}\right)$ is

$$\frac{a^2x}{a/\sqrt2}-\frac{b^2y}{b/\sqrt2}=a^2-b^2$$
$$\Rightarrow \quad \sqrt2ax-\sqrt2\,by=a^2-b^2 \qquad \ldots(ii)$$

$\therefore P_2$ = Length of the perpendicular from (0, 0) on line (ii)

$$\Rightarrow \quad P_2=\frac{a^2-b^2}{\sqrt{(\sqrt2a)^2+(-\sqrt2b)^2}}=\frac{a^2-b^2}{\sqrt{2\,(a^2+b^2)}}$$

$\therefore$ Area of the rectangle $=P_1P_2$

$$=\frac{\sqrt2\,ab}{\sqrt{a^2+b^2}}\times\frac{a^2-b^2}{\sqrt{2\,(a^2+b^2)}}=\left(\frac{a^2-b^2}{a^2+b^2}\right)ab$$

22. (b) We have, $a^2=\frac14, b^2=\frac19, m=\frac89$

The required points are

$$\left(\pm\frac{a^2m}{\sqrt{a^2m+b^2}},\pm\frac{b^2}{\sqrt{a^2m^2+b^2}}\right)=\left(\pm\frac25,\pm\frac15\right)$$

23. (d) Let P (2 cos θ, 3 sin θ) be the point on the ellipse at which the normal is parallel to $2x-y=1$. The equation of the normal at P is $2x\sec\theta-3y\,\text{cosec}\,\theta=-5$.

If it is parallel to $2x-y=1$, then $\frac23\tan\theta=2$

$$\Rightarrow \quad \tan\theta=3$$
$$\Rightarrow \quad \sin\theta=\frac{3}{\sqrt{10}} \text{ and } \cos\theta=\frac{1}{\sqrt{10}}$$
$$\text{or} \quad \sin\theta=-\frac{3}{\sqrt{10}} \text{ and } \cos\theta=-\frac{1}{\sqrt{10}}$$

Hence, the required points are $\left(\pm\frac{2}{\sqrt{10}},\pm\frac{3}{\sqrt{10}}\right)$.

24. (b) The equation of the tangent at $(3\sqrt3\cos\theta, \sin\theta)$ to the ellipse $\frac{x^2}{27}+\frac{y^2}{1}=1$ is

$$\left(\frac{3\sqrt3\cos\theta}{27}\right)x+(\sin\theta)\,y=1$$
$$\Rightarrow \quad \frac{x\cos\theta}{3\sqrt3}+y\sin\theta=1$$

Let S be the sum of the intercepts made by this tangent on the coordinates axes. Then,

$$S=3\sqrt3\sec\theta+\text{cosec}\,\theta$$

Now, $$\frac{dS}{d\theta}=0$$
$$\Rightarrow \quad 3\sqrt3\,\frac{\sin\theta}{\cos^2\theta}=\frac{\cos\theta}{\sin^2\theta}$$
$$\Rightarrow \quad \tan^3\theta=\frac{1}{3\sqrt3}$$
$$\Rightarrow \quad \tan\theta=\frac{1}{\sqrt3}$$
$$\Rightarrow \quad \theta=\frac{\pi}{6}$$

Clearly, $$\left[\frac{d^2S}{d\theta^2}\right]_{\theta=\frac{\pi}{6}}>0$$

Hence, S is minimum for $\theta=\frac{\pi}{6}$.

25. (a) Let P (4 cos θ, 3 sin θ) be a point on the given ellipse such that its distance from the centre (0, 0) of the ellipse is equal to the mean of the lengths of the semi-major and semi-minor axes, i.e.

$$OP=\frac{4+3}{2}$$
$$\Rightarrow \quad \sqrt{16\cos^2\theta+9\sin^2\theta}=\frac72$$
$$\Rightarrow \quad 7\cos^2\theta+9=\frac{49}{4}$$
$$\Rightarrow \quad \cos^2\theta=\frac{13}{28}$$
$$\Rightarrow \quad \cos\theta=\pm\sqrt{\frac{13}{28}} \text{ and } \sin\theta=\pm\sqrt{\frac{15}{28}}$$
$$\therefore \quad \cos\theta=\pm\frac{\sqrt{91}}{14} \text{ and } \sin\theta=\pm\frac{\sqrt{105}}{14}$$

Hence, the required points are given by $P\left(\pm\frac{2\sqrt{91}}{7},\frac{\pm3\sqrt{105}}{14}\right)$.

26. (c) Let P (3 cos θ, 2 sin θ) be a point on $4x^2+9y^2=36$. The equation of the tangent at P is

$$2x\cos\theta+3y\sin\theta=6$$

This meets the coordinate axes at T (3 sec θ, 0) and T' (0, 2 cosec θ). The equation of the circle with TT' as diameter is $(x-3\sec\theta)(x-0)+(y-0)(y-2\,\text{cosec}\theta)=0$

or $$x^2+y^2-3x\sec\theta-2y\,\text{cosec}\theta=0$$

Clearly, it passes through (0, 0).

27. (d) We have, $\frac{x^2}{9}+\frac{y^2}{5}=1$

Let e be the eccentricity of this ellipse. Then,

$$e^2=1-\frac59 \Rightarrow e=\frac23$$

The coordinates of the end-points of latusreta are

$$L\left(2,\frac53\right), M\left(-2,\frac53\right), M'\left(-2,-\frac53\right) \text{ and } L'\left(2,-\frac53\right)$$

The equations of tangents at these points are

$$2x+3y-9=0 \qquad \ldots(i)$$
$$-2y+3y-9=0 \qquad \ldots(ii)$$
$$2x+3y+9=0 \qquad \ldots(iii)$$
$$-2x+3y+9=0 \qquad \ldots(iv)$$

Clearly, these tangents form a parallelogram whose area is given by

$$A=\left|\frac{[9-(-9)]\times[9(-9)]}{\begin{vmatrix}2 & 3\\-2 & 3\end{vmatrix}}\right|=\frac{18\times18}{12}=27 \text{ sq units}$$

28. (c) Let $R(h, k)$ be the point of intersection of tangents at P and Q. Then,

$$h = \frac{a\cos\left(\frac{\alpha+\beta}{2}\right)}{\cos\left(\frac{\alpha-\beta}{2}\right)}$$

and
$$k = \frac{b\sin\left(\frac{\alpha+\beta}{2}\right)}{\cos\left(\frac{\alpha-\beta}{2}\right)}$$

$$\Rightarrow \quad \frac{h^2}{a^2} + \frac{k^2}{b^2} = \frac{1}{\cos^2\left(\frac{\alpha-\beta}{2}\right)}$$

Hence, the locus of $R(h, k)$ is

$$\frac{x^2}{a^2} + \frac{y^2}{b^2} = \frac{1}{\cos^2\left(\frac{\alpha-\beta}{2}\right)} \quad [\because \alpha - \beta = \text{constant}]$$

Clearly, it represents an ellipse.

29. (b) The equation of the circle described as SS' as a diameter is

$$(x - ae)(x + ae) + (y - 0)(y - 0) = 0$$

$$\Rightarrow \quad x^2 + y^2 = a^2e^2$$

The abscissae of thue points of intersection of the ellipse and this circle are the roots of the equation

$$\frac{x^2}{a^2} + \frac{a^2e^2 - x^2}{b^2} = 1$$

$$\Rightarrow \quad x^2\left(\frac{1}{a^2} - \frac{1}{b^2}\right) = 1 - \frac{a^2e^2}{b^2}$$

$$\Rightarrow \quad x^2\left(\frac{b^2 - a^2}{a^2b^2}\right) = \frac{b^2 - a^2e^2}{b^2}$$

$$\Rightarrow \quad -\frac{x^2a^2e^2}{a^2} = a^2 - 2a^2e^2$$

$$\Rightarrow \quad x^2 = a^2\left(\frac{2e^2 - 1}{e^2}\right)$$

$$\Rightarrow \quad x = \pm\frac{a}{e}\sqrt{2e^2 - 1}$$

This will give distinct values of x, if $2e^2 - 1 > 0$

$$\Rightarrow \quad e^2 > \frac{1}{2}$$

$$\therefore \quad e \in \left(\frac{1}{\sqrt{2}}, 1\right)$$

30. (b) Let (h, k) be the foot of perpendicular drawn from the focus $S(ae, 0)$ on tangent

$$y = mx + \sqrt{a^2m^2 + b^2} \quad \ldots(\text{i})$$

at point P on the ellipse $\frac{x^2}{a^2} + \frac{y^2}{b^2} = 1$.

Then,
$$k = mh + \sqrt{a^2m^2 + b^2}$$

$$\Rightarrow \quad (k - mh)^2 = a^2m^2 + b^2 \quad \ldots(\text{ii})$$

Also, SP is perpedicular to Eq. (i)

$$\therefore \quad m \times \frac{k - 0}{h - ae} = -1$$

$$\Rightarrow \quad h + mk = ae \quad (\text{iii})$$

From Eqs. (ii) and (iii), we get

$$(k - mh)^2 + (h + mk)^2 = a^2m^2 + b^2 + a^2e^2$$

$$\Rightarrow \quad (h^2 + k^2)(1 + m^2) = a^2(1 + m^2)$$

$$\Rightarrow \quad h^2 + k^2 = a^2$$

Hence, the locus of (h, k) is $x^2 + y^2 = a^2$.

31. (d) The equation of any tangent to the ellipse $\frac{x^2}{a^2} + \frac{y^2}{b^2} = 1$ is

$$y = mx \pm \sqrt{a^2m^2 + b^2} \quad \ldots(\text{i})$$

Let $P(h, k)$ be the point of intersection of tangents.

If Eq. (i) passes throguh $P(h, k)$, then

$$k = mh \pm \sqrt{a^2m^2 + b^2}$$

$$\Rightarrow \quad (k - mh)^2 = a^2m^2 + b^2$$

$$\Rightarrow \quad m^2(h^2 - a^2) - 2mhk + k^2 - b^2 = 0$$

This given two values of m, say m_1 and m_2. These values represent the slopes of the tangents passing through P. If the tangents drawn from P make complementary angles with X-axis, then

$$m_1m_2 = 1$$

$$\Rightarrow \quad \frac{k^2 - b^2}{h^2 - a^2} = 1$$

$$\Rightarrow \quad h^2 - k^2 = a^2 - b^2$$

Hence, the locus (h, k) is $x^2 - y^2 = a^2 - b^2$.

32. (c) Let (h, k) be the mid-point of the chord $2x + 3y + \lambda = 0$. Then, its equation is

$$hx + 4ky = h^2 + 4k^2 \quad \ldots(\text{i})$$

Since, $2x + 3y + \lambda = 0$ and Eq. (i) represent the same line.

$$\therefore \quad \frac{h}{2} = \frac{4k}{3} = \frac{h^2 + 4k^2}{-\lambda} \Rightarrow 3h = 8k$$

Hence, the locus of (h, k) is $3x - 8y = 0$.

33. (c) Equation of any tangent to $\frac{x^2}{a^2} + \frac{y^2}{b^2} = 1$ is

$$y = mx + \sqrt{a^2m^2 + b^2} \quad \ldots(\text{i})$$

The equation of a tangent to $\frac{x^2}{a^2 + \lambda} + \frac{y^2}{b^2 + \lambda} = 1$ perpendicular to Eq. (i) is

$$y = -\frac{1}{m}x + \sqrt{(a^2 + \lambda)\frac{1}{m^2} + (b^2 + \lambda)}$$

$$\Rightarrow \quad my = -x + \sqrt{(a^2 + \lambda) + (b^2 + \lambda)m^2} \quad \ldots(\text{ii})$$

Let $P(h, k)$ be the point of intersection of Eqs. (i) and (ii). Then,

$(k - mh)^2 = a^2m^2 + b^2$ and $(mk + h)^2 = (a^2 + \lambda) + (b^2 + \lambda)m^2$

$$\Rightarrow \quad (k - mh)^2 + (mk + h)^2 = (a^2 + b^2 + \lambda)(m^2 + 1)$$

$$\Rightarrow \quad h^2 + k^2 = a^2 + b^2 + \lambda.$$

Hence, the locus of (h, k) is

$$x^2 + y^2 = a^2 + b^2 + \lambda$$

34. (c) We know that the locus the foot of the perpendiculars drawn from foci on any tangent to the ellipse is its auxiliary circle. Therefore, $(1, -4)$ lies on the auxiliary circle of the ellipse.

The foci are $S(3, 4)$ and $S'(9, 12)$. Therefore, coordinates of the centre of the ellipse are $(6, 8)$.

Also, distance between two foci = 10

$$\Rightarrow \quad 2ae = 10 \Rightarrow ae = 5$$

The equation of the auxiliary circle is $(x - 6)^2 + (y - 8)^2 = a^2$

It passes through $(1, -4)$.

$$\therefore \quad 25 + 144 = a^2 \Rightarrow a = 13$$

$$\therefore \quad ae = 5 \Rightarrow e = \frac{5}{13}$$

35. (a) We have, $x^2 + 2y^2 = 2$ or $\frac{x^2}{(\sqrt{2})^2} + \frac{y^2}{1^2} = 1$

The equation of any tangent to this ellipse in parametric form at point $(\sqrt{2}\cos\theta, \sin\theta)$ is $\frac{x}{\sqrt{2}}\cos\theta + y\sin\theta = 1$.

This cuts the coordinate axes at points $A\left(\frac{\sqrt{2}}{\cos\theta}, 0\right)$ and $B\left(0, \frac{1}{\sin\theta}\right)$.

Let $P(h, k)$ be the mid-point of intercept AB. Then,

$$h = \frac{\sqrt{2}\cos\theta}{2} \text{ and } k = \frac{1}{2\sin\theta}$$

$$\Rightarrow \quad \cos\theta = \frac{1}{\sqrt{2}h} \text{ and } \sin\theta = \frac{1}{2k}$$

$$\therefore \quad \cos^2\theta + \sin^2\theta = 1 \Rightarrow \frac{1}{2h^2} + \frac{1}{4k^2} = 1$$

Hence, the locus of (h, k) is $\frac{1}{2x^2} + \frac{1}{4y^2} = 1$.

36. (c) The coordinates of the extremity of the latusrectum which lies in the first quadrant are $(ae, b^2/a)$.

The equation of the normal at (x_1, y_1) is

$$\frac{a^2x}{x_1} - \frac{b^2y}{y_1} = (a^2 - b^2).$$

Therefore, the equation of the normal at $(ae, b^2/a)$ is

$$\frac{a^2x}{ae} - \frac{b^2y}{b^2/a} = a^2 - b^2$$

$$\Rightarrow \quad ax - aey = e(a^2 - b^2)$$

$$\Rightarrow \quad ax - aey = ea^2e^2 \Rightarrow x - ey = ae^3$$

This passes through the extremity of the minor axis, i.e. $(0, -b)$.

$$\therefore \quad 0 + eb - ae^3 = 0 \Rightarrow b = ae^2$$

$$\Rightarrow \quad b^2 = a^2e^4$$

$$\Rightarrow \quad a^2(1 - e^2) = a^2e^4$$

$$\Rightarrow \quad e^4 + e^2 - 1 = 0$$

37. (c) Let $P(a\cos\theta, b\sin\theta)$ be a point on the ellipse $\frac{x^2}{a^2} + \frac{y^2}{b^2} = 1$.

The equation of the normal at P is $ax\sec\theta - by\operatorname{cosec}\theta = a^2 - b^2$. It meets the axes at $G\left(\frac{a^2 - b^2}{a}\cos\theta, 0\right)$ and $g\left(0, -\frac{a^2 - b^2}{a}\sin\theta\right)$

$$\therefore \quad PG^2 = \left(a\cos\theta - \frac{a^2 - b^2}{a}\cos\theta\right)^2 + b^2\sin^2\theta$$

$$= \frac{b^2}{a^2}(b^2\cos^2\theta + a^2\sin^2\theta)$$

and, $$Pg^2 = \frac{a^2}{b^2}(b^2\cos^2\theta + a^2\sin^2\theta)$$

$$\Rightarrow \quad PG : Pg = b^2 : a^2$$

38. (a) The equation of the tangent at any point ϕ the ellipse is $\frac{x}{a}\cos\phi + \frac{y}{b}\sin\phi = 1$.

It meets the coordinate axes at the points $A(a\sec\phi, 0)$ and $B(0, b\operatorname{cosec}\phi)$.

Let (h, k) be the mid-point of AB. Then,

$$h = \frac{a\sec\phi}{2}, k = \frac{b\operatorname{cosec}\phi}{2}$$

$$\Rightarrow \quad \cos\phi = \frac{a}{2h} \text{ and } \sin\phi = \frac{b}{2k}$$

$$\Rightarrow \quad \frac{a^2}{4h^2} + \frac{b^2}{4k^2} = 1$$

Hence, the locus of (h, k) is $\frac{a^2}{x^2} + \frac{b^2}{y^2} = 4$

$$\Rightarrow \quad a^2y^2 + b^2x^2 = 4x^2y^2$$

39. (a) Let (h, k) be the mid-point of a focal chord of the ellipse $\frac{x^2}{a^2} + \frac{y^2}{b^2} = 1$. Then, its equation is

$$\frac{hx}{a^2} + \frac{ky}{b^2} = \frac{h^2}{a^2} + \frac{k^2}{b^2}$$

It passes through the focus $S(ae, 0)$.

$$\therefore \quad \frac{he}{a} = \frac{h^2}{a^2} + \frac{k^2}{b^2}$$

Hence, the locus of (h, k) is $\frac{x^2}{a^2} + \frac{y^2}{b^2} = \frac{xe}{a}$.

40. (b) If $y = m_1x$ and $y = m_2x$ are conjugate diameters with respect to the ellipse $\frac{x^2}{a^2} + \frac{y^2}{b^2} = 1$. Then, $y = m_2x$ bisects the system of chords parallel to $y = m_1x$. So, its equation is

$$y = \frac{-b^2}{a^2m_1}x \quad \ldots\text{(i)}$$

Clearly, Eq. (i) and $y = m_2x$ represent the same line.

$$\therefore \quad m_2 = -\frac{b^2}{a^2m_1} \Rightarrow m_1m_2 = -\frac{b^2}{a^2}$$

41. (d) $\frac{x^2}{16} + \frac{y^2}{9} = 1$

Directrix $\quad x = \pm\frac{a}{e}$

$$\Rightarrow \quad x = \pm\frac{16}{\sqrt{7}}$$

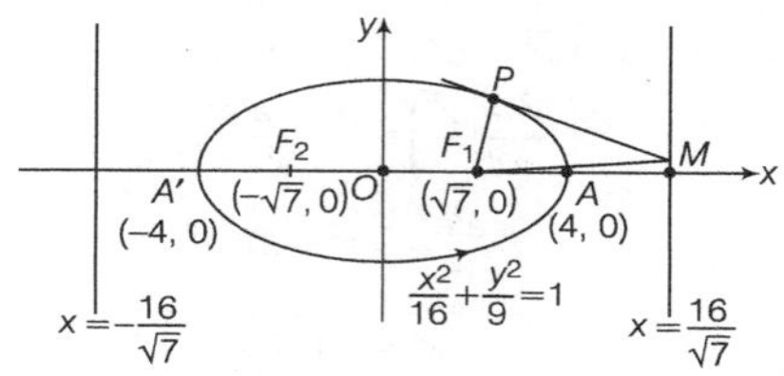

Focus

$(\pm ae, 0) = (\pm\sqrt{7}, 0)$

Since, angle PM subtends right angle at $F_1(\sqrt{7}, 0)$

i.e. $\angle PF_1M = 90°$

$\Rightarrow$ Statement I is incorrect and Statement II is correct.

42. (c) Equation of ellipse $\frac{x^2}{4} + \frac{y^2}{16} = 1$. Condition for tangency $c^2 = a^2m^2 + b^2$

$\Rightarrow$ $2\sqrt{3}x + y = 8$ is tangent. Foot of perpendicular of focii on any tangent lies on auxiliary circle.

$\Rightarrow$ Statement I is correct and statement II is wrong.

43. (a) $x^2 + y^2 + xy = 1$

Replacing x by $-x$ and y by $-y$ we get the same equation.

$\therefore$ Centre of conic is $(0, 0)$ and every chord passing through the centre is bisected by the point. Hence Statement I is correct and Statement II is correct explanation of statement I.

44. (a) Since, the circle drawn with focal radius as diameter touches auxilliary circle.

$\therefore P\left(\frac{3\sqrt{3}}{2}, 1\right)$ and $A(-\sqrt{5}, 0)$ represents focal radius.

$\Rightarrow$ Circle drawn AP as diameter touches another circle $x^2 + y^2 = 9$.

45. (a) As, $P(-\sqrt{3}, \sqrt{2})$ lies on directors circle $x^2 + y^2 = 5$

$\therefore$ tangents are mutually perpendicular.

Thus, Statement I and Statement II both are correct and Statement II is correct explanation of statement I.

Revisal Problems (JEE Advanced)

1. (b) Here, P lies on $x^2 + y^2 = 5$

Let, $P(\sqrt{5}\cos\theta, \sqrt{5}\sin\theta)$

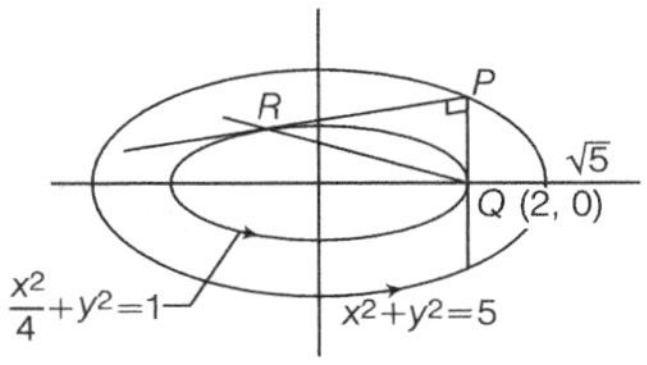

Comparing chord of contact with middle point

$$\frac{xh}{h^2+4k^2} + \frac{yk}{h^2+4k^2} = 1$$

and $$\frac{x\sqrt{5}\cos\theta}{4} + \frac{y\sqrt{5}\sin\theta}{1} = 1$$

$\Rightarrow$ $$\frac{\frac{\sqrt{5}\cos\theta}{4}}{\frac{h}{h^2+4k^2}} = \frac{\frac{\sqrt{5}\sin\theta}{1}}{\frac{k}{h^2+4k^2}} = \frac{1}{1}$$

eliminating θ, we get

$$x^2 + y^2 = \frac{5}{16}(x^2+4y^2)^2$$

2. (d) Any normal to the ellipse is

$$\frac{\sqrt{5}x}{\cos\theta} - \frac{2y}{\sin\theta} = 1$$

i.e. $$y = x\frac{\sqrt{5}}{2}\tan\theta - \frac{\sin\theta}{2}$$

It should be of the form $y = mx + \frac{1}{m}$

$\Rightarrow$ $m = \frac{\sqrt{5}}{2}\tan\theta,\ \frac{1}{m} = -\frac{\sin\theta}{2}$

$\Rightarrow$ $\frac{\sqrt{5}}{4}\cdot\frac{\sin^2\theta}{\cos\theta} = -1$

$\therefore$ $\cos\theta = -\frac{1}{\sqrt{5}}$

$\therefore$ $\theta = \pi - \cos^{-1}\left(\frac{1}{\sqrt{5}}\right)$ or $\pi + \cos^{-1}\left(\frac{1}{\sqrt{5}}\right)$

3. (d) Clearly, origin will be a point lying on the director circle of the ellipse.

Thus, equation of the locus of centre of ellipse will be

$$x^2 + y^2 = a^2 + b^2$$

$\Rightarrow$ $$x^2 + y^2 = 100$$

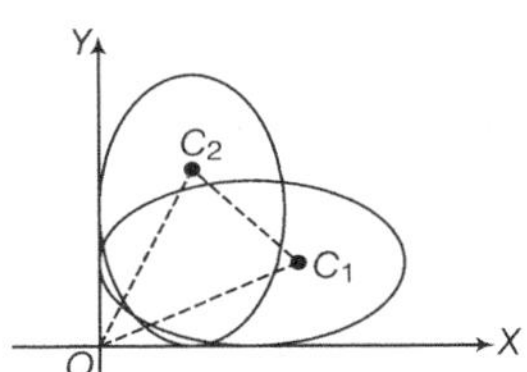

The adjacent figure shows the position of centre of ellipse in two extreme position

$$Q = \angle C_1OC_2$$

$$= \frac{\pi}{2} - 2\tan^{-1}\left(\frac{5}{5\sqrt{3}}\right)$$

$$= \frac{\pi}{6}$$

Thus, arc length $C_1C_2 = 10\cdot\frac{\pi}{6} = \frac{5\pi}{3}$

4. (a) $\frac{ar\cdot\Delta PQR}{ar\cdot\Delta P'Q'R'} = \frac{\sqrt{5}}{3}$

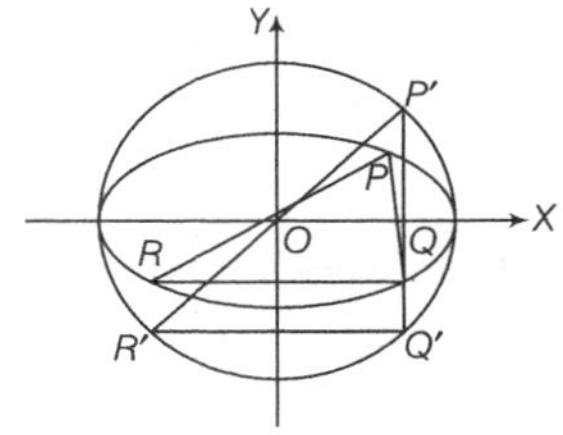

$\Rightarrow$ $\frac{b}{a} = \frac{\sqrt{5}}{3}$

$e^2 = 1 - \frac{b^2}{a^2}$

$\Rightarrow$ $e^2 = 1 - \frac{5}{9}$

$\therefore$ $e = \frac{2}{3}$

5. (a) Let $A(x_1, x_1 - 5)$ be a point on $x - y - 5 = 0$, then chord of contact of $x^2 + 4y^2 = 4$ w.r.t. A is $xx_1 + 4y(x_1 - 5) = 4$

$\Rightarrow$ $(x + 4y)x_1 - (20y + 4) = 0$

It passes through a fixed pt.

$x + 4y = 0$ and $20y + 4 = 0$ ($\because P + \lambda\theta = 0$ form)

$\Rightarrow$ $y = -\frac{1}{5}$ and $x = \frac{4}{5}$

Then coordinates of fixed point is $\left(\frac{4}{5}, -\frac{1}{5}\right)$

6. (c) Here, $a = 3, b = 2$

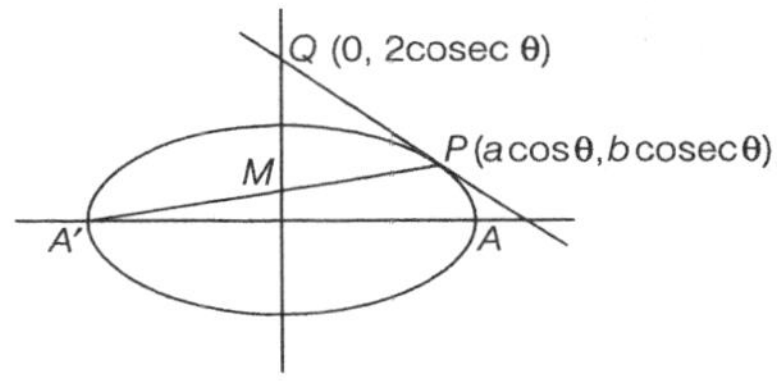

$$T: \frac{x\cos\theta}{3} + \frac{y\sin\theta}{2} = 1$$

$$x = 0;\ y = 2\operatorname{cosec}\theta$$

chord $A'P$

$\Rightarrow$ $$y = \frac{2\sin\theta}{3(\cos\theta+1)}\cdot(x+3)$$

put $$x = 0,\ y = \frac{2\sin\theta}{1+\cos\theta} = OM$$

Now, $OQ^2 - MQ^2 = OQ^2 - (OQ - OM)^2$

$= 2(OQ)(OM) - (OM)^2$

$= OM\{2(OQ) - (OM)\}$

$= \frac{2\sin\theta}{1+\cos\theta}\left\{\frac{4}{\sin\theta} - \frac{2\sin\theta}{1+\cos\theta}\right\} = 4$

7. (a) $a^2e^2 = 36 \Rightarrow a^2 - b^2 = 36$ ...(i)

using $r = (s-a)\tan\frac{A}{2}$ in ΔOCF

$\Rightarrow$ $1 = (s - a)\tan 45°$, when $a = CF$

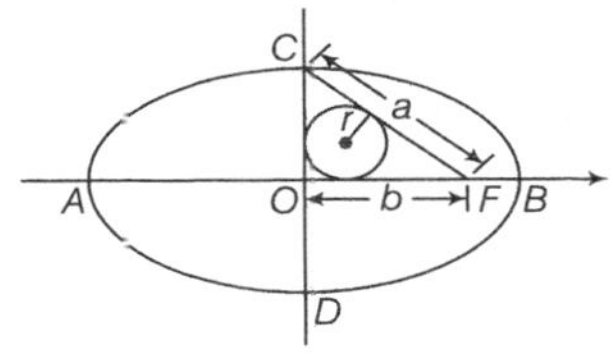

$\Rightarrow$ $2 = 2(s-a)$

$\therefore$ $2 = 2s - 2a$

$\Rightarrow$ $2 = 2s - AB = (OF + OC + CF) - AB$

$\Rightarrow$ $2 = 6 + \frac{CD}{2} + \frac{AB}{2} - AB$

$$\Rightarrow \quad \frac{AB - CD}{2} = 4$$

$$\Rightarrow \quad 2(a - b) = 8$$

$$\Rightarrow \quad a - b = 4 \quad \ldots (ii)$$

from Eqs. (i) and (ii), $\quad a + b = 9 \quad \ldots (iii)$

$$\therefore \quad 2a = 13, 2b = 5$$

$$\Rightarrow \quad (AB)(CD) = 65$$

8. (d) Since, Y-axis is major axis

$$\Rightarrow \quad f(4a) < f(a^2 - 5)$$

$$\Rightarrow \quad 4a > a^2 - 5 \quad (\because f \text{ is decreasing})$$

$$\Rightarrow \quad a^2 - 4a - 5 < 0 \text{ or } (a - 5)(a + 1) < 0$$

$$\Rightarrow \quad a \in (-1, 5)$$

9. (c) Let the equation of circle, $(x - 1)^2 + y^2 = r^2$ and ellipse $\frac{x^2}{16} + \frac{y^2}{4} = 1$ touch

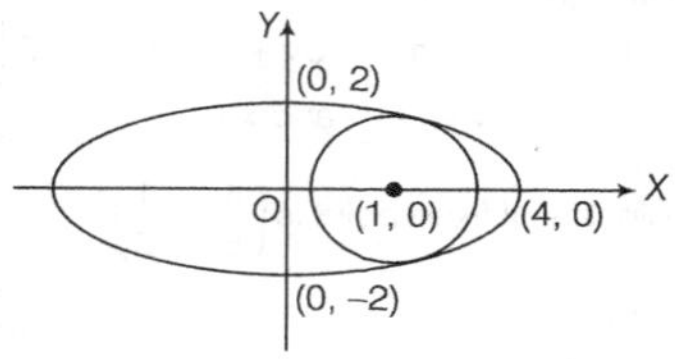

$\therefore$ eliminating y^2, we get

$$\Rightarrow \quad 3x^2 - 8x + 20 - 4r^2 = 0$$

Circle lies inside ellipse

$$\Rightarrow \quad \Delta \le 0 \Rightarrow r \le \sqrt{\frac{11}{3}}$$

$\Rightarrow$ Equation of circle $3(x^2 + y^2) - 6x - 8 = 0$

10. (a) $R = \frac{abc}{4\Delta} = \frac{a(1 - e\cos\theta)\cdot a(1 + e\cos\theta)\cdot 2ae}{4 \times \frac{1}{2} \times 2ae \times b\sin\theta}$

$$= \frac{a^2(1 - e^2\cos^2\theta)}{2b\sin\theta} = \frac{a^2}{2b}\left(\frac{b^2}{a^2}\operatorname{cosec}\theta + e^2\sin\theta\right)$$

$$\Rightarrow \quad \frac{a^2}{2b} \times \frac{2be}{a} = ae$$

11. (c) If PSQ is focal chord then $\frac{1}{SP} + \frac{1}{SQ} = \frac{2}{l}$, l = semi latus rectum.

The product of perpendiculars from foci to any tangent is b^2.
The points A, B lie on auxilliary circle,

$$\therefore \quad \frac{SP + SQ}{SP \cdot SQ} = \frac{2a}{b^2} = \frac{8}{9}$$

and $\quad SA \cdot S'B = 9, \; OA \cdot OB = a^2 = 16$

$$\Rightarrow \quad \frac{SP + SQ}{SP \cdot SQ} + \frac{OA \cdot OB}{SA \cdot S'B} = \frac{8}{9} + \frac{16}{9} = \frac{24}{9}$$

12. (b) $\frac{ax}{\cos\theta} - \frac{by}{\sin\theta} = (a^2 - b^2)$

$$\frac{ah}{\cos\theta} = a^2 - b^2 \quad \text{[as normals pass through } (h, 0)]$$

$$\Rightarrow \quad \cos\theta = \frac{ah}{a^2 - h^2}$$

$$\Rightarrow \quad \left|\frac{ah}{a^2 - b^2}\right| \le 1 \text{ or } |h| \le he^2$$

13. (b) Circle C is $x^2 + y^2 + \lambda y - 16 = 0$, it intersects the circle $x^2 + y^2 + \lambda y - 16x - 10y + 9 = 0$, orthogonally

Then, $\lambda = \frac{7}{5}$

Now, $\quad |r_1 - r_2| < C_1C_2 < r_1 + r_2$

$\therefore$ Number of common tangents is 2.

14. (c) Using property $S_1P_i \times S_2Q_i = b^2$

$$\Rightarrow \quad 20b^2 = 5120$$

$$\Rightarrow \quad b = 16 \text{ and } a = 20$$

$$\therefore \quad e = \frac{3}{5}.$$

15. (d) Sides of the square will be perpendicular tangents to the ellipse so vertices of square will lie on director circle. So diameter of director circle

$$\Rightarrow \quad 2\sqrt{(a^2 - 3) + (a + 4)} = \sqrt{2a^2 + 2a^2}$$

$$\Rightarrow \quad 2\sqrt{a^2 + a + 1} = 2a$$

$\Rightarrow \; a = -1$, but $a^2 > 3$ and $a > -4$.

So 'a' cannot take value -1.

16. (b,c) Any tangent at $P(a\cos\theta, b\sin\theta)$ is

$$\frac{x}{a}\cos\theta + \frac{y}{b}\sin\theta = 1 \quad \ldots (i)$$

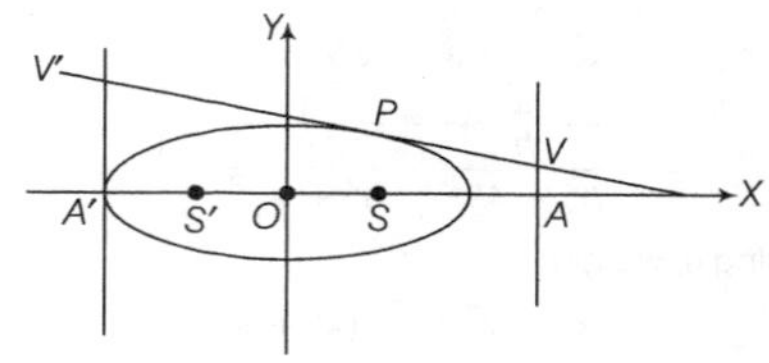

line AV is $x = a$ and $A'V'$ is $x = -a$

So points $\quad V = \left(a, \frac{b(1 - \cos\theta)}{\sin\theta}\right)$

and $\quad V' = \left(-a, \frac{b(1 + \cos\theta)}{\sin\theta}\right)$

So, $\quad l(AV)\cdot l(A'V') = \frac{b(1 - \cos\theta)}{\sin\theta} \cdot \frac{b(1 + \cos\theta)}{\sin\theta} = b^2$

Slope of line $V'S$ is $\quad = \frac{b(1 + \cos\theta)}{\sin\theta(-a - ae)} = m_1$

also, slope of VS is $\quad = \frac{b(1 - \cos\theta)}{\sin\theta(a - ae)} = m_2$

Now, $\quad m_1m_2 = \frac{b^2(1 - \cos^2\theta)}{\sin^2\theta(-b^2)} = -1$

$\Rightarrow \angle V'SV = 90°$ and $V'S'VS$ is not a cyclic quadrilateral.

17. (a,b) $h = \pm ae$, $k = \pm \frac{b^2}{a}$

$$\therefore \quad k = \pm a(1 - e^2) = \pm a\left(1 - \frac{h^2}{a^2}\right) = \pm\left(a - \frac{h^2}{a}\right)$$

On taking + ve sign, we get

$$k = a - \frac{h^2}{a} \Rightarrow h^2 = a(a - k)$$

$\therefore$ locus of (h, k) is $x^2 = a(a - y) \quad \ldots (i)$

On taking – ve sign, we get

$$k = -a + \frac{h^2}{a}$$

$$\Rightarrow \quad h^2 = a(a + h)$$

$\therefore$ locus of (h, k), is $x^2 = a(a + y) \quad \ldots (ii)$

18. (a,b,c) If PSS' form a Δ, then ellipse

$$PS + PS' > SS'$$

$$\Rightarrow \quad K \in (-2, 2)$$

If PSS' are collinear the pair of coincident lines $\Rightarrow K = \pm 2$

If PSS' neither form triangle nor collinear

Then $\quad PS + PS' < SS'$

$$\Rightarrow \quad k \in (-\infty, -2) \cup (2, \infty)$$

19. (a,b,c) Focal property of ellipse

$$PS + PS' = 2a, a > b$$
$$PS + PS' = 2b, a > b$$
$$PS \cos\alpha + PS' \cos\beta = 2ae \quad \ldots(i)$$
$$PS \sin\alpha + PS' \sin\beta = 0 \quad \ldots(ii)$$
$$PS + PS' = 2a \quad \ldots(iii)$$

from Eqs. (i) and (ii), we get

$$PS = \frac{2ae \cdot \sin\beta}{\sin(\alpha+\beta)}, \; PS' = \frac{2ae \cdot \sin\alpha}{\sin(\alpha+\beta)} \quad \ldots(iv)$$

from Eqs. (iii) and (iv)

$$e(\sin\alpha + \sin\beta) = \sin(\alpha+\beta)$$

$$\therefore \; e\left[2\sin\left(\frac{\alpha+\beta}{2}\right)\cdot\cos\left(\frac{\alpha-\beta}{2}\right)\right] = 2\sin\left(\frac{\alpha+\beta}{2}\right)\cdot\cos\left(\frac{\alpha+\beta}{2}\right)$$

$$\therefore \; e\left(\cos\frac{\alpha}{2}\cdot\cos\frac{\beta}{2} + \sin\frac{\alpha}{2}\cdot\sin\frac{\beta}{2}\right) = \cos\frac{\alpha}{2}\cdot\cos\frac{\beta}{2} - \sin\frac{\alpha}{2}\cdot\sin\frac{\beta}{2}$$

$$\Rightarrow \; \tan\frac{\alpha}{2}\cdot\tan\frac{\beta}{2} = \frac{1-e}{1+e}$$

20. (b,d) Clearly (α, β) on curve will be of the form $\left(\frac{2m^2-1}{2m^2+1}, \frac{-2m}{2m^2+1}\right)$

21. (a)

(i) The product of perpendicular, from foci to any tangent is $b^2 = 9$

(ii) The sum of square of the distances from the points at a distance 'ae' from centre, on minor axis is constant equal to $2a^2 = 32$

(iii) The tangent at any point P on the ellipse meets the tangent at A, A' at L, M then $AL \cdot A'M = b^2 = 9$.

(iv) The circle on TT' as diameter will pass through foci of the ellipse hence the intercept is

$$2ae = 2\sqrt{a^2 - b^2} = 2\sqrt{16-9} = 2\sqrt{7}$$

$\therefore$ Square = 28

22. (a)

(i) Any point on ellipse at a distance 'r' from origin is $(r\cos\theta, r\sin\theta)$

$\Rightarrow r^2 = \frac{8}{5 - 3\sin 2\theta}$ is maximum, if $5 - 3\sin 2\theta$ is minimum

$$\Rightarrow \; r^2 = 4$$

r^2 is minimum, if $(5 - 3\sin 2\theta)$ is maximum = 8

$$\Rightarrow \; r^2 = 1$$

$$\therefore \; r_1 + r_2 = 2 + 1 = 3$$

(ii) $r_1 = \frac{4a\cos\theta}{\sin^2\theta}$ and $16a^2(r_1^{2/3} + r_2^{2/3}) = (r_1r_2)^{4/3}$

$$\Rightarrow \; a^2 = \frac{1}{5}$$

(iii) Equation of tangent to $y^2 = 8x$, is $y = mx + \frac{2}{m}$ is also tangent to $x^2 + y^2 = 8$

$$\Rightarrow \; \frac{\frac{2}{m}}{\sqrt{m^2+1}} = \frac{\sqrt{8}}{\sqrt{2}} \Rightarrow m^4 + m^2 - 2 = 0$$

$$\therefore \; m = \pm 1$$

Equation of common tangents $y = x + 2$, $y = -x - 2$ meet at $A(-2, 0)$ chord of contact of $A(-2, 0)$ is $x = -1$, $x = 2$

$\therefore$ Area of trapeziam = 15

(iv) PQ is focal chord $\Rightarrow t_1t_2 = -1$

Centroid $(x, y) = \left(3\left(t_1^2 + \frac{1}{t_1^2}\right), 6\left(t_1 - \frac{1}{t_1}\right)\right)$,

$$\Rightarrow \; y^2 = 12(x + 6)$$

Explanation for Q. Nos. (23-25)

$$2x + 3y + 9 = 0$$
$$P(-1, 2) \text{ and } Q(-2, 3)$$

Hence, image of point P form line $2x + 3y + 9 = 0$

$$\Rightarrow \; \frac{x+1}{2} = \frac{y-2}{3} = \frac{-2(-2+6+9)}{4+9}$$

$$\Rightarrow \; x = -5, y = -4$$

$$\Rightarrow \; P'(-5, -4)$$

Now length of $P'Q = \sqrt{9 + 49} = \sqrt{58} = 2a$

$$\Rightarrow \; a = \frac{\sqrt{58}}{2}$$

We know that, $2ae = PQ \Rightarrow ae = \frac{1}{\sqrt{2}}$

$$\therefore \; a^2e^2 = \frac{1}{2}$$

$$b^2 = a^2 - a^2e^2 = \frac{58}{4} - \frac{1}{2} = \frac{56}{4}$$

$$\therefore \; b^2 = 14$$

$\Rightarrow$ length of minor axis $= 2\sqrt{14}$

Area of ellipse $= \pi ab = \pi\frac{\sqrt{58}}{2}\sqrt{14} = \pi\sqrt{203}$

Equation of line $P'Q$ is

$$\Rightarrow \; (y - 3) = \frac{4-3}{-3}(x + 2)$$

$$\Rightarrow \; 3y - 9 = 7x + 14$$

or $\quad 7x - 3y + 23 = 0$

On solving $7x - 3y + 23 = 0$ and $2x + 3y + 9 = 0$, we get

$$A\left(-\frac{32}{9}, -\frac{17}{27}\right)$$

23. (c) **24.** (d) **25.** (b)

Explanation for Q. Nos. (26-28)

Here, $b^2 = \left|\frac{2+1-9}{\sqrt{2}}\right|\left|\frac{4+1-9}{\sqrt{2}}\right| = 12$

$$2ae = 2 \Rightarrow ae = 1$$
$$b^2 = a^2 - a^2e^2 \Rightarrow a^2 = 13$$

$$\Rightarrow \; e = \frac{1}{\sqrt{13}}$$

length of major axis $= 2\sqrt{13} = \sqrt{52}$

length of latus-rectum $= \frac{2b^2}{a} = \frac{2 \times 12}{\sqrt{13}} = \frac{24}{\sqrt{13}}$

26. (b) **27.** (c) **28.** (c)

29. (2) Here, $(2\alpha, \alpha)$ has between $x^2 + y^2 = 9 + 3$ and $\frac{x^2}{9} + \frac{y^2}{3} = 1$

$$\Rightarrow \; 5\alpha^2 - 12 < 0 \text{ and } \frac{4\alpha^2}{9} + \frac{\alpha^2}{3} - 1 > 0$$

$$\Rightarrow \; \frac{9}{7} < \alpha^2 < \frac{12}{5}$$

$\therefore$ possible integral value of $\alpha^2 = 2$.

30. (4) The director circle of $\frac{x^2}{2} + \frac{y^2}{1} = 1$, is $x^2 + y^2 = 3$

Let, $P(\sqrt{3}\cos\theta, \sqrt{3}\sin\theta)$ chord of contact is

$$x(\sqrt{3}\cos\theta) + y(2\sqrt{3}\sin\theta) = 2$$

It touches $x^2 + y^2 = r^2$

$$\Rightarrow \; r = \frac{2}{\sqrt{3\cos^2\theta + 12\sin^2\theta}} = \frac{2}{\sqrt{3 + 9\sin^2\theta}}$$

$$r_{max} = \frac{2}{\sqrt{3}}, \; r_{min} = \frac{2}{\sqrt{12}}$$

So, $\quad \frac{A_{max}}{A_{min}} = 4$

JEE Main & AIEEE Archive

1. (d) Given equation of ellipse is $\frac{x^2}{9}+\frac{y^2}{5}=1$ ...(i)

$\therefore \quad a^2=9, b^2=5$

$\Rightarrow \quad a=3, b=\sqrt{5}$

Now, $\quad e=\sqrt{1-\frac{b^2}{a^2}}=\sqrt{1-\frac{5}{9}}=\frac{2}{3}$

foci $=(\pm ae, 0)=(\pm 2, 0)$

and $\quad \frac{b^2}{a}=\frac{5}{3}$

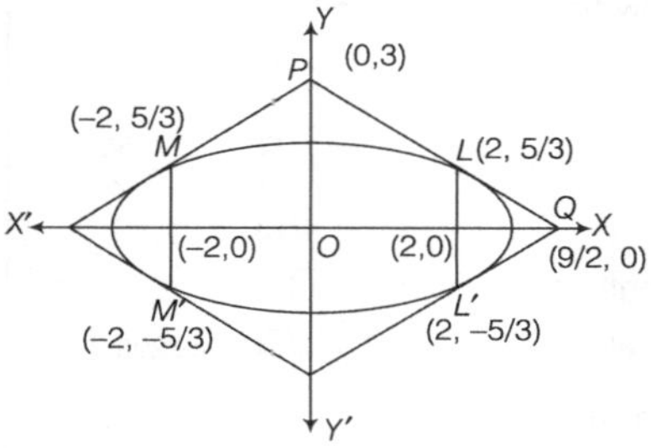

$\therefore$ Extremities of one of latusrectum are

$$\left(2, \frac{5}{3}\right) \text{ and } \left(2, \frac{-5}{3}\right).$$

$\therefore$ Equation of tangent at $\left(2, \frac{5}{3}\right)$ is,

$$\frac{x(2)}{9}+\frac{y(5/3)}{5}=1 \text{ or } 2x+3y=9 \quad \text{...(ii)}$$

Eq.(ii) intersects X and Y-axes at $\left(\frac{9}{2}, 0\right)$ and (0, 3), respectively.

$\therefore$ Area of quadrilateral = 4 × Area of ΔPOQ

$$=4\times\left(\frac{1}{2}\times\frac{9}{2}\times 3\right)=27 \text{ sq units}$$

2. (c) Equation of ellipse is $x^2+3y^2=6$.

Equation of the tangent is $\frac{x\cos\theta}{a}+\frac{y\sin\theta}{b}=1$.

Let (h, k) be any point on the locus, then

$$\frac{h}{a}\cos\theta+\frac{k}{b}\sin\theta=1 \quad \text{...(i)}$$

Slope of the tangent line is $\frac{-b}{a}\cot\theta$.

Slope of perpendicular drawn from centre (0, 0) to (h, k) is $\frac{k}{h}$.

Since, both the lines are perpendicular.

$\therefore \quad \left(\frac{k}{h}\right)\times\left(-\frac{b}{a}\cot\theta\right)=-1$

$\Rightarrow \quad \frac{\cos\theta}{ha}=\frac{\sin\theta}{kb}=\alpha$ [say]

$\Rightarrow \quad \cos\theta=\alpha ha, \sin\theta=\alpha kb$

From Eq. (i), $\quad \frac{h}{a}(\alpha ha)+\frac{k}{b}(\alpha kb)=1$

$\Rightarrow \quad h^2\alpha+k^2\alpha=1 \Rightarrow \alpha=\frac{1}{h^2+k^2}$

Also, $\quad \sin^2\theta+\cos^2\theta=1$

$\Rightarrow \quad (\alpha kb)^2+(\alpha ha)^2=1$

$\Rightarrow \quad \alpha^2k^2b^2+\alpha^2h^2a^2=1$

$\Rightarrow \quad \frac{k^2b^2}{(h^2+k^2)^2}+\frac{h^2a^2}{(h^2+k^2)^2}=1$

$\Rightarrow \quad \frac{2k^2}{(h^2+k^2)^2}+\frac{6h^2}{(h^2+k^2)^2}=1 \quad [\because a^2=6 \text{ and } b^2=2]$

$\Rightarrow \quad 6x^2+2y^2=(x^2+y^2)^2$ [replacing k by y and h by x]

3. (a) Given equation of ellipse is

$$\frac{x^2}{16}+\frac{y^2}{9}=1$$

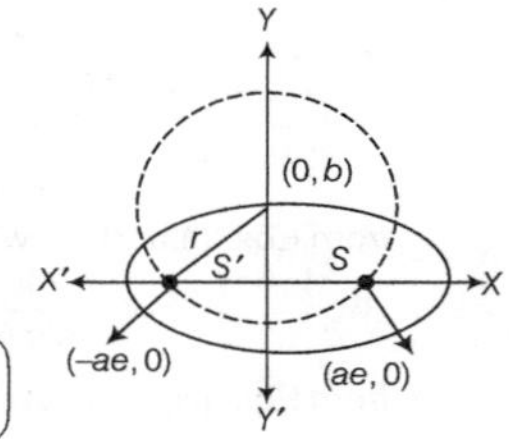

Here, $a=4, b=3$,

$$e=\sqrt{1-\frac{9}{16}} \Rightarrow \frac{\sqrt{7}}{4}$$

$\therefore$ Foci is $(\pm ae, 0)=\left(\pm 4\times\frac{\sqrt{7}}{4}, 0\right)$

$$=(\pm\sqrt{7}, 0)$$

and radius of the circle,

$$r=\sqrt{(ae)^2+b^2}$$
$$=\sqrt{7+9}$$
$$=\sqrt{16}=4$$

Then equation of circle is

$$(x-0)^2+(y-3)^2=16$$

$\therefore \quad x^2+y^2-6y-7=0$

4. (d) **Given**

(i) An ellipse whose semi-minor axis coincides with one of the diameters of the circle $(x-1)^2+y^2=1$.

(ii) The semi-major axis of the ellipse coincides with one of the diameters of circle $x^2+(y-2)^2=4$.

(iii) The centre of the ellipse is at origin.

(iv) The axes of the ellipse are coordinate axes.

To find The equation of the ellipse.

Diameter of circle $(x-1)^2+y^2=1$ is 2 units and that of circle $x^2+(y-2)^2=4$ is 4 units.

$\Rightarrow$ Semi-minor axis of ellipse, $b=2$ units and semi-major axis of ellipse, $a=4$ units.

Hence, the equation of ellipse is

$$\frac{x^2}{a^2}+\frac{y^2}{b^2}=1$$

$\Rightarrow \quad \frac{x^2}{16}+\frac{y^2}{4}=1$

$\therefore \quad x^2+4y^2=16$

5. (d) Given, $\frac{x^2}{a^2}+\frac{y^2}{b^2}=1$

Passes through $P(-3, 1)$ and $e=\sqrt{\frac{2}{5}}$

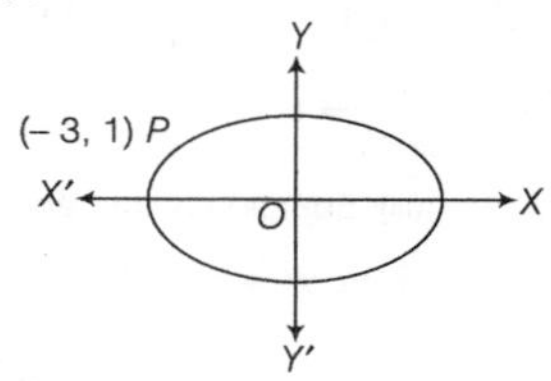

$\therefore \quad \frac{9}{a^2}+\frac{1}{b^2}=1$

and $\quad e^2=1-\frac{b^2}{a^2}$

$\Rightarrow \quad \frac{9}{a^2}+\frac{5}{3a^2}=1$

and $\quad \frac{2}{5}=1-\frac{b^2}{a^2}$

$\Rightarrow \quad \frac{27+5}{3a^2}=1$ and $\frac{b^2}{a^2}=\frac{3}{5}$

$\Rightarrow \quad a^2=\frac{32}{3}$

and $\quad b^2=\frac{32}{5}$

$\therefore$ Equation on ellipse

$$\frac{3x^2}{32}+\frac{5y^2}{32}=1$$

$\Rightarrow \quad 3x^2+5y^2=32$

6. (a) The shortest distance between $y - x = 1$ and $x = y^2$ along the common normal.

$\therefore$ Tangent at P is parallel to $y = x + 1$ ...(i)

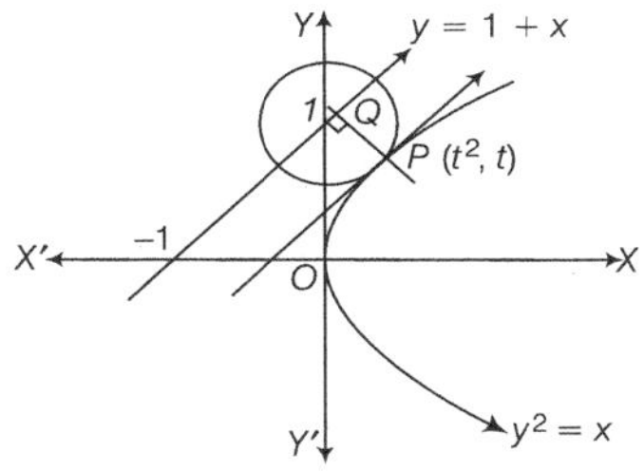

$\therefore$ Slope of tangent at $P(t^2, t)$.

$$\Rightarrow \quad \frac{dy}{dx} = \left(\frac{1}{2y}\right)_{(t^2, t)} = \frac{1}{2t} \quad \text{...(ii)}$$

$$\Rightarrow \quad \frac{1}{2t} = 1 \quad \text{[Eqs. (i) and (ii) are parallel]}$$

$$\Rightarrow \quad t = \frac{1}{2}$$

$$\therefore \quad P\left(\frac{1}{4}, \frac{1}{2}\right)$$

$$\text{Shortest distance} = |PQ| = \frac{\left|\frac{1}{4} - \frac{1}{2} + 1\right|}{\sqrt{1+1}} = \frac{3}{4\sqrt{2}}$$

Hence, shortest distance is $\frac{3\sqrt{2}}{8}$.

7. (a) Let the equation of required ellipse be $\frac{x^2}{16} + \frac{y^2}{b^2} = 1$.

But the ellipse passes through the point (2, 1).

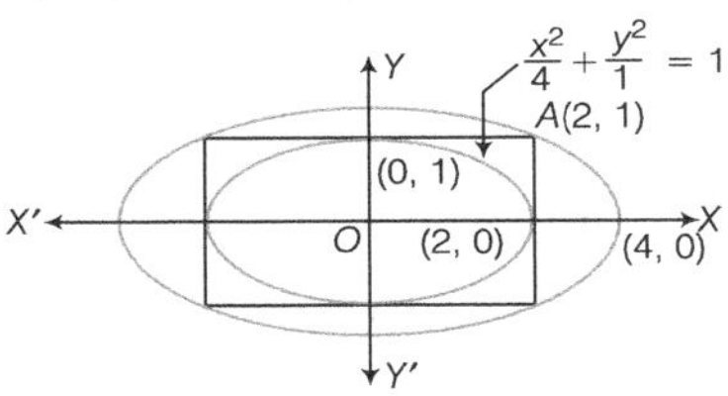

$$\Rightarrow \quad \frac{1}{4} + \frac{1}{b^2} = 1 \Rightarrow \frac{1}{b^2} = \frac{3}{4}$$

$$\therefore \quad b^2 = \frac{4}{3}$$

Hence, equation is

$$\frac{x^2}{16} + \frac{3y^2}{4} = 1$$

$$\Rightarrow \quad x^2 + 12y^2 = 16$$

8. (b) $\because \quad \frac{a}{e} - ae = 4$ and $e = \frac{1}{2}$

$$\therefore \quad 2a - \frac{a}{2} = 4$$

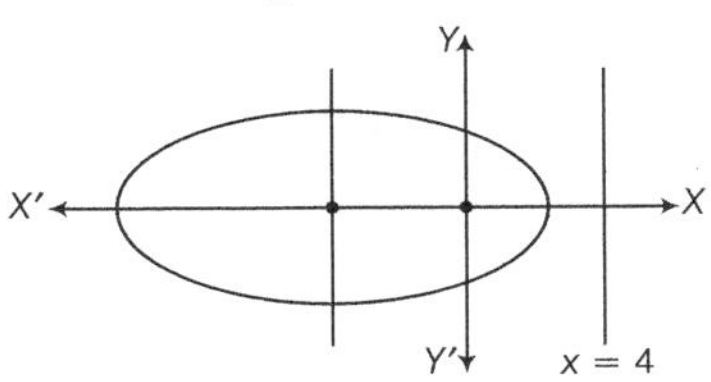

$$\Rightarrow \quad \frac{3a}{2} = 4 \Rightarrow a = \frac{8}{3}$$

9. (d) Given that, $\quad 2ae = 6$ and $2b = 8$

$$\Rightarrow \quad ae = 3 \text{ and } b = 4$$

$$\Rightarrow \quad \frac{ae}{b} = \frac{3}{4} \Rightarrow \frac{b^2}{a^2} = \frac{16e^2}{9}$$

We know that, $\quad \frac{b^2}{a^2} = 1 - e^2 \Rightarrow \frac{16e^2}{9} = 1 - e^2$

$$\Rightarrow \quad \left(\frac{16+9}{9}\right)e^2 = 1 \quad \Rightarrow \quad e^2 = \frac{9}{25}$$

$$\therefore \quad e = \frac{3}{5}$$

10. (d) Let the coordinates of the vertices of rectangle $ABCD$ be

$A(a\cos\theta, b\sin\theta)$, $B(-a\cos\theta, b\sin\theta)$,

$C(-a\cos\theta, -b\sin\theta)$ and $D(a\cos\theta, -b\sin\theta)$

Then, length of rectangle, $AB = 2a\cos\theta$

and breadth of rectangle, $AD = 2b\sin\theta$

$\therefore$ Area of rectangle, $A = AB \times AD$

$$= 2a\cos\theta \times 2b\sin\theta$$

$$\Rightarrow \quad A = 2ab\sin 2\theta \quad \text{...(i)}$$

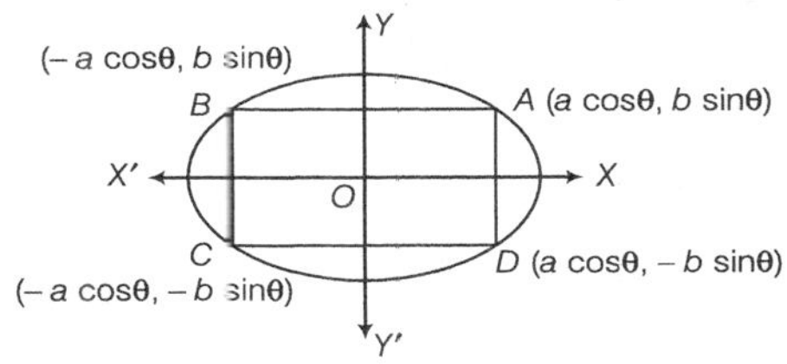

On differentiating Eq. (i) w.r.t. θ, we get

$$\frac{dA}{d\theta} = 2 \times 2ab\cos 2\theta$$

For maxima or minima, put $\frac{dA}{d\theta} = 0$

$$\Rightarrow \quad 4ab\cos 2\theta = 0 \quad \Rightarrow 2\theta = \frac{\pi}{2}$$

$$\therefore \quad \theta = \frac{\pi}{4}$$

Now, $\quad \frac{d^2A}{d\theta^2} = -8ab\sin 2\theta$

At $\theta = \frac{\pi}{4}$, $\quad \left(\frac{d^2A}{d\theta^2}\right) < 0$

$\therefore$ Area is maximum at $\theta = \frac{\pi}{4}$.

So, maximum area of rectangle $= 2ab$ sq units [from Eq. (i)]

Alternate Solution

From Eq. (i), Area of rectangle, $A = 2ab\sin 2\theta$

$\therefore$ A is maximum when $\sin 2\theta = 1$

$\Rightarrow$ Maximum area of rectangle $= 2ab$ sq units

11. (d) Since, $\angle FBF' = 90°$, then $\angle OBF' = 45°$ and $\angle BF'O = 45°$

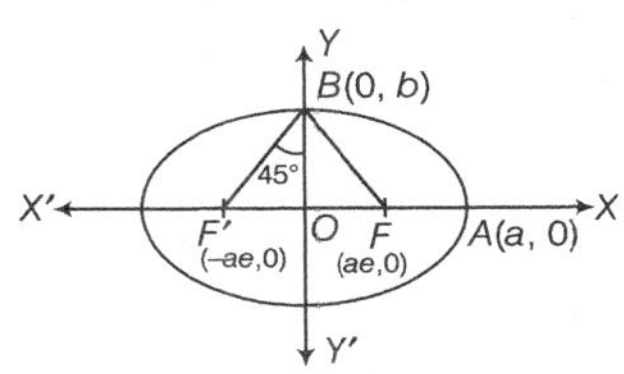

$$\Rightarrow \quad ae = b \quad [\because \Delta BOF' \text{ is an isosceles triangle}]$$

and $\quad e^2 = 1 - \frac{b^2}{a^2} \quad \Rightarrow e^2 = 1 - \frac{a^2e^2}{a^2}$

$$\Rightarrow \quad e^2 = 1 - e^2 \quad \Rightarrow 2e^2 = 1$$

$$\therefore \quad e = \frac{1}{\sqrt{2}} \quad [\because e \text{ cannot be negative}]$$

Alternate Solution

Since, F and F' are foci of an ellipse, whose coordinates are $(ae, 0)$ and $(-ae, 0)$ respectively and coordinates of B are $(0, b)$.

$\therefore$ Slope of $BF = \dfrac{b}{-ae}$

and slope of $BF' = \dfrac{b}{ae}$

$\because \angle FBF' = 90°$

$\therefore -\dfrac{b}{ae} \cdot \left(\dfrac{b}{ae}\right) = -1$

$\Rightarrow b^2 = a^2e^2$

$\therefore e^2 = 1 - \dfrac{b^2}{a^2} = 1 - \dfrac{a^2e^2}{a^2}$

$\Rightarrow 2e^2 = 1$

$\therefore e = \dfrac{1}{\sqrt{2}}$

12. (b) Since, equation of directrix is $x = 4$, then major axis of an ellipse is along X-axis.

$\therefore \dfrac{a}{e} = 4$

$\Rightarrow a = 4 \times \dfrac{1}{2}$ $\left[\because e = \dfrac{1}{2}\right]$

$\Rightarrow a = 2$

Now, $b^2 = a^2(1 - e^2)$

$\therefore b^2 = 4\left(1 - \dfrac{1}{4}\right) = 4 \times \dfrac{3}{4}$

$\Rightarrow b^2 = 3$

Hence, equation of ellipse is

$$\frac{x^2}{4} + \frac{y^2}{3} = 1$$

or $3x^2 + 4y^2 = 12$

13. (a) The equation of an ellipse is

$$\frac{x^2}{16} + \frac{y^2}{9} = 1$$

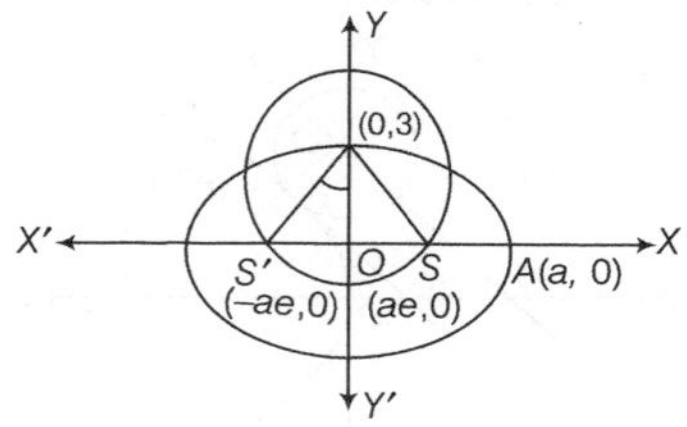

Here, $a = 4, \quad b = 3$

Eccentricity, $e = \sqrt{1 - \dfrac{b^2}{a^2}}$

$= \sqrt{1 - \dfrac{9}{16}} = \dfrac{\sqrt{7}}{4}$

Foci of an ellipse are $(\pm\, ae, 0)$ i.e. $(\pm\, \sqrt{7}, 0)$.

$\therefore$ Radius of required circle

$= \sqrt{(\sqrt{7} - 0)^2 + (0 - 3)^2}$

$= \sqrt{7 + 9} = \sqrt{16}$

$= 4$ units

14. (b) Given that, $e = \dfrac{1}{2}$, $ae = 2$

$\Rightarrow a = 4$

$\therefore b^2 = a^2(1 - e^2)$

$= 16\left(1 - \dfrac{1}{4}\right) = 12$

Hence, equation of an ellipse is

$$\frac{x^2}{16} + \frac{y^2}{12} = 1.$$

JEE Advanced & IIT JEE Archive

1. (4)

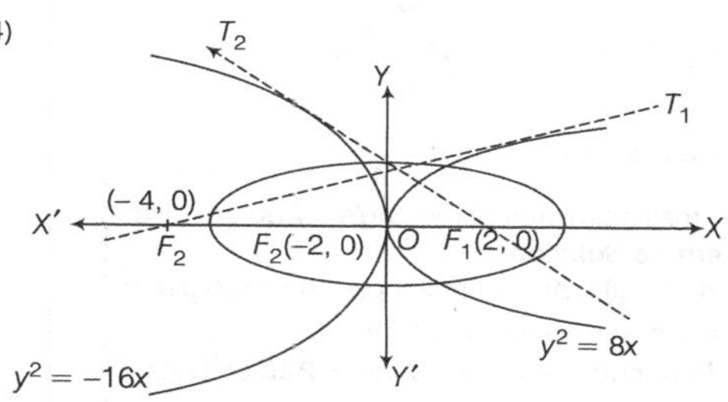

Tangent to P_1 passes through $(2f_2, 0)$ i.e. $(-4, 0)$.

$\therefore T_1 : y = m_1 x + \dfrac{2}{m_1}$

$\Rightarrow 0 = -4m_1 + \dfrac{2}{m_1}$

$\Rightarrow m_1^2 = 1/2$ ...(i)

Also, tangent to P_2 passes through $(f_1, 0)$ i.e. $(2, 0)$.

$\Rightarrow T_2 : y = m_2 x + \dfrac{(-4)}{m_2}$

$\Rightarrow 0 = 2m_2 - \dfrac{4}{m_2}$

$\Rightarrow m_2^2 = 2$ ...(ii)

$\therefore \dfrac{1}{m_1^2} + m_2^2 = 2 + 2 = 4$

2. (a,b) Here, $E_1 : \dfrac{x^2}{a^2} + \dfrac{y^2}{b^2} = 1, \ (a > b)$

$E_2 : \dfrac{x^2}{c^2} + \dfrac{y^2}{d^2} = 1, \ (c < d)$ and $S : x^2 + (y-1)^2 = 2$

as tangent to E_1, E_2 and S is $x + y = 3$.

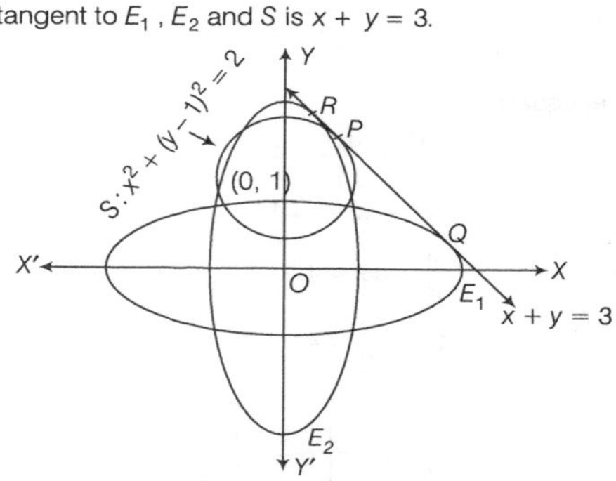

Let the point of contact of tangent be (x_1, y_1) to S.

$\therefore x \cdot x_1 + y \cdot y_1 - (y + y_1) + 1 = 2$

or $x x_1 + y y_1 - y = (1 + y_1)$, same as $x + y = 3$.

$\Rightarrow \dfrac{x_1}{1} = \dfrac{y_1 - 1}{1} = \dfrac{1 + y_1}{3}$

i.e. $x_1 = 1$ and $y_1 = 2$

$\therefore P = (1, 2)$

Since, $PR = PQ = \frac{2\sqrt{2}}{3}$

Thus, by parametric form

$$\frac{x-1}{-1/\sqrt{2}} = \frac{y-2}{1/\sqrt{2}} = \pm\frac{2\sqrt{2}}{3}$$

$$\Rightarrow \left(x = \frac{5}{3}, y = \frac{4}{3}\right) \text{ and } \left(x = \frac{1}{3}, y = \frac{8}{3}\right)$$

$$\therefore \quad Q = \left(\frac{5}{3}, \frac{4}{3}\right) \text{ and } R = \left(\frac{1}{3}, \frac{8}{3}\right)$$

Now, equation of tangent at Q on ellipse E_1 is

$$\frac{x\cdot 5}{a^2\cdot 3} + \frac{y\cdot 4}{b^2\cdot 3} = 1$$

On comparing with $x + y = 3$, we get

$$a^2 = 5 \text{ and } b^2 = 4$$

$$\therefore \quad e_1^2 = 1 - \frac{b^2}{a^2} = 1 - \frac{4}{5} = \frac{1}{5} \quad \ldots(i)$$

Also, equation of tangent at R on ellipse E_2 is

$$\frac{x\cdot 1}{a^2\cdot 3} + \frac{y\cdot 8}{b^2\cdot 3} = 1$$

On comparing with $x + y = 3$, we get

$$a^2 = 1, b^2 = 8$$

$$\therefore \quad e_2^2 = 1 - \frac{a^2}{b^2} = 1 - \frac{1}{8} = \frac{7}{8} \quad \ldots(ii)$$

Now, $e_1^2 \cdot e_2^2 = \frac{7}{40}$

$$\Rightarrow \quad e_1 e_2 = \frac{\sqrt{7}}{2\sqrt{10}}$$

and $e_1^2 + e_2^2 = \frac{1}{5} + \frac{7}{8} = \frac{43}{40}$

Also, $\left|e_1^2 - e_2^2\right| = \left|\frac{1}{5} - \frac{7}{8}\right| = \frac{27}{40}$

3. (a) Given equation of ellipse is $\frac{x^2}{16} + \frac{y^2}{9} = 1$

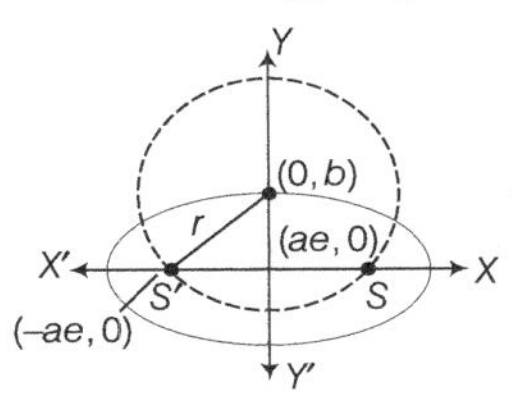

Here, $a = 4, b = 3, e = \sqrt{1 - \frac{9}{16}} \Rightarrow \frac{\sqrt{7}}{4}$

$\therefore$ Foci is $(\pm\, ae, 0) = \left(\pm\, 4 \times \frac{\sqrt{7}}{4}, 0\right) = (\pm\sqrt{7}, 0)$

$\therefore$ Radius of the circle, $r = \sqrt{(ae)^2 + b^2}$

$$= \sqrt{7 + 9} = \sqrt{16} = 4$$

Now, equation of circle is

$$(x - 0)^2 + (y - 3)^2 = 16$$

$$\therefore \quad x^2 + y^2 - 6y - 7 = 0$$

4. (c) Equation of ellipse is $x^2 + 3y^2 = 6$

or $\frac{x^2}{6} + \frac{y^2}{2} = 1$

Equation of the tangent is

$$\frac{x\cos\theta}{a} + \frac{y\sin\theta}{b} = 1$$

Let (h, k) be any point on the locus.

$$\therefore \quad \frac{h}{a}\cos\theta + \frac{k}{b}\sin\theta = 1 \quad \ldots(i)$$

Slope of the tangent line is $\frac{-b}{a}\cot\theta$.

Slope of perpendicular drawn from centre (0,0) to (h, k) is k/h.

Since, both the lines are perpendicular.

$$\therefore \quad \left(\frac{k}{h}\right) \times \left(-\frac{b}{a}\cot\theta\right) = -1$$

$$\Rightarrow \quad \frac{\cos\theta}{ha} = \frac{\sin\theta}{kb} = \alpha \quad \text{[Say]}$$

$$\Rightarrow \quad \cos\theta = \alpha ha, \ \sin\theta = \alpha kb$$

From Eq. (i), we get

$$\frac{h}{a}(\alpha ha) + \frac{k}{b}(\alpha kb) = 1$$

$$\Rightarrow \quad h^2\alpha + k^2\alpha = 1$$

$$\Rightarrow \quad \alpha = \frac{1}{h^2 + k^2}$$

Also, $\sin^2\theta + \cos^2\theta = 1$

$$\Rightarrow \quad (\alpha kb)^2 + (\alpha ha)^2 = 1$$

$$\Rightarrow \quad \alpha^2 k^2 b^2 + \alpha^2 h^2 a^2 = 1$$

$$\Rightarrow \quad \frac{k^2 b^2}{(h^2 + k^2)^2} + \frac{h^2 a^2}{(h^2 + k^2)^2} = 1$$

$$\Rightarrow \quad \frac{2k^2}{(h^2 + k^2)^2} + \frac{6h^2}{(h^2 + k^2)^2} = 1 \quad [\because a^2 = 6, b^2 = 2]$$

$$\Rightarrow \quad 6x^2 + 2y^2 = (x^2 + y^2)^2$$

Replacing k by y and h by x.

5. (c) **Concept Involved** Equation of an ellipse is

$$\frac{x^2}{a^2} + \frac{y^2}{b^2} = 1 \ (a > b)$$

$$\text{Eccentricity } e^2 = 1 - \frac{b^2}{a^2} \ (a > b)$$

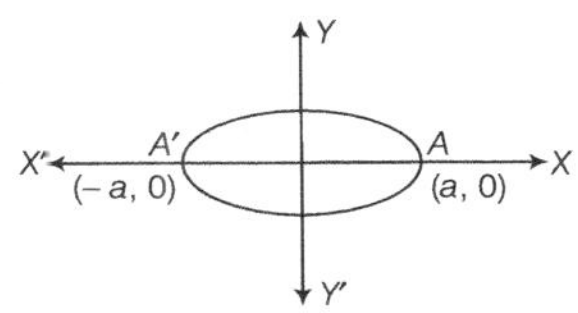

Let the equation of Ellipse E_2 be

$$\frac{x^2}{a^2} + \frac{y^2}{b^2} = 1, \text{ where } a < b \text{ and } b = 4.$$

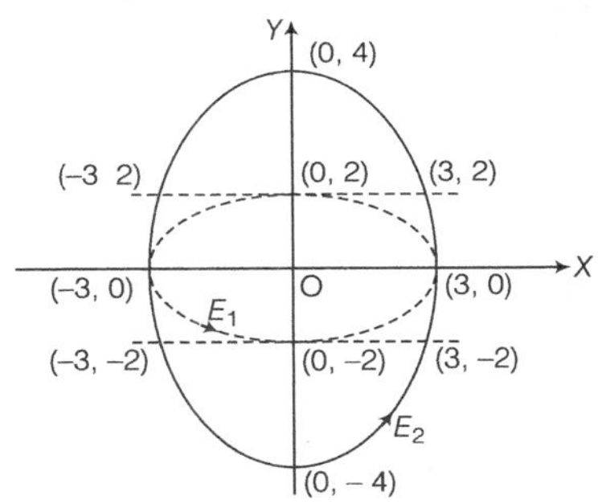

Also, it passes through (3, 2).

$$\Rightarrow \quad \frac{9}{a^2} + \frac{4}{b^2} = 1 \quad [\because b = 4]$$

$$\Rightarrow \quad \frac{9}{a^2} + \frac{1}{4} = 1 \quad \text{or} \quad a^2 = 12$$

Eccentricity of $E_2 \Rightarrow e^2 = 1 - \frac{a^2}{b^2} = 1 - \frac{12}{16} = \frac{1}{4} \quad [\because a < b]$

$$\therefore \quad e = \frac{1}{2}$$

6. (d) Equation of auxiliary circle is

$$x^2 + y^2 = 9 \qquad \dots \text{(i)}$$

Equation of AM is $\dfrac{x}{3} + \dfrac{y}{1} = 1 \qquad \dots \text{(ii)}$

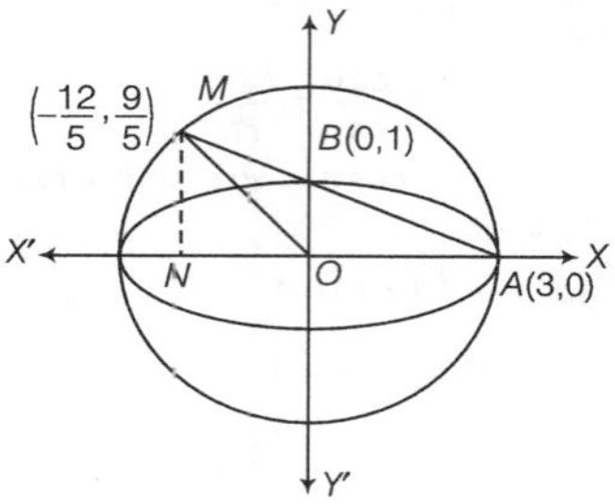

On solving Eqs. (i) and (ii), we get

$$M\left(-\frac{12}{5}, \frac{9}{5}\right)$$

Now, area of $\Delta\, AOM = \dfrac{1}{2} \cdot OA \times MN = \dfrac{27}{10}$ sq units

7. (c) Given, $\dfrac{x^2}{16} + \dfrac{y^2}{4} = 1$

Here, $a = 4, b = 2$

Equation of normal

$$4x \sec\theta - 2y \operatorname{cosec}\theta = 12$$

$$M\left(\frac{7\cos\theta}{2}, \sin\theta\right) = (h, k) \qquad \text{[say]}$$

$\therefore \qquad h = \dfrac{7\cos\theta}{2} \Rightarrow \cos\theta = \dfrac{2h}{7}$

and $\qquad k = \sin\theta$

$\Rightarrow \qquad \dfrac{4h^2}{49} + k^2 = 1 \qquad (\because \cos^2\theta + \sin^2\theta = 1)$

Hence, locus is $\dfrac{4x^2}{49} + y^2 = 1 \qquad \dots \text{(i)}$

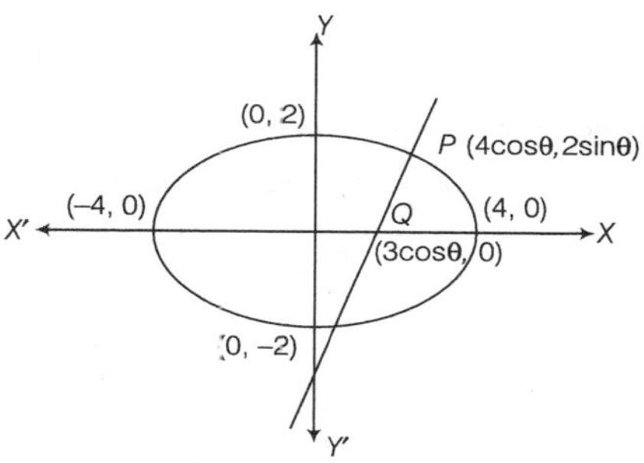

For given ellipse $e^2 = 1 - \dfrac{4}{16} = \dfrac{3}{4}$

$\therefore \qquad e = \dfrac{\sqrt{3}}{2}$

$\therefore \qquad x = \pm 4 \times \dfrac{\sqrt{3}}{2} = \pm 2\sqrt{3} \qquad [\because x = \pm ae] \dots \text{(ii)}$

On solving Eqs. (i) and (ii), we get

$$\frac{4}{49} \times 12 + y^2 = 1$$

$\Rightarrow \qquad y^2 = 1 - \dfrac{48}{49} = \dfrac{1}{49}$

$$y = \pm\frac{1}{7}$$

$\therefore$ Required points $\left(\pm 2\sqrt{3}, \pm\dfrac{1}{7}\right)$.

8. (d) Figure is self explanatory

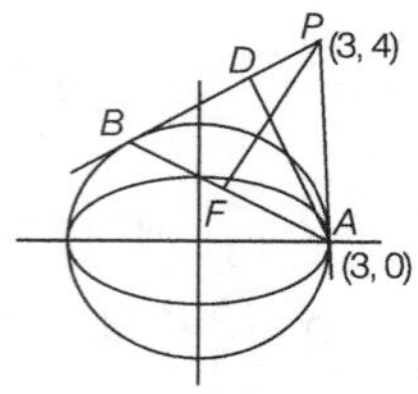

9. (c) Equation of AB is

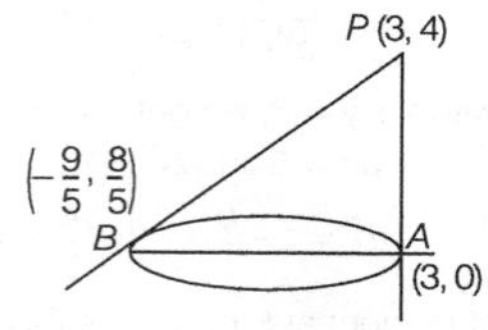

$$y - 0 = \frac{\frac{8}{5}}{-\frac{9}{5} - 3}(x - 3) = \frac{8}{-24}(x - 3)$$

$\Rightarrow \qquad y = -\dfrac{1}{3}(x - 3)$

$\Rightarrow \qquad x + 3y = 3 \qquad \dots \text{(i)}$

Equation of the straight line perpendicular to AB through P is $3x - y = 5$.

Equation of PA is $x - 3 = 0$.

The equation of straight line perpendicular to PA through $B\left(\dfrac{-9}{5}, \dfrac{8}{5}\right)$ is $y = \dfrac{8}{5}$.

Hence, the orthocentre is $\left(\dfrac{11}{5}, \dfrac{8}{5}\right)$.

10. (a) Equation of AB is $y - 0 = -\dfrac{1}{3}(x - 3)$

$$x + 3y - 3 = 0$$

$$|x + 3y - 3|^2 = 10[(x - 3)^2 + (y - 4)^2]$$

(Look at coefficient of x^2 and y^2 in the answers)

11. Let the common tangent to $x^2 + y^2 = 16$ and $\dfrac{x^2}{25} + \dfrac{y^2}{4} = 1$ be

$$y = mx + 4\sqrt{1 + m^2} \qquad \dots \text{(i)}$$

and $\qquad y = mx + \sqrt{25m^2 + 4} \qquad \dots \text{(ii)}$

Since, Eqs. (i) and (ii) are same tangent.

$\therefore \qquad 4\sqrt{1 + m^2} = \sqrt{25m^2 + 4}$

$\Rightarrow \qquad 16(1 + m^2) = 25m^2 + 4$

$\Rightarrow \qquad 9m^2 = 12$

$\Rightarrow \qquad m = \pm 2/\sqrt{3}$

Since, tangent is in Ist quadrant.

$\therefore \qquad m < 0$

$\Rightarrow \qquad m = -2/\sqrt{3}$

So, the equation of the common tangent is

$$y = -\frac{2x}{\sqrt{3}} + 4\sqrt{\frac{7}{3}}$$

which meets coordinate axes at $A(2\sqrt{7}, 0)$ and $\left(0, 4\sqrt{\dfrac{7}{3}}\right)$.

$\therefore \qquad AB = \sqrt{(2\sqrt{7} - 0)^2 + \left(0 - 4\sqrt{\dfrac{7}{3}}\right)^2} = \sqrt{28 + \dfrac{11}{3}} = \sqrt{\dfrac{196}{3}}$

$$= \frac{14}{\sqrt{3}} \times \frac{\sqrt{3}}{\sqrt{3}} = \frac{14\sqrt{3}}{3} \text{ units}$$

3. Hyperbola

DPP-1 Equation of Hyperbola

1. (c) Given, $4x^2 - 9y^2 - 36 = 0$

$\Rightarrow \quad 4x^2 - 9y^2 = 36$

$\Rightarrow \quad \frac{x^2}{9} - \frac{y^2}{4} = 1$

Here, $a^2 = 9$ and $b^2 = 4$

Foci of hyperbola are $(\pm\, ae, 0)$, where $ae = \sqrt{a^2 + b^2}$.

Hence, foci of hyperbola are $(\pm\, \sqrt{9+4}, 0) = (\pm\, \sqrt{13}, 0)$.

2. (a) Given, $\frac{x^2}{16} - \frac{(y-2)^2}{9} = 1$

Now, $a^2 = 16, b^2 = 9$

$ae = \sqrt{16+9} = 5$

Foci of hyperbola of $\frac{(x-h)^2}{a^2} - \frac{(y-k)^2}{b^2} = 1$ are $(\pm\, ae + h, k)$.

$\therefore$ Foci of hyperbola $\frac{x^2}{16} - \frac{(y-2)^2}{9} = 1$ are $(\pm\, 5, 2)$.

3. (a) The difference of focal distance of any point on an hyperbola is constant and equal to the length of transverse axis of hyperbola. Therefore, difference of focal distance is 8.

[$\because$ length of transverse is 8]

4. (d) We have, equation of directrix of hyperbola is $2x + y = 1$.

Focus $\equiv (1, 1)$ and $e = \sqrt{3}$

Now, the equation of hyperbola is

$$(x-1)^2 + (y-1)^2 = \left[\sqrt{3}\,\frac{(2x+y-1)}{\sqrt{5}}\right]^2$$

$\Rightarrow x^2 - 2x + 1 + y^2 - 2y + 1$

$= \frac{3}{5}(4x^2 + y^2 + 1 - 4x - 2y + 4xy)$

$\Rightarrow 5x^2 - 10x + 5y^2 - 10y + 10$

$= 12x^2 + 3y^2 - 12x - 6y + 12xy + 3$

$\therefore \quad 7x^2 + 12xy - 2y^2 - 2x + 4y - 7 = 0$

5. (b) Foci of ellipse of $\frac{x^2}{25} + \frac{y^2}{9} = 1$ is $(\pm\, \sqrt{25-9}, 0)$, i.e. $(\pm\, 4, 0)$.

Now, foci of ellipse is foci of hyperbola.

$\therefore$ Foci of hyperbola is $(\pm\, 4, 0)$

$\therefore \quad ae = 4$ and $e = 2$ [given]

$\therefore \quad a = \frac{4}{e} \Rightarrow a = \frac{4}{2} = 2$, i.e. $a^2 = 4$

Now, $(ae)^2 = a^2 + b^2$

$\Rightarrow \quad (4)^2 = (2)^2 + b^2 \Rightarrow b^2 = 12$

$\therefore$ Equation of hyperbola is

$$\frac{x^2}{4} - \frac{y^2}{12} = 1.$$

6. (b) Given, eccentricity of an ellipse $= \frac{2}{3}$

and length of latusrectum = 5

We know that, length of latusrectum of an ellipse is $\frac{2b^2}{a}$.

$\therefore \quad \frac{2b^2}{a} = 5$ and $e = \frac{2}{3}$

$\Rightarrow \quad 2b^2 = 5a$ and $e^2 = \frac{4}{9}$

$\Rightarrow \quad 2b^2 = 5a$ and $\frac{a^2 - b^2}{a^2} = \frac{4}{9}$ $\quad \left[\because e = \sqrt{\frac{a^2 - b^2}{a^2}}\right]$

$\Rightarrow \quad 9a^2 - 9b^2 = 4a^2$

$\Rightarrow \quad 5a^2 = 9b^2$

$\Rightarrow \quad 5a^2 = \frac{9 \times 5a}{2}$

Thus, $a = \frac{9}{2}$ and $b = \frac{\sqrt{45}}{2}$

$\therefore$ Equation of ellipse is $\frac{x^2}{\left(\frac{9}{2}\right)^2} + \frac{y^2}{\left(\frac{\sqrt{45}}{2}\right)^2} = 1$

i.e. $\frac{4x^2}{81} + \frac{4y^2}{45} = 1$

7. (b) The centre of hyperbola is the mid-point of vertices.

So, coordinates of centre are $\left(\frac{0+10}{2}, \frac{0+0}{2}\right)$, i.e. $(5, 0)$.

One of the foci is $(18, 0)$.

$\therefore$ Other foci of hyperbola is $(-8, 0)$.

[$\because$ centre of hyperbola is also mid-point of foci of hyperbola]

Let $2a$ and $2b$ be the lengths of transverse and conjugate axes and e be the eccentricity. Then, the equation of hyperbola is

$$\frac{(x-5)^2}{a^2} - \frac{y^2}{b^2} = 1$$

Distance between two vertices $= 2a$

i.e. $2a = \sqrt{(10-0)^2 + (0-0)^2} = 10$

$\Rightarrow \quad a = 5$

Now, distance between two foci $= 2ae$

i.e. $2ae = \sqrt{(18+8)^2 + (0-0)^2} = 26$

$\Rightarrow \quad ae = \frac{26}{2} = 13$

In hyperbola, $(ae)^2 = a^2 + b^2$

$\Rightarrow \quad b^2 = (ae)^2 - a^2$

$\Rightarrow \quad b^2 = (13)^2 - (5)^2 = 144$

$\therefore$ Equation of hyperbola is

$$\frac{(x-5)^2}{25} - \frac{y^2}{144} = 1.$$

8. (a) Given eccentricity of hyperbola, i.e. $e = \frac{3}{2}$

and foci $(\pm\, 2, 0)$ i.e. $ae = 2$.

$\therefore \quad a = \frac{2}{e} \Rightarrow a = \frac{2 \times 2}{3} = \frac{4}{3}$

In hyperbola, $(ae)^2 = a^2 + b^2$

$\Rightarrow \quad b^2 = (ae)^2 - a^2$

$\Rightarrow \quad b^2 = (2)^2 - \left(\frac{4}{3}\right)^2 = \frac{20}{9}$

$\therefore$ Equation of hyperbola is

$$\frac{x^2}{\frac{16}{9}} - \frac{y^2}{\frac{20}{9}} = 1$$

$\Rightarrow \quad \frac{x^2}{4} - \frac{y^2}{5} = \frac{4}{9}$

9. (c) Given, centre (0, 0), vertex (4, 0) and focus (6, 0).

$\therefore \quad a = 4, ae = 6$

Now, $b^2 = (ae)^2 - a^2 = (6)^2 - (4)^2 = 36 - 16 = 20$

Thus, equation of hyperbola is $\frac{x^2}{a^2} - \frac{y^2}{b^2} = 1 \Rightarrow \frac{x^2}{16} - \frac{y^2}{20} = 1$

$\therefore \quad 5x^2 - 4y^2 = 80$

10. (a) Given, length of latusrectum = 8

and eccentricity of hyperbola = $\frac{3}{\sqrt{5}}$

i.e. $\frac{2b^2}{a} = 8$ and $e = \frac{3}{\sqrt{5}}$

$\Rightarrow \quad b^2 = 4a$ and $e^2 = \frac{9}{5}$

$\Rightarrow \quad b^2 = 4a$ and $\frac{a^2 + b^2}{a^2} = \frac{9}{5}$

On solving these equations, we get

$$a^2 = 25 \text{ and } b^2 = 20$$

$\therefore$ Equation of hyperbola is $\frac{x^2}{25} - \frac{y^2}{20} = 1$

i.e. $\quad 4x^2 - 5y^2 = 100$

11. (c) We have, $\quad \frac{x^2}{12 - \lambda} + \frac{y^2}{8 - \lambda} = 1$

If $\lambda < 8$, then this equation represents an ellipse.

If $8 < \lambda < 12$, then this equation represents a hyperbola.

12. (d) $\frac{x^2}{1 - k} - \frac{y^2}{1 + k} = 1$

Ellipse, when $1 - k > 0$ and $1 + k < 0$

$\Rightarrow \quad k < 1$ and $k < -1$

$\therefore \quad k < -1 \Rightarrow$ Ellipse

Hyperbola, when $(1 - k)(1 + k) > 0$

$\Rightarrow \quad -1 < k < 1 \Rightarrow$ Hyperbola

When $k > 1 \Rightarrow$ no figure

13. (c) Equation of conic with focus (1, – 1), directrix along $x - y + 1 = 0$ and eccentricity $\sqrt{2}$ is

$$(x - 1)^2 + (y + 1)^2 = \left[\sqrt{2}\, \frac{(x - y + 1)}{\sqrt{2}}\right]^2$$

$\Rightarrow x^2 - 2x + 1 + y^2 + 2y + 1 = x^2 + y^2 + 1 + 2x - 2y - 2xy$

$\therefore \quad 2xy - 4x + 4y + 1 = 0$

14. (d) We have, $\quad -\frac{x^2}{a^2} + \frac{y^2}{b^2} = 1$

Now, foci of this hyperbola lie on Y-axis.

$\therefore \quad$ Eccentricity $= \sqrt{\frac{a^2 + b^2}{b^2}}$

15. (c) Given, $\quad e^2 + e'^2 = 3$

We know that, the eccentricity of hyperbola is greater than one.

$\therefore$ S and S' both are hyperbola.

16. (a) Centre of hyperbola is mid-point of line joining the foci.

So, coordinates of centre are $\left(\frac{6 - 4}{2}, \frac{5 + 5}{2}\right)$, i.e. (1, 5).

Equation of hyperbola is $\frac{(x - 1)^2}{a^2} - \frac{(y - 5)^2}{b^2} = 1$

$\therefore$ Distance between two foci $= 2ae = \sqrt{(6 + 4)^2 + (5 - 5)^2} = 10$

$\Rightarrow \quad 2ae = 10 \Rightarrow a = \frac{5}{e} \Rightarrow a = 4 \quad [\because e = 5/4]$

$\Rightarrow \quad b^2 = (ae)^2 - a^2 = 25 - 16 = 9$

Hence, the equation of hyperbola is

$$\frac{(x - 1)^2}{16} - \frac{(y - 5)^2}{9} = 1$$

17. (c) Given, length of latusrectum is 8

i.e. $\quad \frac{2b^2}{a} = 8$

$\Rightarrow \quad b^2 = 4a \quad$ ...(i)

and $\quad 2b = \frac{1}{2}(2ae)$

$\Rightarrow \quad 2b = ae$

Now, $\quad (ae)^2 = b^2 + a^2$

$\Rightarrow \quad 4b^2 = b^2 + a^2$

$\Rightarrow \quad 3b^2 = a^2 \quad$ ...(ii)

From Eqs. (i) and (ii), we get

$$a = 12 \text{ and } b = 4\sqrt{3}$$

$\therefore \quad e = \frac{2b}{a} = \frac{2 \times 4\sqrt{3}}{12} = \frac{2}{\sqrt{3}}$

18. (d) Given, $\quad \frac{x}{a} + \frac{y}{b} = \lambda$ and $\frac{x}{a} - \frac{y}{b} = \frac{1}{\lambda}$

On multiplying both equations, we get

$$\left(\frac{x}{a} + \frac{y}{b}\right)\left(\frac{x}{a} - \frac{y}{b}\right) = \lambda \times \frac{1}{\lambda}$$

$\Rightarrow \quad \frac{x^2}{a^2} - \frac{y^2}{b} = 1$

which represents the locus of a hyperbola.

19. (c) We have,

$$9x^2 - 16y^2 - 18x + 32y - 151 = 0$$

$\Rightarrow \quad 9(x^2 - 2x + 1) - 16(y^2 - 2y + 1) = 151 + 9 - 16$

$\Rightarrow \quad 9(x - 1)^2 - 16(y - 1)^2 = 144$

$\Rightarrow \quad \frac{(x - 1)^2}{16} - \frac{(y - 1)^2}{9} = 1$

(a) Length of transverse axes $= 2a = 2 \times 4 = 8$

(b) Length of latusrectum $= \frac{2b^2}{a} = \frac{2 \times 9}{4} = \frac{9}{2}$

(c) Equation of directrix is $x = \pm \frac{a}{e} + 1$

$\Rightarrow \quad x = \pm \frac{4}{5} \times 4 + 1 \quad \left[\because e = \sqrt{\frac{a^2 + b^2}{a^2}} = \frac{5}{4}\right]$

$\therefore \quad x = \frac{21}{5}$

and $\quad x = -\frac{11}{5}$

20. (b) We have, $\quad x = t^2 + 1$

and $\quad y = 2t$

On eliminating t, we get

$$y^2 = 4x - 4$$

and $\quad x = 2s$ and $y = \frac{2}{s}$

On substituting this $y^2 = 4x - 4$, we get

$$2s^3 - s^2 - 1 = 0$$

$\Rightarrow \quad (s - 1)(2s^2 + s + 1) = 0$

$\Rightarrow \quad s = 1$

Putting $s = 1$ in $x = 2s$, $y = \frac{2}{s}$, we get

$$x = 2, y = 2$$

So, point is (2, 2).

21. (b) We have,

$$\frac{x^2}{\cos^2\alpha} - \frac{y^2}{\sin^2\alpha} = 1$$

Vertices of this curves are

$(\pm \cos \alpha, 0)$ and $(0, \pm \sin \alpha)$.

Foci of this curves are $(\pm 1, 0)$.

Eccentricity = $\sec \alpha$

∴ Equation of directrix is

$$x = \pm \cos^2\alpha$$

If we change α, then foci are remain unchanged.

22. (d) We have,

$$\sqrt{3}x - y - 4\sqrt{3}k = 0 \quad \text{...(i)}$$

and $$\sqrt{3}xk + ky - 4\sqrt{3} = 0 \quad \text{...(ii)}$$

On eliminating k from these equations, we get

$$(\sqrt{3}x - y)(\sqrt{3}x + y) = (4\sqrt{3})^2$$

$$3x^2 - y^2 = 48$$

Which represents a hyperbola.

23. (d) Graph of conic $x^2 - (y - 1)^2 = 1$ is

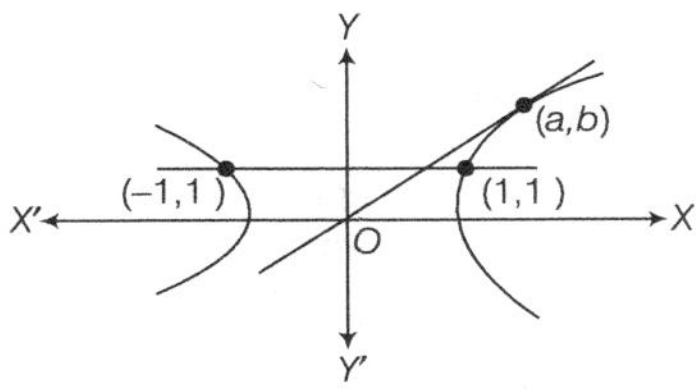

We have,

$$x^2 - (y - 1)^2 = 1$$

On differentiating the curve

$$2x - 2(y - 1)\frac{dy}{dx} = 0$$

$$\Rightarrow \quad \frac{dy}{dx} = \frac{x}{y - 1}$$

$$\Rightarrow \quad \left(\frac{dy}{dx}\right)_{(a,\, b)} = \frac{a}{b - 1}$$

Now, $$\text{slope of } OP = \frac{b}{a}$$

$$\therefore \quad \frac{b}{a} = \frac{a}{b - 1}$$

$$\Rightarrow \quad a^2 = b^2 - b$$

Also, (a, b) satisfy the curve

$$a^2 - (b - 1)^2 = 1$$

$$\Rightarrow \quad a^2 - b^2 + 2b = 2$$

$$\Rightarrow \quad -b + 2b = 2 \quad [\because a^2 - b^2 = -b]$$

$$\Rightarrow \quad b = 2$$

$$\Rightarrow \quad a = \sqrt{2}$$

$$\therefore \quad \sin^{-1}\left(\frac{a}{b}\right) = \sin^{-1}\left(\frac{\sqrt{2}}{2}\right) = \frac{\pi}{4}$$

24. (c) Length of latusrectum $= \dfrac{2b^2}{a} = 2a$

Distance between the vertices = 2

25. (d) Curve is a rectangular hyperbola.

$$\Rightarrow \quad e = \sqrt{2}$$

26. (2) Foci of hyperbola $\dfrac{x^2}{a^2} - \dfrac{y^2}{b^2} = 1$ and $\dfrac{x^2}{a^2} - \dfrac{y^2}{b^2} = -1$ are $(\pm\sqrt{a^2 + b^2}, 0)$ and $(0, \pm\sqrt{a^2 + b^2})$.

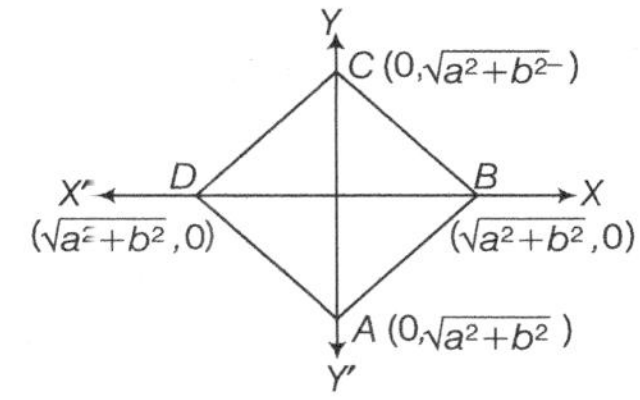

Foci of the hyperbola formed a rhombus.

∴ Area of rhombus $= \dfrac{1}{2} AC \times BD$

$$= \frac{1}{2} \times 2\sqrt{a^2 + b^2} \times 2\sqrt{a^2 + b^2}$$

$$= 2(a^2 + b^2)$$

$$\therefore \quad \lambda = 2$$

27. (6) We have,

$$(10x - 5)^2 + (10y - 2)^2 = 9(3x + 4y - 7)^2$$

$$\left(x - \frac{1}{2}\right)^2 + \left(y - \frac{1}{5}\right)^2 = \frac{9}{100} \frac{(3x + 4y - 7)^2 \times 25}{25}$$

$$\left(x - \frac{1}{2}\right)^2 + \left(y - \frac{1}{5}\right)^2 = \left(\frac{3}{2}\right)^2 \left(\frac{3x + 4y - 7}{5}\right)^2$$

From this equation,

foci of hyperbola is $\left(\dfrac{1}{2}, \dfrac{1}{5}\right)$

eccentricity of hyperbola $= \dfrac{3}{2}$

equation of directrix is $3x + 4y - 7 = 0$

equation of latusrectum is $3x + 4y = \lambda$.

[∵ latusrectum is parallel to directrix]

Since, latusrectum is passes through the foci

i.e. $$\frac{3}{2} + \frac{4}{5} = \lambda$$

$$\Rightarrow \quad \lambda = \frac{23}{10}$$

∴ Equation of latusrectum is

$$30x + 40y - 23 = 0.$$

Now, $$|b + 2c| = |40 - 46| = 6$$

28. (1) Foci of the hyperbola are (5, 12) and (24, 7).

Let $P(x, y)$ be any point on the hyperbola and S and S' be the foci of parabola.

By definition,

$$|PS - PS'| = \text{Length of transverse axis}$$

$$\therefore \left|\sqrt{(x - 5)^2 + (y - 12)^2} - \sqrt{(x - 24)^2 + (y - 7)^2}\right| = 2a$$

Since, hyperbola is passing through the origin.

$$\therefore \quad 2a = |\sqrt{25 + 144} - \sqrt{5 + 6 + 49}|$$

$$= |13 - 25| = 12$$

Now, distance between two foci = $2ae$

i.e. $$\sqrt{(5 - 24)^2 + (12 - 7)^2} = 2ae$$

$$\sqrt{19^2 + 5^2} = 2ae$$

$$\Rightarrow \quad \sqrt{368} = 12e$$

$$\therefore \quad \frac{12e}{\sqrt{368}} = 1$$

29. (2) Let the length of each side of the equilateral ΔOPQ be l units.

Then, the coordinates of P are $\left(\frac{\sqrt{3}l}{2}, \frac{l}{2}\right)$.

Point $P\left(\frac{\sqrt{3}l}{2}, \frac{l}{2}\right)$ lies on the hyperbola $\frac{x^2}{a^2} - \frac{y^2}{b^2} = 1$

$\therefore \qquad \frac{3l^2}{4a^2} - \frac{l^2}{4b^2} = 1$

$\Rightarrow \qquad (3b^2 - a^2)l^2 = 4a^2b^2$

$\Rightarrow \qquad (3e^2 - 4)l^2 = 4a^2(e^2 - 1)$

$\Rightarrow \qquad l = 2a\sqrt{\frac{e^2 - 1}{3e^2 - 4}}$

Since, l is real and $e > 1$.

$\therefore \qquad e^2 > \frac{4}{3} \Rightarrow e > \frac{2}{\sqrt{3}}$

$= \sqrt{3}\, e > 2$

$\Rightarrow \qquad k = 2$

30. (2) Here, $\qquad e^2 = 1 + \frac{16}{9}$

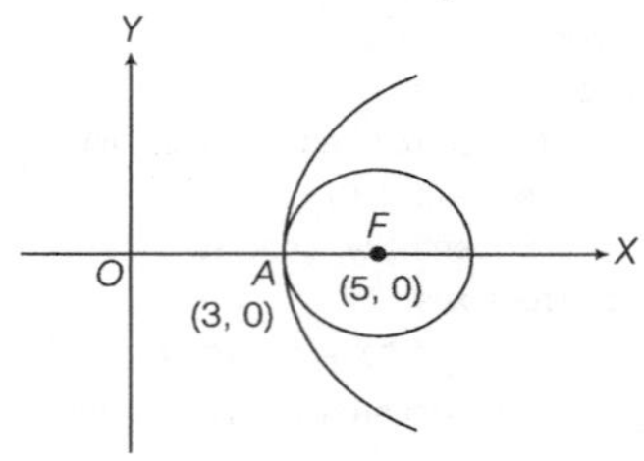

$\Rightarrow \qquad e = \frac{5}{3}$

$\therefore$ Focus (5, 0)

Use reflection property to prove that circle cannot touch at two points. It can only be tangent at vertex.

$\therefore \qquad r = 5 - 3 = 2$

DPP-2 Tangent and Its Properties

1. (c) We have,

$y = x + 2$ touches the hyperbola $5x^2 - 9y^2 = 45$

We put the value of y in equations of hyperbola.

$5x^2 - 9(x + 2)^2 = 45$

$\Rightarrow \qquad 5x^2 - 9(x^2 + 4x + 4) = 45$

$\Rightarrow \qquad 5x^2 - 9x^2 - 36x - 36 = 45$

$\Rightarrow \qquad 4x^2 + 36x + 81 = 0$

$\Rightarrow \qquad (2x + 9)^2 = 0$

$\Rightarrow \qquad x = -9/2$

and $\qquad y = -\frac{9}{2} + 2 = -5/2$

Hence, points of contact are $\left(-\frac{9}{2}, -\frac{5}{2}\right)$.

2. (a) The equation of line is

$lx + my + n = 0$

$\Rightarrow \qquad y = \left(-\frac{l}{m}\right)x + \left(\frac{-n}{m}\right)$

We know that the line $y = mx + c$ touches the hyperbola $\frac{x^2}{a^2} - \frac{y^2}{b^2} = 1$, if $c^2 = a^2m^2 - b^2$.

So, line will touch $\frac{x^2}{a^2} - \frac{y^2}{b^2} = 1$, if

$\left(-\frac{n}{m}\right)^2 = a^2\left(-\frac{l}{m}\right)^2 - b^2$

$\Rightarrow \qquad n^2 = a^2l^2 - b^2m^2$

3. (b) We know that equation of tangent of hyperbola $\frac{x^2}{a^2} - \frac{y^2}{b^2} = 1$ at (x_1, y_1) is

$\frac{xx_1}{a^2} - \frac{yy_1}{b^2} = 1$

Hence, equation of tangent of hyperbola $2x^2 - 3y^2 = 6$ at (3, 2) is

$2(3x) - 3(2y) = 6$

$\Rightarrow \qquad x - y = 1$

$\therefore$ Slope of the tangent of $x - y = 1$ is 1.

4. (b) We have, length of intercepts of tangent is unity.

$\therefore$ Equation of tangent of hyperbola is

$x + y = 1$

$\Rightarrow \qquad y = -x + 1$

We know that the line $y = mx + c$ touches the hyperbola $\frac{x^2}{a^2} - \frac{y^2}{b^2} = 1$, if $c^2 = a^2m^2 - b^2$.

So, line touch $\frac{x^2}{a^2} - \frac{y^2}{b^2} = 1$, if $1 = a^2 - b^2$

$\therefore$ (a, b) line is in curve $x^2 - y^2 = 1$.

5. (b) We have,

$x\cos\alpha + y\sin\alpha = p$

$\Rightarrow \qquad y = -\frac{\cos\alpha}{\sin\alpha}x + \frac{p}{\sin\alpha}$

This will touch $\frac{x^2}{a^2} - \frac{y^2}{b^2} = 1$, if

$\left(\frac{p}{\sin\alpha}\right)^2 = a^2\left(-\frac{\cos\alpha}{\sin\alpha}\right)^2 - b^2$

$\therefore \qquad p^2 = a^2\cos^2\alpha - b^2\sin^2\alpha$

6. (c) We have,

$\frac{x^2}{a^2} + \frac{y^2}{b^2} = 1$

$\Rightarrow \qquad \frac{dy}{dx} = -\frac{b^2x}{a^2y}$

and $\qquad x^2 - y^2 = c^2$

$\Rightarrow \qquad \frac{dy}{dx} = x/y$

The two curves will cut at right angles if

$\left(\frac{dy}{dx}\right)_{C_1} \times \left(\frac{dy}{dx}\right)_2 = -1$

$\Rightarrow \qquad \left(-\frac{b^2x}{a^2y}\right)\left(\frac{x}{y}\right) = -1$

$\Rightarrow \qquad \frac{x^2}{a^2} = \frac{y^2}{b^2} = \frac{1}{2} \qquad \left[\text{using } \frac{x^2}{a^2} + \frac{y^2}{b^2} = 1\right]$

On subsituting these values in $x^2 - y^2 = c^2$, we get

$$\frac{a^2}{2} - \frac{b^2}{2} = c^2$$

$\therefore \quad a^2 - b^2 = 2c^2$

7. (b) Let $y = mx + c$ be a common tangent to $9x^2 - 16y^2 = 144$ and $x^2 + y^2 = 9$

Since, $y = mx + c$ is a tangent to $9x^2 - 16y^2 = 144$

$\therefore \quad c^2 = a^2m^2 - b^2$

$\Rightarrow \quad c^2 = 16m^2 - 9 \quad \ldots(i)$

Now, $y = mx + c$ is a tangent to $x^2 + y^2 = 9$

$\therefore \quad \frac{c}{\sqrt{m^2+1}} = 3$

$\Rightarrow \quad c^2 = 9(m^2 + 11) \quad \ldots(ii)$

From Eqs. (i) and (ii), we get

$$16m^2 - 9 = 9 + 9m^2$$

$\Rightarrow \quad m = \pm 3\sqrt{\frac{2}{7}}$

Putting the value of m in Eq. (ii), we get

$$c = \pm \frac{15}{\sqrt{7}}$$

Hence, equation of common tangent is

$$y = 3\sqrt{\frac{2}{7}}x + \frac{15}{\sqrt{7}}$$

8. (a) The line $y = mx + 6$ is a tangent to the hyperbola

$$\frac{x^2}{100} - \frac{y^2}{49} = 1$$

$\therefore \quad 6^2 = 100m^2 - 49$

$\Rightarrow \quad 36 = 100m^2 - 49$

$\Rightarrow \quad m = \sqrt{\frac{17}{20}}$

9. (b) The equation of any tangent to the hyperbola $\frac{x^2}{a^2} - \frac{y^2}{b^2} = 1$ is

$$y = mx + \sqrt{a^2m^2 - b^2} \quad \ldots(i)$$

Let $S(ae, 0)$ and $S'(-ae, 0)$ be two foci of the hyperbola and p and q be the lengths of perpendicular from $S(ae, 0)$ and $S'(-ae, 0)$ on Eq. (i)

$$p = \left|\frac{mae - 0 + \sqrt{a^2m^2 - b^2}}{\sqrt{1+m^2}}\right|$$

$$= \left|\frac{mae + \sqrt{a^2m^2 - b^2}}{\sqrt{1+m^2}}\right|$$

Similarly,

$$q = \left|\frac{-mae + \sqrt{a^2m^2 - b^2}}{\sqrt{1+m^2}}\right|$$

Hence, $p \cdot q = \left(\frac{mae + \sqrt{a^2m^2 - b^2}}{\sqrt{1+m^2}}\right)\left(\frac{-mae + \sqrt{a^2m^2 - b^2}}{\sqrt{1+m^2}}\right)$

$$= \left(\frac{-m^2a^2e^2 + a^2m^2 - b^2}{1+m^2}\right)$$

$$= \left|\frac{-m^2a^2(e^2-1) - b^2}{m^2+1}\right|$$

$$= \frac{(m^2+1)}{m^2+1}b^2 = b^2 \quad [\because b^2 = a^2(e^2-1)]$$

10. (b) Let $P(h, k)$ be the point of intersection of tangents

$\Rightarrow y = mx \pm \sqrt{a^2m^2 - b^2}$ passes through (h, k)

$\Rightarrow \quad k - mh = \pm\sqrt{a^2m^2 - b^2}$

$\Rightarrow \quad (k - mh)^2 = a^2m^2 - b^2$

$\Rightarrow \quad k^2 + m^2h^2 - 2khm = a^2m^2 - b^2$

$\Rightarrow m^2(h^2 - a^2) - 2khm + (k^2 + b^2) = 0$

Given, $m_1m_2 = c^2$

$\Rightarrow \quad \frac{k^2+b^2}{h^2-a^2} = c^2$

$\Rightarrow \quad k^2 + b^2 = c^2(h^2 - a^2)$

$\therefore$ Locus of (h, k) is

$$y^2 + b^2 = c^2(x^2 - a^2)$$

11. (d) Since, the tangent is parallel to the line

$$2x + y + 8 = 0$$

$\therefore \quad m =$ slope of tangent $= -2$

Now, $\quad x^2 - y^2 = 3$

$\Rightarrow \quad \frac{x^2}{3} - \frac{y^2}{3} = 1$

This is the form $\quad \frac{x^2}{a^2} - \frac{y^2}{b^2} = 1$

where $a^2 = 3$ and $b^2 = 3$

So, the equation of the tangents are

$$y = mx \pm \sqrt{a^2m^2 - b^2}$$

$\Rightarrow \quad y = -2x \pm \sqrt{3(2^2) - 3}$

$\Rightarrow \quad y = -2x \pm 3$

$\Rightarrow \quad 2x + y = \pm 3$

The coordinates of point of contact are

$$\left(\pm \frac{a^2m}{\sqrt{a^2m^2 - b^2}}, \pm \frac{b^2}{\sqrt{a^2m^2 - b^2}}\right)$$

i.e. $\quad (-2, 1)$

12. (b) Let $y = mx + c$ be a common tangent of $y^2 = 8x$ and $xy = -1$

Since, $y = mx + c$ is a tangent to $y^2 = 8x$

$\therefore \quad y = mx + \frac{2}{m}$

Since, $y = mx + \frac{2}{m}$ is also tangent of $x - y = -1$

$\therefore \quad x\left(mx + \frac{2}{m}\right) = -1$

$\Rightarrow \quad m^2x^2 + 2x + m = 0$

$\Rightarrow \quad 4 - 4m^3 = 0$

$m^3 = 1$

$\Rightarrow \quad m = 1$

$\therefore$ Equation of common tangent is $y = x + 2$

13. (a) The line $y = ax + b$ is tangent to the hyperbola

$$\frac{x^2}{a^2} - \frac{y^2}{b^2} = 1$$

$\therefore \quad b^2 = a^2(a^2) - b^2$

$\Rightarrow \quad y^2 = a^2x^2 - b^2$

[a, b replaced by x and y]

$\Rightarrow \quad a^2x^2 - y^2 = b^2$

Which represents the equation of hyperbola.

14. (a) Equation of tangent to $y^2 = 8x$ is

$$y = mx + \frac{2}{m} \quad \ldots(i)$$

and tangent to $x^2 - \frac{y^2}{3} = 1$ is

$$y = mx \pm \sqrt{1 \cdot m^2 - 3} \quad \ldots(ii)$$

Since, Eqs. (i) and (ii) are same, then

$$\frac{2}{m} = \sqrt{m^2 - 3}$$

$$\Rightarrow \quad \frac{4}{m^2} = (m^2 - 3)$$

$$\Rightarrow \quad m^4 - 3m^2 - 4 = 0$$

$$(m^2 - 4)(m^2 + 1) = 0$$

$$\Rightarrow \quad m = \pm 2$$

$\therefore$ Common tangent is $y = \pm 2x \pm 1$

$$\Rightarrow \quad 2x \pm y + 1 = 0$$

15. (a) Let equation of tangent of hyperbola $x^2 - y^2 = a^2$ at point $(a \sec \theta, a \tan \theta)$ is $x \sec \theta - y \tan \theta = a$.

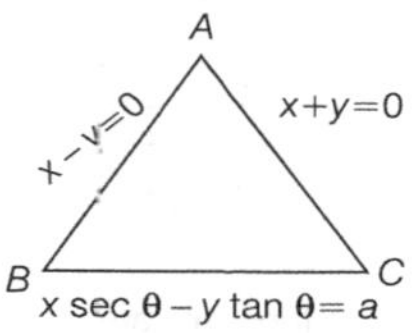

Area of triangle formed by line $x - y = 0$, $x + y = 0$ and $x \sec \theta - y \tan \theta = a$.

Coordinates of A, B and C are

$(0, 0)$, $\{a(\sec \theta = \tan \theta), a(\sec \theta + \tan \theta)\}$

$\{a(\sec \theta - \tan \theta) - a(\sec - \tan \theta)\}$

$$\therefore \text{ Area of } \Delta ABC = \frac{1}{2}\begin{vmatrix} 0 & 0 & 1 \\ a(\sec \theta + \tan \theta) & a(\sec \theta + \tan \theta) & 1 \\ a(\sec \theta - \tan \theta) & -a(\sec \theta - \tan \theta) & 1 \end{vmatrix}$$

$$= \frac{1}{2}| - a^2(\sec^2 \theta - \tan^2 \theta) - a^2(\sec^2 \theta - \tan^2 \theta)|$$

$$= \frac{1}{2}| - 2a^2 | = a^2 \text{ sq units}$$

16. (b) Any tangent to the hyperbola $\frac{x^2}{a^2} - \frac{y^2}{b^2} = 1$ is

$$y = mx \pm \sqrt{a^2m^2 - b^2}$$

or $\quad y = mx + c$, where $c = \sqrt{a^2m^2 - b^2}$

This will touch the hyperbola $\frac{y^2}{a^2} - \frac{x^2}{b^2} = 1$, if the equation

$\frac{(mx + c)^2}{a^2} - \frac{x^2}{b^2} = 1$ has equal roots

or $\quad x^2(b^2m^2 - a^2) + 2b^2mcx + (c^2 - a^2)b^2 = 0$

has equal roots

$$\therefore \quad 4b^4m^2c^2 - 4(b^2m^2 - a^2)(c^2 - a^2)b^2 = 0$$

$$\Rightarrow \quad c^2 = a^2 - b^2m^2$$

$$\therefore \quad a^2m^2 - b^2 = a^2 - b^2m^2 \quad [\because c = \sqrt{a^2m^2 - b^2}]$$

$$\Rightarrow \quad m^2(a^2 + b^2) = a^2 + b^2$$

$$\Rightarrow \quad m = \pm 1$$

Hence, the equation of common tangent is $y = \pm x \pm \sqrt{a^2 - b^2}$.

17. (c) Let $P(h, k)$ be the point of intersection of tangents.

$\Rightarrow y = mx \pm \sqrt{a^2m^2 - b^2}$ passes through (h, k)

$$\Rightarrow \quad (k - mh)^2 = a^2m^2 + b^2$$

$$\Rightarrow \quad m^2(h^2 - a^2) - 2khm + (k^2 + b^2) = 0$$

Given, $\quad m_1 + m_2 = c^2$

$$\Rightarrow \quad \frac{2hk}{h^2 - a^2} = c^2$$

$$\Rightarrow \quad 2xy = c^2(x^2 - a^2)$$

18. (a) We have, $P(a \sec \alpha, b \tan \alpha)$ and $\theta(a \sec \beta, b \tan \beta)$ are points lie on the hyperbola $\frac{x^2}{a^2} - \frac{y^2}{b^2} = 1$

$\therefore$ PQ is a chord of hyperbola $\frac{x^2}{a^2} - \frac{y^2}{b^2} = 1$

Equation of chord of hyperbola joining $P(a \sec \alpha, b \tan \alpha)$ and $\theta(a \sec \beta, b \tan \beta)$ is

$$\frac{x}{a}\cos\left(\frac{\alpha - \beta}{2}\right) - \frac{y}{b}\sin\left(\frac{\alpha + \beta}{2}\right) = \cos\left(\frac{\alpha + \beta}{2}\right)$$

$$\Rightarrow \quad \frac{x \sec\left(\frac{\alpha + \beta}{2}\right)}{a \sec\left(\frac{\alpha - \beta}{2}\right)} - \frac{y}{b}\tan\left(\frac{\alpha + \beta}{2}\right) = 1$$

$$\Rightarrow \quad \frac{x \sec\left(\frac{\alpha + \beta}{2}\right)}{a \sec \theta} - \frac{y}{b}\tan\left(\frac{\alpha + \beta}{2}\right) = 1 \quad [\because \alpha - \beta = 2\theta]$$

This equation of line touches the hyperbola

$$\frac{x^2}{a^2 \sec^2 \theta} - \frac{y^2}{b^2} = 1$$

19. (a)

(i) Equation of chord of contact of tangent from point ϕ $P(h, k)$ to the parabola $y^2 = 4ax$ is

$$yk = 2a(x + h) \Rightarrow y = \frac{2ax}{k} + \frac{2ah}{k}$$

This line touches the parabola $x^2 = 4by$

$\therefore x^2 = 4b\left(\frac{2ax}{k} + \frac{2ah}{k}\right)$ has equal roots

$kx^2 - 8abx - 8abh = 0$ has equal roots

$$\therefore \quad 64a^2b^2 - 32abhk = 0$$

$$2ab + hk = 0$$

Locus of $P(h, k)$ is $xy + 2ab = 0$

which represents the hyperbola.

(ii) We have,

$$x^2 + y^2 - 2(t^2 - 3t + 1)x - 2(t^2 + 2t)y + t = 0$$

Centre of circle is $(t^2 - 3t + 1, t^2 + 2t)$.

$$\therefore \quad h = t^2 - 3t + 1 \text{ and } k = t^2 + 2t$$

(iii) Equation of tangent at $(a \cos \alpha, b \sin \alpha)$ of ellipse

$$\frac{x^2}{a^2} + \frac{y^2}{b^2} = 1 \text{ is } \frac{x}{a}\cos \alpha + \frac{y}{b}\sin \alpha = 1 \quad \ldots(i)$$

Similarly, at $(a \cos \beta, b \sin \beta)$ is $\frac{x \cos \beta}{a} = \frac{y}{b}\sin \beta = 1 \quad \ldots(ii)$

From Eqs. (i) and (ii), we get

$$x = \frac{a \cos\left(\frac{\alpha + \beta}{2}\right)}{\cos\left(\frac{\alpha - \beta}{2}\right)} \text{ and } y = \frac{b \sin\left(\frac{\alpha + \beta}{2}\right)}{\cos\left(\frac{\alpha - \beta}{2}\right)}$$

or $\quad \frac{a \cos\left(\frac{\alpha + \beta}{2}\right)}{x} = \frac{b \sin\left(\frac{\alpha + \beta}{2}\right)}{y}$

$$\Rightarrow \quad ay \cos\left(\frac{\alpha + \beta}{2}\right) = bx \sin\left(\frac{\alpha + \beta}{2}\right)$$

Which represents equation of line.

(iv) Let the two given lines be taken as the X and Y-axes respectively and $c(h, k)$ be the centre of the ellipse any point.

Consider a and b be the lengths of semi-major and semi-minor axes of the ellipse.

Equation of director circle of the ellipse is

$$(x-h)^2 + (y-k)^2 = a^2 + b^2$$

Since, OX and OY are mutually perpendicular tangents sliding ellipse of all its position.

$\therefore$ $O(0, 0)$ will lie on circle.

$$\therefore \quad h^2 + k^2 = a^2 + b^2$$

Hence, locus of $c(h, k)$ is $x^2 + y^2 = a^2 + b^2$ which represents the circle.

Hence, (i) $\rightarrow$ (s), (ii) $\rightarrow$ (r), (iii) $\rightarrow$ (p), (iv) $\rightarrow$ (q)

Explanations for Q. Nos. (20-22)

$Y - y = m(X - x)$, if $Y = 0$. Then,

$X = x - \dfrac{y}{m}$ and if $X = 0$, then $Y = y - mx$

Hence, $\quad x - \dfrac{y}{m} = 2x$

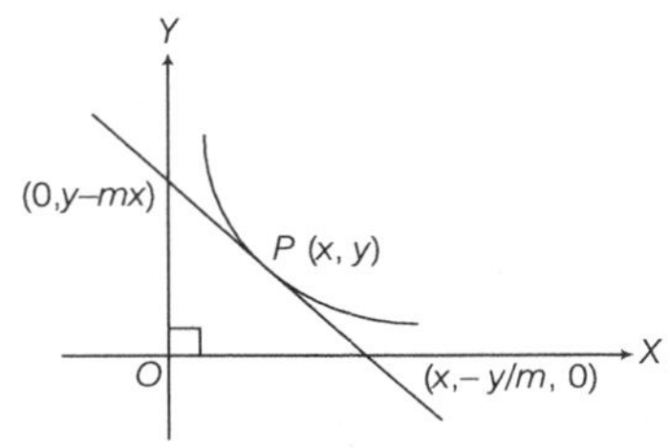

$$\Rightarrow \quad \frac{dy}{dx} = -\frac{y}{x}$$

$$\int \frac{dy}{y} + \int \frac{dx}{x} = c$$

$$\Rightarrow \quad xy = c$$

Passes through (2, 4)

$\Rightarrow$ Equation of conic is $xy = 8$

which is a rectangular hyperbola with $e = \sqrt{2}$

Hence, the two vertices are $(2\sqrt{2}, 2\sqrt{2})$ and $(-2\sqrt{2}, -2\sqrt{2})$

Focii are (4, 4) and (– 4, – 4)

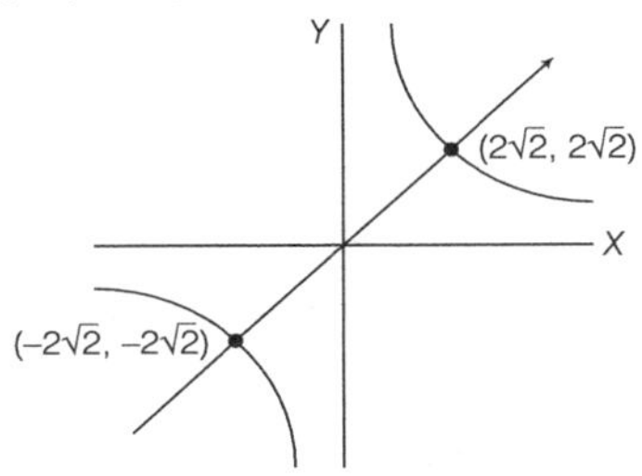

$\therefore$ Equation of S is $x^2 + y^2 = 32$

20. (b) **21.** (d) **22.** (c)

23. (6) Director circle of $\dfrac{x^2}{a^2} + \dfrac{y^2}{b^2} = 1$ is

$$x^2 + y^2 = a^2 + b^2$$

and director circle of $\dfrac{x^2}{a^2} - \dfrac{y^2}{b^2}$ is

$$x^2 + y^2 = a^2 - b^2$$

It is given that

$$a^2 + b^2 = 4r^2$$

and $\quad a^2 - b^2 = r^2$

On solving the equations, we get

$$a = \sqrt{\frac{5}{2}}r \text{ and } b = \frac{\sqrt{3}}{\sqrt{2}}r$$

e_e is the eccentricity of ellipse

$$\therefore \quad e_e = \sqrt{\frac{a^2 - b^2}{a^2}} = \sqrt{\frac{r^2}{\frac{5}{2}r^2}} = \sqrt{\frac{2}{5}}$$

and e_h is the eccentricity of hyperbola

$$\therefore \quad e_h = \sqrt{\frac{a^2 + b^2}{a^2}} = \sqrt{\frac{4r^2}{\frac{5}{2}r^2}} = \sqrt{\frac{8}{5}}$$

Then, $\quad 4e_h^2 - e_e^2 = 4 \times \dfrac{8}{5} - \dfrac{2}{5} = \dfrac{30}{5} = 6$

24. (2) Equation of tangent at point (h, k) on the hyperbola $\dfrac{x^2}{a^2} - \dfrac{y^2}{b^2} = 1$ is

$$\frac{xh}{a^2} - \frac{yk}{b^2} = 1$$

$$\Rightarrow \quad x = \frac{(a^2b^2 + yka^2)}{b^2h}$$

This tangent cuts the circle $x^2 + y^2 = a^2$

Putting the value of x in equation of circle, we get

$$\left(\frac{a^2b^2 + yka^2}{b^2h}\right)^2 + y^2 = a^2$$

$$a^4b^4 + 2yka^4b^4 + y^2k^2a^4 + y^2b^4h^2 = a^2b^4h^2$$

$$\Rightarrow \quad (k^2a^4 + b^4h^2)y^2 + 2y\,ka^4b^2 + a^4b^4 - a^2b^4h^2 = 0$$

Now, sum of roots, i.e.

$$y_1 + y_2 = \frac{-2ka^4b^2}{k^2a^4 + b^4h^2}$$

and product of roots i.e.

$$y_1y_2 = \frac{a^4b^4 - a^2b^4h^2}{k^2a^4 + b^4h^2}$$

Now, $\quad \dfrac{1}{y_1} + \dfrac{1}{y_2} = \dfrac{y_1 + y_2}{y_1y_2}$

$$= \frac{-2ka^4b^2}{a^4b^4 - a^2b^4h^2} = \frac{-2ka^4b^2}{a^2b^4(a^2 - h^2)}$$

$$= -\frac{2ka^2}{b^2(a^2 - h^2)} = \frac{-2k}{b^2\left(1 - \frac{h^2}{a^2}\right)}$$

$$= -\frac{2k}{b^2\left[-\frac{k^2}{b^2}\right]} = \frac{2}{k} \quad \left[\because \frac{h^2}{a^2} - \frac{k^2}{a^2} = 1\right]$$

$$\therefore \quad \lambda = 2$$

25. (1) Let the tangent of point $P(a \sec \theta, b \tan \theta)$ on the hyperbola $\dfrac{x^2}{a^2} - \dfrac{y^2}{b^2} = 1$ be $\dfrac{x \sec \theta}{a} - \dfrac{y \tan \theta}{b} = 1$.

this line meets the coordinate axes at A and B

$\therefore$ Coordinates of point $A(a \cos \theta, 0)$ and $B(0, -b\cot \theta)$.

Since, $OACB$ is formed a rectangle.

$\therefore \quad C$ is $(a \cos \theta, -b\cot \theta)$

Now, C lies on $\quad \dfrac{a^2}{x^2} - \dfrac{b^2}{y^2} = \lambda$

Then, $\quad \dfrac{a^2}{a^2\cos^2\theta} - \dfrac{b^2}{b^2\cot^2\theta} = \lambda$

$$\Rightarrow \quad \sec^2\theta - \tan^2\theta = \lambda$$

$$\therefore \quad \lambda = 1$$

DPP-3 **Normal, Chord Bisected and Product of Pair of Tangents**

1. (b) Given equation of hyperbola is $25x^2 - 16y^2 = 400$

and mid-point of the chord is (5, 3).

We know that, equation of chord bisected at given point is

$$T = S_1$$

$\because \quad 25xx_1 - 16yy_1 - 400 = 25x_1^2 - 16y_1^2 - 400$

$\Rightarrow \quad 25x(5) - 16y(3) = 25(5)^2 - 16(3)^2$

$\Rightarrow \quad 125x - 48y = 625 - 144$

$\therefore \quad 125x - 48y = 481$

2. (d) Given equation of hyperbola is

$$25x^2 - 16y^2 = 400$$

and mid-point of the chord is (6, 2).

We know that, the equation of mid-point of the chord is

$$T = S_1$$

Now, $\quad 25xx_1 - 16yy_1 - 400 = 25x_1^2 - 16y_1^2 - 400$

$\Rightarrow \quad 25x(6) - 16y(2) = 25(6)^2 - 16(2)^2$

$\Rightarrow \quad 150x - 32y = 900 - 64 = 836$

$\therefore \quad 75x - 16y = 418$

3. (a) Let the mid-point of the chord be $P(h, k)$.

Equation of chord bisected at $P(h, k)$ is

$3xh - 2yk + 2(x + h) - 3(y + k)$

$$= 3h^2 - 2k^2 + 4h - 6k$$

$\Rightarrow x(3h + 2) + y(-2k - 3) + 2h - 3k$

$$= 3h^2 - 2k^2 + 4h - 6k$$

$\Rightarrow \quad x(3h + 2) + y(-2k - 3) = 3h^2 - 2k^2 + 2h - 3k$

$\therefore \quad$ Slope of this chord $= \dfrac{3h + 2}{2k + 3}$

Since, this chord is parallel to the line $y = 2x$

[slope of this line is 2]

$\because \quad \dfrac{3h + 2}{2k + 3} = 2$

[$\because$ two lines are parallel, if $m_1 = m_2$]

$\Rightarrow \quad 3h + 2 = 4k + 6$

$$3h - 4k = 4$$

Now, taking locus of $P(h, k)$, we get

$$3x - 4y = 4$$

4. (c) We know that, the equation of normal to the hyperbola $\dfrac{x^2}{a^2} - \dfrac{y^2}{b^2} = 1$ at (x_1, y_1) is $\dfrac{a^2x}{x_1} + \dfrac{b^2y}{y_1} = a^2 + b^2$.

$\because \quad \dfrac{a^2x}{a \sec \theta} + \dfrac{b^2y}{b \tan \theta} = a^2 + b^2$

$\Rightarrow \quad ax \cos \theta + by \cot \theta = a^2 + b^2$

5. (b) We have, $x = 9$ is the chord of contact, hence $(9, 6\sqrt{2})$ and $(9, -6\sqrt{2})$ is point of contact of the hyperbola.

Let $P(h, k)$ be the point for which pair of tangents are drawn.

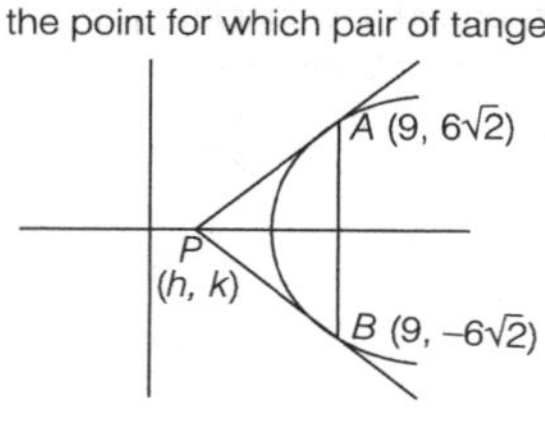

$\because$ Equation of pair of tangents is $SS_1 = T^2$.

i.e. $\quad (x^2 - y^2 - 9)(h^2 - k^2 - 9) = (xh - yk - 9)^2 \quad$ ...(i)

Since, pair of tangents passes through $A(9, 6\sqrt{2})$ and $B(9, -6\sqrt{2})$.

Hence, $\quad 9h - 6\sqrt{2}k - 9 = 0 \quad$ ...(ii)

$9h + 6\sqrt{2} - 9 = 0 \quad$ ...(iii)

On solving Eqs. (ii) and (iii), we get

$$h = 1 \text{ and } k = 0$$

Now, substituting $h = 1$ and $k = 0$ in Eq. (i), we get

$$-8(x^2 - y^2 - 9) = x^2 - 18x + 81$$

$\Rightarrow \quad 9x^2 - 8y^2 - 18x + 9 = 0$

6. (d) The equation of normal to the hyperbola $\dfrac{x^2}{a^2} - \dfrac{y^2}{b^2} = 1$ at $P(a \sec \theta, b \tan \theta)$ is

$ax \cos \theta + by \cot \theta = a^2 + b^2 \quad$ ...(i)

Similarly, the equation at $Q(a \sec \phi, b \tan \phi)$ is

$ax \cos \phi + by \cot \phi = a^2 + b^2 \quad$ ...(ii)

We have, $\quad \theta + \phi = \dfrac{\pi}{2}$

$$ax \cos\left(\frac{\pi}{2} - \theta\right) + by \cot\left(\frac{\pi}{2} - \theta\right) = a^2 + b^2$$

$ax \sin \theta + by \tan \theta = a^2 + b^2 \quad$ ...(iii)

Now, multiplying in Eq. (i) by $\sin \theta$ and Eqs. (iii) by $\cos \theta$ and then subtracting, we get

$$by(\cos \theta - \sin \theta) = (a^2 + b^2)(\sin \theta - \cos \theta)$$

$\because \quad y = k = -\left(\dfrac{a^2 + b^2}{b}\right)$

7. (c) Let the mid-point of the chord of the circle be (h, k).

$\because$ Equation of chord is $T = S_1$

$\Rightarrow \quad xh + yk - 16 = h^2 + k^2 - 16$

$\Rightarrow \quad xh + yk = h^2 + k^2$

$\Rightarrow \quad y = -\dfrac{xh}{k} + \dfrac{h^2 + k^2}{k}$

Comparing with $y = mx + c$, we get

$$m = -\frac{h}{k}$$

and $\quad c = \dfrac{h^2 + k^2}{k}$

Since, this chord of the circle is the tangent to the hyperbola

$\dfrac{x^2}{16} - \dfrac{y^2}{9} = 1 \quad [a^2 = 16 \text{ and } b^2 = 9]$

By condition of tangency

$$c^2 = a^2m^2 - b^2$$

$$\left(\frac{h^2 + k^2}{k^2}\right)^2 = 16\left(-\frac{h}{k}\right)^2 - 9$$

$\Rightarrow \quad (h^2 + k^2)^2 = k^2\left(\dfrac{16h^2}{k^2} - 9\right)$

$\Rightarrow \quad (h^2 + k^2)^2 = 16h^2 - 9k^2$

Now, on taking locus of (h, k), we get

$$(x^2 + y^2)^2 = 16x^2 - 9y^2$$

8. (b) Equation normals of slope m to the hyperbola $\frac{x^2}{a^2}-\frac{y^2}{b^2}=1$ is

$$y = mx \pm \frac{m(a^2+b^2)}{\sqrt{a^2-b^2m^2}}$$

which is biquadratic equation in m.

$\therefore$ 4 normals can be drawn from an external point of a hyperbola.

9. (a) We have, $x\cos\alpha + y\sin\alpha = p$

$$\Rightarrow \quad y = -\cot\alpha\cdot x + p\,\text{cosec}\,\alpha$$

$$\because \quad m = -\cot\alpha \text{ and } c = p\,\text{cosec}\,\alpha$$

We know that

$y = mx + c$ is normal to the hyperbola $\frac{x^2}{a^2}-\frac{y^2}{b^2}=1$, if

$$c^2 = \frac{m^2(a^2+b^2)^2}{a^2-b^2m^2}$$

$$\Rightarrow \quad p^2\,\text{cosec}^2\alpha = \frac{\cot^2\alpha\,(a^2+b^2)^2}{a^2-b^2\cot^2\alpha}$$

$$\Rightarrow p^2(a^2\,\text{cosec}^2\alpha - b^2\,\text{cosec}^2\alpha\cdot\cot^2\alpha) = \cot^2\alpha\,(a^2+b^2)^2$$

$$\Rightarrow \quad p^2\sec^2\alpha\,(a^2-b^2\cot^2\alpha) = (a^2+b^2)^2$$

$$\therefore \quad a^2\sec^2\alpha - b^2\,\text{cosec}^2\alpha = \frac{(a^2+b^2)^2}{p^2}$$

10. (c) Let point on the hyperbola $x^2 - y^2 = 4$ be $(2\sec\theta, 2\tan\theta)$.

Equation of tangent at $(2\sec\theta, 2\tan\theta)$

$$\frac{x}{2}\sec\theta - \frac{y}{2}\tan\theta = 1$$

X-intercept, i.e. $\quad a_1 = 2\cos\theta$

Y-intercept, i.e. $\quad b_1 = -2\cot\theta$

Now, the equation of normal at $(2\sec\theta, 2\tan\theta)$ is

$$2x\cos\theta + 2y\cot\theta = a^2+b^2$$

X-intercept, i.e. $\quad a_2 = \frac{a^2+b^2}{2\cos\theta}$

Y-intercept, i.e. $\quad b_2 = \frac{a^2+b^2}{2\cot\theta}$

$$\therefore \quad a_1a_2 + b_1b_2 = 2\cos\theta\times\frac{a^2+b^2}{2\cos\theta} + (-2\cot\theta)\left(\frac{a^2+b^2}{2\cot\theta}\right)$$

$$= a^2+b^2-a^2-b^2 = 0$$

11. (c) Let $P(a\sec\theta, b\tan\theta)$ be a point on the hyperbola

$$\frac{x^2}{a^2}-\frac{y^2}{b^2}=1,$$

where $e = \sqrt{\frac{a^2+b^2}{a^2}}$

Equation of normal is

$$ax\cos\theta + by\cot\theta = a^2+b^2$$

$\therefore$ X-intercept i.e. $\quad L = \left(\frac{a^2+b^2}{a\cos\theta}, 0\right)$

and Y-intercept i.e. $\quad M = \left(0, \frac{a^2+b^2}{b\cot\theta}\right)$

Let mid-point of L and M be $Q(h, k)$.

$$\therefore \quad h = \frac{a^2+b^2}{2a\cos\theta} \quad\text{and}\quad k = \frac{a^2+b^2}{2b\cot\theta}$$

$$\text{or} \quad \frac{2ha}{a^2+b^2} = \sec\theta \quad\text{and}\quad \frac{2bk}{a^2+b^2} = \tan\theta$$

$$\because \quad \sec^2\theta - \tan^2\theta = 1$$

$$\Rightarrow \quad \frac{4h^2a^2}{(a^2+b^2)^2} - \frac{4b^2k^2}{(a^2+b^2)^2} = 1$$

On taking locus of $Q(h, k)$, we get

$$\frac{4x^2a^2}{(a^2+b^2)^2} - \frac{4b^2y^2}{(a^2+b^2)^2} = 1$$

$$\Rightarrow \quad \frac{x^2}{\left(\frac{a^2+b^2}{2a}\right)^2} - \frac{y^2}{\left(\frac{a^2+b^2}{2b}\right)^2} = 1$$

which is a hyperbola.

And its eccentricity is

$$e_1^2 = \frac{\left(\frac{a^2+b^2}{2a}\right)^2 + \left(\frac{a^2+b^2}{2b}\right)^2}{\left(\frac{a^2+b^2}{2a}\right)^2}$$

$$e_1^2 = \frac{\frac{1}{4a^2}+\frac{1}{4b^2}}{\frac{1}{4a^2}} = \frac{b^2+a^2}{b^2}$$

$$\because \quad e_1^2 = \frac{a^2+b^2}{b^2} = \frac{a^2e^2}{b^2}$$

$$\because \quad e_1^2 = e^2\cdot\frac{1}{e^2-1} \qquad \left[\because \frac{a^2}{b^2} = \frac{1}{e^2-1}\right]$$

$$\therefore \quad e_1 = \frac{e}{\sqrt{e^2-1}}$$

12. (b) We have,

$$lx + my + n = 0$$

$$\text{or} \quad y = -\frac{lx}{m} - \frac{n}{m}$$

Now, $\quad$ slope $m' = -\frac{l}{m}$ and $c = -\frac{n}{m}$

Then, condition of normal is

$$c^2(a^2 - b^2m'^2) = m'^2(a^2+b^2)^2$$

$$\frac{n^2}{m^2}\left(a^2 - b^2\frac{l^2}{m^2}\right) = \frac{l^2}{m^2}(a^2+b^2)^2$$

$$\Rightarrow \quad \frac{n^2}{m^2}(a^2m^2 - b^2l^2) = l^2(a^2+b^2)^2$$

$$\therefore \quad \frac{a^2}{l^2} - \frac{b^2}{m^2} = \frac{(a^2+b^2)^2}{n^2}$$

13. (a) Given equation ot hyperbola is $\frac{x^2}{a^2}-\frac{y^2}{b^2}=1$.

Its vertices $A(a, 0)$ and $A' = (-a, 0)$

Normal at any point θ i.e. $(a\sec\theta, b\tan\theta)$ is

$$ax\cos\theta + by\cot\theta = a^2+b^2$$

Now, G is it intersection on transverse axis i.e. X-axis.

$$\therefore \quad G = \left(\frac{(a^2+b^2)\sec\theta}{a}, 0\right)$$

$$\Rightarrow \quad AG = a - \frac{(a^2+b^2)\sec\theta}{a}$$

$$\Rightarrow \quad A'G = -a - \frac{(a^2+b^2)\sec\theta}{a}$$

$$\therefore \quad AG\times A'G = a^2 - \frac{(a^2+b^2)^2\cdot\sec^2\theta}{a^2}$$

$$= a^2 - \frac{a^4e^4\cdot\sec^2\theta}{a^2} \qquad [\because a^2+b^2 = a^2e^2]$$

$$= a^2 - a^2e^4\cdot\sec^2\theta$$

$$= a^2(1 - e^4\sec^2\theta)$$

$$= a^2(e^4\sec^2\theta - 1)$$

14. (c) Let the normal points be $(x_1, y_1), (x_2, y_2), (x_3, y_3)$ and (x_4, y_4).

Equation of normal is

$$\frac{a^2(x - x_i)}{x_i} = \frac{b^2(y - y_i)}{y_i} \quad [\text{where, } i = 1, 2, 3, 4]$$

Let these normal passes through the point (h, k).

So, $$\frac{a^2(h - x)}{x} = \frac{b^2(k - y)}{y}$$

$\Rightarrow$ $$a^2y(h - x) = b^2x(k - y)$$

i.e. $$a^2y(x - h) + b^2x(y - k) = 0$$

15. (a) Let point on the circle $x^2 + y^2 = a^2$ is $P(a\cos\theta, a\sin\theta)$.

Now, the equation of chord of contact of $P(a\cos\theta, a\sin\theta)$ with respect to hyperbola $x^2 - y^2 = a^2$ is

$$x \cdot a\cos\theta - y \cdot a\sin\theta = a^2$$

$\Rightarrow$ $$x\cos\theta - y\sin\theta = a \quad \text{...(i)}$$

Let mid-point of chord of contact $x^2 - y^2 = 4$ be (h, k).

Equation of chord of contact is $T = S_1$

$$xh - yk - 4 = h^2 - k^2 - 4$$

$\therefore$ $$xh - yk = h^2 - k^2 \quad \text{...(ii)}$$

Since, Eqs. (i) and (ii) are identical, then

$$\frac{\cos\theta}{h} = \frac{\sin\theta}{k} = \frac{a}{h^2 - k^2}$$

$\Rightarrow$ $$\cos\theta = \frac{ah}{h^2 - k^2} \text{ and } \sin\theta = \frac{ak}{h^2 - k^2}$$

$\because$ $$\sin^2\theta + \cos^2\theta = 1$$

$\Rightarrow$ $$\left(\frac{ak}{h^2 - k^2}\right)^2 + \left(\frac{ah}{h^2 - k^2}\right)^2 = 1$$

$\Rightarrow$ $$a^2k^2 + a^2h^2 = (h^2 - k^2)^2$$

Now, on taking locus of (h, k), we get

$$a^2y^2 + a^2x^2 = (x^2 - y^2)^2$$

or $$(x^2 - y^2)^2 = a^2(x^2 + y^2)$$

16. (a) Given the pair of lines $lx^2 + 2mxy + ny^2 = 0$

which represents two straight lines passing through origin having equation $y = Mx$ and $y = M_1x$.

$\therefore$ $$MM_1 = \frac{l}{n} \quad \text{...(i)}$$

Let $y = M_x$ and $y = M_1x$ be two conjugate diameter of hyperbola $\frac{x^2}{a^2} - \frac{y^2}{b^2} = 1$, then

$$M_1M = \frac{b^2}{a^2} \quad \text{...(ii)}$$

From Eqs. (i) and (ii), we get

$$\frac{l}{n} = \frac{b^2}{a^2}$$

$\Rightarrow$ $$la^2 = nb^2$$

17. (c) Let the pole be (x_1, y_1).

$\because$ Equation of polar is $$\frac{xx_1}{a^2} - \frac{yy_1}{b^2} = 1 \quad \text{...(i)}$$

[equation of polar is $T = 0$]

We have the polar is $lx + my = -n$ ...(ii)

Since, Eqs. (i) and (ii) are identical, then

$$\frac{x_1}{a^2l} = -\frac{y_1}{b^2m} = -\frac{1}{n}$$

$\Rightarrow$ $$x_1 = -\frac{a^2l}{n} \text{ and } y_1 = \frac{b^2m}{n}$$

$\therefore$ $$\text{Pole} = \left(-\frac{a^2l}{n}, \frac{b^2m}{n}\right)$$

18. (a) Let (h, k) be the pole of a normal chord

$$ax\cos\theta + by\cot\theta = a^2 + b^2 \quad \text{...(i)}$$

of the hyperbola $\frac{x^2}{a^2} - \frac{y^2}{b^2} = 1$

Now, the equation of polar of (h, k) with the given hyperbola is

$$\frac{xh}{a^2} - \frac{yk}{b^2} = 1 \quad \text{...(ii)}$$

Since, Eqs. (i) and (ii) are identical, then

$$\frac{a\cos\theta}{\frac{h}{a^2}} = \frac{b\cot\theta}{-\frac{k}{b^2}} = a^2 + b^2$$

$\Rightarrow$ $$a\cos\theta = \frac{h}{a^2(a^2 + b^2)}$$

and $$b\cot\theta = -\frac{k}{b^2(a^2 + b^2)}$$

$\Rightarrow$ $$\sec\theta = \frac{a^3}{h(a^2 + b^2)}$$

and $$\tan\theta = -\frac{b^3}{k(a^2 + b^2)}$$

$\because$ $$\sec^2\theta - \tan^2\theta = 1$$

$\Rightarrow$ $$\frac{a^6}{(a^2 + b^2)^2 h^2} - \frac{b^6}{k^2(a^2 + b^2)^2} = 1$$

$\Rightarrow$ $$k^2a^6 - b^6h^2 = h^2k^2(a^2 + b^2)^2$$

Now, on taking locus of (h, k), we get

$$a^6y^2 - b^6x^2 = x^2y^2(a^2 + b^2)^2$$

19. (a)

(i) Property of ellipse and hyperbola

Hence, **(i) → (r, s)**

(ii) Length of subnormal $= |y| \frac{dy}{dx}$

$\because$ $$y \cdot \frac{dy}{dx} = 2$$

$\therefore$ $$\int y\,dy = \int 2dx$$

$\Rightarrow$ $$\frac{y^2}{2} = 2x + C$$

Since, the curve passes through $(1, 2)$.

$\Rightarrow$ $$\frac{4}{2} = 2 + C$$

$\Rightarrow$ $$C = 0$$

$\because$ $$\frac{y^2}{2} = 2x$$

$\therefore$ $$y^2 = 4x,$$

which is a parabola.

Hence, **(ii) → (q)**

(iii) Let $P(x_1, y_1)$

Equation of normal at P is

$$y - y_1 = -\frac{1}{p}(x - x_1),$$

where $p = \left(\frac{dy}{dx}\right)_{\text{at }(x_1, y_1)}$.

Its X-intercept $(Q) = (py_1 + x_1, 0)$

Now, the mid-point of PQ is

$$\left(\frac{py_1 + 2x_1}{2}, \frac{y_1}{2}\right)$$

Since, mid-point of PQ is bisected by Y-axis, then

$$\frac{py_1 + 2x_1}{2} = 0$$

$\therefore$ $$py_1 = -2x_1$$

$$\Rightarrow \quad y \cdot \frac{dy}{dx} = -2x$$

$$\Rightarrow \quad \int y\,dy = \int -2x\,dx$$

$$\Rightarrow \quad \frac{y^2}{2} = -x^2 + C$$

Since, the curve passes through the point (1, 4).

$$\therefore \quad \frac{16}{2} = -1 + C$$

$$\Rightarrow \quad C = 9$$

The curve is $\quad x^2 + \dfrac{y^2}{2} = 9$

$$\Rightarrow \quad \frac{x^2}{9} + \frac{y^2}{18} = 1$$

Which is an ellipse.

Hence, **(iii) → (r)**

(iv) We know that length of normal at any point (x_1, y_1) is $|y_1| \sqrt{1 + p^2}$, where $p = \left(\dfrac{dy}{dx}\right)_{\text{at }(x_1, y_1)}$

$$\because \quad y_1 \sqrt{1 + p^2} = 2$$

$$\Rightarrow \quad 1 + p^2 = \frac{4}{y^2}$$

$$\Rightarrow \quad p^2 = \frac{4}{y^2} - 1 = \frac{4 - y^2}{y^2}$$

$$\Rightarrow \quad p = \frac{\sqrt{4 - y^2}}{y}$$

$$\Rightarrow \quad \frac{dy}{dx} = \frac{\sqrt{4 - y^2}}{y}$$

$$\Rightarrow \quad \int \frac{y}{\sqrt{4 - y^2}} dx = \int dx + C$$

$$\Rightarrow \quad -\sqrt{4 - y^2} = x + C$$

Since, the curve passes through the point (1, 2),

$$1 + C = 0$$

$$\Rightarrow \quad C = -1$$

$$\Rightarrow \quad -\sqrt{4 - y^2} = x - 1$$

$$\Rightarrow \quad 4 - y^2 = (x - 1)^2$$

$$\Rightarrow \quad (x - 1)^2 + y^2 = 4,$$

Which is a circle.

Hence, **(iv) → (p)**

Explanations for Q. Nos. (20 and 21)

Let $P(a \sec \theta, b \tan \theta)$ be any point on the hyperbola $\dfrac{x^2}{a^2} - \dfrac{y^2}{b^2} = 1$.

Then, equation of tangent and normal at P are

$$\frac{x}{a} \sec \theta - \frac{y}{b} \tan \theta = 1 \quad \text{...(i)}$$

and $\quad ax \cos \theta + by \cot \theta = a^2 + b^2 \quad$ ...(ii)

X-intercept of Eq.(ii), i.e. $G = \left[\dfrac{(a^2 + b^2)}{a} \sec \theta, 0\right]$

Y-intercept of Eq.(ii), i.e. $g = \left[0, \dfrac{(a^2 + b^2)}{b} \tan \theta\right]$

$$\therefore \quad PG^2 = \left[a \sec \theta - \frac{(a^2 + b^2)}{a} \sec \theta\right]^2 + (b \tan \theta - 0)^2$$

$$= \frac{b^4}{a^2} \sec^2\theta + b^2 \tan^2\theta$$

$$= \frac{b^2}{a^2} (b^2 \sec^2\theta + a^2 \tan^2\theta) \quad \text{...(iii)}$$

and $\quad Pg^2 = (a \sec \theta - 0)^2 + \left[b \tan \theta - \dfrac{(a^2 + b^2)}{b} \tan \theta\right]^2$

$$= a^2 \sec^2\theta + \frac{a^4}{b^2} \tan^2\theta$$

$$= \frac{a^2}{b^2} (b^2 \sec^2\theta + a^2 \tan^2\theta) \quad \text{...(iv)}$$

20. (b) Clearly, from figure PF is equal to CL.

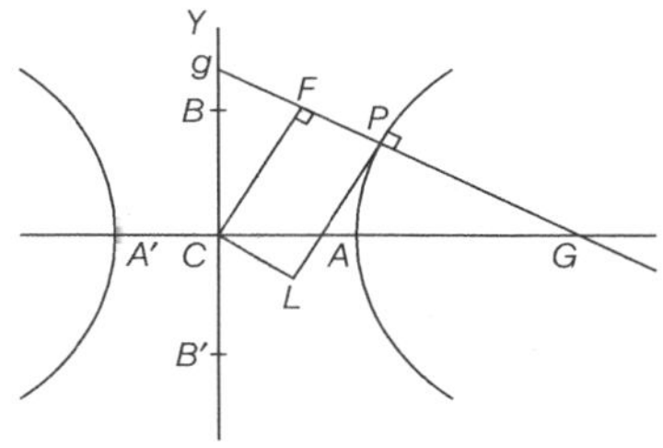

i.e. Length of perpendicular from $C(0, 0)$ to Eq. (i) is

$$PF = CL = \frac{1}{\sqrt{\dfrac{\sec^2\theta}{a^2} + \dfrac{\tan^2\theta}{b^2}}}$$

$$= \frac{ab}{\sqrt{b^2 \sec^2\theta + a^2 \tan^2\theta}}$$

$$\Rightarrow \quad PF = \frac{ab}{\frac{a}{b} PG} \quad \text{[from Eq. (iii)]}$$

$$\because \quad PF \cdot FG = b^2$$

21. (a) Again, $PF = \dfrac{ab}{\sqrt{b^2 \sec^2\theta + a^2 \tan^2\theta}} = \dfrac{ab}{\frac{b}{a} \cdot Pg}$

$$\therefore \quad PF \cdot Pg = a^2$$

22. (b) We have, $\quad L : x = 9$

and $\quad H : x^2 - y^2 = 9$

In Q. No. 5 of this DPP it is solved.

23. (c) In Q. No. 5, we solve the point of intersection of tangents at the extremities of the hyperbola H.

Clearly, $R = (1, 0)$

Now, chord of contact of parabola P

$$y^2 = 4x - 20 \text{ is}$$

$$\Rightarrow \quad yy_1 = 2(x + x_1) - 20$$

$$\Rightarrow \quad 0 = 2(x + 1) - 20$$

$$\Rightarrow \quad 2(x + 1) = 20$$

$$\Rightarrow \quad x + 1 = 10$$

$$\therefore \quad x = 9$$

24. (c) We have the parabola $y^2 = 4(x - 5)$

$$\because \quad T = (9, 4) \text{ and } T' = (9, -4)$$

Focus S of the parabola is (6, 0).

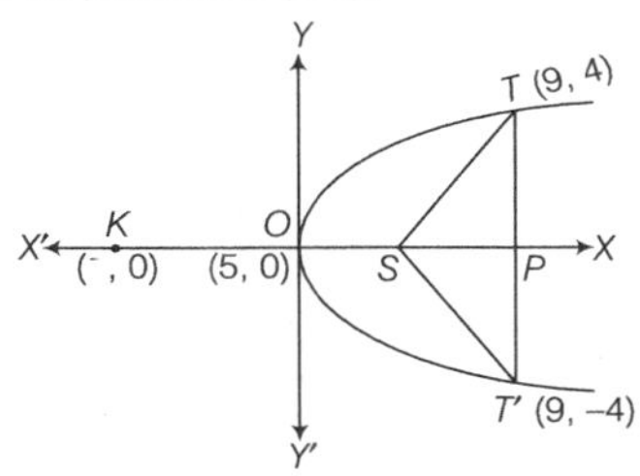

$\therefore$ Area of $\Delta STT'$ is $= \frac{1}{2} \times TT' \times SP$

$$= \frac{1}{2} \times 8 \times 3 = 4 \times 3$$

$$= 12 \text{ sq units}$$

25. (c)In Q. No. 20 to 21 of passage 1, we have

$$D = \left(\frac{a^2 + b^2}{a} \sec\theta, 0\right),$$

$$E = \left(0 \; \frac{a^2 + b^2}{b} \tan\theta\right)$$

and $\quad AD^2 = \frac{b^2}{a^2}(b^2 \sec^2\theta + a^2 \tan^2\theta)$

26. (a) We have,

$$AD^2 = \frac{b^2}{a^2}(b^2 \sec^2\theta + a^2 \tan^2\theta)$$

$$= \frac{b^2}{a^2}[b^2 \sec^2\theta + a^2(\sec^2\theta - 1)]$$

$$= \frac{b^2}{a^2}[\sec^2\theta (a^2 + b^2) - a^2]$$

We know, that $\sec^2\theta \ge 1$

$$(a^2 + b^2)\sec^2\theta \ge a^2 + b^2$$

$\Rightarrow \quad (a^2 + b^2)\sec^2\theta - a^2 \ge b^2$

$\Rightarrow \quad \frac{b^2}{a^2}[(a^2 + b^2)\sec^2\theta - a^2] \ge \frac{b^2}{a^2} \times b^2$

$\Rightarrow \quad AD^2 \ge \frac{b^4}{a^2}$

$\Rightarrow \quad AD \ge \frac{b^2}{a}$

$\therefore$ Minimum value of $AD = \frac{b^2}{a}$

27. (a)From Q. 20 to 21, clearly

$$AE^2 = \frac{a^2}{b^2}(b^2 \tan^2\theta + a^2 \sec^2\theta)$$

28. (4) Equation of normal to the hyperbola $x^2 - y^2 = a^2$ at the point $(a \sec\theta, b \tan\theta)$.

$$ax \cos\theta + ay \cot\theta = 2a^2$$

$\because \quad x \cos\theta + y \cot\theta = 2a \quad$ ...(i)

Again, equation of chord to the given hyperbola whose mid-point is (h, k), is

$$hx - yk = h^2 - k^2 \quad \text{...(i)}$$

$[\because T = S_1]$

Since, Eqs. (i) and (ii) are identical, then

$$\frac{\cos\theta}{h} = \frac{\cot\theta}{-k} = \frac{2a}{h^2 - k^2}$$

$\Rightarrow \quad \sec\theta = \frac{h^2 - k^2}{2ah}$

$$\tan\theta = \frac{h^2 - k^2}{-2ak}$$

$\because \quad \sec^2\theta - \tan^2\theta = 1$

i.e. $\quad \frac{(h^2 - k^2)^2}{4a^2h^2} - \frac{(h^2 - k^2)^2}{4a^2k^2} = 1$

$\Rightarrow \quad \frac{(h^2 - k^2)^2}{4a^2}\left[\frac{1}{h^2} - \frac{1}{k^2}\right] = 1$

$\Rightarrow \quad (h^2 - k^2)^2(k^2 - h^2) = 4a^2h^2k^2$

$\because \quad -(h^2 - k^2)^3 = 4a^2h^2k^2$

On taking locus of (h, k), we get

$$-(x^2 - y^2)^3 = 4a^2x^2y^2 = \lambda a^2x^2y^2$$

$\Rightarrow \quad (y^2 - x^2)^3 = 4a^2x^2y^2$

$\therefore \quad \lambda = 4$

29. (0)Polar of (x_1, y_1) and (x_2, y_2) are

$$\frac{xx_1}{a^2} - \frac{yy_1}{b^2} = 1 \quad \text{...(i)}$$

and $\quad \frac{xx_2}{a^2} - \frac{yy_2}{b^2} = 1 \quad$ ...(ii)

Since, Eqs. (i) and (ii) are right angle.

Hence, $\quad m_1 = \frac{b^2x_1}{a^2y_1}$

and $\quad m_2 = \frac{b^2}{a^2}\frac{x_2}{y_2}$

Now, $\quad m_1m_2 = -1$

$\Rightarrow \quad \frac{b^4x_1x_2}{a^4y_1y_2} = -1$

$\Rightarrow \quad a^4y_1y_2 + b^4x_1x_2 = 0$

$\therefore \quad \lambda = 0$

30. (2) We have, CP and CD are part of conjugate diameters of the hyperbola $x^2 - y^2 = a^2$.

Hence, $P = (a \sec\theta, a \tan\theta)$

and $\quad D = (a \tan\theta, b \sec\theta)$

$\quad C = (0, 0)$

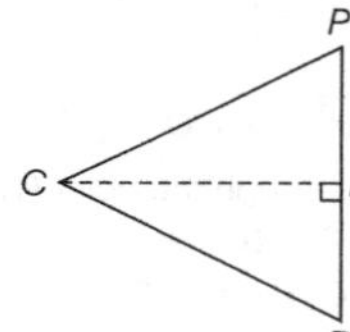

Clearly, ΔCPD is an isosceles triangle.

$\because M$ is the mid-point of PD.

i.e. $\quad M = \left[\frac{a}{2}(\sec\theta + \tan\theta), \frac{a}{2}(\tan\theta + \sec\theta)\right]$

$\Rightarrow \quad CM = \left|\frac{\sec\theta + \tan\theta}{2}\right|\sqrt{a^2 + a^2}$

$\Rightarrow \quad PD = \sqrt{[a(\sec\theta - \tan\theta)]^2 + [a(\sec\theta - \tan\theta)]^2}$

$$= |\sec\theta - \tan\theta|\sqrt{a^2 + a^2}$$

$\therefore$ Areas of $\Delta CPD = \frac{1}{2}PD \times CM$

$$= \frac{1}{4}(\sec\theta - \tan\theta)\sqrt{2a^2} \times \frac{(\sec\theta + \tan\theta)}{2}\sqrt{2a^2}$$

$$= \frac{1}{4}a^2(\sec^2\theta - \tan^2\theta)$$

$$= \frac{a^2}{2} \text{ but it is } \lambda a^2$$

$\therefore \quad \lambda = \frac{1}{2}$

Hence, $\quad |4\lambda| = 2$

DPP- 4 Asymptotes and Rectangular Hyperbola

1. (a) Since, $\frac{x}{a} \pm \frac{y}{b}$ are the two asymptotes of the hyperbola $\frac{x^2}{a^2} - \frac{y^2}{b^2} = 1.$

Now, $m_1 = -\frac{b}{a}$ and $m_2 = \frac{b}{a}$

$$\therefore \quad \tan\theta = \frac{\frac{b}{a} + \frac{b}{a}}{1 - \frac{b}{a} \times \frac{b}{a}} = \frac{2ab}{a^2 - b^2}$$

$$\Rightarrow \quad \theta = \tan^{-1}\frac{2ab}{a^2 - b^2} = 2\tan^{-1}\frac{b}{a}$$

2. (b) We have, hyperbola $x^2 - 2y^2 - 2 = 0$

$$\Rightarrow \quad \frac{x^2}{2} - y^2 = 1$$

Let any point on it is $(\sqrt{2}\sec\theta, \tan\theta)$ and its asymptotes are

$$\frac{x}{\sqrt{2}} + y = 0 \text{ and } \frac{x}{\sqrt{2}} - y = 0.$$

Let p_1 and p_2 be the length of perpendicular from $(\sqrt{2}\sec\theta, \tan\theta)$ to the asymptotes $\frac{x}{\sqrt{2}} + y = 0$ and $\frac{x}{\sqrt{2}} - y = 0$, respectively.

Now, $p_1 = \frac{\sec\theta + \tan\theta}{\sqrt{\frac{3}{2}}}$ and $p_2 = \frac{\sec\theta - \tan\theta}{\sqrt{\frac{3}{2}}}$

Hence, $p_1 \cdot p_2 = \frac{\sec^2\theta - \tan^2\theta}{\frac{3}{2}} = \frac{2}{3}$

3. (c) Equation of hyperbola is $\frac{x^2}{64} - \frac{y^2}{64} = 1$

or $$x^2 - y^2 = 64$$

Its eccentricity $= \sqrt{2}$ and directrices $= \pm\frac{8}{\sqrt{2}} = \pm 4\sqrt{2}$

$\therefore$ Distance between directrices is $8\sqrt{2}$.

4. (d) Equation of the asymptotes are

$$2x^2 - 5xy - 3y^2 - 5x - 3y + \lambda = 0 \quad \text{...(i)}$$

Comparing it with $ax^2 + 2hxy + by^2 + 2gx + 2fy + c = 0$

we get,

$$a = 2, h = -\frac{5}{2}, b = -3,$$

$$g = -\frac{5}{2}, f = -\frac{3}{2} \text{ and } c = \lambda.$$

Eq. (i) represents a pair of straight line, if

$$abc + 2fgh - af^2 - bg^2 - ch^2 = 0$$

$$\Rightarrow \quad 2\times(-3)\lambda + 2\left(-\frac{3}{2}\right)\left(-\frac{5}{2}\right)\left(-\frac{5}{2}\right) - 2\left(-\frac{3}{2}\right)^2 + \left(-\frac{5}{2}\right)^2 3 - \lambda\left(-\frac{5}{2}\right)^2 = 0$$

$$\Rightarrow \quad -6\lambda - \frac{75}{4} - \frac{9}{2} + 3\times\frac{25}{4} - \frac{25}{4}\lambda = 0$$

$$\Rightarrow \quad -6\lambda - \frac{9}{2} - \frac{25}{4}\lambda = 0$$

$$\Rightarrow \quad \frac{49}{4}\lambda = -\frac{9}{2}$$

$$\therefore \quad \lambda = -\frac{18}{49}$$

$\therefore$ Required asymptotes is

$$2x^2 - 5xy - 3y^2 - 5x - 3y - \frac{18}{49} = 0.$$

5. (b) The equation of hyperbola having asymptotes are $3x - 4y + 7 = 0$ and $4x + 3y + 1 = 0$ is

$$(3x - 4y + 7)(4x + 3y + 1) + \lambda = 0$$

Since, their hyperbola passes through the origin,

i.e. $$7 + \lambda = 0 \Rightarrow \lambda = -7$$

$\therefore$ Education hyperbola is

$$(3x - 4y + 7)(4x + 3y + 1) - 7 = 0$$

6. (c) Asymptotes are parallel to the lines

$$2x + 3y = 0 \text{ and } 3x + 2y = 0$$

$\therefore$ Equation of asymptotes are

$$2x + 3y = \lambda \text{ and } 3x + 2y = \mu$$

Since, this passes through the centre of the hyperbola (1, 2).

$$\lambda = +8 \text{ and } \mu = 7$$

Thus, the equation of asymptotes are

$$2x + 3y - 8 = 0 \text{ and } 3x + 2y - 7 = 0$$

Let the equation of hyperbola be

$$(2x + 3y - 8)(3x + 2y - 7) + k = 0$$

Since, this passes through (5, 3).

Hence, $$(10 + 9 - 8)(15 + 6 - 7) + k = 0$$

$$(11)(14) + k = 0$$

$$k = -154$$

$$\therefore \quad (2x + 3y - 8)(3x + 2y - 7) - 154 = 0$$

$$\Rightarrow \quad \lambda = 154$$

7. (b) We have, $$xy = c^2$$

This equation is obtained from one hyperbola

$$x^2 - y^2 = c^2$$

Its eccentricity $= \sqrt{2}$

$$\therefore \quad \text{Foci} = (\pm\sqrt{2}c, 0)$$

8. (d) Let the equation of circle be

$$x^2 + y^2 + 2gx + 2fy + k = 0 \quad \text{...(i)}$$

We have, $$\text{radius} = g^2 + f^2 - k = r^2$$

Let $P(x_1, y_1)$, $Q(x_2, y_2)$, $R(x_3, y_3)$ and $S(x_4, y_4)$ be the point of intersection of Eq. (i) and $xy = c^2$.

$$x^2 + \frac{c^4}{x^2} + 2gx + \frac{2fc^2}{x} + k = 0$$

$$\Rightarrow \quad x^4 + 2gx^3 + kx^2 + 2fc^2x + c^4 = 0$$

$$\Rightarrow \quad x_1 + x_2 + x_3 + x_4 = -2g$$

and $$\Sigma x_1x_2 = k$$

$$\therefore \quad x_1^2 + x_2^2 + x_3^2 + x_4^2 = (x_1 + x_2 + x_3 + x_4)^2 - 2\Sigma x_1x_2$$

$$= 4g^2 - 2k$$

Similarly, $y_1^2 + y_2^2 + y_3^2 + y_4^2 = 4f^2 - 2k$

$$\therefore \quad CP^2 + CQ^2 + CR^2 + CS^2$$

$$= x_1^2 + x_2^2 + x_3^2 + x_4^2 + y_1^2 + y_2^2 + y_3^2 + y_4^2$$

$$= 4g^2 - 2k + 4f^2 - 2k$$

$$= 4(g^2 + f^2 - k) = 4r^2$$

9. (a) We have, $$x^2 + y^2 = a^2 \quad \text{...(i)}$$

and $$xy = c^2 \quad \text{...(ii)}$$

Point of intersection of Eqs. (i) and (ii), we get

$$x^2 + \left(\frac{c^2}{x^2}\right)^2 = a^2 \Rightarrow x^2 + \frac{c^4}{x^2} = a^2$$

$$\Rightarrow \quad x^4 - a^2x^2 + c^4 = 0$$

$$\therefore \quad x_1 + x_2 + x_3 + x_4 = 0$$

10. (b) Let $P(x_1, y_1)$ be a point on the hyperbola $\frac{x^2}{a^2} - \frac{y^2}{b^2} = 1$, then

$$\frac{x_1^2}{a^2} - \frac{y_1^2}{b^2} = 1$$

The chord of contact of tangents from P to the hyperbola $\frac{x^2}{a^2} - \frac{y^2}{b^2} = 2$ is $\frac{xx_1}{a^2} - \frac{yy_1}{b^2} = 2$ ...(i)

The equation of asymptotes are $\frac{x}{a} + \frac{y}{b} = 0$ and $\frac{x}{a} - \frac{y}{b} = 0$

Now, the points of intersection of Eq. (i) with the two asymptotes are given by

and $$x_1 = \frac{2a}{\frac{x_1}{a} - \frac{y_1}{b}}, \; y_1 = \frac{2b}{\frac{x_1}{a} - \frac{y_1}{b}}$$

and $$x_2 = \frac{2a}{\frac{x_1}{a} + \frac{y_1}{b}}, \; y_2 = \frac{-2b}{\frac{x_1}{a} + \frac{y_1}{b}}$$

$$\therefore \quad \text{Area of triangle} = \frac{1}{2}(x_1y_2 - y_1x_2) = \frac{1}{2}\left(\frac{4ab \times 2}{\frac{x_1^2}{a^2} - \frac{y_1^2}{b^2}}\right) = 4ab$$

11. (d) Let rectangular hyperbola is $xy = c^2$ and PQR be a triangle inscribed in the $xy = c^2$.

$$\because \quad P\left(ct_1, \frac{c}{t_1}\right), Q = \left(ct_2, \frac{c}{t_2}\right), R = \left(ct_3, \frac{c}{t_3}\right)$$

Let orthocentre be $H(h, k)$.

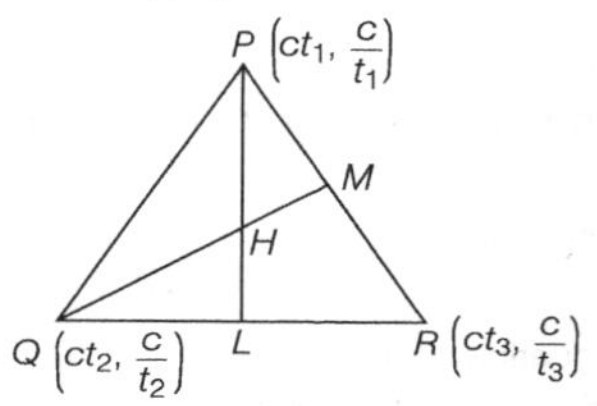

$$\therefore \quad \text{Slope of } QR = \frac{\frac{c}{t_3} - \frac{c}{t_2}}{ct_3 - ct_2} = \frac{t_2 - t_3}{t_2t_3(t_3 - t_2)} = -\frac{1}{t_2t_3}$$

$$\text{Slope of } PL = t_2t_3$$

Equation of PL, $y - \frac{c}{t_1} = t_2t_3(x - ct_1)$ ...(i)

Similarly, equation of QM is

$$y - \frac{c}{t_2} = t_1t_3(x - ct_2) \quad \text{...(ii)}$$

Now, on subtracting Eqs. (i) and (ii), we get

$$\frac{c}{t_2} - \frac{c}{t_1} = t_2t_3(x - ct_1) - t_1t_3(x - ct_2)$$

$$\Rightarrow \quad \frac{c}{t_2} - \frac{c}{t_1} = x(t_2t_3 - t_1t_3)$$

$$\Rightarrow \quad \frac{c(t_1 - t_2)}{t_1t_2} = t_3x(t_2 - t_1)$$

$$\Rightarrow \quad x = -\frac{c}{t_1t_2t_3}$$

On substituting x in Eq. (i), we get

$$y - \frac{c}{t_1} = t_2t_3\left(-\frac{c}{t_1t_2t_3} - ct_1\right) = -\frac{c}{t_1} - ct_1t_2t_3$$

$$y = -ct_1t_2t_3$$

Orthocentre is $h = -\frac{c}{t_1t_2t_3}$ and $k = -ct_1t_2t_3$

$$h = -\frac{c}{-k}c$$

$$\therefore \quad hk = c^2$$

Taking locus, we get

$$xy = c^2$$

Which is a rectangular hyperbola, i.e. curve.

12. (c) We have, $$\frac{x^2}{a^2} - \frac{y^2}{b^2} = 1$$

$$\because \quad e^2 = 1 + \frac{b^2}{a^2}$$

and asymmetrical are $\frac{x}{a} \pm \frac{y}{b} = 0$

Angle between asymptotes are

$$\theta = 2\tan^{-1}\frac{b}{a} \Rightarrow \tan\frac{\theta}{2} = \frac{b}{a}$$

$$\therefore \quad \sec\frac{\theta}{2} = \frac{\sqrt{a^2 + b^2}}{a} = \sqrt{1 + \frac{b^2}{a^2}} = e$$

13. (c) The equation of normal at $\left(ct, \frac{c}{t}\right)$ to the hyperbola $xy = c^2$ is

$$xt^3 - yt - ct^4 + c = 0 \quad \text{...(i)}$$

Since, Eq. (i) intersect at hyperbola $\left(ct_1, \frac{c}{t_1}\right)$.

$$ct_1t^3 - \frac{c}{t_1}t - ct^4 + c = 0$$

$$\Rightarrow \quad t^3t_1 = \frac{t}{t_1} + t^4 - 1$$

$$\Rightarrow \quad t^3t_1^2 - t - t^4t_1 + t_1 = 0$$

$$\Rightarrow \quad (t^3t_1^2 - t^4t_1) + (t_1 - t) = 0$$

$$\Rightarrow \quad t^3t_1(t_1 - t) + 1(t_1 - t) = 0$$

$$\Rightarrow \quad (t_1 - t)(t^3t_1 + 1) = 0$$

$$\Rightarrow \quad t_1 \neq t$$

$$\text{or} \quad t^3t_1 + 1 = 0$$

$$\therefore \quad t^3t_1 = -1$$

14. (c) Let (h, k) be the point of intersection of tangent at the end of normal chord. The equation of normal to the hyperbola $x^2 - y^2 = a^2$ at the point $(a\cos\theta, a\sin\theta)$ is

$$x\cos\theta + y\cot\theta = 2a \quad \text{...(i)}$$

Equation of chord of contact from the point (h, k) is

$$xh - yk = a^2 \quad \text{...(ii)}$$

Since, Eqs. (i) and (ii) are identical, then

$$\frac{\cos\theta}{h} = \frac{\cot\theta}{-k} = \frac{2}{a}$$

$$\Rightarrow \quad \sec\theta = \frac{a}{2h} \text{ and } \tan\theta = \frac{a}{-2k}$$

$$\because \quad \sec^2\theta - \tan^2\theta = 1$$

$$\Rightarrow \quad \frac{a^2}{4h^2} - \frac{a^2}{4k^2} = 1$$

$$\Rightarrow \quad \frac{k^2 - h^2}{h^2k^2} = \frac{4}{a^2}$$

$$\Rightarrow \quad a^2(k^2 - h^2) = 4h^2k^2$$

On taking locus of (h, k), we get

$$a^2(y^2 - x^2) = 4x^2y^2$$

15. (d) We have, (x_i, y_i), $i = 1, 2, 3, 4$ are the points on the rectangular hyperbola $xy = c^2$.

Equation of normal to the hyperbola $xy = c^2$ at point $\left(ct, \frac{c}{t}\right)$ is

$$ct^4 - t^3x + ty - c = 0$$

It passes through (h, k), then

$$ct^4 - t^3h + tk - c = 0$$

Which is a biquadratic in t.

Let its roots be t_1, t_2, t_3 and t_4, then

$$\Sigma t_1 = \frac{h}{c}, \Sigma t_1 t_2 = 0$$

$$\Sigma t_1 t_2 t_3 = -\frac{k}{c}$$

and $\quad t_1 t_2 t_3 t_4 = -1$

Now, $\quad \Sigma x_i = c\,\Sigma t_1 = c \cdot \frac{h}{c} = h$

and
$$\Sigma y_i = c\left(\Sigma \frac{1}{t_1}\right) = c\left(\frac{\Sigma t_1 t_2 t_3}{t_1 t_2 t_3 t_4}\right) = \frac{c \cdot \left(-\frac{k}{c}\right)}{-1} = k$$

$$= \frac{(x_1 + x_2 + x_3 + x_4)(y_1 + y_2 + y_3 + y_4)}{x_1 x_2 x_3 x_4}$$

$$= \frac{hk}{-c^4} \quad [x_1 x_2 x_3 x_4 = c^4 t_1 t_2 t_3 t_4 = -c^4]$$

$$= -\frac{hk}{c^4}$$

16. (d) We have, rectangular hyperbola

$$xy - y - 2x - 2 = 0 \quad \text{...(i)}$$

On differentiating w.r.t. x, we get

$$xy' + y' - y' - 2 = 0$$

$$\Rightarrow \quad y' = \frac{2 - y}{x - 1}$$

Slope of tangent at (5, 3) is $-\frac{1}{4}$.

Slope of normal at (5, 3) is 4.

$\therefore$ Equation of normal is $y - 3 = 4(x - 5)$.

$$y = 4x - 17 \quad \text{...(ii)}$$

Since, this normal intersect again at hyperbola

$$x(4x - 17) - 4x + 17 - 2x - 2 = 0$$

$$\Rightarrow \quad 4x^2 - 23x + 15 = 0$$

$$\Rightarrow \quad (x - 5)(4x - 3) = 0$$

$$\Rightarrow \quad x = \frac{3}{4} \text{ and } x = 5$$

$\therefore$ Point is $\left(\frac{3}{4}, -14\right)$.

17. (b) Clearly, given parabolas $2y^2 = 2x - 1$ and $2x^2 = 2y - 1$ are symmetrical about the line $y = x$.

$\because$ Shortest distance occurs along the common normal which is perpendicular to the line $y = x$.

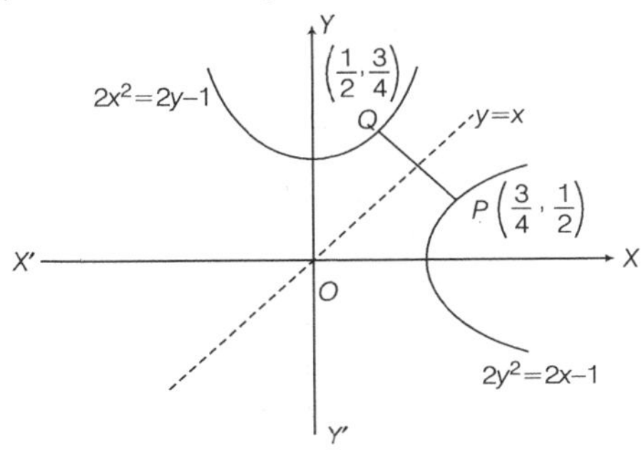

On differentiating $2y^2 = 2x - 1$ w.r.t. x, we get

$$2 \cdot 2y \frac{dy}{dx} = 2$$

$$\therefore \quad \frac{dy}{dx} = \frac{1}{2y} = 1$$

$$\Rightarrow \quad y = \frac{1}{2}$$

Now, the points are

$$P\left(\frac{3}{4}, \frac{1}{2}\right) \text{ and } Q\left(\frac{1}{2}, \frac{3}{4}\right).$$

Hence, the shortest distance $= \sqrt{\frac{1}{16} + \frac{1}{16}} = \frac{1}{2\sqrt{2}}$

18. (a) Given, equation of parabola is

$$y^2 = 2x - 1$$

$$\Rightarrow \quad y^2 = 2\left(x - \frac{1}{2}\right)$$

$$y^2 = 4aX$$

$$\because \quad a = \frac{1}{2}, X = x - \frac{1}{2}$$

Equation of normal of the parabola $y^2 = 4aX$ is

$$y = mX - 2am - am^3$$

$$\Rightarrow \quad y = m\left(x - \frac{1}{2}\right) - m - \frac{1}{2}m^3$$

$$\Rightarrow \quad y = mx - \frac{3m}{2} - \frac{m^3}{2}$$

$$\Rightarrow \quad 2y = 2mx - 3m - m^3$$

Since, this normal passes through $\left(\frac{7}{6}, 4\right)$.

$$2 \times 4 = 2m\left(\frac{7}{6}\right) - 3m - m^3$$

$$8 = \frac{7m}{3} - 3m - m^3$$

$$\Rightarrow \quad m^3 + \frac{2m}{3} + 8 = 0$$

$$\Rightarrow \quad 3m^3 + 2m + 24 = 0$$

which gives only one real value of m.

Hence, one normal can be drawn.

19. (d) $xy = (15)! = 2^{11} \cdot 3^6 \cdot 5^3 \cdot 7^2 \cdot 11^1 \cdot 13^1$

Number of positive integral solutions

= number of ways of fixing x

= the number of factors of (15)!

$= (1 + 11)(1 + 6)(1 + 3)(1 + 2)(1 + 1)(1 + 1)$

$= 4032$

Number of negative integral solutions = 4032

$\therefore$ Number of integral solutions $= 2 \times 4032$

$= 8064$

20. (a) HCF $(\alpha, \beta) = 1$, α and β will not have common factor other than 1, so identical prime numbers should not be separated e.g., 2^{11} will completely go with either α or β.

So, the number of solutions $= 2^6 = 64$

21. (a) The largest number whose perfect square can be made with

$$(15)! = 2^5 \cdot 3^3 \cdot 5^1 \cdot 7^1$$

So, that number of ways of selecting x will be

$$(1 + 5)(1 + 3)(1 + 1)(1 + 1) = 96$$

Explanations for Q. Nos. (22-24)

$$(OM)(OQ) = 2C^2$$

$$\Rightarrow \quad 2C^2 = 4$$

$$\Rightarrow \quad C = \sqrt{2}$$

So, equation of hyperbola is

$$(x-1)(y-2) = 2$$

Its foci are $(-1, 0)$, $(3, 4)$ and equation of directrix is $x + y - 5 = 0$, $x + y - 7 = 0$ is the equation of latusrectum of ellipse and tangents at end points of latusrectum always intersects on corresponding directrix. Also, circle with diameter SP, touches auxiliary circle.

22. (d)

23. (c)

24. (d)

25. (b) Let any point on the hyperbola $xy = 16$ be $\left(4t, \frac{4}{t}\right)$.

Normal at this point is $y - \frac{4}{t} = t^2(x - 4t)$

If this normal passes through $P(h, k)$, then

$$k - \frac{4}{t} = t^2(h - 4t)$$

$$\Rightarrow \quad 4t^4 - t^3h + tk - 4 = 0$$

this equation has roots t_1, t_2, t_3, t_4, which are parameters of the four normals on the hyperbola.

$$\therefore \quad \Sigma t_1 = \frac{h}{4}$$

$$\Sigma t_1 t_2 = 0$$

$$\Sigma t_1 t_2 t_3 = -\frac{k}{4}$$

and $\quad t_1 t_2 t_3 t_4 = -1$

$$\therefore \quad \frac{1}{t_1} + \frac{1}{t_2} + \frac{1}{t_3} + \frac{1}{t_4} = \frac{k}{4}$$

$$\Rightarrow \quad y_1 + y_2 + y_3 + y_4 = k$$

We have,

$$t_1^2 + t_2^2 + t_3^2 + t_4^2 = \frac{h^2}{16} = k$$

$$\therefore \quad h^2 = 16k$$

On taking locus of (h, k), we get

$$x^2 = 16y$$

26. (a) We have, the curve $c \equiv x^2 = 16y$

Equal tangent at point $(8t, 4t^2)$ to the curve $x^2 = 16y$

$$xt = y + 4t^2$$

Its X-intercept $= A = (4t, 0)$

Y-intercept $= B = (0, -4t^2)$

Let mid-point of AB be i.e. (h, k)

$$h = 2t \text{ and } k = -2t^2$$

$$\Rightarrow \quad k = -2\left(\frac{h}{2}\right)^2$$

$$\Rightarrow \quad 2k = -h^2$$

$$\Rightarrow \quad h^2 + 2k = 0$$

On taking locus of (h, k), we get

$$x^2 + 2y = 0$$

27. (c) Let l is length of side of equilateral triangle.

$\because$ Coordinates of $P = (l \cos 60°, l \sin 60°)$

i.e. $\quad P\left(\frac{l}{2}, \frac{l\sqrt{3}}{2}\right)$

Since, P lies on the curve $x^2 = 16y$

$$\therefore \quad l = 32\sqrt{3}$$

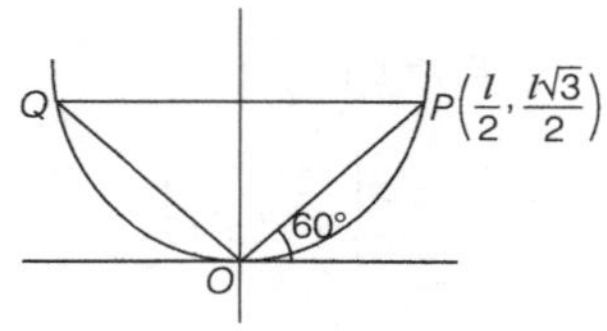

We know that area of equilateral triangle is $= \frac{\sqrt{3}}{4} l^2$

$$\therefore \quad \text{Area} = \frac{\sqrt{3}}{4} \times (32\sqrt{3})^2 = \frac{\sqrt{3}}{4} \times 32 \times 32 \times 3$$

$$= 768\sqrt{3} \text{ sq units}$$

28. (8) We have, the equation of asymptotes is

$$xy + 2x + 4y + c = 0$$

On comparing it with

$$ax^2 + 2hxy + by^2 + 2gx + 2fy + k = 0$$

$$a = 0, h = \frac{1}{2}, b = 0, g = 1, f = 2 \text{ and } k = c$$

It represents a pair of straight lines, if

$$abk - 2fgh - af^2 - bg^2 - ch^2 = 0$$

$$\Rightarrow \quad 0 - 2(2)(1)\left(\frac{1}{2}\right) - 0 - 0 - c\frac{1}{4} = 0$$

$$\Rightarrow \quad -2 - \frac{c}{4} = 0 \Rightarrow \frac{c}{4} = -2$$

$$\therefore \quad c = 8$$

29. (1) Let $P = \left(ct_1, \frac{c}{t_1}\right)$, $Q = \left(ct_2, \frac{c}{t_2}\right)$, $R = \left(ct_3, \frac{c}{t_3}\right)$ and $S = \left(ct_4, \frac{c}{t_4}\right)$

Now, equation of chord PQ is

$$x + yt_1t_2 - c(t_1 + t_2) = 0$$

Equation of chord RS is

$$x + yt_3t_4 - c(t_3 + t_4) = 0$$

$$\text{Slope of } PQ = -\frac{1}{t_1t_2}$$

and $\quad \text{Slope of } RS = -\frac{1}{t_3t_4}$

Since, PQ and RS are perpendicular, then

$$\frac{1}{t_1t_2t_3t_4} = -1$$

$$\therefore \quad t_1t_2t_3t_4 = -1$$

Now, $\quad \text{slope of } CP = \frac{1}{t_1^2}$

$\therefore$ Product of slopes of CP, CQ, CR and CS are

$$\frac{1}{t_1^2} \cdot \frac{1}{t_2^2} \cdot \frac{1}{t_3^2} \cdot \frac{1}{t_4^2} = \left(\frac{1}{-1}\right)^2 = 1$$

30. (2) Since, the asymptotes are perpendicular.

Hence, hyperbola is rectangular.

$$\text{Eccentricity} = \sqrt{2} = e$$

$$\therefore \quad e^2 = 2$$

Revisal Problems (JEE Main)

1. (c) Let two fixed points $A(-ae, 0)$ and $B(ae, 0)$ and $P(x, y)$ be moving point such that

$$PA - PB = k$$

$$\sqrt{(x + ae)^2 + y^2} - \sqrt{(x - ae)^2 + y^2} = k$$

or $$\sqrt{x^2 + y^2 + 2aex + a^2e^2} = k$$

$$-\sqrt{x^2 + y^2 - 2aex + y^2 + a^2e^2} = k \quad \text{...(i)}$$

Consider $$x^2 + y^2 + 2aex + a^2e^2 = l \quad \text{...(ii)}$$

$$x^2 + y^2 - 2aex + a^2e^2 = m \quad \text{...(iii)}$$

Eq. (i) can be rewritten as

$$\sqrt{l} - \sqrt{m} = k \quad \text{...(iv)}$$

From Eqs. (ii) and (iii), we get

$$l - m = 4aex$$

$$(\sqrt{l} + \sqrt{m})(\sqrt{l} - \sqrt{m}) = 4aex$$

$$K(\sqrt{l} + \sqrt{m}) = 4aex$$

$$\sqrt{l} + \sqrt{m} = \frac{4aex}{k} \quad \text{...(v)}$$

On adding Eqs. (iv) and (v), we get

$$2\sqrt{l} = k + \frac{4aex}{k}$$

$$\Rightarrow \sqrt{l} = \frac{k}{2} + \frac{2aex}{k}$$

$$\Rightarrow l = \left(\frac{k}{2} + \frac{2aex}{k}\right)^2$$

On replacing K by $2a$, we get

$$l = (a + ex)^2$$

$$\Rightarrow x^2 + y^2 + 2aex + a^2e^2 = a^2 + e^2x^2 + 2aex$$

$$\Rightarrow x^2(1 - e^2) + y^2 = a^2(1 - e^2)$$

$$\Rightarrow x^2(e^2 - 1) - y^2 = a^2(e^2 - 1)$$

$$\Rightarrow \frac{x^2}{a^2} - \frac{y^2}{a^2(e^2 - 1)} = 1$$

Put $a^2(e^2 - 1) = b^2$, we get

$$\frac{x^2}{a^2} - \frac{y^2}{b^2} = 1$$

Which represents a hyperbola.

2. (d) In previous question, we see that

$PA - PB$ = constant represent a hyperbola

Clearly, option (d) when $k_2 = -1$

Then, $k_1PA - PB = k_3$

k_1 and k_3 are positive.

Then, it represents a hyperbola.

3. (a) We know that, e and e' are the eccentricity of the hyperbola and conjugate hyperbola, then

$$\frac{1}{e} + \frac{1}{e'^2} = 1$$

We have, $$x^2 - \frac{y^2}{\frac{1}{3}} = 1$$

$$\Rightarrow e^2 - 1 = \frac{1}{3}$$

$$\Rightarrow e^2 = 1 + \frac{1}{3} = \frac{4}{3}$$

$$\because \frac{1}{e'^2} = 1 - \frac{1}{e^2} = 1 - \frac{3}{4} = \frac{1}{4}$$

$$\Rightarrow e'^2 = 4$$

$$\therefore e' = 2$$

4. (b) We have,

$$x^2 - y^2 - 4x + 4y + 16 = 0$$

this can be rewritten as

$$x^2 - 4x + 4 - y^2 + 4y + 12 = 0$$

$$\Rightarrow x^2 - 4x + 4 - (y^2 - 4y + 4) + 16 = 0$$

$$\Rightarrow (x - 2)^2 - (y - 2)^2 = -16$$

$$\Rightarrow (y - 2)^2 - (x - 2)^2 = 16$$

$$\Rightarrow Y^2 - X^2 = 16$$

Clearly, this is a rectangular hyperbola, hence its eccentricity is $\sqrt{2}$.

5. (b) We have,

transverse axis $= x + 2y - 3 = 0$

and conjugate axis $= 2x - y + 4 = 0$

both are perpendicular,

and $$2a = \sqrt{2} \text{ and } 2b = \frac{2}{\sqrt{3}}$$

$$\Rightarrow a = \frac{1}{\sqrt{2}} \text{ and } b = \frac{1}{\sqrt{3}}$$

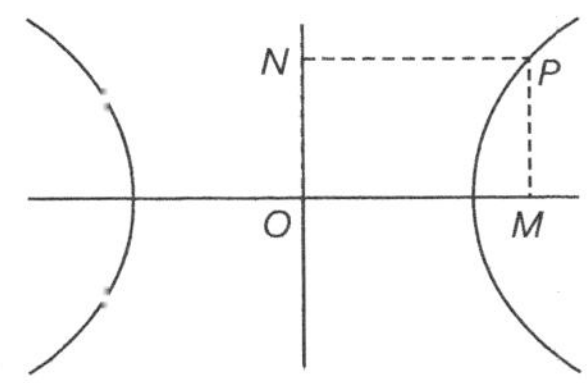

We know that,

Equation of the hyperbola referred to two perpendicular lines,

i.e. $$\frac{PN^2}{a^2} - \frac{PM^2}{b^2} = 1$$

$$\Rightarrow \frac{\left(\frac{2x - y + 4}{\sqrt{5}}\right)^2}{\frac{1}{2}} - \frac{\left(\frac{x + 2y - 3}{\sqrt{5}}\right)^2}{\frac{1}{3}} = 1$$

$$\therefore \frac{2}{5}(2x - y + 4)^2 - \frac{3}{5}(x + 2y - 3)^2 = 1$$

6. (c) Let three circles having centres P, A and B. We have to find the locus of circle having centre P touches the circle having centres A and B externally, then

$$PA = r_1 + r_2 \quad \text{...(i)}$$

$$PB = r_1 + r_3 \quad \text{...(ii)}$$

Now, $$PA - PB = r_2 - r_3 \quad \text{(constant)}$$

From question 1,

It represents a hyperbola.

7. (d) Let the hyperbola be $\frac{x^2}{a^2} - \frac{y^2}{b^2} = 1$.

We have, OLL' is an equilateral triangle.

$$OL = OL' = LL' = \frac{2b^2}{a}$$

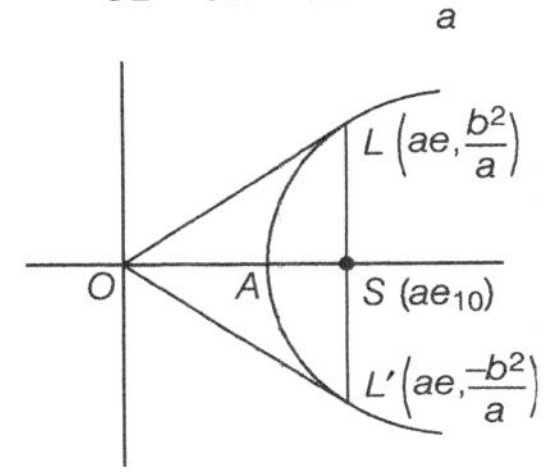

In ΔOSL,

$$OL^2 = OS^2 + SL^2$$

$$\Rightarrow \left(\frac{2b^2}{a}\right)^2 = a^2e^2 + \left(\frac{b^2}{a}\right)^2$$

$$\Rightarrow 3\left(\frac{b^2}{a}\right)^2 = a^2e^2$$

$$\because e^2 = 3\frac{b^4}{a^4} = 3(e^2 - 1)^2$$

$$\Rightarrow e^2 = 3(e^4 + 1 - 2e^2)$$

$$\Rightarrow 3e^4 - 6e^2 + 3 = e^2$$

$$\Rightarrow 3e^4 - 7e^2 + 3 = 0$$

$$\Rightarrow e^2 = \frac{7 \pm \sqrt{49 - 36}}{6}$$

$$\Rightarrow e^2 = \frac{7 \pm \sqrt{13}}{6}$$

$$\Rightarrow e^2 = \frac{14 + 2\sqrt{13}}{12} = \frac{13 + 1 + 2\sqrt{13}}{12}$$

$$\Rightarrow e^2 = \left(\frac{\sqrt{13} + 1}{2\sqrt{3}}\right)^2$$

$$\therefore e = \frac{\sqrt{13} + 1}{2\sqrt{3}}$$

8. (b) Let hyperbola be $\frac{x^2}{a^2} - \frac{y^2}{b^2} = 1$

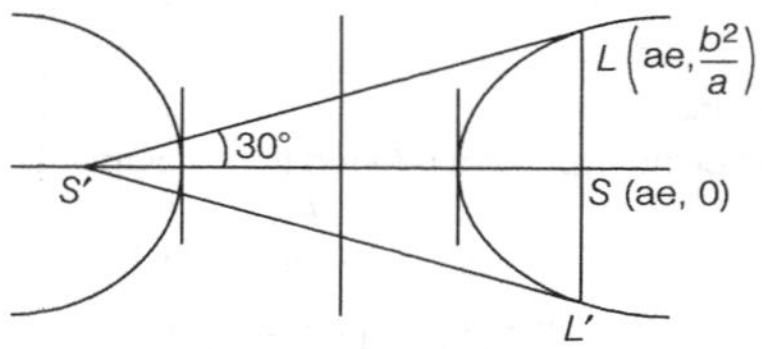

Since, $LRLL'$ subtends $60°$ at S'

$$\because \angle LS'S = 30°$$

In $\Delta LSS'$,

$$\tan 30° = \frac{\frac{b^2}{a}}{2ae} = \frac{1}{\sqrt{3}}$$

$$\Rightarrow \frac{b^2}{2a^2e} = \frac{1}{\sqrt{3}}$$

$$\Rightarrow \sqrt{3}\frac{b^2}{a^2} = 2e$$

$$\Rightarrow \sqrt{3}(e^2 - 1) = 2e$$

$$\Rightarrow \sqrt{3}e^2 - 2e - \sqrt{3} = 0$$

$$\Rightarrow e = \frac{2 \pm \sqrt{4 + 4\cdot\sqrt{3}\sqrt{3}}}{2\sqrt{3}} = \frac{2 + 4}{2\sqrt{3}} = \frac{6}{2\sqrt{3}} = \frac{3}{\sqrt{3}}$$

$$\therefore e = \sqrt{3}$$

9. (b) Let the equation of hyperbola be

$$\frac{x^2}{a^2} - \frac{y^2}{b^2} = 1 \quad \text{...(i)}$$

$$\text{Eccentricity } e^2 = 1 + \frac{b^2}{a^2} = \frac{a^2 + b^2}{a^2}$$

$$\Rightarrow \frac{1}{e^2} = \frac{a^2}{a^2 + b^2}$$

$\therefore$ Now, equation of conjugate hyperbola to Eq. (i) is

$$\frac{x^2}{a^2} - \frac{y^2}{b^2} = -1$$

$$\text{Eccentricity } e'^2 = 1 + \frac{a^2}{b^2} = \frac{a^2 + b^2}{a^2}$$

$$\Rightarrow \frac{1}{e'^2} = \frac{b^2}{a^2 + b^2}$$

$$\therefore \frac{1}{e^2} + \frac{1}{e'^2} = \frac{a^2 + b^2}{a^2 + b^2} = 1$$

10. (d) We have, $A'LL'$ is an equilateral triangle.

$$\because A'S = a + ae$$

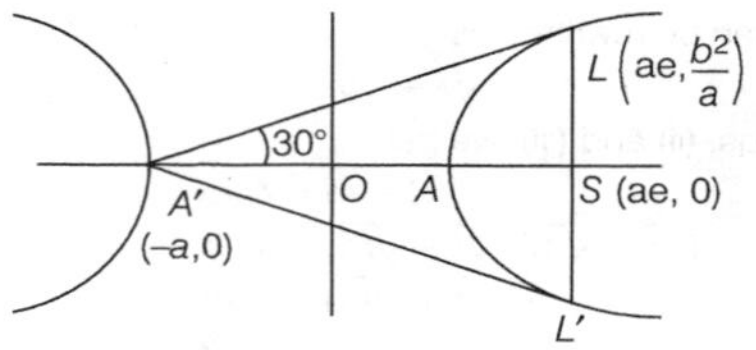

$\therefore$ In $\Delta A'SL$,

$$\tan 30° = \frac{SL}{A'S} = \frac{\frac{b^2}{a}}{a + ae} = \frac{b^2}{a^2(1 + e)}$$

$$\Rightarrow \frac{1}{\sqrt{3}} = \frac{1}{1 + e}\frac{b^2}{a^2} = \frac{1}{1 + e}(e^2 - 1) = \frac{(e + 1)(e - 1)}{e + 1} = e - 1$$

$$\because e - 1 = \frac{1}{\sqrt{3}}$$

$$\Rightarrow e = 1 + \frac{1}{\sqrt{3}} = \frac{\sqrt{3} + 1}{\sqrt{3}}$$

11. (d) Let the line $2x + \sqrt{6}y = 2$ touches the hyperbola $x^2 - 2y^2 = 4$ at (x_1, y_1), i.e. point of contact.

$\because$ Equation of tangent at (x_1, y_1) is

$$xx_1 - 2yy_1 = 4 \quad \text{...(i)}$$

$$2x + \sqrt{6}y = 2 \quad \text{...(ii)}$$

Since, Eqs. (i) and (ii) are identical, we get

$$\frac{x_1}{2} = \frac{-2y_1}{\sqrt{6}} = 2$$

$$\Rightarrow x_1 = 4 \text{ and } y_1 = -\sqrt{6}$$

$\therefore$ Point of contact $= (4, -\sqrt{6})$

12. (b) We know that, the equation of tangent to the hyperbola $\frac{x^2}{a^2} - \frac{y^2}{b^2} = 1$ in slope form is

$$y = mx \pm \sqrt{a^2m^2 - b^2}$$

$\because$ Equation of tangent to the hyperbola $x^2 - y^2 = 3$ is $y = mx \pm \sqrt{3m^2 - 3}$, since this tangent is parallel to the line $2x + y + 8 = 0$.

$$\because m = -2$$

$$\Rightarrow y = -2x \pm 3$$

$$\text{or} \quad 2x + y \pm 3 = 0 \quad \text{...(i)}$$

Let (x_1, y_1) be the point of contact

$$xx_1 - yy_1 = 3 \quad \text{...(ii)}$$

Since, Eqs. (i) and (ii) are identical, then

$$\frac{x_1}{2} = -\frac{y_1}{1} = \pm 1$$

$$\Rightarrow x_1 = \pm 2 \text{ and } y_1 = \pm 1$$

$\therefore$ Point of contact $= (2, -1)$

13. (a) We have hyperbola $16x^2 - 25y^2 - 96x + 100y - 356 = 0$

This can be rewritten as

$$\frac{(x-3)^2}{25} - \frac{(y-2)^2}{16} = 1$$

or $$\frac{X^2}{25^2} - \frac{Y^2}{16} = 1 \quad \text{...(i)}$$

Equation of tangent to (i) is

$$Y = mX + \sqrt{a^2m^2 - b^2}$$

Clearly, $m = \tan 45° = 1, a^2 = 25$ and $b^2 = 16$

$\because$ $$Y = X + 3$$

$\Rightarrow$ $$y - 2 = x - 3 + 3$$

i.e. $y = x + 2$ is the required equation of tangent.

14. (c) Equation of chord bisected at given point is

$$T = S_1$$

and $$xx_1 - yy_1 - 9 = x_1^2 - y_1^2 - 9$$

$\Rightarrow$ $$5x + 3y = 25 - 9$$

$\Rightarrow$ $$5x + 3y = 16$$

15. (c) We know that, the line $y = mx + c$ touches the hyperbola $\frac{x^2}{a^2} - \frac{y^2}{b^2} = 1$, if $c = a^2m^2 - b^2$.

We have, $$y = 3x + \lambda,$$

then $$\frac{x^2}{5} - \frac{y^2}{9} = 1$$

$\therefore$ $$\lambda^2 = 5 \times 9 - 9 = 36$$

$\Rightarrow$ $$\lambda = \pm 6 \Rightarrow \lambda = 6$$

16. (b) Tangent at $(a \sec \phi, b \tan \phi)$ on the I st hyperbola is

$$\frac{x}{a} \sec \phi - \frac{y}{b} \tan \phi = 1 \quad \text{...(i)}$$

Similarly, tangent at any point $(b \tan \theta, a \sec \theta)$ on IInd hyperbola is

$$\frac{y}{a} \sec \theta - \frac{x}{b} \tan \theta = 1 \quad \text{...(ii)}$$

Since, Eqs. (i) and (ii) are common tangents, then they should be identical.

$\Rightarrow$ $$\frac{\sec \theta}{a} = -\frac{\tan \phi}{b}$$

$\Rightarrow$ $$\sec \theta = -\frac{a}{b} \tan \phi \quad \text{...(iii)}$$

$\Rightarrow$ $$-\frac{\tan \theta}{b} = \frac{\sec \phi}{a}$$

$\Rightarrow$ $$\tan \theta = -\frac{b}{a} \sec \phi \quad \text{...(iv)}$$

$\because$ $$\sec^2\theta - \tan^2\theta = 1$$

$\Rightarrow$ $$\frac{a^2}{b^2} \tan^2 \phi - \frac{b^2}{a^2} \sec^2\phi = 1 \quad \text{[from Eqs. (iii) and (iv)]}$$

$\Rightarrow$ $$\frac{a^2}{b^2} \tan^2 \phi - \frac{b^2}{a^2} (1 + \tan^2 \phi) = 1$$

$\Rightarrow$ $$\left(\frac{a^2}{b^2} - \frac{b^2}{a^2}\right) \tan^2 \phi = 1 + \frac{b^2}{a^2}$$

$\Rightarrow$ $$\tan^2 \phi = \frac{b^2}{a^2 - b^2}$$

$\therefore$ $$\sec^2\phi = 1 + \tan^2 \phi = \frac{a^2}{a^2 - b^2}$$

Now, substituting $\sec \phi$ and $\tan \phi$ in Eq. (i), we get

$$\pm \frac{x}{\sqrt{a^2 - b^2}} \mp \frac{y}{\sqrt{a^2 - b^2}} = 1$$

$\therefore$ $$x \mp y = \pm \sqrt{a^2 - b^2}$$

or $$y = \pm x \pm \sqrt{a^2 - b^2}$$

17. (c) The equation of tangent having slope n to the hyperbola is

$$y = mx \pm \sqrt{a^2m^2 - b^2}$$

Since, this tangent passes through (c, d), hence

$$d = mc \pm \sqrt{a^2m^2 - b^2}$$

$\Rightarrow$ $$(d - mc)^2 = a^2m^2 - b^2$$

$\Rightarrow$ $$d^2 + m^2c^2 - 2mcd = a^2m^2 - b^2$$

$\Rightarrow$ $$m^2(c^2 - a^2) - 2cdm + d^2 + b^2 = 0$$

$\Rightarrow$ $$m_1m_2 = \frac{d^2 + b^2}{c^2 - a^2} \quad [\because m_1 = \tan \alpha, m_2 = \tan \beta]$$

We have,

$$m_1m_2 = 1$$

$\therefore$ $$\frac{d^2 + b^2}{c^2 - a^2} = 1$$

$\Rightarrow$ $$d^2 + b^2 = c^2 - a^2$$

$\Rightarrow$ $$d^2 - c^2 = -a^2 - b^2$$

$\therefore$ $$c^2 - d^2 = a^2 + b^2$$

18. (d) Let the point on the given hyperbola (α, β) be $(a \sec \theta, b \tan \theta)$.

Equation of tangent at the point $(a \sec \theta, b \tan \theta)$ is

$$\frac{x}{a} \sec \theta - \frac{y}{b} \tan \theta = 1 \quad \text{...(i)}$$

Equation of auxiliary circle is

$$x^2 + y^2 = a^2 \quad \text{...(ii)}$$

On eliminating x from Eqs. (i) and (ii), we get

$$x = \left(1 + \frac{y}{b} \tan \theta\right) a \cos \theta$$

On substituting x in Eq. (ii), we get

$$a^2 \cos^2\theta \left(1 + \frac{y}{b} \tan \theta\right)^2 + y^2 = a^2$$

$$\Rightarrow a^2 \cos^2\theta \left(1 + \frac{y^2}{b^2} \tan^2\theta + \frac{2 \tan \theta}{b} y\right) + y^2 = a^2$$

$$\Rightarrow y^2 \left(a^2 \cos^2\theta \cdot \frac{\tan^2\theta}{b^2} + 1\right) + 2 \tan \theta \cdot \cos^2\theta \cdot \frac{a^2}{b} y + a^2 \cos^2\theta - a^2 = 0$$

$$\Rightarrow y^2 \left(\frac{a^2 \sin^2\theta}{b^2} + 1\right) + 2 \frac{a^2}{b} \sin \theta \cos \theta \, y - a^2 \sin^2\theta = 0$$

Now, $$y_1 + y_2 = -\frac{2 \frac{a^2}{b} \sin \theta \cos \theta}{\frac{a^2 \sin^2\theta}{b^2} + 1},$$

$\Rightarrow$ $$y_1y_2 = -\frac{a^2 \sin^2\theta}{\frac{a^2 \sin^2\theta}{b^2} + 1}$$

$\therefore$ $$\frac{y_1 + y_2}{y_1y_2} = \frac{-\frac{2a^2}{b} \sin \theta \cos \theta}{-a^2 \sin^2\theta}$$

$$= \frac{2 \cos \theta}{b \sin \theta}$$

$$= \frac{2}{b \tan \theta} = \frac{2}{\beta}$$

19. (a) Let the point be (h, k) on the hyperbola

$$x^2 - y^2 = a^2 - b^2$$

$\therefore$ $$h^2 - k^2 = a^2 - b^2 \quad \text{...(i)}$$

Equation of tangent to ellipse $\frac{x^2}{a^2} + \frac{y^2}{b^2} = 1$ is

$$y = mx \pm \sqrt{a^2m^2 + b^2}$$

this tangent passes through (h, k), hence

$$k = mh \pm \sqrt{a^2m^2 + b^2}$$

$$\Rightarrow \quad (k - mh)^2 = a^2m^2 + b^2$$

$$\Rightarrow \quad k^2 + m^2h^2 - 2hkm = a^2m^2 + b^2$$

$$\Rightarrow \quad m^2(h^2 - a^2) - 2hkm + k^2 - b^2 = 0$$

$$\Rightarrow \quad m_1m_2 = \frac{k^2 - b^2}{h^2 - a^2} = \frac{h^2 - a^2}{h^2 - a^2} = 1 \quad \text{[from Eq. (i)]}$$

$$\therefore \quad \tan\alpha \cdot \tan\beta = 1$$

20. (a) Given hyperbola is

$$\frac{x^2}{a^2} - \frac{y^2}{b^2} = 1$$

Equation of chord PQ to the given hyperbola is

$$\frac{x}{a}\cos\left(\frac{\alpha - \beta}{2}\right) - \frac{y}{b}\sin\left(\frac{\alpha + \beta}{2}\right) = \cos\left(\frac{\alpha + \beta}{2}\right)$$

Given, $\alpha - \beta = 2\theta$

$$\frac{x}{a}\cos\theta - \frac{y}{b}\sin\left(\frac{\alpha + \beta}{2}\right) = \cos\left(\frac{\alpha + \beta}{2}\right)$$

i.e. $$y = \frac{b}{a}\frac{\cos\theta}{\sin\left(\frac{\alpha + \beta}{2}\right)}x + \frac{b\cos\left(\frac{\alpha + \beta}{2}\right)}{\sin\left(\frac{\alpha + \beta}{2}\right)}$$

On comparing this line with $y = mx + c$, then

$$m = \frac{b}{a}\frac{\cos\theta}{\sin\left(\frac{\alpha + \beta}{2}\right)}$$

and $$c = \frac{b\cos\left(\frac{\alpha + \beta}{2}\right)}{\sin\left(\frac{\alpha + \beta}{2}\right)}$$

For line $y = mx + c$ to be a tangent on $\frac{x^2}{a^2}\cos^2\theta - \frac{y^2}{b} = 1$, we have

$$c^2 = \frac{a^2}{\cos^2\theta}m^2 - b^2$$

$$\Rightarrow \quad c^2 = \frac{b^2\cos^2\left(\frac{\alpha + \beta}{2}\right)}{\sin^2\left(\frac{\alpha + \beta}{2}\right)}$$

Take option (a),

and $$a^2m^2 - b^2 = \frac{a^2}{\cos^2\theta} \times \frac{b^2\cos^2\theta}{a^2\sin^2\left(\frac{\alpha + \beta}{2}\right)} - b^2$$

$$= \frac{b^2}{\sin^2\left(\frac{\alpha + \beta}{2}\right)} - b^2 = \frac{b^2\cos^2\left(\frac{\alpha + \beta}{2}\right)}{\sin^2\left(\frac{\alpha + \beta}{2}\right)}$$

21. (a) We have, the points (x_1, y_1) and (x_2, y_2) on the curve $xy = c^2$.

Now, the mid-point of the chord is

$$\left(\frac{x_1 + x_2}{2}, \frac{y_1 + y_2}{2}\right)$$

The equation of the mid-point of the chord is

$$x\left(\frac{y_1 + y_2}{2}\right) + y\left(\frac{x_1 + x_2}{2}\right) - c^2 = 2\left(\frac{x_1 + x_2}{2}\right)\left(\frac{y_1 + y_2}{2}\right) - c^2$$

$$\Rightarrow \quad x(y_1 + y_2) + y(x_1 - x_2) = (x_1 + x_2)(y_1 + y_2)$$

$$\therefore \quad \frac{x}{x_1 + x_2} + \frac{y}{y_1 + y_2} = 1$$

22. (c) Let the point be $(a\sec\theta, b\tan\theta)$ on the hyperbola

$$\frac{x^2}{a^2} - \frac{y^2}{b^2} = 1$$

Its asymptotes is

$$\frac{x}{a} + \frac{y}{b} = 0 \text{ and } \frac{x}{a} - \frac{y}{b} = 0.$$

$$\Rightarrow \quad p_1 = \frac{\sec\theta + \tan\theta}{\sqrt{\frac{1}{a^2} + \frac{1}{b^2}}}, p_2 = \frac{\sec\theta - \tan\theta}{\sqrt{\frac{1}{a^2} + \frac{1}{b^2}}}$$

$$\therefore \quad p_1p_2 = \frac{\sec^2\theta - \tan^2\theta}{\frac{1}{a^2} + \frac{1}{b^2}} = \frac{a^2b^2}{a^2 + b^2}$$

23. (c) Equation of normal to hyperbola $x^2 - y^2 = a^2$ at the point $(a\sec\theta, a\tan\theta)$ is

$$x\cos\theta + y\cot\theta = 2a \quad \text{...(i)}$$

Let (h, k) be the point of intersection of tangent at ends of normal chord.

Equation of chord of contact is

$$xh - yk = a^2 \quad \text{...(ii)}$$

Since, Eqs. (i) and (ii) are identical, then

$$\frac{\cos\theta}{h} = \frac{\cot\theta}{-k} = \frac{2}{a}$$

$$\Rightarrow \quad \cos\theta = \frac{2h}{a}, \cot\theta = -\frac{2k}{a}$$

$$\Rightarrow \quad \sec\theta = \frac{a}{2h}, \tan\theta = -\frac{a}{2k}$$

$$\because \quad \sec^2\theta - \tan^2\theta = 1$$

$$\Rightarrow \quad \frac{a^2}{4h^2} - \frac{a^2}{4k^2} = 1$$

$$\Rightarrow \quad \frac{k^2 - h^2}{h^2k^2} = \frac{4}{a^2}$$

$$\therefore \quad a^2(k^2 - h^2) = 4h^2k^2$$

On taking locus of (h, k), we get

$$a^2(y^2 - x^2) = 4x^2y^2$$

24. (b)

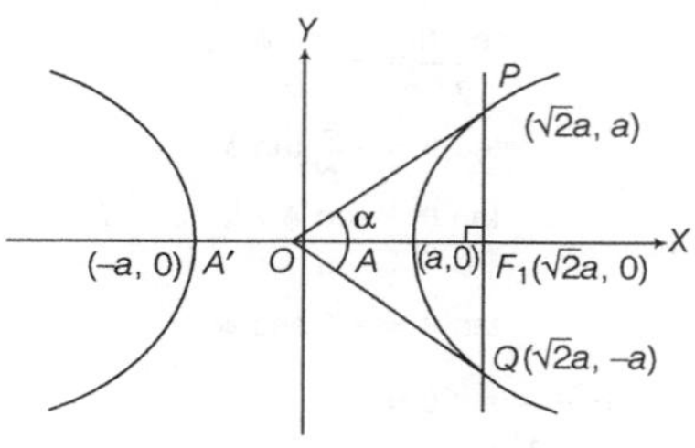

$$m_{OP} = \frac{1}{\sqrt{2}}, m_{OQ} = -\frac{1}{\sqrt{2}}$$

$$\tan\alpha = \frac{-\frac{1}{\sqrt{2}} - \frac{1}{\sqrt{2}}}{1 - \frac{1}{2}} = \frac{-\sqrt{2}}{1/2} = -2\sqrt{2} \quad \text{...(i)}$$

Also, equation of tangent at P and Q are respectively

$$x \cdot (\sqrt{2}a) - y(a) = a^2$$

$$\Rightarrow \quad \sqrt{2}x - y = a$$

and $$x(\sqrt{2}a) + y(a) = a^2$$

$$\Rightarrow \quad \sqrt{2}x + y = a$$

$$\therefore \quad \tan\beta = \frac{-(\sqrt{2} + \sqrt{2})}{1 - 2} = +2\sqrt{2} \quad \text{...(ii)}$$

From Eqs. (i) and (ii), we get

$$\tan\alpha = \tan(\pi - \beta)$$

$$\Rightarrow \quad \alpha + \beta = \pi$$

25. (c) Since, axes of the hyperbola are bisectors of pair of asymptotes.

Transverse axis is the bisector which contains origin

$$\because \quad \frac{3x-4y-1}{5} = + \frac{4x-3y-6}{5}$$

$$\Rightarrow \quad x + y - 5 = 0$$

Conjugate axis is the bisectors which does not containing origin

$$\frac{3x-4y-1}{5} = - \frac{4x-3y-6}{5}$$

$$\therefore \quad x - y - 1 = 0$$

26. (d) Equation of asymptotes is

$$2x^2 + 5xy + 2y^2 - 11x - 7\lambda + \lambda = 0 \quad \text{...(i)}$$

Since, Eq. (i) represents a pair of straight lines.

Comparing with $ax^2 + 2hxy + by^2 + 2gx + 2fy + c = 0$,

we get $a = 2, h = \frac{5}{2}, b = 2, g = -\frac{11}{2}, f = -\frac{7}{2}, c = \lambda$

Now, $abc + 2fgh - af^2 - bg^2 - ch^2 = 0$

$$2 \cdot 2 \cdot \lambda + 2 \cdot \left(-\frac{7}{2}\right)\left(-\frac{11}{2}\right)\left(\frac{5}{2}\right) - 2 \cdot \frac{49}{4} - 2 \cdot \frac{121}{4} - \lambda \cdot \frac{25}{4} = 0$$

$$\Rightarrow \quad 4\lambda - \frac{25\lambda}{4} + \frac{385}{4} - 85 = 0$$

$$\Rightarrow \quad -\frac{9\lambda}{4} + 45 = 0 \Rightarrow \lambda = 20$$

$\therefore$ Equation of asymptotes are

$$2x^2 + 5xy + 2y^2 - 11x - 7y + 20 = 0$$

27. (c) We know that ellipse and a hyperbola cut at right angle, if they confocal (i.e. having same focus).

Now, focus of ellipse = (ae, 0)

Focus of hyperbola = $(c\sqrt{2}, 0)$

[eccentricity of hyperbola = $\sqrt{2}$]

$$\because \quad ae = c\sqrt{2}$$

$$\Rightarrow \quad a^2e^2 = 2c^2$$

$$\Rightarrow \quad a^2\left(1 - \frac{b^2}{a^2}\right) = 2c^2$$

$$\therefore \quad a^2 - b^2 = 2c^2$$

28. (c) We have a variable chord

$$x \cos \alpha + y \sin \alpha = p$$

and this chord intersect the hyperbola at A and B, then the combined equation of OA and OB is given by

$$\frac{x^2}{a^2} - \frac{y^2}{b^2} = 1\left(\frac{x \cos \alpha + y \sin \alpha}{p}\right)^2$$

$$x^2\left(\frac{1}{a^2} - \frac{\cos^2 \alpha}{p^2}\right) + y^2\left(-\frac{1}{b^2} - \frac{\sin^2 \alpha}{p^2}\right) - \frac{2xy \sin \alpha \cos \alpha}{p} = 0$$

Since, this chords subtends right angle at centre.

$\therefore$ Coefficient of x^2 + Coefficient of $y^2 = 0$

$$\frac{1}{a^2} - \frac{\cos^2 \alpha}{p^2} - \frac{1}{b^2} - \frac{\sin^2 \alpha}{p^2} = 0$$

$$\frac{1}{a^2} - \frac{1}{b^2} = \frac{1}{p^2}$$

$$\Rightarrow \quad p^2 = \frac{a^2b^2}{b^2 - a^2}$$

$$\therefore \quad p = \frac{ab}{\sqrt{b^2 - a^2}}$$

Hence, p is constant, i.e. it touches a fixed circle whose radius is $\frac{ab}{\sqrt{b^2 - a^2}}$.

29. (a) We have, one asymptote of a hyperbola is

$$3x - 4y - 6 = 0 \quad \text{...(i)}$$

Other asymptotes s $4x + 3y + \lambda = 0$ ...(ii)

[$\because$ asymptotes are perpendicular]

We have, equation of line passing through the centre

$$x - y - 1 = 0 \quad \text{...(iii)}$$

Eqs. (i), (ii) and (iii) are concurrent, then

$$\begin{vmatrix} 3 & -4 & -6 \\ 4 & 3 & \lambda \\ 1 & -1 & -1 \end{vmatrix} = 0$$

$$\Rightarrow \quad \lambda = 17$$

$\therefore$ Other asymptotes is

$$4x + 3y + 17 = 0$$

30. (c) We know that, a rectangular hyperbola circumscribing a triangle also passes through its orthocentre.

i.e. $\left(ct_i, \frac{c}{t_i}\right)$, where $i = 1, 2, 3$ are the vertices of the triangle, then

orthocentre is $\left(-\frac{c}{t\, t_2 t_3}, -ct_1 t_2 t_3\right)$,

where $t_1 t_2 t_3 t_4 = 1$.

Hence, orthocentre is $\left(-ct_4, -\frac{c}{t_4}\right)$

$$= (-x_4, -y_4)$$

31. (a) We have ellipse $\frac{x^2}{25} + \frac{y^2}{16} = 1$

$$\Rightarrow \quad e = \sqrt{1 - \frac{16}{25}} = \frac{3}{5}$$

$\therefore$ Foci are $(\pm 3, 0)$

and hyperbola $12x^2 - 4y^2 = 27$

or $\frac{x^2}{\frac{27}{12}} - \frac{y^2}{\frac{27}{4}} = 1 \Rightarrow e = \sqrt{1 + \frac{12}{4}} = 2$

The foci are $\pm (\pm 3, 0)$.

Therefore, the two conics are confocal.

Hence, curves are intersect each other orthogonally.

32. (b) We have, $e = \frac{5}{3}$ and $e' = \frac{5}{4}$

$$\Rightarrow \quad \frac{1}{e^2} + \frac{1}{e'^2} = \frac{9}{25} + \frac{16}{25} = \frac{25}{25} = 1 \quad \left[\because \frac{1}{e^2} + \frac{1}{e'^2} = 1\right]$$

$\because e$ and e' are eccentricities of the hyperbola and its conjugate hyperbola.

Thus Statement I is correct.

Clearly, $e \cdot e'$

i.e. $\frac{5}{3} \times \frac{5}{4} = \frac{25}{12} > 1$

Hence, Statement II is also correct but Statement II is not correct explanation of Statement I.

33. (a) Let P be the position of gun and Q be the position of the target, and u be the velocity of the sound, v be the velocity of the bullet and R be the position of the man. Then, we have

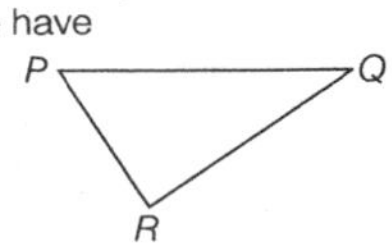

$$\frac{PR}{u} = \frac{QR}{u} + \frac{PQ}{v}$$

$$\Rightarrow \quad \frac{PR}{u} - \frac{QR}{u} = \frac{PQ}{v}$$

$$\Rightarrow \quad PR - QR = \frac{u}{v} PQ = \text{constant}$$

and $\frac{u}{v} PQ < PQ$

Therefore, the locus of R is a hyperbola.

34. (d) We know that if a circle and the rectangular hyperbola $xy = c^2$ meet at the 4 points t_1, t_2, t_3, t_4, then

$$t_1 t_2 t_3 t_4 = 1$$

So, Statement II is correct

For the points (2, 2), (4, 1) and $\left(6, \frac{2}{3}\right)$.

$\therefore t_1 = 2, t_2 = 2, t_3 = 3$ respectively.

For the point $\left(\frac{1}{4}, 16\right), t_4 = \frac{1}{8}$

Now, $t_1 t_2 t_3 t_4 = 1 \times 2 \times 3 \times \frac{1}{8} = \frac{3}{4} \neq 1$

Hence, Statement I is not correct.

35. (d) Statement I is incorrect because the foci of $xy = c^2$ is $(\pm c\sqrt{2}, \pm c\sqrt{2})$.

Statement II is correct.

In $x^2 - y^2 = a^2$, if axes are rotated clockwise through an angle 45°, we obtained $xy = \frac{a^2}{2}$.

Revisal Problems (JEE Advanced)

1. (d) We have, $x + iy = \sqrt{\phi + i\psi}$

On squaring both sides, we get

$$x^2 - y^2 + 2ixy = \phi + i\psi$$

On equating real and imaginary parts on both sides, we get

$$x^2 - y^2 = \phi$$

$$2xy = \psi$$

We have, ϕ and ψ are constants.

$\because x^2 - y^2 = a^2$ and $xy = c^2$ are represent two hyperbola and they intersect each other at right angle.

$\therefore$ Angle $= \frac{\pi}{2}$

2. (c) We have,

ellipse $x^2 + 9y^2 = 9$

$\Rightarrow \frac{x^2}{9} + y^2 = 1$

Let $P(3\cos\theta, \sin\theta)$ and $Q(3\cos\theta, -\sin\theta)$ be the ends of double ordinate of the given ellipse.

Now, the equation of normal at $P(3\cos\theta, \sin\theta)$ to the given ellipse is

$$ax \sec\theta - by \operatorname{cosec}\theta = a^2 - b^2 \quad \text{...(i)}$$

and the equation of diameter passing through Q is

$$y = -\frac{x}{3}\tan\theta \quad \text{...(ii)}$$

On solving Eqs. (i) and (ii), we get

$$R\left(\frac{12\cos\theta}{5}, -\frac{4\sin\theta}{5}\right)$$

Let mid-point of P and R be (h, k),

i.e. $(h, k) = \left(\frac{1}{2}\left(3\cos\theta + \frac{12\cos\theta}{5}\right), \frac{1}{2}\left(\sin\theta - \frac{4\sin\theta}{5}\right)\right)$

$= \left(\frac{1}{2}\cdot\frac{27\cos\theta}{5}, \frac{1}{2}\cdot\frac{\sin\theta}{5}\right)$

$\because h = \frac{27\cos\theta}{10}$ and $k = \frac{\sin\theta}{10}$

$\Rightarrow \cos\theta = \frac{10h}{27}$

and $\sin\theta = 10k$

$\because \cos^2\theta + \sin^2\theta = 1$

$\Rightarrow \left(\frac{10h}{27}\right)^2 + (10k)^2 = 1$

$\Rightarrow \frac{100h^2}{27} + 100k^2 = 1$

On taking locus of (h, k), we get

$$\frac{x^2}{27} + y^2 = \frac{1}{100}$$

Which represents the equation of an ellipse.

3. (c) Let standard equation of ellipse be $\frac{x^2}{a^2} - \frac{y^2}{b^2} = 1$

and hyperbola be $\frac{x^2}{a_1^2} - \frac{y^2}{b_1^2} = 1$

We have, major axis = transverse axis

$$2a = 2a_1$$

$\Rightarrow a = a_1$ ...(i)

and radius of director circle of ellipse

$$a^2 + b^2 = 4R^2 \quad \text{...(ii)}$$

and radius of director circle of hyperbola

$$a_1^2 - b_1^2 = R \quad \text{...(iii)}$$

On dividing Eqs. (ii) by (iii), we get

$$\frac{a^2 + b^2}{a_1^2 - b_1^2} = 4$$

$\Rightarrow a^2 + b^2 = 4a_1^2 - 4b_1^2$

$\Rightarrow b^2 = 3a^2 - 4b_1^2$ [from Eq. (i)]

or $b_1^2 = \frac{1}{4}(3a^2 - b^2)$ ...(iv)

We have, $1 - e_1^2 = \frac{b^2}{a^2}$ ...(v)

and $e_2^2 - 1 = \frac{b_1^2}{a_1^2}$ ...(vi)

Now, on dividing Eqs. (v) by (vi), we get

$$\frac{1 - e_1^2}{e_2^2 - 1} = \frac{b^2}{b_1^2}$$

$\Rightarrow \frac{1 - e_1^2}{e_2^2 - 1} = \frac{4b^2}{3a^2 - b^2}$

$\Rightarrow \frac{4}{3\frac{a^2}{b^2} - 1} = \frac{4}{\frac{3}{1 - e_1^2} - 1}$

$\Rightarrow \frac{1 - e_1^2}{e_2^2 - 1} = \frac{4(1 - e_1^2)}{3 - 1 + e_1^2}$

$\Rightarrow e_1^2 + 2 = 4(e_2^2 - 1)$

$\therefore 4e_2^2 - e_1^2 = 6$

4. (a) We have, the curve $y^2 - x^2 = 1$

It abscissa $= n$ and ordinate $= \sqrt{1 + n^2}$

$\therefore P$ is $(n, \sqrt{1 + n^2})$

d_n = Shortest distance = Perpendicular distance of P from the line $y = x$

$\Rightarrow dx = \frac{\sqrt{1 + n^2} - n}{\sqrt{2}} = \lim_{n \to \infty} \frac{n(\sqrt{1 + n^2} - n)}{\sqrt{2}} \times \frac{\sqrt{1 + n^2} + n}{\sqrt{1 + n^2} + n}$

$\therefore \lim_{n \to \infty} \frac{n}{\sqrt{2}\,[\sqrt{1 + n^2} + n]} = \frac{1}{2\sqrt{2}}$

5. (d) We have, the hyperbola

$$\frac{x^2}{16} - \frac{y^2}{9} = 1$$

and eccentricity $e = \frac{5}{4}$

$\therefore$ Foci are (5, 0) and (−5, 0).

Clearly, the given line $3x + (5 - 4\sqrt{2})\,y = 15$ passes through the one focus (5, 0).

Hence, the reflected ray must be passes through other focus (−5, 0)

and clearly only option (d) is passes through (−5, 0).

$\therefore$ Reflected ray is $3x - y(4\sqrt{2} + 5) + 15 = 0$.

6. (b) For hyperbola,

$$\frac{x^2}{5} - \frac{y^2}{5\cos^2\alpha} = 1$$

$$e_1^2 = 1 + \frac{b^2}{a^2} = 1 + \frac{5\cos^2\alpha}{5} = 1 + \cos^2\alpha$$

For ellipse, $\frac{x^2}{25\cos^2\alpha} + \frac{y^2}{25} = 1$

$$e_2 = 1 - \frac{a^2}{b^2} = 1 - \frac{25\cos^2\alpha}{25} = \sin^2\alpha$$

Given that $e_1 = \sqrt{3}\, e_2 \Rightarrow e_1^2 = 3e_2^2$

$\Rightarrow$ $1 + \cos^2\alpha = 3\sin^2\alpha$

$\Rightarrow$ $4\sin^2\alpha = 2$

$\Rightarrow$ $\sin\alpha = \frac{1}{\sqrt{2}}$

$\therefore$ $\alpha = \frac{\pi}{4}$

7. (c) We have, the hyperbola $\frac{x^2}{4} - y^2 = 1$

Equation of normal at point $P(2\sec\theta, \tan\theta)$ to hyperbola is

$$2x\cos\theta + y\cot\theta = 5 \quad \text{...(i)}$$

X-intercept $= \frac{5}{2}\sec\theta$

and Y-intercept $= 5\tan\theta$

Given, $\frac{5}{2}\sec\theta = 5\tan\theta$

$\Rightarrow$ $\frac{1}{2\cos\theta} = \frac{\sin\theta}{\cos\theta}$

$\Rightarrow$ $\sin\theta = \frac{1}{2}$

$\therefore$ $\theta = \frac{\pi}{6}$

$\therefore$ Equation of normal (i) becomes

$$2x\frac{\sqrt{3}}{2} + y\sqrt{3} = 5$$

$\Rightarrow$ $x + y = \frac{5}{\sqrt{3}}$

$\Rightarrow$ $y = -x + \frac{5}{\sqrt{3}}$ ...(ii)

The line (ii) touches the ellipse, if

$$\frac{25}{3} = a^2 + b^2 \quad [\because c^2 = a^2m^2 + b^2]$$

$\therefore$ $a^2 + b^2 = \frac{25}{3}$

8. (d) Now, the slope of tangent at $B\left(ct_2, \frac{c}{t_2}\right)$ is $-\frac{1}{t_2^2}$,

and the slope of chord joining A and C is $-\frac{1}{t_1t_3}$.

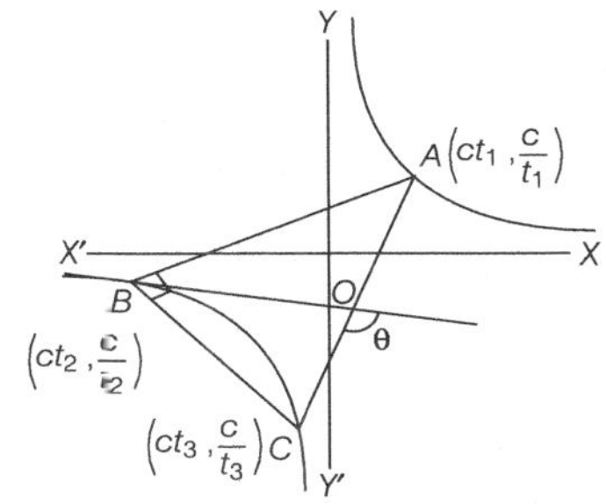

We have, $AB \perp BC$

$\Rightarrow$ $-\frac{1}{t_1t_2} \times -\frac{1}{t_2t_3} = -1$

$\Rightarrow$ $t_1t_2^2t_3 = -1$ ...(i)

Let θ be the angle between the tangent at B and AC

i.e. $$\tan\theta = \frac{-\frac{1}{t_1t_3} + \frac{1}{t_2}}{1 + \frac{1}{t_2^2\, t_1t_3}}$$

From Eq. (i), we get $\tan\theta = \infty$

$\therefore$ $\theta = 90°$

9. (b) We know that $C + H = 2A$

We have,

$$H: x^2 + 3xy + 2y^2 + 2x + 3y = 0$$

Its asymptotes is

$$x^2 + 3xy + 2y^2 + 2x + 3y + \lambda = 0$$

Compare it with

$$ax^2 + 2hxy + by^2 + 2gx + 2fy + c = 0$$

We get, $a = 1, h = \frac{3}{2}, b = 2, g = 1, f = \frac{3}{2}, c = \lambda$

It represents a pair of straight lines, if

$$abc + 2fgh - af^2 - bg^2 - ch^2 = 0$$

$\Rightarrow$ $1\cdot 2\cdot\lambda + 2\cdot\frac{3}{2}\cdot 1\cdot\frac{3}{2} - 1\cdot\frac{9}{4} - 2 - \lambda\cdot\frac{9}{4} = 0$

$\Rightarrow$ $2\lambda + \frac{9}{2} - \frac{9}{4} - 2 - \frac{9\lambda}{4} = 0$

$\lambda = 1$

$$A = x^2 + 3xy + 2y^2 + 2x + 3y + 1$$

$$C = 2x^2 + 6xy + 4y^2 + 4x + 6y + 2$$
$$- x^2 - 3xy - 2y^2 - 2x - 3y$$

$$C = x^2 + 3xy + 2y^2 + 2x + 3y + 2 = 0$$

10. (a) We know that,

any tangent to the hyperbola forms a triangle with the asymptotes has constant area $= ab$

Given, $ab = a^2\tan\lambda$

$\Rightarrow$ $\tan\lambda = \frac{b}{a}$

$\Rightarrow$ $\tan^2\lambda = \frac{b^2}{a^2} = e^2 - 1$

$\because$ $e^2 = 1 + \tan^2\lambda = \sec^2\lambda$

$\therefore$ $e = \sec\lambda$

11. (d) Let $P(x_1, y_1)$ lie on $x^2 - y^2 = a^2$ and $Q(h, k)$ be the mid-point of PN.

Then, the coordinates of Q are $\left(x_1, \frac{y_1}{2}\right)$

i.e. $x_1 = h$ and $\frac{y_1}{2} = k$

$\Rightarrow$ $x_1 = h$ and $y_1 = 2k$

$\therefore$ $x_1^2 - y_1^2 = a^2$

$\Rightarrow$ $h^2 - 4k^2 = a^2$

On taking locus of (h, k), we get

$$x^2 - 4y^2 = a^2$$

Which is a hyperbola.

12. (c) Let $y = f(x)$ be a conic

and (x_1, y_1) be a point on the conic such that the segment of the tangent between the coordinate axes is bisected by the point.

$\because$ Equation of tangent at (x_1, y_1),

$y - y_1 = p(x - x_1)$, where $p = \left(\frac{dy}{dx}\right)_{(x_1, y_1)}$

$$A \equiv X\text{-intercept} = \left(x_1 - \frac{y_1}{p}, 0\right)$$

$$B \equiv Y\text{-intercept} = (0, y_1 - px_1)$$

Mid-point of $AB = \left(\frac{x_1 - \frac{y_1}{p}}{2}, \frac{y_1 - px_1}{2}\right) = (x_1, y_1)$

$\Rightarrow$ $x_1 - \frac{y_1}{p} = 2x_1$ and $\frac{y_1 - px_1}{2} = y_1$

$\Rightarrow$ $\frac{y_1}{p} = -x_1$ and $-px_1 = y_1$

$\therefore$ $p = -\frac{y_1}{x_1}$ and $p = -\frac{y_1}{x_1}$

$\because$ $\left(\frac{dy}{dx}\right)_{(x_1, y_1)} = -\frac{y_1}{x_1}$

$\Rightarrow$ $\frac{dy}{dx} = -\frac{y}{x}$

$\Rightarrow$ $\int \frac{1}{y} dy = -\int \frac{1}{x} dx + C$

$\Rightarrow$ $\ln y = -\ln x + \ln C$

$\Rightarrow$ $y = \frac{C}{x}$

$\therefore$ $xy = C$

Since, this curve passes through (2, 4).

$xy = 8$ is equation of conic.

$xy = (2\sqrt{2})^2$ is rectangular hyperbola having eccentricity $= \sqrt{2}$

We know that, foci of the rectangular hyperbola is $(\pm C\sqrt{2}, \pm C\sqrt{2})$

$\therefore$ (4, 4) and (–4, – 4) are two foci of the conic.

13. (a) Let point of intersection be (x_1, y_1).

We have, rectangular hyperbola

$$xy = c^2$$

Equation of tangent at (x_1, y_1)

$\Rightarrow$ $xy_1 - yx_1 = 2c^2$

$\therefore$ $\tan\theta = -\frac{y_1}{x_1}$ ...(i)

and $y^2 = 4ax$

$\Rightarrow$ $yy_1 = 2a(x + x_1)$

$\Rightarrow$ $\tan\phi = \frac{2a}{y_1}$

$\because (x_1, y_1)$ lies on both the curve.

$\therefore$ $\tan\phi = \frac{y_1}{2x_1}$

or $\frac{y_1}{x_1} = 2\tan\phi$

$\Rightarrow$ $\tan\theta = -(2\tan\phi)$

$\therefore$ $\theta = \tan^{-1}(-2\tan\phi)$

14. (b) Equation of circle is

$$x^2 + y^2 = r^2 \quad \text{...(i)}$$

and hyperbola $xy = 1$ ...(ii)

On solving Eqs. (i) and (ii), we get

$$x^2 + \frac{1}{x^2} = r^2$$

$$x^4 + r^2x^2 + 1 = 0$$

Since, circle does not intersect the hyperbola, then $\Delta < 0$

i.e. $r^4 - 4 < 0$

$\Rightarrow$ $(r^2 + 2)(r^2 - 2) < 0$

$\Rightarrow$ $r^2 < 2$

$\Rightarrow$ $-\sqrt{2} < r < \sqrt{2}$

$\Rightarrow$ $0 < r < \sqrt{2}$

$\therefore$ $r \in (0, \sqrt{2})$

15. (b) Let the hyperbola is $\frac{x^2}{a^2} - \frac{y^2}{b^2} = 1$

$\therefore$ $e = \sqrt{1 + \frac{b^2}{a^2}} = \frac{\sqrt{a^2 + b^2}}{a}$

and $e' = \frac{\sqrt{a^2 + b^2}}{b}$

Equation of line is $\frac{x}{e} + \frac{y}{e'} = 1$

i.e. $\frac{ax}{\sqrt{a^2 + b^2}} + \frac{by}{\sqrt{a^2 + b^2}} = 1$

If $\frac{a}{\sqrt{a^2 + b^2}} = \cos\theta, \frac{b}{\sqrt{a^2 + b^2}} = \sin\theta$

$$x\cos\theta + y\sin\theta = 1$$

Since, this line touching fixed circle.

Let fixed circle be $x^2 + y^2 = r^2$

$\therefore$ $r = 1$

$\therefore$ Area of circle $= \pi r^2 = \pi$

16. (a) We have, $f(x) = ax^3 + bx^2 + cx + d$

The eccentricity of parabola = 1

and eccentricity of rectangular hyperbola $= \sqrt{2}$

$\because$ 1 and $\sqrt{2}$ are the roots of $f(x) = 0$

$\therefore$ $f(1) = 0$

$\Rightarrow$ $a + b + c + d = 0$

17. (c) Equation of normal at $\left(ct, \frac{c}{t}\right)$ to hyperbola $xy = c^2$

$$xt^3 - yt - ct^4 + c = 0 \quad \text{...(i)}$$

It touches its conjugate.

Equation of tangent, $x + yt^2 - 2ct = 0$ ...(ii)

Since, Eqs. (i) and (ii) are identical, then

$$\frac{t^3}{1} = \frac{-t}{t^2} = \frac{c - ct^4}{2ct} = \frac{1 - t^4}{2t}$$

$\Rightarrow$ $2t^4 = 1 - t^4$

$\Rightarrow$ $3t^4 = 1$ [it is 4 degree]

So, there are four normals.

18. (a) On solving $\quad y = \frac{x}{2}$ and $y = x^2$

$\Rightarrow \quad x = 0$ and $x = \frac{1}{2}$

$\therefore \quad -\frac{1}{2} < \alpha < \frac{1}{2}$

19. (a) Given, $P(x, y)$ lies on $x^2 + y^2 = 1$.

Let $P = (\cos\theta, \sin\theta)$ and $(x + y)^2 \le \lambda$

$$\Rightarrow \quad x^2 + y^2 + 2xy < \lambda$$

$$\Rightarrow \quad 2xy \le \lambda - 1$$

$$\Rightarrow \quad 2\sin\theta\cos\theta \le \lambda - 1$$

$$\Rightarrow \quad \sin 2\theta \le \lambda - 1 \qquad [\because |\sin\theta| \le 1]$$

$$\therefore \quad -1 \le \lambda - 1 \le 1$$

$$\Rightarrow \quad 0 \le \lambda \le 2$$

Now, check the position of $(\lambda, 0)$ w.r.t. the hyperbola,

$$(x-2)^2 - y^2 = 1$$

$$(\lambda - 2)^2 - 1 = \text{negative}$$

Since, $(\lambda, 0)$ lies inside the hyperbola.

Hence, no tangent can be drawn.

20. (a) We have,

$$R = (2, 0)$$

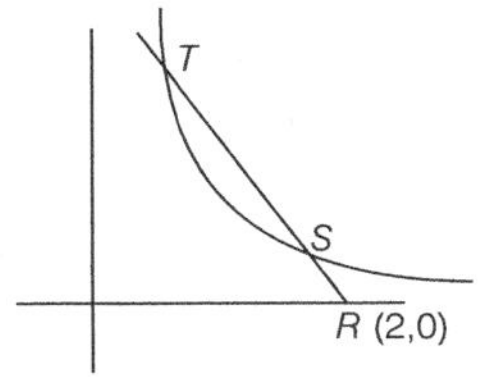

Let $\quad S = \left(2t_1, \frac{2}{t_1}\right)$ and $T = \left(2t_2, \frac{2}{t_2}\right)$

Equation of chord joining S and T is

$$x + yt_1t_2 - 2(t_1 + t_2) = 0$$

this chord passes through the point $R(2, 0)$, i.e.

$$2 - 2(t_1 + t_2) = 0$$

$$t_1 + t_2 = 1 \qquad \text{...(i)}$$

Now, $\quad RS^2 = (2t_1 - 2)^2 + \frac{4}{t_1^2}$

and $\quad RT^2 = (2t_2 - 2)^2 + \frac{4}{t_2^2}$

Then, $\quad RS^2 \times RT^2 = 4\left[(t_1 - 1)^2 + \frac{1}{t_1^2}\right] 4\left[(t_2 - 1)^2 + \frac{1}{t_2^2}\right]$

$$= 16\left[\left(t_2^2 + \frac{1}{t_1^2}\right)\left(t_1^2 + \frac{1}{t_2^2}\right)\right] \qquad \text{[from Eq. (i)]}$$

$$= 16\left[t_1^2t_2^2 + 1 + 1 + \frac{1}{t_1^2t_2^2}\right]$$

$$= 16\left[(t_1t_2)^2 + \frac{1}{(t_1t_2)^2} + 2\right]$$

Since, $\quad \text{AM} \ge \text{GM}$

$$\Rightarrow \quad (t_1t_2)^2 + \frac{1}{(t_1t_2)^2} \ge 2$$

$$\Rightarrow \quad 16\left[(t_1t_2)^2 + \frac{1}{(t_1t_2)^2} + 2\right] \ge 4 \times 16$$

$$\Rightarrow \quad RS^2 \times RT^2 \ge 64$$

$$\therefore \quad RS \times RT \ge \sqrt{64} = 8$$

Hence, minimum value of $RS \times RT$ is 8.

21. (a,c,d) Verify each option in $\frac{x^2}{a^2} - \frac{y^2}{b^2} = 1$ or $\frac{x^2}{a^2} - \frac{y^2}{b^2} = -1$ and $xy = c^2$.

$$\frac{2x}{a} = t + \frac{1}{t} \text{ and } \frac{2y}{b} = t - \frac{1}{t}$$

$$\frac{x}{a} + \frac{y}{b} = t \qquad \text{...(i)}$$

and $\quad \frac{x}{a} - \frac{y}{b} = \frac{1}{t} \qquad$ ...(ii)

On multiplying Eqs. (i) and (ii), we get

$$\frac{x^2}{a^2} - \frac{y^2}{b^2} = 1$$

Which is a hyperbola.

(b) $\quad \frac{tx}{a} - \frac{y}{b} = 1 \qquad$ ...(i)

and $\quad \frac{x}{a} + \frac{yt}{b} - 1 = 0 \qquad$ ...(ii)

On solving Eqs. (i) and (ii), we get

$$\frac{x}{a} = \frac{1-t^2}{1+t^2} \text{ and } \frac{y}{b} = \frac{2t}{1+t^2}$$

which is not a hyperbola.

(c) $\quad x = e^t + e^{-t} \qquad$ ...(i)

and $\quad y = e^t - e^{-t} \qquad$ ...(ii)

$$\Rightarrow \quad x + y = 2e^t \text{ and } x - y = 2e^{-t}$$

$$\therefore \quad (x + y)(x - y) = 4$$

$$\Rightarrow \quad x^2 - y^2 = 4$$

which is a hyperbola.

(d) $\quad x^2 - 6 = 2\cos t \qquad$ ...(i)

$$\Rightarrow \quad y^2 + 2 = 4\cos^2\frac{t}{2} = 2(1 + \cos t)$$

$\therefore \quad y^2 = 2\cos t \qquad$ ...(ii)

$$\Rightarrow \quad x^2 - y^2 - 6 = 0$$

$$\Rightarrow \quad x^2 - y^2 = 6$$

Which is a hyperbola.

22. (a,b,d) We have,

$$(px^2 + qy^2 + r)(4x^2 + 4y^2 - 8x - 4) = 0$$

$$S_1 \equiv px^2 + qy^2 + r = 0$$

and $\quad S_2 \equiv 4x^2 + 4y^2 - 8x - 4 = 0$

(a) S_2 represents a circle

and if $r = 0$ and p and q are of the opposite signs, then S_1 represents a pair of straight lines.

(b) Clearly, if $p = q$ and r is negative, then S_1 represents a circle. S_2 is a circle.

(c) When p and q are of opposite sign $r \ne 0$, then S_1 represents a hyperbola not an ellipse.

(d) When p and q are of same sign but unequal and r is of sign opposite to that of p.

S_1 represents an ellipse.

S_2 represents a circle.

23. (a,b,d) We have, $\quad (1 - x^2)\frac{dy}{dx} + xy = ax$

$$\frac{dy}{dx} + \frac{x}{1 - x^2}\cdot y = \frac{ax}{1 - x^2}$$

Now, $\quad \text{IF} = e^{\int \frac{x}{1-x^2}dx} = e^{-\frac{1}{2}\log(1-x^2)} = (1 - x^2)^{-\frac{1}{2}} = \frac{1}{\sqrt{1 - x^2}}$

$$\Rightarrow \quad y \cdot \frac{1}{\sqrt{1 - x^2}} = \int \frac{ax}{(1 - x^2)^{3/2}}\,dx$$

Let $1 - x^2 = t^2$, then $-x\,dx = t\,dt$

$$\Rightarrow \quad \frac{y}{\sqrt{1-x^2}} = -a\int \frac{t}{t^3}dt = -a\int \frac{1}{t^2}dt = -\frac{at^{-1}}{-1} = \frac{a}{1}$$

$$\therefore \quad \frac{y}{\sqrt{1-x^2}} = \frac{a}{\sqrt{1-x^2}} + C$$

$$\therefore \quad y = a + C\sqrt{1-x^2}$$

$$\Rightarrow \quad (y-a)^2 = c^2(1-x^2)$$

$$\Rightarrow \quad c^2x^2 + (y-a)^2 = c^2$$

$$\Rightarrow \quad x^2 + \frac{(y-a)^2}{c^2} = 1 \quad \text{...(i)}$$

(a) Eq. (i) represents a conic which is either an ellipse or a hyperbola. Principle axes parallel to coordinate axes.

(b) Centre = $(0, a)$ and length of the one principle axis is 2.

24. (c,d) We have, $a\sec\theta - x\tan\theta - y = 0$ ...(i)

and $b\sec\theta + y\tan\theta - x = 0$ ...(ii)

By cross-multiplication, we get

$$\frac{\sec\theta}{x^2+y^2} = -\frac{\tan\theta}{-ax+by} = \frac{1}{ay+bx}$$

$$\Rightarrow \quad \sec\theta = \frac{x^2+y^2}{ay+bx} \text{ and } \tan\theta = \frac{by-ax}{ay+bx}$$

$$\because \quad \sec^2\theta - \tan^2\theta = 1$$

$$\Rightarrow \quad \frac{(x^2+y^2)^2}{(ay+bx)^2} - \frac{(by-ax)^2}{(ay+bx)^2} = 1$$

$$\Rightarrow \quad (x^2+y^2)^2 - (by-ax)^2 = (ay+bx)^2$$

$$\Rightarrow x^4 + y^4 + 2x^2y^2 - b^2y^2 - a^2x^2 + 2abxy = a^2y^2 + b^2x^2 + 2abxy$$

$$\Rightarrow \quad (x^2+y^2)^2 - b^2(x^2+y^2) - a^2(x^2+y^2) = 0$$

$$\Rightarrow \quad x^2 + y^2 - a^2 - b^2 = 0$$

$$\Rightarrow \quad x^2 + y^2 = a^2 + b^2$$

It is the director circle of $\frac{x^2}{a^2} + \frac{y^2}{b^2} = 1$

and director circle of circle $x^2 + y^2 = \frac{a^2+b^2}{2}$.

25. (b,c) We have, the hyperbola $x^2 - 3y^2 = 3$

$$\Rightarrow \quad \frac{x^2}{3} - y^2 = 1$$

Equation of tangent at $(\sqrt{3}, 0)$ to the given hyperbola is

$$\sqrt{3}x = 1$$

$$\therefore \quad x = \frac{1}{\sqrt{3}}$$

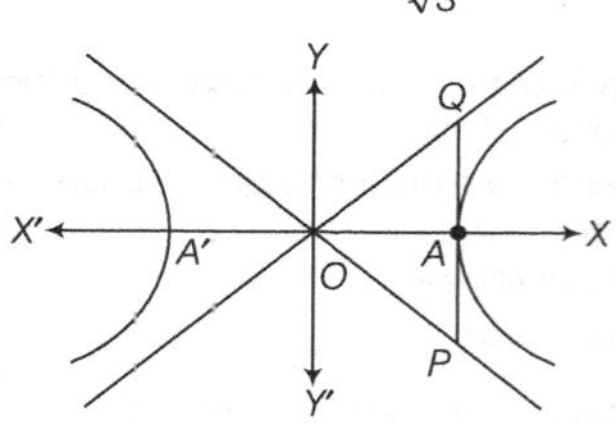

and asymptotes of the hyperbola

$$\Rightarrow \quad \frac{x}{\sqrt{3}} \pm y = 0$$

$$\therefore \quad y = \pm\frac{x}{\sqrt{3}} \text{ i.e. } \left(\frac{1}{\sqrt{3}}, \pm\frac{1}{0}\right)$$

Clearly, ΔOPQ forms an equilateral triangle, then

its area $= \frac{\sqrt{3}}{4} \times (2)^2 = \frac{\sqrt{3}\times 4}{4} = \sqrt{3}$ sq units

26. (a) If the circle

$$x^2 + y^2 + 2gx + 2fy + d = 0 \quad \text{...(i)}$$

cut the rectangular hyperbola

$$xy = c^2 \quad \text{...(ii)}$$

at 4 points, then from Eqs. (i) and (ii), we get

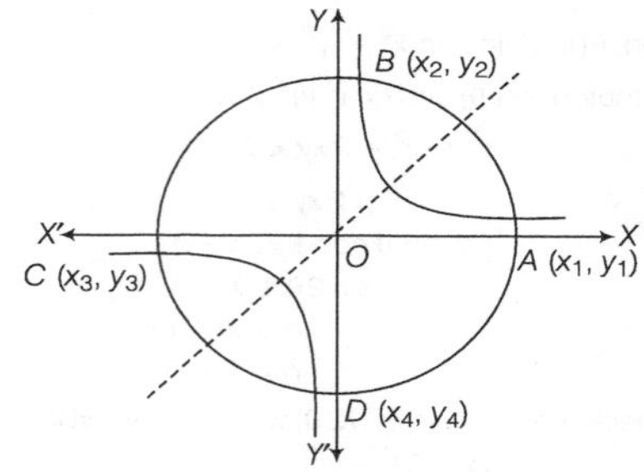

$$x^2 + \frac{c^4}{x^2} + 2gx + 2f\frac{c^2}{x} + d = 0$$

$$\Rightarrow \quad x^4 + 2gx^3 + dx^2 + 2fc^2x + c^4 = 0$$

$$\Rightarrow \quad \Sigma x_1 = -2g$$

$$\Rightarrow \quad \Sigma x_1x_2 = d$$

$$\Rightarrow \quad \Sigma x_1x_2x_3 = -2fc^2$$

$$\Rightarrow \quad x_1x_2x_3x_4 = c^4$$

We have, the circle

$$(x-1)^2 + (y-2)^2 = 25$$

$$\Rightarrow \quad x^2 + y^2 - 2x - 4y - 20 = 0$$

$$\therefore \quad g = -1, f = -2 \text{ and } d = -20$$

(i) $x_1 + x_2 + x_3 + x_4 = -2g = -2(-1) = 2$

(ii) $x_1^2 + x_2^2 + x_3^2 + x_4^2 = (x_1 + x_2 + x_3 + x_4)^2 - 2\,\Sigma x_1x_2$

$$= 4g^2 - 2d = 4(-1)^2 - 2(-20)$$

$$= 4 + 40 = 44$$

Similarly, $\Sigma y_1 = -2f$

and $\Sigma y_1y_2 = d$

(iii) $OA^2 + OB^2 + OC^2 + OD^2$

$$= x_1^2 + y_1^2 + x_2^2 + y_2^2 + x_3^2 + y_3^2 + x_4^2 + y_4^2$$

$$\therefore \quad \Sigma x_1^2 + \Sigma y_1^2 = 44 + 56 = 100$$

(iv) $\Sigma y_1^2 = (\Sigma y_1)^2 - 2\,\Sigma y_1y_2 = 4f^2 - 2d$

$$= 4(-2)^2 - 2(-20)$$

$$= 16 + 40 = 56$$

Hence, (i) → (p), (ii) → (r), (iii) → (t), (iv) → (s)

27. (a) Equation of normal at point $\left(ct, \frac{c}{t}\right)$ to the rectangular hyperbola

$$xt^3 - yt - ct^4 + c = 0$$

Since, this normal passes through (p, q).

$$\therefore \quad pt^3 - qt - ct^4 + c = 0$$

$$\Rightarrow \quad ct^4 - pt^3 + qt - c = 0 \quad \text{...(i)}$$

Eq. (i) has 4 roots t_1, t_2, t_3 and t_4.

$$\Sigma t_1 = \frac{p}{c}$$

$$\Sigma t_1t_2 = 0$$

$$\Sigma t_1t_2t_3 = -\frac{q}{c}$$

$$t_1t_2t_3t_4 = -1$$

$$p = c\Sigma t_1 = ct_1 + ct_2 + ct_3 + ct_4$$

$$\therefore \quad p = x_1 + x_2 + x_3 + x_4$$

Take $y_1 + y_2 + y_3 + y_4 = \frac{c}{t_1} + \frac{c}{t_2} + \frac{c}{t_3} + \frac{c}{t_4}$

$$= c\,\frac{\Sigma t_1 t_2 t_3}{t_1 t_2 t_3 t_4}$$

$$= \frac{c\cdot\left(-\frac{g}{c}\right)}{-1} = \frac{-q}{-1} = q$$

For $\Sigma x_1^2 = (\Sigma x_1)^2 - 2\,\Sigma x_1 x_2$

$\Rightarrow \quad c^2\,\Sigma t_1^2 = (c\,\Sigma t_1)^2 - 2c^2\,\Sigma t_1 t_2$

$\Rightarrow \quad \Sigma t_1^2 = \frac{c^2 p^2}{c^2} - 2c^2 \times 0 = p^2$

$\therefore \quad \Sigma y_1^2 = (\Sigma y_1)^2 - 2\,\Sigma y_1 y_2$

$$= q^2 - 2 \times 0$$

$$= q^2$$

Hence, (i) → (r), (ii) → (s), (iii) → (p), (iv) → (q)

Explanations for Q. Nos. (28-30)

Let three points be $P(a\sec\alpha, b\tan\alpha)$, $R(a\sec\beta, b\tan\beta)$, $T(a\sec\gamma, b\tan\gamma)$.

Equation of the chord PR, PT and RT are

$$\frac{x}{\sqrt{2}}\cos\left(\frac{\alpha-\beta}{2}\right) - y\cdot\sin\left(\frac{\alpha+\beta}{2}\right) = \cos\left(\frac{\alpha+\beta}{2}\right) \quad \ldots\text{(i)}$$

$$\frac{x}{\sqrt{2}}\cos\left(\frac{\alpha-\gamma}{2}\right) - y\sin\left(\frac{\alpha+\gamma}{2}\right) = \cos\left(\frac{\alpha+\gamma}{2}\right) \quad \ldots\text{(ii)}$$

$$\frac{x}{\sqrt{2}}\cos\left(\frac{\beta-\gamma}{2}\right) - y\sin\left(\frac{\beta+\gamma}{2}\right) = \cos\left(\frac{\beta+\gamma}{2}\right) \quad \ldots\text{(iii)}$$

$\because PR$ and PT are perpendicular to each other

$$\Rightarrow \frac{\cos\left(\frac{\alpha-\beta}{2}\right)}{\sqrt{2}\sin\left(\frac{\alpha+\beta}{2}\right)}\cdot\frac{\cos\left(\frac{\alpha-\beta}{2}\right)}{\sqrt{2}\sin\left(\frac{\alpha+\gamma}{2}\right)} = -1$$

$$\Rightarrow \frac{1}{2}\left[\cos\left(\frac{\beta-\gamma}{2}\right) + \cos\left(\alpha - \left(\frac{\beta+\gamma}{2}\right)\right)\right] + \cos\left(\frac{\beta+\gamma}{2}\right) - \cos\left(\frac{\alpha+\beta+\gamma}{2}\right) = 0$$

$$\Rightarrow \frac{3}{2}\left[\cos\left(\frac{\beta-\gamma}{2}\right)\right] + \frac{1}{2}\left[\cos\alpha\cdot\cos\left(\frac{\beta+\gamma}{2}\right) + \sin\alpha\cdot\sin\left(\frac{\beta+\gamma}{2}\right)\right] - \cos\alpha\cdot\cos\left(\frac{\beta+\gamma}{2}\right) + \sin\alpha\cdot\sin\left(\frac{\beta+\gamma}{2}\right)\cdot\cos\left(\frac{\beta+\gamma}{2}\right)$$

$$= \frac{3}{\cos\alpha}\left[\sin\alpha\cdot\sin\left(\frac{\beta+\gamma}{2}\right) + \cos\left(\frac{\beta-\gamma}{2}\right)\right] \quad \ldots\text{(iv)}$$

If $\alpha = \frac{\pi}{4}$, then

$$\cos\left(\frac{\beta+\gamma}{2}\right) = 3\left[\sin\left(\frac{\beta+\gamma}{2}\right) + \sqrt{2}\cos\left(\frac{\beta-\gamma}{2}\right)\right]$$

Putting in Eq, (iii), we get

$$(x-6) - \frac{\sqrt{2}\sin\left(\frac{\beta+\gamma}{2}\right)}{\cos\left(\frac{\beta-\gamma}{2}\right)}\cdot(y+3) = 0$$

Hence, for all values of β and γ, RT passes through the point $(6, -3)$.

Slope of $PC = \frac{1}{2}$ $\quad$ [$\because C$ is centre (0, 0)]

Slope of $CQ = -\frac{3}{6} = -\frac{1}{2}$

Thus, sum of slopes is zero.

$\therefore$ One of the bisectors is $y = 0$

Putting the value of $\cos\left(\frac{\beta+\gamma}{2}\right)$ from Eq. (iv) in Eq. (iii), we get

$$(x - 3\sqrt{2}\sec\alpha) - \sqrt{2}\,\frac{\sin\left(\frac{\beta+\gamma}{2}\right)}{\cos\left(\frac{\beta-\gamma}{2}\right)}\cdot(y + 3\tan\alpha) = 0$$

$\therefore RT$ is concurrent at $Q(3\sqrt{2}\sec\alpha, -3\tan\alpha)$

Let $h = 3\sqrt{2}\sec\alpha$, $k = -3\tan\alpha$

$\Rightarrow \frac{h^2}{18} - \frac{k^2}{9} = 1$

$\therefore$ Locus of $Q(h, k)$ is $\frac{x^2}{18} - \frac{y^2}{9} = 1$

28. (b) **29.** (c) **30.** (a)

31. (2) We have, $\frac{e}{2}$ and $\frac{e'}{2}$ are eccentricities of a hyperbola and its conjugate hyperbola.

$\because \quad \frac{4}{e^2} + \frac{4}{e'^2} = 1$

$\Rightarrow \quad 4 = \frac{e^2 e'^2}{e^2 + e'^2}$

The line passes through the point $(e, 0)$ and $(0, e')$.

$$\frac{x}{e} + \frac{y}{e'} = 1$$

$\Rightarrow \quad e'x + ey - ee' = 0$

It is tangent to the circle

$$x^2 + y^2 = r^2$$

$\because \quad r = \frac{ee'}{\sqrt{e^2 + e'^2}}$

$\Rightarrow \quad r^2 = \frac{e^2 e'^2}{e^2 + e'^2} = 4$

$\therefore \quad r = 2$

32. (9) Circle, hyperbola intersect at 4 points.

Given, point $= (-3, 12) =$ Centre of circle

Radius $= \sqrt{9 + 144 - 72} = 9$

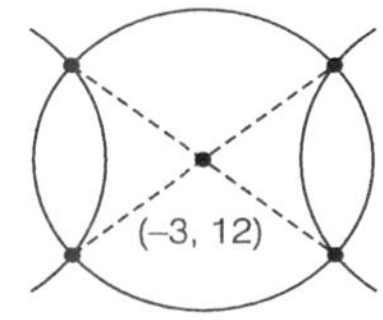

$\therefore \quad$ Required sum $= 36 = 2^2\cdot 3^2$

Number of divisors $= (1 + 2)(1 + 2) = 9$

33. (4) Clearly, $C(0, 0)$ let the equation of circle be $x^2 + y^2 - ax - by = 0$

On homogenising with chord $11x + 3y - 14 = 0$ and comparing with pair of asymptotes, we get equation of circle as $x^2 + y^2 - \frac{112}{23}x - \frac{64}{23}y = 0$

$$\therefore \quad \frac{23(a - b + c)}{12} = \frac{23\left(+\frac{112}{23} - \frac{64}{23}\right)}{12}$$

$$= 4$$

34. (1) Clearly, $\angle SOL = 45°$

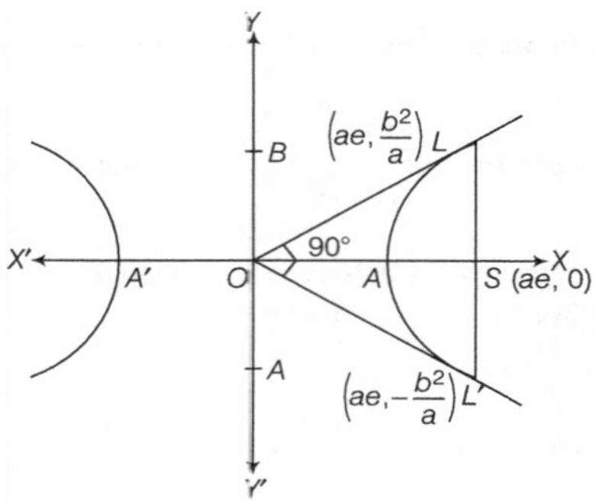

In ΔOSL, $\tan 45° = \frac{\frac{b^2}{a}}{ae} = 1$

$\because \quad \frac{b^2}{a^2e} = 1 \Rightarrow e = \frac{b^2}{a^2} = e^2 - 1$

$\Rightarrow \quad e^2 - e - 1 = 0 \Rightarrow e = \frac{1 + \sqrt{1+4}}{2}$

$\therefore \quad e = \frac{\sqrt{5}+1}{2}$ i.e. $[e] = 1$

35. (4) $x = \frac{55 - y}{2y - 5}, y = \frac{55 + 5x}{2x + 1}$

Satisfied by 16 pairs of (x, y).

$\therefore \quad \lambda = 16$

$[\sqrt{\lambda}] = 4$

JEE Main & AIEEE Archive

1. (c)

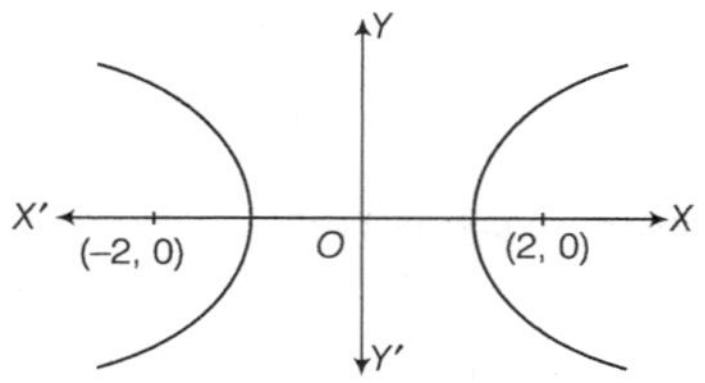

Let equation of hyperbola be

$$\frac{x^2}{a^2} - \frac{y^2}{b^2} = 1$$

where, $2ae = 4$ and $e = 2$

$\Rightarrow \quad a = 1$

$\because \quad a^2e^2 = a^2 + b^2$

$\Rightarrow \quad 4 = 1 + b^2$

$\therefore \quad b^2 = 3$

Thus, equation of hyperbola is

$$\frac{x^2}{1} - \frac{y^2}{3} = 1 \text{ or } 3x^2 - y^2 = 3$$

2. (d) The given equation of hyperbola is

$$\frac{x^2}{\cos^2 \alpha} - \frac{y^2}{\sin^2 \alpha} = 1$$

Here, $a^2 = \cos^2 \alpha$

and $b^2 = \sin^2 \alpha$

Now, $e = \sqrt{1 + \frac{b^2}{a^2}}$

$\Rightarrow \quad e = \sqrt{1 + \frac{\sin^2 \alpha}{\cos^2 \alpha}} = \sqrt{1 + \tan^2 \alpha}$

$\Rightarrow \quad e = \sec \alpha$

Coordinates of foci are $(\pm ae, 0)$ i.e. $(\pm 1, 0)$.

Hence, abscissae of foci remains constant, when α varies.

3. (a) Since, the line $y = \alpha x + \beta$ is tangent to the hyperbola $\frac{x^2}{a^2} - \frac{y^2}{b^2} = 1$.

$\therefore \quad \beta^2 = a^2\alpha^2 - b^2$

So, the locus of (α, β) is

$$y^2 = a^2x^2 - b^2$$

$\Rightarrow \quad a^2x^2 - y^2 - b^2 = 0$

Since, this equation represents a hyperbola, so locus of a point $P(\alpha, \beta)$ is a hyperbola.

4. (c) Given equation of hyperbola can be rewritten as

$$\frac{x^2}{\left(\frac{12}{5}\right)^2} - \frac{y^2}{\left(\frac{9}{5}\right)^2} = 1$$

$\therefore \quad$ Eccentricity, $e'^2 = 1 + \frac{b'^2}{a'^2}$

$\Rightarrow \quad e'^2 = 1 + \frac{9}{16} = \frac{25}{16}$

$\therefore \quad e' = \frac{5}{4}$

The foci of a hyperbola are

$$(\pm a'e', 0) = \left(\pm \frac{12}{5} \times \frac{5}{4}, 0\right) = (\pm 3, 0)$$

Given equation of ellipse is

$$\frac{x^2}{16} + \frac{y^2}{b^2} = 1$$

Foci of an ellipse are $(\pm ae, 0) = (\pm 4e, 0)$ but given focus of ellipse and hyperbola coincide, then

$$4e = 3 \Rightarrow e = \frac{3}{4}$$

Also, $b^2 = a^2(1 - e^2)$

$$= 16\left(1 - \frac{9}{16}\right) = 16 - 9 = 7$$

5. (a) The mid-point of the chord is $\left(\frac{x_1 + x_2}{2}, \frac{y_1 + y_2}{2}\right)$.

The equation of the chord $T = S_1$.

$\therefore \quad x\left(\frac{y_1 + y_2}{2}\right) + y\left(\frac{x_1 + x_2}{2}\right) = 2\left(\frac{x_1 + x_2}{2}\right)\left(\frac{y_1 + y_2}{2}\right)$

$\Rightarrow \quad x(y_1 + y_2) + y(x_1 + x_2) = (x_1 + x_2)(y_1 + y_2)$

$\Rightarrow \quad \frac{x}{x_1 + x_2} + \frac{y}{y_1 + y_2} = 1$

JEE Advanced & IIT JEE Archive

1. (a,b,d)

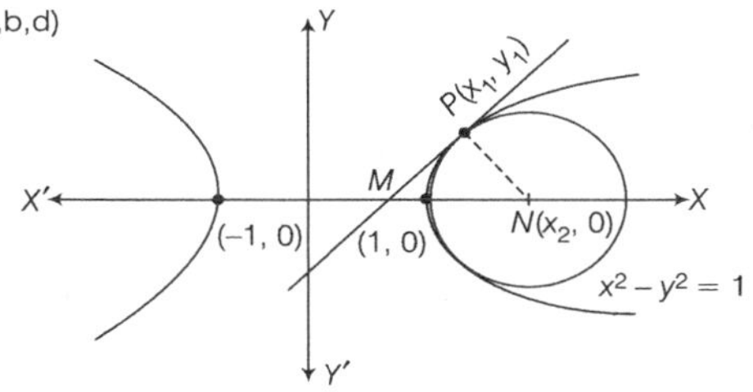

Equation of family of circles touching hyperbola at (x_1, y_1) is

$$(x-x_1)^2+(y-y_1)^2+\lambda(xx_1-yy_1-1)=0$$

Now, its centre is $(x_2, 0)$.

$$\therefore \left[\frac{-(\lambda x_1-2x_1)}{2}, \frac{-(-2y_1-\lambda y_1)}{2}\right]=(x_2,0)$$

$\Rightarrow \quad 2y_1+\lambda y_1=0 \Rightarrow c\lambda=-2$

and $\quad 2x_1-\lambda x_1=2x_2 \Rightarrow x_2=2x_1$

$\therefore \quad P(x_1, \sqrt{x_1^2-1})$ and $N(x_2, 0)=(2x_1, 0)$

As tangent intersect X-axis at $M\left(\frac{1}{x}, 0\right)$.

Centroid of $\Delta PMN=(l, m)$

$$\Rightarrow \left[\frac{3x_1+\frac{1}{x_1}}{3}, \frac{y_1+0+0}{3}\right]=(l,m) \Rightarrow l=\frac{3x_1+\frac{1}{x_1}}{3}$$

On differentiating w.r.t. x_1, we get

$$\frac{dl}{dx_1}=\frac{3-\frac{1}{x_1^2}}{3}$$

$$\Rightarrow \frac{dl}{dx_1}=1-\frac{1}{3x_1^2} \text{ for } x_1>1 \text{ and } m=\frac{\sqrt{x_1^2-1}}{3}$$

On differentiating w.r.t. x_1, we get

$$\frac{dm}{dx_1}=\frac{2x_1}{2\times3\sqrt{x_1^2-1}}=\frac{x_1}{3\sqrt{x_1^2-1}} \text{ for } x_1>1$$

Also, $\quad m=\frac{y_1}{3}$

On differentiating w.r.t. y_1, we get

$$\frac{dm}{dy_1}=\frac{1}{3} \text{ for } y_1>0$$

2. (b) Equation of normal to hyperbola at (x_1, y_1) is

$$\frac{a^2x}{x_1}+\frac{b^2y}{y_1}=(a^2+b^2)$$

$$\therefore \text{ At } (6, 3) \Rightarrow \frac{a^2x}{6}+\frac{b^2y}{3}=(a^2+b^2)$$

It passes through (9, 0).

$$\Rightarrow \frac{a^2\cdot9}{6}=a^2+b^2$$

$$\Rightarrow \frac{3a^2}{2}-a^2=b^2 \Rightarrow \frac{a^2}{b^2}=2$$

$$\therefore e^2=1+\frac{b^2}{a^2}=1+\frac{1}{2} \Rightarrow e=\sqrt{\frac{3}{2}}$$

3. (b,d) Here, equation of ellipse $\frac{x^2}{4}+\frac{y^2}{1}=1$

$$\Rightarrow e^2=1-\frac{b^2}{a^2}=1-\frac{1}{4}=\frac{3}{4}$$

$$\therefore e=\frac{\sqrt{3}}{2} \text{ and focus } (\pm ae, 0) \Rightarrow (\pm\sqrt{3}, 0)$$

For hyperbola $\frac{x^2}{a^2}-\frac{y^2}{b^2}=1, e_1^2=1+\frac{b^2}{a^2}$, where, $e_1^2=\frac{1}{e^2}=\frac{4}{3}$

$$\Rightarrow 1+\frac{b^2}{a^2}=\frac{4}{3}$$

$$\therefore \frac{b^2}{a^2}=\frac{1}{3} \quad \text{...(i)}$$

and hyperbola passes through $(\pm\sqrt{3}, 0)$

$$\Rightarrow \frac{3}{a^2}=1$$

$$\Rightarrow a^2=3 \quad \text{...(ii)}$$

From Eqs. (i) and (ii), we get

$$b^2=1 \quad \text{...(iii)}$$

$\therefore$ Equation of hyperbola is $\frac{x^2}{3}-\frac{y^2}{1}=1$

Focus is $(\pm ae_1, 0)$ i.e. $\left(\pm\sqrt{3}\cdot\frac{2}{\sqrt{3}}, 0\right) \Rightarrow (\pm 2, 0)$

4. (b) Equation of tangent to hyperbola having slope m is

$$y=mx+\sqrt{9m^2-4} \quad \text{...(i)}$$

Equation of tangent to circle is

$$y=m(x-4)+\sqrt{16m^2+16} \quad \text{...(ii)}$$

Eqs. (i) and (ii) will be identical for $m=\frac{2}{\sqrt{5}}$ satisfy.

$\therefore$ Equation of common tangent is $2x-\sqrt{5}y+4=0$.

5. (a) Equation of the hyperbola is $\frac{x^2}{9}-\frac{y^2}{4}=1$ and that of circle is

$$x^2+y^2-8x=0$$

For their points of intersection

$$\frac{x^2}{9}+\frac{x^2-8x}{4}=1$$

$$\Rightarrow 4x^2+9x^2-72x=36$$

$$\Rightarrow 13x^2-72x-36=0$$

$$\Rightarrow 13x^2-78x+6x-36=0$$

$$\Rightarrow 13x(x-6)+6(x-6)=0$$

$$\Rightarrow x=6, x=-\frac{13}{6}$$

$$\because x=-\frac{13}{6} \text{ is not acceptable.}$$

Now, for $x=6, y=\pm 2\sqrt{3}$

Required equation is $(x-6)^2+(y+2\sqrt{3})(y-2\sqrt{3})=0$.

$$\Rightarrow x^2-12x+y^2+24=0$$

$$\therefore x^2+y^2-12x+24=0$$

6. (2) On substituting $\left(\frac{a}{e}, 0\right)$

in $y=-2x+1$,

we get

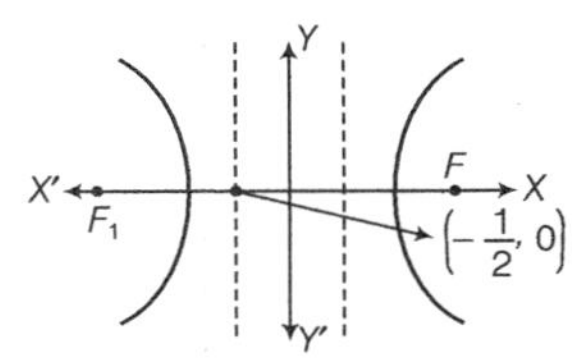

$$0=-\frac{2a}{e}+1$$

$$\Rightarrow \frac{a}{e}=\frac{1}{2}$$

Also, $y=-2x+1$ is tangent to hyperbola.

$$\therefore 1=4a^2-b^2$$

$$\Rightarrow \frac{1}{a^2}=4-(e^2-1) \Rightarrow \frac{4}{e^2}=5-e^2$$

$$\Rightarrow e^4-5e^2+4=0$$

$$\Rightarrow (e^2-4)(e^2-1)=0 \Rightarrow e=2, e=1$$

$e=1$ gives the conic as parabola. But conic is given as hyperbola, hence $e=2$.

7. (a,b) Given, $2x^2 - 2y^2 = 1$

$$\Rightarrow \quad \frac{x^2}{\left(\frac{1}{2}\right)} - \frac{y^2}{\left(\frac{1}{2}\right)} = 1 \quad \text{... (i)}$$

Eccentricity of hyperbola $= \sqrt{2}$

So, eccentricity of ellipse $= 1/\sqrt{2}$

Let equation of ellipse be

$$\frac{x^2}{a^2} + \frac{y^2}{b^2} = 1 \; (a > b)$$

$$\therefore \quad \frac{1}{\sqrt{2}} = \sqrt{1 - \frac{b^2}{a^2}}$$

$$\Rightarrow \quad \frac{b^2}{a^2} = \frac{1}{2} \Rightarrow a^2 = 2b^2$$

$$\therefore \quad x^2 + 2y^2 = 2b^2 \quad \text{... (ii)}$$

Let ellipse and hyperbola intersect at $A\left(\frac{1}{\sqrt{2}} \sec\theta, \frac{1}{\sqrt{2}} \tan\theta\right)$.

On differentiating Eq. (i),we get

$$4x - 4y\frac{dy}{dx} = 0 \Rightarrow \frac{dy}{dx} = \frac{x}{y}$$

$$\left.\frac{dy}{dx}\right|_{\text{at } A} = \frac{\sec\theta}{\tan\theta} = \operatorname{cosec}\theta$$

and differentiating Eq. (ii),

$$2x + 4y\frac{dy}{dx} = 0$$

$$\left.\frac{dy}{dx}\right|_{\text{at } A} = -\frac{x}{2y} = -\frac{1}{2}\operatorname{cosec}\theta$$

Since, ellipse and hyperbola are orthogonal.

$$\therefore \quad -\frac{1}{2}\operatorname{cosec}^2\theta = -1$$

$$\Rightarrow \quad \operatorname{cosec}^2\theta = 2 \Rightarrow \theta = \pm\frac{\pi}{4}$$

$$\therefore \quad A\left(1, \frac{1}{\sqrt{2}}\right) \text{ or } \left(1, -\frac{1}{\sqrt{2}}\right)$$

From Eq. (i), $\quad 1 + 2\left(\frac{1}{\sqrt{2}}\right)^2 = 2b^2$

$$\Rightarrow \quad b^2 = 1$$

Equation of ellipse is $x^2 + 2y^2 = 2$ and

coordinates of foci $(\pm ae, 0) = \left(\pm\sqrt{2} \cdot \frac{1}{\sqrt{2}}, 0\right) = (\pm 1, 0)$

Hence, options (a) and (b) are correct.

If major axis is along Y-axis, then

$$\frac{1}{\sqrt{2}} = \sqrt{1 - \frac{a^2}{b^2}}$$

$$\Rightarrow \quad b^2 = 2a^2$$

$$\therefore \quad 2x^2 + y^2 = 2a^2$$

$$\Rightarrow \quad y' = -\frac{2x}{y}$$

$$\Rightarrow \quad y'_{\left(\frac{1}{\sqrt{2}}\sec\theta, \frac{1}{\sqrt{2}}\tan\theta\right)} = -\frac{2}{\sin\theta}$$

As ellipse and hyperbola are orthogonal.

$$\therefore \quad -\frac{2}{\sin\theta} \cdot \operatorname{cosec}\theta = -1$$

$$\Rightarrow \quad \operatorname{cosec}^2\theta = 1 \Rightarrow \theta = \pm\frac{\pi}{4}$$

$$\therefore \quad 2x^2 + y^2 = 2a^2$$

$$\Rightarrow \quad 2 + \frac{1}{2} = 2a^2 \Rightarrow a^2 = \frac{5}{4}$$

$\therefore \quad 2x^2 + y^2 = \frac{5}{2}$, corresponding foci are $(1, \pm 0)$.

8. (b) Given equation can be rewritten as $\frac{(x - \sqrt{2})^2}{4} - \frac{(y + \sqrt{2})^2}{2} = 1$

For point $A(x, y), e = \sqrt{1 + \frac{2}{4}} = \sqrt{\frac{3}{2}}$

$$x - \sqrt{2} = 2 \quad \Rightarrow \quad x = 2 + \sqrt{2}$$

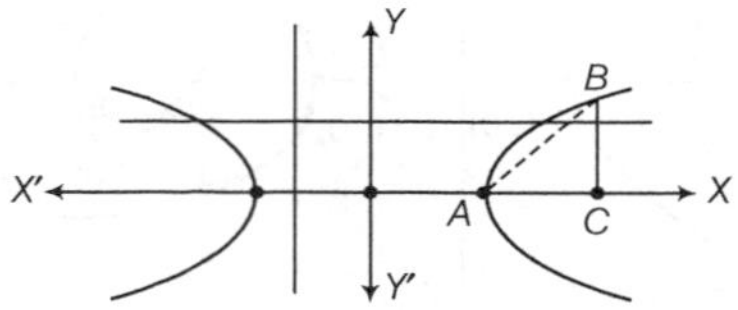

For point $C(x, y)$, $x - \sqrt{2} = ae = \sqrt{6}$

$/ \; x = \sqrt{6} + \sqrt{2}$

Now, $\quad AC = \sqrt{6} + \sqrt{2} - 2 - \sqrt{2} = \sqrt{6} - 2$

and $\quad BC = \frac{b^2}{a} = \frac{2}{2} = 1$

$\therefore$ Area of $\Delta ABC = \frac{1}{2} \times (\sqrt{6} - 2) \times 1 = \sqrt{\frac{3}{2}} - 1$ sq unit

9. (a) The given ellipse is

$$\frac{x^2}{4} + \frac{y^2}{3} = 1 \Rightarrow a = 2, b = \sqrt{3}$$

$$\therefore \quad 3 = 4(1 - e^2) \Rightarrow e = \frac{1}{2}$$

$$\therefore \quad ae = 2 \times \frac{1}{2} = 1$$

Hence, eccentricity e_1, of the hyperbola is given by

$$1 = e_1 \sin\theta \Rightarrow e_1 = \operatorname{cosec}\theta$$

$$\Rightarrow \quad b^2 = \sin^2\theta(\operatorname{cosec}^2\theta - 1) = \cos^2\theta$$

Hence, equation of hyperbola is $\frac{x^2}{\sin^2\theta} - \frac{y^2}{\cos^2\theta} = 1$

or $\quad x^2 \operatorname{cosec}^2\theta - y^2 \sec^2\theta = 1$

10. (b) The eccentricity of $\frac{x^2}{16} + \frac{y^2}{25} = 1$ is $e_1 = \sqrt{1 - \frac{16}{25}} = \frac{3}{5}$.

$$\Rightarrow \quad e_2 = \frac{5}{3} \qquad [\because e_1 e_2 = 1]$$

$\Rightarrow$ Foci of ellipse $(0, \pm 3)$.

$\therefore$ Equation of hyperbola is $\frac{x^2}{16} - \frac{y^2}{9} = -1$.

11. Let any point on the hyperbola be $(3\sec\theta, 2\tan\theta)$.

$\therefore$ Chord of contact of the circle $x^2 + y^2 = 9$ with respect to the point $(3\sec\theta, 2\tan\theta)$ is,

$$(3\sec\theta)x + (2\tan\theta)y = 9. \quad \text{...(i)}$$

Let (x_1, y_1) be the mid-point of the chord of contact.

$\Rightarrow$ Equation of chord in mid-point form is

$$xx_1 + yy_1 = x_1^2 + y_1^2 \quad \text{...(ii)}$$

Since, Eqs. (i) and (ii) are identically equal.

$$\therefore \quad \frac{3\sec\theta}{x_1} = \frac{2\tan\theta}{y_1} = \frac{9}{x_1^2 + y_1^2}$$

$$\Rightarrow \quad \sec\theta = \frac{9x_1}{3(x_1^2 + y_1^2)}$$

and $\quad \tan\theta = \frac{9y_1}{2(x_1^2 + y_1^2)}$

Thus, eliminating θ from the above equation, we get

$$\frac{81x_1^2}{9(x_1^2 + y_1^2)^2} - \frac{81y_1^2}{4(x_1^2 + y_1^2)^2} = 1 \qquad [\because \sec^2\theta - \tan^2\theta = 1]$$

$\therefore$ Required locus is $\frac{x^2}{9} - \frac{y^2}{4} = \frac{(x^2 + y^2)^2}{81}$.

4. Vectors

DPP-1 Introduction to Vectors

1. (c) It is given that **b** is collinear with the vector **a**.

$$\therefore \quad \mathbf{b} = \lambda \mathbf{a} \quad \ldots \text{(i)}$$
$$= 2\sqrt{2}\lambda \hat{\mathbf{i}} - \lambda \hat{\mathbf{j}} + 4\lambda \hat{\mathbf{k}}$$

Also, $|\mathbf{b}| = 10$

$$\Rightarrow \sqrt{(2\sqrt{2}\lambda)^2 + (-\lambda)^2 + (4\lambda)^2} = 10$$
$$\Rightarrow 25\lambda^2 = 100$$
$$\Rightarrow \lambda = \pm 2 \quad \ldots \text{(ii)}$$

From Eqs. (i) and (ii), we have

$$\mathbf{b} = \pm 2\mathbf{a}$$
$$\Rightarrow 2\mathbf{a} \pm \mathbf{b} = 0$$

2. (a) If **a**, **b**, **c** and **d** are linearly independent, if

$$k_1 \mathbf{a} + k_2 \mathbf{b} + k_3 \mathbf{c} + k_4 \mathbf{d} = 0$$

then, $k_1 = k_2 = k_3 = k_4 = 0$

$$\Rightarrow k_1 + k_2 + k_3 + k_4 = 0$$

3. (a) Let the position vector of point C is **c**. It is given that C divides AB in the ratio 3 : 2.

$$\therefore \quad \mathbf{c} = \frac{3(\mathbf{a} - 3\mathbf{b}) + 2(6\mathbf{a} + 2\mathbf{b})}{3 + 2}$$
$$= \frac{3\mathbf{a} - 9\mathbf{b} + 12\mathbf{a} + 4\mathbf{b}}{5} = \frac{15\mathbf{a} - 5\mathbf{b}}{5}$$
$$= 3\mathbf{a} - \mathbf{b}$$

4. (d) In ΔADB and ΔCBD, we have $\mathbf{AB} + \mathbf{AD} = 2\mathbf{AF}$ and $\mathbf{CB} + \mathbf{CD} = 2\mathbf{CF}$

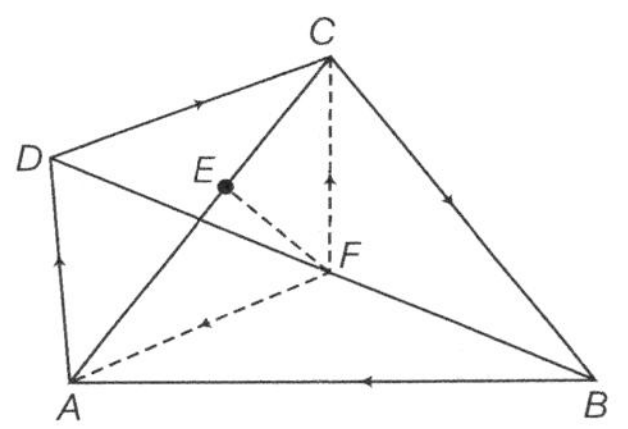

$$\therefore \quad \mathbf{AB} + \mathbf{AD} + \mathbf{CB} + \mathbf{CD} = 2(\mathbf{AF} + \mathbf{CF}) = -2(\mathbf{FA} + \mathbf{FC})$$
$$= -2(2\mathbf{FE}) = 4\mathbf{EF}$$

5. (d) We have,

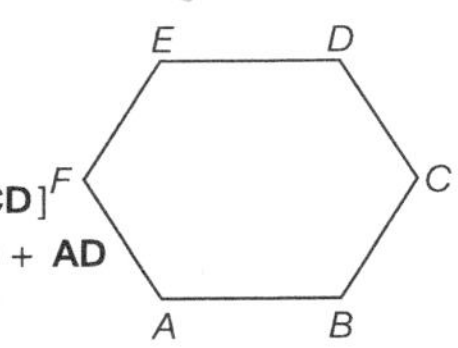

$$\mathbf{AB} + \mathbf{AC} + \mathbf{AD} + \mathbf{AE} + \mathbf{AF}$$
$$= \mathbf{ED} + \mathbf{AC} + \mathbf{AD} + \mathbf{CD}$$
$$[\because \mathbf{AB} = \mathbf{ED}, \mathbf{AF} = \mathbf{CD}]$$
$$= (\mathbf{AC} + \mathbf{CD}) + (\mathbf{AE} + \mathbf{ED}) + \mathbf{AD}$$
$$= \mathbf{AD} + \mathbf{AD} + \mathbf{AD} = 3\mathbf{AD}$$
$$\therefore \quad \lambda = 3$$

6. (c)

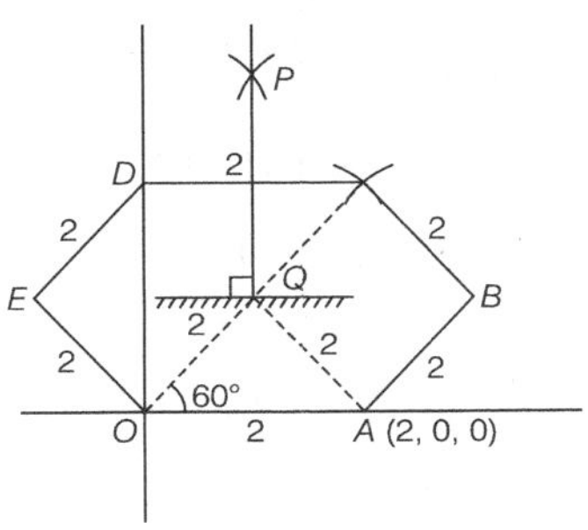

Here, coordinate of Q are $(2\cos 60°, 2\sin 60°)$

$$\Rightarrow Q(1, \sqrt{3}, 0)$$
$$\therefore P(1, \sqrt{3}, z)$$
$$OP = 3$$
$$\Rightarrow \sqrt{1 + 3 + z^2} = 3$$

or
$$z^2 = 5$$
$$z = \sqrt{5}$$
$$\therefore P(1, \sqrt{3}, \sqrt{5}) \Rightarrow \overrightarrow{OP} = \hat{\mathbf{i}} + \sqrt{3}\hat{\mathbf{j}} + \sqrt{5}\hat{\mathbf{k}}$$

Now,
$$\overrightarrow{AP} = \overrightarrow{OP} - \overrightarrow{OA} = \hat{\mathbf{i}} + \sqrt{3}\hat{\mathbf{j}} + \sqrt{5}\hat{\mathbf{k}} - 2\hat{\mathbf{i}}$$
$$= -\hat{\mathbf{i}} + \sqrt{3}\hat{\mathbf{j}} + \sqrt{5}\hat{\mathbf{k}}$$

7. (c) From the figure, we have

$$\mathbf{b} = \mathbf{AC} = \hat{\mathbf{i}} + 2\hat{\mathbf{j}} + 3\hat{\mathbf{k}}$$

and
$$\mathbf{c} = \mathbf{AB} = 3\hat{\mathbf{i}} + 2\hat{\mathbf{j}} - \hat{\mathbf{k}}$$

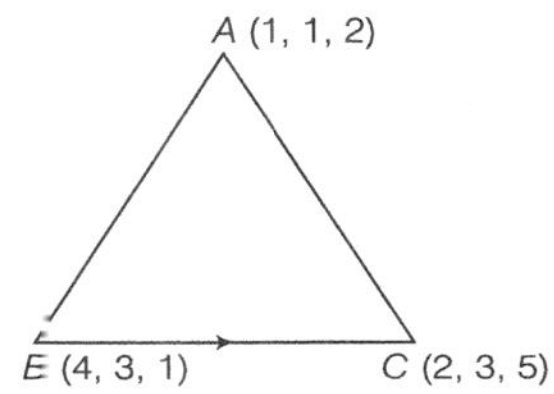

$\therefore$ Unit vector along the bisector of $\angle A$ is given by

$$= \frac{\mathbf{b} + \mathbf{c}}{2} = \frac{(\hat{\mathbf{i}} + 2\hat{\mathbf{j}} + 3\hat{\mathbf{k}}) + (3\hat{\mathbf{i}} + 2\hat{\mathbf{j}} - \hat{\mathbf{k}})}{\sqrt{14}}$$
$$= \frac{2\hat{\mathbf{i}} + 2\hat{\mathbf{j}} + \hat{\mathbf{k}}}{\sqrt{14}}$$

$\therefore$ Any vector along the angle bisector of $\angle A = 2\hat{\mathbf{i}} + 2\hat{\mathbf{j}} + \hat{\mathbf{k}}$.

8. (a) It is given that **c** is along the vector which bisects the angle between **a** and **b**. Therefore, **c** is parallel $\frac{\mathbf{a} + \mathbf{b}}{2}$.

So, $\mathbf{c} = \lambda(\mathbf{a} + \mathbf{b})$

$$= \lambda\left[\frac{7\hat{\mathbf{i}} - 4\hat{\mathbf{j}} - 4\hat{\mathbf{k}}}{9} + \frac{-2\hat{\mathbf{i}} - \hat{\mathbf{j}} + 2\hat{\mathbf{k}}}{3}\right] = \lambda\left[\frac{\hat{\mathbf{i}} - 7\hat{\mathbf{j}} + 2\hat{\mathbf{k}}}{9}\right]$$

Also, we have $|\mathbf{c}| = 5\sqrt{6}$

$$\Rightarrow \lambda = \pm 15$$
$$\therefore \quad \mathbf{c} = \pm \frac{5}{3}(\hat{\mathbf{i}} - 7\hat{\mathbf{j}} + 2\hat{\mathbf{k}})$$

9. (a) Let the forces P and Q be represented by **OL** and **OM**, so that the diagonal **ON** represents the resultant R.

$$\therefore \quad \mathbf{P} + \mathbf{Q} = \mathbf{R} \quad \ldots \text{(i)}$$

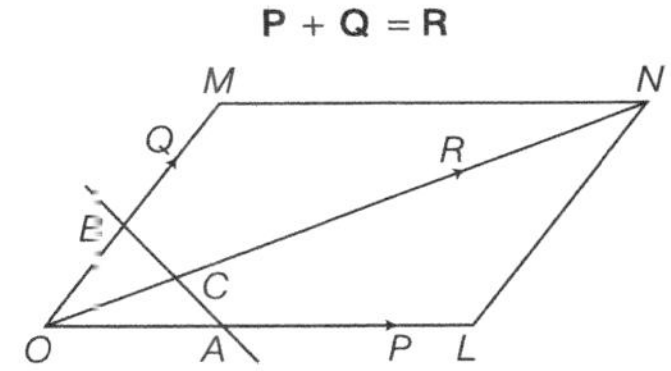

Again, let
$$\mathbf{P} = \mathbf{OL} = m\mathbf{OA}$$
$$\mathbf{Q} = \mathbf{OM} = n\mathbf{OB}$$
$$\mathbf{R} = \mathbf{ON} = t\mathbf{OC}$$
$$\therefore \quad m = \frac{\mathbf{P}}{\mathbf{OA}}; n = \frac{\mathbf{Q}}{\mathbf{OB}}; t = \frac{\mathbf{R}}{\mathbf{OC}} \quad \ldots \text{(ii)}$$

Hence, from Eq. (i), we get

$$m\cdot\mathbf{OA} + n\cdot\mathbf{OB} = t\cdot\mathbf{OC}$$

$$\Rightarrow \quad m\cdot\mathbf{OA} + n\cdot\mathbf{OB} - t\cdot\mathbf{OC} = 0 \qquad \ldots\text{(iii)}$$

But A, B, C are collinear and we know that, if there exists a relation of the form $x\mathbf{a} + y\mathbf{b} + z\mathbf{c} = 0$ between the position vectors of three collinear points, then $x + y + z = 0$.

Hence, from Eq. (iii), we must have

$$m + n - t = 0$$

$$\therefore \quad m + n = t$$

Now, putting the values of m, n and t from Eq. (ii) is $m + n = t$, we get

$$\left|\frac{\mathbf{P}}{\mathbf{OA}}\right| + \left|\frac{\mathbf{Q}}{\mathbf{OB}}\right| = \left|\frac{\mathbf{R}}{\mathbf{OC}}\right|$$

10. (c) From the given data, we have the following relations

If $x, y = x\hat{\mathbf{i}} + y\hat{\mathbf{j}}$

$$A(0) = 2, 3 \Rightarrow f_1(0) = 2, \quad f_2(0) = 3$$
$$A(1) = 6, 2 \Rightarrow f_1(1) = 6, \quad f_2(1) = 2$$
$$B(0) = 3, 2 \Rightarrow g_1(0) = 3, \ g_2(0) = 2$$
$$B(1) = 2, 6 \Rightarrow g_1(1) = 2, \ g_2(1) = 6$$

If two vectors are parallel, then their cross-product is zero.

$$\mathbf{A}(t) \times \mathbf{B}(t) = 0$$

$$\Rightarrow \quad \begin{vmatrix} \hat{\mathbf{i}} & \hat{\mathbf{j}} & \hat{\mathbf{k}} \\ f_1(t) & f_2(t) & 0 \\ g_1(t) & g_2(t) & 0 \end{vmatrix} = 0$$

$$\Rightarrow \quad [f_1(t)\, g_2(t) - f_2(t)\, g_1(t)]\,\hat{\mathbf{k}} = 0$$

$$\Rightarrow \quad f_1(t)\, g_2(t) - f_2(t)\, g_1(t) = 0 \text{ for some } t.$$

$$\Rightarrow \quad F(t) = f_1(t)\, g_2(t) - f_2(t) g_1(t)$$

Put $t = 0, 1$

$$F(0) = 2\cdot 2 - 3\cdot 3 = -5$$
$$F(1) = 6\cdot 6 - 2\cdot 2 = 32$$

Since, $F(0)$ and $F(1)$ are of opposite signs, we have $F(t) = 0$ for some $t \in [0, 1]$. For such $A(t)$ and $B(t)$ will be parallel.

11. (b) Here, $\bar{c} = t(\hat{\mathbf{a}} + \hat{\mathbf{b}}) = t\left(\dfrac{7\hat{\mathbf{i}} - 4\hat{\mathbf{j}} - 4\hat{\mathbf{k}}}{9} + \dfrac{(-2\hat{\mathbf{i}} - \hat{\mathbf{j}} + 2\hat{\mathbf{k}})}{3}\right)$

$$\Rightarrow \quad \bar{c} = t\left(\frac{\hat{\mathbf{i}} - 7\hat{\mathbf{j}} + 2\hat{\mathbf{k}}}{9}\right)$$

also, $\quad |\bar{c}| = 5\sqrt{6}$

$$\Rightarrow \quad \frac{t}{9}\cdot\sqrt{1 + 49 + 4} = 5\sqrt{6}$$

$$\therefore \quad t = 15$$

$$\Rightarrow \quad \bar{c} = \frac{15}{9}(\hat{\mathbf{i}} - 7\hat{\mathbf{j}} + 2\hat{\mathbf{k}})$$

or $\quad \dfrac{5}{3}(\hat{\mathbf{i}} - 7\hat{\mathbf{j}} + 2\hat{\mathbf{k}})$

12. (c) Here, $\dfrac{AB}{AC} = \dfrac{BD}{DC}$ and $\dfrac{AB}{AC} = \dfrac{BE}{CE}$

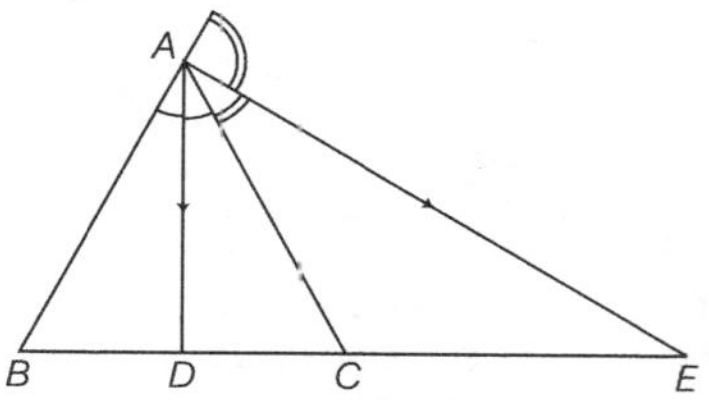

$$\Rightarrow \quad \frac{BD}{DC} = \frac{BE}{CE}$$

$$\Rightarrow \quad \frac{BD}{BC - BD} = \frac{BE}{BE - BC}$$

$$\Rightarrow \quad BD\cdot BE - BD\cdot BC = BC\cdot BE - BD\cdot BE$$

$$\Rightarrow \quad 2BD\cdot BE = (BD + BE)\cdot BC$$

or $\quad \dfrac{2}{BC} = \dfrac{1}{BD} + \dfrac{1}{BE}$

13. (1) Let G be the centroid of ΔABC. Then, O', G, O are collinear and G divides $O'O$ in the ratio $2 : 1$, i.e. $\dfrac{O'G}{OG} = \dfrac{2}{1}$.

Replacing S by O in illustration 1, we have

$$\mathbf{OA} + \mathbf{OB} + \mathbf{OC} = 3\mathbf{OG}$$

$$\Rightarrow \quad \mathbf{OA} + \mathbf{OB} + \mathbf{OC} = 2\mathbf{OG} + \mathbf{OG}$$

$$\Rightarrow \quad \mathbf{OA} + \mathbf{OB} + \mathbf{OC} = \mathbf{GO'} + \mathbf{OG} \qquad [\because 2OG = GO']$$

$$\Rightarrow \quad \mathbf{OA} + \mathbf{OB} + \mathbf{OC} = \mathbf{OO'}$$

$$\therefore \quad \lambda = 1$$

14. (3) Let O be the origin and A, B, C, P, Q are five coplanar points.

Now, $\mathbf{PA} + \mathbf{PB} + \mathbf{PC} + \mathbf{AQ} + \mathbf{BQ} + \mathbf{CQ}$

$$= \mathbf{OA} - \mathbf{OP} + \mathbf{OB} - \mathbf{OP} + \mathbf{OC} - \mathbf{OP} + \mathbf{OQ} - \mathbf{OA} + \mathbf{OQ} - \mathbf{OB} + \mathbf{OQ} - \mathbf{OC}$$

$$= 3(\mathbf{OQ} - \mathbf{OP})$$

$$= 3\mathbf{QP}$$

On comparing, we get $\lambda = 3$.

15. (0) It is given that $\mathbf{a} + 2\mathbf{b}$ is collinear with $\mathbf{c}$ and $\mathbf{b} + 3\mathbf{c}$ with $\mathbf{a}$.

$$\therefore \quad \mathbf{a} + 2\mathbf{b} = \mu_1\mathbf{c}$$

and $\quad \mathbf{b} + 3\mathbf{c} = \mu_2\mathbf{a}$ for some μ_1 and μ_2.

$$\Rightarrow \quad \mathbf{a} + 2(\mu_2\mathbf{a} - 3\mathbf{c}) = \mu_1\mathbf{c}$$

$$\Rightarrow \quad \mathbf{a} + 2\mu_2\mathbf{a} - 6\mathbf{c} - \mu_1\mathbf{c} = 0$$

$$\Rightarrow \quad (1 + 2\mu_2)\mathbf{a} - (\mu_1 + 6)\mathbf{c} = 0$$

$$\Rightarrow \quad 1 + 2\mu_2 = 0 \text{ and } \mu_1 + 6 = 0$$

$$\Rightarrow \quad \mu_2 = -\frac{1}{2}, \quad \mu_1 = -6$$

$$\therefore \quad \mathbf{a} + 2\mathbf{b} = -6\mathbf{c}$$

$$\Rightarrow \quad \mathbf{a} + 2\mathbf{b} + 6\mathbf{c} = 0$$

or $\quad \mathbf{a} + 2\mathbf{b} + 6\mathbf{c} = 0\cdot\mathbf{a}$

$$\therefore \quad \lambda = 0$$

16. (1) Since, angle bisector of $\bar{a}$ and $\bar{b}$

$$\Rightarrow \quad h(\hat{\mathbf{a}} + \hat{\mathbf{b}}) = h\left(\frac{\mathbf{a}}{|\mathbf{a}|} + \frac{\mathbf{b}}{|\mathbf{b}|}\right) \qquad \ldots\text{(i)}$$

given, $\bar{a} + \bar{b}$ is along angle bisector.

$$\Rightarrow \quad \mu\left(\frac{\mathbf{a}}{|\mathbf{a}|} + \frac{\mathbf{b}}{|\mathbf{b}|}\right) = \bar{a} + \bar{b}$$

only, when $|\mathbf{a}| = |\mathbf{b}| = \mu$

$$\therefore \quad |\mathbf{a}| = |\mathbf{b}|$$

$$\Rightarrow \quad \lambda = 1$$

17. (3) We have,

$$\mathbf{AB} = \hat{\mathbf{i}} + 2\hat{\mathbf{j}} + \hat{\mathbf{k}}, \ \mathbf{AC} = -6\hat{\mathbf{i}} + 3\hat{\mathbf{j}} - 3\hat{\mathbf{k}}$$

$$\Rightarrow \quad |\mathbf{AB}| = \sqrt{6} \text{ and } |\mathbf{AC}| = 3\sqrt{6}$$

Clearly, point D divides BC in the ratio $AB : AC$, i.e. $1 : 3$.

$\therefore$ Position vector of $D = \dfrac{(-5\hat{\mathbf{i}} + 2\hat{\mathbf{j}} - 6\hat{\mathbf{k}}) + 3(2\hat{\mathbf{i}} + \hat{\mathbf{j}} - 2\hat{\mathbf{k}})}{1 + 3}$

$$= \frac{1}{4}(\hat{\mathbf{i}} + 5\hat{\mathbf{j}} - 12\hat{\mathbf{k}})$$

$$\therefore \quad \mathbf{AD} = \frac{1}{4}(\hat{\mathbf{i}} + 5\hat{\mathbf{j}} - 12\hat{\mathbf{k}}) - (\hat{\mathbf{i}} - \hat{\mathbf{j}} - 3\hat{\mathbf{k}})$$

$$= \frac{3}{4}(-\hat{\mathbf{i}} + 3\hat{\mathbf{j}})$$

$$\Rightarrow \quad |\mathbf{AD}| = AD = \frac{3}{4}\sqrt{10}$$

$$\therefore \quad \lambda = 3$$

18. (0) Since, X and Y divide $\mathbf{PQ}$ internally and externally in the ratio $2:1$, then $X = \dfrac{2\mathbf{b} + \mathbf{a}}{3}$ and $y = 2\mathbf{b} - \mathbf{a}$

$\therefore \quad \mathbf{XY}$ = Position vector of $\mathbf{y}$ – Position vector of x

$$= 2\mathbf{b} - \mathbf{a} - \frac{2\mathbf{b} + \mathbf{a}}{3} = \frac{4\mathbf{b}}{3} - \frac{4\mathbf{a}}{3}$$

On comparing it with $\lambda\mathbf{a} + \mu\mathbf{b}$, we get

$$\lambda = -\frac{4}{3} \text{ and } \mu = \frac{4}{3}$$

$$\therefore \quad |\lambda + \mu| = \left|\frac{-4}{3} + \frac{4}{3}\right| = 0$$

19. (3)

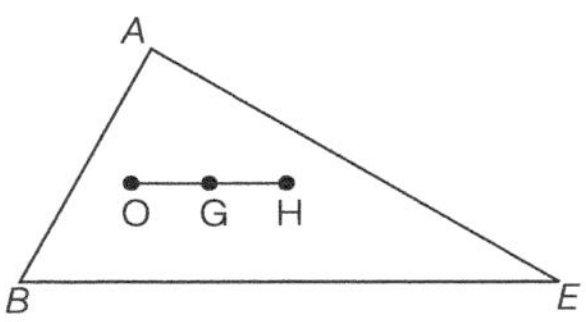

Here, θ is circum centre = 0, G is centroid = $\mathbf{g}$

H is orthocentre = $\mathbf{p}$

Since, $\dfrac{\mathbf{OG}}{\mathbf{GH}} = \dfrac{1}{2}$

$$\Rightarrow \quad \frac{\mathbf{g}}{\mathbf{p} - \mathbf{g}} = \frac{1}{2}$$

$$\Rightarrow \quad 2\mathbf{g} = \mathbf{p} - \mathbf{g}$$

or $\quad \mathbf{p} = 3\mathbf{g}$

$\therefore \quad k = 3$

20. (2) We have,

$$\mathbf{a} = 2p\hat{\mathbf{i}} + \hat{\mathbf{j}}$$

Let $\mathbf{b}$ be the vector obtained from $\mathbf{a}$ by rotating the axes. Then, the components of $\mathbf{b}$ are $p + 1$ and 1. Therefore,

$$\mathbf{b} = (p + 1)\,\hat{\alpha} + \hat{\beta}$$

where $\hat{\alpha}$ and $\hat{\beta}$ are unit vectors along the new axes.

But $\quad |\mathbf{b}| = |\mathbf{a}|$

$$\Rightarrow \quad 4p^2 + 1 = (p + 1)^2 + 1$$

$$\Rightarrow \quad 3p^2 - 2p - 1 = 0$$

$$\Rightarrow \quad p = 1, -\frac{1}{3}$$

$$\Rightarrow \quad p_1 = 1 \text{ and } p_2 = -\frac{1}{3}$$

$$\therefore \quad 3\,|p_1 + p_2| = 3\left|1 - \frac{1}{3}\right| = 2$$

DPP-2 **Scalar or Dot Product**

1. (i) We know that

$$|\hat{\mathbf{a}} - \hat{\mathbf{b}}|^2 = (\hat{\mathbf{a}} - \hat{\mathbf{b}})\cdot(\hat{\mathbf{a}} - \hat{\mathbf{b}})$$

$$= \hat{\mathbf{a}}\cdot\hat{\mathbf{a}} - \hat{\mathbf{a}}\cdot\hat{\mathbf{b}} - \hat{\mathbf{b}}\cdot\hat{\mathbf{a}} + \hat{\mathbf{b}}\cdot\hat{\mathbf{b}}$$

$$= |\hat{\mathbf{a}}|^2 + |\hat{\mathbf{b}}|^2 - 2\hat{\mathbf{a}}\cdot\hat{\mathbf{b}} \qquad [\because \hat{\mathbf{a}}\cdot\hat{\mathbf{b}} = \hat{\mathbf{b}}\cdot\hat{\mathbf{a}}]$$

$$= |\hat{\mathbf{a}}|^2 + |\hat{\mathbf{b}}|^2 - 2|\hat{\mathbf{a}}|\,|\hat{\mathbf{b}}|\cos\theta$$

$$= 1 + 1 - 2\cos\theta = 2\,(1 - \cos\theta)$$

$$= 2 \times 2\sin^2\frac{\theta}{2} = 4\sin^2\frac{\theta}{2}$$

$$\therefore \quad \sin\frac{\theta}{2} = \frac{1}{2}|\hat{\mathbf{a}} - \hat{\mathbf{b}}|$$

(ii) Let θ be the angle between vectors $\mathbf{a}$ and $\mathbf{b}$. Then,

$$\mathbf{a}\cdot\mathbf{b} = |\mathbf{a}|\,|\mathbf{b}|\cos\theta$$

$$\Rightarrow \quad (\mathbf{a}\cdot\mathbf{b})^2 = |\mathbf{a}|^2|\mathbf{b}|^2\cos^2\theta$$

Now, $\quad \cos^2\theta \le 1$

$$\Rightarrow \quad |\mathbf{a}|^2|\mathbf{b}|^2\cos^2\theta \le |\mathbf{a}|^2|\mathbf{b}|^2$$

$$\therefore \quad (\mathbf{a}\cdot\mathbf{b})^2 \le |\mathbf{a}|^2|\mathbf{b}|^2$$

(iii) For any two vectors we know that

$$|\mathbf{a} + \mathbf{b}|^2 + |\mathbf{a} - \mathbf{b}|^2 = 2\,(|\mathbf{a}|^2 + |\mathbf{b}|^2)$$

$$\Rightarrow \quad |\hat{\mathbf{a}} + \hat{\mathbf{b}}|^2 + |\hat{\mathbf{a}} - \hat{\mathbf{b}}|^2 = 2\,(|\hat{\mathbf{a}}|^2 + |\hat{\mathbf{b}}|^2)$$

$$\Rightarrow \quad (1)^2 + |\hat{\mathbf{a}} - \hat{\mathbf{b}}|^2 = 2\,(1 + 1) \qquad [\because |\hat{\mathbf{a}} + \hat{\mathbf{b}}| = 1]$$

$$\Rightarrow \quad |\hat{\mathbf{a}} - \hat{\mathbf{b}}|^2 = 3$$

$$\Rightarrow \quad |\hat{\mathbf{a}} - \hat{\mathbf{b}}| = \sqrt{3}$$

(iv) Let $|\mathbf{a}| = |\mathbf{b}| = |\mathbf{c}| = \lambda$ (say)

Since, $\mathbf{a}$, $\mathbf{b}$, $\mathbf{c}$ are mutually perpendicular vectors, therefore

$$\mathbf{a}\cdot\mathbf{b} = \mathbf{b}\cdot\mathbf{c} = \mathbf{c}\cdot\mathbf{a} = 0$$

We know that

$$|\mathbf{x}|^2 = \mathbf{x}\cdot\mathbf{x}$$

$$\therefore |\mathbf{a} + \mathbf{b} + \mathbf{c}|^2 = (\mathbf{a} + \mathbf{b} + \mathbf{c})\cdot(\mathbf{a} + \mathbf{b} + \mathbf{c})$$

$$= \mathbf{a}\cdot\mathbf{a} + \mathbf{b}\cdot\mathbf{b} + \mathbf{c}\cdot\mathbf{c} + 2\mathbf{a}\cdot\mathbf{b} + 2\mathbf{b}\cdot\mathbf{c} + 2\mathbf{c}\cdot\mathbf{a}$$

$$= |\mathbf{a}|^2 + |\mathbf{b}|^2 + |\mathbf{c}|^2 = 3\lambda^2$$

$$\Rightarrow |\mathbf{a} + \mathbf{b} + \mathbf{c}| = \sqrt{3}\lambda$$

Suppose $\mathbf{a} + \mathbf{b} + \mathbf{c}$ makes angles θ_1, θ_2, θ_3 with $\mathbf{a}$, $\mathbf{b}$ and $\mathbf{c}$ respectively. Then,

$$\cos\theta_1 = \frac{\mathbf{a}\cdot(\mathbf{a} + \mathbf{b} + \mathbf{c})}{|\mathbf{a}|\,|\mathbf{a} + \mathbf{b} + \mathbf{c}|} = \frac{\mathbf{a}\cdot\mathbf{a} + \mathbf{a}\cdot\mathbf{b} + \mathbf{a}\cdot\mathbf{c}}{|\mathbf{a}|\,|\mathbf{a} + \mathbf{b} + \mathbf{c}|}$$

$$= \frac{\mathbf{a}}{|\mathbf{a}|\,|\mathbf{a} + \mathbf{b} + \mathbf{c}|} = \frac{\lambda}{\lambda \times \sqrt{3}\lambda} = \frac{1}{\sqrt{3}}$$

$$\Rightarrow \quad \theta_1 = \cos^{-1}\left(\frac{1}{\sqrt{3}}\right)$$

Similarly, we obtain $\theta_2 = \cos^{-1}\left(\dfrac{1}{\sqrt{3}}\right)$

and $\quad \theta_3 = \cos^{-1}\left(\dfrac{1}{\sqrt{3}}\right)$

$$\therefore \quad \theta_1 = \theta_2 = \theta_3$$

(v) We know that

$$|\mathbf{a} + \mathbf{b}|^2 = (\mathbf{a} + \mathbf{b})\cdot(\mathbf{a} + \mathbf{b})$$

$$\Rightarrow \quad |\mathbf{a} - \mathbf{b}|^2 = \mathbf{a}\cdot\mathbf{a} + \mathbf{b}\cdot\mathbf{b} + 2\mathbf{a}\cdot\mathbf{b}$$

$$\Rightarrow \quad |\mathbf{a} + \mathbf{b}|^2 + |\mathbf{a}|^2 + |\mathbf{b}|^2 + 2\mathbf{a}\cdot\mathbf{b}$$

$$\Rightarrow \quad 2\mathbf{a}\cdot\mathbf{b} = 0 \qquad [\because |\mathbf{a} + \mathbf{b}|^2 = |\mathbf{a}|^2 + |\mathbf{b}|^2]$$

$$\Rightarrow \quad \mathbf{a}\cdot\mathbf{b} = 0$$

If scalar product of two non-zero vectors is zero, then both vectors are perpendicular.

2. (i) Let the point of intersection O of two altitudes BQ and CR be taken as origin and the position vectors of the vertices A, B, C be $\mathbf{a}$, $\mathbf{b}$, $\mathbf{c}$ respectively. Let AO produced meet BC at P. We will show that AP is perpendicular to BC, showing thereby that the three altitudes are concurrent.

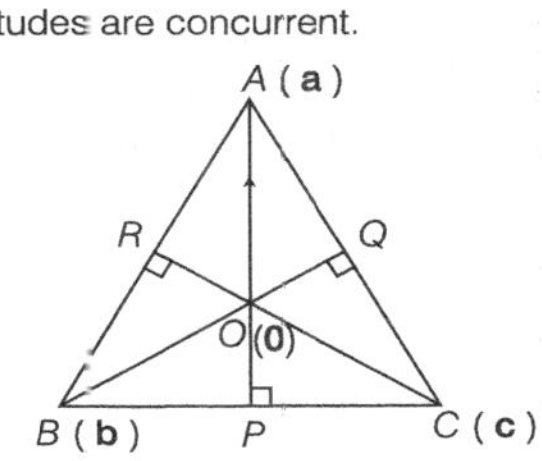

$\therefore \qquad \mathbf{OB} = \mathbf{b}, \quad \mathbf{BQ} = \mu\mathbf{b}$

as it is collinear with **OB**.

Similarly, since $\mathbf{OC} = \mathbf{C}$

$\therefore \qquad \mathbf{CR} = \nu\mathbf{C}$

Now, $\qquad \mathbf{AC} = \mathbf{c} - \mathbf{a}$

and $\qquad \mathbf{AB} = \mathbf{b} - \mathbf{a}$

Since, $BQ \perp AC$, we have $\mu\mathbf{b} \cdot (\mathbf{c} - \mathbf{a})$ and so $\mathbf{a} \cdot \mathbf{b} = \mathbf{b} \cdot \mathbf{c}$

Again, since $CR \perp AB$,

$$\nu\mathbf{c} \cdot (\mathbf{b} - \mathbf{a}) = 0$$

$\therefore \qquad \mathbf{b} \cdot \mathbf{c} = \mathbf{c} \cdot \mathbf{a}$

$\therefore \qquad \mathbf{a} \cdot \mathbf{b} = \mathbf{b} \cdot \mathbf{c} = \mathbf{c} \cdot \mathbf{a}$

or $\qquad \mathbf{a} \cdot (\mathbf{c} - \mathbf{b}) = 0$

$\Rightarrow \qquad \lambda\mathbf{a} \cdot (\mathbf{c} - \mathbf{b}) = 0$

$\therefore \qquad \mathbf{AP} \cdot \mathbf{BC} = 0$

$\Rightarrow \qquad AP \perp BC$

(ii) Take the centre O as origin and AB is the diameter, so that $OA = OB$.

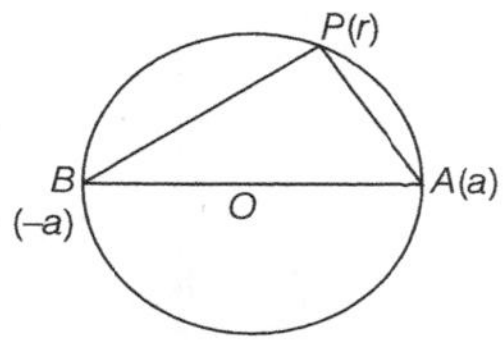

If the point A is a, then B is $-a$ and $|a| = r =$ radius.

Let P be any point r on the circumference, so that $|r| = OP = r$

Then, $\quad \mathbf{AP}$ = Position vector of P – Position vector of A
$= r - a$

and $\quad \mathbf{BP}$ = Position vector of P – Position vector of B
$= r + a$

$$\therefore \quad \mathbf{AP} \cdot \mathbf{BP} = (r - a) \cdot (r + a) = r^2 - a^2$$
$$= r^2 - r^2 = 0$$

(iii) Let the vectors represented by sides BC, CA and AB be denoted by $\mathbf{a}$, $\mathbf{b}$ and $\mathbf{c}$ respectively and their modulus be denoted by a, b and c.

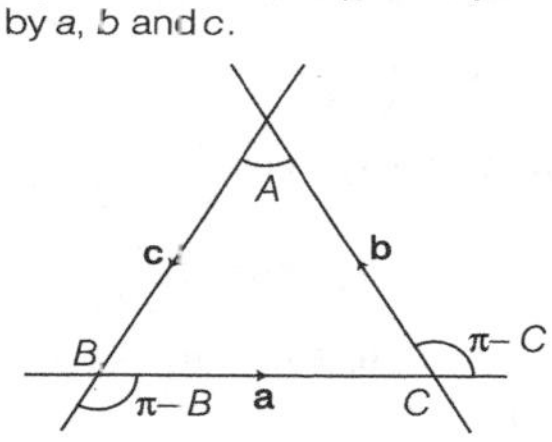

Also, we know that

$$\mathbf{BC} + \mathbf{CA} + \mathbf{AB} = \mathbf{BB} = 0$$

or $\qquad \mathbf{a} + \mathbf{b} + \mathbf{c} = 0$

$\therefore \qquad \mathbf{BC} = -(\mathbf{CA} + \mathbf{AB}) \qquad \ldots(i)$

On squaring both sides, we get

$$\mathbf{BC}^2 = \mathbf{CA}^2 + \mathbf{AB}^2 + 2\mathbf{CA} \cdot \mathbf{AB}$$

$\Rightarrow \qquad a^2 = b^2 + c^2 + 2bc\cos(\pi - A)$

$\Rightarrow \qquad a^2 = b^2 + c^2 - 2bc\cos A$

$\therefore \qquad \cos A = \dfrac{b^2 + c^2 - a^2}{2bc}$

(iv) In a ΔABC, we know that

$$\mathbf{CA} = -(\mathbf{AB} + \mathbf{BC})$$

On multiplying both sides by **CA**, we get

$$\mathbf{CA} \cdot \mathbf{CA} = -(\mathbf{AB} + \mathbf{BC}) \cdot \mathbf{CA}$$
$$b^2 = -\mathbf{AB} \cdot \mathbf{CA} - \mathbf{BC} \cdot \mathbf{CA}$$

$\Rightarrow \qquad b^2 = -bc\cos(\pi - A) - ab\cos(\pi - C)$

$\Rightarrow \qquad b^2 = bc\cos A + ab\cos C$

$\therefore \qquad b = c\cos A + a\cos C$

(v) Let there be two unit vectors $\hat{\mathbf{i}}$ and $\hat{\mathbf{j}}$ along OX and OY, two perpendicular lines in the plane of the paper. If OP and OR be any two lines in the same plane making angles α and β with OX respectively, then $\angle POR = \alpha + \beta$. Again, let **OA** and **OB** represent unit vectors along OP and OR respectively.

$\angle POR = \alpha + \beta$. Again, let **OA** and **OB′** represent with unit vectors along **OP** and **OR** respectively.

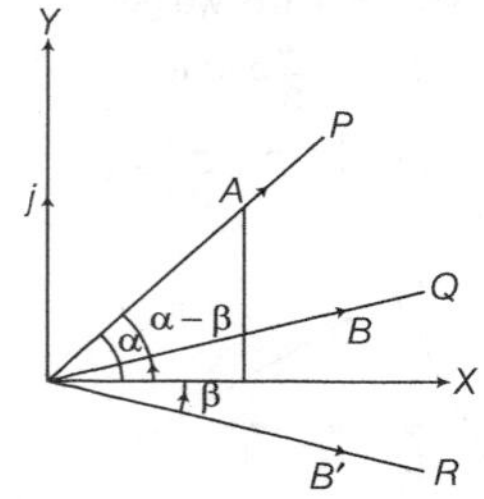

$\therefore \qquad \mathbf{OA} \cdot \mathbf{OB}' = 1 \cdot 1\cos(\alpha + \beta) = \cos(\alpha + \beta) \qquad \ldots(i)$

Again, $\qquad \mathbf{OA} = \mathbf{OM} + \mathbf{MA}$

$$= OA\cos\alpha\hat{\mathbf{i}} + OA\sin\alpha\hat{\mathbf{j}}$$
$$= \cos\alpha\,\hat{\mathbf{i}} + \sin\alpha\hat{\mathbf{j}}$$

$\because \qquad OA = 1 \qquad \ldots(ii)$

Similarly, $\quad \mathbf{OB}' = \cos\beta\hat{\mathbf{i}} - \sin\beta\hat{\mathbf{j}} \qquad \ldots(iii)$

$\therefore \qquad \mathbf{OA} \cdot \mathbf{OB}' = (\cos\alpha\hat{\mathbf{i}} + \sin\alpha\hat{\mathbf{j}}) \cdot (\cos\beta\hat{\mathbf{i}} - \sin\beta\hat{\mathbf{j}})$

$$= \cos\alpha\cos\beta - \sin\alpha\sin\beta$$

$\Rightarrow \cos(\alpha + \beta) = \cos\alpha\cos\beta - \sin\alpha\sin\beta$

3. (b) We know, that, for any two vectors

$$|\mathbf{a} + \mathbf{b}|^2 + |\mathbf{a} - \mathbf{b}|^2 = 2(|\mathbf{a}|^2 + |\mathbf{b}|^2)$$

$\Rightarrow \qquad (10)^2 + (8)^2 = 2[(5)^2 + |\mathbf{b}|^2]$

$\Rightarrow \qquad 100 + 64 = 50 + 2|\mathbf{b}|^2$

$\Rightarrow \qquad |\mathbf{b}|^2 = 57$

$\therefore \qquad |\mathbf{b}| = \sqrt{57}$

4. (a) Let **c** and **d** be the diagonals of parallelogram.

Then, $\qquad \mathbf{c} = \mathbf{a} + \mathbf{b}$ and $\mathbf{d} = \mathbf{a} - \mathbf{b}$

$\Rightarrow \qquad \mathbf{c} = 3\hat{\mathbf{i}}$ and $\mathbf{d} = \hat{\mathbf{i}} + 2\hat{\mathbf{j}} + 2\hat{\mathbf{k}}$

Let θ be the angle between **c** and **d**.

Then, $\qquad \cos\theta = \dfrac{\mathbf{c} \cdot \mathbf{d}}{|\mathbf{c}|\,|\mathbf{d}|} = \dfrac{3\hat{\mathbf{i}} \cdot (\hat{\mathbf{i}} + 2\hat{\mathbf{j}} + 2\hat{\mathbf{k}})}{\sqrt{(3)^2}\sqrt{1^2 + 2^2 + 2^2}}$

$$= \frac{3}{3 \times 3} = \frac{1}{3}$$

$\therefore \qquad \theta = \cos^{-1}\left(\dfrac{1}{3}\right)$

5. (c) We have,

$$(\mathbf{a} - \mathbf{d}) \cdot (\mathbf{b} - \mathbf{c}) = (\mathbf{b} - \mathbf{d}) \cdot (\mathbf{c} - \mathbf{a}) = 0$$

$\Rightarrow \qquad \mathbf{DA} \cdot \mathbf{BC} = 0$ and $\mathbf{DB} \cdot \mathbf{AC} = 0$

$\Rightarrow AD \perp BC$ and $DB \perp AC$.

$\therefore D$ is the orthocentre of ΔABC.

6. (d) We have, $|\mathbf{a}| = 3, |\mathbf{b}| = 4$ and $|\mathbf{c}| = 5$. It is given that

$$\mathbf{a} \perp (\mathbf{b} + \mathbf{c}), \mathbf{b} \perp (\mathbf{c} + \mathbf{a}) \text{ and } \mathbf{c} \perp (\mathbf{a} + \mathbf{b})$$

$\Rightarrow \qquad \mathbf{a} \cdot (\mathbf{b} + \mathbf{c}) = 0, \mathbf{b} \cdot (\mathbf{c} + \mathbf{a}) = 0$ and $\mathbf{c} \cdot (\mathbf{a} + \mathbf{b}) = 0$

$\Rightarrow \qquad \mathbf{a} \cdot \mathbf{b} + \mathbf{a} \cdot \mathbf{c} = \mathbf{b} \cdot \mathbf{c} + \mathbf{b} \cdot \mathbf{a} = \mathbf{c} \cdot \mathbf{a} + \mathbf{c} \cdot \mathbf{b} = 0$

or $\mathbf{a} \cdot \mathbf{b} + \mathbf{b} \cdot \mathbf{c} + \mathbf{c} \cdot \mathbf{a} = 0$ (adding all the above equations)

Now, $\quad |\mathbf{a} + \mathbf{b} + \mathbf{c}|^2 = |\mathbf{a}|^2 + |\mathbf{b}|^2 + |\mathbf{c}|^2 + 2(\mathbf{a} \cdot \mathbf{b} + \mathbf{b} \cdot \mathbf{c} + \mathbf{c} \cdot \mathbf{a})$

$$= 3^2 + 4^2 + 5^2 = 50$$

$\therefore \qquad |\mathbf{a} + \mathbf{b} + \mathbf{c}| = 5\sqrt{2}$

7. (a) It is given that $\mathbf{e}_1$ and $\mathbf{e}_2$ are two unit vectors inclined at an angle 2θ and $|e_1 - e_2| < 1$.

$\therefore \quad |e_1 - e_2| < 1$

$\Rightarrow \quad |e_1 - e_2|^2 < 1$

$\Rightarrow \quad 4\sin^2\theta < 1 \quad [\because |\mathbf{e}_1 - \mathbf{e}_2| = 4\sin^2\theta]$

$\Rightarrow \quad \sin^2\theta < \frac{1}{4}$

$\therefore \quad \theta \in \left[0, \frac{\pi}{6}\right)$

8. (c) Let the required vector be $\mathbf{a} = x\hat{i} + y\hat{j}$. It is given that the vector $\mathbf{c}$ is in XY-plane and is perpendicular to $\mathbf{b} = 4\hat{i} + 3\hat{j}$.

Therefore, $\mathbf{c} = \lambda(3\hat{i} - 4\hat{j})$

$\Rightarrow \quad \hat{\mathbf{c}} = \frac{1}{5}(3\hat{i} - 4\hat{j})$

Now, $\mathbf{a} \cdot \hat{\mathbf{b}} = 1$ and $\mathbf{a} \cdot \hat{\mathbf{c}} = 2$

$\Rightarrow \quad (4x + 3y = 5 \text{ and } 3x - 4y = 10);$

$(4x + 3y = 5 \text{ and } 3x - 4y = -10)$

$\Rightarrow \quad (x = 2 \text{ and } y = 1);\ x = -\frac{2}{5} \text{ and } y = \frac{11}{5}$

Hence, $\mathbf{a} = 2\hat{i} - \hat{j};\ a = \frac{1}{5}(-2\,\hat{i} + 11\,\hat{j})$

9. (c)

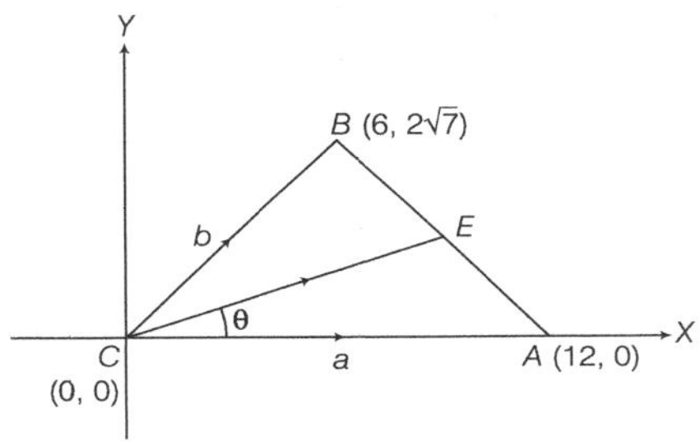

Here, $\mathbf{CA} = 12\,\hat{i} = \mathbf{a}$ and $\mathbf{CB} = 6\hat{i} + 2\sqrt{7}\,\hat{j} = \mathbf{b}$

$$\mathbf{CE} = \frac{3\bar{a} + \bar{b}}{4}$$

$\therefore \quad \cos\theta = \frac{\mathbf{CE}\,.\,\mathbf{CA}}{|\mathbf{CE}||\mathbf{CA}|} \Rightarrow \cos\theta = \frac{3\sqrt{7}}{8}$

10. (a) $A\mathbf{u}\,.\,\mathbf{u} = 0$

$\Rightarrow |A||\mathbf{u}|^2 = 0$, Since $|\mathbf{u}| \neq 0$

$\Rightarrow |A| = 0$

$\therefore$ A is singular.

11. (b) It is given that

$2\mathbf{p} + \mathbf{q} = \hat{i} + \hat{j}$

and $\quad \mathbf{p} + 2\mathbf{q} = \hat{i} - \hat{j}$

$\Rightarrow \quad \mathbf{p} = \frac{1}{3}\hat{i} + \hat{j}$ and $\mathbf{q} = \frac{1}{3}\hat{i} - \hat{j}$

Let θ be the angle between $\mathbf{p}$ and $\mathbf{q}$. Then,

$$\cos\theta = \frac{\mathbf{p}\cdot\mathbf{q}}{|\mathbf{p}|\,|\mathbf{q}|} = \frac{\left(\frac{1}{3}\hat{i} + \hat{j}\right)\cdot\left(\frac{1}{3}\hat{i} - \hat{j}\right)}{\sqrt{\left(\frac{1}{3}\right)^2 + (1)^2}\sqrt{\left(\frac{1}{3}\right)^2 + (-1)^2}}$$

$$= \frac{\frac{1}{9} - 1}{\sqrt{\frac{1}{9} + 1}\sqrt{\frac{1}{9} + 1}} = \frac{-\frac{8}{9}}{\frac{10}{9}} = -\frac{8}{10} = -\frac{4}{5}$$

$\therefore \quad \lambda = \cos\theta = -\frac{4}{5}$

12. (b) Let, $\mathbf{a} = 3\hat{i} + 4\hat{j} + 5\hat{k}$, $\mathbf{b} = \hat{i} + 7\hat{k}$ and $\mathbf{c} = 5\hat{i} + 5\hat{j}$

Then, $\quad |\mathbf{a}| = |\mathbf{b}| = |\mathbf{c}|$

So, origin is circum centre.

and the orthocentre, is $\mathbf{a} + \mathbf{b} + \mathbf{c} = |9\hat{i} + 9\hat{j} + 12\hat{k}|$

$\therefore$ The distance between circum centre and ortho is

$$|\mathbf{a} + \mathbf{b} + \mathbf{c}| = |9\hat{i} + 9\hat{j} + 12\hat{k}| = \sqrt{306}$$

13. (a) We have, $\quad \mathbf{a} + \mathbf{b} + \mathbf{c} = 0$

$\Rightarrow \quad |\mathbf{a} + \mathbf{b} + \mathbf{c}| = 0$

$\Rightarrow \quad |\mathbf{a} + \mathbf{b} + \mathbf{c}|^2 = 0$

$\Rightarrow \quad |\mathbf{a}|^2 + |\mathbf{b}|^2 + |\mathbf{c}|^2 + 2(\mathbf{a}\cdot\mathbf{b} + \mathbf{b}\cdot\mathbf{c} + \mathbf{c}\cdot\mathbf{a}) = 0$

$\Rightarrow \quad 1 + 4 + 4 + 2(\mathbf{a}\cdot\mathbf{b} + \mathbf{b}\cdot\mathbf{c} + \mathbf{c}\cdot\mathbf{a}) = 0$

$\therefore \quad \mathbf{a}\cdot\mathbf{b} + \mathbf{b}\cdot\mathbf{c} + \mathbf{c}\cdot\mathbf{a} = -\frac{9}{2}$

14. (a) We have,

Projection of $\mathbf{v}$ and $\mathbf{u}$ = Projection of $\mathbf{w}$ along $\mathbf{u}$

$\Rightarrow \quad \frac{\mathbf{v}\cdot\mathbf{u}}{|\mathbf{u}|} = \frac{\mathbf{w}\cdot\mathbf{u}}{|\mathbf{u}|} \Rightarrow \mathbf{v}\cdot\mathbf{u} = \mathbf{w}\cdot\mathbf{u} \quad \ldots\text{(i)}$

Also, $\mathbf{v}$ and $\mathbf{w}$ are perpendicular to each other.

$\therefore \quad \mathbf{v}\cdot\mathbf{w} = 0 \quad \ldots\text{(ii)}$

Now, $\quad \mathbf{u} - \mathbf{v} + \mathbf{w}|^2 = |\mathbf{u}|^2 + |\mathbf{v}|^2 + |\mathbf{w}|^2 - 2(\mathbf{u}\cdot\mathbf{v}) - 2(\mathbf{v}\cdot\mathbf{w}) + 2(\mathbf{u}\cdot\mathbf{w})$

$\Rightarrow \quad |\mathbf{u} - \mathbf{v} + \mathbf{w}|^2 = 1 + 4 + 9$

$\Rightarrow \quad |\mathbf{u} - \mathbf{v} + \mathbf{w}| = \sqrt{14}$

15. (b) We have,

$|\mathbf{a} - \mathbf{b}|^2 + |\mathbf{b} - \mathbf{c}|^2 + |\mathbf{c} - \mathbf{a}|^2$

$= |\mathbf{a}|^2 + |\mathbf{b}^2 - 2(\mathbf{a}\cdot\mathbf{b}) + |\mathbf{b}|^2 + |\mathbf{c}|^2 - 2(\mathbf{b}\cdot\mathbf{c}) + |\mathbf{c}|^2 + |\mathbf{a}|^2 - 2(\mathbf{c}\cdot\mathbf{a})$

$= 2\,[|\mathbf{a}|^2 + |\mathbf{b}|^2 + |\mathbf{c}|^2 - (\mathbf{a}\cdot\mathbf{b} + \mathbf{b}\cdot\mathbf{c} + \mathbf{c}\cdot\mathbf{a})]$

$= 2\,[3 - (\mathbf{a}\cdot\mathbf{b} + \mathbf{b}\cdot\mathbf{c} + \mathbf{c}\cdot\mathbf{a})]$

$= 6 - 2(\mathbf{a}\cdot\mathbf{b} + \mathbf{b}\cdot\mathbf{c} + \mathbf{c}\cdot\mathbf{a}) \quad \ldots\text{(i)}$

Now, $\quad |\mathbf{a} + \mathbf{b} + \mathbf{c}|^2 \geq 0$

$\Rightarrow \quad |\mathbf{a}|^2 + |\mathbf{b}|^2 + |\mathbf{c}|^2 + 2(\mathbf{a}\cdot\mathbf{b} + \mathbf{b}\cdot\mathbf{c} + \mathbf{c}\cdot\mathbf{a}) \geq 0$

$\Rightarrow \quad 3 + 2(\mathbf{a}\cdot\mathbf{b} + \mathbf{b}\cdot\mathbf{c} + \mathbf{c}\cdot\mathbf{a}) \geq 0$

$\Rightarrow \quad \mathbf{a}\cdot\mathbf{b} + \mathbf{b}\cdot\mathbf{c} + \mathbf{c}\cdot\mathbf{a} \geq -\frac{3}{2}$

$\Rightarrow \quad -2(\mathbf{a}\cdot\mathbf{b} + \mathbf{b}\cdot\mathbf{c} + \mathbf{c}\cdot\mathbf{a}) \leq 3 \quad \ldots\text{(iii)}$

From Eqs. (i) and (ii), we obtain

$|\mathbf{a} - \mathbf{b}|^2 + |\mathbf{b} - \mathbf{c}|^2 + |\mathbf{c} - \mathbf{a}|^2 \leq 6 + 3$

$\Rightarrow \quad |\mathbf{a} - \mathbf{b}|^2 + |\mathbf{b} - \mathbf{c}|^2 + |\mathbf{c} - \mathbf{a}|^2 \leq 9$

16. (a) We have, $\mathbf{OP} = \hat{\mathbf{a}}\cos t + \hat{\mathbf{b}}\sin t$

$\Rightarrow \quad |\mathbf{OP}|^2 = (\hat{\mathbf{a}}\cos t + \hat{\mathbf{b}}\sin t)\cdot(\hat{\mathbf{a}}\cos t + \hat{\mathbf{b}}\sin t)$

$= \cos^2 t(\hat{\mathbf{a}}\cdot\hat{\mathbf{a}}) + \sin^2 t\,(\hat{\mathbf{b}}\cdot\hat{\mathbf{b}}) + 2\cos t\sin t\,(\hat{\mathbf{a}}\cdot\hat{\mathbf{b}})$

$= \cos^2 t + \sin^2 t + \sin 2t\,(\hat{\mathbf{a}}\cdot\hat{\mathbf{b}})$

$\Rightarrow \quad |\mathbf{OP}| = \sqrt{1 + \sin 2t(\hat{\mathbf{a}}\cdot\hat{\mathbf{b}})}$

Clearly, $\hat{\mathbf{a}}\cdot\hat{\mathbf{b}}$ is constant. So, $|\mathbf{OP}|$ is maximum, when $\sin 2t$ is maximum, i.e. at $t = \frac{\pi}{4}$.

For $t = \frac{\pi}{4}$, we have $|\mathbf{OP}| = (1 + \hat{\mathbf{a}}\cdot\hat{\mathbf{b}})^{1/2}$

$\Rightarrow \quad M = (1 + \hat{\mathbf{a}}\cdot\hat{\mathbf{b}})^{1/2}$

Also, $\quad \hat{\jmath} = \frac{\mathbf{OP}}{|\mathbf{OP}|} = \frac{\hat{\mathbf{a}}\cos\frac{\pi}{4} + \hat{\mathbf{b}}\sin\frac{\pi}{4}}{\left|\hat{\mathbf{a}}\cos\frac{\pi}{4} + \hat{\mathbf{b}}\sin\frac{\pi}{4}\right|} = \frac{\hat{\mathbf{a}} + \hat{\mathbf{b}}}{|\hat{\mathbf{a}} + \hat{\mathbf{b}}|}$

17. (b) Clearly, AD' is perpendicular to **AB** and lies in the plane of **AB** and **AD**. Therefore,

$$\mathbf{AD'} = \mathbf{AB} \times (\mathbf{AB} \times \mathbf{AD})$$
$$= (\mathbf{AB} \cdot \mathbf{AD})\,\mathbf{AB} - (\mathbf{AB} \cdot \mathbf{AB})\,\mathbf{AD}$$
$$= 5\,(61\hat{i} - 10\hat{j} - 2\hat{k})$$

$\therefore$ $$\cos\alpha = \frac{\mathbf{AD} \cdot \mathbf{AD'}}{|\mathbf{AD}|\,|\mathbf{AD'}|}$$
$$= \frac{(-\hat{i} + 2\hat{j} + 2\hat{k}) \cdot (61\hat{i} - 10\hat{j} - 2\hat{k})\,5}{\sqrt{(-1)^2 + (2)^2 + (2)^2} \times 5\sqrt{(61)^2 + (10)^2 + (-2)^2}}$$
$$= \frac{\sqrt{17}}{9}$$

18. (a,b,c) $|z_1| = |z_2| = r$ and $z_1, z_2 = 0$

$\Rightarrow$ $a^2 + b^2 = c^2 + d^2 = r$ ...(i)

and $ac + bd = 0$

as, $ac = -bd \Rightarrow \dfrac{a}{b} = \dfrac{c}{-d} = \lambda$ ...(ii)

From Eqs. (i) and (ii), $a^2(1 + \lambda^2) = d^2(1 + \lambda^2)$

$\Rightarrow$ $a^2 = d^2$

and $b^2 = c^2$

Now, $|w_1| = a^2 + c^2 = a^2 + b^2 = h = |w_2|$

$w_1 \,.\, w_2 = ab + bd = 0$

19. (b,d) We have,

$$\mathbf{a} \cdot \mathbf{b} = \mathbf{b} \cdot \mathbf{c} = \mathbf{c} \cdot \mathbf{a} = 0 \quad \text{...(i)}$$

Again, their magnitudes are equal, i.e.

$$|\mathbf{a}| = |\mathbf{b}| = |\mathbf{c}| = \lambda \quad \text{...(ii)}$$

Now, $$|\mathbf{a} + \mathbf{b} + \mathbf{c}|^2 = (\mathbf{a} + \mathbf{b} + \mathbf{c}) \cdot (\mathbf{a} + \mathbf{b} + \mathbf{c})$$
$$= |\mathbf{a}|^2 + |\mathbf{b}|^2 + |\mathbf{c}| + 2\mathbf{a} \cdot \mathbf{b} + 2\mathbf{b} \cdot \mathbf{c} + 2\mathbf{c} \cdot \mathbf{a}$$
$$= \lambda^2 + \lambda^2 + \lambda^2 = 3\lambda^2$$

$\Rightarrow$ $|\mathbf{a} + \mathbf{b} + \mathbf{c}| = \sqrt{3}\lambda$

Let θ be the angle between $(\mathbf{a} + \mathbf{b} + \mathbf{c})$ and $\mathbf{a}$.

$\therefore$ $$\cos\theta = \frac{(\mathbf{a} + \mathbf{b} + \mathbf{c}) \cdot \mathbf{a}}{|\mathbf{a} + \mathbf{b} + \mathbf{c}|\,|\mathbf{a}|}$$
$$= \frac{|\mathbf{a}|^2}{|\mathbf{a} + \mathbf{b} + \mathbf{c}|\,|\mathbf{a}|}$$
$$= \frac{\lambda^2}{\sqrt{3}\lambda \times \lambda} = \frac{1}{\sqrt{3}}$$

$\Rightarrow$ $$\theta = \cos^{-1}\frac{1}{\sqrt{3}} \text{ or } \tan^{-1}\sqrt{2}$$

20. (a,b,d) We have, $\mathbf{p} = 3\hat{i} + 4\hat{j}$ and $\mathbf{q} = 5\hat{i}$

Also, $$4\mathbf{r} = \mathbf{p} + \mathbf{q}$$
$$= 3\hat{i} + 4\hat{j} + 5\hat{i}$$
$$= 8\hat{i} + 4\hat{j}$$

$\Rightarrow$ $\mathbf{r} = 2\hat{i} + \hat{j}$

and $$2\mathbf{s} = \mathbf{p} - \mathbf{q}$$
$$= 3\hat{i} + 4\hat{j} - 5\hat{i}$$
$$= -2\hat{i} + 4\hat{j}$$

$\Rightarrow$ $\mathbf{s} = -\hat{i} + 2\hat{j}$

Now, $|\mathbf{r} + k\mathbf{s}| = |\mathbf{r} - k\mathbf{s}|$

$\Rightarrow$ $|2\hat{i} + \hat{j} - k\hat{i} + 2k\hat{j}|^2 = |2\hat{i} + \hat{j} + k\hat{i} - 2k\hat{j}|^2$

$\Rightarrow$ $(2 - k)^2 + (1 + 2k)^2 = (2 + k)^2 + (1 - 2k)^2$

which is true for all values of k.

Now, $$\mathbf{r} \cdot \hat{\mathbf{s}} = (2\hat{i} + \hat{j})(-\hat{i} + 2\hat{j})$$
$$= -2 + 2 = 0$$

$\therefore$ $\mathbf{r} \perp \mathbf{s}$

Also, $$(\mathbf{r} + \mathbf{s}) \cdot (\mathbf{r} - \mathbf{s}) = (\hat{i} + 3\hat{j}) \cdot (3\hat{i} - \hat{j})$$
$$= 3 - 3 = 0$$

$\therefore$ $(\mathbf{r} + \mathbf{s}) \perp (\mathbf{r} - \mathbf{s})$

Also, $$|\mathbf{r}| = \sqrt{(2)^2 + 1} = \sqrt{5}$$
$$|\mathbf{s}| = \sqrt{(-1)^2 + (2)^2} = \sqrt{5}$$
$$|\mathbf{p}| = \sqrt{3^2 + 4^2} = 5$$
$$|\mathbf{q}| = \sqrt{5^2} = 5$$

$\therefore$ $|\mathbf{r}| = |\mathbf{s}|$ and $|\mathbf{p}| = |\mathbf{q}|$.

21. (c) Let $\mathbf{r} = x\hat{i} + y\hat{j}$

Now, $$\mathbf{r} \cdot (10\hat{j} - 8\hat{i} - \mathbf{r}) = 40$$

$\Rightarrow$ $(x\hat{i} + y\hat{j}) \cdot (10\hat{j} - 8\hat{i} - x\hat{i} - y\hat{j}) = 40$

$\Rightarrow$ $x^2 + y^2 + 8x - 10y + 40 = 0$

$\Rightarrow$ $(x + 4)^2 + (y - 5)^2 = 1$

which is an equation of circle with radius 1.

22. (d) $p_1 = (cp + r)^2$, $p_2 = (cp - r)^2$

$\therefore$ $p_1 = (2\sqrt{2} + 1)^2$

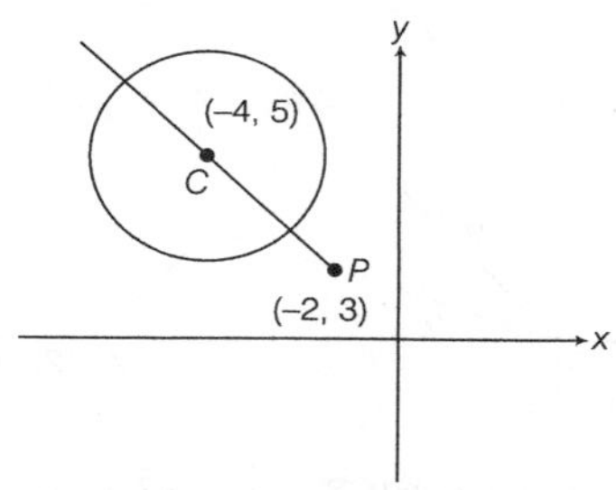

and $p_2 = (2\sqrt{2} - 1)^2$

$\Rightarrow$ $p_1 + p_2 = 18$

23. (a) Here, $D\,(2, 2)$

$\therefore$ Tangent at D for $y = 8/x^2$ is

$$2x + y = 6$$

$\therefore$ $E\,(3, 0) \Rightarrow \mathbf{OE} = 3\hat{i}$

$$\mathbf{OD} = 2\hat{i} + 2\hat{j},$$
$$\mathbf{DE} = \hat{i} - 2\hat{j}$$

$\Rightarrow$ $\mathbf{DE} \,.\, \mathbf{OE} = 3$

24. (9) $\mathbf{r} \,.\, \mathbf{a} = 20$

$\Rightarrow x + 2y + z = 20;\ x, y, z \in N$

$\therefore$ number of non-negative integer solution are

$\Rightarrow {}^{17}C_1 + {}^{15}C_1 + {}^{13}C_1 + \ldots\ldots +$

${}^{1}C_1 = 81 = 9\lambda$

$\Rightarrow$ $\lambda = 9$

25. (1) $\alpha \,.\, \beta = \dfrac{b}{a} + \dfrac{4a}{b} + 1$

as, $\dfrac{b}{a} + \dfrac{4a}{b} + 1, \geq 5$ (using AM $\geq$ G.M)

So $$\left(\frac{10}{5 + \bar{\alpha} \,.\, \bar{\beta}}\right)_{max} = 1$$

DPP-3 Vector or Cross Product

1. (a) We have,

$$|\mathbf{a}\times\mathbf{b}| = |\mathbf{a}||\mathbf{b}|\sin\theta$$

$$\Rightarrow \quad |\mathbf{a}\times\mathbf{b}|^2 = |\mathbf{a}|^2|\mathbf{b}|^2\sin^2\theta$$

$$= |\mathbf{a}|^2|\mathbf{b}|^2(1-\cos^2\theta)$$

$$= |\mathbf{a}|^2|\mathbf{b}|^2 - |\mathbf{a}|^2|\mathbf{b}|^2\cos^2\theta$$

$$= |\mathbf{a}|^2|\mathbf{b}|^2 - (|\mathbf{a}||\mathbf{b}|\cos\theta)^2$$

$$= |\mathbf{a}|^2|\mathbf{b}|^2 - (\mathbf{a}\cdot\mathbf{b})^2$$

$$\therefore \quad |\mathbf{a}\times\mathbf{b}|^2 + (\mathbf{a}\cdot\mathbf{b})^2 = |\mathbf{a}|^2|\mathbf{b}|^2$$

2. (c) Any vector perpendicular to both **a** and **b** $= \lambda(\mathbf{a}\times\mathbf{b})$.

Now, $$\mathbf{a}\times\mathbf{b} = \begin{vmatrix} \hat{i} & \hat{j} & \hat{k} \\ 1 & 1 & 1 \\ 1 & 1 & -1 \end{vmatrix}$$

$$= -2\hat{i} + 2\hat{j}$$

$$= -2(\hat{i} - \hat{j})$$

$\therefore$ Required vector $= \lambda(\hat{i} - \hat{j})$

3. (c) We have,

$$\mathbf{a}\times\mathbf{b} = \mathbf{b}\times\mathbf{c}$$

$$\Rightarrow \quad \mathbf{a}\times\mathbf{b} - \mathbf{b}\times\mathbf{c} = 0$$

$$\Rightarrow \quad \mathbf{a}\times\mathbf{b} + \mathbf{c}\times\mathbf{b} = 0$$

$$\Rightarrow \quad (\mathbf{a}+\mathbf{c})\times\mathbf{b} = 0$$

$$\therefore \quad (\mathbf{a}+\mathbf{c}) \,||\, \mathbf{b}$$

[$\because$ if vector product of two vectors is zero, then both vectors are parallel to each other]

4. (b) We have,

$$\mathbf{BC} + \mathbf{CA} = \mathbf{AB}$$

$$\Rightarrow \quad \mathbf{a} + \mathbf{b} = -\mathbf{c}$$

$$\Rightarrow \quad \mathbf{a} + \mathbf{b} + \mathbf{c} = \mathbf{0}$$

$$\Rightarrow \quad \mathbf{a}\times(\mathbf{a}+\mathbf{b}+\mathbf{c}) = \mathbf{a}\times 0$$

$$\Rightarrow \quad \mathbf{a}\times\mathbf{b} + \mathbf{a}\times\mathbf{c} = 0$$

$$\Rightarrow \quad \mathbf{a}\times\mathbf{b} = \mathbf{c}\times\mathbf{a}$$

Similarly, we can say that

$$\mathbf{b}\times\mathbf{c} = \mathbf{c}\times\mathbf{a}$$

$$\therefore \quad \mathbf{a}\times\mathbf{b} = \mathbf{b}\times\mathbf{c} = \mathbf{c}\times\mathbf{a}$$

5. (a) We have,

$$\mathbf{a} = \hat{i} + 2\hat{j} + 3\hat{k}$$

$$\mathbf{b} = -\hat{i} + 2\hat{j} + \hat{k}$$

and $$\mathbf{c} = 3\hat{i} + \hat{j}$$

Since, $(\mathbf{a} + t\,\mathbf{b})$ is perpendicular to **c**.

$$\therefore \quad (\mathbf{a} + t\,\mathbf{b})\cdot\mathbf{c} = 0$$

$$[(1-t)\hat{i} + (2+2t)\hat{j} + (3+t)\hat{k}]\cdot[3\hat{i} + \hat{j}] = 0$$

$$\Rightarrow \quad 3(1-t) + (2+2t) = 0$$

$$\Rightarrow \quad t = 5$$

6. (c) We have,

$$\mathbf{a} = 2\hat{i} - 3\hat{j} + \hat{k}$$

$$\mathbf{b} = -\hat{i} + \hat{k}$$

and $$\mathbf{c} = 2\hat{j} - \hat{k}$$

Since, $(\mathbf{a} + \mathbf{b})$ and $(\mathbf{b} + \mathbf{c})$ are the diagonals of the parallelogram.

Now, $$\mathbf{a} + \mathbf{b} = \hat{i} - 3\hat{j} + 2\hat{k}$$

and $$\mathbf{b} + \mathbf{c} = -\hat{i} + 2\hat{j}$$

$\therefore$ Area of parallelogram $= \frac{1}{2}|(\mathbf{a}+\mathbf{b})\times(\mathbf{b}+\mathbf{c})|$

$$= \frac{1}{2}|(\hat{i} - 3\hat{j} + 2\hat{k})\times(-\hat{i} + 2\hat{j})|$$

$$= \left|\frac{1}{2}\begin{vmatrix} \hat{i} & \hat{j} & \hat{k} \\ 1 & -3 & 2 \\ -1 & 2 & 0 \end{vmatrix}\right|$$

$$= \left|\frac{1}{2}(-4\hat{i} - 2\hat{j} - \hat{k})\right|$$

$$= \frac{1}{2}\sqrt{(-4)^2 + (-2)^2 + (-1)^2}$$

$$= \frac{\sqrt{21}}{2} \text{ sq units}$$

7. (d) Let the points be $A(10\hat{i} + 3\hat{j})$, $B(12\hat{i} - 15\hat{j})$ and $C(a\hat{i} + 11\hat{j})$.

$$\therefore \quad \mathbf{AB} = 12\hat{i} - 18\hat{j} \text{ and } \mathbf{AC} = (a-10)\hat{i} + 8\hat{j}$$

Since, A, B and C are collinear, then

$$\frac{2}{a-10} = -\frac{18}{8}$$

$$\therefore \quad a = \frac{82}{9}$$

8. (b) We have,

$$\mathbf{OA} = \mathbf{a},\ \mathbf{OB} = 10\mathbf{a} + 2\mathbf{b} \text{ and } \mathbf{OC} = \mathbf{b}$$

Since, p denotes the area of quadrilateral $OABC$.

$$\therefore \quad p = \frac{1}{2}|\mathbf{AC}\times\mathbf{OB}|$$

$$= \frac{1}{2}|(\mathbf{b}-\mathbf{a}) - (10\mathbf{a} + 2\mathbf{b})|$$

$$= \frac{1}{2}|10\,\mathbf{b}\times\mathbf{a} - 2\mathbf{a}\times\mathbf{b}|$$

$$= \frac{1}{2}|10\,\mathbf{b}\times\mathbf{a} + 2\mathbf{b}\times\mathbf{a}| = 6|\mathbf{b}\times\mathbf{a}|$$

Also, q denotes the area of parallelogram with OA and OC as adjacent sides.

$$\therefore \quad q = |\mathbf{OC}\times\mathbf{OA}| = |\mathbf{b}\times\mathbf{a}|$$

$$\therefore \quad p = 6q$$

$$\Rightarrow \quad k = 6$$

9. (c) We have, $\mathbf{AB} = \hat{i} + \hat{j} + \hat{k}$ and $\mathbf{AC} = \hat{i} - \hat{j} + \hat{k}$

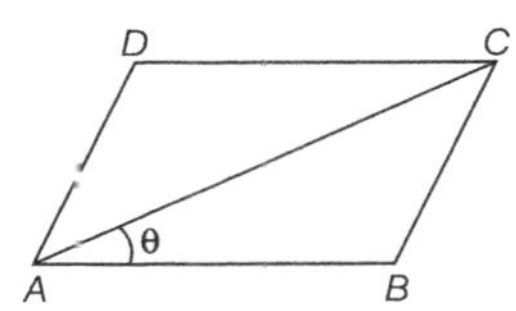

Let θ be the ∠BAC. Then,

$$\sin\theta = \frac{|\mathbf{AB}\times\mathbf{AC}|}{|\mathbf{AB}||\mathbf{AC}|}$$

Now, $$\mathbf{AB}\times\mathbf{AC} = \begin{vmatrix} \hat{i} & \hat{j} & \hat{k} \\ 1 & 1 & 1 \\ 1 & -1 & 1 \end{vmatrix} = 2\hat{i} - 2\hat{k}$$

$$\therefore \quad |\mathbf{AB}\times\mathbf{AC}| = \sqrt{8}$$

Hence, $$\sin\theta = \frac{\sqrt{8}}{\sqrt{3}\times\sqrt{3}}$$

$$\Rightarrow \quad \theta = \sin^{-1}\left(\frac{\sqrt{8}}{3}\right)$$

10. (b) Since, $|2\hat{\mathbf{u}} \times 3\hat{\mathbf{v}}| = 1$

$\Rightarrow \quad 6|\hat{\mathbf{u}}||\hat{\mathbf{v}}| \sin\theta = 1$

$\Rightarrow \quad \sin\theta = \dfrac{1}{6} \qquad [\because |\hat{\mathbf{u}}| = |\hat{\mathbf{v}}| = 1]$

Since, θ is an acute angle, then there is exactly one value of θ for which $(2\hat{\mathbf{u}} \times 3\hat{\mathbf{v}})$ is a unit vector.

11. (c) We have, $\mathbf{u} = \hat{\mathbf{i}} + \hat{\mathbf{j}}$, $\mathbf{v} = \hat{\mathbf{i}} - \hat{\mathbf{j}}$ and $\mathbf{w} = \hat{\mathbf{i}} + 2\hat{\mathbf{j}} + 3\hat{\mathbf{k}}$

$\hat{\mathbf{n}}$ is a unit vector such that $\mathbf{u} \cdot \hat{\mathbf{n}} = 0$ and $\mathbf{v} \cdot \hat{\mathbf{n}} = 0$.

$\Rightarrow \quad \hat{\mathbf{n}} \perp \mathbf{u}$ and $\hat{\mathbf{n}} \perp \mathbf{v}$

$\therefore \quad \mathbf{n} = \mathbf{u} \times \mathbf{v} = \begin{vmatrix} \hat{\mathbf{i}} & \hat{\mathbf{j}} & \hat{\mathbf{k}} \\ 1 & 1 & 0 \\ 1 & -1 & 0 \end{vmatrix} = -2\hat{\mathbf{k}}$

$\therefore \quad \hat{\mathbf{n}} = \dfrac{\mathbf{n}}{|\mathbf{n}|} = \pm\hat{\mathbf{k}}$

$\therefore \quad |\mathbf{w} \cdot \hat{\mathbf{n}}| = |(\hat{\mathbf{i}} + 2\hat{\mathbf{j}} + 3\hat{\mathbf{k}}) \cdot (\pm\hat{\mathbf{k}})|$

$= |\pm 3| = 3$

12. (b) Let $\mathbf{a} = a_1\hat{\mathbf{i}} + a_2\hat{\mathbf{j}} + a_3\hat{\mathbf{k}}$

$|\mathbf{a}|^2 = a_1^2 + a_2^2 + a_3^2$

and $\quad \mathbf{a} \times \hat{\mathbf{i}} = (a_1\hat{\mathbf{i}} + a_2\hat{\mathbf{j}} + a_3\hat{\mathbf{k}}) \times \hat{\mathbf{i}}$

$= -a_2\hat{\mathbf{k}} + a_3\hat{\mathbf{j}}$

$\Rightarrow \quad |\mathbf{a} \times \hat{\mathbf{i}}|^2 = a_2^2 + a_3^2$

Similarly, $\quad |\mathbf{a} \times \hat{\mathbf{j}}|^2 = a_3^2 + a_1^2$

and $\quad |\mathbf{a} \times \hat{\mathbf{k}}|^2 = a_1^2 + a_2^2$

Now, $|\mathbf{a} \times \hat{\mathbf{i}}|^2 + |\mathbf{a} \times \hat{\mathbf{j}}|^2 + |\mathbf{a} \times \hat{\mathbf{k}}|^2$

$= a_2^2 + a_3^2 + a_3^2 + a_1^2 + a_1^2 + a_2^2$

$= 2(a_1^2 + a_2^2 + a_3^2) = 2|\mathbf{a}|^2$

13. (b) The equation of two lines are

$(\mathbf{r} - \mathbf{b}) \times \mathbf{a} = 0$ and $(\mathbf{r} - \mathbf{a}) \times \mathbf{b} = 0$

$\therefore$ $(\mathbf{r} - \mathbf{b})$ is parallel to $\mathbf{a}$ and $(\mathbf{r} - \mathbf{a})$ is parallel to $\mathbf{b}$.

$\therefore \quad \mathbf{r} = \mathbf{b} + p\mathbf{a}$

and $\quad \mathbf{r} = \mathbf{a} + q\mathbf{b}$

For their point of intersection, we have identical values of r.

$\therefore \quad p = q = 1$ and hence, $\mathbf{r} = \mathbf{a} + \mathbf{b} = 3\hat{\mathbf{i}} + \hat{\mathbf{j}} - \hat{\mathbf{k}}$

14. (b) Let A be the first term and D be the common difference of the corresponding AP. Then,

$\dfrac{1}{a} = A + (p-1)D, \quad \dfrac{1}{b} = A + (q-1)D, \quad \dfrac{1}{c} = A + (r-1)D$

$\Rightarrow \quad a^{-1}(q - r) + b^{-1}(r - p) + c^{-1}(p - q) = 0$

$\Rightarrow \quad \mathbf{v} \cdot \mathbf{u} = 0$

$\Rightarrow \quad \mathbf{u} \perp \mathbf{v}$

Hence, $\mathbf{u}$ and $\mathbf{v}$ are orthogonal vectors.

15. (c) Let $A(\alpha, \beta)$ point be given and O be taken as the origin.

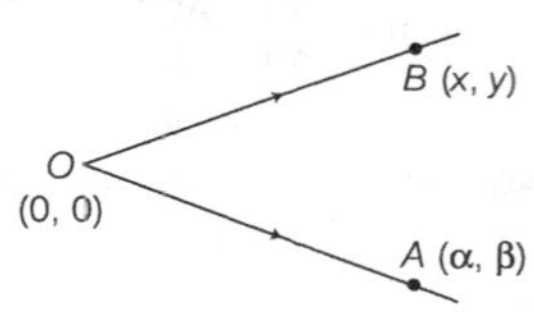

We have, $\quad \mathbf{OA} = \alpha\hat{\mathbf{i}} + \beta\hat{\mathbf{j}}$

and $\quad \mathbf{OB} = xi + yj$

Now, $\quad |\mathbf{OA} \times \mathbf{OB}| = |(\alpha y - \beta x)\hat{\mathbf{k}}| = \text{constant}$

$\Rightarrow \quad \alpha y - \beta x = \text{constant}$

$\therefore$ Locus of $B(x, y)$ is a line parallel to OA because slope of $OA = \dfrac{\beta}{\alpha}$.

16. (b) The required vector is given by

$\hat{\mathbf{n}} = \dfrac{\mathbf{AB} \times \mathbf{AC}}{|\mathbf{AB} \times \mathbf{AC}|}$

Now, $\quad \mathbf{AB} \times \mathbf{AC} = (\mathbf{b} \times \mathbf{a}) \times (\mathbf{c} - \mathbf{a})$

$= \mathbf{b} \times \mathbf{c} - \mathbf{b} \times \mathbf{a} - \mathbf{a} \times \mathbf{c} + \mathbf{a} \times \mathbf{a}$

$= \mathbf{b} \times \mathbf{c} + \mathbf{a} \times \mathbf{b} + \mathbf{c} \times \mathbf{a} \qquad [\because \mathbf{a} \times \mathbf{a} = 0]$

We also know that,

$\text{Area of } \Delta ABC = \dfrac{1}{2}|\mathbf{a} \times \mathbf{b} + \mathbf{b} \times \mathbf{c} + \mathbf{c} \times \mathbf{a}|$

$\therefore \quad \hat{\mathbf{n}} = \dfrac{\mathbf{a} \times \mathbf{b} + \mathbf{b} \times \mathbf{c} + \mathbf{c} \times \mathbf{a}}{|\mathbf{a} \times \mathbf{b} + \mathbf{b} \times \mathbf{c} + \mathbf{c} \times \mathbf{a}|}$

$= \dfrac{\mathbf{a} \times \mathbf{b} + \mathbf{b} \times \mathbf{c} + \mathbf{c} \times \mathbf{a}}{2\Delta}$

$\left[\because \Delta = \dfrac{1}{2}|\mathbf{a} \times \mathbf{b} + \mathbf{b} \times \mathbf{c} + \mathbf{c} \times \mathbf{a}|\right]$

Explanations for Q. Nos. (17-19)

Here, $|\mathbf{a}| = 2, |\mathbf{b}| = 3$

$\mathbf{c} \times \mathbf{a} = \mathbf{b} \Rightarrow |\mathbf{c}||\mathbf{a}| \sin\alpha = |\mathbf{b}|$

$\Rightarrow \quad |\mathbf{c}| = \dfrac{3}{2} \operatorname{cosec}\alpha$

consider, $|\mathbf{c} - \mathbf{a}|^2 = |\mathbf{c}|^2 + |\mathbf{a}|^2 - 2(\mathbf{a} \cdot \mathbf{c})$

$= \dfrac{9}{4}\operatorname{cosec}^2\alpha + 4 - 2(2) \cdot \left(\dfrac{3}{2}\right) \cdot \operatorname{cosec}\alpha \cdot \cos\alpha$

$= \dfrac{25}{4} + \dfrac{9}{4} \cdot \cot^2\alpha - 6\cot\alpha$

$= \dfrac{25}{4} + \left(\dfrac{3}{2}\cot\alpha - 2\right)^2 - 4$

$= \dfrac{9}{4} + \left(\dfrac{3}{2}\cot\alpha - 2\right)^2 \cdot \geq \dfrac{9}{4}$

$|\mathbf{c} - \mathbf{a}| \geq \dfrac{3}{2}$ and least possible when

$\dfrac{3}{2}\cot\alpha = 2$

$\Rightarrow \quad \tan\alpha = \dfrac{3}{4}$

also, $\quad |\vec{c}| = \dfrac{3}{2\sin\alpha} = \dfrac{3}{2\left(\dfrac{3}{5}\right)} = \dfrac{5}{2}$

17. (d) **18.** (b) **19.** (c)

20. (d) Required vector is given by

$\hat{\mathbf{n}} = \dfrac{(\mathbf{x} + \mathbf{y}) \times (\mathbf{y} + \mathbf{z})}{|(\mathbf{x} + \mathbf{y}) \times (\mathbf{y} + \mathbf{z})|}$

Now, $(\mathbf{x} + \mathbf{y}) \times (\mathbf{y} + \mathbf{z})$

$= (\mathbf{a} + \mathbf{b} - \mathbf{c} - \mathbf{a} + \mathbf{b} - 2\mathbf{c}) \times (-\mathbf{a} + \mathbf{b} - 2\mathbf{c} - \mathbf{a} + 2\mathbf{b} - \mathbf{c})$

$= (2\mathbf{b} - 3\mathbf{c}) \times (-2\mathbf{a} + 3\mathbf{b} - 3\mathbf{c})$

$= -4\mathbf{b} \times \mathbf{a} - 6\mathbf{b} \times \mathbf{c} + 6\mathbf{c} \times \mathbf{a} - 9\mathbf{c} \times \mathbf{b}$

$= 4\mathbf{a} \times \mathbf{b} + 3\mathbf{b} \times \mathbf{c} + 6\mathbf{c} \times \mathbf{a} = 3\mathbf{a} + 6\mathbf{b} + 4\mathbf{c}$

$\therefore \quad \hat{\mathbf{n}} = \dfrac{3\mathbf{a} + 3\mathbf{b} + 4\mathbf{c}}{|3\mathbf{a} + 3\mathbf{b} + 4\mathbf{c}|}$

21. (c) The equation of the two lines are $(\mathbf{r} - \mathbf{y}) \times \mathbf{x} = 0$ and $(\mathbf{r} - \mathbf{x}) \times \mathbf{y} = 0$.

$\therefore \quad \mathbf{r} - \mathbf{y} \,||\, \mathbf{a}$ and $\mathbf{r} - \mathbf{x} \,||\, \mathbf{y}$

$\therefore \quad \mathbf{r} = \mathbf{y} + \lambda\mathbf{x}$ and $\mathbf{r} = \mathbf{x} + \mu\mathbf{y}$

For their point of intersection, we have identical values of r

$\therefore \quad \lambda = \mu = 1$

$\therefore \quad \mathbf{r} = \mathbf{x} + \mathbf{y}$

$= (\mathbf{a} + \mathbf{b}) + (2\mathbf{a} - \mathbf{b})$

$= 3\mathbf{a}$

22. (3) We have, $|\mathbf{a}| = 3, |\mathbf{b}| = 1$ and $|\mathbf{a} + \mathbf{b}| = |\mathbf{a} - \mathbf{b}|$

Now, $|\mathbf{a} + \mathbf{b}|^2 = |\mathbf{a} - \mathbf{b}|^2$

$$\Rightarrow |\mathbf{a}|^2 + |\mathbf{b}|^2 + 2|\mathbf{a}||\mathbf{b}|\cos\theta = |\mathbf{a}|^2 + |\mathbf{b}|^2 - 2|\mathbf{a}||\mathbf{b}|\cos\theta$$

$$\Rightarrow 4|\mathbf{a}||\mathbf{b}|\cos\theta = 0$$

$$\Rightarrow \cos\theta = 0 \Rightarrow \theta = 90°$$

Now, $|\mathbf{a} \times \mathbf{b}| = |\mathbf{a}||\mathbf{b}|\sin\theta = 3 \times 1 \times \sin 90° = 3$

23. (3) $\mathbf{c} \times \mathbf{a} = \mathbf{b} \Rightarrow |\mathbf{c} \times \mathbf{a}| = |\mathbf{b}| \Rightarrow |\mathbf{c}||\mathbf{a}|\sin\theta = 3$

$$|\mathbf{c}| = \frac{3}{2\sin\theta} \quad |\mathbf{c} - \mathbf{a}|^2 = |\mathbf{c}|^2 + |\mathbf{a}|^2 - 2\,\mathbf{c}\,.\,\mathbf{a}$$

$$\Rightarrow |\mathbf{c} - \mathbf{a}|^2 = |\mathbf{c}|^2 + 4 - 2|\mathbf{c}||\mathbf{a}|\cos\theta$$

$$= \frac{9}{4\sin^2\theta} + 4 - 6\cot\theta = 4 + \frac{9}{4}\operatorname{cosec}^2\theta - 6\cot\theta$$

$$= \frac{9}{4} + \left(\frac{3}{3}\cot\theta - 2\right)^2$$

$$|\mathbf{c} - \mathbf{a}|^2 \geq \frac{9}{4}$$

$$\Rightarrow 2|\mathbf{c} - \mathbf{a}| \geq 3$$

24. (6) The vector **c** is coplanar with vectors **a** and **b**

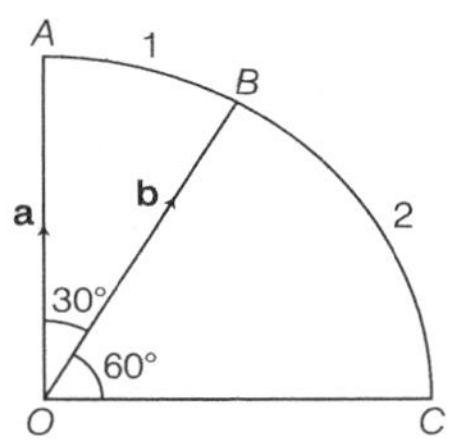

$\therefore \quad \mathbf{c} = x\mathbf{a} + y\mathbf{b}$ ...(i)

The point B divides arc AC in the ratio 1 : 2, so that $\angle AOB = 30°$, $\angle BOC = 60°$. We have to find the values of x and y when we are given $|\mathbf{a}| = |\mathbf{b}| = |\mathbf{c}| = \gamma$ (say)

$$\mathbf{a} \cdot \mathbf{b} = r^2 \cos 30° = r^2 \frac{\sqrt{3}}{2}, \mathbf{a} \cdot \mathbf{c} = 0$$

$$\mathbf{b} \cdot \mathbf{c} = r^2 \cos 60° = \frac{r^2}{2}$$

Multiply both sides of Eq. (i) scalarly by **c** and **a**.

$$\therefore \quad \mathbf{c} \cdot \mathbf{c} = x\mathbf{a} \cdot \mathbf{c} = y\mathbf{b} \cdot \mathbf{c}$$

$$\mathbf{c} \cdot \mathbf{a} = x\mathbf{a} \cdot \mathbf{a} + y\mathbf{b} \cdot \mathbf{a}$$

or $\quad r^2 = 0 + \frac{r^2}{2}y$

$\therefore \quad y = 2$

$$0 = xr^2 + yr^2 \frac{\sqrt{3}}{2}$$

Put $y = 2$

$\therefore \quad x = -\sqrt{3}$

Now, $\quad \mathbf{c} = -\sqrt{3}a + 2b$

$\therefore \quad |\sqrt{3}xy| = |\sqrt{3}(-\sqrt{3})(2)| = 6$

25. (1) We have, $\mathbf{a} = \hat{\mathbf{i}} + \hat{\mathbf{j}} + \hat{\mathbf{k}} \Rightarrow |\mathbf{a}| = \sqrt{3}$

$$\mathbf{a} \cdot \mathbf{b} = 1$$

and $\quad \mathbf{a} \times \mathbf{b} = \hat{\mathbf{j}} - \hat{\mathbf{k}} \Rightarrow |\mathbf{a} \times \mathbf{b}| = \sqrt{2}$

We know that

$$|\mathbf{a} \times \mathbf{b}|^2 + |\mathbf{a} \cdot \mathbf{b}|^2 = |\mathbf{a}|^2|\mathbf{b}|^2$$

$$\Rightarrow (\sqrt{2})^2 + (1)^2 = (\sqrt{3})^2|\mathbf{b}|^2$$

$$\Rightarrow 2 + 1 = 3|\mathbf{b}|^2$$

$$\therefore |\mathbf{b}| = 1$$

26. (4) We have,

$$\mathbf{a} \times (2\hat{\mathbf{i}} + 3\hat{\mathbf{j}} + 4\hat{\mathbf{k}}) = (2\hat{\mathbf{i}} + 3\hat{\mathbf{j}} + 4\hat{\mathbf{k}}) \times \mathbf{b}$$

$$= \mathbf{a} \times (2\hat{\mathbf{i}} + 3\hat{\mathbf{j}} + 4\hat{\mathbf{k}}) + \mathbf{b} \times (2\hat{\mathbf{i}} + 3\hat{\mathbf{j}} + 4\hat{\mathbf{k}}) = 0$$

$$= (\mathbf{a} + \mathbf{b}) \times (2\hat{\mathbf{i}} + 3\hat{\mathbf{j}} + 4\hat{\mathbf{k}}) = 0$$

$\Rightarrow (\mathbf{a} + \mathbf{b})$ is parallel to $2\hat{\mathbf{i}} + 3\hat{\mathbf{j}} + 4\hat{\mathbf{k}}$.

$$\Rightarrow \mathbf{a} + \mathbf{b} = \pm(2\hat{\mathbf{i}} + 3\hat{\mathbf{j}} + 4\hat{\mathbf{k}}) \quad [\because |\mathbf{a} + \mathbf{b}| = \sqrt{29}]$$

$$\therefore |(\mathbf{a} + \mathbf{b}) \cdot (-7\hat{\mathbf{i}} + 2\hat{\mathbf{j}} + 3\hat{\mathbf{k}})| = \pm(2\hat{\mathbf{i}} + 3\hat{\mathbf{j}} + 4\hat{\mathbf{k}})| (-7\hat{\mathbf{i}} + 2\hat{\mathbf{j}} + 3\hat{\mathbf{k}})$$

$$= \pm 4$$

27. (9) We have,

$$\mathbf{a} = -\hat{\mathbf{i}} - \hat{\mathbf{k}},$$

$$\mathbf{b} = -\hat{\mathbf{i}} + \hat{\mathbf{j}}$$

and $\quad \mathbf{c} = \hat{\mathbf{i}} + 2\hat{\mathbf{j}} + 3\hat{\mathbf{k}}$

Also, we have

$$\mathbf{r} \times \mathbf{b} = \mathbf{c} \times \mathbf{b}$$

$$\Rightarrow (\mathbf{r} - \mathbf{c}) \times \mathbf{b} = 0$$

$$\Rightarrow \mathbf{r} - \mathbf{c} \,||\, \mathbf{b}$$

$$\therefore \mathbf{r} = \mathbf{c} + \lambda\mathbf{b}$$

$$\Rightarrow \mathbf{r} \cdot \mathbf{a} = \mathbf{c} \cdot \mathbf{a} + \lambda\mathbf{b} \cdot \mathbf{a}$$

$$\Rightarrow \lambda = -\frac{(\mathbf{c} \cdot \mathbf{a})}{(\mathbf{b} \cdot \mathbf{a})} \quad [\because \mathbf{r} \cdot \mathbf{a} = 0]$$

$$\Rightarrow \lambda = 4$$

$$\therefore \mathbf{r} = \mathbf{c} + 4\mathbf{b}$$

$$\Rightarrow \mathbf{r} = -3\hat{\mathbf{i}} + 6\hat{\mathbf{j}} + 3\hat{\mathbf{k}}$$

Hence, $\quad \mathbf{r} \cdot \mathbf{b} = (-3\hat{\mathbf{i}} + 6\hat{\mathbf{j}} + 3\hat{\mathbf{k}}) \cdot (-\hat{\mathbf{i}} + \hat{\mathbf{j}}) = 9$

28. (4) Let **a**, **b**, **c**, **d** be the position vectors of points P, Q, R and S respectively. Then,

$$\mathbf{PQ} \times \mathbf{RS} = (\mathbf{b} - \mathbf{a}) \times (\mathbf{d} - \mathbf{c}) = \mathbf{b} \times \mathbf{d} - \mathbf{b} \times \mathbf{c} + \mathbf{d} \times \mathbf{a} - \mathbf{c} \times \mathbf{a}$$

$$\mathbf{QR} \times \mathbf{PS} = (\mathbf{c} - \mathbf{b}) \times (\mathbf{d} - \mathbf{a}) = \mathbf{c} \times \mathbf{d} - \mathbf{c} \times \mathbf{a} - \mathbf{b} \times \mathbf{d} - \mathbf{a} \times \mathbf{b}$$

$$\mathbf{RP} \times \mathbf{QS} = (\mathbf{a} - \mathbf{c}) \times (\mathbf{d} - \mathbf{b}) = \mathbf{a} \times \mathbf{d} - \mathbf{a} \times \mathbf{b} - \mathbf{c} \times \mathbf{d} - \mathbf{b} \times \mathbf{c}$$

$$\therefore \mathbf{PQ} \times \mathbf{RS} - \mathbf{QR} \times \mathbf{PS} + \mathbf{RP} \times \mathbf{QS}$$

$$= \mathbf{b} \times \mathbf{d} - \mathbf{b} \times \mathbf{c} + \mathbf{d} \times \mathbf{a} - \mathbf{c} \times \mathbf{a} - \mathbf{c} \times \mathbf{d} + \mathbf{c} \times \mathbf{a} + \mathbf{b} \times \mathbf{d} + \mathbf{a} \times \mathbf{b} + \mathbf{a} \times \mathbf{d} - \mathbf{a} \times \mathbf{b} - \mathbf{c} \times \mathbf{d} - \mathbf{b} \times \mathbf{c}$$

$$= 2(\mathbf{b} \times \mathbf{d}) - 2(\mathbf{b} \times \mathbf{c}) - 2(\mathbf{c} \times \mathbf{d})$$

$$= 2[\mathbf{b} \times \mathbf{d} + \mathbf{c} \times \mathbf{b} + \mathbf{d} \times \mathbf{c}]$$

$$= 2(\mathbf{QS} \times \mathbf{QR}) = 4 \text{ Area of } \Delta QSR$$

$$\therefore \lambda = 4$$

29. (1) $\mathbf{x} \times |\mathbf{a} - \mathbf{x}| = \mathbf{b}$

$$\therefore \mathbf{x} \times \mathbf{a} = \mathbf{b}$$

$$\Rightarrow \mathbf{a} \times (\mathbf{x} \times \mathbf{a}) = \mathbf{a} \times \mathbf{b}$$

$$\Rightarrow |\mathbf{a}|^2\,\mathbf{x} - (\mathbf{a}\,.\,\mathbf{x})\,\mathbf{a} = \mathbf{a} \times \mathbf{b}$$

$$\therefore \mathbf{x} = \frac{\mathbf{a} + |\mathbf{a} \times \mathbf{b}|}{|\mathbf{a}|^2} \Rightarrow \lambda = 1$$

30. (1) We have, $\quad \mathbf{a} \times \mathbf{b} = 3\mathbf{a} \times \mathbf{c}$

$$\Rightarrow \mathbf{a} \times (\mathbf{b} - 3\mathbf{c}) = 0$$

$\Rightarrow$ **a** is parallel to $\mathbf{b} - 3\mathbf{c}$.

Now, $\quad \mathbf{b} - 3\mathbf{c} = \lambda\mathbf{a}$

$$\Rightarrow |\mathbf{b} - 3\mathbf{c}|^2 = \lambda^2|\mathbf{a}|^2$$

$$\Rightarrow |\mathbf{b}|^2 + 9|\mathbf{c}|^2 = 6(\mathbf{b} \cdot \mathbf{c}) = \lambda^2|\mathbf{a}|^2$$

$$\Rightarrow 1 + 9 \times \frac{1}{9} - 6 \times 1 \times \frac{1}{3} \times \frac{1}{2} = \lambda^2 (1)^2$$

$$\Rightarrow 1 + 1 - 1 = \lambda^2 \Rightarrow \lambda^2 = 1$$

$$\therefore \lambda = \pm 1$$

DPP-4 Scalar Triple Product

1. (a) We have,

$\hat{i}\cdot(\hat{j}\times\hat{k})+\hat{j}\cdot(\hat{k}\times\hat{i})+\hat{k}\cdot(\hat{i}\times\hat{j})$

$=\hat{i}\cdot\hat{i}+\hat{j}\cdot\hat{j}+\hat{k}\cdot\hat{k}$ $\quad[\because \hat{i}\times\hat{j}=\hat{k}, \hat{j}\times\hat{k}=\hat{i}, \hat{k}\times\hat{i}=\hat{j}]$

$=|\hat{i}|^2+|\hat{j}|^2+|\hat{k}|^2=1+1+1$ $\quad[\because \hat{i}, \hat{j}, \hat{k}$ are unit vectors$]$

$=3$

2. (c) Dot product of two vectors is a saclar quantity while cross product of two vectors is a vector quantity. Hence, only $(\mathbf{u}\cdot\mathbf{v})\mathbf{w}$ is meaningful.

3. (a) We have,

$(\mathbf{u}+\mathbf{v}-\mathbf{w})\cdot(\mathbf{u}-\mathbf{v})\times(\mathbf{v}-\mathbf{w})$

$=[\mathbf{u}+\mathbf{v}-\mathbf{w}\;\; \mathbf{u}-\mathbf{v}\;\; \mathbf{v}-\mathbf{w}]$

$=[\mathbf{u}+\mathbf{v}-\mathbf{w}\;\; \mathbf{u}-\mathbf{v}+0\mathbf{w}\;\; 0\mathbf{u}+\mathbf{v}-\mathbf{w}]$

$$=\begin{vmatrix}1 & 1 & -1\\ 1 & -1 & 0\\ 0 & 1 & -1\end{vmatrix}[\mathbf{u}\;\mathbf{v}\;\mathbf{w}]$$

$=[\mathbf{u}\;\mathbf{v}\;\mathbf{w}]=\mathbf{u}\cdot(\mathbf{v}\times\mathbf{w})$

4. (d) Let $\mathbf{a}=2\hat{i}-\hat{j}+\hat{k}$,

$\mathbf{b}=\hat{i}+2\hat{j}-3\hat{k}$

and $\mathbf{c}=3\hat{i}+p\hat{j}+5\hat{k}$

Three vectors **a**, **b** and **c** are coplanar, if

$[a\;b\;c]=0$

$$\Rightarrow \begin{vmatrix}2 & -1 & 1\\ 1 & 2 & -3\\ 3 & p & 5\end{vmatrix}=0$$

$\Rightarrow 2(10+3p)+1(5+9)+1(p-6)=0$

$\Rightarrow 7p+28=0$

$\therefore p=-4$

5. (a) The given vectors are coplanar, if

$$\begin{vmatrix}a & 1 & 1\\ 1 & b & 1\\ 1 & 1 & c\end{vmatrix}=0$$

Applying $C_2\to C_2-C_1$ and $C_3\to C_3-C_1$, we get

$$\Delta=\begin{vmatrix}a & 1-a & 1-a\\ 1 & b-1 & 0\\ 1 & 0 & c-1\end{vmatrix}=0$$

$\Rightarrow a(b-1)(c-1)-(1-a)(c-1)-(1-a)(b-1)=0$

On dividing by $(1-a)(1-b)(1-c)$, we get

$$\frac{a}{1-a}+\frac{1}{1-b}+\frac{1}{1-c}=0$$

$$\Rightarrow \frac{1}{1-b}+\frac{1}{1-c}=-\frac{a}{1-a}$$

On adding $\frac{1}{1-a}$ to both sides, we get

$$\therefore \frac{1}{1-a}+\frac{1}{1-b}+\frac{1}{1-c}=\frac{1-a}{1-a}=1$$

6. (a) Volume $=[a\;b\;c]=\begin{vmatrix}a_1 & a_2 & a_3\\ b_1 & b_2 & b_3\\ c_1 & c_2 & c_3\end{vmatrix}$

$$\therefore \text{Volume}=\begin{vmatrix}2 & -2 & 0\\ 1 & 1 & -1\\ 3 & 0 & -1\end{vmatrix}$$

$=2(-1+0)+2(-1+3)+0$

$=-2+4=2$

7. (d) We have, $|(\mathbf{a}\times\mathbf{b})\cdot\mathbf{c}|=|\mathbf{a}||\mathbf{b}||\mathbf{c}|$

$\Rightarrow ||\mathbf{a}||\mathbf{b}||\mathbf{c}|\sin\theta\cos\alpha|=|\mathbf{a}||\mathbf{b}||\mathbf{c}|$

$\Rightarrow |\sin\theta||\cos\alpha|=1$

$\Rightarrow \theta=\frac{\pi}{2}$ and $\alpha=0$

$\Rightarrow \mathbf{a}\perp\mathbf{b}$ and $\mathbf{c}\,||\,\hat{\mathbf{n}}$

$\Rightarrow \mathbf{a}\perp\mathbf{b}$ and $\mathbf{c}\perp$ both **a** and **b**.

$\Rightarrow$ **a**, **b**, **c** are mutually perpendicular.

$\Rightarrow \mathbf{a}\cdot\mathbf{b}=\mathbf{b}\cdot\mathbf{c}=\mathbf{c}\cdot\mathbf{a}=0$

8. (d) We have, $[\mathbf{a}+\mathbf{b}\;\; \mathbf{b}+\mathbf{c}\;\; \mathbf{c}+\mathbf{a}]$

$$=\begin{vmatrix}1 & 1 & 0\\ 0 & 1 & 1\\ 1 & 0 & 1\end{vmatrix}[\mathbf{a}\;\mathbf{b}\;\mathbf{c}]$$

$=[1(1-0)-1(0-1)+0(0-1)][\mathbf{a}\;\mathbf{b}\;\mathbf{c}]$

$=2[\mathbf{a}\;\mathbf{b}\;\mathbf{c}]$

9. (d) We have,

$A(6\hat{i}+0\hat{j}+0\hat{k}), B(0\hat{i}+6\hat{j}+0\hat{k}), C(0\hat{i}+0\hat{j}+\hat{k})$

and $O(0\hat{i}+0\hat{j}+0\hat{k})$.

$\therefore$ $\mathbf{OA}=6\hat{i}+0\hat{j}+0\hat{k}$,

$\mathbf{OB}=0\hat{i}+6\hat{j}+0\hat{k}$

and $\mathbf{OC}=0\hat{i}+0\hat{j}+\hat{k}$

$\therefore$ Volume of tetrahedron $=\frac{1}{6}[\mathbf{OA}\;\mathbf{OB}\;\mathbf{OC}]=\frac{1}{6}[6\hat{i}\;6\hat{j}\;\hat{k}]$

$$=\frac{1}{6}\begin{vmatrix}6 & 0 & 0\\ 0 & 6 & 0\\ 0 & 0 & 1\end{vmatrix}=\frac{1}{6}[6(6-0)]$$

$=6$ cu units

10. (a) Since, **a**, **b**, **c** are unit coplanar vectors, therefore $2\mathbf{a}-\mathbf{b}$, $2\mathbf{b}-\mathbf{c}$, $2\mathbf{c}-\mathbf{a}$ are also coplanar vectors.
Hence, their scalar triple product is zero.

11. (a) $[\mathbf{a}-\mathbf{b}\;\; \mathbf{b}-\mathbf{c}\;\; \mathbf{c}-\mathbf{a}]+\begin{bmatrix}1 & -1 & 0\\ 0 & 1 & -1\\ -1 & 0 & 1\end{bmatrix}[\mathbf{a}\;\mathbf{b}\;\mathbf{c}]$

$=[1(1-0)+1(0-1)+0(0+1)][\mathbf{a}\;\mathbf{b}\;\mathbf{c}]$

$=0\times[\mathbf{a}\;\mathbf{b}\;\mathbf{c}]=0$

12. (d) Let $\alpha=3\mathbf{a}-7\mathbf{b}-4\mathbf{c}$, $\beta=3\mathbf{a}-2\mathbf{b}+\mathbf{c}$ and $\gamma=\mathbf{a}+\mathbf{b}+\lambda\mathbf{c}$

For α, β and γ to be coplanar $[\alpha\;\beta\;\gamma]=0$

$$\Rightarrow \begin{vmatrix}3 & -7 & -4\\ 3 & -2 & 1\\ 1 & 1 & \lambda\end{vmatrix}[\mathbf{a}\;\mathbf{b}\;\mathbf{c}]=0$$

$\Rightarrow [3(-2\lambda-1)+7(3\lambda-1)-4(3+2)][\mathbf{a}\;\mathbf{b}\;\mathbf{c}]=0$

$\Rightarrow (15\lambda-30)[\mathbf{a}\;\mathbf{b}\;\mathbf{c}]=0$

Since, **a**, **b**, **c** are non-coplanar.

$\therefore 15\lambda-30=0$

$\Rightarrow \lambda=2$

13. (a) Let $\alpha=\mathbf{a}+2\mathbf{b}+3\mathbf{c}$, $\beta=\lambda\mathbf{b}+4\mathbf{c}$ and $\gamma=(2\lambda-1)\mathbf{c}$

$$\text{Then, } [\alpha\;\beta\;\gamma]=\begin{vmatrix}1 & 2 & 3\\ 0 & \lambda & 4\\ 0 & 0 & (2\lambda-1)\end{vmatrix}$$

$[\mathbf{a}\;\mathbf{b}\;\mathbf{c}]=\lambda(2\lambda-1)[\mathbf{a}\;\mathbf{b}\;\mathbf{c}]$

$\Rightarrow [\alpha\;\beta\;\gamma]=0$, if $\lambda=0, \frac{1}{2}$

Hence, α, β, γ are non-coplanar for all values of λ except two values 0 and $\frac{1}{2}$.

14. (a) $[\mathbf{a}\ \mathbf{b}\ \mathbf{c}] = \begin{vmatrix} 1 & 0 & -1 \\ x & 1 & 1-x \\ y & x & 1+x-y \end{vmatrix}$

$$= 1(1 + x - y - x + x^2) - 1(x^2 - y)$$
$$= 1 + x^2 - y - x^2 + y = 1$$

Hence, $[\mathbf{a}\ \mathbf{b}\ \mathbf{c}]$ does not depend upon neither x nor y.

15. (d) Since, the given points lie in a plane.

$$\therefore \quad \begin{vmatrix} a & a & c \\ 1 & 0 & 1 \\ c & c & b \end{vmatrix} = 0$$

On applying $C_1 \to C_1 - C_2$

$$\Rightarrow \quad \begin{vmatrix} 0 & a & c \\ 1 & 0 & 1 \\ 0 & c & b \end{vmatrix} = 0$$

$$\Rightarrow \quad -1(ab - c^2) = 0$$
$$\Rightarrow \quad c^2 = ab$$

Hence, c is GM of a and b.

16. (c) Given, $[\lambda(\mathbf{a} + \mathbf{b})\lambda^2\mathbf{b}\lambda\mathbf{c}] = [\mathbf{a}\ \mathbf{b} + \mathbf{c}\,\mathbf{b}]$

$$\Rightarrow \begin{vmatrix} \lambda(a_1 + b_1) & \lambda(a_2 + b_2) & \lambda(a_3 + b_3) \\ \lambda^2 b_1 & \lambda^2 b_2 & \lambda^2 b_3 \\ \lambda c_1 & \lambda c_2 & \lambda c_3 \end{vmatrix} = \begin{vmatrix} a_1 & a_2 & a_3 \\ b_1 + c_1 & b_2 + c_2 & b_3 + c_3 \\ b_1 & b_2 & b_3 \end{vmatrix}$$

$$\Rightarrow \lambda^4 \begin{vmatrix} a_1 + b_1 & a_2 + b_2 & a_3 + b_3 \\ b_1 & b_2 & b_3 \\ c_1 & c_2 & c_3 \end{vmatrix} = \begin{vmatrix} a_1 & a_2 & a_3 \\ b_1 + c_1 & b_2 + c_2 & b_3 + c_3 \\ b_1 & b_2 & b_3 \end{vmatrix}$$

[applying $R_1 \to R_1 - R_2$ in LHS and $R_2 \to R_2 - R_3$ in RHS]

$$\Rightarrow \quad \lambda^4 \begin{vmatrix} a_1 & a_2 & a_3 \\ b_1 & b_2 & b_3 \\ c_1 & c_2 & c_3 \end{vmatrix} = -\begin{vmatrix} a_1 & a_2 & a_3 \\ b_1 & b_2 & b_3 \\ c_1 & c_2 & c_3 \end{vmatrix}$$

$$\Rightarrow \quad \lambda^4 = 1$$

Hence, no real value of λ exists.

17. (b) Since, $\mathbf{c}$ lies in the plane of $\mathbf{a}$ and $\mathbf{b}$, then

$$\begin{vmatrix} 1 & 1 & 1 \\ 1 & -1 & 2 \\ x & x-2 & -1 \end{vmatrix} = 0$$

$$\Rightarrow 1(1 - 2x + 4) - 1(-1 - 2x) + 1(x - 2 + x) = 0$$
$$\Rightarrow \quad 5 - 2x + 1 + 2x + 2x - 2 = 0$$
$$\Rightarrow \quad 2x + 4 = 0$$
$$\therefore \quad x = -2$$

18. (a) We have, $[3\mathbf{u}\ p\mathbf{v}\ p\mathbf{w}] - [p\mathbf{v}\ \mathbf{w}\ q\mathbf{u}] - [2\mathbf{w}\ q\mathbf{v}\ q\mathbf{u}] = 0$

$$\Rightarrow \quad 3p^2[\mathbf{u}\ \mathbf{v}\ \mathbf{w}] - pq[\mathbf{v}\ \mathbf{w}\ \mathbf{u}] - 2q^2[\mathbf{w}\ \mathbf{v}\ \mathbf{u}] = 0$$
$$\Rightarrow \quad 3p^2[\mathbf{u}\ \mathbf{v}\ \mathbf{w}] - pq\,[\mathbf{u}\ \mathbf{v}\ \mathbf{w}] + 2q\,[\mathbf{u}\ \mathbf{v}\ \mathbf{w}] = 0$$
$$\Rightarrow \quad (3p^2 - pq + 2q^2)\,[\mathbf{u}\ \mathbf{v}\ \mathbf{w}] = 0$$
$$\Rightarrow \quad 3p^2 - pq + 2q^2 = 0 \qquad [\because [\mathbf{u}\ \mathbf{v}\ \mathbf{w}] \neq 0]$$
$$\Rightarrow \quad p^2 - \frac{1}{3}pq + \frac{2}{3}q^2 = 0 \Rightarrow \left(p - \frac{q}{6}\right)^2 + \frac{23}{36}q^2 = 0$$
$$\Rightarrow \quad p - \frac{q}{6} = 0, \quad q = 0 \quad \Rightarrow \quad p = 0, q = 0$$

Hence, there is exactly one value of (p, q).

19. (d) $\mathbf{a}$, $\mathbf{b}$, $\mathbf{c}$ are linearly dependent vectors.

$$\therefore \quad [\mathbf{a}\ \mathbf{b}\ \mathbf{c}] = 0$$

$$\Rightarrow \quad \begin{vmatrix} 1 & 1 & 1 \\ 4 & 3 & 4 \\ 1 & \alpha & \beta \end{vmatrix} = 0$$

$$\Rightarrow 1(3\beta - 4\alpha) - 1(43 - 4) + 1(4\alpha - 3) = 0$$
$$\Rightarrow \quad 3\beta - 4\alpha - 4\beta + 4 + 4\alpha - 3 = 0$$
$$\Rightarrow \quad -\beta + 1 = 0 \Rightarrow \beta = 1$$

Also,
$$|\mathbf{c}| = \sqrt{3}$$
$$\Rightarrow \quad \sqrt{1 + \alpha^2 + \beta^2} = \sqrt{3}$$
$$\Rightarrow \quad 1 + \alpha^2 + 1 = 3$$
$$\Rightarrow \quad \alpha^2 = 1$$
$$\therefore \quad \alpha = \pm 1$$

20. (b) We have, $\Delta = \begin{vmatrix} a_1 & a_2 & a_3 \\ b_1 & b_2 & b_3 \\ c_1 & c_2 & c_3 \end{vmatrix}^2$

$$= [\mathbf{a}\ \mathbf{b}\ \mathbf{c}]^2 = [(\mathbf{a} \times \mathbf{b}) \cdot \mathbf{c}]^2$$
$$= |\mathbf{a} \times \mathbf{b}|^2 |\mathbf{c}|^2 = |\mathbf{a}|^2 |\mathbf{b}| \sin^2 \frac{\pi}{6} |\mathbf{c}|^2$$
$$= \frac{1}{4}|\mathbf{a}||\mathbf{b}|^2$$
$$= \frac{1}{4}(a_1^2 + a_2^2 + a_3^2)(b_1^2 + b_2^2 + b_3^2)$$

21. (c) We have,
$$\mathbf{a} = \hat{\mathbf{i}} + \hat{\mathbf{j}} + \hat{\mathbf{k}}, \quad \mathbf{b} = \hat{\mathbf{i}} - \hat{\mathbf{j}} + \hat{\mathbf{k}}$$
and
$$\mathbf{c} = \hat{\mathbf{i}} + 2\hat{\mathbf{j}} - \hat{\mathbf{k}}$$

We know that

$$\begin{vmatrix} \mathbf{a}\cdot\mathbf{a} & \mathbf{a}\cdot\mathbf{b} & \mathbf{a}\cdot\mathbf{c} \\ \mathbf{b}\cdot\mathbf{a} & \mathbf{b}\cdot\mathbf{b} & \mathbf{b}\cdot\mathbf{c} \\ \mathbf{c}\cdot\mathbf{a} & \mathbf{c}\cdot\mathbf{b} & \mathbf{c}\cdot\mathbf{c} \end{vmatrix} = [\mathbf{a}\ \mathbf{b}\ \mathbf{c}]^2$$

$$= \begin{vmatrix} 1 & 1 & 1 \\ 1 & -1 & 1 \\ 1 & 2 & -1 \end{vmatrix}^2$$

$$= [1(1 - 2) - 1(-1 - 1) + 1(2 + 1)]^2$$
$$= [-1 + 2 + 3]^2 = [4]^2 = 16$$

22. (a)

(i) Given, $\mathbf{a} = 2\hat{\mathbf{i}} - 3\hat{\mathbf{j}} + 6\hat{\mathbf{k}}$

and $\mathbf{b} = -2\hat{\mathbf{i}} + 2\hat{\mathbf{j}} - \hat{\mathbf{k}}$

Now,
$$\lambda = \frac{\text{Projection of } \mathbf{a} \text{ on } \mathbf{b}}{\text{Projection of } \mathbf{b} \text{ on } \mathbf{a}} = \frac{\left(\dfrac{\mathbf{a}\cdot\mathbf{b}}{|\mathbf{b}|}\right)}{\left(\dfrac{\mathbf{b}\cdot\mathbf{a}}{|\mathbf{a}|}\right)} = \frac{|\mathbf{a}|}{|\mathbf{b}|}$$

$$= \frac{\sqrt{2^2 + (-3)^2 + 6^2}}{\sqrt{(-2)^2 + 2^2 + (-1)^2}} = \frac{\sqrt{4 + 9 + 36}}{\sqrt{4 + 4 + 1}} = \frac{7}{3}$$

$$\therefore \quad 3\lambda = 7$$

(ii) Given,
$$\mathbf{a} = \hat{\mathbf{i}} + 2\hat{\mathbf{j}} + 3\hat{\mathbf{k}}, \quad \mathbf{b} = -\hat{\mathbf{i}} + 2\hat{\mathbf{j}} + \hat{\mathbf{k}}$$
and
$$\mathbf{c} = 3\hat{\mathbf{i}} + \hat{\mathbf{j}}$$

Since, $\mathbf{a} + p\mathbf{b}$ is normal to $\mathbf{c}$.

$$\Rightarrow \quad (\mathbf{a} + p\mathbf{b}) \cdot \mathbf{c} = 0$$
$$\Rightarrow [(1 - p)\hat{\mathbf{i}} - (2 + 2p)\hat{\mathbf{j}} + (3 + p)\hat{\mathbf{k}}] \cdot (3\hat{\mathbf{i}} + \hat{\mathbf{j}}) = 0$$
$$\Rightarrow \quad 3(1 - p) + 2 + 2p = 0$$
$$\Rightarrow \quad 3 - 3p + 2 + 2p = 0$$
$$\therefore \quad p = 5$$

(iii) Given, $\mathbf{a} + \mathbf{b} + \mathbf{c} = 0$

$\Rightarrow \quad \mathbf{a} \times \mathbf{a} + \mathbf{a} \times \mathbf{b} + \mathbf{a} \times \mathbf{c} = \mathbf{a} \times 0$

$\Rightarrow \quad \mathbf{a} \times \mathbf{b} + \mathbf{a} \times \mathbf{c} = 0 \quad \ldots(\text{i})$

Also, $\mathbf{b} \times \mathbf{a} + \mathbf{b} \times \mathbf{b} + \mathbf{b} \times \mathbf{c} = \mathbf{b} \times 0$

$\Rightarrow \quad \mathbf{b} \times \mathbf{a} + \mathbf{b} \times \mathbf{c} = 0$

$\Rightarrow \quad -\mathbf{a} \times \mathbf{b} - \mathbf{c} \times \mathbf{b} = 0$

$\Rightarrow \quad \mathbf{a} \times \mathbf{b} + \mathbf{c} \times \mathbf{b} = 0 \quad \ldots(\text{ii})$

On adding Eqs. (i) and (ii), we get

$2(\mathbf{a} \times \mathbf{b}) + \mathbf{c} \times \mathbf{b} + \mathbf{a} \times \mathbf{c} = 0$

On comparing, we get

$\lambda = 2$

(iv) $\begin{vmatrix} p & q & r \\ q & r & p \\ r & p & q \end{vmatrix} = 0$

$\Rightarrow (p + q + r)(p^2 + q^2 + r^2 - pq - qr - rp) = 0$

$\therefore p^2 + q^2 + r^2 - pq - qr - rp = 0$

$\Rightarrow$ (iv) $\rightarrow$ (p)

Explanations for Q. Nos. (23-25)

Volume of parallelopiped $= [\mathbf{a}\ \mathbf{b}\ \mathbf{c}] = \mathbf{a} \cdot (\mathbf{b} \times \mathbf{c})$

Now, we have

$\mathbf{a} \times \mathbf{b} + \mathbf{b} \times \mathbf{c} = p\mathbf{a} + q\mathbf{b} + r\mathbf{c}$

$\Rightarrow \quad \mathbf{a} \cdot (\mathbf{a} \times \mathbf{b}) + \mathbf{a} \cdot (\mathbf{b} \times \mathbf{c}) = p(\mathbf{a} \cdot \mathbf{a}) + q(\mathbf{a} \cdot \mathbf{b}) + r(\mathbf{a} \cdot \mathbf{c})$

$\Rightarrow [\mathbf{a}\ \mathbf{a}\ \mathbf{b}] + [\mathbf{a}\ \mathbf{b}\ \mathbf{c}] = p|\mathbf{a}|^2 + q|\mathbf{a}||\mathbf{b}|\cos\theta + r|\mathbf{a}||\mathbf{c}|\cos\theta$

$\Rightarrow \quad [\mathbf{a}\ \mathbf{b}\ \mathbf{c}] = p + q\cos\theta + r\cos\theta = p + (q + r)\cos\theta$

$\therefore$ Volume of parallelopiped $= p + (q + r)\cos\theta$

Taking dot products with **a**, **b**, **c** respectively with given equation

$[\mathbf{a}\ \mathbf{b}\ \mathbf{c}] = p + (q + r)\cos\theta \quad \ldots(\text{i})$

$0 = (p + r)\cos\theta + q \quad \ldots(\text{ii})$

$[\mathbf{a}\ \mathbf{b}\ \mathbf{c}] = (p + q)\cos\theta + r \quad \ldots(\text{iii})$

also, $[\mathbf{a}\ \mathbf{b}\ \mathbf{c}]^2 = \begin{vmatrix} 1 & \cos\theta & \cos\theta \\ \cos\theta & 1 & \cos\theta \\ \cos\theta & \cos\theta & 1 \end{vmatrix}$

$\Rightarrow \quad [\mathbf{a}\ \mathbf{b}\ \mathbf{c}]^2 = (1 - \cos\theta)^2(1 + 2\cos\theta)$

$\therefore \quad v = |[\mathbf{a}\ \mathbf{b}\ \mathbf{c}]| = |1 - \cos\theta|\,|\sqrt{1 + 2\cos\theta}|$

$= 2\sin^2\frac{\theta}{2} \cdot |\sqrt{1 + 2\cos\theta}|$

From (i) and (iii), $p = r$ substituting in (ii), we get

$2p\cos\theta + q = 0$

$\Rightarrow \frac{q}{p} + 2\cos\theta = 0$

23. (a) **24.** (c) **25.** (b)

26. (3) We have,

$\mathbf{a} \cdot \mathbf{p} = \mathbf{a} \cdot \frac{(\mathbf{b} \times \mathbf{c})}{[\mathbf{a}\ \mathbf{b}\ \mathbf{c}]} = \frac{1}{[\mathbf{a}\ \mathbf{b}\ \mathbf{c}]}\mathbf{a} \cdot (\mathbf{b} \times \mathbf{c}) = \frac{[\mathbf{a}\ \mathbf{b}\ \mathbf{c}]}{[\mathbf{a}\ \mathbf{b}\ \mathbf{c}]} = 1$

and $\mathbf{a} \cdot \mathbf{q} = \mathbf{a} \cdot \frac{\mathbf{c} \times \mathbf{a}}{[\mathbf{a}\ \mathbf{b}\ \mathbf{c}]} = \frac{1}{[\mathbf{a}\ \mathbf{b}\ \mathbf{c}]}\mathbf{a} \cdot (\mathbf{c} \times \mathbf{a}) = 0$

Similarly, we have

$\mathbf{b} \cdot \mathbf{q} = \mathbf{c} \cdot \mathbf{r} = 1$

and $\mathbf{a} \cdot \mathbf{r} = \mathbf{b} \cdot \mathbf{p} = \mathbf{c} \cdot \mathbf{q} = \mathbf{c} \cdot \mathbf{p} = \mathbf{a} \cdot \mathbf{r} = 0$

$\therefore (\mathbf{a} + \mathbf{b}) \cdot \mathbf{p} + (\mathbf{b} + \mathbf{c}) \cdot \mathbf{q} + (\mathbf{c} + \mathbf{a}) \cdot \mathbf{r}$

$= \mathbf{a} \cdot \mathbf{p} + \mathbf{b} \cdot \mathbf{p} + \mathbf{b} \cdot \mathbf{q} + \mathbf{c} \cdot \mathbf{q} + \mathbf{c} \cdot \mathbf{r} + \mathbf{a} \cdot \mathbf{r}$

$= 1 + 1 + 1 = 3$

27. (8) We have, $[\mathbf{u}\ \mathbf{v}\ \mathbf{w}] = \mathbf{u} \cdot (\mathbf{v} \times \mathbf{w})$

$\Rightarrow \quad [\mathbf{u}\ \mathbf{v}\ \mathbf{w}] \le |\mathbf{u}||\mathbf{v} \times \mathbf{w}| \quad [\because \mathbf{a} \cdot \mathbf{b} \le |\mathbf{a}||\mathbf{b}|]$

$\Rightarrow \quad [\mathbf{u}\ \mathbf{v}\ \mathbf{w}] \le [\mathbf{v} \times \mathbf{w}] \quad [\because |\mathbf{u}| = 1]$

Now, $\mathbf{v} \times \mathbf{w} = \begin{vmatrix} \hat{\mathbf{i}} & \hat{\mathbf{j}} & \hat{\mathbf{k}} \\ 2 & 1 & -1 \\ 1 & 0 & 3 \end{vmatrix} = 3\hat{\mathbf{i}} - 7\hat{\mathbf{j}} - \hat{\mathbf{k}}$

$\therefore \quad |\mathbf{v} \times \mathbf{w}| = \sqrt{9 + 49 + 1} = \sqrt{59}$

Hence, maximum value of $[\mathbf{u}\ \mathbf{v}\ \mathbf{w}] = \sqrt{59}$

On comparing, we get

$\lambda = 59$

$\therefore \quad \lambda - 51 = 8$

28. (3) Let the volume of parallelopiped be V.

$\therefore \quad V = \begin{vmatrix} 1 & a & 1 \\ 0 & 1 & a \\ a & 0 & 1 \end{vmatrix} = (1 - 0) - a(0 - a^2) + 1(0 - a)$

$= a^3 - a + 1$

Now, $\frac{dV}{da} = 3a^2 - 1$

and $\frac{d^2V}{da^2} = 6a$

For minimum, $\frac{dV}{da} = 0 \Rightarrow 3a^2 - 1 = 0$

$\Rightarrow \quad a = \pm\frac{1}{\sqrt{3}}$

At $a = \frac{1}{\sqrt{3}}$, $\frac{d^2V}{da^2} = \frac{6}{\sqrt{3}}$ = Positive

Hence, V is minimum at $a = \frac{1}{\sqrt{3}}$.

$\therefore \quad \lambda = 3$

29. (5) In ΔPQR and ΔPSQ, we have

$\mathbf{PQ} + \mathbf{QR} = \mathbf{PR}$ and $\mathbf{PS} + \mathbf{SQ} = \mathbf{PQ}$

$\Rightarrow \quad \mathbf{PQ} + \mathbf{QR} = \mathbf{PR}$ and $\mathbf{PQ} - \mathbf{PS} = \mathbf{SQ}$

$\Rightarrow \quad \mathbf{PQ} + \mathbf{PS} = \mathbf{PR}$ and $\mathbf{PQ} - \mathbf{PS} = \mathbf{SQ}$

$\Rightarrow \quad \mathbf{PQ} = \frac{\mathbf{PR} + \mathbf{SQ}}{2}$ and $\mathbf{PS} = \frac{\mathbf{PR} - \mathbf{SQ}}{2}$

$\Rightarrow \quad \mathbf{PQ} = 2\hat{\mathbf{i}} - \hat{\mathbf{j}} - 3\hat{\mathbf{k}}$ and $\mathbf{PS} = \hat{\mathbf{i}} + 2\hat{\mathbf{j}} + \hat{\mathbf{k}}$

Thus, we obtain $\mathbf{PQ} = 2\hat{\mathbf{i}} - \hat{\mathbf{j}} - 3\hat{\mathbf{k}}$,

$\mathbf{PS} = \hat{\mathbf{i}} + 2\hat{\mathbf{j}} + \hat{\mathbf{k}}$

and $\mathbf{PT} = \hat{\mathbf{i}} + 2\hat{\mathbf{j}} + 3\hat{\mathbf{k}}$

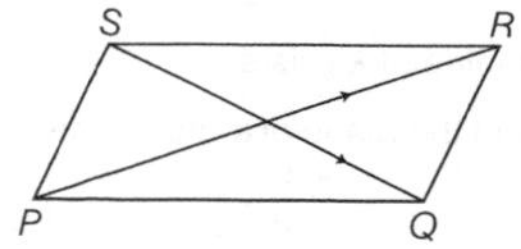

Now, $[\mathbf{PT}\ \mathbf{PQ}\ \mathbf{PS}] = \begin{vmatrix} 1 & 2 & 3 \\ 2 & -1 & -3 \\ 1 & 2 & 1 \end{vmatrix} = 5 - 10 + 15 = 10$

$\Rightarrow \quad \lambda + 5 = 10$

Thus, $(\lambda - 5) = 10 - 5$

$\therefore \quad \lambda = 5$

30. Let (1, 1, 1), (–1 , 1, 1), (1, – 1, 1),, (–1, – 1, 1) be vectors **a**, **b**, **c**, **d** and rest of vectors –**a**, – **b**, – **c**, – **d** and let as find the number of ways of selecting coplanar vectors.

Observe that out of any three coplanar vectors two will be collinear (antiparallel)

Number of ways of selecting the antiparallel pair = 4.

Number of ways of selecting the third vectors = 6 Total = 24.

Total number of ways of selecting the vector = 24

Number of non-coplanar selections $= {}^8C_3 - 24 = 32 = 2^5$

$\Rightarrow \quad p = 5$

DPP-5 Vector Triple Product

1. (a) Let $\mathbf{a} = a_1\hat{\mathbf{i}} + a_2\hat{\mathbf{j}} + a_3\hat{\mathbf{k}}$.

Then, $\hat{\mathbf{i}} \times (\mathbf{a} \times \hat{\mathbf{i}}) + \hat{\mathbf{j}} \times (\mathbf{a} \times \hat{\mathbf{j}}) + \hat{\mathbf{k}} \times (\mathbf{a} \times \hat{\mathbf{k}})$

$= \{(\hat{\mathbf{i}}\cdot\hat{\mathbf{i}})\mathbf{a} - (\hat{\mathbf{i}}\cdot\mathbf{a})\hat{\mathbf{i}}\} + \{(\hat{\mathbf{j}}\cdot\hat{\mathbf{j}})\mathbf{a} - (\hat{\mathbf{j}}\cdot\mathbf{a})\hat{\mathbf{j}}\} + \{(\hat{\mathbf{k}}\cdot\hat{\mathbf{k}})\mathbf{a} - (\hat{\mathbf{k}}\cdot\mathbf{a})\hat{\mathbf{k}}\}$

$= \{\mathbf{a} - (\hat{\mathbf{i}} - \mathbf{a})\hat{\mathbf{i}}\} + \{\mathbf{a} - (\hat{\mathbf{j}}\cdot\mathbf{a})\hat{\mathbf{j}}\} + \{\mathbf{a} - (\hat{\mathbf{k}} - \mathbf{a})\hat{\mathbf{k}}\}$

$= 3\mathbf{a} - \{(\hat{\mathbf{i}} - \mathbf{a})\hat{\mathbf{i}} + (\hat{\mathbf{j}}\cdot\mathbf{a})\hat{\mathbf{j}} + (\hat{\mathbf{k}}\cdot a)\hat{\mathbf{k}}\}$

$[\because \hat{\mathbf{i}}\cdot\mathbf{a} = a_1,\ \hat{\mathbf{j}}\cdot\mathbf{a} = a_2 \text{ and } \hat{\mathbf{k}}\cdot\mathbf{a} = a_3]$

$= 3\mathbf{a} - (a_1\hat{\mathbf{i}} + a_2\hat{\mathbf{j}} + a_3\hat{\mathbf{k}})$

$= 3\mathbf{a} - \mathbf{a} = 2\mathbf{a}$

2. (c) Given, $|\mathbf{a}| = 1, |\mathbf{b}| = 1$ and $|\mathbf{c}| = 2$

Also, $\mathbf{a} \times (\mathbf{a} \times \mathbf{c}) + \mathbf{b} = 0$

$\Rightarrow (\mathbf{a}\cdot\mathbf{c})\mathbf{a} - (\mathbf{a}\cdot\mathbf{a})\mathbf{c} + \mathbf{b} = 0$

$\Rightarrow (\mathbf{a}\cdot\mathbf{c})\mathbf{a} - \mathbf{c} + \mathbf{b} = 0 \quad [\because \mathbf{a}\cdot\mathbf{a} = |\mathbf{a}|^2 = 1]$

$\Rightarrow (\mathbf{a}\cdot\mathbf{c})\mathbf{a} - \mathbf{c} = -\mathbf{b}$

$\Rightarrow |(\mathbf{a}\cdot\mathbf{c})\mathbf{a} - \mathbf{c}| = |-\mathbf{b}|$

$\Rightarrow |(\mathbf{a}\cdot\mathbf{c})\mathbf{a} - \mathbf{c}|^2 = |\mathbf{b}|^2$

$\Rightarrow |(\mathbf{a}\cdot\mathbf{c})\mathbf{a}|^2 + |\mathbf{c}|^2 - 2\{(\mathbf{a}\cdot\mathbf{c})\mathbf{a}\cdot\mathbf{c}\} = |\mathbf{b}|^2$

$\Rightarrow (\mathbf{a}\cdot\mathbf{c})|\mathbf{a}|^2 + |\mathbf{c}|^2 - 2(\mathbf{a}\cdot\mathbf{c})(\mathbf{a}\cdot\mathbf{c}) = |\mathbf{b}|^2$

$\Rightarrow (\mathbf{a}\cdot\mathbf{c})^2\{|\mathbf{a}|^2 - 2\} + |\mathbf{c}|^2 = |\mathbf{b}|^2$

$\Rightarrow -(\mathbf{a}\cdot\mathbf{c})^2 + 4 = 1$

$\Rightarrow \mathbf{a}\cdot\mathbf{c} = \pm\sqrt{3}$

$\Rightarrow |\mathbf{a}|\cdot|\mathbf{c}|\cos\theta = \sqrt{3}$

where, θ is an acute angle between $\mathbf{a}$ and $\mathbf{c}$.

$\Rightarrow \cos\theta = \frac{\sqrt{3}}{2}$

$\therefore \theta = \frac{\pi}{6}$

3. (a) $\mathbf{a} \times (\mathbf{a} \times \mathbf{b}) = (\mathbf{a}\cdot\mathbf{b})\mathbf{a} - (\mathbf{a}\cdot\mathbf{a})\mathbf{b}$

$= 3(2\hat{\mathbf{i}} + \hat{\mathbf{j}} + \hat{\mathbf{k}}) - 6(\hat{\mathbf{i}} + 2\hat{\mathbf{j}} - \hat{\mathbf{k}})$

$= -9\hat{\mathbf{j}} + 9\hat{\mathbf{k}}$

$\therefore$ Required unit vector $= \frac{\mathbf{a} \times (\mathbf{a} \times \mathbf{b})}{|\mathbf{a} \times (\mathbf{a} \times \mathbf{b})|}$

$= \pm\frac{1}{\sqrt{2}}(-\hat{\mathbf{j}} + \hat{\mathbf{k}})$

4. (b) $|\mathbf{c} - \mathbf{a}| = 2\sqrt{2}$

$\Rightarrow |\mathbf{c}|^2 + |\mathbf{a}|^2 - 2\mathbf{a}\cdot\mathbf{c} = 8$

$\Rightarrow |\mathbf{c}|^2 + (\sqrt{9})^2 - 2|\mathbf{c}| = 8$

$\Rightarrow |\mathbf{c}|^2 - 2|\mathbf{c}| + 1 = 0$

$\Rightarrow (|\mathbf{c}| - 1)^2 = 0$

$\Rightarrow |\mathbf{c}| = 1$

Now, $\mathbf{a} \times \mathbf{b} = 2\hat{\mathbf{i}} + 2\hat{\mathbf{j}} + \hat{\mathbf{k}}$

$\therefore |\mathbf{a} \times \mathbf{b}| = \sqrt{4 + 4 + 1} = 3$

$\therefore |(\mathbf{a} \times \mathbf{b}) \times \mathbf{c}| = |\mathbf{a} \times \mathbf{b}||\mathbf{c}|\sin 30^\circ$

$= 3 \times 1 \times \frac{1}{2} = \frac{3}{2}$

5. (d) Given, $\frac{1}{3}|\mathbf{b}||\mathbf{c}|\mathbf{a} = (\mathbf{a} \times \mathbf{b}) \times \mathbf{c} = (\mathbf{c}\cdot\mathbf{a})\mathbf{b} - (\mathbf{b}\cdot\mathbf{c})\mathbf{a}$

$\therefore \mathbf{b}\cdot\mathbf{c} = -\frac{1}{3}|\mathbf{b}||\mathbf{c}|$

$\Rightarrow \cos\theta = -\frac{1}{3}$

$\Rightarrow \sin\theta = \frac{2\sqrt{2}}{3}$

6. (c) We have,

$(\mathbf{a} \times \mathbf{b}) \times \mathbf{c} = \mathbf{a} \times (\mathbf{b} \times \mathbf{c})$

$\Rightarrow -\{\mathbf{c} \times (\mathbf{a} \times \mathbf{b})\} = \mathbf{a} \times (\mathbf{b} \times \mathbf{c})$

$\Rightarrow -\{(\mathbf{c}\cdot\mathbf{b})\mathbf{a} - (\mathbf{c}\cdot\mathbf{a})\mathbf{b}\} = (\mathbf{a}\cdot\mathbf{c})\mathbf{b} - (\mathbf{a}\cdot\mathbf{b})\mathbf{c}$

$\Rightarrow -(\mathbf{c}\cdot\mathbf{b})\mathbf{a} + (\mathbf{a}\cdot\mathbf{c})\mathbf{b} = (\mathbf{a}\cdot\mathbf{c})\mathbf{b} - (\mathbf{a}\cdot\mathbf{b})\mathbf{c}$

$\Rightarrow -(\mathbf{c}\cdot\mathbf{b})\mathbf{a} = (\mathbf{a}\cdot\mathbf{b})\mathbf{c}$

$\Rightarrow (\mathbf{c}\cdot\mathbf{b})\mathbf{a} = (\mathbf{a}\cdot\mathbf{b})\mathbf{c}$

$\Rightarrow \mathbf{a} = \frac{(\mathbf{a}\cdot\mathbf{b})}{(\mathbf{c}\cdot\mathbf{b})}\mathbf{c}$

$\Rightarrow \mathbf{a} = \lambda\mathbf{c}$,

where $\lambda = \frac{\mathbf{a}\cdot\mathbf{b}}{\mathbf{c}\cdot\mathbf{b}}$.

$\therefore$ $\mathbf{a}$ and $\mathbf{c}$ are parallel.

7. (a) We have,

$\mathbf{a} \times \mathbf{b} + \mathbf{c} = 0$

$\Rightarrow \mathbf{a} \times (\mathbf{a} \times \mathbf{b}) + \mathbf{a} \times \mathbf{c} = 0$

$\Rightarrow (\mathbf{a}\cdot\mathbf{b})\mathbf{a} - (\mathbf{a}\cdot\mathbf{a})\mathbf{b} + \mathbf{a} \times \mathbf{c} = 0$

$\Rightarrow 3\mathbf{a} - 2\mathbf{b} + \mathbf{a} \times \mathbf{c} = 0$

$\Rightarrow 2\mathbf{b} = 3\mathbf{a} + \mathbf{a} \times \mathbf{c}$

$\Rightarrow 2\mathbf{b} = 3\hat{\mathbf{j}} - 3\hat{\mathbf{k}} - 2\hat{\mathbf{i}} - \hat{\mathbf{j}} - \hat{\mathbf{k}}$

$= -2\hat{\mathbf{i}} + 2\hat{\mathbf{j}} - 4\hat{\mathbf{k}}$

$\therefore \mathbf{b} = -\hat{\mathbf{i}} + \hat{\mathbf{j}} - 2\hat{\mathbf{k}}$

8. (a) Since, the given vectors are mutually orthogonal, therefore

$\mathbf{a}\cdot\mathbf{b} = 2 - 4 + 2 = 0$

$\mathbf{a}\cdot\mathbf{c} = \lambda - 1 + 2\mu = 0 \quad \ldots(i)$

$\mathbf{b}\cdot\mathbf{c} = 2\lambda + 4 + \mu = 0 \quad \ldots(ii)$

On solving Eqs (i) and (ii), we get

$\mu = 2$

and $\lambda = -3$

Hence, $(\lambda, \mu) = (-3, 2)$

9. (d) Given, $\mathbf{a} = \frac{1}{\sqrt{10}}(3\hat{\mathbf{i}} + \hat{\mathbf{k}})$

and $\mathbf{b} = \frac{1}{7}(2\hat{\mathbf{i}} + 3\hat{\mathbf{j}} - 6\hat{\mathbf{k}})$

$\therefore (2\mathbf{a} - \mathbf{b})\cdot\{(\mathbf{a} \times \mathbf{b}) \times (\mathbf{a} + 2\mathbf{b})\}$

$= (2\mathbf{a} - \mathbf{b})\cdot\{(\mathbf{a} \times \mathbf{b}) \times \mathbf{a} + (\mathbf{a} \times \mathbf{b}) \times 2\mathbf{b}\}$

$= (2\mathbf{a} - \mathbf{b})\{(\mathbf{a}\cdot\mathbf{a})\mathbf{b} + (\mathbf{b}\cdot\mathbf{a})\mathbf{a} + 2(\mathbf{a}\cdot\mathbf{b})\mathbf{b} - 2(\mathbf{b}\cdot\mathbf{b})\mathbf{a}\}$

$= (2\mathbf{a} - \mathbf{b})\cdot\{1(\mathbf{b})\cdot(0)\mathbf{a} + 2(0)\mathbf{b} - 2(1)\mathbf{a}\}$

$[\because \mathbf{a}\cdot\mathbf{b} = 0 \text{ and } \mathbf{a}\cdot\mathbf{a} = \mathbf{b}\cdot\mathbf{b} = 1]$

$= (2\mathbf{a} - \mathbf{b})(\mathbf{b} - 2\mathbf{a}) = -(4|\mathbf{a}|^2 + 4\mathbf{a}\cdot\mathbf{b} + |\mathbf{b}|^2$

$= -\{4 + 0 + 1\} = -5$

10. (c) Given, $\mathbf{a}\cdot\mathbf{b} \neq 0$, $\mathbf{a}\cdot\mathbf{d} = 0 \quad \ldots(i)$

and $\mathbf{b} \times \mathbf{c} = \mathbf{b} \times \mathbf{d}$

$\Rightarrow \mathbf{b} \times (\mathbf{c} - \mathbf{d}) = 0$

$\Rightarrow \mathbf{b} \,||\, (\mathbf{c} - \mathbf{d})$

$\Rightarrow \mathbf{c} - \mathbf{d} = \lambda\mathbf{b}$

$\Rightarrow \mathbf{d} = \mathbf{c} - \lambda\mathbf{b} \quad \ldots(ii)$

On taking dot product with $\mathbf{a}$, we get

$\mathbf{a}\cdot\mathbf{d} = \mathbf{a}\cdot\mathbf{c} - \lambda\mathbf{a}\cdot\mathbf{b}$

$\Rightarrow 0 = \mathbf{a}\cdot\mathbf{c} - \lambda(\mathbf{a}\cdot\mathbf{b}) \quad$ [from Eq. (i)]

$\Rightarrow \lambda = \frac{\mathbf{a}\cdot\mathbf{c}}{\mathbf{a}\cdot\mathbf{b}}$

$\therefore \mathbf{d} = \mathbf{c} - \left(\frac{\mathbf{a}\cdot\mathbf{c}}{\mathbf{a}\cdot\mathbf{b}}\right)\mathbf{b}$

11. (a) Given, $\mathbf{a} = 2\hat{i} - \hat{j} + \hat{k}, \mathbf{b} = \hat{i} + 2\hat{j} - \hat{k}$ and $\mathbf{c} = \hat{i} + \hat{j} - 2\hat{k}$

Now, we have $\mathbf{b} + \lambda\mathbf{c} = (1+\lambda)\hat{i} + (2+\lambda)\hat{j} + (-1-2\lambda)\hat{k}$

$\therefore$ Projection of $(\mathbf{b} + \lambda\mathbf{c})$ on $\mathbf{a}$

$$= \left|\frac{(\mathbf{b}+\lambda\mathbf{c})\cdot\mathbf{a}}{|\mathbf{a}|}\right| = \sqrt{\frac{2}{3}} \quad \text{[given]}$$

$$\Rightarrow \left|\frac{2(1-\lambda) - (2+\lambda) + (-1-2\lambda)}{\sqrt{4+1+1}}\right| = \sqrt{\frac{2}{3}}$$

$$\Rightarrow \left|\frac{-\lambda-1}{\sqrt{6}}\right| = \frac{\sqrt{2}}{\sqrt{3}}$$

$$\Rightarrow \lambda + 1 = 2$$

$$\Rightarrow \lambda = 1$$

$$\therefore \mathbf{b} + \lambda\mathbf{c} = 2\hat{i} + 3\hat{j} - 3\hat{k}$$

12. (c) Given, $\mathbf{a} = \hat{i} + \hat{j} + \hat{k}, \mathbf{b} = \hat{i} - \hat{j} + \hat{k}$ and $\mathbf{c} = \hat{i} - \hat{j} - \hat{k}$

Now, we have $\mathbf{a} + \lambda\mathbf{b} = (1+\lambda)\hat{i} + (1-\lambda)\hat{j} + (1+\lambda)\hat{k}$

$\therefore$ Projection of $(\mathbf{a} + \lambda\mathbf{b})$ on $\mathbf{c}$

$$= \left|\frac{(\mathbf{a}+\lambda\mathbf{b})\cdot\mathbf{c}}{|\mathbf{c}|}\right| = \frac{1}{\sqrt{3}} \quad \text{[given]}$$

$$\Rightarrow \left|\frac{(1+\lambda) - (1-\lambda) - (1+\lambda)}{\sqrt{1+1+1}}\right| = \frac{1}{\sqrt{3}}$$

$$\Rightarrow \left|\frac{1-\lambda-1+\lambda-1-\lambda}{\sqrt{3}}\right| = \frac{1}{\sqrt{3}}$$

$$\Rightarrow \left|\frac{\lambda-1}{\sqrt{3}}\right| = \frac{1}{\sqrt{3}}$$

$$\Rightarrow \lambda - 1 = 1$$

$$\Rightarrow \lambda = 2$$

$$\therefore \mathbf{a} + \lambda\mathbf{b} = 3\hat{i} - \hat{j} + 3\hat{k}$$

13. (a) We have, $\mathbf{a}\cdot\hat{\mathbf{d}} = 0$ and $[\mathbf{b}\ \mathbf{c}\ \hat{\mathbf{d}}] = 0$

$\Rightarrow \mathbf{a} \perp \hat{\mathbf{d}}$ and $\mathbf{b}, \mathbf{c}, \hat{\mathbf{d}}$ are coplanar.

$\Rightarrow \hat{\mathbf{d}} \perp \mathbf{a}$ and $\hat{\mathbf{d}}$ lies in the plane of $\mathbf{b}$ and $\mathbf{c}$, we know that the vector $\mathbf{r} = \mathbf{a} \times (\mathbf{b} \times \mathbf{c})$ is perpendicular to $\mathbf{a}$ and lies in the plane of $\mathbf{b}$ and $\mathbf{c}$.

$$\therefore \hat{\mathbf{d}} = \pm\frac{\mathbf{r}}{|\mathbf{r}|}$$

Now, $\mathbf{r} = \mathbf{a} \times (\mathbf{b} \times \mathbf{c})$

$$\Rightarrow \mathbf{r} = (\mathbf{a}\cdot\mathbf{c})\mathbf{b} - (\mathbf{a}\cdot\mathbf{b})\mathbf{c}$$

$$\Rightarrow \mathbf{r} = -(\hat{j} - \hat{k}) + (\hat{k} - \hat{i}) = -\hat{i} - \hat{j} + 2\hat{k}$$

$$\therefore \hat{\mathbf{d}} = \pm\frac{\mathbf{r}}{|\mathbf{r}|} = \pm\frac{(-\hat{i} - \hat{j} + 2\hat{k})}{\sqrt{1+1+4}} = \pm\frac{\hat{i} + \hat{j} - 2\hat{k}}{\sqrt{6}}$$

14. (a) We have, $\mathbf{a} \times (\mathbf{b} \times \mathbf{c}) = \dfrac{\mathbf{b}+\mathbf{c}}{\sqrt{2}}$

$$\Rightarrow (\mathbf{a}\cdot\mathbf{c})\mathbf{b} - (\mathbf{a}\cdot\mathbf{b})\mathbf{c} = \frac{1}{\sqrt{2}}\mathbf{b} + \frac{1}{\sqrt{2}}\mathbf{c}$$

$$\Rightarrow \left\{\mathbf{a}\cdot\mathbf{c} - \frac{1}{\sqrt{2}}\right\}\mathbf{b} - \left\{\mathbf{a}\cdot\mathbf{b} + \frac{1}{\sqrt{2}}\right\}\mathbf{c} = \mathbf{0}$$

$$\Rightarrow o\mathbf{a} + \left\{\mathbf{a}\cdot\mathbf{c} - \frac{1}{\sqrt{2}}\right\}\mathbf{b} + \left\{-(\mathbf{a}\cdot\mathbf{b}) - \frac{1}{\sqrt{2}}\right\}\mathbf{c} = \mathbf{o}$$

$$\Rightarrow \mathbf{a}\cdot\mathbf{c} - \frac{1}{\sqrt{2}} = 0$$

and $-(\mathbf{a}\cdot\mathbf{b}) - \dfrac{1}{\sqrt{2}} = 0$ [$\because$ **a**, **b**, **c** are non-coplanar]

Now, $\mathbf{a}\cdot\mathbf{b} = -\dfrac{1}{\sqrt{2}}$

$$\Rightarrow |\mathbf{a}||\mathbf{b}|\cos\theta = -\frac{1}{\sqrt{2}} \Rightarrow \cos\theta = -\frac{1}{\sqrt{2}}$$

$$\therefore \theta = \frac{3\pi}{4}$$

15. (b) We have,

$$\mathbf{p} \times \{(\mathbf{x}-\mathbf{q}) \times \mathbf{p}\} + \mathbf{q}\{(\mathbf{x}-\mathbf{r}) \times \mathbf{q}\} + \mathbf{r} \times \{(\mathbf{x}-\mathbf{p}) \times \mathbf{r}\} = 0$$

$$\Rightarrow (\mathbf{p}\cdot\mathbf{p})(\mathbf{x}-\mathbf{q}) - \{\mathbf{p}\cdot(\mathbf{x}-\mathbf{q})\}\mathbf{p} + (\mathbf{q}\cdot\mathbf{q})(\mathbf{x}-\mathbf{r})$$
$$- \{\mathbf{q}\cdot(\mathbf{x}-\mathbf{r})\}\mathbf{q} + (\mathbf{r}\cdot\mathbf{r})(\mathbf{x}-\mathbf{p}) - \{\mathbf{r}\cdot(\mathbf{x}-\mathbf{p})\}\mathbf{r} = \mathbf{0}$$

$$\Rightarrow (|\mathbf{p}|^2 + |\mathbf{q}|^2 + |\mathbf{r}|^2)\mathbf{x} - (|\mathbf{p}|^2\mathbf{q} + |\mathbf{q}|^2\mathbf{r}$$
$$+ |\mathbf{r}|^2\mathbf{p}) - \{(\mathbf{p}\cdot\mathbf{x})\mathbf{p} + (\mathbf{q}\cdot\mathbf{x})\mathbf{q} + (\mathbf{r}\cdot\mathbf{x})\mathbf{r}\} = 0$$

$$\Rightarrow 3\lambda^2\mathbf{x} - \lambda^2(\mathbf{p}+\mathbf{q}+\mathbf{r}) - \{(\mathbf{p}\cdot\mathbf{x})\mathbf{p} + (\mathbf{q}\cdot\mathbf{x})\mathbf{q} + (\mathbf{r}\cdot\mathbf{x})\mathbf{r}\} = 0$$

where $\lambda = |\mathbf{p}| = |\mathbf{q}| = |\mathbf{r}|$ ...(i)

$$\Rightarrow 3\lambda^2\mathbf{x} - \lambda^2(\mathbf{p}+\mathbf{q}+\mathbf{r}) - \lambda^2\mathbf{x} = 0$$

$$\left[\because \mathbf{x} = \frac{(\mathbf{p}\cdot\mathbf{x})\mathbf{p}}{|\mathbf{p}|^2} + \frac{(\mathbf{q}\cdot\mathbf{x})\mathbf{q}}{|\mathbf{q}|^2} + \frac{(\mathbf{r}\cdot\mathbf{x})\mathbf{r}}{|\mathbf{r}|^2}\right]$$

$$\Rightarrow \mathbf{x} = \frac{1}{2}(\mathbf{p}+\mathbf{q}+\mathbf{r})$$

16. (c) Refer to solution of Q. 4.

17. (a,b,c)

$$\mathbf{a} \times (\mathbf{b} \times \mathbf{c}) = (\mathbf{a}\cdot\mathbf{c})\mathbf{b} - (\mathbf{a}\cdot\mathbf{b})\mathbf{c}$$
$$= (xz + yx + yz)(y\hat{i} + z\hat{j} + x\hat{k})$$
$$- (xy + yz + zx)(z\hat{i} + x\hat{j} + y\hat{k})$$
$$= (xy + yz + zx)[(y-z)\hat{i} + (z-x)\hat{j} + (x-y)\hat{k}]$$

Clearly 1 to $\hat{i} + \hat{j} + \hat{k}$ and also to $(y+z)\hat{i} + (z+x)\hat{j} + (x+y)\hat{k}$, as dot product are zero. clearly parallel to $(y-z)\hat{i} + (z-x)\hat{j} + (x-y)\hat{k}$

18. (b,c,d)

(a) $\mathbf{a} \times (\mathbf{b} \times \mathbf{c}) + \mathbf{b} \times (\mathbf{c} \times \mathbf{a}) + \mathbf{c} \times (\mathbf{a} \times \mathbf{b}) = 0$

$\Rightarrow$ vector are coplanar, so do not form RHS.

(b) $(\mathbf{a} \times \mathbf{b}) \times \mathbf{c}, \mathbf{a} \times \mathbf{b}, \mathbf{c}$ in that order form RHS.

$\Rightarrow \mathbf{c}, (\mathbf{a} \times \mathbf{b}) \times \mathbf{c}, \mathbf{a} \times \mathbf{b}$ also forms RHS as they are in same cyclic order.

(c) $\mathbf{a} + \mathbf{b} + \mathbf{c} = 0 \Rightarrow |\mathbf{a} + \mathbf{b} + \mathbf{c}|^2 = 0$

$\Rightarrow a^2 + b^2 + c^2 = -2(\mathbf{a}\cdot\mathbf{b} + \mathbf{b}\cdot\mathbf{c} + \mathbf{c}\cdot\mathbf{a})$

$\Rightarrow \mathbf{a}\cdot\mathbf{b} + \mathbf{b}\cdot\mathbf{c} + \mathbf{c}\cdot\mathbf{a} < 0$

(d) $a + b + c = 0$

$\Rightarrow \mathbf{a} \times \mathbf{b} = \mathbf{b} \times \mathbf{c} = \mathbf{c} \times \mathbf{a}$

19. (a, b, c, d)

$$\mathbf{a}\cdot\mathbf{b} = 0, \mathbf{c}\cdot\mathbf{a} = \mathbf{c}\cdot\mathbf{b} = \cos\alpha.$$

Take dot products with **a**, **b** and **c** respectively

$$\Rightarrow l = m, l^2 + m^2 + n^2 = 1,$$
$$n^2 = -\cos 2\alpha,$$
$$m^2 = \frac{1 + \cos 2\alpha}{2}$$

20. (a,b,c,d) Since **a**, **b** and $\mathbf{a} \times \mathbf{b}$ are non-coplanar vectors.

$\therefore$ $\mathbf{r} = x\mathbf{a} + y\mathbf{b} + z(\mathbf{a} \times \mathbf{b})$ for some scalars x, y, z ...(i)

Now,
$$\mathbf{a} = \mathbf{r} \times \mathbf{b}$$
$$\mathbf{a} = \{x\mathbf{a} + y\mathbf{b} + z(\mathbf{a} \times \mathbf{b})\} \times \mathbf{b}$$
$$\mathbf{a} = \{x(\mathbf{a} \times \mathbf{b}) + z[(\mathbf{a} \times \mathbf{b}) \times \mathbf{b})]\}$$
$$\mathbf{a} = x\cdot(\mathbf{a} \times \mathbf{b}) - z[\mathbf{b} \times (\mathbf{a} \times \mathbf{b})]$$
$$\mathbf{a} = x(\mathbf{a} \times \mathbf{b}) - z[(\mathbf{b}\cdot\mathbf{b})\mathbf{a} - (\mathbf{b}\cdot\mathbf{a})\cdot\mathbf{b}]$$
$$\mathbf{a} = x(\mathbf{a} \times \mathbf{b}) - z(\mathbf{b}\cdot\mathbf{b})\mathbf{a} \quad [\because \mathbf{a}\cdot\mathbf{b} = 0]$$

Comparing the coefficients, we get

$$x = 0, z = -\frac{1}{\mathbf{b}\cdot\mathbf{b}} = -\frac{1}{|\mathbf{b}|^2}$$

Putting the values of x and z in Eq. (i), we get

$$\mathbf{r} = y\mathbf{b} - \frac{1}{|\mathbf{b}|^2}(\mathbf{a} \times \mathbf{b})$$

Now, y can take any value. Hence, all the options are true.

Explanations for Q. Nos. (21-23)

We have $|\mathbf{x}| = |\mathbf{y}| = |\mathbf{z}| = \sqrt{2}$ and $\mathbf{x}, \mathbf{y}\ \mathbf{z}$ make angle 60° with each other.

$\therefore \quad \mathbf{x}\cdot\mathbf{y} = |\mathbf{x}|\,|\mathbf{y}|\cos 60^\circ = 1$

$\mathbf{y}\cdot\mathbf{z} = 1$ and $\mathbf{x}\cdot\mathbf{z} = 1$

$\mathbf{x}\cdot\mathbf{x} = |\mathbf{x}|^2 = 2 = |\mathbf{y}|^2 = |\mathbf{z}|^2$

Now, $\quad \mathbf{x} \times |\mathbf{y} \times \mathbf{z}| = \mathbf{a}$ and $\mathbf{y} \times (\mathbf{z} \times \mathbf{x}) = \mathbf{b}$

$\Rightarrow \quad (\mathbf{x}\cdot\mathbf{z})\,\mathbf{y} - (\mathbf{x}\cdot\mathbf{y})\,\mathbf{z} = \mathbf{a}$ and $(\mathbf{x}\cdot\mathbf{y})\,\mathbf{z} - (\mathbf{y}\cdot\mathbf{z})\,\mathbf{x} = \mathbf{b}$

$\Rightarrow \quad \mathbf{y} - \mathbf{z} = \mathbf{a}$ and $\mathbf{z} - \mathbf{x} = \mathbf{b}$

$\Rightarrow \quad \mathbf{y} - \mathbf{x} = \mathbf{a} + \mathbf{b}$

Thus, $\quad \mathbf{y} - \mathbf{z} = \mathbf{a},\ \mathbf{z} - \mathbf{x} = \mathbf{b},\ \mathbf{y} - \mathbf{x} = \mathbf{a} + \mathbf{b}$

Now, $\mathbf{x} \times \mathbf{y} = \mathbf{c}$, given

$\Rightarrow \quad \mathbf{x} \times (\mathbf{x} \times \mathbf{y}) = \mathbf{x} \times \mathbf{c}$

$\Rightarrow \quad \mathbf{x} - 2\mathbf{y} = \mathbf{x} \times \mathbf{c} \quad \ldots(\text{i})$

Again, $\mathbf{x} \times \mathbf{y} = \mathbf{c}$

$\Rightarrow \quad \mathbf{y} \times (\mathbf{x} \times \mathbf{y}) = \mathbf{y} \times \mathbf{c}$

$\Rightarrow \quad 2\mathbf{x} - \mathbf{y} = \mathbf{y} \times \mathbf{c} \quad \ldots(\text{ii})$

On subtracting (i) and (ii), we get

$\mathbf{x} + \mathbf{y} = (\mathbf{y} \times \mathbf{c}) - (\mathbf{x} \times \mathbf{c})$

$\Rightarrow \quad (\mathbf{x} + \mathbf{y}) = (\mathbf{y} - \mathbf{x}) \times \mathbf{c}$

$\Rightarrow \quad \mathbf{x} + \mathbf{y} = (\mathbf{a} + \mathbf{b}) \times \mathbf{c} \quad \ldots(\text{iii})$

From above equations

$$\mathbf{x} = \frac{1}{2}[(\mathbf{a} + \mathbf{b}) \times \mathbf{c} - (\mathbf{a} + \mathbf{b})],$$

$$\mathbf{y} = \frac{1}{2}[(\mathbf{a} + \mathbf{b}) + (\mathbf{a} + \mathbf{b}) \times \mathbf{c}]$$

and $\quad \mathbf{z} = \dfrac{1}{2}[(\mathbf{b} - \mathbf{a}) + (\mathbf{a} + \mathbf{b}) \times \mathbf{c}]$

21. (c) **22.** (a) **23.** (b)

24. (9) $\mathbf{a} \times \mathbf{b} = p\,\mathbf{b} + q\,\mathbf{b} + r\,\mathbf{c}$

Taking dot product with $\mathbf{a}, \mathbf{b}, \mathbf{c}$ respectively, we get

$$0 = p + \frac{3}{5}q + \frac{3}{5}r \quad \ldots(\text{i})$$

$$0 = \frac{3}{5}p + q + \frac{3}{5}r \quad \ldots(\text{ii})$$

$$[\mathbf{a}\ \mathbf{b}\ \mathbf{c}] = \frac{3}{5}p + \frac{3}{5}q + r \quad \ldots(\text{iii})$$

$$\therefore \quad [\mathbf{a}\ \mathbf{b}\ \mathbf{c}]^2 = \begin{vmatrix} \mathbf{a}\cdot\mathbf{a} & \mathbf{a}\cdot\mathbf{b} & \mathbf{a}\cdot\mathbf{c} \\ \mathbf{b}\cdot\mathbf{a} & \mathbf{b}\cdot\mathbf{b} & \mathbf{b}\cdot\mathbf{c} \\ \mathbf{c}\cdot\mathbf{a} & \mathbf{c}\cdot\mathbf{b} & \mathbf{c}\cdot\mathbf{c} \end{vmatrix}$$

$$= \frac{44}{125}$$

$$\Rightarrow \quad \frac{5}{11}[\mathbf{a}\ \mathbf{b}\ \mathbf{c}] = -\frac{2}{3}q$$

$$\Rightarrow \quad \frac{25}{121}\cdot\frac{44}{125} = \frac{4}{49}q^2$$

$$\Rightarrow \quad 55q^2 = 9$$

25. (5) $(\mathbf{a} \times \mathbf{b}) \times (\mathbf{a} - 2\mathbf{b})$

$= (\mathbf{a} \times \mathbf{b}) \times \mathbf{a} - 2(\mathbf{a} \times \mathbf{b}) \times \mathbf{b}$

$= -\mathbf{a} \times (\mathbf{a} \times \mathbf{b}) + 2\mathbf{b}(\mathbf{a} \times \mathbf{b})$

$= -\{(\mathbf{a}\cdot\mathbf{b})\mathbf{a} - (\mathbf{a}\cdot\mathbf{a})\mathbf{b}\} + 2\{(\mathbf{b}\cdot\mathbf{a})\mathbf{a} - (\mathbf{b}\cdot\mathbf{a})\mathbf{b}\}$

$= 2\mathbf{a} + \mathbf{b} \quad [\because \mathbf{a}\cdot\mathbf{b} = 0, \mathbf{a}\cdot\mathbf{a} = 1, \mathbf{b}\cdot\mathbf{b} = 1]$

$\therefore (2\mathbf{a} + \mathbf{b})\cdot[(\mathbf{a} \times \mathbf{b}) \times (\mathbf{a} - 2\mathbf{b})]$

$= (2\mathbf{a} + \mathbf{b})\cdot(2\mathbf{a} + \mathbf{b})$

$= 4|\mathbf{a}|^2 + |\mathbf{b}|^2 + 4\mathbf{a}\cdot\mathbf{b} = 4 + 1 + 0 = 5$

Revisal Problems (JEE Main)

1. (c) Let $\mathbf{c}$ be $x\hat{\mathbf{i}} + y\hat{\mathbf{j}} + z\hat{\mathbf{k}}$.

By given condition,

$$|a| = |b| = \sqrt{2}$$

$\therefore \quad x^2 + y^2 + z^2 = 2 \quad \ldots(\text{i})$

Also, $\mathbf{c}$ makes obtuse angle with X-axis.

$\therefore \quad \mathbf{c}\hat{\mathbf{i}} = x = \text{negative}$

Again, $\quad \cos\theta = \dfrac{a}{\sqrt{2}}\cdot\dfrac{b}{\sqrt{2}} = \dfrac{b\cdot c}{\sqrt{2}\,\sqrt{2}} = \dfrac{c\cdot a}{\sqrt{2}\,\sqrt{2}}$

$$\frac{1}{2} = \frac{y+z}{2} = \frac{x+y}{2}$$

$\therefore \quad y + z = 1$ and $x = z$

$\therefore \quad y = 1 - z = 1 - x$

$\because \quad z = x$

Putting in Eq. (i), we get

$\therefore \quad x^2 + (1 - x)^2 + x^2 = 2$

or $\quad 3x^2 - 2x - 1 = 0$

$(3x + 1)(x - 1) = 0$

$\therefore \quad x = -\dfrac{1}{3}$ as x is negative.

$\therefore \quad y = \dfrac{4}{3}$ and $z = -\dfrac{1}{3}$

$\therefore \quad \mathbf{c} = -\dfrac{1}{3}\hat{\mathbf{i}} + \dfrac{4}{3}\hat{\mathbf{j}} - \dfrac{1}{3}\hat{\mathbf{k}}$

$= \dfrac{1}{3}(-\hat{\mathbf{i}} + 4\hat{\mathbf{j}} - \hat{\mathbf{k}})$

2. (d) Let $x\hat{\mathbf{i}} + y\hat{\mathbf{j}} + z\hat{\mathbf{k}}$ be the unit vector along $\mathbf{c}$.

Since, $-\hat{\mathbf{i}} + \hat{\mathbf{j}} - \hat{\mathbf{k}}$ bisects the angle between $\mathbf{c}$ and $3\hat{\mathbf{i}} + \hat{\mathbf{j}}$.

$\therefore \quad \lambda(-\hat{\mathbf{i}} + \hat{\mathbf{j}} - \hat{\mathbf{k}}) = x\hat{\mathbf{i}} + y\hat{\mathbf{j}} + z\hat{\mathbf{k}} + \dfrac{3\hat{\mathbf{i}} + 4\hat{\mathbf{j}}}{5}$

$\Rightarrow \quad x + \dfrac{3}{5} = -\lambda,\ y + \dfrac{4}{5} = \lambda$ and $z = -\lambda$

Now, $\quad x^2 + y^2 + z^2 = 1 \quad [\because x\hat{\mathbf{i}} + y\hat{\mathbf{j}} + z\hat{\mathbf{k}}$ is a unit vector$]$

$\Rightarrow \quad \left(-\lambda - \dfrac{3}{5}\right)^2 + \left(\lambda - \dfrac{4}{5}\right)^2 + \lambda^2 = 1$

$\Rightarrow \quad \lambda = 0, \dfrac{2}{15}$

But $\lambda \neq 0$ because $\lambda = 0$ implies that the given vectors are parallel.

$\therefore \quad \lambda = \dfrac{2}{15} \Rightarrow x = -\dfrac{11}{15}, y = -\dfrac{10}{15}, z = -\dfrac{2}{15}$

Hence, $x\hat{\mathbf{i}} + y\hat{\mathbf{j}} + z\hat{\mathbf{k}} = -\dfrac{1}{15}(11\hat{\mathbf{i}} + 10\hat{\mathbf{j}} + 2\hat{\mathbf{k}})$

3. (a) Let $\mathbf{a} = 3\hat{\mathbf{i}} + 2\hat{\mathbf{j}} + 2\hat{\mathbf{k}}$ and $\mathbf{b} = 18\hat{\mathbf{i}} - 22\hat{\mathbf{j}} - 5\hat{\mathbf{k}}$.

Then, the required vector $\mathbf{r}$ is given by $\mathbf{r} = \lambda(\mathbf{a} \times \mathbf{b})$

$\Rightarrow \quad \mathbf{r} = \lambda(34\hat{\mathbf{i}} + 51\hat{\mathbf{j}} - 102\hat{\mathbf{k}})$

$= 17\lambda(2\hat{\mathbf{i}} + 3\hat{\mathbf{j}} - 3\hat{\mathbf{k}})$

Now, $\quad |\mathbf{r}| = 14$

$\Rightarrow \quad 119|\lambda| = 14$

$\Rightarrow \quad |\lambda| = \dfrac{2}{17}$

Since, $\mathbf{r}$ makes an obtuse angle with Y-axis. Therefore, $\lambda = -\dfrac{2}{17}$.

Hence, $\quad \mathbf{r} = -2(2\hat{\mathbf{i}} + 3\hat{\mathbf{j}} - 6\hat{\mathbf{k}})$

or $\quad \mathbf{r} = 2(-2\hat{\mathbf{i}} - 3\hat{\mathbf{j}} + 6\hat{\mathbf{k}})$

4. (a) Taking A as the origin, let the position vectors of B and C be $\mathbf{b}$ and $\mathbf{c}$ respectively. Then, the position vector of a is $\dfrac{\mathbf{b}+\mathbf{c}}{3}$.

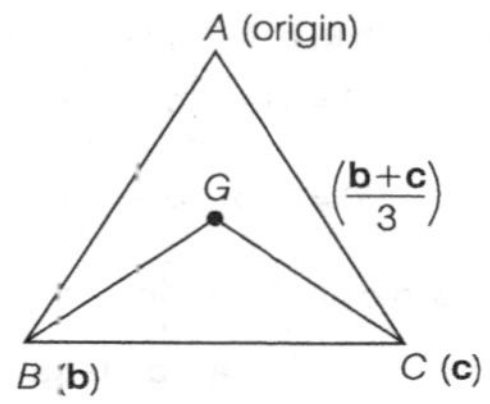

Since, **GB** and **GC** are inclined at an obtuse angle.

$$\therefore \quad \mathbf{GB}\cdot\mathbf{GC} < 0$$

$$\Rightarrow \quad \left(\mathbf{b} - \frac{\mathbf{b}+\mathbf{c}}{3}\right)\cdot\left(\mathbf{c} - \frac{\mathbf{b}+\mathbf{c}}{3}\right) < 0$$

$$\Rightarrow \quad (2\mathbf{b}-\mathbf{c})\cdot(2\mathbf{c}-\mathbf{b}) < 0$$

$$\Rightarrow \quad 5(\mathbf{b}\cdot\mathbf{c}) - 2|\mathbf{b}|^2 - 2|\mathbf{c}|^2 < 0 \quad \ldots(i)$$

We have,

$$c = AB = |\mathbf{b}|$$
$$b = AC = |\mathbf{c}|$$
$$a = BC = |\mathbf{c}-\mathbf{b}|$$
$$\therefore \quad a^2 = |\mathbf{c}-\mathbf{b}|^2$$
$$= |\mathbf{c}|^2 + |\mathbf{b}|^2 - 2(\mathbf{b}\cdot\mathbf{c})$$
$$= \mathbf{b} + \mathbf{c} - 2(\mathbf{b}\cdot\mathbf{c})$$
$$\Rightarrow \quad \mathbf{b}\cdot\mathbf{c} = \frac{b^2+c^2-a^2}{2} \quad \ldots(ii)$$

From Eqs. (i) and (ii), we get

$$\frac{5(b^2+c^2-a^2)}{2} < 2c^2 + 2b^2$$
$$\Rightarrow \quad b^2 + c^2 < 5a^2$$

5. (d) Let $|\mathbf{a}| = |\mathbf{b}| = |\mathbf{c}| = \lambda$

We have,

$$\mathbf{a}\cdot\mathbf{b} = |\mathbf{a}|\,|\mathbf{b}|\cos\alpha = \lambda^2\cos\alpha$$
$$\mathbf{b}\cdot\mathbf{c} = |\mathbf{b}|\,|\mathbf{c}|\cos\beta = \lambda^2\cos\beta$$
$$\mathbf{c}\cdot\mathbf{a} = |\mathbf{c}|\,|\mathbf{a}|\cos\gamma = \lambda^2\cos\gamma$$

Now,

$$|\mathbf{a}+\mathbf{b}+\mathbf{c}|^2 \ge 0$$
$$\Rightarrow \quad |\mathbf{a}|^2 + |\mathbf{b}|^2 + \mathbf{c}|^2 = 2(\mathbf{a}\cdot\mathbf{b} + \mathbf{b}\cdot\mathbf{c} + \mathbf{c}\cdot\mathbf{a}) \ge 0$$
$$\Rightarrow \quad 3\lambda^2 + 2\lambda^2(\cos\alpha + \cos\beta + \cos\gamma) \ge 0$$
$$\Rightarrow \quad \cos\alpha + \cos\beta + \cos\gamma \ge -\frac{3}{2}$$

Hence, the minimum value of $\cos\alpha + \cos\beta + \cos\gamma$ is $-\dfrac{3}{2}$.

6. (b) Since, $\mathbf{u}$, $\mathbf{v}$ and $\mathbf{u}\times\mathbf{b}$ are non-coplanar vectors.

$$\therefore \quad (a-b)\,\mathbf{u} + (b-c)\mathbf{v} + (c-a)(\mathbf{u}\times\mathbf{v}) = 0$$
$$\Rightarrow \quad a-b = 0 = b-c = c-a$$
$$\Rightarrow \quad a = b = c$$

So, the triangle is equilateral.

7. (a) We have,

$$\mathbf{r}\times\mathbf{b} = \mathbf{c}\times\mathbf{b}$$
$$\Rightarrow \quad (\mathbf{r}-\mathbf{c})\times\mathbf{b} = 0$$
$$\Rightarrow \quad \mathbf{r}-\mathbf{c} = \lambda\mathbf{b}$$
$$\Rightarrow \quad \mathbf{r} = \mathbf{c} + \lambda\mathbf{b}$$

Now, $\mathbf{r}\perp\mathbf{a} \Rightarrow \mathbf{r}\cdot\mathbf{a} = 0$

$$\Rightarrow \quad (\mathbf{c}+\lambda\mathbf{b})\cdot\mathbf{a} = 0$$
$$\Rightarrow \quad \lambda = -\frac{\mathbf{c}\cdot\mathbf{a}}{\mathbf{b}\cdot\mathbf{a}}$$
$$\therefore \quad \mathbf{r} = \mathbf{c} - \frac{\mathbf{c}\cdot\mathbf{a}}{\mathbf{b}\cdot\mathbf{a}}\cdot\mathbf{b} = \frac{(\mathbf{a}\cdot\mathbf{b})\mathbf{c} - (\mathbf{a}\cdot\mathbf{c})\,\mathbf{b}}{\mathbf{a}\cdot\mathbf{b}} = \frac{\mathbf{a}\times(\mathbf{c}\times\mathbf{b})}{\mathbf{a}\cdot\mathbf{b}}$$

8. (b) Let $\mathbf{c} = x\hat{\mathbf{i}} + y\hat{\mathbf{i}} + z\hat{\mathbf{k}}$,

Then,

$$\mathbf{a}\cdot\mathbf{c} = 2 \text{ and } \mathbf{a}\times\mathbf{c} = \mathbf{b}$$
$$\Rightarrow \quad \mathbf{a}\cdot\mathbf{c} = 2 \text{ and } (\mathbf{a}\times\mathbf{c})\cdot\mathbf{c} = \mathbf{b}\cdot\mathbf{c}$$
$$\Rightarrow \quad \mathbf{a}\cdot\mathbf{c} = 2 \text{ and } \mathbf{b}\cdot\mathbf{c} = 0 \qquad [\because (\mathbf{a}\times\mathbf{c})\cdot\mathbf{c} = 0]$$
$$\Rightarrow \quad x+y+z = 2 \text{ and } x-2y+z = 0$$
$$\Rightarrow \quad y = \frac{2}{3} \text{ and } x+z = \frac{4}{3}$$

Clearly, option (b) satisfies the above conditions.

9. (a) We have, $\mathbf{r}\times\mathbf{a} = \mathbf{b}\times\mathbf{a}$

and $\mathbf{r}\times\mathbf{b} = \mathbf{a}\times\mathbf{b}$

$$\Rightarrow \quad \mathbf{r}\times\mathbf{a} = -(\mathbf{r}\times\mathbf{b})$$
$$\Rightarrow \quad \mathbf{r}\times(\mathbf{a}+\mathbf{b}) = 0$$

$\Rightarrow$ **r** is parallel to $\mathbf{a}+\mathbf{b}$.

$$\Rightarrow \quad \mathbf{r} = \lambda(\mathbf{a}+\mathbf{b})$$
$$\Rightarrow \quad \mathbf{r} = \lambda(\hat{\mathbf{i}} + 3\hat{\mathbf{j}} - \hat{\mathbf{k}})$$
$$\Rightarrow \quad |\mathbf{r}| = \sqrt{11}\lambda$$
$$\therefore \quad \text{Required vector} = \frac{\mathbf{r}}{|\mathbf{r}|}$$
$$= \frac{1}{\sqrt{11}}(\hat{\mathbf{i}} + 3\hat{\mathbf{j}} - \hat{\mathbf{k}})$$

10. (b) Since, **a**, **b**, **c** are non-coplanar vectors. Therefore, vectors are $\mathbf{a}\times\mathbf{b}$, $\mathbf{b}\times\mathbf{c}$, $\mathbf{c}\times\mathbf{a}$.

We know that, any vector in space is expressible as the linear combination of three non-coplanar vectors. So, let

$$\mathbf{a} = x(\mathbf{b}\times\mathbf{c}) + y(\mathbf{c}\times\mathbf{a}) + z(\mathbf{a}\times\mathbf{b}) \quad \ldots(i)$$

Taking dot products successively with **a**, **b**, **c**, we get

$$x = \frac{\mathbf{a}\cdot\mathbf{a}}{[\mathbf{a}\ \mathbf{b}\ \mathbf{c}]},\ y = \frac{\mathbf{a}\cdot\mathbf{b}}{[\mathbf{a}\ \mathbf{b}\ \mathbf{c}]},\ z = \frac{\mathbf{a}\cdot\mathbf{c}}{[\mathbf{a}\ \mathbf{b}\ \mathbf{c}]}$$

Substituting these values in Eq. (i), we obtain that the given expression is equal to $[\mathbf{a}\ \mathbf{b}\ \mathbf{c}]\,\mathbf{a}$.

11. (d) Clearly,

$$[\mathbf{a}+\mathbf{b}+\mathbf{c}\quad \mathbf{a}-\mathbf{c}\quad \mathbf{a}-\mathbf{b}] = \begin{vmatrix} 1 & 1 & 1 \\ 1 & 0 & -1 \\ 1 & -1 & 0 \end{vmatrix}[\mathbf{a}\ \mathbf{b}\ \mathbf{c}]$$
$$= [1(0-1) - 1(0+1) + 1(-1-0)]\,[\mathbf{a}\ \mathbf{b}\ \mathbf{c}]$$
$$= -3\,[\mathbf{a}\ \mathbf{b}\ \mathbf{c}]$$

12. (b) We have,

$$[\mathbf{u}\ \mathbf{v}\ \mathbf{w}] = \begin{vmatrix} al + a_1l_1 & am + a_1m_1 & an + a_1n_1 \\ bl + b_1l_1 & bm + b_1m_1 & bn + b_1n_1 \\ cl + c_1l_1 & cm + c_1m_1 & cn + c_1n_1 \end{vmatrix}$$

$$\Rightarrow \quad [\mathbf{u}\ \mathbf{v}\ \mathbf{w}] = \begin{vmatrix} a & a_1 & 0 \\ b & b_1 & 0 \\ c & c_1 & 0 \end{vmatrix}\begin{vmatrix} l & l_1 & 0 \\ m & m_1 & 0 \\ n & n_1 & 0 \end{vmatrix} = 0$$

Hence, the given vectors are coplanar.

13. (b) Let

$$\mathbf{a} = \hat{\mathbf{i}} + \hat{\mathbf{j}},$$
$$\mathbf{b} = \hat{\mathbf{j}} + \hat{\mathbf{k}}$$
$$\text{and} \quad \mathbf{c} = \hat{\mathbf{k}} + \hat{\mathbf{i}}$$

Let $\mathbf{n}_1$, $\mathbf{n}_2$, $\mathbf{n}_3$ be the normals to the given planes. Then,

$$\mathbf{n}_1 = \mathbf{a}\times\mathbf{b}$$
$$\mathbf{n}_2 = \mathbf{b}\times\mathbf{c}$$
$$\text{and} \quad \mathbf{n}_3 = \mathbf{c}\times\mathbf{a}$$

$\therefore$ Volume of the given parallelopiped is given by

$$[\mathbf{n}_1\ \mathbf{n}_2\ \mathbf{n}_3] = [\mathbf{a}\ \mathbf{b}\ \mathbf{c}]^2 = \begin{vmatrix} 1 & 1 & 0 \\ 0 & 1 & 1 \\ 1 & 0 & 1 \end{vmatrix}^2 = 4$$

14. (d) 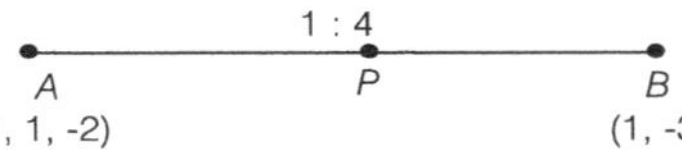

$\therefore \quad P = \left(\frac{1+4}{5}, \frac{-3+4}{54}, \frac{1-8}{5}\right) = \left(1, \frac{1}{5}, \frac{-7}{5}\right)$

$\therefore \quad \mathbf{OP} = \left(\hat{\mathbf{i}} + \frac{\hat{\mathbf{j}}}{5} - \frac{7\hat{\mathbf{k}}}{5}\right)$ or $\frac{1}{5}(5\hat{\mathbf{i}} + \hat{\mathbf{j}} - 7\hat{\mathbf{k}})$

15. (d) We have, work done = $\mathbf{F}\cdot(\mathbf{AB} + \mathbf{AC}) = \mathbf{F}\cdot\mathbf{AC}$

Now, $\quad \mathbf{F} = \frac{24\sqrt{3}}{\sqrt{6}}(\hat{\mathbf{i}} + 2\hat{\mathbf{j}} + \hat{\mathbf{k}})$

$= 12\sqrt{2}(\hat{\mathbf{i}} + 2\hat{\mathbf{j}} + \hat{\mathbf{k}})$

and $\quad \mathbf{AC} = \hat{\mathbf{i}} + 4\hat{\mathbf{j}} - \hat{\mathbf{k}}$

$\therefore \quad$ Work done $= \mathbf{F}\cdot\mathbf{AC} = 72\sqrt{2}$

16. (a) Let the required vector be $\mathbf{r} = l\hat{\mathbf{i}} + m\hat{\mathbf{j}} + n\hat{\mathbf{k}}$, where l, m, n are the direction cosines of $\mathbf{r}$ such that $l = m$.

It is given that $\mathbf{r}$ is prependicular to $-\hat{\mathbf{i}} + 2\hat{\mathbf{j}} + 2\hat{\mathbf{k}}$. Therefore,

$\mathbf{r}\cdot(-\hat{\mathbf{i}} + 2\hat{\mathbf{k}} + 2\hat{\mathbf{k}}) = 0$

$\Rightarrow \quad -l + 2m + 2n = 0$

$\Rightarrow \quad l + 2n = 0 \qquad [\because l = m]$

$\Rightarrow \quad l = -2n$

Now, $\quad l^2 + m^2 + n^2 = 1$

$\Rightarrow \quad 4n^2 + 4n^2 + n^2 = 1$

$\Rightarrow \quad n = \pm\frac{1}{3}$

$\therefore \quad l = \mp\frac{2}{3}, m = \mp\frac{2}{3}, n = \mp\frac{1}{3}$

Hence, $\quad \mathbf{r} = \mp\frac{1}{3}(2\hat{\mathbf{i}} + 2\hat{\mathbf{j}} - \hat{\mathbf{k}})$

17. (a) A vector parallel to the bisector of the angle between the vectors **a** and **b** is

$$\frac{\mathbf{a}}{|\mathbf{a}|} + \frac{\mathbf{b}}{|\mathbf{b}|} = \hat{\mathbf{a}} + \hat{\mathbf{b}}.$$

$\therefore$ Unit vector along the bisector $= \frac{\hat{\mathbf{a}} + \hat{\mathbf{b}}}{|\hat{\mathbf{a}} + \hat{\mathbf{b}}|} = \frac{1}{\sqrt{2}}(\hat{\mathbf{a}} + \hat{\mathbf{b}})$

$(\hat{\mathbf{a}} + \hat{\mathbf{b}})^2 = |\hat{\mathbf{a}}|^2 + |\hat{\mathbf{b}}| + 2\hat{\mathbf{a}}\cdot\hat{\mathbf{b}}$

$= 1 + 1 + 0 = 2$

$\therefore \quad$ Required projection

$= \left[l\frac{\mathbf{a}}{|\mathbf{a}|} + m\frac{\mathbf{b}}{|\mathbf{b}|} + n\frac{\mathbf{a}\times\mathbf{b}}{|\mathbf{a}\times\mathbf{b}|}\right]\cdot\frac{1}{\sqrt{2}}(\hat{\mathbf{a}} + \hat{\mathbf{b}})$

$= \frac{1}{\sqrt{2}}(l + m)$

$\left[\because \frac{\mathbf{a}\cdot\hat{\mathbf{a}}}{|\mathbf{a}|} = \frac{\mathbf{b}\cdot\hat{\mathbf{b}}}{|\mathbf{b}|} = 1 \text{ and } \hat{\mathbf{a}}\cdot\hat{\mathbf{b}} = \hat{\mathbf{a}}\cdot(\mathbf{a}\times\mathbf{b}) = \hat{\mathbf{b}}\cdot(\hat{\mathbf{a}}\times\mathbf{b}) = 0\right]$

18. (a) We have, $\quad \mathbf{a}\times(\mathbf{a}\times\mathbf{c}) + 3\mathbf{b} = 0$

$\Rightarrow \quad (\mathbf{a}\cdot\mathbf{c})\mathbf{a} - (\mathbf{a}\cdot\mathbf{a})\mathbf{c} + 3\mathbf{b} = 0$

$\Rightarrow \quad (2\sqrt{3}\cos\theta)\,\mathbf{a} - 3\mathbf{c} + 3\mathbf{b} = 0$

$\Rightarrow \quad 2\cos\theta\,\mathbf{a} - \sqrt{3}\mathbf{c} + \sqrt{3}\mathbf{b} = 0$

$\Rightarrow \quad |2\cos\theta\,\mathbf{a} - \sqrt{3}\mathbf{c}| = |-\sqrt{3}\,\mathbf{b}|$

$\Rightarrow \quad |2\cos\theta\,\mathbf{a} - \sqrt{3}\mathbf{c}|^2 = |-\sqrt{3}\mathbf{b}|^2$

$\Rightarrow \quad 4\cos^2\theta\,|\mathbf{a}|^2 + 3|\mathbf{c}|^2 - 4\sqrt{3}\cos\theta\,(\mathbf{a}\cdot\mathbf{c}) = 3|\mathbf{b}|^2$

$\Rightarrow \quad 12\cos^2\theta + 12 - 4\sqrt{3}\cos\theta \times \sqrt{3} \times 2\cos\theta = 3$

$\Rightarrow \quad 12\cos^2\theta + 9 - 24\cos^2\theta = 0$

$\Rightarrow \quad 12\cos^2\theta = 9 \quad \Rightarrow \cos^2\theta = \frac{9}{12}$

$\therefore \quad \cos^2\theta = \frac{3}{4}$

19. (b) We have,

$|a\hat{\mathbf{i}} + b\hat{\mathbf{j}} + c\hat{\mathbf{k}}| = |a| + |b| + |c|$

$\Rightarrow \quad \sqrt{a^2 + b^2 + c^2} = |a| + |b| + |c|$

$\Rightarrow \quad a^2 + b^2 + c^2 = a^2 + b^2 + c^2 + 2(|a|\,|b| + |b|\,|c| + |c|\,|a|)$

$\Rightarrow \quad |a|\,|b| + |b|\,|c| + |c|\,|a| = 0$

$\Rightarrow \quad ab = bc = ca = 0$

$\therefore$ Any two of a, b, c are zero.

20. (a) We have,

$\{(\mathbf{a} + \mathbf{b})\times(\mathbf{a} + \mathbf{c})\}\times(\mathbf{b}\times\mathbf{c})\cdot(\mathbf{b} + \mathbf{c})$

$= (\mathbf{a}\times\mathbf{c} + \mathbf{b}\times\mathbf{a} + \mathbf{b}\times\mathbf{c})\times(\mathbf{b}\times\mathbf{c})\cdot(\mathbf{b} + \mathbf{c})$

$= \{(\mathbf{a}\times\mathbf{c})\times(\mathbf{b}\times\mathbf{c}) + (\mathbf{b}\times\mathbf{a})\times(\mathbf{b}\times\mathbf{c})\}\cdot(\mathbf{b} + \mathbf{c})$

$= [\{(\mathbf{a}\times\mathbf{c})\cdot\mathbf{c}\}\mathbf{b} - \{(\mathbf{a}\times\mathbf{c})\cdot\mathbf{b}\}\mathbf{c} + \{(\mathbf{b}\times\mathbf{a})\cdot\mathbf{c}\}\,\mathbf{b}$
$\qquad - \{(\mathbf{b}\times\mathbf{a})\cdot\mathbf{b}\}\mathbf{c}]\cdot(\mathbf{b} + \mathbf{c})$

$= \{-[\mathbf{a}\ \mathbf{c}\ \mathbf{b}]\mathbf{c} + [\mathbf{b}\ \mathbf{a}\ \mathbf{c}]\mathbf{b}\}\cdot(\mathbf{b} + \mathbf{c})$

$= [\mathbf{a}\ \mathbf{b}\ \mathbf{c}]\{\mathbf{c}\cdot(\mathbf{b} + \mathbf{c})\} - [\mathbf{a}\ \mathbf{b}\ \mathbf{c}]\{\mathbf{b}\cdot(\mathbf{b} + \mathbf{c})\}$

$= [\mathbf{a}\ \mathbf{b}\ \mathbf{c}]\,[\mathbf{c}\cdot\mathbf{b} + \mathbf{c}\cdot\mathbf{c} - \mathbf{b}\cdot\mathbf{b} - \mathbf{b}\cdot\mathbf{c}]$

$= [\mathbf{a}\ \mathbf{b}\ \mathbf{c}]\{\,|\mathbf{c}|^2 - |\mathbf{b}|^2\} \qquad [\because |\mathbf{b}| = |\mathbf{c}|]$

$= 0$

21. (c) We have, $\quad |[\mathbf{a}\ \mathbf{b}\ \mathbf{c}]| = 27$cu units

Now, $\quad [\alpha\,\beta\,\gamma] = \begin{vmatrix} 1 & 2 & -1 \\ 1 & -1 & 0 \\ 1 & -1 & -1 \end{vmatrix}[\mathbf{a}\ \mathbf{b}\ \mathbf{c}] = 3\,[\mathbf{a}\ \mathbf{b}\ \mathbf{c}]$

$\therefore$ Required volume $= |[\alpha\,\beta\,\gamma]|$

$= 3|[\mathbf{a}\ \mathbf{b}\ \mathbf{c}]|$

$= 3 \times 27 = 81$ cu units

22. (a) We have, $\quad |[\mathbf{a}\ \mathbf{b}\ \mathbf{c}]| = V$

Let V_1 be the volume of the parallelepiped formed by the vectors α, β and γ. Then,

$V_1 = |[\alpha\,\beta\,\gamma]|$

Now, $\quad [\alpha\,\beta\,\gamma] = \begin{vmatrix} \mathbf{a}\cdot\mathbf{a} & \mathbf{a}\cdot\mathbf{b} & \mathbf{a}\cdot\mathbf{c} \\ \mathbf{a}\cdot\mathbf{b} & \mathbf{b}\cdot\mathbf{b} & \mathbf{b}\cdot\mathbf{c} \\ \mathbf{a}\cdot\mathbf{c} & \mathbf{b}\cdot\mathbf{c} & \mathbf{c}\cdot\mathbf{c} \end{vmatrix}[\mathbf{a}\ \mathbf{b}\ \mathbf{c}]$

$\Rightarrow \quad [\alpha\,\beta\,\gamma] = [\mathbf{a}\ \mathbf{b}\ \mathbf{c}]^2[\mathbf{a}\ \mathbf{b}\ \mathbf{c}]$

$\Rightarrow \quad [\alpha\,\beta\,\gamma] = [\mathbf{a}\ \mathbf{b}\ \mathbf{c}]^3$

$\because \quad V_1 = |[\alpha\,\beta\,\gamma]| = |[\mathbf{a}\ \mathbf{b}\ \mathbf{c}]^3| = V^3$

23. (a) Since, **a** is perpendicular to both **b** and **c**. Therefore, **a** is parallel to $\mathbf{b}\times\mathbf{c}$.

Now, $\quad |[\mathbf{a}\ \mathbf{b}\ \mathbf{c}]| = |\mathbf{a}\cdot(\mathbf{b}\times\mathbf{c})|$

$\Rightarrow \quad |[\mathbf{a}\ \mathbf{b}\ \mathbf{c}]| = |\mathbf{a}|\,|\mathbf{b}\times\mathbf{c}| \qquad [\because \mathbf{a} \parallel \mathbf{b}\times\mathbf{c}]$

$= |\mathbf{a}|\,|\mathbf{b}|\,|\mathbf{c}|\sin\frac{5\pi}{6}$

$\therefore \quad |[\mathbf{a}\ \mathbf{b}\ \mathbf{c}]| = 5 \times 3 \times 4 \times \frac{1}{2}$

$= 30$ cu units

24. (a) If **a**, **b**, **c** are three non-coplanar vectors, then any vector **r** can be expressed as

$$\mathbf{r} = \left\{\frac{\mathbf{r}\cdot\mathbf{a}}{|\mathbf{a}|^2}\right\}\mathbf{a} + \left\{\frac{\mathbf{r}\cdot\mathbf{b}}{|\mathbf{b}|^2}\right\}\mathbf{b} + \left\{\frac{\mathbf{r}\cdot\mathbf{c}}{|\mathbf{c}|^2}\right\}\mathbf{c}$$

Clearly, **a**, **b** and $\mathbf{a}\times\mathbf{b}$ are three non-coplanas vectors.

$\therefore \quad \mathbf{v} = \left\{\frac{\mathbf{v}\cdot\mathbf{a}}{|\mathbf{a}|^2}\right\}\mathbf{a} + \left\{\frac{\mathbf{v}\cdot\mathbf{b}}{|\mathbf{b}|^2}\right\}\mathbf{b} + \left\{\frac{\mathbf{v}\cdot(\mathbf{a}\times\mathbf{b})}{|\mathbf{a}\times\mathbf{b}|^2}\right\}(\mathbf{a}\times\mathbf{b})$

$= 0\,\mathbf{a} + \frac{1}{|\mathbf{b}|^2}\mathbf{b} + \frac{1}{|\mathbf{a}+\mathbf{b}|^2}(\mathbf{a}\times\mathbf{b})$

$= \frac{\mathbf{b}}{|\mathbf{b}|^2} + \frac{\mathbf{a}\times\mathbf{b}}{|\mathbf{a}\times\mathbf{b}|^2}$

25. (a) We have, $\mathbf{r}\cdot\mathbf{a} = 10$

$\Rightarrow \quad x + y + z = 10$

The total number of possible positions of P is same as the number of solutions of the above equation in the set of natural numbers. The total number of solution is

$$^{10-1}C_{3-1} = {}^9C_2 = 36$$

26. (d) Clearly, **b**, **c** and **b** × **c** are non-coplanar vectors.

$$\therefore \quad \mathbf{a} = \left\{\frac{\mathbf{a}\cdot\mathbf{b}}{|\mathbf{b}|^2}\right\}\mathbf{b} + \left\{\frac{\mathbf{a}\cdot\mathbf{c}}{|\mathbf{c}|^2}\right\}\mathbf{c} + \frac{\mathbf{a}\cdot(\mathbf{b}\times\mathbf{c})}{|\mathbf{b}\times\mathbf{c}|^2}(\mathbf{b}\times\mathbf{c})$$

Hence, the magnitude of the given vector

$$= |\mathbf{a}| = \sqrt{14}$$

27. (b) Let the required vector be $\mathbf{r} = x\hat{\mathbf{i}} + z\hat{\mathbf{k}}$, since **r** is a unit vector.

$$\therefore \quad x^2 + y^2 = 1$$

It is given that **r** makes 45° and 60° angles with **a** and **b** respectively.

$$\therefore \quad \cos 45° = \frac{\mathbf{r}\cdot\mathbf{a}}{|\mathbf{a}||\mathbf{r}|} \text{ and } \cos 60° = \frac{\mathbf{r}\cdot\mathbf{b}}{|\mathbf{r}||\mathbf{b}|}$$

$$\Rightarrow \quad \frac{1}{\sqrt{2}} = \frac{2x - y}{3} \text{ and } \frac{1}{2} = -\frac{-y}{\sqrt{2}}$$

$$\Rightarrow \quad 2x - y = \frac{3}{\sqrt{2}} \text{ and } y = -\frac{1}{\sqrt{2}}$$

$$\Rightarrow \quad x = \frac{1}{\sqrt{2}}$$

$$\therefore \quad \mathbf{r} = \frac{1}{\sqrt{2}}(\hat{\mathbf{i}} - \hat{\mathbf{k}})$$

28. (a) We have, $\mathbf{a} = 2\hat{\mathbf{i}} + \hat{\mathbf{j}} + \hat{\mathbf{k}}$,

$$\mathbf{b} = \hat{\mathbf{i}} + 2\hat{\mathbf{j}} + 2\hat{\mathbf{k}} \text{ and } \mathbf{c} = \hat{\mathbf{i}} + \hat{\mathbf{j}} + 2\hat{\mathbf{k}}$$

Now, $\mathbf{a}\times(\mathbf{b}\times\mathbf{c}) = (\mathbf{a}\cdot\mathbf{c})\mathbf{b} - (\mathbf{a}\cdot\mathbf{b})\mathbf{c} = 5\mathbf{b} - 6\mathbf{c}$

$$= 5\hat{\mathbf{j}} + 10\hat{\mathbf{j}} + 10\hat{\mathbf{k}} - 6\hat{\mathbf{i}} - 6\hat{\mathbf{j}} - 12\hat{\mathbf{k}}$$

$$= -\hat{\mathbf{i}} + 4\hat{\mathbf{j}} - 2\hat{\mathbf{k}}$$

Now, $\mathbf{a}\times(\mathbf{b}\times\mathbf{c}) = (1+\alpha)\hat{\mathbf{i}} + \beta(1+\alpha)\hat{\mathbf{j}} + \gamma(1+\alpha) + (1+\beta)\hat{\mathbf{k}}$

$$\Rightarrow \quad 1 + \alpha = -1, \beta(1+\alpha) = 4, \gamma(1+\alpha)(1+\beta) = -2$$

$$\therefore \quad \alpha = -2, \beta = -4, \gamma = -\frac{2}{3}$$

29. (a) We have,

$$(\mathbf{a}\times\mathbf{b})\times(\mathbf{r}\times\mathbf{c}) = [\mathbf{a}\ \mathbf{b}\ \mathbf{c}]\mathbf{r} - [\mathbf{a}\ \mathbf{b}\ \mathbf{r}]\mathbf{c}$$

$$(\mathbf{b}\times\mathbf{c})\times(\mathbf{r}\times\mathbf{a}) = [\mathbf{b}\ \mathbf{c}\ \mathbf{a}]\mathbf{c} - [\mathbf{b}\ \mathbf{c}\ \mathbf{r}]\mathbf{a}$$

$$(\mathbf{c}\times\mathbf{a})\times(\mathbf{r}\times\mathbf{b}) = [\mathbf{c}\ \mathbf{a}\ \mathbf{b}]\mathbf{r} - [\mathbf{c}\ \mathbf{a}\ \mathbf{r}]\mathbf{b}$$

$$\therefore (\mathbf{a}\times\mathbf{b})\times(\mathbf{r}\times\mathbf{c}) + (\mathbf{b}+\mathbf{c})\times(\mathbf{r}\times\mathbf{a}) + (\mathbf{c}\times\mathbf{a})\times(\mathbf{r}\times\mathbf{b})$$

$$= 3[\mathbf{a}\ \mathbf{b}\ \mathbf{c}]\mathbf{r} - [\mathbf{a}\ \mathbf{b}\ \mathbf{r}]\mathbf{c} + [\mathbf{b}\ \mathbf{c}\ \mathbf{r}]\mathbf{a} + [\mathbf{c}\ \mathbf{a}\ \mathbf{r}]\mathbf{b}$$

$$= 3[\mathbf{a}\ \mathbf{b}\ \mathbf{c}]\mathbf{r} - [\mathbf{a}\ \mathbf{b}\ \mathbf{c}]\mathbf{r}$$

$$= 2[\mathbf{a}\ \mathbf{b}\ \mathbf{c}]\mathbf{r}$$

30. (c) It is given that **r** perpendicular (**a** + **b** + **c**).

$$\therefore \quad \mathbf{r}\cdot(\mathbf{a}+\mathbf{b}+\mathbf{c}) = 0$$

$$\Rightarrow l[\mathbf{a}\ \mathbf{b}\ \mathbf{c}] + m[\mathbf{c}\ \mathbf{a}\ \mathbf{b}] + n[\mathbf{a}\ \mathbf{b}\ \mathbf{c}] = 0$$

$$\Rightarrow \quad 2(l + m + n) = 0 \qquad [\because [\mathbf{a}\ \mathbf{b}\ \mathbf{c}] = 2]$$

$$\Rightarrow \quad l + m + n = 0$$

31. (b) We have,

$$[l\mathbf{a} + m\mathbf{b} + n\mathbf{c} \quad l\mathbf{b} + m\mathbf{c} + n\mathbf{a} \quad l\mathbf{c} + m\mathbf{a} + n\mathbf{b}] = 0$$

$$\Rightarrow \quad [l\mathbf{a} + m\mathbf{b} + n\mathbf{c} \quad n\mathbf{a} + l\mathbf{b} + m\mathbf{c} \quad m\mathbf{a} + n\mathbf{b} + l\mathbf{c}] = 0$$

$$\Rightarrow \quad \begin{vmatrix} l & m & n \\ n & l & m \\ m & n & l \end{vmatrix}[\mathbf{a}\ \mathbf{b}\ \mathbf{c}] = 0$$

$$\Rightarrow \quad \begin{vmatrix} l & m & n \\ n & l & m \\ m & n & l \end{vmatrix} = 0 \qquad [\because [\mathbf{a}\ \mathbf{b}\ \mathbf{c}] = 0]$$

$$\Rightarrow \quad l^3 + m^3 + n^2 - 3lmn = 0$$

$$\Rightarrow \quad (l + m + n)(l^2 + m^2 + n^2 - lm - mn - nl) = 0$$

$$\therefore \quad l + m + n = 0$$

32. (a) We have, $\mathbf{a}\cdot\mathbf{b} = \mathbf{a}\cdot\mathbf{c} = 0$

$$\Rightarrow \quad \mathbf{a}\perp\mathbf{b}, \mathbf{a}\perp\mathbf{c}$$

$$\Rightarrow \quad \mathbf{a} \parallel \mathbf{b}\times\mathbf{c}$$

Also, $|\mathbf{b}\times\mathbf{c}| = |\mathbf{b}||\mathbf{c}|\sin\frac{\pi}{3}$

$$= \frac{\sqrt{3}}{2} \qquad [\because |\mathbf{b}| = |\mathbf{c}| = 1]$$

$$\therefore \quad |[\mathbf{a}\ \mathbf{b}\ \mathbf{c}]| = |\mathbf{a}\cdot(\mathbf{b}\times\mathbf{c})|$$

$$= |\mathbf{a}||\mathbf{b}\times\mathbf{c}| = \frac{\sqrt{3}}{2}$$

33. (b) We have,

$$(\mathbf{a}\times\mathbf{b})\times(\mathbf{a}\times\mathbf{c}) = \{(\mathbf{a}\times\mathbf{b})\cdot\mathbf{c}\}\mathbf{a} - \{(\mathbf{a}\times\mathbf{b})\cdot\mathbf{a}\}\mathbf{c}$$

$$\Rightarrow \quad (\mathbf{a}\times\mathbf{b})\times(\mathbf{a}\times\mathbf{c}) = [\mathbf{a}\ \mathbf{b}\ \mathbf{c}]\mathbf{a} - [\mathbf{a}\ \mathbf{b}\ \mathbf{a}]\mathbf{c} = [\mathbf{a}\ \mathbf{b}\ \mathbf{c}]\mathbf{a}$$

Similarly, we have

$$(\mathbf{b}\times\mathbf{c})(\mathbf{b}\times\mathbf{a}) = [\mathbf{b}\ \mathbf{c}\ \mathbf{a}]\mathbf{b}$$

and $(\mathbf{c}\times\mathbf{a})\times(\mathbf{c}\times\mathbf{b}) = [\mathbf{c}\ \mathbf{a}\ \mathbf{b}]\mathbf{c}$

So, the given expression is equal to $[\mathbf{a}\ \mathbf{b}\ \mathbf{c}](\mathbf{a} + \mathbf{b} + \mathbf{c})$.

34. (b) We have,

$$[\mathbf{a}\ \mathbf{b}\ \hat{\mathbf{i}}]\hat{\mathbf{i}} + [\mathbf{a}\ \mathbf{b}\ \hat{\mathbf{j}}]\hat{\mathbf{j}} + [\mathbf{a}\ \mathbf{b}\ \hat{\mathbf{k}}]\hat{\mathbf{k}}$$

$$= \{(\mathbf{a}\times\mathbf{b}\cdot\hat{\mathbf{i}}\}\hat{\mathbf{i}} + \{(\mathbf{a}\times\mathbf{b})\cdot\hat{\mathbf{j}}\}\hat{\mathbf{j}} + \{(\hat{\mathbf{a}}\times\hat{\mathbf{b}})\cdot\hat{\mathbf{k}}]\hat{\mathbf{k}}$$

$$= \mathbf{a}\times\mathbf{b}$$

35. (b) $(\mathbf{a}\times\mathbf{b})\times(\mathbf{a}\times\mathbf{c}) = \{(\mathbf{a}\times\mathbf{b})\cdot\mathbf{c}\}\mathbf{a} - \{(\mathbf{a}\times\mathbf{b})\cdot\mathbf{a}\}\mathbf{c}$

$$= [\mathbf{abc}]\mathbf{a}$$

Similarly, $(\mathbf{b}\times\mathbf{c})\times(\mathbf{c}\times\mathbf{a}) = [\mathbf{bca}]\mathbf{b}$

and $(\mathbf{c}\times\mathbf{a})\times(\mathbf{c}\times\mathbf{b}) = [\mathbf{cab}]\mathbf{c}$

⇒ Given expression is equal to

$$(\mathbf{a} + \mathbf{b} + \mathbf{c})[\mathbf{abc}]$$

36. (c) Clearly, **RS** + **ST** = **RT**, which is not parallel to **PQ**.

$$\therefore \quad \mathbf{PQ}\times(\mathbf{RS} + \mathbf{ST}) \neq 0$$

So, Statement I is correct.

Also, PQ is not parallel to **RS**.

$$\therefore \quad \mathbf{PQ}\times\mathbf{RS} \neq 0$$

So, Statement II is not correct.

37. (b) Both the statements are true and statement 2 is not the correct explanation of statement 1.

Because **b**, **c**, **d** is statement-1 are coplanar.

38. (a) $3\mathbf{a} - 2\mathbf{b} + 5\mathbf{c} - 6\mathbf{d} = (2\mathbf{a} - 2\mathbf{b}) + (-5\mathbf{a} + 5\mathbf{c}) + (6\mathbf{a} - 6\mathbf{d})$

$$= -2\mathbf{AB} + 5\mathbf{AC} - 6\mathbf{AD} = 0$$

∴ **AB**, **AC** and **AD** are linearly dependent, hence by statement 2, the statement 1 is true.

39. (d) $(\mathbf{a}\times(\mathbf{b}\times\mathbf{c}))\cdot(\mathbf{a}\times\mathbf{c}) = 5$

$$\Rightarrow \quad (\mathbf{a}\cdot\mathbf{c})[\mathbf{b}\mathbf{a}\ \mathbf{c}] = 5$$

$$\Rightarrow \quad [\mathbf{b}\mathbf{a}\ \mathbf{c}] = 10$$

$$\therefore \quad [\mathbf{a}\mathbf{b}\ \mathbf{c}] = -10$$

Statement 1 is false.

40. (a) Let $\alpha = \hat{i} + a\hat{j}$, $\beta = a\hat{i} + \hat{j} + \hat{k}$ and $\gamma = \hat{j} + a\hat{k}$

$\therefore$ Volume of the parallelopiped with α, β, γ as its conterminous sides is

$$V = [\alpha\,\beta\,\gamma] = \begin{vmatrix} 1 & a & 0 \\ a & 1 & 1 \\ 0 & 1 & a \end{vmatrix}$$

$$= 1(a-1) - a(a^2 - 0) + 0(a - 0)$$

$$= a - 1 - a^3$$

Now, $\frac{dV}{da} = 1 - 3a^2$

and $\frac{d^2V}{da^2} = -6a$

For extremum, $\frac{dV}{da} = 0$

$\Rightarrow \quad 1 - 3a^2 = 0$

$\Rightarrow \quad a = \pm \frac{1}{\sqrt{3}}$

$\therefore$ At $a = \frac{1}{\sqrt{3}}$,

$$\frac{d^2V}{da^2} = -\frac{6}{\sqrt{3}} \text{ (negative)}$$

Hence, at $a = \frac{1}{\sqrt{3}}$, V is maximum.

The volume of the parallelopiped having three conterminous edges **a**, **b** and **c** is [**a b c**].

Revisal Problems (JEE Advanced)

1. (d) Since, **a**, **b** and **c** are linearly independent,

$$[\mathbf{a\ b\ c}] \neq 0$$

We know that

$$[\mathbf{a}x_1 + \mathbf{b}x_2 + \mathbf{c}x_3 \;\; \mathbf{a}y_1 + \mathbf{b}y_2 + \mathbf{c}y_3 \;\; \mathbf{a}z_1 + \mathbf{b}z_2 + \mathbf{c}z_3]$$

$$= [\mathbf{a\ b\ c}] = \begin{vmatrix} x_1 & x_2 & x_3 \\ y_1 & y_2 & y_3 \\ z_1 & z_2 & z_3 \end{vmatrix}$$

Hence, the given system can be written as

$$[\mathbf{a\ b\ c}] \begin{vmatrix} \cos\theta & \sin\theta & \cos 2\theta \\ \cos\left(\frac{2\pi}{3} + \theta\right) & \sin\left(\frac{2\pi}{3} + \theta\right) & \cos\left(\theta + \frac{2\pi}{3}\right) \\ \cos\left(\theta - \frac{2\pi}{3}\right) & \sin\left(\theta - \frac{2\pi}{3}\right) & \cos 2\left(\theta - \frac{2\pi}{3}\right) \end{vmatrix}$$

On applying $R_2 \to R_3 + R_2 + R_1$, we get

$$[\mathbf{a\ b\ c}] \begin{vmatrix} \cos\theta & \sin\theta & \cos 2\theta \\ \cos\left(\frac{2\pi}{3} + \theta\right) & \sin\left(\frac{2\pi}{3} + \theta\right) & \cos\left(\theta + \frac{2\pi}{3}\right) \\ 0 & 0 & -2\cos\left(3\theta + \frac{\pi}{3}\right)\cdot\cos\theta \end{vmatrix}$$

$$\Rightarrow \quad [\mathbf{a\ b\ c}] - 2\cos\theta \cdot \cos\left(3\theta + \frac{\pi}{3}\right) \cdot \frac{\sqrt{3}}{2}$$

2. (a) $\because$ $\quad \mathbf{AB} + \mathbf{BC} = \mathbf{AC}$

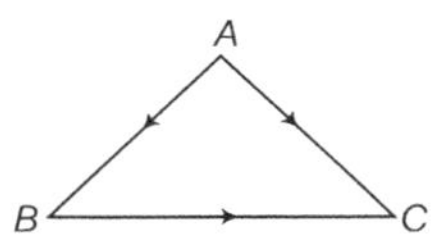

$\therefore \quad \mathbf{AB} = \mathbf{AC} - \mathbf{BC}$

$$= \frac{2\mathbf{e}}{|\mathbf{e}|} - \frac{\mathbf{e}}{|\mathbf{e}|} + \frac{\mathbf{f}}{|\mathbf{f}|} = \frac{\mathbf{e}}{|\mathbf{e}|} + \frac{\mathbf{f}}{|\mathbf{f}|}$$

Now, $\quad \mathbf{AB}\cdot\mathbf{BC} = \left(\frac{\mathbf{e}}{|\mathbf{e}|} + \frac{\mathbf{f}}{|\mathbf{f}|}\right)\cdot\left(\frac{\mathbf{e}}{|\mathbf{e}|} - \frac{\mathbf{f}}{|\mathbf{f}|}\right)$

$$= 1 - 1 = 0$$

$\because \quad \mathbf{AB} \perp \mathbf{BC}$

$$\angle B = 90^\circ$$

We know that in a $\Delta\, ABC$

$$\cos 2A + \cos 2B + \cos 2C = -1 + 4\cos A \cos B \cos C$$

$\therefore \quad \cos 2A + \cos 2B + \cos 2C = -1$

3. (c) Given, $\quad |\mathbf{a}| = |\mathbf{b}| = |\mathbf{c}| = 1$

and **c** is equally inclined to **a** and **b**,

$\because$ θ be the angle between **a** and **c**

$\therefore \quad \cos\theta = \mathbf{a}\cdot\mathbf{c} = \mathbf{b}\cdot\mathbf{c}$

We have, $\quad \mathbf{c} = \alpha\mathbf{a} + \beta\mathbf{b} + \gamma(\mathbf{a}\times\mathbf{b}) \quad \ldots(i)$

On taking dot product by **a**, we get

$$\mathbf{a}\cdot\mathbf{c} = \cos\theta = \alpha|\mathbf{a}|^2 + \beta\mathbf{a}\cdot\mathbf{b} + \gamma[\mathbf{a}\cdot\mathbf{ab}]$$

$\because \quad \alpha = \cos\theta$

Similarly taking dot product by **b** of Eq. (i), we get

$$\beta = \cos\theta$$

On taking modulus of Eq. (i), we get

$$|\mathbf{c}| = |\cos\theta\,\mathbf{a} + \cos\theta\,\mathbf{b} + \gamma(\mathbf{a}\times\mathbf{b})|$$

On squaring both sides, we get

$$|\mathbf{c}|^2 = 1 = \cos^2\theta|\mathbf{a}|^2 + \cos^2\theta|\mathbf{b}|^2 + \gamma^2(\mathbf{a}\times\mathbf{b})^2$$

$$1 = \cos^2\theta + \cos^2\theta + \gamma^2|\mathbf{a}|\,|\mathbf{b}|^2\sin^2 90^\circ$$

$\Rightarrow \quad 1 = 2\cos^2\theta + \gamma^2$

$\Rightarrow \quad \gamma^2 = 1 - 2\cos^2\theta = -\cos 2\theta$

$\Rightarrow \quad \alpha = \beta = \cos\theta$ and $\gamma^2 = -\cos 2\theta$

4. (a) We have, $\quad (\mathbf{a} + 3\mathbf{b})\cdot(7\mathbf{a} - 5\mathbf{b}) = 0$

$\Rightarrow \quad 7|\mathbf{a}|^2 + 16\mathbf{a}\cdot\mathbf{b} - 15|\mathbf{b}|^2 = 0 \quad \ldots(i)$

and $\quad (\mathbf{a} - 4\mathbf{b})\cdot(7\mathbf{a} - 2\mathbf{b}) = 0$

$\Rightarrow \quad 7|\mathbf{a}|^2 - 30\,\mathbf{a}\cdot\mathbf{b} + 8|\mathbf{b}|^2 = 0 \quad \ldots(ii)$

From Eqs. (i) and (ii), we get

$$\mathbf{a}\cdot\mathbf{b} = \frac{15|\mathbf{b}|^2 - 7|\mathbf{a}|^2}{16}$$

$$\mathbf{a}\cdot\mathbf{b} = \frac{1}{30}(7|\mathbf{a}|^2 + 8|\mathbf{b}|^2)$$

$\Rightarrow \quad \frac{15|\mathbf{b}|^2 - 7|\mathbf{a}|^2}{8} = \frac{1}{15}(7|\mathbf{a}|^2 + 8|\mathbf{b}|)^2$

$\Rightarrow \quad 15(15|\mathbf{b}|^2 - 7|\mathbf{a}|^2 = 8(7|\mathbf{a}|^2 + 8|\mathbf{b}|^2)$

$\Rightarrow \quad 225|\mathbf{b}|^2 - 105|\mathbf{a}|^2 = 56|\mathbf{a}|^2 + 64|\mathbf{b}|^2$

$\Rightarrow \quad 161|\mathbf{b}|^2 = 161|\mathbf{a}|^2$

$\Rightarrow \quad |\mathbf{b}|^2 = |\mathbf{a}|^2$

From Eq. (i), we get

$$16\mathbf{a}\cdot\mathbf{b} = 15|\mathbf{b}|^2 - 7|\mathbf{a}|^2 = 15|\mathbf{b}|^2 - 7|\mathbf{b}|^2$$

$\Rightarrow \quad 16\mathbf{a}\cdot\mathbf{b} = 8|\mathbf{b}|^2$

$\Rightarrow \quad \mathbf{a}\cdot\mathbf{b} = \frac{1}{2}|\mathbf{b}|^2$

$\Rightarrow \quad |\mathbf{a}|\,|\mathbf{b}|\cos\theta = \frac{1}{2}|\mathbf{b}|^2$

$\Rightarrow \quad \cos\theta = \frac{1}{2}$

$\therefore \quad \theta = 60^\circ$

5. (b) The coordinates of O (0, 0, 0), A (1, 0, 0), B (0, 1, 0), C (0, 0, 1), D (1, 1, 0), E (0, 1, 1), F (1, 0, 1), G (1, 1, 1)

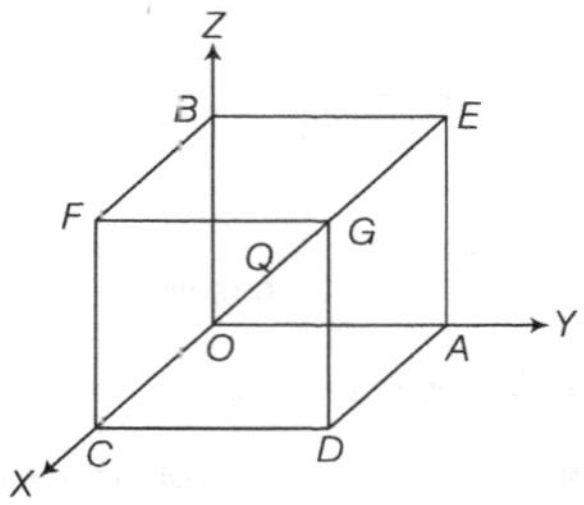

Since, OG is a diagonal of the cube.

Equation of OG is $\frac{x-1}{1} = \frac{y-1}{1} = \frac{z-1}{1} = \lambda$

Let the foot of perpendicular drawn from D (1, 1, 0) be the point Q (λ).

$\because \quad Q = (\lambda + 1, \lambda + 1, \lambda + 1)$

$\because$ DR's of DQ is $< \lambda, \lambda, \lambda + 1 >$ and DR's of OG is $< 1, 1, 1 >$

$\Rightarrow DQ \perp OG$, hence $\lambda + \lambda + \lambda + 1 = 0$

$$\lambda = -\frac{1}{3}$$

$$\Rightarrow \quad Q = \left(\frac{2}{3}, \frac{2}{3}, \frac{2}{3}\right)$$

$$\therefore \quad DQ = \sqrt{\left(1-\frac{2}{3}\right)^2 + \left(1-\frac{2}{3}\right)^2 + \left(\frac{2}{3}\right)^2}$$

$$= \sqrt{\frac{1}{9} + \frac{1}{9} + \frac{4}{9}} = \sqrt{\frac{6}{9}} = \sqrt{\frac{2}{3}}$$

6. (c) $\mathbf{p} \times \mathbf{q}$, $\mathbf{p}$, $\mathbf{q}$ are non-coplanar vectors

$\Rightarrow \quad b - c = 0, c - a = 0, a - b = 0$

$\Rightarrow \quad a = b = c$

$\Rightarrow \quad < 1$ is equilateral.

7. (d) Translating the axes through $A(1, 2, 3)$. A changes to (0, 0, 0), B changes to (8,6,2).

$\therefore$ Coterminus edges are of length 8, 6, 2.

Volume of parallelopiped

$$= 8 \times 6 \times 2 = 96 \text{ cu units}$$

8. (d) Given, $\mathbf{d} \cdot (\mathbf{a} + \mathbf{b} + \mathbf{c}) = 0$

and $\quad \mathbf{d} = \sin x\,(\mathbf{a} \times \mathbf{b}) + \cos y\,(\mathbf{b} \times \mathbf{c}) + 2\,(\mathbf{c} \times \mathbf{a})$ ...(i)

$\because \quad \mathbf{a} \cdot \mathbf{d} = \cos y\,[\mathbf{a}\ \mathbf{b}\ \mathbf{c}]$ ...(ii)

$\mathbf{b} \cdot \mathbf{d} = 2\,[\mathbf{b}\ \mathbf{c}\ \mathbf{a}]$ ...(iii)

$\mathbf{c} \cdot \mathbf{d} = \sin x\,[\mathbf{a}\ \mathbf{b}\ \mathbf{c}]$ ...(iv)

On adding Eqs. (ii), (iii) and (iv), we get

$$\mathbf{a} \cdot \mathbf{d} + \mathbf{b} \cdot \mathbf{d} + \mathbf{c} \cdot \mathbf{d} = (\cos y + 2 + \sin x)\,[\mathbf{a}\ \mathbf{b}\ \mathbf{c}]$$

$\because \quad \sin x + \cos y + 2 = 0$

$\Rightarrow \quad \sin x + \cos y = -2$

$\Rightarrow \quad \sin x = -1 \quad [\because -1 < \sin x \le 1 \text{ and } -1 \le \cos y \le 1]$

and $\quad \cos y = -1$

Since, we have to find the minimum value of

$$x^2 + y^2;\ x = -\frac{\pi}{2}, y = \pi$$

$$\therefore \quad x^2 + y^2 = \frac{\pi^2}{4} + \pi^2 = \frac{5\pi^2}{4}$$

9. (a)

$\alpha x + \beta y = 0$ ...(i)

$\beta y + \gamma z = 0$ ...(ii)

$\gamma z + \alpha x = 0$ ...(iii)

$\alpha x + \beta y + \gamma z = p$ ...(iv)

(iv) cuts (ii) at $\left(0, 0, \frac{p}{\gamma}\right)$, (iv) cuts (ii) at $\left(\frac{p}{\gamma}, 0, 0\right)$

(iv) cuts (iii) at $\left(0, \frac{p}{\beta}, 0\right)$,

from (i), (ii), (iii) are coordinate planes intersecting at (0, 0, 0). So lengths of the coterminus edges are $\left|\frac{p}{\alpha}\right|, \left|\frac{p}{\beta}\right|, \left|\frac{p}{\gamma}\right|$ so volume of tetrahedron is $\left(\frac{p^3}{6\alpha\beta\gamma}\right)$ cu. units

10. (b) We have, $\mathbf{p} \cdot \mathbf{q} = 0$ and $\mathbf{r} \cdot \mathbf{s} = 0$

$$\mathbf{x} = \frac{1}{3}(\mathbf{p} + \mathbf{r} + \mathbf{s}) = \frac{1}{3}(5\mathbf{a} - 3\mathbf{b} - 4\mathbf{a} - \mathbf{b} - \mathbf{a} + \mathbf{b})$$

$$= \frac{1}{3}(-3\mathbf{b}) = -\mathbf{b}$$

$$\mathbf{y} = \frac{1}{5}(\mathbf{r} + \mathbf{s}) = \frac{1}{5}(-4\mathbf{a} - \mathbf{b} - \mathbf{a} + \mathbf{b}) = \frac{1}{5}(-5\mathbf{a}) = -\mathbf{a}$$

Now, let θ is the angle between $\mathbf{x}$ and $\mathbf{y}$, then

$$\cos\theta = \frac{\mathbf{x} \cdot \mathbf{y}}{|\mathbf{x}|\,|\mathbf{y}|} = \frac{(-\mathbf{b}) \cdot (-\mathbf{a})}{|\mathbf{b}|\,|\mathbf{a}|} = \frac{\mathbf{a} \cdot \mathbf{b}}{|\mathbf{a}|\,|\mathbf{b}|}$$

Now, $\quad \mathbf{p} \cdot \mathbf{q} = (5\mathbf{a} - 3\mathbf{b}) \cdot (-\mathbf{a} - 2\mathbf{b})$

$= -5\mathbf{a}^2 + 6\mathbf{b}^2 - 7\mathbf{a} \cdot \mathbf{b} = 0$ ...(i)

and $\mathbf{r} \cdot \mathbf{s} = (-4\mathbf{a} - \mathbf{b}) \cdot (-\mathbf{a} + \mathbf{b}) = 4\mathbf{a}^2 - (\mathbf{b}^2) - 3\mathbf{a} \cdot \mathbf{b} = 0$ ...(ii)

From Eqs. (i) and (ii), we get

$$\mathbf{a} \cdot \mathbf{b} = \frac{6|\mathbf{b}|^2 - 5|\mathbf{a}|^2}{7} = \frac{4|\mathbf{a}|^2 - |\mathbf{b}|^2}{3}$$

$\Rightarrow \quad 18|\mathbf{b}|^2 - 15|\mathbf{a}|^2 = 28|\mathbf{a}|^2 - 7|\mathbf{b}|^2$

$\Rightarrow \quad 25|\mathbf{b}|^2 = 43|\mathbf{a}|^2$

$\Rightarrow \quad |\mathbf{b}|^2 = \frac{43}{25}|\mathbf{a}|^2$

On substituting the value of $|\mathbf{b}|^2$ in Eq. (i)

$$7\mathbf{a} \cdot \mathbf{b} = -5|\mathbf{a}|^2 + 6 \cdot \frac{43}{25}|\mathbf{a}|^2$$

$$\Rightarrow \quad 7\mathbf{a} \cdot \mathbf{b} = \frac{-125|\mathbf{a}|^2 + 258|\mathbf{a}|^2}{25}$$

$$\Rightarrow \quad 7\mathbf{a} \cdot \mathbf{b} = \frac{133}{25}|\mathbf{a}|^2 \Rightarrow \mathbf{a} \cdot \mathbf{b} = \frac{133}{175}|\mathbf{a}|^2$$

$$\therefore \quad \cos\theta = \frac{\mathbf{a} \cdot \mathbf{b}}{|\mathbf{a}|\,|\mathbf{b}|} = \frac{133|\mathbf{a}|^2}{175|\mathbf{a}|\,|\mathbf{b}|} = \frac{133\,|\mathbf{a}|}{175\,|\mathbf{b}|}$$

$$= \frac{133}{175} \times \sqrt{\frac{25}{43}} = \frac{133 \times 5}{175\sqrt{43}}$$

$$= \frac{133}{35\sqrt{43}} = \frac{19}{5\sqrt{43}}$$

$$\Rightarrow \quad \cos\theta = \frac{19}{5\sqrt{43}}$$

$$\therefore \quad \theta = \cos^{-1}\left(\frac{19}{5\sqrt{43}}\right)$$

11. (d) $\mathbf{a} \cdot \mathbf{b} = \mathbf{a} \cdot \mathbf{c} = 0, |\mathbf{a}| = |\mathbf{c}| = 1, |\mathbf{b}| = 2$ and $\mathbf{b} \cdot \mathbf{c} = 1$

let $\quad \mathbf{d} = x(\mathbf{a} + \mathbf{b}) + y(2\mathbf{b} - \mathbf{c})$

But $\quad \mathbf{d} \cdot \mathbf{a} = 1$

$\Rightarrow \quad x(1 + 0) + 0 = 1$

$\Rightarrow \quad x = 1$

$\Rightarrow \quad \mathbf{d} = \mathbf{a} + \mathbf{b} + y(2\mathbf{b} - \mathbf{c})$

$\Rightarrow \quad |\mathbf{d}|^2 = |\mathbf{a}|^2 + |\mathbf{b}| + 2\mathbf{a} \cdot \mathbf{b} + y^2(2\mathbf{b} - \mathbf{c})^2 + 2y(\mathbf{a} + \mathbf{b}) \cdot (2\mathbf{b} - \mathbf{c})$

$\Rightarrow \quad |\mathbf{d}|^2 = 1 + 4 + y^2(16 + 1 - 4) + 2y\,(8 - 1)$

$= 13y^2 - 14y + 5$

$$\therefore \quad |d|_{\min} = \sqrt{\frac{4 \times 13 \times 5 - 14 \times 14}{4 \times 13}} = \frac{4}{\sqrt{13}}$$

12. (b) We have, **P** and **Q** are parallel i.e. $\frac{3ax^2}{b\,(x-1)} = -\frac{2\,(x-1)}{x}$

$\Rightarrow \quad 3ax^3 = -2b\,(x-1)^2$

$\Rightarrow \quad 3ax^3 = -2b\,(x^2+1-2x)$

$\Rightarrow \quad 3ax^3 + 2bx^2 - 4bx + 2b = 0$

Let polynomial be

$f(x) = 3ax^3 + 2bx^2 - 4bx + 2b$

$\Rightarrow \quad f(0) = -2b$

$\therefore \quad f(1) = 3a + 2b - 4b + 2b = 3a$

Clearly, we have $ab < 0$

i.e. a and b is of opposite in sign

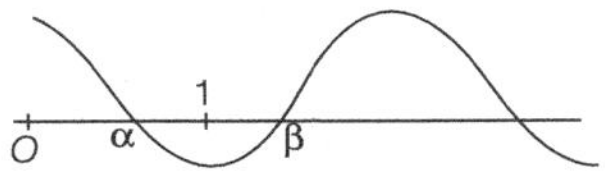

$f(0)$ and $f(1)$ is of opposite in sign, this shows atleast one of the roots lies in the interval $0 < x < 1$.

13. (c) Given, $\mathbf{r}\cdot\mathbf{a} = 0$

$|\mathbf{r}\times\mathbf{b}| = |\mathbf{r}|\,|\mathbf{b}|$ and $|\mathbf{r}\times\mathbf{c}| = |\mathbf{r}|\,|\mathbf{c}|$

This shows **r** is perpendicular to both **b** and **c**.

$\Rightarrow$ **r** is perpendicular **a**, **b** and **c**.

$\therefore \quad [\mathbf{a}\ \mathbf{b}\ \mathbf{c}] = 0$

14. (c) We have,

$\{(\mathbf{a}\times\mathbf{b})\times(\mathbf{c}\times\mathbf{d})\}\cdot(\mathbf{a}\times\mathbf{d}) = 0$

take, $\mathbf{a}\times\mathbf{b} = \mathit{l}$

$\Rightarrow \quad l\times(\mathbf{c}\times\mathbf{d}) = (l\cdot\mathbf{d})\,\mathbf{c} - (l\cdot\mathbf{c})\,\mathbf{d}$

$= \{(\mathbf{a}\times\mathbf{b})\cdot\mathbf{d}\}\,\mathbf{c} - \{(\mathbf{a}\times\mathbf{b})\cdot\mathbf{c}\}\,\mathbf{d}$

$= [\mathbf{a}\ \mathbf{b}\ \mathbf{d}]\,\mathbf{c} - [\mathbf{a}\ \mathbf{b}\ \mathbf{c}]\,\mathbf{d}$

$\Rightarrow \quad ([\mathbf{a}\ \mathbf{b}\ \mathbf{d}]\,\mathbf{c} - [\mathbf{a}\ \mathbf{b}\ \mathbf{c}]\,\mathbf{d})\cdot(\mathbf{a}\times\mathbf{d})$

$\Rightarrow \quad [\mathbf{a}\ \mathbf{b}\ \mathbf{d}]\,[\mathbf{c}\ \mathbf{a}\ \mathbf{d}] - [\mathbf{a}\ \mathbf{b}\ \mathbf{c}]\cdot[\mathbf{d}\ \mathbf{a}\ \mathbf{d}] = 0$

$\Rightarrow \quad [\mathbf{a}\ \mathbf{b}\ \mathbf{d}]\,[\mathbf{c}\ \mathbf{a}\ \mathbf{d}] = 0$

$\Rightarrow \quad [\mathbf{a}\ \mathbf{b}\ \mathbf{d}] = 0$ or $[\mathbf{c}\ \mathbf{a}\ \mathbf{d}] = 0$

$\Rightarrow \quad [\mathbf{a}\ \mathbf{d}\ \mathbf{b}] = 0$ or $[\mathbf{a}\ \mathbf{d}\ \mathbf{c}] = 0$

$\Rightarrow \quad (\mathbf{a}\times\mathbf{d})\cdot\mathbf{b} = 0$ or $(\mathbf{a}\times\mathbf{d})\cdot\mathbf{c} = 0$

which shows either **b** or **c** must lie in plane of $\mathbf{a} \perp \mathbf{d}$,

15. (a,d) We have, $|\mathbf{a}| = |\mathbf{b}| = 1$ and $\mathbf{a}\cdot\mathbf{b} = 0$

and $\quad \mathbf{c} = \lambda_1\mathbf{a} + \lambda_2\mathbf{b} + \lambda_3(\mathbf{a}\times\mathbf{b}) \quad \ldots(\text{i})$

Taking dot product by **a** of Eq. (i) we get

$\mathbf{a}\cdot\mathbf{c} = \lambda_1|\mathbf{a}|^2$

$\therefore \quad \lambda_1 = \mathbf{a}\cdot c$

Similarly, $\quad \mathbf{b}\cdot\mathbf{c} = \lambda_2|\mathbf{b}|^2 = \lambda_2$

$\therefore \quad \lambda_2 = \mathbf{b}\cdot\mathbf{c}$

Now, taking dot product by $(\mathbf{a}\times\mathbf{b})$ of Eq. (i), we get

$\mathbf{c}\cdot(\mathbf{a}\times\mathbf{b}) = \lambda_3|\mathbf{a}\times\mathbf{b}|^2 = \lambda_3|\mathbf{a}|^2|\mathbf{b}|^2\sin^2 90^\circ$

$\Rightarrow \quad \lambda_3 = \mathbf{c}\cdot(\mathbf{a}\times\mathbf{b})$

$\therefore \quad \lambda_1 + \lambda_2 + \lambda_3 = \mathbf{a}\cdot\mathbf{c} + \mathbf{b}\cdot\mathbf{c} + (\mathbf{a}\times\mathbf{b})\cdot\mathbf{c}$

$= (\mathbf{a} + \mathbf{b} + \mathbf{a}\times\mathbf{b})\cdot\mathbf{c}$

16. (a,b,c) We have, S be the centre of sphere.

$\mathbf{SO} = |\mathbf{SA}| = |\mathbf{SB}| = |\mathbf{SC}| = |\mathbf{x}|$

$\Rightarrow \quad |\mathbf{x} - \mathbf{a}| = x$

$\Rightarrow \quad |\mathbf{x} - \mathbf{a}|^2 = x^2$

$\Rightarrow \quad x^2 - 2\mathbf{x}\cdot\mathbf{a} + a^2 = x^2$

$\Rightarrow \quad 2\mathbf{x}\cdot\mathbf{a} = a^2$

Similarly, $2\mathbf{x}\cdot\mathbf{b} = b^2$ and $2\mathbf{x}\cdot\mathbf{c} = c^2$

We have **a**, **b**, **c** are non-coplanar, then

$\mathbf{x} = l\mathbf{a} + m\mathbf{b} + n\mathbf{c}$

i.e. linear combination of **a**, **b** and **c**, also $\mathbf{a}\times\mathbf{b}$, $\mathbf{b}\times\mathbf{c}$ and $\mathbf{c}\times\mathbf{a}$ are non-coplanar.

$x = l\,(\mathbf{b}\times\mathbf{c}) + m\,(\mathbf{c}\times\mathbf{a}) + n\,(\mathbf{a}\times\mathbf{b}) \quad \ldots(\text{i})$

i.e. linear combination of $\mathbf{b}\times\mathbf{c}$, $\mathbf{c}\times\mathbf{a}$ and $\mathbf{a}\times\mathbf{b}$.

Taking dot product by **a** in Eq. (i), we get

$\mathbf{a}\cdot\mathbf{x} = l\,[\mathbf{a}\ \mathbf{b}\ \mathbf{c}]$

$\Rightarrow \quad \frac{a^2}{2} = l\,[\mathbf{a}\ \mathbf{b}\ \mathbf{c}]$

$\Rightarrow \quad l = \frac{a^2}{2\,[\mathbf{a}\ \mathbf{b}\ \mathbf{c}]}$

Similarly, $\quad m = \frac{b^2}{2\,[\mathbf{a}\ \mathbf{b}\ \mathbf{c}]}$

and $\quad n = \frac{c^2}{2\,[\mathbf{a}\ \mathbf{b}\ \mathbf{c}]}$

$\therefore \quad \mathbf{x} = \frac{a^2(\mathbf{b}\times\mathbf{c}) + b^2(\mathbf{c}\times\mathbf{a}) + c^2(\mathbf{a}\times\mathbf{b})}{2\,[\mathbf{a}\ \mathbf{b}\ \mathbf{c}]}$

17. (a, c) Given, $\mathbf{a} = \hat{i} + \hat{j} + \hat{k}$ and $\mathbf{b} = \hat{i} - \hat{j}$

$(a\cdot\hat{i})\hat{i} + (a\cdot\hat{j})\hat{j} + (\mathbf{a}\cdot\hat{k})\,\hat{k} = \hat{i} + \hat{j} + \hat{k} = \mathbf{x} \quad$ (say)

and $\quad (b\cdot\hat{i})\hat{i} + (b\cdot\hat{j})\hat{j} + (b\cdot\hat{k})\hat{k} = \hat{i} - \hat{j} = \mathbf{y} \quad$ (say)

and $\quad \hat{i} + \hat{j} - 2\hat{k} = \hat{z} \quad$ (say)

Clearly, $\quad \mathbf{x}\cdot\mathbf{y} = \mathbf{yz} = \mathbf{z}\cdot\mathbf{x} = 0$

$\therefore$ **x**, **y** and **z** are mutually perpendicular

Volume of parallelopiped $= \begin{vmatrix} 1 & 1 & 1 \\ 1 & -1 & 0 \\ 1 & 1 & -2 \end{vmatrix}$

$= 1\,(2-0) - 1\,(-2-0) + 1\,(1+1)$

$= 2 + 2 + 2 = 6$

$\because$ **x**, **y** and **z** are not coplanar, i.e. $[\mathbf{x}\ \mathbf{y}\ \mathbf{z}] \neq 0$.

$\therefore$ Volume of parallelopiped formed by **x**, **y** and **z** is 6 cu units.

18. (b,c,d) We have, $|\mathbf{a}| = |\mathbf{b}| = 1$

Let θ be the angle between **a** and **b**

$|\mathbf{a} + \mathbf{b}| = 2\cos\frac{\theta}{2}$ and $|\mathbf{a} - \mathbf{b}| = 2\sin\frac{\theta}{2}$

$\because \quad 3\cos\frac{\theta}{2} + 4\sin\frac{\theta}{2} : \theta \in [0, \pi]$

$-5 \le 3\cos\frac{\theta}{2} + 4\sin\frac{\theta}{2} \le 5$

$\therefore$ The possible range are 3, 4 or 5.

19. (a, d) Given, $\quad \mathbf{a}\cdot\mathbf{b} = 0$

and $\quad pr + (\mathbf{r}\cdot\mathbf{b})\,\mathbf{a} = \mathbf{c} \quad \ldots(\text{i})$

On taking dot product by **b**, we get

$p\,(\mathbf{r}\cdot\mathbf{b}) + (\mathbf{r}\cdot\mathbf{b})\,\mathbf{a}\cdot\mathbf{b} = \mathbf{b}\cdot\mathbf{c}$

$\Rightarrow \quad p\,(\mathbf{r}\cdot\mathbf{b}) = \mathbf{b}\cdot\mathbf{c}$

$\Rightarrow \quad p\left(\frac{\mathbf{c} - pr}{\mathbf{a}}\right) = \mathbf{b}\cdot\mathbf{c}$

$\Rightarrow \quad p\mathbf{c} - p^2\mathbf{r} = (\mathbf{b}\cdot\mathbf{c})\,\mathbf{a}$

$\therefore \quad p^2 r = p\mathbf{c} - (\mathbf{b}\cdot\mathbf{c})\,\mathbf{a}$

$\Rightarrow \quad \mathbf{r} = \frac{\mathbf{c}}{p} - \frac{(\mathbf{b}\cdot\mathbf{c})}{p^2}\,\mathbf{a}$

$\Rightarrow \quad [\mathbf{r}\ \mathbf{a}\ \mathbf{c}] = 0$

20. (c,d) Given, $[\mathbf{a}\ \mathbf{b}\ \mathbf{c}] = 0$ **a** **b** and **c** are coplanar vectors.

Further **d** is equally inclined to **a**, **b** and **c**, then

$\mathbf{d}\cdot\mathbf{a} = \mathbf{d}\cdot\mathbf{b} = \mathbf{d}\cdot\mathbf{c} = 0$

$\therefore \quad \mathbf{d}\cdot\mathbf{x} = \mathbf{d}\cdot\mathbf{y} = \mathbf{d}\cdot\mathbf{z} = 0$

If $\mathbf{r} = \lambda\mathbf{x} + \mu\mathbf{y} - \delta\mathbf{z}$, then

$\mathbf{d}\cdot\mathbf{r} = 0$

21. (a)

(i) Clearly, $\mathbf{a}\cdot\mathbf{b}+\mathbf{b}\cdot\mathbf{c}+\mathbf{c}\cdot\mathbf{a}=0$

$\Rightarrow |\mathbf{a}+\mathbf{b}+\mathbf{c}|^2=|\mathbf{a}|^2+|\mathbf{b}|^2+|\mathbf{c}|^2+2(\mathbf{a}\cdot\mathbf{b}+\mathbf{b}\cdot\mathbf{c}+\mathbf{c}\cdot\mathbf{a})$

$=1+16+64+0=81$

$\therefore\quad \mathbf{a}+\mathbf{b}+\mathbf{c}=9$

(ii) We have reciprocal of $\mathbf{b}_1$, $\mathbf{b}_2$ and $\mathbf{b}_3$ and $\mathbf{a}_1\,\mathbf{a}_2$ and $\mathbf{a}_3$

$\therefore\quad [\mathbf{b}_1\ \mathbf{b}_2\ \mathbf{b}_3]\,[b_1^1\ b_2^1\ b_3^1]=1$

(iii) Area of quadrilateral $ABCD$ is

$=\frac{1}{2}|\mathbf{a}\times(2\mathbf{a}+3\mathbf{b})|+\frac{1}{2}|(2\mathbf{a}+3\mathbf{b})\times\mathbf{b}|$

$\Rightarrow\quad \frac{1}{2}[3(\mathbf{a}\times\mathbf{b})+2(\mathbf{a}\times\mathbf{b})]=\frac{5}{2}(\mathbf{a}\times\mathbf{b})$

Area of quadrilateral $=\frac{5}{2}$ area of parallelogram

$\therefore\quad \alpha=\frac{5}{2}$

(iv) We have, $\quad \mathbf{d}=x(\mathbf{a}\times\mathbf{b})+y(\mathbf{b}\times\mathbf{c})+z(\mathbf{c}\times\mathbf{a})\quad$ ...(i)

$\because\quad \mathbf{d}\cdot\mathbf{a}=y\{\mathbf{a}\cdot(\mathbf{b}\times\mathbf{c})\}=y[\mathbf{a}\ \mathbf{b}\ \mathbf{c}]\quad$ ...(ii)

$\therefore\quad \mathbf{b}\cdot\mathbf{d}=z[\mathbf{a}\ \mathbf{b}\ \mathbf{c}]\quad$ ...(iii)

$\mathbf{c}\cdot\mathbf{d}=x[\mathbf{a}\ \mathbf{b}\ \mathbf{c}]\quad$ ...(iv)

On adding Eqs. (ii), (iii) and (iv), we get

$[\mathbf{a}\ \mathbf{b}\ \mathbf{c}](x+y+z)=\mathbf{a}\cdot\mathbf{d}+\mathbf{b}\cdot\mathbf{d}+\mathbf{c}\cdot\mathbf{d}$

$\frac{1}{8}(x+y+z)=(\mathbf{a}+\mathbf{b}+\mathbf{c})\cdot\mathbf{d}$

$\because\quad x+y+z=8(\mathbf{a}+\mathbf{b}+\mathbf{c})\cdot\mathbf{d}$

$\therefore\quad R=8$

Hence, (i) → (q), (ii) → (s), (iii) → (p), (iv) → (r).

22. (b)

(i) We know that 3 vectors are coplanar, if $x\mathbf{a}+y\mathbf{b}+z\mathbf{c}=0$

Clearly, $-3\hat{i}+3\hat{j}+4\hat{k}$ and $\hat{i}+\hat{j}$ are two vectors lie in the plane.($\mathbf{a}+\mathbf{b}$ and $\mathbf{a}-\mathbf{b}$).

(ii) $\mathbf{a}\times\mathbf{b}$ is a vector which perpendicular to both $\mathbf{a}$ and $\mathbf{b}$.

$\therefore\ \mathbf{a}\times\mathbf{b}=\begin{vmatrix}\hat{i} & \hat{j} & \hat{k}\\ -1 & 2 & 2\\ -2 & 1 & 2\end{vmatrix}=\hat{i}(4-2)-\hat{j}(-2+4)+\hat{k}(-1+4)$

$=2\hat{i}-2\hat{j}+3\hat{k}$

(iii) $\mathbf{c}$ is a vector which is equally inclined to $\mathbf{a}$ and $\mathbf{b}$.

$\therefore\quad \mathbf{c}\cdot\mathbf{a}=\mathbf{c}\cdot\mathbf{b}$

Clearly, $\hat{i}-\hat{j}+5\hat{k}$ satisfies the condition.

(iv) $\mathbf{a}$, $\mathbf{b}$ and $\mathbf{c}$ are form the triangle

$\Rightarrow\quad \mathbf{AB}+\mathbf{BC}=\mathbf{AC}$

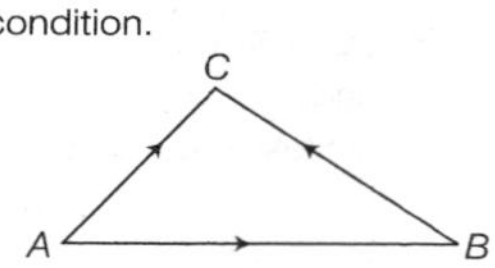

Hence,

(i) → (p,r), (ii) → (q), (iii) → (s), (iv) → (p).

23. (a)

(i) We have,

$|a\hat{i}+b\hat{j}+\hat{k}|=\sqrt{2}$

$\Rightarrow\quad \sqrt{a^2+b^2+1}=\sqrt{2}$

$\Rightarrow\quad a^2+b^2=1\quad$ ...(i)

Since, $a\hat{i}+b\hat{j}+\hat{k}$ is l nearly dependent with $\hat{i}+2\hat{j}$ and $\hat{j}-2\hat{k}$.

$\therefore$ They are coplanar $\begin{vmatrix}a & b & 1\\ 1 & 2 & 0\\ 0 & 1 & -2\end{vmatrix}=0$

$a(-4)-b(-2)+1=0$

$-4a+2b+1=0$

$2b=4a-1\quad$ ...(ii)

On solving Eqs. (i) and (ii), we get

$a^2+\left(\frac{4a-1}{2}\right)^2=1$

$\Rightarrow\quad a^2+\frac{16a^2+1-8a}{4}=1$

$\Rightarrow\quad 4a^2+16a^2+1-8a=4$

$\Rightarrow\quad 20a^2-8a-3=0$

$\Rightarrow\quad a_1+a_2=\frac{8}{20}$ and $a_1a_2=-\frac{3}{20}$

$\therefore\quad \frac{1}{a_1}+\frac{1}{a_2}=\frac{a_1+a_2}{a_1a_2}=\frac{\frac{8}{20}}{-\frac{3}{20}}=-\frac{8}{3}$

$\Rightarrow\quad \left|\frac{1}{a_1}+\frac{1}{a_2}\right|=\frac{8}{3}$

$\Rightarrow\quad \left[\left|\frac{1}{a_1}+\frac{1}{a_2}\right|=\left[\frac{8}{3}\right]\right]=2$

(ii) $\because\ \mathbf{AB}=\hat{i}(1-\alpha)-\hat{j}-\hat{k}$

Projection of $\mathbf{AB}$ on $\hat{i}+\hat{j}+\hat{k}$ is $\frac{1-\alpha-1-1}{\sqrt{3}}$

i.e. $\quad \frac{-\alpha-1}{\sqrt{3}}=\frac{8}{\sqrt{3}}$

$\Rightarrow\quad -\alpha-1=8\Rightarrow-\alpha=9$

$\Rightarrow\quad \alpha=-9$

and $\quad y=x^2-(\alpha+1)x+\beta^2$

$\Rightarrow\ y=x^2+8x+\beta^2$, it does not intersect the X-axis.

Now, $\Delta<0\ \Rightarrow\ 64-4\beta^2<0$

$\Rightarrow\quad 16-\beta^2<0\Rightarrow\beta^2>16$

$\Rightarrow\quad \beta<-4$ or $\beta>4\Rightarrow|\beta|=4$

$\therefore$ Minimum value of $|\beta|$ is 4.

(iii) Given, $\mathbf{a}\cdot\mathbf{b}=0$

We have to find the value of

$\frac{(3c+a)(3c-a)}{b^2}=\frac{9c^2-a^2}{b^2}$

and we have $\quad \mathbf{a}+2\mathbf{b}-3\mathbf{c}=0$

$\Rightarrow\quad 3\mathbf{c}=\mathbf{a}+2\mathbf{b}$

On squaring both sides, we get

$9|\mathbf{c}|^2=|\mathbf{a}|^2+4|\mathbf{b}|^2+4\mathbf{a}\cdot\mathbf{b}$

$\Rightarrow\quad 9c^2=a^2+4b^2+0$

$\Rightarrow\quad 9c^2-a^2=4b^2$

$\therefore\quad \frac{9c^2-a^2}{b^2}=4$

(iv) We have, $[\mathbf{a}\ \mathbf{b}\ \mathbf{c}]=1$

$(\mathbf{a}+2\mathbf{b}+3\mathbf{c})\cdot(\mathbf{a}\times\mathbf{b}+\mathbf{b}\times\mathbf{c}+\mathbf{c}\times\mathbf{a})$

$=[\mathbf{a}\ \mathbf{b}\ \mathbf{c}]+2[\mathbf{b}\ \mathbf{c}\ \mathbf{a}]+3[\mathbf{c}\ \mathbf{a}\ \mathbf{b}]$

$=1+2+3=6$

Hence, (i) → (q), (ii) → (r), (iii) → (r), (iv) → (s).

24. (a)

(i) $x\times y=(\mathbf{a}\times\mathbf{b})\times(\mathbf{a}-\mathbf{b})$

$=\mathbf{a}\times\mathbf{a}-\mathbf{a}\times\mathbf{b}+\mathbf{b}\times\mathbf{a}-\mathbf{b}\times\mathbf{b}=-2(\mathbf{a}\times\mathbf{b})$

LHS $=|x\times y|=|-2(\mathbf{a}\times\mathbf{b})|=2|\mathbf{a}||\mathbf{b}|\sin\theta$

RHS $=2\{\lambda-(\mathbf{a}\cdot\mathbf{b})^2\}^{\frac{1}{2}}=2\{\lambda-|\mathbf{a}|^2|\mathbf{b}|^2\cos^2\theta\}^{\frac{1}{2}}$

$\because\quad 8\sin\theta=2(\lambda-16\cos^2\theta)^{\frac{1}{2}}$

$\Rightarrow\quad 16\sin^2\theta=\lambda-16\cos^2\theta$

$\Rightarrow\quad 16(\sin^2\theta+\cos^2\theta)=\lambda$

$\therefore\quad \lambda=16$

(ii) We have, $\cos\frac{\pi}{3}=\frac{1}{2}=\frac{(\lambda\hat{i}+\hat{j}+\hat{k})\cdot(\hat{i}+\lambda\hat{j}+\hat{k})}{\sqrt{\lambda^2+2}\sqrt{\lambda^2+2}}$

$\because \quad \frac{1}{2}=\frac{\lambda+\lambda+1}{\lambda^2+2}$

$\Rightarrow \quad \lambda^2+2=2(2\lambda+1)$

$\Rightarrow \quad \lambda^2-4\lambda=0 \Rightarrow \lambda(\lambda-4)=0$

$\Rightarrow \quad \lambda=0$ and 4

$\therefore \quad \lambda=4$

(iii) We have,

$|\mathbf{a}-2\mathbf{b}|^2+|\mathbf{b}-2\mathbf{c}|^2+|\mathbf{c}-2\mathbf{a}|^2$

$=|\mathbf{a}|^2+4|\mathbf{b}|^2-4\mathbf{a}\cdot\mathbf{b}+|\mathbf{b}|^2+4|\mathbf{c}|^2-4\mathbf{b}\cdot\mathbf{c}+|\mathbf{c}|^2+4|\mathbf{a}|^2-4\mathbf{c}\cdot\mathbf{a}$

$\Rightarrow 1+4+1+16+4+4-4(\mathbf{a}\cdot\mathbf{b}+\mathbf{b}\cdot\mathbf{c}+\mathbf{c}\cdot\mathbf{a})$

$\Rightarrow \quad 30-4(\mathbf{a}\cdot\mathbf{b}+\mathbf{b}\cdot\mathbf{c}+\mathbf{c}\cdot\mathbf{a})$

We know that

$(\mathbf{a}+\mathbf{b}+\mathbf{c})^2=\Sigma\mathbf{a}^2+2\Sigma\mathbf{a}\cdot\mathbf{b}\geq 0$

i.e. $\quad 1+1+4+2(\mathbf{a}\cdot\mathbf{b}+\mathbf{b}\cdot\mathbf{c}+\mathbf{c}\cdot\mathbf{a})\geq 0$

$\Rightarrow \quad \mathbf{a}\cdot\mathbf{b}+\mathbf{b}\cdot\mathbf{c}+\mathbf{c}\cdot\mathbf{a}\geq -3$

$\Rightarrow \quad \geq 30-4(-3)\geq 30+12\geq 42$

Hence, (i) $\rightarrow$ (r), (ii) $\rightarrow$ (p), (iii) $\rightarrow$ (q)

Explanations for Q. Nos. (25-27)

Since, sum of three vectors $9\mathbf{AP}, 4\mathbf{BQ}$ and $7\mathbf{CR}$ is zero, there is a Δ whose sides have lengths $9|\mathbf{AP}|, 4|\mathbf{BQ}|, 7|\mathbf{CR}|$ and are parallel to the corresponding vectors.

H is orthocentre.

$\Rightarrow \angle BHP=90^\circ-\angle QBC=\angle ACB$

$\Rightarrow$ Angle between $\mathbf{AP}$ and $\mathbf{BQ}$ equal to $\angle ACB$, similarly angle between $\mathbf{BQ}$ and $\mathbf{CR}$ be $\angle BAC$

$\Rightarrow \quad \frac{AB}{7|\mathbf{CR}|}=\frac{BC}{9|\mathbf{AP}|}=\frac{AC}{4|\mathbf{BQ}|}$

$2.\ ar\cdot(\Delta ABC)=|\mathbf{AB}\times\mathbf{CR}|=|\mathbf{BC}\times\mathbf{AP}|=|\mathbf{CA}\times\mathbf{BQ}|$

$\Rightarrow \quad \frac{c^2}{7}=\frac{a^2}{9}=\frac{b^2}{4}$

$\therefore \quad a:b:c=3:2:\sqrt{7}$

$\Rightarrow \quad \cos C=\frac{a^2+b^2-c^2}{2ab}=\frac{1}{2}$

$\Rightarrow \quad \angle C=60^\circ$

and $\quad \cos B=\frac{a^2+b^2-c^2}{2ac}=\frac{2}{\sqrt{7}}$

$\cos A=\frac{c^2+b^2-a^2}{2bc}=\frac{1}{2\sqrt{7}}$

25. (b) **26.** (a) **27.** (b)

28. (a) We have, $\mathbf{n}_1, \mathbf{n}_2$ and $\mathbf{n}_3$ are non-coplanar vectors.

$\therefore\ \mathbf{n}_1\times\mathbf{n}_2, \mathbf{n}_2\times\mathbf{n}_3$ and $\mathbf{n}_3\times\mathbf{n}_1$ are non-coplanar vectors, so every vector can be written as

$\mathbf{r}=a(\mathbf{n}_1\times\mathbf{n}_2)+b(\mathbf{n}_2\times\mathbf{n}_3)+c(\mathbf{n}_3\times\mathbf{n}_1)$

Now, substituting this value in $\mathbf{r}\cdot\mathbf{n}_1=d_1$, **a**, **b**, we get

$a(\mathbf{n}_1\times\mathbf{n}_2)\cdot\mathbf{n}_1+b(\mathbf{n}_2\times\mathbf{n}_3)\cdot\mathbf{n}_1+c(\mathbf{n}_3\times\mathbf{n}_1)\cdot\mathbf{n}_1=d_1$

$\Rightarrow \quad b[\mathbf{n}_1\ \mathbf{n}_2\ \mathbf{n}_3]=d_1$

$\Rightarrow \quad b=\frac{d_1}{[\mathbf{n}_1\ \mathbf{n}_2\ \mathbf{n}_3]}$

Similarly, $\quad a=\frac{d_3}{[\mathbf{n}_1\ \mathbf{n}_2\ \mathbf{n}_3]}$ and $c=\frac{d_2}{[\mathbf{n}_1\ \mathbf{n}_2\ \mathbf{n}_3]}$

Hence, the required point of intersection will have the position vector

$\mathbf{r}=\frac{1}{[\mathbf{n}_1\ \mathbf{n}_2\ \mathbf{n}_3]}[d_3(\mathbf{n}_1\times\mathbf{n}_2)+d_1(\mathbf{n}_2\times\mathbf{n}_3)+d_2(\mathbf{n}_3\times\mathbf{n}_1)]$

29. (d) Equation of plane passing through the line of intersection of first two planes is

$\mathbf{r}\cdot(\mathbf{x}_1+\lambda\mathbf{x}_2)=d_1-\lambda d_2$, where λ is parameter.

Since, three planes have a common line of intersection, the above equation should be identical with the $\mathbf{r}\cdot\mathbf{x}_3=d_3$ for some λ

i.e. $\quad \mathbf{x}_1+\lambda\mathbf{x}_2=k\mathbf{x}_3 \quad$ ...(i)

$d_1+\lambda d_2=kd_3 \quad$ ...(ii)

From Eq. (i), we get

$\mathbf{x}_1\times\mathbf{x}_3+\lambda(\mathbf{x}_2\times\mathbf{x}_3)=0$

and $\quad \mathbf{x}_1\times\mathbf{x}_2=k(\mathbf{x}_3\times\mathbf{x}_2) \quad$ ...(iii)

From Eq. (ii), we get

$(d_1+\lambda d_2)(\mathbf{x}_2\times\mathbf{x}_3)=kd_3(\mathbf{x}_2\times\mathbf{x}_3)=d_3(\mathbf{x}_2\times\mathbf{x}_1) \quad$ ...(iv)

$d_1(\mathbf{x}_2\times\mathbf{x}_3)+\lambda d_2(\mathbf{x}_2\times\mathbf{x}_3)+d_3(\mathbf{x}_1+\mathbf{x}_2)=0$

$\Rightarrow \quad d_1(\mathbf{x}_2\times\mathbf{x}_3)+d_2(\mathbf{x}_3\times\mathbf{x}_1)+d_3(\mathbf{x}_1+\mathbf{x}_2)=0$

30. (d) Since, **a**, **b**, **c** are non-coplanar vectors, then

$[\mathbf{a}\ \mathbf{b}\ \mathbf{c}]\neq 0$

$\Rightarrow \quad [\mathbf{a}\ \mathbf{b}\ \mathbf{c}]^2\neq 0$

$\Rightarrow \quad [\mathbf{a}\ \mathbf{b}\ \mathbf{c}]^2=\begin{vmatrix}\mathbf{a}\cdot\mathbf{a} & \mathbf{b}\cdot\mathbf{a} & \mathbf{c}\cdot\mathbf{a}\\ \mathbf{a}\cdot\mathbf{b} & \mathbf{b}\cdot\mathbf{b} & \mathbf{c}\cdot\mathbf{b}\\ \mathbf{a}\cdot\mathbf{c} & \mathbf{b}\cdot\mathbf{c} & \mathbf{c}\cdot\mathbf{c}\end{vmatrix}\neq 0$

Since, any vector **r** in space can be expressed as a linear combination of three non-coplanar vectors.

So, let $\quad \mathbf{r}=l\mathbf{a}+m\mathbf{b}+n\mathbf{c} \quad$ ...(i)

taking dot product by **a**, **b**, **c** successively, we get

$\mathbf{r}\cdot\mathbf{a}=l\cdot\mathbf{a}\cdot\mathbf{a}+m\mathbf{a}\cdot\mathbf{b}+n\mathbf{c}\cdot\mathbf{a} \quad$ ...(ii)

$\mathbf{r}\cdot\mathbf{b}=l\mathbf{a}\cdot\mathbf{b}+m\mathbf{b}\cdot\mathbf{b}+n\mathbf{c}\cdot\mathbf{b} \quad$ ...(iii)

$\mathbf{r}\cdot\mathbf{c}=l\,\mathbf{a}\cdot\mathbf{c}+m\mathbf{b}\cdot\mathbf{c}+n\mathbf{a}\cdot\mathbf{c} \quad$ ...(iv)

Now, eliminating, l, m and n from above 4 relations, we get

$$\begin{vmatrix}\mathbf{r} & \mathbf{a} & \mathbf{b} & \mathbf{c}\\ \mathbf{r}\cdot\mathbf{a} & \mathbf{a}\cdot\mathbf{a} & \mathbf{b}\cdot\mathbf{a} & \mathbf{c}\cdot\mathbf{a}\\ \mathbf{r}\cdot\mathbf{b} & \mathbf{a}\cdot\mathbf{b} & \mathbf{b}\cdot\mathbf{b} & \mathbf{c}\cdot\mathbf{b}\\ \mathbf{r}\cdot\mathbf{c} & \mathbf{a}\cdot\mathbf{c} & \mathbf{b}\cdot\mathbf{c} & \mathbf{c}\cdot\mathbf{c}\end{vmatrix}=0$$

Now, expanding along first row, we get

$$\mathbf{r}=\left(\frac{\Delta_1}{\Delta}\right)\mathbf{a}+\left(\frac{\Delta_2}{\Delta}\right)\mathbf{b}+\left(\frac{\Delta_3}{\Delta}\right)\mathbf{c}$$

31. (c) Since, **a**, **b**, **c** are three non-coplanar vectors, then

On three exists scalars x, y, z, such that

$\mathbf{r}=x\mathbf{a}+y\mathbf{b}+z\mathbf{c} \quad$ ...(i)

Taking dot product by $\mathbf{b}\times\mathbf{c}$, $\mathbf{c}\times\mathbf{a}$ and $\mathbf{a}\times\mathbf{b}$ successively, we get

$\mathbf{r}\cdot(\mathbf{b}\times\mathbf{c})=(x\mathbf{a}+y\mathbf{b}+z\mathbf{c})\cdot(\mathbf{b}\times\mathbf{c})=x[\mathbf{a}\ \mathbf{b}\ \mathbf{c}]$

$\mathbf{r}\cdot(\mathbf{c}\times\mathbf{a})=y[\mathbf{b}\ \mathbf{c}\ \mathbf{a}]$

$\mathbf{r}\cdot(\mathbf{a}\times\mathbf{b})=z[\mathbf{c}\ \mathbf{a}\ \mathbf{b}]$

$\because \quad x=\frac{[\mathbf{r}\ \mathbf{b}\ \mathbf{c}]}{[\mathbf{a}\ \mathbf{b}\ \mathbf{c}]}, y=\frac{[\mathbf{r}\ \mathbf{c}\ \mathbf{a}]}{[\mathbf{a}\ \mathbf{b}\ \mathbf{c}]}$ and $z=\frac{[\mathbf{r}\ \mathbf{a}\ \mathbf{b}]}{[\mathbf{a}\ \mathbf{b}\ \mathbf{c}]}$

On substituting the values of x, y, z in Eq. (i), we get

$$\mathbf{r}=\left\{\frac{[\mathbf{r}\ \mathbf{b}\ \mathbf{c}]}{[\mathbf{a}\ \mathbf{b}\ \mathbf{c}]}\right\}\mathbf{a}+\left\{\frac{[\mathbf{r}\ \mathbf{c}\ \mathbf{a}]}{[\mathbf{a}\ \mathbf{b}\ \mathbf{c}]}\right\}\mathbf{b}+\left\{\frac{[\mathbf{r}\ \mathbf{a}\ \mathbf{b}]}{[\mathbf{a}\ \mathbf{b}\ \mathbf{c}]}\right\}\mathbf{c}$$

or $\quad \mathbf{r}=\frac{1}{[\mathbf{a}\ \mathbf{b}\ \mathbf{c}]}\{[\mathbf{r}\ \mathbf{b}\ \mathbf{c}]\mathbf{a}+[\mathbf{r}\ \mathbf{c}\ \mathbf{a}]\mathbf{b}+[\mathbf{r}\ \mathbf{a}\ \mathbf{b}]\mathbf{c}\}$

32. (b) We know that,

$[\mathbf{a}\times\mathbf{b}\ \ \mathbf{b}\times\mathbf{c}\ \ \mathbf{c}\times\mathbf{a}]=[\mathbf{a}\ \mathbf{b}\ \mathbf{c}]^2$

Clearly, $\quad [\mathbf{a}\times\mathbf{b}\ \ \mathbf{b}\times\mathbf{c}\ \ \mathbf{c}\times\mathbf{a}]\neq 0\{[\mathbf{a}\ \mathbf{b}\ \mathbf{c}]\neq 0\}$

$\Rightarrow \mathbf{a}\times\mathbf{b}\ \mathbf{b}\times\mathbf{c}\ \mathbf{c}\times\mathbf{a}$ are non-coplanar.

We also know that any vector in space can be expressed as a linear combination of any three non-coplanar vectors, so let

$\mathbf{a}=l(\mathbf{b}\times\mathbf{c})+m(\mathbf{c}\times\mathbf{a})+n(\mathbf{a}\times\mathbf{b}) \quad$ ...(i)

On taking dot product on both sides by **a**, **b**, **c** successively, we get

$$\mathbf{a}\cdot\mathbf{a} = l\,[\mathbf{a\ b\ c}]$$
$$\mathbf{a}\cdot\mathbf{b} = m\,[\mathbf{c\ a\ b}]$$
$$\mathbf{c}\cdot\mathbf{a} = n\,[\mathbf{c\ a\ b}]$$

$\because \quad l = \dfrac{\mathbf{a}\cdot\mathbf{a}}{[\mathbf{a\ b\ c}]}, m = \dfrac{\mathbf{a}\cdot\mathbf{b}}{[\mathbf{a\ b\ c}]}$ and $n = \dfrac{\mathbf{a}\cdot\mathbf{c}}{[\mathbf{a\ b\ c}]}$

On substituting these values in Eq. (i), we get

$$\mathbf{a} = \frac{\mathbf{a}\cdot\mathbf{a}}{[\mathbf{a\ b\ c}]}(\mathbf{b}\times\mathbf{c}) + \frac{\mathbf{a}\cdot\mathbf{b}}{[\mathbf{a\ b\ c}]}(\mathbf{c}\times\mathbf{a}) + \frac{\mathbf{a}\cdot\mathbf{c}}{[\mathbf{a\ b\ c}]}(\mathbf{a}\times\mathbf{b})$$

or $\mathbf{a} = \dfrac{1}{[\mathbf{a\ b\ c}]}\{\mathbf{a}\cdot\mathbf{a}\,(\mathbf{b}\times\mathbf{c}) + \mathbf{a}\cdot\mathbf{b}\,(\mathbf{c}\times\mathbf{a}) + \mathbf{a}\cdot\mathbf{c}\,(\mathbf{a}\times\mathbf{b})\}$

33. (d) Let $\mathbf{a} = a_1\hat{\mathbf{i}} + a_2\hat{\mathbf{j}} + a_3\hat{\mathbf{k}}, \mathbf{b} = b_1\hat{\mathbf{i}} + b_2\hat{\mathbf{j}} + b_3\hat{\mathbf{k}}$

$$\mathbf{c} = c_1\hat{\mathbf{i}} + c_2\hat{\mathbf{j}} + c_3\hat{\mathbf{k}}, \quad \mathbf{p} = p_1\hat{\mathbf{i}} + p_2\hat{\mathbf{j}} + p_3\hat{\mathbf{k}}$$

and $\mathbf{q} = q_1\hat{\mathbf{i}} + q_2\hat{\mathbf{j}} + q_3\hat{\mathbf{k}}$

$$\begin{vmatrix} a_1\hat{\mathbf{i}} + a_2\hat{\mathbf{j}} + a_3\hat{\mathbf{k}} & b_1\hat{\mathbf{i}} + b_2\hat{\mathbf{j}} + b_3\hat{\mathbf{k}} & c_1\hat{\mathbf{i}} + c_2\hat{\mathbf{j}} + c_3\hat{\mathbf{k}} \\ a_1p_1 + a_2p_2 + a_3p_3 & b_1p_1 + p_2p_2 + b_3p_3 & c_1p_1 + c_2p_2 + c_3p_3 \\ a_1q_1 + a_2q_2 + a_3q_3 & b_1q_1 + b_2q_2 + b_3q_3 & c_1q_1 + c_2q_2 + c_3q_3 \end{vmatrix}$$

$$\Rightarrow \quad \begin{vmatrix} \hat{\mathbf{i}} & \hat{\mathbf{j}} & \hat{\mathbf{k}} \\ p_1 & p_2 & p_3 \\ q_1 & q_2 & q_3 \end{vmatrix}\begin{vmatrix} a_1 & a_2 & a_3 \\ b_1 & b_2 & b_3 \\ c_1 & c_2 & c_3 \end{vmatrix}$$

$$\Rightarrow \quad (\mathbf{p}\times\mathbf{q})\,[\mathbf{a\ b\ c}] = \sqrt{[\mathbf{a\ b\ c}]^2}\,(\mathbf{p}\times\mathbf{q})$$
$$= \sqrt{[\mathbf{a}\times\mathbf{b}\ \mathbf{b}\times\mathbf{c}\ \mathbf{c}\times\mathbf{a}]}\,(\mathbf{p}\times\mathbf{q})$$

$\therefore \quad (\mathbf{p}\times\mathbf{q})\sqrt{|[\mathbf{a}\times\mathbf{b}\ \mathbf{b}\times\mathbf{c}\ \mathbf{c}\times\mathbf{a}]|}$

34. (a) We have,

$$\mathbf{b}\times\mathbf{c} = \mathbf{a},$$
$$\mathbf{c}\times\mathbf{a} = \mathbf{b} \text{ and } \mathbf{a}\times\mathbf{b} = \mathbf{c}$$

this shows a right handed system

$$\hat{\mathbf{i}}\times\hat{\mathbf{j}} = \hat{\mathbf{k}}$$
$$\hat{\mathbf{j}}\times\hat{\mathbf{k}} = \hat{\mathbf{i}}$$
$$\hat{\mathbf{k}}\times\hat{\mathbf{i}} = \hat{\mathbf{j}}$$

Hence, $|\mathbf{a}| = 1$

35. (c) $|\mathbf{a}| = |\mathbf{b}| = |\mathbf{c}| = 1$

$|\mathbf{a}| + |\mathbf{b}| + |\mathbf{c}| = 3$

36. (4) Median through A is

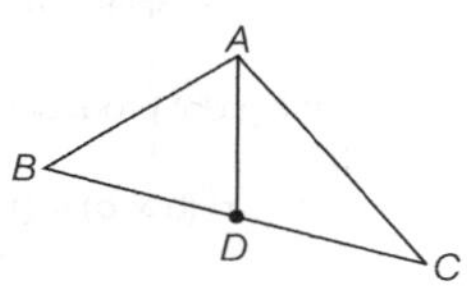

$$AD = \frac{\mathbf{AB} + \mathbf{AC}}{2}$$

$$\mathbf{AB} = -3\hat{\mathbf{i}} - 3\hat{\mathbf{k}}$$

$$\mathbf{AC} = (\lambda - 2)\,\hat{\mathbf{i}} + 2\hat{\mathbf{j}} + (\mu - 5)\,\hat{\mathbf{k}}$$

$$\mathbf{AD} = \frac{1}{2}\,[(\lambda - 5)\,\hat{\mathbf{i}} + 2\hat{\mathbf{j}} + (\mu - 8)\hat{\mathbf{k}}]$$

We have, AD is equally inclined to the positive direction of axes

$$\cos\theta = \frac{\frac{1}{2}[(\lambda - 5)\hat{\mathbf{i}} + 2\hat{\mathbf{j}} + (\mu - 8)\,\hat{\mathbf{k}}]\cdot\hat{\mathbf{i}}}{|\mathbf{AD}|\,|\hat{\mathbf{i}}|}$$

$$= \frac{\frac{1}{2}[(\lambda - 5)\hat{\mathbf{i}} + 2\hat{\mathbf{j}} + (\mu - 8)\,\hat{\mathbf{k}}]\cdot\hat{\mathbf{j}}}{|\mathbf{AD}|\,|\hat{\mathbf{j}}|}$$

$$= \frac{\frac{1}{2}[(\lambda - 5)\,\hat{\mathbf{i}} + 2\hat{\mathbf{j}} + (\mu - 8)\,\hat{\mathbf{k}}]\cdot\hat{\mathbf{k}}}{|\mathbf{AD}|\,|\hat{\mathbf{k}}|}$$

$$\Rightarrow \quad \frac{\lambda - 5}{2} = 1 = \frac{\mu - 8}{2}$$

$$\Rightarrow \quad \lambda = 7 \text{ and } \mu = 10$$

$$\therefore \quad 2\lambda - \mu = 14 - 10 = 4$$

37. (3) We know that,

Volume of parallelopiped having coterminous edges **a**, **b** and **c** is $[\mathbf{a\ b\ c}]$

$$\because \quad V = [\mathbf{a\ b\ c}]$$

Now, volume of parallelopiped having coterminqus edges α, β and γ are

$$V^\lambda = [\alpha\ \beta\ \gamma]$$

$$V^\lambda = [(\mathbf{a}\cdot\mathbf{a})\mathbf{a} + (\mathbf{a}\cdot\mathbf{b})\,\mathbf{b} + (\mathbf{a}\cdot\mathbf{c})\mathbf{c}\ (\mathbf{b}\cdot\mathbf{a})\,\mathbf{a} + (\mathbf{b}\cdot\mathbf{b})\mathbf{b} + (\mathbf{b}\cdot\mathbf{c})\,\mathbf{c}\ (\mathbf{c}\cdot\mathbf{a})\,\mathbf{a} + (\mathbf{c}\cdot\mathbf{b})\,\mathbf{b} + (\mathbf{c}\cdot\mathbf{c})\mathbf{c}]$$

$$= [\mathbf{a\ b\ c}]\begin{vmatrix} \mathbf{a}\cdot\mathbf{a} & \mathbf{a}\cdot\mathbf{b} & \mathbf{a}\cdot\mathbf{c} \\ \mathbf{b}\cdot\mathbf{a} & \mathbf{b}\cdot\mathbf{b} & \mathbf{b}\cdot\mathbf{c} \\ \mathbf{c}\cdot\mathbf{a} & \mathbf{c}\cdot\mathbf{b} & \mathbf{c}\cdot\mathbf{c} \end{vmatrix}$$

$$= [\mathbf{a\ b\ c}]\,[\mathbf{a\ b\ c}]\,[\mathbf{a\ b\ c}]$$

$$V^\lambda = [\mathbf{a\ b\ c}]^3 = V^3$$

By comparing, we get

$$\lambda = 3$$

38. (2) We have, **a**, **b** and **c** forming right handed system .

Hence, take $\mathbf{a} = \hat{\mathbf{i}}, \mathbf{b} = \hat{\mathbf{j}}$ and $\mathbf{c} = \hat{\mathbf{k}}$

$$\therefore \text{ Volume of tetrahedron} = \frac{1}{6}\,[\mathbf{a\ b\ c}] = \frac{1}{6}\begin{vmatrix} 3 & -2 & 2 \\ -1 & 0 & -2 \\ 2 & -3 & 4 \end{vmatrix}$$

$$= \frac{1}{6}\,[3\,(0 - 6) + 2\,(-4 + 4) + 2\,(3 - 0)]$$

$$= \frac{1}{6}\,[-18 + 6] = \frac{1}{6}\,(-12) = |-2|$$

$= 2$ cu units

39. (3) $\mathbf{c}\times\mathbf{a} = \mathbf{b}$

$$\Rightarrow \quad |\mathbf{c}\times\mathbf{a}| = |\mathbf{b}|$$

$$\Rightarrow \quad |\mathbf{c}|\,|\mathbf{a}|\sin\theta = 3$$

$$\Rightarrow \quad |\mathbf{c}| = \frac{3}{\sin\theta},$$

$$|\mathbf{c} - \mathbf{a}|^2 = |\mathbf{c}|^2 + |\mathbf{a}|^2 - 2\mathbf{a}\cdot\mathbf{c}$$

$$\therefore \quad |\mathbf{c} - \mathbf{a}|^2 = |\mathbf{c}|^2 + 4 - 2|\mathbf{a}|\,|\mathbf{c}|\cos\theta$$

$$= \frac{9}{4\sin^2\theta} + 4 - 2\cdot\frac{3}{2\sin\theta}\cdot 2\cos\theta$$

$$= \frac{9}{4} + \left(\frac{3}{2}\cot\theta - 2\right)^2$$

$$\Rightarrow \quad |\mathbf{c} - \mathbf{a}|^2 \geq \frac{9}{4}$$

$$\Rightarrow \quad |\mathbf{c} - \mathbf{a}| \geq \frac{3}{2}$$

$$\Rightarrow \quad 2|\mathbf{c} - \mathbf{a}| \geq 3$$

$\therefore$ min $(2|\mathbf{c} - \mathbf{a}|) = 3$

40. (9) $\mathbf{a}\times\mathbf{b} = p\mathbf{a}\times q\mathbf{b} + r\mathbf{c}$

Taking dot product with **a**, **b** and **c**

$$0 = p + \frac{3}{5}q + \frac{3}{5}r \qquad \text{...(i)}$$

$$0 = \frac{3}{5}p + q + \frac{3}{5}r \qquad \text{...(ii)}$$

$$[\mathbf{abc}] = \frac{3}{5}p + \frac{3}{5}q + r \qquad \text{...(iii)}$$

$$[\mathbf{abc}]^2 = \begin{vmatrix} \mathbf{a}\cdot\mathbf{a} & \mathbf{a}\cdot\mathbf{b} & \mathbf{a}\cdot\mathbf{c} \\ \mathbf{b}\cdot\mathbf{a} & \mathbf{b}\cdot\mathbf{b} & \mathbf{b}\cdot\mathbf{c} \\ \mathbf{c}\cdot\mathbf{a} & \mathbf{c}\cdot\mathbf{b} & \mathbf{c}\cdot\mathbf{c} \end{vmatrix} = \frac{44}{125}$$

$$\therefore \quad \frac{5}{11}[\mathbf{abc}] = -\frac{2}{3}q \quad \Rightarrow \quad \frac{25}{121}\cdot\frac{44}{125} = \frac{4}{9}q^2$$

$$\Rightarrow \quad 55\,q^2 = 9$$

JEE Main & AIEEE Archive

1. (a) Given, $(\mathbf{a} \times \mathbf{b}) \times \mathbf{c} = \frac{1}{3}|\mathbf{b}||\mathbf{c}|\mathbf{a}$

$\Rightarrow \quad -\mathbf{c} \times (\mathbf{a} \times \mathbf{b}) = \frac{1}{3}|\mathbf{b}||\mathbf{c}|\mathbf{a}$

$\Rightarrow \quad -(\mathbf{c} \cdot \mathbf{b}) \cdot \mathbf{a} + (\mathbf{c} \cdot \mathbf{a})\mathbf{b} = \frac{1}{3}|\mathbf{b}||\mathbf{c}|\mathbf{a}$

$$\left[\frac{1}{3}|\mathbf{b}||\mathbf{c}| + (\mathbf{c} \cdot \mathbf{b})\right]\mathbf{a} = (\mathbf{c} \cdot \mathbf{a})\mathbf{b}$$

Since, **a** and **b** are non-collinear.

$$\mathbf{c} \cdot \mathbf{b} + \frac{1}{3}|\mathbf{b}||\mathbf{c}| = 0 \text{ and } \mathbf{c} \cdot \mathbf{a} = 0$$

$\Rightarrow \quad |\mathbf{c}||\mathbf{b}|\cos\theta + \frac{1}{3}|\mathbf{b}||\mathbf{c}| = 0$

$\Rightarrow \quad |\mathbf{b}||\mathbf{c}|\left(\cos\theta + \frac{1}{3}\right) = 0$

$\Rightarrow \quad \cos\theta + \frac{1}{3} = 0 \qquad [\because |\mathbf{b}| \neq 0, |\mathbf{c}| \neq 0]$

$\Rightarrow \quad \cos\theta = -\frac{1}{3}$

$\therefore \quad \sin\theta = \frac{\sqrt{8}}{3} = \frac{2\sqrt{2}}{3}$

2. (b) $\mathbf{a} \times (\mathbf{b} \times \mathbf{c}) = (\mathbf{a} \cdot \mathbf{c})\mathbf{b} - (\mathbf{a} \cdot \mathbf{b})\mathbf{c}, [\mathbf{a}\ \mathbf{b}\ \mathbf{c}] = [\mathbf{b}\ \mathbf{c}\ \mathbf{a}] = [\mathbf{c}\ \mathbf{a}\ \mathbf{b}]$

and $[\mathbf{a}\ \mathbf{a}\ \mathbf{b}] = [\mathbf{a}\ \mathbf{b}\ \mathbf{b}] = [\mathbf{a}\ \mathbf{c}\ \mathbf{c}] = 0$

Further simplify it and get the result.

Now, $[\mathbf{a} \times \mathbf{b}\ \ \mathbf{b} \times \mathbf{c}\ \ \mathbf{c} \times \mathbf{a}]$

$= \mathbf{a} \times \mathbf{b} \cdot [(\mathbf{b} \times \mathbf{c}) \times (\mathbf{c} \times \mathbf{a})]$

$= \mathbf{a} \times \mathbf{b} \cdot [(\mathbf{k} \times \mathbf{c} \times \mathbf{a})] \qquad [\because \mathbf{k} = \mathbf{b} \times \mathbf{c}]$

$= \mathbf{a} \times \mathbf{b} \cdot [(\mathbf{k} \cdot \mathbf{a})\mathbf{c} - (\mathbf{k} \cdot \mathbf{c})\mathbf{a}]$

$= (\mathbf{a} \times \mathbf{b}) \cdot ((\mathbf{b} \times \mathbf{c} \cdot \mathbf{a})\mathbf{c} - (\mathbf{b} \times \mathbf{c} \cdot \mathbf{c})\mathbf{a})$

$= (\mathbf{a} \times \mathbf{b}) \cdot ([\mathbf{b}\ \mathbf{c}\ \mathbf{a}]\mathbf{c}) - 0 \qquad (\because [\mathbf{b} \times \mathbf{c} \cdot \mathbf{c}] = 0)$

$= \mathbf{a} \times \mathbf{b} \cdot \mathbf{c}\ [\mathbf{b}\ \mathbf{c}\ \mathbf{a}] = [\mathbf{a}\ \mathbf{b}\ \mathbf{c}]\ [\mathbf{b}\ \mathbf{c}\ \mathbf{a}]$

$= [\mathbf{a}\ \mathbf{b}\ \mathbf{c}]^2 \qquad (\because [\mathbf{a}\ \mathbf{b}\ \mathbf{c}] = [\mathbf{b}\ \mathbf{c}\ \mathbf{a}])$

Hence, $[\mathbf{a} \times \mathbf{b}\ \mathbf{b} \times \mathbf{c}\ \mathbf{c} \times \mathbf{a}] = \lambda\ [\mathbf{a}\ \mathbf{b}\ \mathbf{c}]^2$

$\Rightarrow \quad [\mathbf{a}\ \mathbf{b}\ \mathbf{c}]^2 = \lambda\ [\mathbf{a}\ \mathbf{b}\ \mathbf{c}]^2$

$\Rightarrow \quad \lambda = 1$

3. (c) We know that, the sum of three vectors of a triangle is zero.

$\therefore \quad \mathbf{AB} + \mathbf{BC} + \mathbf{CA} = 0$

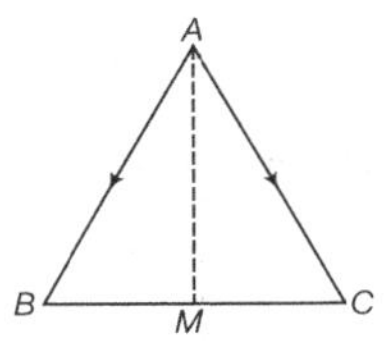

$\Rightarrow \quad \mathbf{BC} = \mathbf{AC} - \mathbf{AB} \qquad [\because \mathbf{AC} = -\mathbf{CA}]$

$\Rightarrow \quad \mathbf{BM} = \frac{\mathbf{AC} - \mathbf{AB}}{2} \qquad [\because M \text{ is a mid-point of } BC]$

Also, $\mathbf{AB} + \mathbf{BM} + \mathbf{MA} = 0$ [by properties of a triangle]

$\Rightarrow \quad \mathbf{AB} + \frac{\mathbf{AC} - \mathbf{AB}}{2} = \mathbf{AM} \qquad [\because \mathbf{AM} = -\mathbf{MA}]$

$\Rightarrow \quad \mathbf{AM} = \frac{\mathbf{AB} + \mathbf{AC}}{2}$

$= \frac{3\hat{\mathbf{i}} + 4\hat{\mathbf{k}} + 5\hat{\mathbf{i}} - 2\hat{\mathbf{j}} + 4\hat{\mathbf{k}}}{2}$

$= 4\hat{\mathbf{i}} - \hat{\mathbf{j}} + 4\hat{\mathbf{k}}$

$\Rightarrow \quad |\mathbf{AM}| = \sqrt{4^2 + 1^2 + 4^2} = \sqrt{33}$

4. (c) Given that,

(i) **a** and **b** are unit vectors,
i.e. $|\mathbf{a}| = |\mathbf{b}| = 1$

(ii) $\mathbf{c} = \mathbf{a} + 2\mathbf{b}$ and $\mathbf{d} = 5\mathbf{a} - 4\mathbf{b}$

(iii) **c** and **d** are perpendicular to each other, i.e. $\mathbf{c} \cdot \mathbf{d} = 0$

To find Angle between **a** and **b**.

Now, $\quad \mathbf{c} \cdot \mathbf{d} = 0 \Rightarrow (\mathbf{a} + 2\mathbf{b}) \cdot (5\mathbf{a} - 4\mathbf{b}) = 0$

$\Rightarrow \quad 5\mathbf{a} \cdot \mathbf{a} - 4\mathbf{a} \cdot \mathbf{b} + 10\mathbf{b} \cdot \mathbf{a} - 8\mathbf{b} \cdot \mathbf{b} = 0$

$\Rightarrow \quad 6\mathbf{a} \cdot \mathbf{b} = 3$

$\Rightarrow \quad \mathbf{a} \cdot \mathbf{b} = \frac{1}{2}$

So, the angle between **a** and **b** is $\frac{\pi}{3}$.

5. (b) **Given**

(i) A parallelogram *ABCD* such that $\mathbf{AB} = \mathbf{q}$ and $\mathbf{AD} = \mathbf{p}$.

(ii) The altitude from vertex *B* to side *AD* coincides with a vector **r**.

To find The vector **r** in terms of **p** and **q**.

Let *E* be the foot of perpendicular from *B* to side *AD*.

AE = Projection of vector **q** on $\mathbf{p} = \mathbf{q} \cdot \mathbf{p} = \frac{\mathbf{q} \cdot \mathbf{p}}{|\mathbf{p}|}$

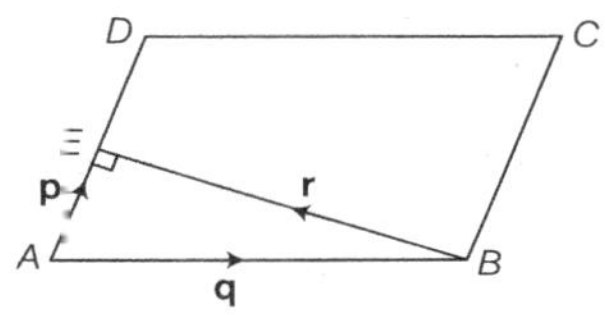

$\mathbf{AE}$ = Vector along *AE* of length *AE*

$= |\mathbf{AE}|\ \mathbf{AE} = \left(\frac{\mathbf{q} \cdot \mathbf{p}}{|\mathbf{p}|}\right)\mathbf{p} = \frac{(\mathbf{q} \cdot \mathbf{p})\mathbf{p}}{|\mathbf{p}|^2}$

Now, applying triangles law in ΔABE, we get

$$\mathbf{AB} + \mathbf{BE} = \mathbf{AE}$$

$\Rightarrow \quad \mathbf{q} + \mathbf{r} = \frac{(\mathbf{q} \cdot \mathbf{p})\mathbf{p}}{|\mathbf{p}|^2}$

$\Rightarrow \quad \mathbf{r} = \frac{(\mathbf{q} \cdot \mathbf{p})\mathbf{p}}{|\mathbf{p}|^2} - \mathbf{q}$

$\therefore \quad \mathbf{r} = -\mathbf{q} + \left(\frac{\mathbf{q} \cdot \mathbf{p}}{\mathbf{p} \cdot \mathbf{p}}\right)\mathbf{p}$

6. (d) Given, $\mathbf{a} = \frac{1}{\sqrt{10}}(3\hat{\mathbf{i}} + \hat{\mathbf{k}})$

and $\quad \mathbf{b} = \frac{1}{7}(2\hat{\mathbf{i}} + 3\hat{\mathbf{j}} - 6\hat{\mathbf{k}})$

$\therefore (2\mathbf{a} - \mathbf{b}) \cdot \{(\mathbf{a} \times \mathbf{b}) \times (\mathbf{a} + 2\mathbf{b})\}$

$= (2\mathbf{a} - \mathbf{b}) \cdot \{(\mathbf{a} \times \mathbf{b}) \times \mathbf{a} + (\mathbf{a} \times \mathbf{b}) \times 2\mathbf{b}\}$

$= (2\mathbf{a} - \mathbf{b}) \cdot \{(\mathbf{a} \cdot \mathbf{a})\mathbf{b} - (\mathbf{b} \cdot \mathbf{a})\mathbf{a} + 2(\mathbf{a} \cdot \mathbf{b})\mathbf{b} - 2(\mathbf{b} \cdot \mathbf{b})\mathbf{a}\}$

$= (2\mathbf{a} - \mathbf{b}) \cdot \{1(\mathbf{b}) - (0)\mathbf{a} + 2(0)\mathbf{b} - 2(1)\mathbf{a}\}$

$[\because \mathbf{a} \cdot \mathbf{b} = 0 \text{ and } \mathbf{a} \cdot \mathbf{a} = \mathbf{b} \cdot \mathbf{b} = 1]$

$= (2\mathbf{a} - \mathbf{b})(\mathbf{b} - 2\mathbf{a})$

$= -(4|\mathbf{a}|^2 - 4\mathbf{a} \cdot \mathbf{b} + |\mathbf{b}|^2)$

$= -(4 - 0 + 1) = -5$

7. (c) Given, $\mathbf{a} \cdot \mathbf{b} \neq 0, \mathbf{a} \cdot \mathbf{d} = 0 \qquad \ldots(i)$

and $\quad \mathbf{b} \times \mathbf{c} = \mathbf{b} \times \mathbf{d}$

$\Rightarrow \quad \mathbf{b} \times (\mathbf{c} - \mathbf{d}) = 0$

$\therefore \quad \mathbf{b} \parallel (\mathbf{c} - \mathbf{d})$

$\Rightarrow \quad \mathbf{c} - \mathbf{d} = \lambda\mathbf{b}$

$\Rightarrow \quad \mathbf{d} = \mathbf{c} - \lambda\mathbf{b} \qquad \ldots(ii)$

Taking dot product with **a**, we get

$$\mathbf{a}\cdot\mathbf{d}=\mathbf{a}\cdot\mathbf{c}-\lambda\mathbf{a}\cdot\mathbf{b}$$

$\Rightarrow \quad 0=\mathbf{a}\cdot\mathbf{c}-\lambda(\mathbf{a}\cdot\mathbf{b})$

$\therefore \quad \lambda=\dfrac{\mathbf{a}\cdot\mathbf{c}}{\mathbf{a}\cdot\mathbf{b}}$ ...(iii)

$\therefore \quad \mathbf{d}=\mathbf{c}-\dfrac{(\mathbf{a}\cdot\mathbf{c})}{(\mathbf{a}\cdot\mathbf{b})}\mathbf{b}$

8. (a) Given, $\mathbf{a}=p\hat{i}+\hat{j}+\hat{k}, \mathbf{b}=\hat{i}+q\hat{j}+\hat{k}$ and $\mathbf{c}=\hat{i}+\hat{j}+r\hat{k}$ are coplanar and $p\neq q\neq r\neq 1$.

Since, **a**, **b** and **c** are coplanar.

$\Rightarrow \quad [\mathbf{a}\ \mathbf{b}\ \mathbf{c}]=0$

$\Rightarrow \quad \begin{vmatrix} p & 1 & 1\\ 1 & q & 1\\ 1 & 1 & r\end{vmatrix}=0$

$\Rightarrow \quad p(qr-1)-1(r-1)+1(1-q)=0$

$\Rightarrow \quad pqr-p-r+1+1-q=0$

$\therefore \quad pqr-(p+q+r)=-2$

9. (d) As, $\mathbf{a}+3\mathbf{b}$ is collinear with **c**.

$\therefore \quad \mathbf{a}+3\mathbf{b}=\lambda\mathbf{c}$ ...(i)

Also, $\mathbf{b}+2\mathbf{c}$ is collinear with **a**.

$\Rightarrow \quad \mathbf{b}+2\mathbf{c}=\mu\mathbf{a}$ ...(ii)

From Eq. (i), we get

$\mathbf{a}+3\mathbf{b}+6\mathbf{c}=(\lambda+6)\mathbf{c}$ ...(iii)

From Eq. (ii), we get

$\mathbf{a}+3\mathbf{b}+6\mathbf{c}=(1+3\mu)\mathbf{a}$ ...(iv)

From Eqs. (iii) and (iv), we get

$\therefore \quad (\lambda+6)\mathbf{c}=(1+3\mu)\mathbf{a}$

Since, **a** is not collinear with **c**.

$\Rightarrow \quad \lambda+6=1+3\mu=0$

From Eq. (iv), we get

$\mathbf{a}+3\mathbf{b}+6\mathbf{c}=0$

10. (a) We have, $\mathbf{a}\times\mathbf{b}+\mathbf{c}=0$

$\Rightarrow \quad \mathbf{a}\times(\mathbf{a}\times\mathbf{b})+\mathbf{a}\times\mathbf{c}=0$

$\Rightarrow \quad (\mathbf{a}\cdot\mathbf{b})\mathbf{a}-(\mathbf{a}\cdot\mathbf{a})\mathbf{b}+\mathbf{a}\times\mathbf{c}=0$

$\Rightarrow \quad 3\mathbf{a}-2\mathbf{b}+\mathbf{a}\times\mathbf{c}=0$

$\Rightarrow \quad 2\mathbf{b}=3\mathbf{a}+\mathbf{a}\times\mathbf{c}$

$\Rightarrow \quad 2\mathbf{b}=3\hat{j}-3\hat{k}-2\hat{i}-\hat{j}-\hat{k}=-2\hat{i}+2\hat{j}-4\hat{k}$

$\therefore \quad \mathbf{b}=-\hat{i}+\hat{j}-2\hat{k}$

11. (a) Since, the given vectors are mutually orthogonal, therefore

$\mathbf{a}\cdot\mathbf{b}=2-4+2=0$

$\mathbf{a}\cdot\mathbf{c}=\lambda-1+2\mu=0$ ...(i)

and $\quad \mathbf{b}\cdot\mathbf{c}=2\lambda+4+\mu=0$ ...(ii)

On solving Eqs. (i) and (ii), we get

$\mu=2$ and $\lambda=-3$

Hence, $\quad (\lambda,\mu)=(-3,2)$

12. (d) Since, $\quad [3\mathbf{u}\ p\mathbf{v}\ p\mathbf{w}]-[p\mathbf{v}\ \mathbf{w}\ q\mathbf{u}]-[2\mathbf{w}\ q\mathbf{v}\ q\mathbf{u}]=0$

$\therefore\ 3p^2[\mathbf{u}\cdot(\mathbf{v}\times\mathbf{w})]-pq\,[\mathbf{v}\cdot(\mathbf{w}\times\mathbf{u})]-2q^2\,[\mathbf{w}\cdot(\mathbf{v}\times\mathbf{u})]=0$

$\Rightarrow \quad (3p^2-pq+2q^2)\,[\mathbf{u}\cdot(\mathbf{v}\times\mathbf{w})]=0$

But $\quad [\mathbf{u}\ \mathbf{v}\ \mathbf{w}]\neq 0$

$\Rightarrow \quad 3p^2-pq+2q^2=0$

$\therefore \quad p=q=0$

13. (a) Given that, $\mathbf{b}=\hat{i}+\hat{j}$ and $\mathbf{c}=\hat{j}+\hat{k}$.

The equation of bisector of **b** and **c** is

$$\mathbf{r}=\lambda(\mathbf{b}+\mathbf{c})=\lambda\left(\frac{\hat{i}+\hat{j}}{\sqrt{2}}+\frac{\hat{j}+\hat{k}}{\sqrt{2}}\right)$$

$=\dfrac{\lambda}{\sqrt{2}}(\hat{i}+2\hat{j}+\hat{k})$ ...(i)

Since, vector **a** lies in plane of **b** and **c**.

$\therefore \quad \mathbf{a}=\mathbf{b}+\mu\mathbf{c}$

$\Rightarrow \quad \dfrac{\lambda}{\sqrt{2}}(\hat{i}+2\hat{j}+\hat{k})=(\hat{i}+\hat{j})+\mu(\hat{j}+\hat{k})$

On equating the coefficient of $\hat{i}$ both sides, we get

$\dfrac{\lambda}{\sqrt{2}}=1$

$\Rightarrow \quad \lambda=\sqrt{2}$

On putting $\lambda=\sqrt{2}$ in Eq. (i), we get

$\mathbf{r}=\hat{i}+2\hat{j}+\hat{k}$

Since, the given vector **a** represents the same bisector equation **r**.

$\therefore \quad \alpha=1$ and $\beta=1$

Alternate Method

Since, **a**, **b** and **c** are coplanar.

$\Rightarrow \quad \begin{vmatrix}\alpha & 2 & \beta\\ 1 & 1 & 0\\ 0 & 1 & 1\end{vmatrix}=0$

$\Rightarrow \quad \alpha(1-0)-2(1-0)+\beta(1-0)=0$

$\Rightarrow \quad \alpha+\beta=2$

which is possible for $\alpha=1,\beta=1$.

14. (a) Since, $\mathbf{a}=8\mathbf{b}$ and $\mathbf{c}=-7\mathbf{b}$

So, **a** is parallel to **b** and **c** is anti-parallel to **b**.

$\Rightarrow$ **a** and **c** are anti-parallel.

So, the angle between **a** and **c** is π.

15. (d) Since, $(2\mathbf{u}\times 3\mathbf{v})$ is a unit vector.

$\Rightarrow \quad |2\mathbf{u}\times 3\mathbf{v}|=1$

$\Rightarrow \quad 6|\mathbf{u}||\mathbf{v}||\sin\theta|=1$

$\Rightarrow \quad \sin\theta=\dfrac{1}{6}$ $\quad [\because |\mathbf{u}|=|\mathbf{v}|=1]$

Since, θ is an acute angle, then there is exactly one value of θ for which $(2\mathbf{u}\times 3\mathbf{v})$ is a unit vector.

16. (d) Since, given vectors **a**, **b** and **c** are coplanar.

$\therefore \quad \begin{vmatrix}1 & 1 & 1\\ 1 & -1 & 2\\ x & x-2 & -1\end{vmatrix}=0$

$\Rightarrow \quad 1\{1-2(x-2)\}-1(-1-2x)+1(x-2+x)=0$

$\Rightarrow \quad 1-2x+4+1+2x+2x-2=0$

$\Rightarrow \quad 2x=-4$

$\Rightarrow \quad x=-2$

17. (c) Since, $\quad (\mathbf{a}\times\mathbf{b})\times\mathbf{c}=\mathbf{a}\times(\mathbf{b}\times\mathbf{c})$

$\therefore \quad (\mathbf{a}\cdot\mathbf{c})\mathbf{b}-(\mathbf{b}\cdot\mathbf{c})\mathbf{a}=(\mathbf{a}\cdot\mathbf{c})\mathbf{b}-(\mathbf{a}\cdot\mathbf{b})\mathbf{c}$

$\Rightarrow \quad (\mathbf{b}\cdot\mathbf{c})\mathbf{a}=(\mathbf{a}\cdot\mathbf{b})\mathbf{c}$

$\Rightarrow \quad \mathbf{a}=\dfrac{(\mathbf{a}\cdot\mathbf{b})}{(\mathbf{b}\cdot\mathbf{c})}\cdot\mathbf{c}$

Hence, **a** is parallel to **c**.

18. (d) Since, position vectors of A, B, C are $2\hat{i}-\hat{j}+\hat{k}$, $\hat{i}-3\hat{j}-5\hat{k}$ and $a\hat{i}-3\hat{j}+\hat{k}$, respectively.

Now, $\quad \mathbf{AC}=(a\hat{j}-3\hat{j}+\hat{k})-(2\hat{i}-\hat{j}+\hat{k})$

$=(a-2)\hat{i}-2\hat{j}$

and $\quad \mathbf{BC}=(a\hat{i}-3\hat{j}+\hat{k})-(\hat{i}-3\hat{j}-5\hat{k})$

$=(a-1)\hat{i}+6\hat{k}$

Since, the ΔABC is right angled at C, then

$\mathbf{AC}\cdot\mathbf{BC}=0$

$\Rightarrow \quad \{(a-2)\hat{i}-2\hat{j}\}\cdot\{(a-1)\hat{i}+6\hat{k}\}=0$

$\Rightarrow \quad (a-2)(a-1)=0$

$\therefore \quad a=1$ and $a=2$

19. (d) Let P be the origin outside of AB and C is mid-point of AB, then

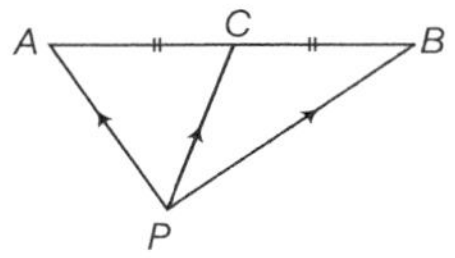

$$\mathbf{PC} = \frac{\mathbf{PA} + \mathbf{PB}}{2}$$

$$\Rightarrow \quad 2\mathbf{PC} = \mathbf{PA} + \mathbf{PB}$$

20. (c) Line is parallel to plane as

$$(\hat{\mathbf{i}} - \hat{\mathbf{j}} + 4\hat{\mathbf{k}}) \cdot (\hat{\mathbf{i}} + 5\hat{\mathbf{j}} + \hat{\mathbf{k}}) = 1 - 5 + 4 = 0$$

General point on the line is $(\lambda + 2, -\lambda - 2, 4\lambda + 3)$.

For $\lambda = 0$, a point on this line is $(2, -2, 3)$ and distance from $\mathbf{r} \cdot (\hat{\mathbf{i}} + 5\hat{\mathbf{j}} + \hat{\mathbf{k}}) = 5$ or $x + 5y + z = 5$ is

$$d = \left| \frac{2 + 5(-2) + 3 - 5}{\sqrt{1 + 25 + 1}} \right|$$

$$\therefore \quad d = \left| \frac{-10}{3\sqrt{3}} \right| = \frac{10}{3\sqrt{3}}$$

21. (b) Let $\mathbf{a} = a_1 \hat{\mathbf{i}} + a_2 \hat{\mathbf{j}} + a_3 \hat{\mathbf{k}}$

Then, $\quad \mathbf{a} \times \hat{\mathbf{i}} = -a_2 \hat{\mathbf{k}} + a_3 \hat{\mathbf{j}}$

$$\mathbf{a} \times \hat{\mathbf{j}} = a_1 \hat{\mathbf{k}} - a_3 \hat{\mathbf{i}}$$

$$\mathbf{a} \times \hat{\mathbf{k}} = -a_1 \hat{\mathbf{j}} + a_2 \hat{\mathbf{i}}$$

$$\therefore \quad (\mathbf{a} \times \hat{\mathbf{i}})^2 + (\mathbf{a} \times \hat{\mathbf{j}})^2 + (\mathbf{a} \times \hat{\mathbf{k}})^2$$

$$= a_2^2 + a_3^2 + a_1^2 + a_3^2 + a_1^2 + a_2^2$$

$$= 2(a_1^2 + a_2^2 + a_3^2)$$

$$= 2\mathbf{a}^2$$

22. (a) Given, vectors are

$$\mathbf{a} = \hat{\mathbf{i}} - \hat{\mathbf{k}}, \quad \mathbf{b} = x\hat{\mathbf{i}} + \hat{\mathbf{j}} + (1 - x)\hat{\mathbf{k}}$$

and $\quad \mathbf{c} = y\hat{\mathbf{i}} + x\hat{\mathbf{j}} + (1 + x - y)\hat{\mathbf{k}}$

$$\therefore \quad [\mathbf{a}\ \mathbf{b}\ \mathbf{c}] = \begin{vmatrix} 1 & 0 & -1 \\ x & 1 & 1 - x \\ y & x & 1 + x - y \end{vmatrix}$$

Applying $C_3 \to C_3 + C_1$, we get

$$= \begin{vmatrix} 1 & 0 & 0 \\ x & 1 & 1 \\ y & x & 1 + x \end{vmatrix}$$

$$= 1(1 + x) - x = 1$$

Thus, $[\mathbf{a}\ \mathbf{b}\ \mathbf{c}]$ depends upon neither x nor y.

23. (d) Since, the given vectors lie in a plane.

$$\therefore \quad \begin{vmatrix} a & a & c \\ 1 & 0 & 1 \\ c & c & b \end{vmatrix} = 0$$

Applying $C_1 \to C_1 - C_2$, we get

$$\begin{vmatrix} 0 & a & c \\ 1 & 0 & 1 \\ 0 & c & b \end{vmatrix} = 0$$

$$\Rightarrow \quad -1(ab - c^2) = 0$$

$$\Rightarrow \quad c^2 = ab$$

Hence, c is GM of a and b.

24. (c) Given that, $[\lambda(\mathbf{a} + \mathbf{b}) \quad \lambda^2\mathbf{b} \quad \lambda\mathbf{c}] = [\mathbf{a} \quad \mathbf{b} + \mathbf{c} \quad \mathbf{b}]$

$$\therefore \quad \begin{vmatrix} \lambda(a_1 + b_1) & \lambda(a_2 + b_2) & \lambda(a_3 + b_3) \\ \lambda^2 b_1 & \lambda^2 b_2 & \lambda^2 b_3 \\ \lambda c_1 & \lambda c_2 & \lambda c_3 \end{vmatrix}$$

$$= \begin{vmatrix} a_1 & a_2 & a_3 \\ b_1 + c_1 & b_2 + c_2 & b_3 + c_3 \\ b_1 & b_2 & b_3 \end{vmatrix}$$

$$\Rightarrow \quad \lambda^4 \begin{vmatrix} a & a_2 & a_3 \\ b_1 & b_2 & b_3 \\ c & c_2 & c_3 \end{vmatrix} = -\begin{vmatrix} a_1 & a_2 & a_3 \\ b_1 & b_2 & b_3 \\ c_1 & c_2 & c_3 \end{vmatrix}$$

$$\Rightarrow \quad \lambda^4 = -1$$

So, no real value of λ exists.

25. (d) If $\mathbf{a} + 2\mathbf{b}$ is collinear with $\mathbf{c}$, then $\mathbf{a} + 2\mathbf{b} = t\mathbf{c}$ $\quad$ ...(i)

Also, $\mathbf{b} + 3\mathbf{c}$ is collinear with $\mathbf{a}$, then

$$\mathbf{b} + 3\mathbf{c} = \lambda\mathbf{a}$$

$$\Rightarrow \quad \mathbf{b} = \lambda\mathbf{a} - 3\mathbf{c} \quad \text{...(ii)}$$

From Eqs. (i) and (ii), we get

$$\mathbf{a} + 2(\lambda\mathbf{a} - 3\mathbf{c}) = t\mathbf{c}$$

$$\Rightarrow \quad (\mathbf{a} - 6\mathbf{c}) = t\mathbf{c} - 2\lambda\mathbf{a}$$

On comparing the coefficients of $\mathbf{a}$ and $\mathbf{c}$, we get

$$1 = -2\lambda$$

$$\Rightarrow \quad \lambda = -\frac{1}{2}$$

and $\quad -6 = t$

$$\Rightarrow \quad t = -6$$

From Eq. (i), we get

$$\mathbf{a} + 2\mathbf{b} = -6\mathbf{c}$$

$$\Rightarrow \quad \mathbf{a} + 2\mathbf{b} + 6\mathbf{c} = 0$$

26. (a) Total force, $\mathbf{F} = (4\hat{\mathbf{i}} + \hat{\mathbf{j}} - 3\hat{\mathbf{k}}) + (3\hat{\mathbf{i}} + \hat{\mathbf{j}} - \hat{\mathbf{k}})$

$$\therefore \quad \mathbf{F} = 7\hat{\mathbf{i}} + 2\hat{\mathbf{j}} - 4\hat{\mathbf{k}}$$

The particle is displaced from

$$A(\hat{\mathbf{i}} + 2\hat{\mathbf{j}} + 3\hat{\mathbf{k}}) \text{ to } B(5\hat{\mathbf{i}} + 4\hat{\mathbf{j}} + \hat{\mathbf{k}}).$$

Now, displacement, $\mathbf{AB} = (5\hat{\mathbf{i}} + 4\hat{\mathbf{j}} + \hat{\mathbf{k}}) - (\hat{\mathbf{i}} + 2\hat{\mathbf{j}} + 3\hat{\mathbf{k}})$

$$= 4\hat{\mathbf{i}} + 2\hat{\mathbf{j}} - 2\hat{\mathbf{k}}$$

$$\therefore \quad \text{Work done} = \mathbf{F} \cdot \mathbf{AB}$$

$$= (7\hat{\mathbf{i}} + 2\hat{\mathbf{j}} - 4\hat{\mathbf{k}}) \cdot (4\hat{\mathbf{i}} + 2\hat{\mathbf{j}} - 2\hat{\mathbf{k}})$$

$$= 28 + 4 + 8 = 40 \text{ units}$$

27. (c) The three vectors $(a + 2b + 3c)$, $(\lambda b + 4c)$ and $(2\lambda - 1)c$ are non-coplanar, if $\begin{vmatrix} 1 & 2 & 3 \\ 0 & \lambda & 4 \\ 0 & 0 & 2\lambda - 1 \end{vmatrix} \neq 0$

$$\Rightarrow \quad (2\lambda - 1)(\lambda) \neq 0 \quad \Rightarrow \quad \lambda \neq 0, \frac{1}{2}$$

So, these three vectors are non-coplanar for all except two values of λ.

28. (c) Since, $\quad |\mathbf{u}| = 1, |\mathbf{v}| = 2, |\mathbf{w}| = 3$

The projection of $\mathbf{v}$ along $\mathbf{u} = \dfrac{\mathbf{v} \cdot \mathbf{u}}{|\mathbf{u}|}$

and the projection of $\mathbf{w}$ along $\mathbf{u} = \dfrac{\mathbf{w} \cdot \mathbf{u}}{|\mathbf{u}|}$

According to the given condition,

$$\frac{\mathbf{v} \cdot \mathbf{u}}{|\mathbf{u}|} = \frac{\mathbf{w} \cdot \mathbf{u}}{|\mathbf{u}|}$$

$$\Rightarrow \quad \mathbf{v} \cdot \mathbf{u} = \mathbf{w} \cdot \mathbf{u} \quad \text{...(i)}$$

Since, **v**, **w** are perpendicular to each other.

$\therefore \quad \mathbf{v} \cdot \mathbf{w} = 0 \qquad \ldots(ii)$

Now, $|\mathbf{u} - \mathbf{v} + \mathbf{w}|^2 = |\mathbf{u}|^2 + |\mathbf{v}|^2 + |\mathbf{w}|^2 - 2\mathbf{u} \cdot \mathbf{v} - 2\mathbf{v} \cdot \mathbf{w} + 2\mathbf{u} \cdot \mathbf{w}$

$\Rightarrow \quad |\mathbf{u} - \mathbf{v} + \mathbf{w}|^2 = 1 + 4 + 9 - 2\mathbf{u} \cdot \mathbf{v} + 2\mathbf{v} \cdot \mathbf{u}$ [from Eqs. (i) and (ii)]

$\Rightarrow \quad |\mathbf{u} - \mathbf{v} + \mathbf{w}|^2 = 1 + 4 + 9$

$\Rightarrow \quad |\mathbf{u} - \mathbf{v} + \mathbf{w}| = \sqrt{14}$

29. (d) Given that, $\frac{1}{3}|\mathbf{b}||\mathbf{c}|\mathbf{a} = (\mathbf{a} \times \mathbf{b}) \times \mathbf{c}$

We know that,

$$(\mathbf{a} \times \mathbf{b}) \times \mathbf{c} = (\mathbf{a} \cdot \mathbf{c})\mathbf{b} - (\mathbf{b} \cdot \mathbf{c})\mathbf{a}$$

$\therefore \quad \frac{1}{3}|\mathbf{b}||\mathbf{c}|\mathbf{a} = (\mathbf{a} \cdot \mathbf{c})\mathbf{b} - (\mathbf{b} \cdot \mathbf{c})\mathbf{a}$

On comparing the coefficients of **a** and **b**, we get

$$\frac{1}{3}|\mathbf{b}||\mathbf{c}| = -\mathbf{b} \cdot \mathbf{c} \text{ and } \mathbf{a} \cdot \mathbf{c} = 0$$

$\Rightarrow \quad \frac{1}{3}|\mathbf{b}||\mathbf{c}| = -|\mathbf{b}||\mathbf{c}| \cos\theta$

$\Rightarrow \quad \cos\theta = -\frac{1}{3}$

$\Rightarrow \quad 1 - \sin^2\theta = \frac{1}{9}$

$\Rightarrow \quad \sin^2\theta = \frac{8}{9}$

$\therefore \quad \sin\theta = \frac{2\sqrt{2}}{3} \qquad \left[\because 0 \le \theta \le \frac{\pi}{2}\right]$

30. (a) Vector perpendicular to face *OAB* is $\mathbf{n}_1$

$$= \mathbf{OA} \times \mathbf{OB} = \begin{vmatrix} \hat{i} & \hat{j} & \hat{k} \\ 1 & 2 & 1 \\ 2 & 1 & 3 \end{vmatrix}$$

$$= 5\hat{i} - \hat{j} - 3\hat{k}$$

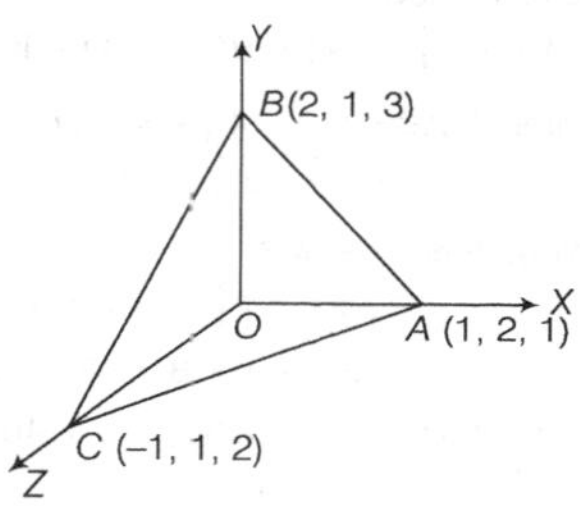

Vector perpendicular to face *ABC* is $\mathbf{n}_2$

$$= \mathbf{AB} \times \mathbf{AC} = \begin{vmatrix} \hat{i} & \hat{j} & \hat{k} \\ 1 & -1 & 2 \\ -2 & -1 & 1 \end{vmatrix}$$

$$= \hat{i} - 5\hat{j} - 3\hat{k}$$

Since, angle between faces is equal to the angle between their normals.

$$\therefore \quad \cos\theta = \frac{\mathbf{n}_1 \cdot \mathbf{n}_2}{|\mathbf{n}_1||\mathbf{n}_2|}$$

$$= \frac{5 \times 1 + (-1) \times (-5) + (-3) \times (-3)}{\sqrt{5^2 + (-1)^2 + (-3)^2}\sqrt{1^2 + (-5)^2 + (-3)^2}}$$

$$= \frac{5 + 5 + 9}{\sqrt{35}\sqrt{35}} = \frac{19}{35}$$

$$\Rightarrow \quad \theta = \cos^{-1}\left(\frac{19}{35}\right)$$

31. (b) Given that, $|\mathbf{a}| = 1, |\mathbf{b}| = 2, |\mathbf{c}| = 3$

and $\mathbf{a} + \mathbf{b} + \mathbf{c} = \mathbf{0}$

Now, $(\mathbf{a} + \mathbf{b} + \mathbf{c})^2 = |\mathbf{a}|^2 + |\mathbf{b}|^2 + |\mathbf{c}|^2 + 2(\mathbf{a} \cdot \mathbf{b} + \mathbf{b} \cdot \mathbf{c} + \mathbf{c} \cdot \mathbf{a})$

$\Rightarrow \quad 0 = 1^2 + 2^2 + 3^2 + 2(\mathbf{a} \cdot \mathbf{b} + \mathbf{b} \cdot \mathbf{c} + \mathbf{c} \cdot \mathbf{a})$

$\Rightarrow \quad 2(\mathbf{a} \cdot \mathbf{b} + \mathbf{b} \cdot \mathbf{c} + \mathbf{c} \cdot \mathbf{a}) = -14$

$\therefore \quad \mathbf{a} \cdot \mathbf{b} + \mathbf{b} \cdot \mathbf{c} + \mathbf{c} \cdot \mathbf{a} = -7$

32. (b) $(\mathbf{u} + \mathbf{v} - \mathbf{w}) \cdot [(\mathbf{u} - \mathbf{v}) \times (\mathbf{v} - \mathbf{w})]$

$= (\mathbf{u} + \mathbf{v} - \mathbf{w}) \cdot [\mathbf{u} \times \mathbf{v} - \mathbf{u} \times \mathbf{w} - \mathbf{v} \times \mathbf{v} + \mathbf{v} \times \mathbf{w}]$

$= \mathbf{u} \cdot (\mathbf{u} \times \mathbf{v}) - \mathbf{u} \cdot (\mathbf{u} \times \mathbf{w}) + \mathbf{u} \cdot (\mathbf{v} \times \mathbf{w}) + \mathbf{v} \cdot (\mathbf{u} \times \mathbf{v}) - \mathbf{v} \cdot (\mathbf{u} \times \mathbf{w}) + \mathbf{v} \cdot (\mathbf{v} \times \mathbf{w}) - \mathbf{w} \cdot (\mathbf{u} \times \mathbf{v}) + \mathbf{w} \cdot (\mathbf{u} \times \mathbf{w}) - \mathbf{w} \cdot (\mathbf{v} \times \mathbf{w})$

$= \mathbf{u} \cdot \mathbf{v} \times \mathbf{w} - \mathbf{v} \cdot \mathbf{u} \times \mathbf{w} - \mathbf{w} \cdot \mathbf{u} \times \mathbf{v} \qquad \{[\mathbf{a}\ \mathbf{a}\ \mathbf{b}] = 0\}$

$= \mathbf{u} \cdot \mathbf{v} \times \mathbf{w} + \mathbf{w} \cdot \mathbf{u} \times \mathbf{v} - \mathbf{w} \cdot \mathbf{u} \times \mathbf{v} = \mathbf{u} \cdot \mathbf{v} \times \mathbf{w}$

33. (d) Given that, $\mathbf{OA} = 7\hat{i} - 4\hat{j} + 7\hat{k}$, $\mathbf{OB} = \hat{i} - 6\hat{j} + 10\hat{k}$, $\mathbf{OC} = -\hat{i} - 3\hat{j} + 4\hat{k}$ and $\mathbf{OD} = 5\hat{i} - \hat{j} + 5\hat{k}$

Now, $AB = \sqrt{(7-1)^2 + (-4+6)^2 + (7-10)^2}$

$= \sqrt{36 + 4 + 9} = 7$

$BC = \sqrt{(1+1)^2 + (-6+3)^2 + (10-4)^2}$

$= \sqrt{4 + 9 + 36} = \sqrt{49} = 7$

$CD = \sqrt{(-1-5)^2 + (-3+1)^2 + (4-5)^2}$

$= \sqrt{36 + 4 + 1} = \sqrt{41}$

and $DA = \sqrt{(5-7)^2 + (-1+4)^2 + (5-7)^2}$

$= \sqrt{4 + 9 + 4} = \sqrt{17}$

Hence, option (d) is correct.

34. (b) Since, $\begin{vmatrix} a & a^2 & 1+a^3 \\ b & b^2 & 1+b^3 \\ c & c^2 & 1+c^3 \end{vmatrix} = \begin{vmatrix} a & a^2 & 1 \\ b & b^2 & 1 \\ c & c^2 & 1 \end{vmatrix} + \begin{vmatrix} a & a^2 & a^3 \\ b & b^2 & b^3 \\ c & c^2 & c^3 \end{vmatrix} = 0$

$\Rightarrow \quad \begin{vmatrix} a & a^2 & 1 \\ b & b^2 & 1 \\ c & c^2 & 1 \end{vmatrix} + abc \begin{vmatrix} a & a^2 & 1 \\ b & b^2 & 1 \\ c & c^2 & 1 \end{vmatrix} = 0$

$\Rightarrow \quad (1 + abc)\begin{vmatrix} a & a^2 & 1 \\ b & b^2 & 1 \\ c & c^2 & 1 \end{vmatrix} = 0 \qquad \left[\because \begin{vmatrix} a & a^2 & 1 \\ b & b^2 & 1 \\ c & c^2 & 1 \end{vmatrix} \ne 0\right]$

$\Rightarrow \quad 1 + abc = 0$

$\Rightarrow \quad abc = -1$

35. (d) Given that, $\mathbf{u} = \hat{i} + \hat{j}$,

$\mathbf{v} = \hat{i} - \hat{j}$,

$\mathbf{w} = \hat{i} + 2\mathbf{j} + 3\hat{k}$,

$\mathbf{u} \cdot \mathbf{n} = 0$

and $\mathbf{v} \cdot \mathbf{n} = 0$

i.e. $\mathbf{n} = \frac{\mathbf{u} \times \mathbf{v}}{|\mathbf{u} \times \mathbf{v}|}$

Now, $\mathbf{u} \times \mathbf{v} = \begin{vmatrix} \hat{i} & \hat{j} & \hat{k} \\ 1 & 1 & 0 \\ 1 & -1 & 0 \end{vmatrix}$

$= 0\hat{i} - 0\hat{j} - 2\hat{k} = -2\hat{k}$

$\therefore \quad |\mathbf{w} \cdot \mathbf{n}| = \frac{|\mathbf{w} \cdot \mathbf{u} \times \mathbf{v}|}{|\mathbf{u} \times \mathbf{v}|} = \frac{|-6\hat{k}|}{|-2\hat{k}|} = 3$

$[\because \mathbf{w} \cdot (\mathbf{u} \times \mathbf{w}) = (\hat{i} + 2\hat{j} + 3\hat{k}) \cdot (-2\hat{k}) = -6\hat{k}]$

Hence, $|\mathbf{w} \cdot \mathbf{n}| = 3$

36. (a) Given two vectors lie in *XY*-plane. So, a vector coplanar with them is

$$\mathbf{a} = x\hat{\mathbf{i}} + y\hat{\mathbf{j}}$$

Since, $\mathbf{a} \perp (\hat{\mathbf{i}} - \hat{\mathbf{j}})$

$\Rightarrow (x\hat{\mathbf{i}} + y\hat{\mathbf{j}}) \cdot (\hat{\mathbf{i}} - \hat{\mathbf{j}}) = 0$

$\Rightarrow x - y = 0$

$\Rightarrow x = y$

$\therefore \mathbf{a} = x\hat{\mathbf{i}} + x\hat{\mathbf{j}}$

and $|\mathbf{a}| = \sqrt{x^2 + x^2} = x\sqrt{2}$

$\therefore$ Required unit vector $= \dfrac{\mathbf{a}}{|\mathbf{a}|} = \dfrac{x(\hat{\mathbf{i}} + \hat{\mathbf{j}})}{x\sqrt{2}} = \dfrac{1}{\sqrt{2}}(\hat{\mathbf{i}} + \hat{\mathbf{j}})$

37. (a) Since, the vector $\hat{\mathbf{i}} + x\hat{\mathbf{j}} + 3\hat{\mathbf{k}}$ is doubled in magnitude, then it becomes $4\hat{\mathbf{i}} + (4x - 2)\hat{\mathbf{j}} + 2\hat{\mathbf{k}}$

$\therefore 2|\hat{\mathbf{i}} + x\hat{\mathbf{j}} + 3\hat{\mathbf{k}}| = |4\hat{\mathbf{i}} + (4x - 2)\hat{\mathbf{j}} + 2\hat{\mathbf{k}}|$

$\Rightarrow 2\sqrt{1 + x^2 + 9} = \sqrt{16 + (4x - 2)^2 + 4}$

$\Rightarrow 40 + 4x^2 = 20 + (4x - 2)^2$

$\Rightarrow 3x^2 - 4x - 4 = 0$

$\Rightarrow (x - 2)(3x + 2) = 0$

$\Rightarrow x = 2, -\dfrac{2}{3}$

38. (b) Since, $\mathbf{a} + \mathbf{b} + \mathbf{c} = \mathbf{0}$

$\Rightarrow \mathbf{a} + \mathbf{b} = -\mathbf{c}$

$\Rightarrow (\mathbf{a} + \mathbf{b}) \times \mathbf{c} = -\mathbf{c} \times \mathbf{c}$

$\Rightarrow \mathbf{b} \times \mathbf{c} = \mathbf{c} \times \mathbf{a}$

Similarly, $\mathbf{a} \times \mathbf{b} = \mathbf{b} \times \mathbf{c}$

Hence, $\mathbf{a} \times \mathbf{b} = \mathbf{b} \times \mathbf{c} = \mathbf{c} \times \mathbf{a}$

39. (a) Since, the vectors $\mathbf{a} = x\hat{\mathbf{i}} + y\hat{\mathbf{j}} + z\hat{\mathbf{k}}$ and $\mathbf{b} = \hat{\mathbf{j}}$ are such that $\mathbf{a}$, $\mathbf{c}$ and $\mathbf{b}$ form a right handed system.

$$\therefore \mathbf{c} = \mathbf{b} \times \mathbf{a} = \begin{vmatrix} \hat{\mathbf{i}} & \hat{\mathbf{j}} & \hat{\mathbf{k}} \\ 0 & 1 & 0 \\ x & y & z \end{vmatrix}$$

$$= z\hat{\mathbf{i}} - x\hat{\mathbf{k}}$$

JEE Advanced & IIT JEE Archive

1. (a,c,d)

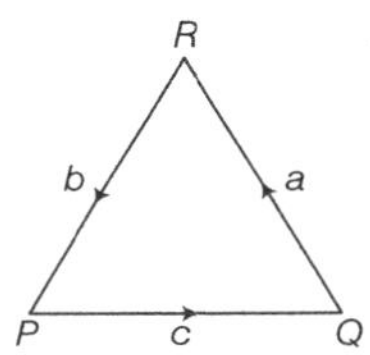

Given, $|\mathbf{a}| = 12, |\mathbf{b}| = 4\sqrt{3}$

$\mathbf{a} + \mathbf{b} + \mathbf{c} = 0$

$\Rightarrow \mathbf{a} = -(\mathbf{b} + \mathbf{c})$

We have, $|\mathbf{a}|^2 = |\mathbf{b} + \mathbf{c}|^2$

$\Rightarrow |\mathbf{a}|^2 = |\mathbf{b}|^2 + |\mathbf{c}|^2 + 2\mathbf{b} \cdot \mathbf{c}$

$\Rightarrow 144 = 48 + |\mathbf{c}|^2 + 48$

$\Rightarrow |\mathbf{c}|^2 = 48$

$|\mathbf{c}| = 4\sqrt{3}$

Also, $|\mathbf{c}|^2 = |\mathbf{a}|^2 + |\mathbf{b}|^2 + 2\,\mathbf{a} \cdot \mathbf{b}$

$\Rightarrow 48 = 144 + 48 + 2\,\mathbf{a} \cdot \mathbf{b}$

$\Rightarrow \mathbf{a} \cdot \mathbf{b} = -72$

$\therefore$ Option (d) is correct.

Also, $\mathbf{a} \times \mathbf{b} = \mathbf{c} \times \mathbf{a}$

$\Rightarrow \mathbf{a} \times \mathbf{b} + \mathbf{c} \times \mathbf{a} = 2\mathbf{a} \times \mathbf{b}$

$\Rightarrow |\mathbf{a} \times \mathbf{b} + \mathbf{c} \times \mathbf{a}| = 2|\mathbf{a} \times \mathbf{b}|$

$= 2\sqrt{|\mathbf{a}|^2|\mathbf{b}|^2 - (\mathbf{a} \cdot \mathbf{b})^2}$

$= 2\sqrt{(144)(48) - (-72)^2}$

$= 2(12)\sqrt{48 - 36}$

$= 48\sqrt{3}$

$\therefore$ Option (c) is correct.

Now, $\dfrac{|\mathbf{c}|^2}{2} - |\mathbf{a}| = 24 - 12 = 12$

$\therefore$ Option (a) is correct.

and $\dfrac{|\mathbf{c}|^2}{2} + |\mathbf{a}| = 24 + 12 = 36$

$\therefore$ Option (b) is not correct.

2. (9) Here, $\mathbf{s} = 4\mathbf{p} + 3\mathbf{q} + 5\mathbf{p}$ ...(i)

and $\mathbf{s} = (-\mathbf{p} + \mathbf{q} + \mathbf{r})x + (\mathbf{p} - \mathbf{q} + \mathbf{r})y + (-\mathbf{p} - \mathbf{q} + \mathbf{r})z$ ...(ii)

$\therefore 4\mathbf{p} + 3\mathbf{q} + 5\mathbf{r} = \mathbf{p}(-x + y - z) + \mathbf{q}(x - y - z) + \mathbf{r}(x + y + z)$

On comparing both sides, we get

$-x + y - z = 4$

$x - y - z = 3$

and $x + y + z = 5$

On solving above equations, we get

$x = 4, y = \dfrac{9}{2}, z = \dfrac{-7}{2}$

$\therefore 2x + y + z = 8 + \dfrac{9}{2} - \dfrac{7}{2} = 9$

3. (c) Let $\mathbf{v} = \mathbf{a} + \lambda\mathbf{b}$

$\mathbf{v} = (1 + \lambda)\hat{\mathbf{i}} + (1 - \lambda)\hat{\mathbf{j}} + (1 + \lambda)\hat{\mathbf{k}}$

Projection of $\mathbf{v}$ on $\mathbf{c} = \dfrac{1}{\sqrt{3}}$.

$\Rightarrow \dfrac{\mathbf{v} \cdot \mathbf{c}}{|\mathbf{c}|} = \dfrac{1}{\sqrt{3}}$

$\Rightarrow \dfrac{(1 + \lambda) - (1 - \lambda) - (1 + \lambda)}{\sqrt{3}} = \dfrac{1}{\sqrt{3}}$

$\Rightarrow 1 + \lambda - 1 + \lambda - 1 - \lambda = 1$

$\Rightarrow \lambda - 1 = 1$

$\Rightarrow \lambda = 2$

$\therefore \mathbf{v} = 3\hat{\mathbf{i}} - \hat{\mathbf{j}} + 3\hat{\mathbf{k}}$

4. (a,d) Let $\mathbf{a} = \hat{\mathbf{i}} + \hat{\mathbf{j}} + 2\hat{\mathbf{k}}$, $\mathbf{b} = \hat{\mathbf{i}} + 2\hat{\mathbf{j}} + \hat{\mathbf{k}}$ and $\mathbf{c} = \hat{\mathbf{i}} + \hat{\mathbf{j}} + \hat{\mathbf{k}}$.

$\therefore$ A vector coplanar to $\mathbf{a}$ and $\mathbf{b}$, and perpendicular to $\mathbf{c}$

$= \lambda(\mathbf{a} \times \mathbf{b}) \times \mathbf{c}$

$= \lambda\{(\mathbf{a} \cdot \mathbf{c})\,\mathbf{v} - (\mathbf{b} \cdot \mathbf{c})\,\mathbf{a}\}$

$= \lambda\{(1 + 1 + 4)(\hat{\mathbf{i}} + 2\hat{\mathbf{j}} + \hat{\mathbf{k}}) - (1 + 2 + 1)(\hat{\mathbf{i}} + \hat{\mathbf{j}} + 2\hat{\mathbf{k}})\}$

$= \lambda\{6\hat{\mathbf{i}} + 12\hat{\mathbf{j}} + 6\hat{\mathbf{k}} - 6\hat{\mathbf{i}} - 6\hat{\mathbf{j}} - 12\hat{\mathbf{k}}\}$

$= \lambda\{6\hat{\mathbf{j}} - 6\hat{\mathbf{k}}\} = 6\lambda\{\hat{\mathbf{j}} - \hat{\mathbf{k}}\}$

for $\lambda = \dfrac{1}{6} \Rightarrow$ (a) is correct.

and $\lambda = -\dfrac{1}{6} \Rightarrow$ (d) is correct.

5. (9) $$\mathbf{r} \times \mathbf{b} = \mathbf{c} \times \mathbf{b}$$

$\Rightarrow$ $$(\mathbf{r} - \mathbf{c}) \times \mathbf{b} = 0$$

$\Rightarrow$ $$\mathbf{r} - \mathbf{c} = \lambda \mathbf{b}$$

or $$\mathbf{r} = \mathbf{c} + \lambda \mathbf{b} \quad \ldots \text{(i)}$$

Given, $\mathbf{r} \cdot \mathbf{a} = 0$, taking dot product with $\mathbf{a}$ for Eq. (i).

$\Rightarrow$ $$\mathbf{r} \cdot \mathbf{a} = \mathbf{a} \cdot \mathbf{c} + \lambda \mathbf{a} \cdot \mathbf{b}$$

$\therefore$ $$\lambda = \frac{-\mathbf{a} \cdot \mathbf{c}}{\mathbf{a} \cdot \mathbf{b}} \quad [\because \mathbf{r} \cdot \mathbf{a} = 0] \ldots \text{(ii)}$$

From Eqs. (i) and (ii), we get

$$\mathbf{r} = \mathbf{c} - \frac{\mathbf{a} \cdot \mathbf{c}}{\mathbf{a} \cdot \mathbf{b}} \mathbf{b}$$, taking dot with $\mathbf{b}$, we get

$$\mathbf{r} \cdot \mathbf{b} = \mathbf{c} \cdot \mathbf{b} - \frac{\mathbf{a} \cdot \mathbf{c}}{\mathbf{a} \cdot \mathbf{b}} (\mathbf{b} \cdot \mathbf{b})$$

$$= (-1 + 2) - \frac{(-1 - 3)}{(1)} (1 + 1)$$

$$\left[\text{where, } \begin{aligned} \mathbf{a} &= -\hat{\mathbf{i}} - \hat{\mathbf{k}}, \\ \mathbf{b} &= -\hat{\mathbf{i}} + \hat{\mathbf{j}}, \\ \mathbf{c} &= \hat{\mathbf{i}} + 2\hat{\mathbf{j}} + 3\hat{\mathbf{k}} \end{aligned}\right]$$

$$= 1 + 8 = 9$$

6. (a) $m_{PQ} = \frac{1}{6}$, $\quad m_{SR} = \frac{1}{6}$,

$m_{RQ} = -3$, $\quad m_{SP} = -3$

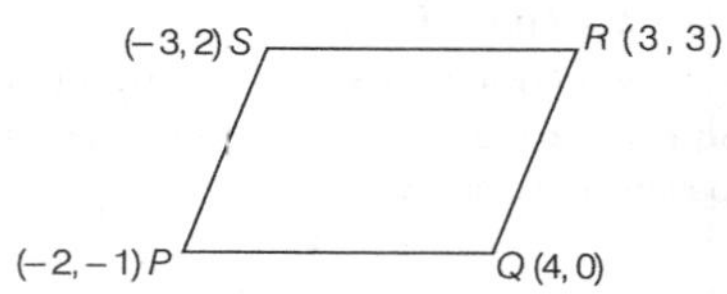

$\Rightarrow$ Parallelogram

But neither $PR = SQ$ nor $PR \perp SQ$.

$\therefore$ Parallelogram, which is neither a rhombus nor a rectangle.

7. (b) Given $$\mathbf{AB} = 2\hat{\mathbf{i}} + 10\hat{\mathbf{j}} + 11\hat{\mathbf{k}}$$

and $$\mathbf{AD} = -\hat{\mathbf{i}} + 2\hat{\mathbf{j}} + 2\hat{\mathbf{k}}$$

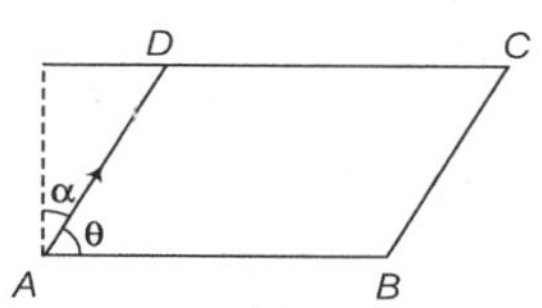

Angle θ between $\mathbf{AB}$ and $\mathbf{AD}$ is

$$\cos\theta = \left|\frac{\mathbf{AB} \cdot \mathbf{AD}}{|\mathbf{AB}|\,|\mathbf{AD}|}\right|$$

$$= \left|\frac{-2 + 20 + 22}{(15)(3)}\right| = \frac{8}{9}$$

$\Rightarrow$ $$\sin\theta = \frac{\sqrt{17}}{9}$$

Since, $$\alpha + \theta = 90^\circ$$

$\therefore$ $$\cos(\alpha) = \cos(90^\circ - \theta)$$

$$= \sin\theta = \frac{\sqrt{17}}{9}$$

8. (5) From the given information, it is clear that

$$\mathbf{a} = \frac{\hat{\mathbf{i}} - 2\hat{\mathbf{j}}}{\sqrt{5}}$$

$\Rightarrow$ $$|\mathbf{a}| = 1, |\mathbf{b}| = 1, \mathbf{a} \cdot \mathbf{b} = 0$$

Now, $(2\mathbf{a} + \mathbf{b}) \cdot [(\mathbf{a} \times \mathbf{b}) \times (\mathbf{a} - 2\mathbf{b})]$

$$= (2\mathbf{a} + \mathbf{b}) \cdot [a^2\mathbf{b} - (\mathbf{a} \cdot \mathbf{b}) \cdot \mathbf{a} + 2\, b^2 \cdot \mathbf{a} - 2(\mathbf{b} \cdot \mathbf{a}) \cdot a]$$

$$= [2\mathbf{a} + \mathbf{b}] \cdot [\mathbf{b} + 2\mathbf{a}] = 4\mathbf{a}_2 + \mathbf{b}_2$$

$$= 4 \cdot 1 + 1 = 5 \quad [\because \mathbf{a} \cdot \mathbf{b} = 0]$$

9. (c) Let angle between $\mathbf{a}$ and $\mathbf{b}$ be θ_1, $\mathbf{c}$ and $\mathbf{d}$ be θ_2 and $\mathbf{a} \times \mathbf{b}$ and $\mathbf{c} \times \mathbf{d}$ be θ.

Since, $$(\mathbf{a} \times \mathbf{b}) \cdot (\mathbf{c} \times \mathbf{d}) = 1$$

$\Rightarrow$ $$\sin\theta_1 \cdot \sin\theta_2 \cdot \cos\theta = 1$$

$\Rightarrow$ $$\theta_1 = 90^\circ, \theta_2 = 90^\circ, \theta = 0^\circ$$

$\Rightarrow$ $$\mathbf{a} \perp \mathbf{b}, \mathbf{c} \perp \mathbf{d}, (\mathbf{a} \times \mathbf{b}) \,||\, (\mathbf{c} \times \mathbf{d})$$

So, $$\mathbf{a} \times \mathbf{b} = k(\mathbf{c} \times \mathbf{d})$$

and $$\mathbf{a} \times \mathbf{b} = k(\mathbf{c} \times \mathbf{b})$$

$\Rightarrow$ $$(\mathbf{a} \times \mathbf{b}) \cdot \mathbf{c} = k(\mathbf{c} \times \mathbf{d}) \cdot \mathbf{c}$$

and $$(\mathbf{a} \times \mathbf{b}) \cdot \mathbf{d} = k(\mathbf{c} \times \mathbf{d}) \cdot \mathbf{d}$$

$\Rightarrow$ $$[\mathbf{a}\ \mathbf{b}\ \mathbf{c}] = 0$$

and $$[\mathbf{a}\ \mathbf{b}\ \mathbf{d}] = 0$$

$\Rightarrow$ $\mathbf{a}, \mathbf{d}, \mathbf{c}$ and $\mathbf{a}, \mathbf{b}, \mathbf{d}$ are coplanar vectors, so options (a) and (b) are incorrect.

Let $\mathbf{b} \,||\, \mathbf{d} \Rightarrow$ $$\mathbf{b} = \pm \mathbf{d}$$

As, $$(\mathbf{a} \times \mathbf{b}) \cdot (\mathbf{c} \times \mathbf{d}) = 1$$

$\Rightarrow$ $$(\mathbf{a} \times \mathbf{b}) \cdot (\mathbf{c} \times \mathbf{b}) = \pm 1$$

$\Rightarrow$ $$[\mathbf{a} \times \mathbf{b}\ \mathbf{c}\ \mathbf{b}] = \pm 1$$

$\Rightarrow$ $$[\mathbf{c}\ \mathbf{b}\ \mathbf{a} \times \mathbf{b}] = \pm 1$$

$\Rightarrow$ $$\mathbf{c} \cdot [\mathbf{b} \times (\mathbf{a} \times \mathbf{b})] = \pm 1$$

$\Rightarrow$ $$\mathbf{c} \cdot [\mathbf{a} - (\mathbf{b} \cdot \mathbf{a})\mathbf{b}] = \pm 1$$

$\Rightarrow$ $$\mathbf{c} \cdot \mathbf{a} = \pm 1 \quad [\because \mathbf{a} \cdot \mathbf{b} = 0]$$

which is a contradiction, so option (c) is correct.

Let option (d) is correct.

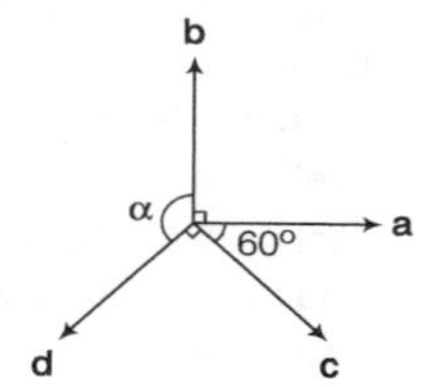

$\Rightarrow$ $$\mathbf{d} = \pm \mathbf{a}$$

and $$\mathbf{c} = \pm \mathbf{b}$$

As $$(\mathbf{a} \times \mathbf{b}) \cdot (\mathbf{c} \times \mathbf{d}) = 1$$

$\Rightarrow$ $$(\mathbf{a} \times \mathbf{b}) \cdot (\mathbf{b} \times \mathbf{a}) = \pm 1$$

which is a contradiction, so option (d) is incorrect.

Alternatively, option (c) and (d) may be observed from the given figure.

10. (a) The volume of the parallelopiped with coterminus edges as $\hat{\mathbf{a}}, \hat{\mathbf{b}}, \hat{\mathbf{c}}$ is given by $[\hat{\mathbf{a}}\ \hat{\mathbf{b}}\ \hat{\mathbf{c}}] = \hat{\mathbf{a}} \cdot (\hat{\mathbf{b}} \times \hat{\mathbf{c}})$

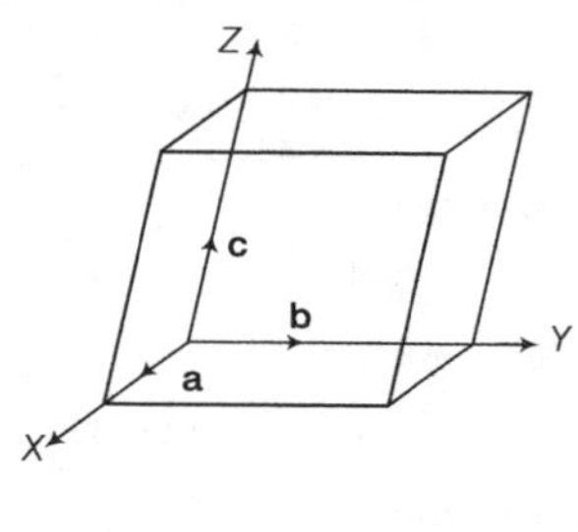

Now, $[\hat{a}\ \hat{b}\ \hat{c}]^2 = \begin{vmatrix} \hat{a}\cdot\hat{a} & \hat{a}\cdot\hat{b} & \hat{a}\cdot\hat{c} \\ \hat{b}\cdot\hat{a} & \hat{b}\cdot\hat{b} & \hat{b}\cdot\hat{c} \\ \hat{c}\cdot\hat{a} & \hat{c}\cdot\hat{b} & \hat{c}\cdot\hat{c} \end{vmatrix}$

$$= \begin{vmatrix} 1 & 1/2 & 1/2 \\ 1/2 & 1 & 1/2 \\ 1/2 & 1/2 & 1 \end{vmatrix}$$

$$\Rightarrow \quad [\hat{a}\ \hat{b}\ \hat{c}]^2 = 1\left(1-\frac{1}{4}\right)-\frac{1}{2}\left(\frac{1}{2}-\frac{1}{4}\right)+\frac{1}{2}\left(\frac{1}{4}-\frac{1}{2}\right)$$

$$= \frac{1}{2}$$

Thus, the required volume of the parallelopiped

$$= \frac{1}{\sqrt{2}} \text{ cu unit.}$$

11. (a) Given, $\mathbf{OP} = \hat{\mathbf{a}}\cos t + \hat{\mathbf{b}}\sin t$

$$\Rightarrow \quad |\mathbf{OP}| = \sqrt{(\hat{\mathbf{a}}\cdot\hat{\mathbf{a}})\cos^2 t + (\hat{\mathbf{b}}\cdot\hat{\mathbf{b}})\sin^2 t + 2\hat{\mathbf{a}}\cdot\hat{\mathbf{b}}\sin t\cos t}$$

$$\Rightarrow \quad |\mathbf{OP}| = \sqrt{1+\hat{\mathbf{a}}\cdot\hat{\mathbf{b}}\sin 2t}$$

$$\Rightarrow \quad |\mathbf{OP}|_{max} = M = \sqrt{1+\hat{\mathbf{a}}\cdot\hat{\mathbf{b}}} \quad \text{at } \sin 2t = 1$$

$$\Rightarrow \quad t = \frac{\pi}{4}$$

At $t = \frac{\pi}{4}$,

$$\mathbf{OP} = \frac{1}{\sqrt{2}}(\hat{\mathbf{a}}+\hat{\mathbf{b}}), \text{ then}$$

unit vector along $\mathbf{OP} = \dfrac{\hat{\mathbf{a}}+\hat{\mathbf{b}}}{|\hat{\mathbf{a}}+\hat{\mathbf{b}}|}$.

12. (c) Since, given vectors are coplanar, then

$$\begin{vmatrix} -\lambda^2 & 1 & 1 \\ 1 & -\lambda^2 & 1 \\ 1 & 1 & -\lambda^2 \end{vmatrix} = 0$$

$$\Rightarrow \quad \lambda^6 - 3\lambda^2 - 2 = 0$$

$$\Rightarrow \quad (1+\lambda^2)^2(\lambda^2-2) = 0$$

$$\therefore \quad \lambda = \pm\sqrt{2}$$

13. (b) Since, **a**, **b**, **c** are unit vectors and $\mathbf{a}+\mathbf{b}+\mathbf{c} = 0$, then **a**, **b**, **c** represent an equilateral triangle.

$$\therefore \quad \mathbf{a}\times\mathbf{b} = \mathbf{b}\times\mathbf{c} = \mathbf{c}\times\mathbf{a} \neq 0.$$

14. (c) Since, **PQ** is not parallel to **TR**.

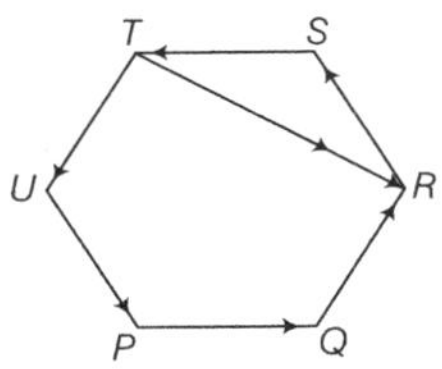

$\because$ **TR** is resultant of **RS** and **ST** vectors.

$$\Rightarrow \quad \mathbf{PQ}\times(\mathbf{RS}+\mathbf{ST}) \neq 0$$

But for Statement II, we have

$$\mathbf{PQ}\times\mathbf{RS} = 0$$

which is not possible as **PQ** not parallel to **RS**.

Hence, Statement I is correct and Statement II is incorrect.

15. (a) Let vector **r** be coplanar to **a** and **b**.

$$\therefore \quad \mathbf{r} = \mathbf{a} + t\mathbf{b}$$

$$\Rightarrow \quad \mathbf{r} = (\hat{i}+2\hat{j}+\hat{k}) + t(\hat{i}-\hat{j}+\hat{k})$$

$$= \hat{i}(1+t) + \hat{j}(2-t) + \hat{k}(1+t)$$

The projection of **r** on **c** $= \dfrac{1}{\sqrt{3}}$. [given]

$$\Rightarrow \quad \frac{\mathbf{r}\cdot\mathbf{c}}{|\mathbf{c}|} = \frac{1}{\sqrt{3}}$$

$$\Rightarrow \quad \frac{|1\cdot(1+t)+1\cdot(2-t)-1\cdot(1+t)|}{\sqrt{3}} = \frac{1}{\sqrt{3}}$$

$$\Rightarrow \quad (2-t) = \pm 1$$

$$\Rightarrow \quad t = 1 \text{ or } 3$$

when $t = 1$, we have $\mathbf{r} = 2\hat{i}+\hat{j}+2\hat{k}$

when $t = 3$, we have $\mathbf{r} = 4\hat{i}-\hat{j}+4\hat{k}$.

16. (b,d) Let vector **AO** be parallel to line of intersection of planes P_1 and P_2 through origin

Normal to plane p_1 is

$$\mathbf{n}_1 = [(2\hat{j}+3\hat{k})\times(4\hat{j}-3\hat{k})] = -18\hat{i}$$

Normal to plane p_2 is

$$\mathbf{n}_2 = (\hat{j}-\hat{k})\times(3\hat{i}+3\hat{j}) = 3\hat{i}-3\hat{j}-3\hat{k}$$

$\therefore$ **OA** is parallel to $\pm(\mathbf{n}_1\times\mathbf{n}_2) = 54\hat{j}-54\hat{k}$

$\therefore$ Angle between $54(\hat{j}-\hat{k})$ and $(2\hat{i}+\hat{j}-2\hat{k})$ is

$$\cos\theta = \pm\left(\frac{54+108}{3\cdot 54\cdot\sqrt{2}}\right) = \pm\frac{1}{\sqrt{2}}$$

$$\therefore \quad \theta = \frac{\pi}{4}, \frac{3\pi}{4}$$

17. (b) Since, $\mathbf{b}_1 = \mathbf{b} - \dfrac{\mathbf{b}\cdot\mathbf{a}}{|\mathbf{a}|^2}\mathbf{a}$,

$$\mathbf{b}_2 = \mathbf{b} + \frac{\mathbf{b}\cdot\mathbf{a}}{|\mathbf{a}|^2}\mathbf{a}$$

and $\quad \mathbf{c}_1 = \mathbf{c} - \dfrac{\mathbf{c}\cdot\mathbf{a}}{|\mathbf{a}|^2}\mathbf{a} - \dfrac{\mathbf{c}\cdot\mathbf{b}}{|\mathbf{b}|^2}\mathbf{b}$,

$$\mathbf{c}_2 = \mathbf{c} - \frac{\mathbf{c}\cdot\mathbf{a}}{|\mathbf{a}|^2}\mathbf{a} - \frac{\mathbf{c}\cdot\mathbf{b}_1}{|\mathbf{b}|^2}\mathbf{b}_1,$$

$$\mathbf{c}_3 = \mathbf{c} - \frac{\mathbf{c}\cdot\mathbf{a}}{|\mathbf{a}|^2}\mathbf{a} - \frac{\mathbf{c}\cdot\mathbf{b}_2}{|\mathbf{b}_2|^2}\mathbf{b}_2,$$

$$\mathbf{c}_4 = \mathbf{a} - \frac{\mathbf{c}\cdot\mathbf{a}}{|\mathbf{a}|^2}\mathbf{a}$$

which shows $\mathbf{a}\cdot\mathbf{b}_1 = 0 = \mathbf{a}\cdot\mathbf{c}_2 = \mathbf{b}_1\cdot\mathbf{c}_2$.

$\therefore$ $\{\mathbf{a}, \mathbf{b}_1, \mathbf{c}_2\}$ are mutually orthogonal vectors.

5. 3D Geometry

DPP-1 Direction Ratios, Direction Cosines and Equation of Straight Line

1. (c) Given direction ratios of two lines are (5, – 12, 13) and (–3, 4, 5).

Let θ be the acute angle between the given vectors.

$$a_1 = 5, b_1 = -12, c_1 = 13$$

and $$a_2 = -3, b_2 = 4, c_2 = 5$$

$\therefore$ $$\cos\theta = \frac{a_1a_2 + b_1b_2 + c_1c_2}{\sqrt{(a_1^2 + b_1^2 + c_1^2)}\sqrt{(a_2^2 + b_2^2 + c_2^2)}}$$

$$= \frac{5\times(-3) + (-12)\times 4 + 13\times 5}{\left(\sqrt{5^2 + (-12)^2 + 13^2}\right)\left(\sqrt{(-3)^2 + (4)^2 + (5)^2}\right)}$$

$$= \frac{-15 - 48 + 65}{\sqrt{25 + 144 + 169}\sqrt{9 + 16 + 25}}$$

$$= \frac{-63 + 65}{\sqrt{338}\sqrt{50}}$$

$$= \frac{2}{\sqrt{338\times 50}}$$

$$= \frac{2}{\sqrt{16900}}$$

$\Rightarrow$ $$\cos\theta = \frac{2}{130} = \frac{1}{65}$$

$\Rightarrow$ $$\theta = \cos^{-1}\left(\frac{1}{65}\right)$$

2. (c) Let each edge of cube be a, then coordinates of the vertices of cube are

$O(0, 0, 0)$, $A(a, 0, 0)$, $B(0, a, 0)$, $C(0, 0, a)$, $N(a, a, 0)$, $P(a, a, a)$, $L(0, a, a)$, $M(a, 0, a)$.

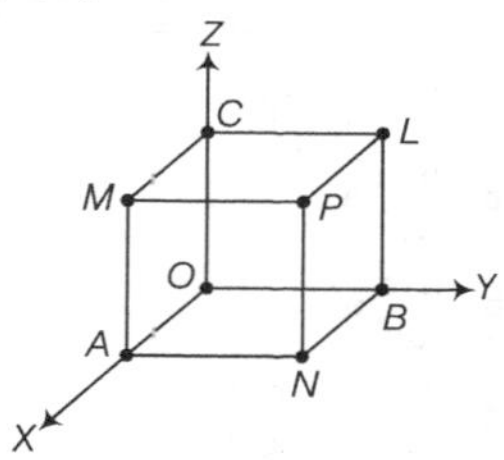

Direction ratios of the diagonals OP, AL, BM and CN are (a, a, a), $(-a, a, a)$, $(a, -a, a)$ and $(a, a, -a)$.

Let θ be the acute angle between diagonals OP and AL.

$\therefore$ $$\cos\theta = \frac{a_1a_2 + b_1b_2 + c_1c_2}{\sqrt{a_1^2 + b_1^2 + c_1^2}\sqrt{a_2^2 + b_2^2 + c_2^2}}$$

$$= \frac{a\times(-a) + a\times a + a\times a}{\sqrt{a^2 + a^2 + a^2}\sqrt{(-a)^2 + a^2 + a^2}}$$

$$= \frac{-a^2 + a^2 + a^2}{\sqrt{3a^2}\sqrt{3a^2}}$$

$$= \frac{a^2}{a\sqrt{3}\times a\sqrt{3}}$$

$\Rightarrow$ $$\cos\theta = \frac{1}{3}$$

$\Rightarrow$ $$\theta = \cos^{-1}\left(\frac{1}{3}\right)$$

3. (c) Let each edge of cube be a, then the coordinates of the vertices of the cube are $O(0, 0, 0)$, $A(a, 0, 0)$, $B(0, a, 0)$, $C(0, 0, a)$, $N(a, a, 0)$ $P(a, a, a)$, $L(0, a, a)$, $M(a, 0, a)$.

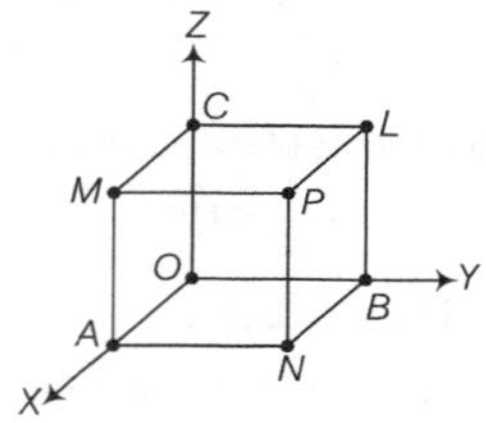

Direction ratios of the diagonals OP, AL, BM and CN are (a, a, a), $(-a, a, a)$, $(a, -a, a)$ and $(a, a, -a)$.

$\therefore$ Direction cosines of OP, AL, BM and CN are $\left(\frac{1}{\sqrt{3}}, \frac{1}{\sqrt{3}}, \frac{1}{\sqrt{3}}\right)$, $\left(\frac{-1}{\sqrt{3}}, \frac{1}{\sqrt{3}}, \frac{1}{\sqrt{3}}\right)$, $\left(\frac{1}{\sqrt{3}}, \frac{-1}{\sqrt{3}}, \frac{1}{\sqrt{3}}\right)$ and $\left(\frac{1}{\sqrt{3}}, \frac{1}{\sqrt{3}}, \frac{-1}{\sqrt{3}}\right)$.

Let (l, m, n) be the direction cosines of a line which makes angle α, β, γ, δ with the four diagonals of the cubes, i.e.

$$\cos\alpha = l\cdot\frac{1}{\sqrt{3}} + m\times\frac{1}{\sqrt{3}} + n\times\frac{1}{\sqrt{3}} = \frac{l + m + n}{\sqrt{3}}$$

$$\cos\beta = l\times\left(\frac{-1}{\sqrt{3}}\right) + m\times\frac{1}{\sqrt{3}} + n\times\frac{1}{\sqrt{3}} = \frac{-l + m + n}{\sqrt{3}}$$

$$\cos\gamma = l\times\left(\frac{1}{\sqrt{3}}\right) + m\times\left(\frac{-1}{\sqrt{3}}\right) + n\times\frac{1}{\sqrt{3}} = \frac{l - m + n}{\sqrt{3}}$$

$$\cos\delta = l\times\left(\frac{1}{\sqrt{3}}\right) + m\times\left(\frac{1}{\sqrt{3}}\right) + n\times\left(-\frac{1}{\sqrt{3}}\right) = \frac{l + m - n}{\sqrt{3}}$$

$$\therefore \cos^2\alpha + \cos^2\beta + \cos^2\gamma + \cos^2\delta$$

$$= \frac{1}{3}[(l + m + n)^2 + (-l + m + n)^2] + (l - m + n)^2 + (l + m - n)^2]$$

$$= \frac{1}{3}\times 4 = \frac{4}{3}$$

4. (a) Given, $\beta = 60^\circ \Rightarrow \cos\beta = \cos 60^\circ = \frac{1}{2}$

and $$\gamma = 60^\circ$$

$\Rightarrow$ $$\cos\gamma = \cos 60^\circ$$

$\Rightarrow$ $$\gamma = \frac{1}{2}$$

and let straight line makes an angle with X-axis be α.

Then, we know that,

$$\cos^2\alpha + \cos^2\beta + \cos^2\gamma = 1$$

$\Rightarrow$ $$\cos^2\alpha + \left(\frac{1}{2}\right)^2 + \left(\frac{1}{2}\right)^2 = 1$$

$\Rightarrow$ $$\cos^2\alpha + \frac{1}{4} + \frac{1}{4} = 1$$

$\Rightarrow$ $$\cos^2\alpha = 1 - \frac{1}{2}$$

$\Rightarrow$ $$\cos^2\alpha = \frac{1}{2}$$

$\Rightarrow$ $$\cos\alpha = \frac{1}{\sqrt{2}}$$

$\therefore$ $$\alpha = 45^\circ$$

5. (c) Let the given points be $A(2, 4, 5)$ and $B(3, 5, -4)$ and the point $P(0, y, 3)$ on YZ-plane divides AB in the ratio $k : 1$, then

$$x\text{-coordinate of point } k = \frac{mx_2 + nx_1}{m + n}$$

$$\Rightarrow \quad \frac{k \times 3 + 1 \times 2}{p + 1} = 0$$

$$\Rightarrow \quad 3k + 2 = 0\,(p + 1)$$

$$\Rightarrow \quad 3k + 2 = 0$$

$$\Rightarrow \quad 2k = -2$$

$$\Rightarrow \quad k = -2/3$$

$$\therefore \quad k : 1 = -2 : 3$$

6. (a) Given, line passes through (3, 4, 5) and (4, 6, 3).

Position vector of point (3, 4, 5),

$$\mathbf{OA} = 3\hat{i} + 4\hat{j} + 5\hat{k}$$

Position vector of point (4, 6, 3)

$$\mathbf{OB} = 4\hat{i} + 6\hat{j} + 3\hat{k}$$

Then, $\mathbf{AB} = (4\hat{i} + 6\hat{j} + 3\hat{k}) - (3\hat{i} + 4\hat{j} + 5\hat{k})$

$$= \hat{i} + 2\hat{j} - 2\hat{k}$$

and given points are $(-1, 2, 4)$ and $(1, 0, 5)$, then

$$\mathbf{CD} = (\hat{i} + 5\hat{k}) - (-\hat{i} + 2\hat{j} + 4\hat{k})$$

$$= 2\hat{i} - 2\hat{j} + \hat{k}$$

$\therefore$ Projection of **AB** on **CD** $= \dfrac{\mathbf{AB} \cdot \mathbf{CD}}{|\mathbf{CD}|}$

$$= \frac{(\hat{i} + 2\hat{j} - 2\hat{k}) \cdot (2\hat{i} - 2\hat{j} + \hat{k})}{\sqrt{(2)^2 + (-2)^2 + (1)^2}}$$

$$= \frac{2 + 2\,(-2) + (-2)\,(1)}{\sqrt{4 + 4 + 1}}$$

$$= \frac{2 - 4 - 2}{\sqrt{9}}$$

$$= -\frac{4}{3}$$

7. (c) Given DR's (1, 1, 2) and $(\sqrt{3} - 1, -\sqrt{3} - 1, 4)$.

Let θ be the acute angle between the given pair of lines, then

$$\cos\theta = \frac{1 \times (\sqrt{3} - 1) + 1 \times (-\sqrt{3} - 1) + 2 \times 4}{(\sqrt{(1)^2 + (1)^2 + (2)^2}\,(\sqrt{(\sqrt{3} - 1)^2 + (-\sqrt{3} - 1)^2 + (4)^2}}$$

$$= \frac{\sqrt{3} - 1 - \sqrt{3} - 1 + 8}{\sqrt{1 + 1 + 4}\,\sqrt{3 + 1 - 2\sqrt{3} + 3 + 1 + 2\sqrt{3} + 16}}$$

$$= \frac{6}{\sqrt{6} \times \sqrt{24}}$$

$$\Rightarrow \quad \cos\theta = \frac{6}{\sqrt{144}} = \frac{6}{12} = \frac{1}{2}$$

$$\therefore \quad \theta = \frac{\pi}{3}$$

8. (a) Given, $\cos\alpha$, $\cos\beta$ and $\cos\gamma$ are DC's, then

$$\cos^2\alpha + \cos^2\beta + \cos^2\gamma = 1$$

$$\frac{1 + \cos 2\alpha}{2} + \frac{1 + \cos 2\beta}{2} + \frac{1 + \cos 2\gamma}{2} = 1$$

$$\Rightarrow \quad 3 + \cos 2\alpha + \cos 2\beta + \cos 2\gamma = 2$$

$$\therefore \quad \cos 2\alpha + \cos 2\beta + \cos 2\gamma = -1$$

9. (a) Given, a line s making α, β, γ angles from X, Y and Z-axes

$\alpha = \theta, \beta = 60°, \gamma = 30°$

$$\therefore \quad \cos^2\alpha + \cos^2\beta + \cos^2\gamma = 1$$

$$\Rightarrow \quad \cos^2\theta + \cos^2 60° + \cos^2 30° = 1$$

$$\Rightarrow \quad \cos^2\theta + \left(\frac{1}{2}\right)^2 + \left(\frac{\sqrt{3}}{2}\right)^2 = 1$$

$$\Rightarrow \quad \cos^2\theta + \frac{1}{4} + \frac{3}{4} = 1$$

$$\Rightarrow \quad \cos^2\theta + 1 = 1$$

$$\Rightarrow \quad \cos^2\theta = 0$$

$$\Rightarrow \quad 1 - \sin^2\theta = 0$$

$$\Rightarrow \quad \sin^2\theta = 1 \Rightarrow \sin\theta = \pm 1$$

10. (a) Given points are $A(-2, 3, 5)$ and $B(a, 2, 5)$.

Let $P(0, y, z)$ be the point which divides YZ-plane and the joining of points in ratio 2 : 3.

$$\frac{2 \times a + 3 \times (-2)}{2 + 3} = 0$$

$$\Rightarrow \quad 2a - 6 = 0 \times 5$$

$$\therefore \quad 2a = 6 \Rightarrow a = 3$$

11. (b) Given, direction cosines are $l + m + n = 0$ ...(i)

and $l^2 + m^2 + n^2 = 0$ ...(ii)

From Eq. (i), $l = -(m + n)$

On putting the value of l in Eq. (ii), we get

$$\{-(m + n)\}^2 = m^2 + n^2$$

$$\Rightarrow \quad m^2 + n^2 + 2mn = m^2 + n^2$$

$$\Rightarrow \quad mn = 0$$

$$\Rightarrow \quad m = 0 \text{ or } n = 0$$

when, $m = 0$, $l = -n$, then direction ratios $-1, 0, 1$

when $n = 0$, $l = -m$, then direction ratios $-1, 0, 1$.

$\therefore$ Angle between lines

$$\cos\theta = \frac{(-1) \cdot (-1) + (0)\,(1) + (1)\,(0)}{\sqrt{(-1)^2 + (0)^2 + (1)^2}\,\sqrt{(-1)^2 + (1)^2 + (0)^2}}$$

$$\Rightarrow \quad \frac{1}{\sqrt{2} \times \sqrt{2}}$$

$$\therefore \quad \cos\theta = \frac{1}{2}$$

$$\Rightarrow \quad \theta = \frac{\pi}{3}$$

12. (b) Given, $\alpha = \dfrac{\pi}{4}$ and $\beta = \dfrac{\pi}{4}$

$$\because \quad \cos^2\alpha + \cos^2\beta + \cos^2\gamma = 1$$

$$\Rightarrow \quad \left(\frac{1}{\sqrt{2}}\right)^2 + \left(\frac{1}{\sqrt{2}}\right)^2 + \cos^2\gamma = 1$$

$$\Rightarrow \quad \frac{1}{2} + \frac{1}{2} + \cos^2\gamma = 1$$

$$\Rightarrow \quad 1 + \cos^2\gamma = 1$$

$$\Rightarrow \quad \cos^2\gamma = 0$$

$$\Rightarrow \quad \cos\gamma = 0$$

$$\therefore \quad \gamma = \frac{\pi}{2}$$

13. (d) Given lines are $\frac{x+1}{2} = \frac{y-1}{5} = \frac{z+3}{4}$

Direction ratios = (2, 5, 4)

and $\frac{x-1}{1} = \frac{y+2}{2} = \frac{z-3}{-3}$

Direction ratios = (1, 2, − 3)

Let θ be the acute angle between given lines is

$$\cos\theta = \frac{2\times 1 + 5\times 2 + 4\times(-3)}{\sqrt{2^2+5^2+4^2}\sqrt{1^2+2^2+(-3)^2}}$$

$$= \frac{2+10-12}{\sqrt{4+25+16}\sqrt{1+4+9}} = \frac{0}{\sqrt{45}\sqrt{14}} = 0$$

$\Rightarrow \quad \cos\theta = \cos\frac{\pi}{2}$

$\therefore \quad \theta = \frac{\pi}{2}$ i.e. 90°

14. (d) Given, lines $\frac{x-1}{-3} = \frac{y-2}{2k} = \frac{z-3}{2}$

Direction ratios (−3, 2k, 2)

and $\frac{x-1}{3k} = \frac{y-5}{1} = \frac{z-6}{-5}$

Direction ratios = (3k, 1, − 5).

Since, lines are perpendicular to each other, then

$$a_1a_2 + b_1b_2 + c_1c_2 = 0$$

$\Rightarrow \quad -3\times 3k + 2k\times 1 + 2\times -5 = 0$

$\Rightarrow \quad -9k + 2k - 10 = 0$

$\therefore \quad k = -\frac{10}{7}$

15. (d) Given lines are $2x = 3y = -z$

$\Rightarrow \quad \frac{x}{1/2} = \frac{y}{1/3} = \frac{z}{-1}$

Direction ratios $= \left(\frac{1}{2}, \frac{1}{3}, -1\right)$

and $6x = -y = -4z$

$\Rightarrow \quad \frac{x}{1/6} = \frac{y}{-1} = \frac{z}{-1/4}$

Direction ratios $= \left(\frac{1}{6}, -1, -\frac{1}{4}\right)$.

Let θ be the acute angle between these lines, i.e.

$$\cos\theta = \frac{\frac{1}{2}\times\frac{1}{6} + \frac{1}{3}\times(-1) + (-1)\times\left(-\frac{1}{4}\right)}{\sqrt{\left(\frac{1}{2}\right)^2+\left(\frac{1}{3}\right)^2+(-1)^2}\sqrt{\left(\frac{1}{6}\right)^2+|(-1)^2+\left(\frac{1}{-4}\right)^2}}$$

$$= \frac{\frac{1}{12} - \frac{1}{3} + \frac{1}{4}}{\sqrt{\frac{1}{4}+\frac{1}{9}+1}\sqrt{\frac{1}{36}+1+\frac{1}{6}}}$$

$$= \frac{\frac{1-4+3}{12}}{\sqrt{\frac{9+4+36}{36}} + \sqrt{\frac{4+144+9}{144}}}$$

$$= \frac{\frac{0}{12}}{\sqrt{\frac{9-4+36}{36}}\sqrt{\frac{4+144+9}{144}}} = 0$$

$\Rightarrow \quad \cos\theta = \cos\frac{\pi}{2}$

$\therefore \quad \theta = \frac{\pi}{2}$

16. (a) Given straight lines $\frac{x-1}{k} = \frac{y-2}{2} = \frac{z-3}{3}$

In vector form, the equation of line

$$r_1 = (\hat{i} + 2\hat{j} + 3\hat{k}) + \lambda(k\hat{i} + 2\hat{j} + 3\hat{k})$$

and $\frac{x-2}{3} = \frac{y-3}{k} = \frac{z-1}{2}$

In vector form, the equation of line

$$r_2 = (2\hat{i} + 3\hat{j} + \hat{k}) + \mu(3\hat{i} + k\hat{j} + 2\hat{k}).$$

Comparing with $r_1 = a_1 + \lambda b_1$ and $r_2 = a_2 + \mu b_2$, we have

$$a_1 = \hat{i} + 2\hat{j} + 3\hat{k},\ b_1 = k\hat{i} + 2\hat{j} + 3\hat{k},$$

$$a_2 = 3\hat{i} + 3\hat{j} + \hat{k},\ b_2 = 2\hat{i} + k\hat{j} + 2\hat{k}$$

Now, $\quad a_2 - a_1 = \hat{i} + \hat{j} - 2\hat{k}$

$\therefore \quad b_1 \times b_2 = \begin{vmatrix} \hat{i} & \hat{j} & \hat{k} \\ k & 2 & 3 \\ 3 & k & 2 \end{vmatrix}$

$$= \hat{i}(4-3k) - \hat{j}(2k-9) + \hat{k}(k^2-6)$$

Since, lines are intersect.

$\therefore$ Shortest distance between these lines is zero, then

$$(b_1 \times b_2)\cdot(a_2 - a_1) = 0$$

$\Rightarrow \quad (4-3k) - (2k-9) - 2(k^2-6) = 0$

$\Rightarrow \quad 4 - 3k - 2k + 9 - 2k^2 + 12 = 0$

$\Rightarrow \quad -2k^2 - 5k + 25 = 0$

$\Rightarrow \quad 2k^2 + 5k - 25 = 0$

$\Rightarrow \quad 2k^2 + 10k - 5k - 25 = 0$

$\Rightarrow \quad 2k(k+5) - 5(k+5) = 0$

$\Rightarrow \quad (2k-5)(k+5) = 0$

$\Rightarrow \quad 2k - 5 = 0$ or $k + 5 = 0$

$\Rightarrow \quad k = 5/2$ or $k = -5$

Since, k is integer, so $k = -5$.

17. (a) Let OP be the normal from O to the plane of ΔABC and l, m, n be the direction cosines of OP.

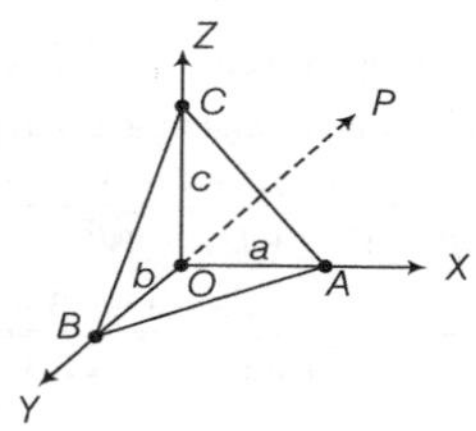

Let s be the area of ΔABC.

$\therefore$ Area of ΔOAC = Projection of ΔABC on ZX-plane = $s\cdot m$

But ΔOAC is a right angled triangle.

Now, the area of $\Delta OAC = \frac{1}{2}ac$

$\Rightarrow \quad \frac{1}{2}ac = s\cdot m$

$\Rightarrow \quad m = \frac{ac}{2s}$

Similarly in ΔOBC, $\quad l = \frac{bc}{2s}$

$\because \quad l^2 + m^2 + n^2 = 1$

$$\left(\frac{ac}{2s}\right)^2 + \left(\frac{bc}{2s}\right)^2 + \left(\frac{ab}{2s}\right)^2 = 1$$

$\Rightarrow \quad a^2c^2 + b^2c^2 + a^2b^2 = 4s^2$

$\therefore \quad s = \frac{1}{2}\sqrt{a^2b^2 + b^2c^2 + a^2c^2}$

18. (a) Given, equation of a line in vector form is

$$\mathbf{r} = (2\hat{i} - \hat{j} + 4\hat{k}) + \lambda(\hat{i} + \hat{j} - 2\hat{k})$$

Cartesian form of the given equation is

$$\frac{x-2}{1} = \frac{y+1}{1} = \frac{z-4}{-2}.$$

19. (a) Equation of line passing through point $A(a, b, c)$ and origin

$$\frac{x - x_1}{x_2 - x_1} = \frac{y - y_1}{y_2 - y_1} = \frac{z - z_1}{z_2 - z_1}$$

$$\Rightarrow \quad \frac{x-0}{a-0} = \frac{y-0}{b-0} = \frac{z-0}{c-0}$$

$$\Rightarrow \quad \frac{x}{a} = \frac{y}{b} = \frac{z}{c}.$$

Since, this line passes through the point $B(a', b', c')$, therefore lies on line.

$$\Rightarrow \quad \frac{a'}{a} = \frac{b'}{b} = \frac{c'}{c}.$$

20. (c) $\cos\theta = \dfrac{lm + mn + nl}{l^2 + m^2 + n^2}$

$$\Rightarrow \quad x^3 + x^2 - 4x - 4 = 0$$

$$l + m + n = -1$$

$$lm + mn + nl = -4$$

$$(l + m + n)^2 = l^2 + m^2 + n^2 + 2(-4)$$

$$\Rightarrow \quad l^2 + m^2 + n^2 = 1 + 8 = 9$$

$$\therefore \quad \cos\theta = \frac{-4}{9}$$

$$\therefore \quad \theta = \cos^{-1}\left(\frac{4}{9}\right)$$

21. (b) Given line, $\dfrac{x - x_1}{0} = \dfrac{y - y_1}{1} = \dfrac{z - z_1}{2}$

lines indicated that not parallel to X-axis and not passing through origin.

Thus, perpendicular to X-axis.

22. (c) Given lines are

$$L_1 : \mathbf{r} = (2\hat{i} + 9\hat{j} + 13\hat{k}) + \lambda(\hat{i} + 2\hat{j} + 3\hat{k})$$

and $$L_2 : \mathbf{r} = (-3\hat{i} + 7\hat{j} + p\hat{k}) + \mu(-\hat{i} + 2\hat{j} - 3\hat{k})$$

Given lines comparing with

$$\mathbf{r} = \mathbf{a} + \lambda\mathbf{b}$$

$$a_1 = 2\hat{i} + 9\hat{j} + 13\hat{k},\ b_1 = \hat{i} + 2\hat{j} + 3\hat{k}$$

$$a_2 = -3\hat{i} + 7\hat{j} + p\hat{k},\ b_2 = -\hat{i} + 2\hat{j} - 3\hat{k}$$

Condition for two given lines to intersects.

$$(\mathbf{a}_2 - \mathbf{a}_1)\cdot(\mathbf{b}_1 \times \mathbf{b}_1) = 0$$

Now, $$\mathbf{b}_1 \times \mathbf{b}_1 = \begin{vmatrix} \hat{i} & \hat{j} & \hat{k} \\ 1 & 2 & 3 \\ -1 & 2 & -3 \end{vmatrix}$$

$$= \hat{i}(-6-6) - \hat{j}(-3+3) + \hat{k}(2+2)$$

$$= \hat{i}(-12) + \hat{k}(4)$$

$$= -12\hat{i} + 4\hat{k}$$

and $$a_2 - a_1 = (-3\hat{i} + 7\hat{j} + p\hat{k}) - (2\hat{i} - 9\hat{j} + 13\hat{k})$$

$$= -5\hat{i} - 2\hat{j} + (p - 13)\hat{k}$$

Now, $$(\mathbf{a}_2 - \mathbf{a}_1)\cdot(\mathbf{b}_1 \times \mathbf{b}_2) = 0$$

$$\Rightarrow \quad [-5\hat{i} - 2\hat{j} + (p-13)\hat{k}]\cdot(-12\hat{i} + 4\hat{k}) = 0$$

$$60 + 4(p - 13) = 0$$

$$\Rightarrow \quad 60 + 4p - 52 = 0$$

$$\Rightarrow \quad 4p = -8$$

$$\Rightarrow \quad p = -2$$

$\therefore$ Intersecting lines for $p = -2$

23. (b) Given lines are

$$L_1 : \frac{x+1}{3} = \frac{y+2}{1} = \frac{z+1}{2}$$

and $$L_2 : \frac{x-2}{1} = \frac{y+2}{2} = \frac{z-3}{3}$$

Now, convert into vector form

$$L_1 : (-\hat{i} - 2\hat{j} - \hat{k}) + \lambda(3\hat{i} + \hat{j} + 2\hat{k})$$

$$L_2 : (2\hat{i} + 2\hat{j} + 3\hat{k}) + \mu(\hat{i} + 2\hat{j} + 3\hat{k})$$

Line L_1 comparing with $a_1 + \lambda b_1$

and L_2 comparing with $a_2 + \mu b_2$ then we have

$$b_1 = 3\hat{i} + \hat{j} + 2\hat{k}$$

and $$b_2 = \hat{i} + 2\hat{j} + 3\hat{k}$$

perpendicular to both b_1 and $b_2 = \mathbf{b}_1 \times \mathbf{b}_2$

$$= \begin{vmatrix} \hat{i} & \hat{j} & \hat{k} \\ 3 & 1 & 2 \\ 1 & 2 & 3 \end{vmatrix}$$

$$= \hat{i}(3-4) - \hat{j}(9-2) + \hat{k}(6-1)$$

$$= \hat{i} - 7\hat{j} + 5\hat{k}$$

$$\therefore \quad \text{Required unit vector} = \frac{\hat{i} - 7\hat{j} + 5\hat{k}}{\sqrt{(-1)^2 + (-7)^2 + (5)^2}}$$

$$= \frac{\hat{i} - 7\hat{j} + 5\hat{k}}{\sqrt{75}}$$

$$= \frac{-\hat{i} - 7\hat{j} + 5\hat{k}}{5\sqrt{3}}$$

24. (d) Shortest distance between lines L_1 and L_2

$$= \left|\frac{(\mathbf{a}_2 - \mathbf{a}_1)\cdot(\mathbf{b}_1 \times \mathbf{b}_2)}{|\mathbf{b}_1 \times \mathbf{b}_2|}\right|$$

Using solution of question no. 23

$$\mathbf{b}_1 \times \mathbf{b}_2 = -\hat{i} - 7\hat{j} + 5\hat{k}$$

$$\Rightarrow \quad |\mathbf{b}_1 \times \mathbf{b}_2| = 5\sqrt{3}$$

Since, $$L_1 : -\hat{i} - 2\hat{j} - \hat{k} + \lambda(3\hat{i} + \hat{j} + 2\hat{k})$$

then, $$a_1 = -\hat{i} - 2\hat{j} - \hat{k},\ b_1 = 3\hat{i} + \hat{j} + 2\hat{k}$$

and $$L_2 : (2\hat{i} - 2\hat{j} - 3\hat{k}) + \mu(\hat{i} + 2\hat{j} + 3\hat{k})$$

then, $$a_2 = 2\hat{i} - 2\hat{j} + 3\hat{k},\ b_2 = \hat{i} + 2\hat{j} + 3\hat{k}$$

Now, $$\mathbf{a}_2 - \mathbf{a}_1 = (2\hat{i} - 2\hat{j} + 3\hat{k}) - (-\hat{i} - 2\hat{j} - \hat{k})$$

$$= 3\hat{i} + 0\hat{j} + 4\hat{k}$$

Then, $$(\mathbf{a}_2 - \mathbf{a}_1)\cdot(\mathbf{b}_1 \times \mathbf{b}_2)$$

$$= (3\hat{i} + 0\hat{j} + 4\hat{k})\cdot(-\hat{i} - 7\hat{j} + 5\hat{k})$$

$$= -3 + 20$$

$$\therefore \quad = 17$$

Shortest distance between lines L_1 and L_2

$$= \left|\frac{17}{5\sqrt{3}}\right|$$

$$= \frac{17}{5\sqrt{3}}$$

25. (c) Equation of plane passing through a given point and parallel to two given lines is

$$(\mathbf{r} - \mathbf{a})\cdot(\mathbf{b} \times \mathbf{c}) = 0$$

Given, passing point $(-1, -2, -1)$

$$\mathbf{a} = -\hat{i} - 2\hat{j} - \hat{k}$$

Required equation of plane

$$[(x\hat{i} + y\hat{j} + z\hat{k}) - (-\hat{i} - 2\hat{j} - \hat{k})]\cdot(-\hat{i} - 7\hat{j} + 5\hat{k}) = 0$$

(where, $\mathbf{b} \times \mathbf{c} = \mathbf{b}_1 \times \mathbf{b}_2$ using solution of question no. 23)

$$\Rightarrow \quad [(x+1)\hat{\mathbf{i}} + (y+2)\hat{\mathbf{j}} + (z+1)\hat{\mathbf{k}}] \cdot (-\hat{\mathbf{i}} - 7\hat{\mathbf{j}} + 5\hat{\mathbf{k}}) = 0$$

$$\Rightarrow \quad (x+1)(-1) + (y+2)(-7) + (z+1)(5) = 0$$

$$\Rightarrow \quad -x - 1 - 7y - 14 + 5z + 5 = 0$$

$$\Rightarrow \quad -x - 7y + 5z - 10 = 0$$

$$\Rightarrow \quad x + 7y - 5z + 10 = 0$$

Again, distance between plane and the point (1, 1, 1) is

$$\frac{1 \times 1 + 7 \times 1 + (-5) \times 1 + 10}{\sqrt{(1)^2 + (7)^2 + (-5)^2}}$$

$$= \frac{1 + 7 - 5 + 10}{\sqrt{1 + 44 + 25}} = \frac{13}{\sqrt{75}}$$

26. (b) Given triangular pyramid $ABCD$ with vertex $A(3, 0, 1)$, $B(-1, 4, 1)$, $C(5, 2, 3)$ and $D(0, -5, 4)$ and G is the point of intersection of medians of ΔBCD.

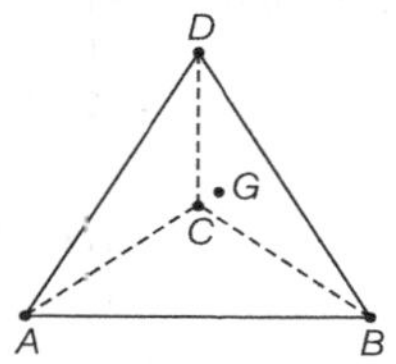

Coordinates of $G\left(\frac{-1+5+0}{3}, \frac{4+2-5}{3}, \frac{1+3+4}{3}\right)$

i.e. $\quad G\left(\frac{4}{3}, \frac{1}{3}, \frac{8}{3}\right)$

$$\therefore \quad \text{Length of } \mathbf{AG} = \sqrt{\left(3-\frac{4}{3}\right)^2 + \left(0-\frac{1}{3}\right)^2 + \left(1-\frac{8}{3}\right)^2}$$

$$= \sqrt{\left(\frac{5}{3}\right)^2 + \left(\frac{1}{3}\right)^2 + \left(\frac{-5}{3}\right)^2}$$

$$= \sqrt{\frac{25 + 1 + 25}{9}} = \frac{\sqrt{51}}{3}$$

27. (c)

$$\mathbf{AB} = -4\hat{\mathbf{i}} + 4\hat{\mathbf{j}} + \hat{\mathbf{k}}$$

$$\mathbf{AC} = 2\hat{\mathbf{i}} + 2\hat{\mathbf{j}} + 2\hat{\mathbf{k}}$$

$$\therefore \quad \mathbf{AB} \times \mathbf{AC} = -8\begin{vmatrix} \hat{\mathbf{i}} & \hat{\mathbf{j}} & \hat{\mathbf{k}} \\ 1 & -1 & 0 \\ 1 & 1 & 1 \end{vmatrix}$$

$$= 8(\hat{\mathbf{i}} + \hat{\mathbf{j}} - 2\hat{\mathbf{k}}) - \mathbf{n}$$

$$\therefore \quad \text{Area of } \Delta ABC = \frac{1}{2}|\mathbf{AB} \times \mathbf{AC}| = 4\sqrt{6}$$

28. (a) $h = |\text{(projection of } \mathbf{AD} \text{ on } \mathbf{n}|$

where, $\quad \mathbf{AD} = -3\hat{\mathbf{i}} - 5\hat{\mathbf{j}} + 3\hat{\mathbf{k}}$

$$\therefore \quad h = \left|\frac{\mathbf{AD} \cdot \mathbf{n}}{|\mathbf{n}|}\right|$$

$$= \left|\frac{(-3\hat{\mathbf{i}} - 5\hat{\mathbf{j}} + 3\hat{\mathbf{k}}) \cdot (\hat{\mathbf{i}} + \hat{\mathbf{j}} - 2\hat{\mathbf{k}})}{\sqrt{6}}\right|$$

$$= \left|\frac{-3 - 5 - 6}{\sqrt{6}}\right| = \frac{14}{\sqrt{6}}$$

29. (6) Let $P(t + 2, -2t + 1, t - 1)$,

$Q\left(2s + \frac{8}{3}, -3 - s, s + 1\right)$

O, P, R are collinear

$$\Rightarrow \quad \frac{t + 2}{2s + \frac{8}{3}} = \frac{-2t + 1}{-s - 3} = \frac{t - 1}{s + 1}$$

On solving $t = 3, s = \frac{4}{3}$

$$\therefore \quad P(5, -5, 2) \text{ and } Q\left(\frac{10}{3}, \frac{-10}{3}, \frac{4}{3}\right)$$

$$\Rightarrow \quad PQ^2 = 6$$

30. (8) Since, point of intersection of the given lines is (0, 0, 0). It must lie on the angle bisector,

$$\therefore \quad \frac{0 - 2}{-1} = \frac{0 + 2}{1} = \frac{0 + k}{4}$$

$$\Rightarrow \quad k = 8$$

31. (5) Given line, $\frac{x - 2}{2} = \frac{y - 1}{3} = \frac{z - 2}{4}$

Let any point on the given line be $L(2r + 2, 3r + 1, 4r + 2)$, then direction ratio of PL are proportional to

$$2r + 2 - 1, 3r + 1 - 2, 4r + 2 - 3$$

i.e. $\quad 2r + 1, 3r - 1, 4r - 1$

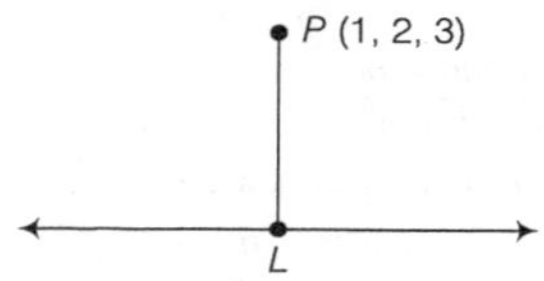

PL is perpendicular to the given line, so

$$2 \times (2r + 1) + 3(3r - 1) + 4(4r - 1) = 0$$

$$\Rightarrow \quad 4r + 2 + 9r - 3 + 16r - 4 = 0$$

$$\Rightarrow \quad 29r - 5 = 0$$

$$\Rightarrow \quad r = \frac{5}{29}$$

So, coordinates of point

$$L\left(2 \times \frac{5}{29} + 2, 3 \times \frac{5}{29} + 1, 4 \times \frac{5}{29} + 2\right)$$

i.e. $\quad L\left(\frac{68}{29}, \frac{44}{29}, \frac{78}{29}\right)$.

According to question (x, y, z) is the foot of perpendicular, so

$$x = \frac{68}{29}, y = \frac{44}{29}, z = \frac{78}{29}$$

$$\therefore \quad \frac{2x + 3y + 4z}{4} = \frac{2 \times \frac{68}{29} + 3 \times \frac{44}{29} + 4 \times \frac{78}{29}}{4}$$

$$= \frac{136 + 132 + 312}{4 \times 29}$$

$$= \frac{580}{4 \times 29} = 5$$

32. (3) Given lines

$$\frac{x - 2}{1} = \frac{y - 3}{1} = \frac{z - 4}{k} \quad \ldots \text{(i)}$$

and $\quad \frac{x - 1}{k} = \frac{y - 4}{-2} = \frac{z - 5}{-1} \quad \ldots \text{(ii)}$

Since, lines are intersect to each other, so shortest distance between these lines will be zero.

$\therefore$ Shortest distance between these two lines, i.e.

$$\frac{\begin{vmatrix} x_2 - x_1 & y_2 - y_1 & z_2 - z_1 \\ a_1 & b_1 & c_1 \\ a_2 & b_2 & c_2 \end{vmatrix}}{\sqrt{(a_1b_2 - a_2b_1)^2 + (b_1c_2 - b_2c_1)^2 + (c_1a_2 - c_2a_1)^2}} = 0$$

$$\Rightarrow \quad \begin{vmatrix} x_2 - x_1 & y_2 - y_1 & z_2 - z_1 \\ a_1 & b_1 & c_1 \\ a_2 & b_2 & c_2 \end{vmatrix} = 0$$

$$\Rightarrow \begin{vmatrix} 1-2 & 4-3 & 5-4 \\ 1 & 1 & k \\ k & -2 & -1 \end{vmatrix} = 0$$

$$\Rightarrow \begin{vmatrix} -1 & 1 & 1 \\ 1 & 1 & k \\ k & -2 & -1 \end{vmatrix} = 0$$

$$\Rightarrow -1(-1+2k) - 1(-1-k^2) + 1(-2-k) = 0$$

$$\Rightarrow 1 - 2k + 1 + k^2 - 2 - k = 0$$

$$\Rightarrow k^2 - 3k = 0$$

$$\Rightarrow k(k-3) = 0$$

$$\Rightarrow k = 0 \text{ or } k - 3 = 0$$

$$k = 3 \qquad [\because k > 0]$$

33. (1) Given lines are

$$\frac{x}{1} = \frac{y-1}{k} = \frac{z}{-1}$$

and
$$\frac{x-k}{2k} = \frac{y-k}{3k-1} = \frac{z-2}{k}$$

since, lines are coplanar.

So, they are intersecting or parallel to each other.

Any point on first line is $(r, kr+1, -r)$ and on second line is

$$(2kr' + k, (3k-1)r' + k, kr' + 2).$$

They are intersects, if point must be coincide for some particular value r and r'.

i.e.
$$r = 2kr' + k \qquad \ldots(\text{i})$$
$$kr + 1 = (3k-1)r' \qquad \ldots(\text{ii})$$
$$-r = kr' + 2 \qquad \ldots(\text{iii})$$

From Eqs. (i) and (iii), we get

$$O = 2kr' + k + kr' + 2$$
$$\Rightarrow 3kr' + k = -2$$
$$\Rightarrow 3kr' = -2 - k$$
$$\Rightarrow r' = \frac{-2-k}{3k} \qquad \ldots(\text{iv})$$

From Eqs. (ii) and (iii), we get

$$1 = (3k-1)r' + k^2r' + 2k$$
$$\Rightarrow 1 = r'(3k - 1 + k^2) + 2k$$
$$\Rightarrow r' = \frac{1-2k}{3k-1+k^2} \qquad \ldots(\text{v})$$

From Eqs. (iv) and (v), we get

$$\frac{-2-k}{3k} = \frac{1-2k}{3k-1+k^2}$$
$$\Rightarrow -(k+2)(k^2+3k-1) = 3k(1-2k)$$
$$\Rightarrow -(k^3 + 3k^2 - k + 2k^2 + 6k - 2) = 3k - 6k^2$$
$$\Rightarrow -(k^3 + 5k^2 + 5k - 2) = -6k^2 + 3k$$
$$\Rightarrow k^3 + 5k^2 + 5k - 2 = 6k^2 - 3k$$
$$\Rightarrow k^3 - k^2 + 8k - 2 = 0$$

Let $f(k) = k^3 - k^2 + 8k - 2$,

$f'(k) = 3k^2 - 2k + 8 > 0$ for $k \in R$

Hence, the equation has only one real root.

34. (2) Since, the given lines are parallel.

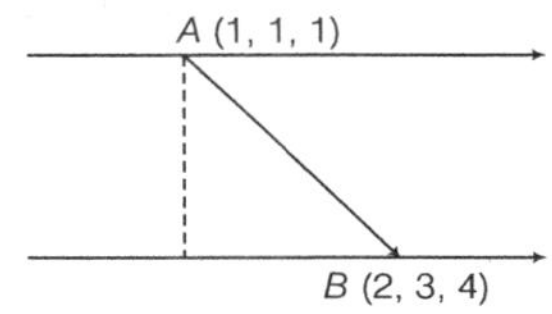

From figure,

$$BC = \frac{(2-1)\cdot 1}{\sqrt{3}} - \frac{(3-1)\cdot 1}{\sqrt{3}} + \frac{(4-1)\cdot 1}{\sqrt{3}} = \frac{1+2+3}{\sqrt{3}} = 2\sqrt{3}$$

$$AB = \sqrt{1+4+9} = \sqrt{14}$$

Shortest distance $= AC = \sqrt{14-12} = \sqrt{2} = \sqrt{k}$

$$\Rightarrow k = 2$$

35. (0) The given equations are

$$al + bm + nc = 0 \qquad \ldots(\text{i})$$

and
$$fmn + gnl + hlm = 0 \qquad \ldots(\text{ii})$$

From Eq. (i), we get

$$nc = -(al + bm)$$
$$n = \frac{-(al+bm)}{c}$$

On putting the value of n in Eq. (ii), we get

$$fm\left[\frac{-(al+bm)}{c}\right] + gl\left[\frac{-(al+bm)}{c}\right] + hlm = 0$$

$$\Rightarrow \frac{-fm(al+bm) - gl(al+bm) + hlmc}{c} = 0$$

$$\Rightarrow afml + fbm^2 + agl^2 + glbm - hlmc = 0$$

On dividing by m^2 both sides, we get

$$\Rightarrow \frac{afl}{m} + fb + \frac{agl^2}{m^2} + \frac{gb}{m} - \frac{hlc}{m} = 0$$

$$\Rightarrow ag\left(\frac{l}{m}\right)^2 + (af + bg - ch)\cdot\left(\frac{l}{m}\right) + bf = 0 \qquad \ldots(\text{iii})$$

Now, Eq. (iii) being a quadratic equation in $\left(\frac{l}{m}\right)$, then let $\frac{l_1}{m_1}$ and $\frac{l_2}{m_2}$ be roots of the equation.

$\therefore$ Product of root $\frac{l_1}{m_1} \times \frac{l_2}{m_2} = \frac{bf}{ag}$

$$\Rightarrow \frac{l_1l_2}{m_1m_2} = \frac{bf}{ag} \Rightarrow \frac{l_1l_2}{bf} = \frac{m_1m_2}{ag}$$

$$\Rightarrow \frac{l_1l_2}{(f/a)} = \frac{m_1m_2}{(g/b)} = \frac{n_1n_2}{(h/c)} = k \qquad \text{[by symmetry]}$$

$$\Rightarrow l_1l_2 = k(f/a)$$
$$m_1m_2 = k(g/b)$$
$$n_1n_2 = k(h/c)$$

$$\therefore l_1l_2 + m_1m_2 + n_1n_2 = k\left(\frac{f}{a} + \frac{g}{b} + \frac{h}{c}\right)$$

Since, given lines are perpendicular to each other,

i.e.
$$l_1l_2 + m_1m_2 + n_1n_2 = 0$$

$$\Rightarrow k\left(\frac{f}{a} + \frac{g}{b} + \frac{h}{c}\right) = 0$$

$$\therefore \frac{f}{a} + \frac{g}{b} + \frac{h}{c} = 0$$

DPP-2 Equation of Plane and Sphere

1. (a) Given equation $xy + yz = 0$

$\Rightarrow \quad y(x + z) = 0$

$\Rightarrow \quad y = 0$ or $x + z = 0$

represents planes which have direction ratio of normal to the plane respectively (0, 1, C) and (1, 0, 1).

$\Rightarrow \quad a_1a_2 + b_1b_2 + c_1c_2 = 0 \cdot 1 + 1 \cdot 0 + 0 \cdot 1 = 0$

So, given equation represents a pair of perpendicular planes.

2. (a) Given, plane passes through point (1, – 3, 1).

Direction ratio line joining two points (3, 4, – 1) and (2, – 1, 5)

$$= 2 - 3, -1 - 4, 5 - (-1)$$
$$= -1, -5, 6$$

Equation of plane

$$A(x - x_1) + B(y - y_1) + C(z - z_1) = 0$$

$\Rightarrow \quad (-1)(x - 2) + (-5)[y - (-3)] + 6(z - 1) = 0$

$\Rightarrow \quad (-1)(x - 2) + (-5)(y + 3) + 6(z - 1) = 0$

$\Rightarrow \quad -x + 2 - 5y - 15 + 6z - 6 = 0$

$\Rightarrow \quad -x - 5y + 6z - 19 = 0$

$\therefore \quad x + 5y - 6z + 19 = 0$

3. (a) Let the equation of plane be

$$\frac{x}{a} + \frac{y}{b} + \frac{z}{c} - 1 = 0$$

Using foot of perpendicular, we get

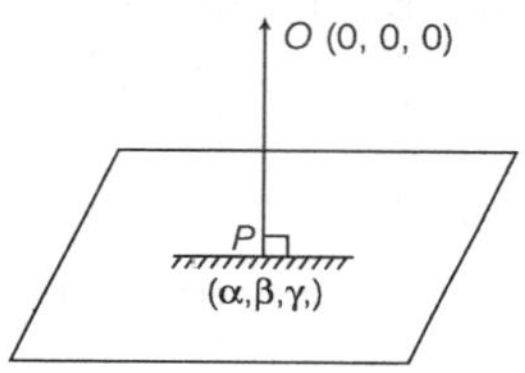

$$\frac{\alpha - 0}{\frac{1}{a}} = \frac{\beta - 0}{\frac{1}{b}} = \frac{\gamma - 0}{\frac{1}{c}} = \frac{-(-1)}{\frac{1}{a^2} + \frac{1}{b^2} + \frac{1}{c^2}} = \lambda$$

$\therefore \quad a = \frac{\lambda}{\alpha}, b = \frac{\lambda}{\beta}, c = \frac{\lambda}{\gamma}$

and $\quad \frac{1}{a^2} + \frac{1}{b^2} + \frac{1}{c^2} = \frac{1}{\lambda}$

$\Rightarrow \quad \frac{\alpha^2}{\lambda^2} + \frac{\beta^2}{\lambda^2} + \frac{\gamma^2}{\lambda^2} = \frac{1}{\lambda}$

$\Rightarrow \quad \lambda = \alpha^2 + \beta^2 + \gamma^2 \quad \ldots(i)$

$\therefore \quad a = \frac{\alpha^2 + \beta^2 + \gamma^2}{\alpha},$

$b = \frac{\alpha^2 + \beta^2 + \gamma^2}{\beta}$

and $\quad c = \frac{\alpha^2 + \beta^2 + \gamma^2}{\gamma}$

Also, $\frac{x}{a} + \frac{y}{b} + \frac{z}{c} = 1$ passes through (1, 2, 3)

$\Rightarrow \quad \frac{1}{a} + \frac{2}{b} + \frac{3}{c} = 1$

$\Rightarrow \quad \frac{\alpha + 2\beta + 3\gamma}{\alpha^2 + \beta^2 + \gamma^2} = 1$

$\Rightarrow$ Locus of (α, β, γ) is $x^2 - y^2 + z^2 = x + 2y + 3z$

4. (a) Any plane through given point (1, – 3, – 2) is

$$a(x - 1) + b[y - (-3)] + c[z - (-2)] = 0$$

$\Rightarrow \quad a(x - 1) + b(y + 3) + c(z + 2) = 0 \quad \ldots(i)$

Plane (i) is at right angle to given plane.

$$x + 2y + 3z = 5$$
$$a(1) + b(2) + c(3) = 0$$

$\Rightarrow \quad a + 2b + 3c = 0 \quad \ldots(ii)$

Plane (i) is at right angle to given plane

$$3x + 3y + 2z = 0$$

$\Rightarrow \quad a \cdot 3 + b \cdot 3 + c \cdot 2 = 0$

$\Rightarrow \quad 3a + 3b + 2c = 0 \quad \ldots(iii)$

From Eqs. (ii) and (iii), we get

$$\frac{a}{2 \times 2 - 3 \times 3} = \frac{b}{3 \times 3 - 1 \times 2} = \frac{c}{1 \times 3 - 2 \times 3}$$

$\Rightarrow \quad \frac{a}{4 - 9} = \frac{b}{9 - 2} = \frac{c}{3 - 6}$

$\Rightarrow \quad \frac{a}{-5} = \frac{b}{7} = \frac{c}{-3} = \lambda$ (let)

$\therefore \quad a = -5\lambda, b = 7\lambda, c = -3\lambda$

Putting the values of a, b, c in Eq. (i), we get

$$(-5\lambda)(x - 1) + (7\lambda)(y + 3) + (-3\lambda)(z + 2) = 0$$

$\Rightarrow \quad (-5)(x - 1) + (7)(y + 3) + (-3)(z + 2) = 0$

$\Rightarrow \quad -5x + 5 + 7y + 21 - 3z - 6 = 0$

$\Rightarrow \quad -5x + 7y - 3z - 20 = 0$

$\therefore \quad 5x - 7y + 3z + 20 = 0$

5. (b) Given point (2, – 1, 3) and plane $3x - 2y - z = 9$.

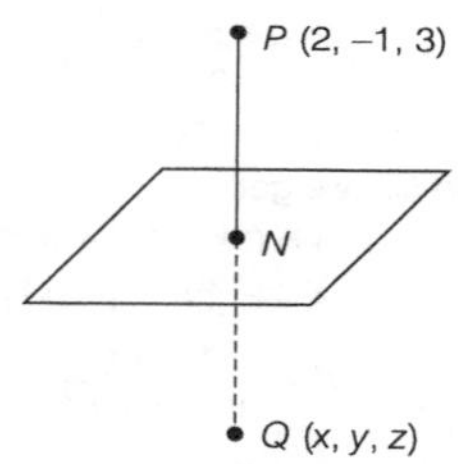

Let $Q(x, y, z)$ be the image of the point $P(2, -1, 3)$ in the given plane.

The equation of the line through $P(2, -1, 3)$ and perpendicular to given plane

$$\frac{x - 2}{3} = \frac{y + 1}{-2} = \frac{z - 3}{-1} = r \quad \text{[say]}$$

The coordinates of a general point on this line plane are

$$N(3r + 2, -2r - 1, -r + 3).$$

$\therefore$ This point lies on the plane, so

$$3(3r + 2) - 2(-2r - 1) - (-r + 3) = 9$$

$\Rightarrow \quad 9r + 6 + 4r + 2 + r - 3 = 9$

$\Rightarrow \quad 14r + 6 + 2 - 3 - 9 = 0$

$\Rightarrow \quad 14r - 4 = 0$

$\therefore \quad r = \frac{2}{7}$

We get, point $N\left(\frac{20}{7}, -\frac{11}{7}, \frac{19}{7}\right)$.

Now, N is the mid-point of PQ.

$\therefore \quad \frac{2 + x}{2} = \frac{20}{7}, \frac{-1 + y}{2} = \frac{-11}{7}, \frac{3 + z}{2} = \frac{19}{7}$

$\Rightarrow \quad 14 + 7x = 40, -7 + 7y = -22, 21 + 7z = 38$

$\Rightarrow \quad 7x = 26 \Rightarrow 7y = -15 \Rightarrow 7z = 17$

$\Rightarrow \quad x = \frac{26}{7} \Rightarrow y = -\frac{15}{7} \Rightarrow z = \frac{17}{7}$

Hence, the required image of $P(2, -1, 3)$ is

$$Q\left(\frac{26}{7}, \frac{-15}{7}, \frac{17}{7}\right).$$

6. (d) Any point on the line

$$\frac{x-2}{3} = \frac{y+1}{4} = \frac{z-2}{12} = r$$

$$x = 3r + 2, \; y = 4r - 1, \; z = 12r + 2$$

Since, line and plane are intersects

So, point lies on the plane $x - y + z = 5$.

$$(3r + 2) - (4r - 1) + (12r + 2) = 5$$

$\Rightarrow \quad 3r + 2 - 4r + 1 + 12r + 2 = 5$

$\Rightarrow \quad 11r + 5 = 5$

$\Rightarrow \quad 11r = 5 - 5$

$\Rightarrow \quad 11r = 0$

$\therefore \quad r = 0$

Intersecting point of plane and line

$(3r + 2, 4r - 1, 12r + 2) = (3 \times 0 + 2, 4 \times 0 - 1, 12 \times 0 + 2)$

$= (2, -1, 2)$

So, distance between points $(-1, -5, -10)$ and $(2, -1, 2)$.

$$d = \sqrt{(2+1)^2 + (-1+5)^2 + (2+10)^2}$$

$$= \sqrt{(3)^2 + (4)^2 + (12)^2}$$

$$= \sqrt{9 + 16 + 144} = \sqrt{169} = 13$$

7. (a) $\because$ Equation of PQ, is

$$\frac{x-1}{2} = \frac{y+2}{3} = \frac{z-3}{-6} = \lambda$$

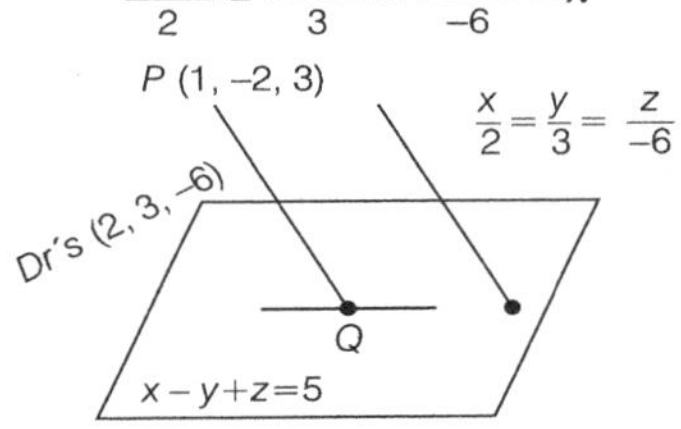

$\Rightarrow Q\,(2\lambda + 1, 3\lambda - 2, -6\lambda + 3)$ lies on $x - y + z = 5$.

$\Rightarrow \quad 2\lambda + 1 - 3\lambda + 2 - 6\lambda + 3 = 5$

$\therefore \quad \lambda = \frac{1}{7}$ ...(i)

Also, distance $PQ = \sqrt{4\lambda^2 + 9\lambda^2 + 36\lambda^2} = 7\lambda = 1$

8. (b) Given sphere $|\hat{\mathbf{r}}| = 5$, Radius of sphere = 5

Centre of sphere = (0, 0, 0)

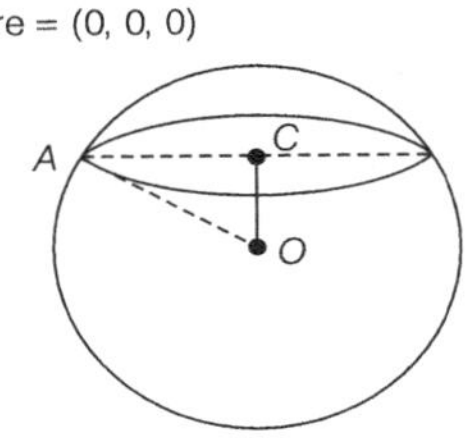

Equation of plane

$$\hat{\mathbf{r}} \cdot (\hat{\mathbf{i}} + \hat{\mathbf{j}} + \hat{\mathbf{k}}) = 3\sqrt{3}$$

$$(x\hat{\mathbf{i}} + y\hat{\mathbf{j}} + z\hat{\mathbf{k}})(\hat{\mathbf{i}} + \hat{\mathbf{j}} + \hat{\mathbf{k}}) = 3\sqrt{3}$$

$\Rightarrow \quad x + y + z = 3\sqrt{3}$

$\therefore$ The length of perpendicular from (0, 0, 0) to the given plane

$$x + y + z = 3\sqrt{3}$$

$\Rightarrow \quad x + y + z - 3\sqrt{3} = 0$

$\therefore \quad OC = \left| \frac{1 \cdot 0 + 1 \cdot 0 + 1 \cdot 0 - 3\sqrt{3}}{\sqrt{1^2 + 1^2 + 1^2}} \right|$

$$OC = \frac{3\sqrt{3}}{\sqrt{3}} = 3$$

Now, radius of the circle is given by

$$AC = \sqrt{(AO)^2 - (OC)^2}$$

$$= \sqrt{(5)^2 - (3)^2} = \sqrt{25 - 9}$$

$$= \sqrt{16} = 4$$

9. (a) Any plane passes through $A(x_1, y_1, z_1)$

$$a(x - x_1) + b(y - y_1) + c(z - z_1) = 0 \quad \text{...(i)}$$

If plane (i) contains the given line, then it must pass through the point (x_2, y_2, z_2) and must be parallel to given line

$$a(x_2 - x_1) + b(y_2 - y_1) + c(z_2 - z_1) = 0 \quad \text{...(ii)}$$

If plane (i) is parallel to the given line

$$ad_1 + bd_2 + cd_3 = 0 \quad \text{...(iii)}$$

Cross multiplying of Eqs. (ii) and (iii), we get

$$\frac{a}{d_3(y_2 - y_1) - d_2(z_2 - z_1)} = \frac{b}{d_1(z_2 - z_1) - d_3(x_2 - x_1)}$$

$$= \frac{c}{d_2(x_2 - x_1) - d_1(y_2 - y_1)} = \lambda \quad \text{[say]}$$

$\Rightarrow \quad a = \lambda \{d_3(y_2 - y_1) - d_2(z_2 - z_1)\}$

$\Rightarrow \quad b = \lambda \{d_1(z_2 - z_1) - d_3(x_2 - x_1)\}$

$\Rightarrow \quad c = \lambda \{d_2(x_2 - x_1) - d_3(x_2 - x_1)\}$

Putting the values of a, b and c in Eq. (i), we get

$\lambda \{d_3(y_2 - y_1) - d_2(z_2 - z_1)\}(x - x_1)$

$+ \lambda \{d_1(z_2 - z_1) - d_3(x_2 - x_1)\}(y - y_1)$

$+ \lambda \{d_1(z_2 - z_1) - d_3(x_2 - x_1)\}(z - z_1) = 0$

$\Rightarrow (x - x_1)\{d_3(y_2 - y_1) - d_2(z_2 - z_1)\}$

$+ (y - y_1)\{d_1(z_2 - z_1) - d_3(x_2 - x_1)\}$

$+ \{z - z_1\}\{d_1(z_2 - z_1) - d_3(x_2 - x_1)(z - z_1)\} = 0$

$$\Rightarrow \quad \begin{vmatrix} x - x_1 & y - y_1 & z - z_1 \\ x_2 - x_1 & y_2 - y_1 & z_2 - z_1 \\ d_1 & d_2 & d_3 \end{vmatrix} = 0$$

10. (a) Given equation of the line

$$x + y + z - 1 = 0$$

and $\quad 4x + y - 2z + 2 = 0$

Let a, b, c be the direction ratios of the required line and required line lies on both the given planes

$$a + b + c - 1 = 0$$

$\Rightarrow \quad a + b + c = 1$

and $\quad 4a + b - 2c + 2 = 0$

$\Rightarrow \quad 4a + b - 2c = -2$

On solving these two equations, we get

$$\frac{a}{1 \times (-2) - (-1) \times 1} = \frac{b}{1 \times 4 - 1 \times (-2)} = \frac{c}{1 \times 1 - 1 \times 4}$$

$\Rightarrow \quad \frac{a}{-2 - 1} = \frac{b}{4 + 2} = \frac{c}{1 - 4}$

$\Rightarrow \quad \frac{a}{-3} = \frac{b}{6} = \frac{c}{-3}$

$\Rightarrow \quad \frac{a}{-1} = \frac{b}{2} = \frac{c}{-1}$

We put $z = 0$, then

$$x + y = 1$$

and $$4x + y = -2$$

Solving these equations

$$-3x = 3$$

$$\Rightarrow \quad x = -1$$

and $$-1 + y = 1$$

$$\Rightarrow \quad y = 2$$

Therefore, coordinates of a point on the required line are (–1, 2, 0).

Hence, the equation of the required line are

$$\frac{x-(-1)}{-1} = \frac{y-2}{2} = \frac{z-0}{-1}$$

$$\Rightarrow \quad \frac{x+1}{-1} = \frac{y-2}{2} = \frac{z}{-1}$$

$$\Rightarrow \quad \frac{x+1}{1} = \frac{y-2}{-2} = \frac{z}{1}$$

11. (d) Let $\mathbf{b}_1 = 2\hat{\mathbf{i}} + 3\hat{\mathbf{j}} - \hat{\mathbf{k}}$

$$\mathbf{b}_2 = \hat{\mathbf{i}} - \hat{\mathbf{j}} + 2\hat{\mathbf{k}}$$

Since, the plane contains the lines which are parallel to the vectors $\mathbf{b}_1$ and $\mathbf{b}_2$ respectively.

The normal to the plane = $\mathbf{b}_1 \times \mathbf{b}_2$

$$= \begin{vmatrix} \hat{\mathbf{i}} & \hat{\mathbf{j}} & \hat{\mathbf{k}} \\ 2 & 3 & -1 \\ 1 & -1 & 2 \end{vmatrix}$$

$$= \hat{\mathbf{i}}\,(6-1) - \hat{\mathbf{j}}\,(4+1) + \hat{\mathbf{k}}\,(-2-3)$$

$$= 5\hat{\mathbf{i}} - 5\hat{\mathbf{j}} - 5\hat{\mathbf{k}}$$

Acute angle between the vector $2\hat{\mathbf{i}} - 2\hat{\mathbf{j}} + \hat{\mathbf{k}}$ and normal to the vector

$$\cos\left(\frac{\pi}{2} - \theta\right) = \frac{\mathbf{b} \cdot \mathbf{n}}{|\mathbf{b}||\mathbf{n}|}$$

$$\sin\theta = \frac{(2\hat{\mathbf{i}} - 2\hat{\mathbf{j}} + \hat{\mathbf{k}}) \cdot (5\hat{\mathbf{i}} - 5\hat{\mathbf{j}} - 5\hat{\mathbf{k}})}{\sqrt{(2)^2 + (-2)^2 + (1)^2}\,\sqrt{(5)^2 + (-5)^2 + (-5)^2}}$$

$$= \frac{10 + 10 - 5}{\sqrt{9}\,\sqrt{75}} = \frac{15}{3 \times 5\sqrt{3}}$$

$$\Rightarrow \quad \sin\theta = \frac{1}{\sqrt{3}}$$

$$\Rightarrow \quad \cot\theta = \sqrt{2}$$

$$\therefore \quad \theta = \cot^{-1}(\sqrt{2})$$

12. (a) Let the equation of plane

$$\frac{x}{a} + \frac{y}{b} + \frac{z}{c} = 1 \quad \ldots\text{(i)}$$

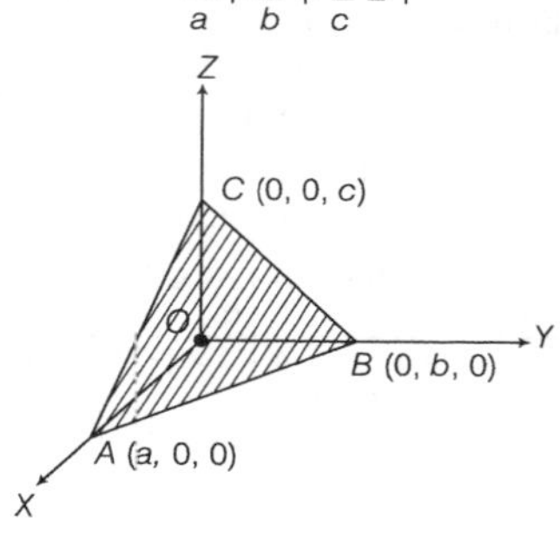

(α, β, γ) be centroid,

$$\Rightarrow \quad \alpha = \frac{a}{3}, \beta = \frac{b}{3}, \gamma = \frac{c}{3} \quad \ldots\text{(ii)}$$

$\therefore$ Required equation of plane is

$$\frac{x}{3\alpha} + \frac{y}{3\beta} + \frac{z}{3\gamma} = 1$$

$$\Rightarrow \quad \frac{x}{\alpha} + \frac{y}{\beta} + \frac{z}{\gamma} = 3$$

13. (b) Here,

$$(1-x)^2 + (1+x)^2 + (1-y)^2 + (1+y)^2 + (1-z)^2 + (z+1)^2 = 10$$

$$\Rightarrow 2[x^2 + 1] + 2[y^2 + 1] + 2[z^2 + 1] = 10$$

$$\therefore \quad x^2 + y^2 + z^2 + 3 = 5$$

$$\Rightarrow \quad x^2 + y^2 + z^2 = 2$$

14. (a) Let the equation of the plane in first system of a rectangular

$$\frac{x}{a} + \frac{y}{b} + \frac{z}{c} = 1$$

It cuts the coordinate axes at $A(a, 0, 0)$, $B(0, b, 0)$ and $C(0, 0, c)$.

Let the distance of plane from the origin p, then

$$p = \left| \frac{\frac{0}{a} + \frac{0}{b} + \frac{0}{c} - 1}{\sqrt{\left(\frac{1}{a}\right)^2 + \left(\frac{1}{b}\right)^2 + \left(\frac{1}{c}\right)^2}} \right|$$

$$\Rightarrow \quad p = \left| \frac{-1}{\sqrt{\frac{1}{a^2} + \frac{1}{b^2} + \frac{1}{c^2}}} \right|$$

$$\Rightarrow \quad \frac{1}{p^2} = \frac{1}{a^2} + \frac{1}{b^2} + \frac{1}{c^2} \quad \ldots\text{(i)}$$

Again, let the equation of plane in second system of rectangular i.e.

$$\frac{x}{a_1} + \frac{y}{b_1} + \frac{z}{c_1} = 1$$

It cuts the coordinate axes at $A_1(a_1, 0, 0)$, $B_1(0, b_1, 0)$ and $C_1(0, 0, c_1)$.

Let the distance of the same plane from the origin p, then

$$p = \left| \frac{\frac{0}{a_1} + \frac{0}{b_1} + \frac{0}{c_1} - 1}{\sqrt{\left(\frac{1}{a_1}\right)^2 + \left(\frac{1}{b_1}\right)^2 + \left(\frac{1}{c_1}\right)^2}} \right|$$

$$p = \left| \frac{-1}{\sqrt{\left(\frac{1}{a_1}\right)^2 + \left(\frac{1}{b_1}\right)^2 + \left(\frac{1}{c_1}\right)^2}} \right|$$

$$= \frac{1}{\sqrt{\left(\frac{1}{a_1}\right)^2 + \left(\frac{1}{b_1}\right)^2 + \left(\frac{1}{c_1}\right)^2}}$$

$$\therefore \quad \frac{1}{p^2} = \frac{1}{(a_1)^2} + \frac{1}{(b_1)^2} + \frac{1}{(c_1)^2} \quad \ldots\text{(ii)}$$

From Eqs. (i) and (ii), we get

$$\frac{1}{a^2} + \frac{1}{b^2} + \frac{1}{c^2} = \frac{1}{(a_1)^2} + \frac{1}{(b_1)^2} + \frac{1}{(c_1)^2}$$

15. (b) Given lines are

$$\frac{x-1}{3}=\frac{y-2}{1}=\frac{z-3}{2} \quad \ldots(\text{i})$$

and $$\frac{x-3}{1}=\frac{y-1}{2}=\frac{z-2}{3} \quad \ldots(\text{ii})$$

Vector form of line (i)

$$\mathbf{r}=\hat{\mathbf{i}}+\hat{\mathbf{j}}+3\hat{\mathbf{k}}+\lambda(3\hat{\mathbf{i}}+\hat{\mathbf{j}}+2\hat{\mathbf{k}})$$
$$=(1+3\lambda)\hat{\mathbf{i}}+(2+\lambda)\hat{\mathbf{j}}+(3+2\lambda)\hat{\mathbf{k}}$$

Vector form of line (ii)

$$\mathbf{r}=(3\hat{\mathbf{i}}+\hat{\mathbf{j}}+2\hat{\mathbf{k}})+\mu(\hat{\mathbf{i}}+2\hat{\mathbf{j}}+3\hat{\mathbf{k}})$$
$$=(3+\mu)\hat{\mathbf{i}}+(1+2\mu)\hat{\mathbf{j}}+(2+3\mu)\hat{\mathbf{k}}$$

Since, lines are intersect to each other, then

$$1+3\lambda=3+\mu,\ 2+\lambda=1+2\mu$$

and $3+2\lambda=2+3\mu$

$\Rightarrow$ $3\lambda-\mu=2,\ \lambda-2\mu=-1$

After solving these equations for the value of λ and μ, we get

$$\lambda=1 \text{ and } \mu=1$$

Thus, position vector of the point of intersection of the given lines is given by

$$\mathbf{r}=(1+3)\hat{\mathbf{i}}+(2+1)\hat{\mathbf{j}}+(3+2)\hat{\mathbf{k}}$$
$$=4\hat{\mathbf{i}}+3\hat{\mathbf{j}}+5\hat{\mathbf{k}}$$

Hence, the point of intersection is $P(4, 3, 5)$.

$\therefore$ Equation of plane is

$$4(x-4)+3(y-3)+5(z-5)=0$$

$\Rightarrow$ $4x+3y+5z=50$

16. (d) Given planes are

$$x+y+2z=7 \quad \ldots(\text{i})$$

and $$x-y=3 \quad \ldots(\text{ii})$$

Let a plane $ax+by+cz+d=0$, which is perpendicular to the both planes (i) and (ii),

i.e. $a\cdot 1+b\cdot 1+c\cdot 2=0$

$\Rightarrow$ $a+b+2c=0 \quad \ldots(\text{iii})$

and $a\cdot 1+b(-1)=0$

$\Rightarrow$ $a-b=0 \quad \ldots(\text{iv})$

From Eqs. (iii) and (iv), we get

$$\frac{a}{1\times 0-2\times(-1)}=\frac{b}{2\times 1-1\times 0}=\frac{c}{1\times(-1)-1\times 1}$$

$\Rightarrow$ $\frac{a}{2}=\frac{b}{2}=\frac{c}{-2}$

$\Rightarrow$ $\frac{a}{1}=\frac{b}{1}=\frac{c}{-1}=r$ [say]

$\Rightarrow$ $a=r, b=r, c=-r$

Required equation of plane which are perpendicular to both planes, is

$$(r)x+(r)y+(-r)z+d=0$$

$\Rightarrow$ $x+y-z+d=0$

Now, angle between two planes

$$2x-y+z=6 \text{ and } x+y-z+d=0 \text{ is}$$

$$\cos\theta=\left|\frac{1\times 2+(-1)\times 1+(1)\times(-1)}{\sqrt{2^2+(-1)^2+(1)^2}\sqrt{(1)^2+(1)^2+(-1)^2}}\right|$$

$\Rightarrow$ $\cos\theta=\left|\frac{2-1-1}{\sqrt{6}\sqrt{3}}\right|$

$\Rightarrow$ $\cos\theta=0$

$\therefore$ $\theta=\frac{\pi}{2}$

17. (a) Dr's of line passing through

$$2x-y+3z+4=0$$

and $$ax+y-z+2=0$$

$$\Rightarrow \begin{vmatrix}\hat{\mathbf{i}} & \hat{\mathbf{j}} & \hat{\mathbf{k}}\\ 2 & -1 & 3\\ a & 1 & -1\end{vmatrix}=-2\hat{\mathbf{i}}+(2+3a)\hat{\mathbf{j}}+(2+a)\hat{\mathbf{k}} \quad \ldots(\text{i})$$

Dr's of straight line passing through

$$x-3y+z=0$$

and $$x+2y+z+1=0$$

$$\Rightarrow \begin{vmatrix}\hat{\mathbf{i}} & \hat{\mathbf{j}} & \hat{\mathbf{k}}\\ 1 & -3 & 1\\ 1 & 2 & 1\end{vmatrix}=\hat{\mathbf{i}}-\hat{\mathbf{k}} \quad \ldots(\text{ii})$$

Let the fixed point in two planes be

$x=0 \Rightarrow y=-5, z=-3$ for $2x-y+3z+4=0$

$=ax+y-z+2$

and $x=0 \Rightarrow y=\frac{-1}{5}, z=-3/5$ for $x-3y+z=0$

$=x+2y+z+1.$

$$\therefore \text{dr's} \Rightarrow \left(0, \frac{24}{5}, \frac{12}{5}\right) \text{ or } (0, 2, 1) \quad \ldots(\text{iii})$$

$$\text{Since, coplanar} \Rightarrow \begin{vmatrix}-2 & 2+3a & 2+a\\ 1 & 0 & -1\\ 0 & 2 & 1\end{vmatrix}=0$$

$\Rightarrow a=-2$

18. (d) Given lines are concurrent

$$\frac{x}{1}=\frac{y}{2}=\frac{z}{3}=r_1 \quad [\text{say}]\ldots(\text{i})$$

Then, $x=r_1, y=2r_1, z=3r_1$

and $$\frac{x-1}{3}=\frac{y-2}{-1}=\frac{z-3}{4}=r_2 \quad [\text{say}]\ldots(\text{ii})$$

Then, $x=3r_2+1, y=-r_2+2, z=4r_2+3$

and $$\frac{x+k}{3}=\frac{y-1}{2}=\frac{z-2}{h}=r_3 \quad [\text{say}]\ldots(\text{iii})$$

Then, $x=3r_3-k, y=2r_3+1, z=hr_3+2$

Given lines are concurrent, so for some values of r_1, r_2 and r_3, Eqs. (i), (ii) and (iii) are same.

From Eqs. (i) and (ii), we get

$$r_1=3r_2+1,\ 2r_1=-r_2+2$$

$\Rightarrow$ $2(3r_2+1)=-r_2+2$

$\Rightarrow$ $6r_2+2=-r_2+2$

$\Rightarrow$ $7r_2=0 \Rightarrow r_2=0$

Then, $r_1=3\times 0+1$

$\Rightarrow$ $r_1=1$

So, concurrent point $(x, y, z)=(r_1, 2r_1, 3r_1)=(1, 2, 3)$

Now, from Eqs. (ii) and (iii) which are

$$x=3r_3-k=1$$
$$y=2r_3+1=2 \Rightarrow r_3=\frac{1}{2}$$
$$z=hr_3+2=3$$

So, $3\times\frac{2}{1}-k=1$

$\Rightarrow$ $k=\frac{3}{2}-1 \Rightarrow k=\frac{1}{2}$

and $h\times\frac{1}{2}+2=3$

$\Rightarrow$ $\frac{h}{2}=1 \Rightarrow h=2$

$\therefore$ $h=2, k=\frac{1}{2}$

19. (a) Dr's of straight line passing through intersection of two planes $x - y + 2z = 3$ and $4x + 3y - z = 1$

$$\Rightarrow \begin{vmatrix} \hat{i} & \hat{j} & \hat{k} \\ 1 & -1 & 2 \\ 4 & 3 & -1 \end{vmatrix} = -5\hat{i} + 9\hat{j} + 7\hat{k} \qquad \text{...(i)}$$

Here, $11x + 10y - 5z = 0$ has Dr's of normal

$$(11, 10, -5) \qquad \text{...(ii)}$$

and Eqs. (i) and (ii) are perpendicular as

$$11(-5) + 10(9) - 5(7)$$

$$\Rightarrow \quad -55 + 90 - 35 \Rightarrow 0$$

$\therefore$ Required equation of plane is $11x + 10y - 5z = 0$

20. (a) Let, given points $A(2 - x, 2, 2)$, $B(2, 2 - y, 2)$, $C(2, 2, 2 - z)$ and $D(1, 1, 1)$ respectively.

Direction ratios of AB, AC and AD are respectively

$$x, -y, 0;\ x, 0, -z;\ -1 + x, -1, -1$$

If they are coplanar so there is a line with direction ratios l, m, n which is perpendicular to AB, AC and AD.

$$lx - my = 0,\ lx - nz = 0,$$

$$l(-1 + x) - m - n = 0$$

Now, $$\begin{vmatrix} x & -y & 0 \\ x & 0 & -z \\ -1 + x & -1 & -1 \end{vmatrix} = 0$$

$$\Rightarrow \quad x(-z) + y\{-x + z(-1 + x)\} = 0$$

$$\Rightarrow \quad -xz - xy - yz(-1 + x) = 0$$

$$\Rightarrow \quad -xz - xy - yz + xyz = 0$$

$$\Rightarrow \quad xz + xy + yz - xyz = 0$$

$$\Rightarrow \quad \frac{xz}{xyz} + \frac{xy}{xyz} + \frac{yz}{xyz} - \frac{xyz}{xyz} = 0 \quad \text{[divide by } xyz \text{ both sides]}$$

$$\Rightarrow \quad \frac{1}{y} + \frac{1}{z} + \frac{1}{x} - 1 = 0$$

$$\Rightarrow \quad \frac{1}{x} + \frac{1}{y} + \frac{1}{z} = 1$$

21. (b) Any plane through $(1, 0, 0)$ is

$$a(x - 1) + b(y - 0) + c(z - 0) = 0 \qquad \text{...(i)}$$

Since, it passes through point $(0, 1, 0)$

$$\Rightarrow \quad a(0 - 1) + b(1 - 0) + c(0 - 0) = 0$$

$$\Rightarrow \quad -a + b = 0$$

$$\Rightarrow \quad a - b = 0 \qquad \text{...(ii)}$$

$$\Rightarrow \quad a = b$$

Angle between plane (i) and plane $x + y = 3$ is $\frac{\pi}{4}$.

i.e. $$\cos\frac{\pi}{4} = \frac{a \cdot 1 + b \cdot 1 + c \cdot 0}{\sqrt{a^2 + b^2 + c^2}\sqrt{1^2 + 1^2}}$$

$$\Rightarrow \quad \frac{1}{\sqrt{2}} = \frac{a + b}{\sqrt{a^2 + b^2 + c^2}\sqrt{2}}$$

$$\Rightarrow \quad \sqrt{a^2 + b^2 + c^2} = a + b$$

$$\Rightarrow \quad a^2 + b^2 + c^2 = (a + b)^2$$

$$\Rightarrow \quad a^2 + b^2 + c^2 = a^2 + b^2 + 2ab$$

$$\Rightarrow \quad 2ab = c^2 \qquad \text{...(iii)}$$

From Eqs. (ii) and (iii), we get,

$$\Rightarrow \quad 2a \times a = c^2 \quad \Rightarrow \quad 2a^2 = c^2$$

$$\Rightarrow \quad \sqrt{2}a = c \quad \Rightarrow \quad \frac{a}{c} = \frac{1}{\sqrt{2}}$$

$$\Rightarrow \quad a : c = 1 : \sqrt{2}$$

So, direction ratios (a, b, c) are $(1, 1, \sqrt{2})$.

22. (b) Any plane through the intersection of the two given planes is

$$(\mathbf{r} \cdot \mathbf{a} - \lambda) + t(\mathbf{r} \cdot \mathbf{b} - \mu) = 0$$

$$\Rightarrow \quad \mathbf{r}\,(\mathbf{a} + t\,\mathbf{b}) + (-\lambda - t\mu) = 0 \qquad \text{...(i)}$$

If plane (i) passes through $(0, 0, 0)$, then $\mathbf{r} = 0 \cdot \hat{i} + 0 \cdot \hat{j} + 0 \cdot \hat{k}$ satisfied it.

$$\Rightarrow (0 \cdot \hat{i} + 0 \cdot \hat{j} + 0 \cdot \hat{k})\,(\mathbf{a} + t\mathbf{b}) + (-\lambda - t\mu) = 0$$

$$\Rightarrow \quad -\lambda - t\mu = 0$$

$$\Rightarrow \quad \lambda + t\mu = 0$$

$$\Rightarrow \quad t = -\frac{\lambda}{\mu}$$

On putting the value of t in Eq. (i), we get

$$\Rightarrow \quad \mathbf{r}\left(\mathbf{a} - \frac{\lambda}{\mu}\mathbf{b}\right) + \left(-\lambda + \frac{\lambda}{\mu}\mu\right) = 0$$

$$\Rightarrow \quad \mathbf{r}\left(\frac{\mu\mathbf{a} - \lambda\mathbf{b}}{\mu}\right) + 0 = 0$$

$$\therefore \quad \mathbf{r}\,(\lambda\mathbf{b} - \mu\mathbf{a}) = 0$$

23. (b) Given equations of the lines are

$$\mathbf{r} = \mathbf{a}_1 + \lambda\mathbf{b} \qquad \text{...(i)}$$

and $$\mathbf{r} = \mathbf{a}_2 + \mu\mathbf{b} \qquad \text{...(ii)}$$

The line (i) passes through a point A with position vector $\mathbf{a}_1$ and is parallel to $\mathbf{b}$, the line (ii) passes through a point B with position vector $\mathbf{a}_2$ and parallel to $\mathbf{b}$.

$$\mathbf{AB} = (\text{Position vector of } B) - (\text{Position vector of } A)$$

$$= (\mathbf{a}_2 - \mathbf{a}_1)$$

$\therefore$ The given lines are coplanar.

$\Rightarrow \mathbf{r}$, $\mathbf{AB}$, $\mathbf{b}$ are coplanar

$$\Rightarrow \quad \mathbf{r} \cdot (\mathbf{AB} \times \mathbf{b}) = 0$$

$$\Rightarrow \quad \mathbf{r} \cdot (\mathbf{a}_2 - \mathbf{a}_1) \times \mathbf{b} = 0$$

$$\Rightarrow \quad [\mathbf{a}_1\mathbf{a}_2\mathbf{b}]$$

24. (a) Given lines are

$$\mathbf{r} = \mathbf{a}_1 + \lambda\mathbf{a}_2$$

and $$\mathbf{r} = \mathbf{a}_2 + \lambda\mathbf{a}_1$$

If given lines are coplanar, then their common plane is perpendicular to the vector $(\mathbf{a}_2 \times \mathbf{a}_1)$, and this plane passes through each of the points $\mathbf{a}_1$ and $\mathbf{a}_2$.

So, its required equation is

$$[\mathbf{r}\ \mathbf{a}_1\mathbf{a}_2] = 0.$$

25. (b) Given plane whose equation is

$$\mathbf{r} \cdot \hat{\mathbf{n}} = 0 \qquad \text{...(i)}$$

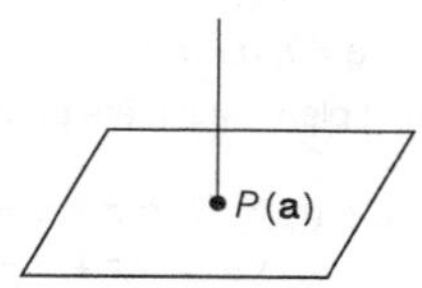

Clearly $\mathbf{n}$ is perpendicular to this plane.

Let, P is the given point with position vector $\mathbf{a}$, equation of line

$$\mathbf{r} = \mathbf{a} + \lambda b \qquad \text{...(ii)}$$

Since, P is point of intersection of plane (i) and line (ii)

$$(\mathbf{a} + \lambda\mathbf{b}) \cdot \hat{\mathbf{n}} = 0$$

$$\Rightarrow \quad \mathbf{a} \cdot \hat{\mathbf{n}} + \lambda\,(\mathbf{b} \cdot \mathbf{n}) = 0$$

$$\Rightarrow \quad \lambda = -\frac{\mathbf{a} \cdot \mathbf{n}}{\mathbf{b} \cdot \mathbf{n}}$$

On putting the value of λ in Eq. (ii)

$$\mathbf{r} = \mathbf{a} + \left(-\frac{\mathbf{a} \cdot \mathbf{n}}{\mathbf{b} \cdot \mathbf{n}}\right)\mathbf{b}$$

$$\mathbf{r} = \mathbf{a} - \left(\frac{\mathbf{a} \cdot \mathbf{n}}{\mathbf{b} \cdot \mathbf{n}}\right)\mathbf{b}$$

26. (c) Here, equation of plane is

$$[\mathbf{ra}-\mathbf{bc}]+[\mathbf{abc}]=0$$

$\therefore$ Distance from origin

$$\Rightarrow \frac{[\mathbf{abc}]}{|(\mathbf{a}-\mathbf{b})\times\mathbf{c}|}=\frac{[\mathbf{abc}]}{[\mathbf{a}\times\mathbf{c}-\mathbf{b}\times\mathbf{c}]} \text{ or } \frac{[\mathbf{abc}]}{[\mathbf{b}\times\mathbf{c}+\mathbf{c}\times\mathbf{a}]}$$

27. (d) Equation of plane passing through the line of intersection of the given plane is

$$(ax+by+cz+d)+\lambda(a'x+b'y+c'z+d')=0$$

$$\Rightarrow (a+\lambda a')x+(b+\lambda b')y+(c+\lambda c')z+(d+\lambda d')=0 \quad \text{...(i)}$$

If this plane parallel to line $y=0$, $z=0$, then normal to the plane perpendicular to this line

$$\frac{x}{1}=\frac{y}{0}=\frac{z}{0}=r$$

$$\Rightarrow (a+\lambda a')\cdot 1+(b+\lambda b')\cdot 0+(c+\lambda c')\cdot 0=0$$

$$\Rightarrow a+\lambda a'=0$$

$$\Rightarrow \lambda=-\frac{a}{a'}$$

Putting the value of λ in Eq. (i)

$$\left(a-\frac{a}{a'}\cdot a'\right)x+\left(b-\frac{ab'}{a'}\right)y+\left(c+\frac{a}{a'}c'\right)z+\left(d-\frac{a}{a'}d'\right)=0$$

$$\Rightarrow \left(\frac{a'b-ab'}{a'}\right)y+\left(\frac{a'c-ac'}{a'}\right)z+\left(\frac{a'd-ad'}{a'}\right)=0$$

$$\Rightarrow (a'b-ab')y+(a'c-ac')z+(a'd-ad')=0$$

28. (b) $\because$ $\mathbf{r}=\mathbf{a}+\lambda(\mathbf{b}\times\mathbf{c})$

and $\mathbf{r}=\mathbf{b}+\mu(\mathbf{c}\times\mathbf{a})$ will intersect.

$$\therefore (\mathbf{a}-\mathbf{b})\cdot[(\mathbf{b}\times\mathbf{c})\times(\mathbf{c}\times\mathbf{a})]=0$$

$$\Rightarrow (\mathbf{a}-\mathbf{b})\cdot[[\mathbf{bca}]\mathbf{c}-[\mathbf{bcc}]\mathbf{a}]=0$$

$$\Rightarrow (\mathbf{a}\cdot\mathbf{c}-\mathbf{b}\cdot\mathbf{c})[\mathbf{abc}]=0$$

$$\Rightarrow \mathbf{a}\cdot\mathbf{c}=\mathbf{b}\cdot\mathbf{c}$$

29. (d) $\mathbf{r}=\mathbf{a}_1+\lambda\mathbf{b}_1$ and $\mathbf{r}=\mathbf{a}_2+\mu\mathbf{b}_2$ are coplanar.

$$\Rightarrow (\mathbf{a}_2-\mathbf{a}_1)\cdot(\mathbf{b}_1\times\mathbf{b}_2)=0$$

$$\Rightarrow [\mathbf{a}_2\mathbf{b}_1\mathbf{b}_2]-[\mathbf{a}_1\mathbf{b}_1\mathbf{b}_2]=0$$

$$\Rightarrow [\mathbf{a}_1\mathbf{b}_1\mathbf{b}_2]=[\mathbf{a}_2\mathbf{b}_1\mathbf{b}_2]$$

30. (d) Given equation of sphere

$$x^2+y^2+z^2+2x-2y-4z-19=0$$

Comparing with

$$x^2+y^2+z^2+2ux+2vy+2wz+d=0$$

$$2u=2\Rightarrow u=1$$

$$2v=-2\Rightarrow v=-1$$

$$2w=-4\Rightarrow w=-2$$

$$d=-19$$

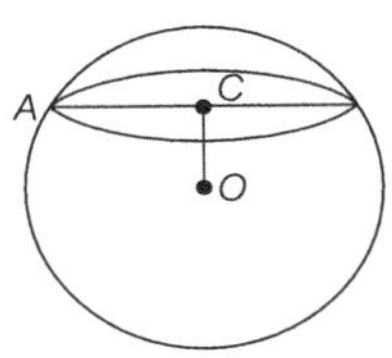

Centre $O(-u,-v,-w)=O(-1,1,2)$

Radius of sphere $=\sqrt{u^2+v^2+w^2-d}$

$$OA=\sqrt{(1)^2+(-1)^2+(-2)^2-(-19)}$$

$$=\sqrt{1+1+4+19}=\sqrt{25}=5$$

$\therefore$ The length of perpendicular from $(-1, 1, 2)$ to the given plane

$$x+2y+2z+7=0$$

$$OC=\frac{1\times(-1)+2\times(1)+2\times(2)+7}{\sqrt{(1)^2+(2)^2+(2)^2}}$$

$$=\frac{-1+2+4+7}{\sqrt{9}}=\frac{12}{3}=4$$

Radius of the circle is given by

$$AC=\sqrt{(OA)^2-(OC)^2}=\sqrt{(5)^2-(4)^2}$$

$$=\sqrt{25-16}=\sqrt{9}=3$$

31. (d) Given sphere

$$x^2+y^2+z+4x-2y-6z=155$$

Here, $2u=4$, $2v=-2$, $2w=-6$, $d=-155$

$$\Rightarrow u=2, v=-1, w=-3$$

Centre of sphere $(-u,-v,-w)=(-2,1,3)$

Distance from $(-2, 1, 3)$ to the plane $12x+4y+3z=327$

$$=\frac{12(-2)+4(1)+3(3)-327}{\sqrt{(12)^2+(4)^2+(3)^2}}$$

$$=\left|\frac{-24+4+9-327}{\sqrt{144+16+9}}\right|$$

$$=\left|\frac{338}{13}\right|=26$$

$\therefore$ Shortest distance $=26-\sqrt{4+1+9+155}$

$$=26-\sqrt{169}$$

$$=26-13$$

$$=13$$

32. (c) Required equation of sphere is

$$(x^2+y^2+z^2-25)+\lambda x=0$$

Passing through $(1, 3, -2)$

$$\Rightarrow (1+9+4-25)+\lambda\cdot(1)=0$$

$$\Rightarrow \lambda=11$$

$\therefore$ Equation of sphere is

$$x^2+y^2+z^2+11x-25=0$$

33. (b) Centre of the given sphere $(-u,-v,-w)$, $d=-19$

$$=(-1,1,2), d=-19$$

Radius of sphere $=\sqrt{u^2+v^2+w^2-d}=5$

Distance from $(-1, 1, 2)$ the plane

$$\mathbf{r}\cdot(\hat{\mathbf{i}}-2\hat{\mathbf{j}}+2\hat{\mathbf{k}})+8=0$$

$$=\frac{1\times(-1)+(-2)(1)+2(2)+8}{\sqrt{(1)^2+(-2)^2+(2)^2}}$$

$$=\left|\frac{-1-2+4+8}{\sqrt{1+4+4}}\right|$$

$$=\left|\frac{9}{3}\right|=3$$

Radius of the circle $=\sqrt{(5)^2-(3)^2}$

$$=\sqrt{25-9}=\sqrt{16}=4$$

34. (b) Equation of line passing through (1, 1, 1) and having Dr's (a, b, c)

$$\Rightarrow \frac{x-1}{a}=\frac{y-1}{b}=\frac{z-1}{c}$$

Parallel to plane $2x+3y+z+5=0$

$$\therefore 2a+3b+c=0$$

Only option (b) $\frac{x-1}{-1}=\frac{y-1}{1}=\frac{z-1}{1}$ satisfy

$$2a+3b+c=0, \quad [\text{as } a=-1, b=1, c=1]$$

So, option (b) is correct.

35. (d) Given line L has the equation

$x = 1 + 3r,\ y = 2 - r$ and $z = 3 + 4r$

$$\Rightarrow \quad \frac{x-1}{3} = r \qquad \Rightarrow \quad \frac{y-2}{-1} = r$$

$$\Rightarrow \quad \frac{z-3}{4} = r$$

$$\Rightarrow \quad \frac{x-1}{3} = \frac{y-2}{-1} = \frac{z-3}{4} = r$$

Direction ratio of AB

$$= 1 + 3r - 1, 2 - r - 2, 3 + 4r + 3$$

$$\Rightarrow \quad 3r, -r, 4r + 6$$

and direction ratio of normal to plane $x + y - z = 1$ is

$$1, 1, -1$$

Since, AB parallel to the plane, then

$$1 \times 3r + 1 \times (-r) + (-1) \times (4r + 6) = 0$$

$$\Rightarrow \quad 3r - r - 4r - 6 = 0$$

$$\Rightarrow \quad -2r - 6 = 0$$

$$\Rightarrow \quad r = -3$$

So, coordinates of point B

$$(1 + 3r, 2 - r, 3 + 4r)$$

$$\Rightarrow \quad \{1 + 3(-3), 2 - (-3), 3 + 4(-3)\}$$

$$\Rightarrow \quad (-8, 5, -9)$$

36. (b) Equation of plane passing through point $A(1, 2, -3)$

$$a(x-1) + b(y-2) + c(z+3) = 0 \quad \text{...(i)}$$

Since, plane (i) contains the given line L, then it passes through point $B(-8, 5, -9)$

$$a(-8-1) + b(5-2) + c(-9+3) = 0$$

$$\Rightarrow \quad a(-9) + b(3) + c(-6) = 0$$

$$\Rightarrow \quad -9a + 3b - 6c = 0$$

$$\Rightarrow \quad 9a - 3b + 6c = 0$$

$$\Rightarrow \quad 3a - b + 2c = 0 \quad \text{...(ii)}$$

If plane (i) contains line L then it parallel to line

$$3a - (-1)b + 4c = 0$$

$$3a - b + 4c = 0 \quad \text{...(iii)}$$

On solving Eqs. (ii) and (iii) by cross-multiplication

$$\frac{a}{(-1)(4) - (2)(-1)} = \frac{b}{(2)(3) - (3)(4)} = \frac{c}{(3)(-1) - (-1)(3)}$$

$$\Rightarrow \quad \frac{a}{-4+2} = \frac{b}{6-12} = \frac{c}{-3+3}$$

$$\Rightarrow \quad \frac{a}{-2} = \frac{b}{-6} = \frac{c}{0} = r \quad \text{[say]}$$

$$\Rightarrow \quad a = -2r, b = -6r, c = 0$$

On putting values of a, b, c in Eq. (i), we get

$$(-2r)(x-1) + (-6r)(y-2) + 0 \times (z+3) = 0$$

$$\Rightarrow \quad (x-1) + 3(y-2) = 0$$

$$\Rightarrow \quad x - 1 + 3y - 6 = 0$$

$$\therefore \quad x + 3y - 7 = 0$$

37. (b) Equation of a plane BC, passing through the points A, B and C respectively

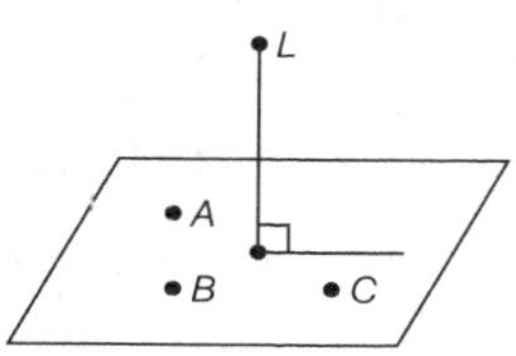

$$\Rightarrow \quad \begin{vmatrix} x - x_1 & y - y_1 & z - z_1 \\ x - x_2 & y - y_2 & z - z_2 \\ x - x_3 & y - y_3 & z - z_3 \end{vmatrix} = 0$$

$$\Rightarrow \quad \begin{vmatrix} x - 2 & y - 1 & z - 0 \\ x - 1 & y - 0 & z - 1 \\ x - 3 & y - 0 & z - 1 \end{vmatrix} = 0$$

$$\Rightarrow \quad \begin{vmatrix} x - 2 & y - 1 & z \\ x - 1 & y & z - 1 \\ x - 3 & y & z - 1 \end{vmatrix} = 0$$

$$\Rightarrow (x-2)\{y(z-1) - (z-1)(y)\} - (y-1)\{(x-1)(z-1) - (x-3) - (z-1)\} + z\{(x-1)y - (x-3)y\} = 0$$

$$\Rightarrow (x-2)(0) - (y-1)(xz - x - z + 1 - xz + x + 3z - 3) + 3(xy - y - xy + 3y) = 0$$

$$\Rightarrow \quad -(y-1)(2z-2) + 3(2y) = 0$$

$$\Rightarrow \quad -(2yz - 2y - 2z + 2) + 2yz = 0$$

$$\Rightarrow \quad -2yz + 2y + 2z - 2 + 2yz = 0$$

$$\Rightarrow \quad 2y + 2z - 2 = 0$$

$$\Rightarrow \quad y + z - 1 = 0$$

38. (c) Equation of line L

Here, direction ratio of line = direction ratio of normal to the plane $BC = (0, 1, 1)$

and line L passes through point $D(0, 0, 2)$ then required equation of line L

$$= 2\hat{\mathbf{k}} + \lambda(\hat{\mathbf{j}} + \hat{\mathbf{k}})$$

39. (d) Perpendicular distance between point D and plane

$$= \left| \frac{1 \times 1 + 1 \times 1 - 1}{\sqrt{(1)^2 + (1)^2}} \right|$$

$$= \left| \frac{1}{\sqrt{2}} \right| = \frac{1}{\sqrt{2}}$$

40. (6) Given equation of plane

$$x + y + z + d = 0 \quad \text{...(i)}$$

Mid-point of line segment joining the points (1, 2, 3) and (3, 2, 1)

$$= \left(\frac{1+3}{2}, \frac{2+2}{2}, \frac{3+1}{2} \right)$$

$$= (2, 2, 2)$$

Since, plane (i) bisects the line segment joining the points (1, 2, 3) and (3, 2, 1), then it passes through the mid-point of the line joining

$$\Rightarrow \quad 2 + 2 + 2 + d = 0$$

$$\Rightarrow \quad 6 + d = 0$$

$$\Rightarrow \quad d = -6$$

$$\Rightarrow \quad |d| = 6$$

41. (1) Given plane $\mathbf{r} \cdot \mathbf{n} = q$

Equation of plane in cartesian form

put $\quad \mathbf{r} = x\hat{\mathbf{i}} + y\hat{\mathbf{j}} + z\hat{\mathbf{k}}$

and $\quad \mathbf{n} = a\hat{\mathbf{i}} + b\hat{\mathbf{j}} + n\hat{\mathbf{k}}$

Now, $\quad (x\hat{\mathbf{i}} + y\hat{\mathbf{j}} + z\hat{\mathbf{k}}) \cdot (a\hat{\mathbf{i}} + b\hat{\mathbf{j}} + n\hat{\mathbf{k}}) = q$

$$\Rightarrow \quad ax + by + cz = q$$

Since, this plane meet to X-axis

$$ax + b \times 0 + c \times 0 = q$$

$$\Rightarrow \quad ax = q,\ x = \frac{q}{a}$$

Given x-intercept is $\lambda\left(\frac{q}{\hat{\mathbf{i}}\cdot\mathbf{n}}\right)$

Now, $\frac{q}{a} = \lambda\left(\frac{q}{\hat{\mathbf{i}}\cdot(a\hat{\mathbf{i}} + b\hat{\mathbf{j}} + c\hat{\mathbf{k}})}\right)$

$\Rightarrow \quad \frac{q}{a} = \lambda\left(\frac{q}{a}\right) \Rightarrow \lambda = 1$

42. (1) Given equation of plane is cartesian form

$$(x\hat{\mathbf{i}} + y\hat{\mathbf{j}} + z\hat{\mathbf{k}})\cdot(2\hat{\mathbf{i}} - 2\hat{\mathbf{j}} + \hat{\mathbf{k}}) + 3 = 0$$

$\Rightarrow \quad 2x - 2y + z + 3 = 0$

$\Rightarrow \quad 2x - 2y + z = -3 \quad \text{...(i)}$

and $(x\hat{\mathbf{i}} + y\hat{\mathbf{j}} + z\hat{\mathbf{k}})\cdot(4\hat{\mathbf{i}} - 4\hat{\mathbf{j}} + 2\hat{\mathbf{k}}) + 5 = 0$

$\Rightarrow \quad 4x - 4y + 2z + 5 = 0$

$\Rightarrow \quad 4x - 4y + 2z = -5$

$\Rightarrow \quad 2(2x - 2y + z) = -5$

$\Rightarrow \quad 2x - 2y + z = -\frac{5}{2} \quad \text{...(ii)}$

Distance between parallel plane (i) and (ii)

$$= \left|\frac{d_1 - d_2}{\sqrt{a^2 + b^2 + c^2}}\right|$$

$\Rightarrow \quad \lambda = \left|\frac{-\frac{5}{2} - (-3)}{\sqrt{(2)^2 + (-2)^2 + (1)^2}}\right|$

$\Rightarrow \quad \lambda = \left|\frac{1/2}{\sqrt{9}}\right| = \left|\frac{1/2}{3}\right|$

$\Rightarrow \quad \lambda = \frac{1}{6}, 6\lambda = 6 \times \frac{1}{6} = 1$

43. (2) Given equation of plane

$$2x - 3y + 6z - 11 = 0$$

$\Rightarrow \quad 2x - 3y + 6z = 11$

$\Rightarrow \quad \frac{2x}{11} - \frac{3}{11}y + \frac{6}{11}z = 1$

$\Rightarrow \quad \frac{x}{11/2} + \frac{y}{(-11/3)} + \frac{z}{(11/6)} = 1$

Here, X-intercept is $\frac{11}{2}$

Equation of X-axis $\frac{x}{a} = \frac{y}{0} = \frac{z}{0}$

We know that angle between line

$$\frac{x - x_1}{a_1} = \frac{y - y_1}{b_1} = \frac{z - z_1}{c_1}$$

and plane $a_2x + b_2y + c_2z + d = 0$

Then, $\sin\theta = \frac{(a_1a_2 + b_1b_2 + c_1c_2)}{\sqrt{a_1^2 + b_1^2 + c_1^2}\sqrt{a_2^2 + b_2^2 + c_2^2}}$

$\Rightarrow \sin\{\sin^{-1}(k)\} = \frac{\frac{11}{2} \times 2}{\sqrt{\left(\frac{11}{2}\right)^2 + (0)^2 + (0)^2}\sqrt{(2)^2 + (-3)^2 + (6)^2}}$

$$= \frac{\frac{11}{2} \times 2}{\frac{11}{2} \times \sqrt{4 + 9 + 36}}$$

$\Rightarrow \quad k = \frac{2}{7}$

$\Rightarrow \quad 7k = 7 \times \frac{2}{7} = 2$

44. (6) Let, the equation of variable plane be

$$lx + my + nz = p \quad \text{...(i)}$$

Also, the equation of coordinate plane are

$$x = 0 \quad \text{...(ii)}$$

$$y = 0 \quad \text{...(iii)}$$

$$z = 0 \quad \text{...(iv)}$$

Solving these equations, we get the coordinates of the vertices of the tetrahedron as $(0, 0, 0)$, $\left(0, 0, \frac{p}{n}\right)$, $\left(\frac{p}{l}, 0, 0\right)$ and $\left(0, \frac{p}{m}, 0\right)$

Centroid of this tetrahedron be (x_1, y_1, z_1)

$\therefore \quad x_1 = \frac{1}{4}\left(0 + 0 + \frac{p}{l} + 0\right) = \frac{p}{4l}$

Similarly, $y_1 = \frac{p}{4m}, \; z_1 = \frac{p}{4n}$

Now, $\frac{p}{l} = 4x_1$

$\frac{p}{m} = 4y_1 \quad \text{...(v)}$

and $\frac{p}{n} = 4z_1$

$\therefore$ Volume of tetrahedron $= 64k^3$

$$\frac{1}{6}\begin{vmatrix} \frac{p}{l} & 0 & 0 \\ 0 & \frac{p}{m} & 0 \\ 0 & 0 & \frac{p}{n} \end{vmatrix} = 64k^3$$

$\Rightarrow \quad \frac{p^3}{6lmn} = 64k^3$

$\Rightarrow \quad \frac{1}{6}\left(\frac{p}{l}\right)\left(\frac{p}{m}\right)\left(\frac{p}{n}\right) = 64k^3$

$\Rightarrow \quad \frac{1}{6}(4x_1)(4y_1)(4z_1) = 64k^3 \quad$ [from Eq. (v)]

$\Rightarrow \quad x_1y_1z_1 = 6k^3$

$\Rightarrow \quad xyz = 6k^3$

and given $\quad xyz = \lambda k^3$

So, $\quad \lambda = 6$

45. (5) Area of $\Delta BCD = \frac{1}{2}|\mathbf{BC} \times \mathbf{BD}|$

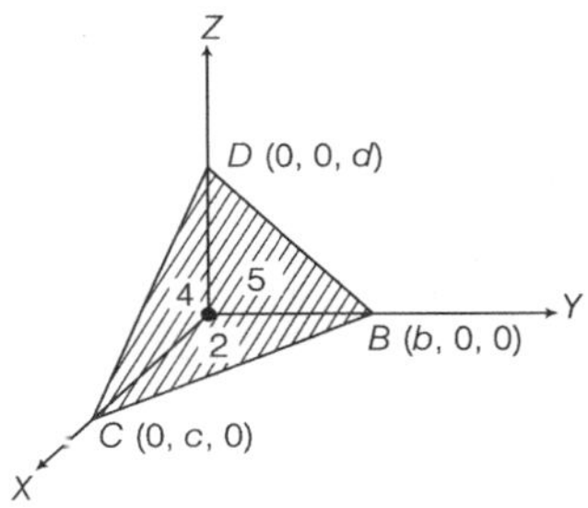

$$= \frac{1}{2}|(b\hat{\mathbf{i}} - c\hat{\mathbf{j}}) \times (b\hat{\mathbf{i}} - d\hat{\mathbf{k}})|$$

$$= \frac{1}{2}|(bd)\hat{\mathbf{j}} + (bc)\hat{\mathbf{k}} + (dc)\hat{\mathbf{i}}|$$

$$= \frac{1}{2}\sqrt{b^2c^2 + c^2d^2 + b^2d^2} \quad \text{...(i)}$$

Now, $6 = bc$, $8 = cd$, $10 = bd$

$b^2c^2 + c^2d^2 + b^2d^2 = 200$

$\therefore \quad A = \frac{1}{2}\sqrt{200} = 5\sqrt{2} \Rightarrow \lambda = 5$

Revisal Problems (JEE Main)

1. (c) A line makes an angle θ with X-axis and Z-axis and β with Y-axis.

$\because \quad l = \cos\theta, m = \cos\beta$ and $n = \cos\theta$

$\therefore \quad l^2 + m^2 + n^2 = 1$

$\Rightarrow \quad \cos^2\theta + \cos^2\beta + \cos^2\theta = 1$

$\Rightarrow \quad 2\cos^2\theta + \cos^2\beta = 1$

$\Rightarrow \quad 2\cos^2\theta = 1 - \cos^2\beta$

$\Rightarrow \quad 2\cos^2\theta = \sin^2\beta \quad \ldots(i)$

But it is given that, $\quad \sin^2\beta = 3\sin^2\theta \quad \ldots(ii)$

From Eqs. (i) and (ii), we get

$$3\sin^2\theta = 2\cos^2\theta$$

$\Rightarrow \quad 3(1 - \cos^2\theta) = 2\cos^2\theta$

$\Rightarrow \quad 3 = 5\cos^2\theta$

$\therefore \quad \cos^2\theta = \dfrac{3}{5}$

2. (a) Let, OA and OB be two lines with direction cosines l_1, m_1, n_1 and l_2, m_2, n_2 and $OA = OB = 1$.

Then, the coordinates of A and B are (l_1, m_1, n_1) and (l_2, m_2, n_2), respectively. Let OC be the bisector of $\angle AOB$. Then, C is the mid-point of AB and so its coordinates are

$$\left(\frac{l_1 + l_2}{2}, \frac{m_1 + m_2}{2}, \frac{n_1 + n_2}{2}\right).$$

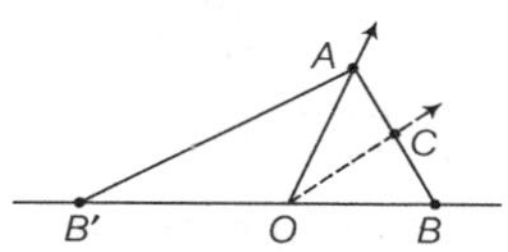

Therefore, the direction ratios of OC are

$$\frac{l_1 + l_2}{2}, \frac{m_1 + m_2}{2}, \frac{n_1 + n_2}{2}.$$

We have,

$$OC = \sqrt{\left(\frac{l_1 + l_2}{2}\right)^2 + \left(\frac{m_1 - m_2}{2}\right)^2 + \left(\frac{n_1 + n_2}{2}\right)^2}$$

$$= \frac{1}{2}\sqrt{(l_1^2 + m_1^2 + n_1^2) + (l_2^2 + m_2^2 + n_2^2) + 2(l_1l_2 + m_1m_2 + n_1n_2)}$$

$$= \frac{1}{2}\sqrt{2 + 2\cos\theta} \quad [\because \cos\theta = l_1l_2 + m_1m_2 + n_1n_2]$$

$$= \frac{1}{2}\sqrt{2(1 + \cos\theta)} = \cos\frac{\theta}{2}$$

Therefore, the direction cosines of **OC** are

$$\frac{l_1 + l_2}{2(OC)}, \frac{m_1 + m_2}{2(OC)}, \frac{n_1 + n_2}{2(OC)}$$

or

$$\frac{l_1 + l_2}{2\cos\frac{\theta}{2}}, \frac{m_1 + m_2}{2\cos\frac{\theta}{2}}, \frac{n_1 + n_2}{2\cos\frac{\theta}{2}}$$

3. (d) Given, equation of line

$$\frac{x - 6}{3} = \frac{y - 7}{2} = \frac{z - 7}{-2} \quad \ldots(i)$$

The given line passes through $A(6, 7, 7)$ and is parallel to the vector $\quad \mathbf{b} = 3\hat{i} + 2\hat{j} - 2\hat{k}$.

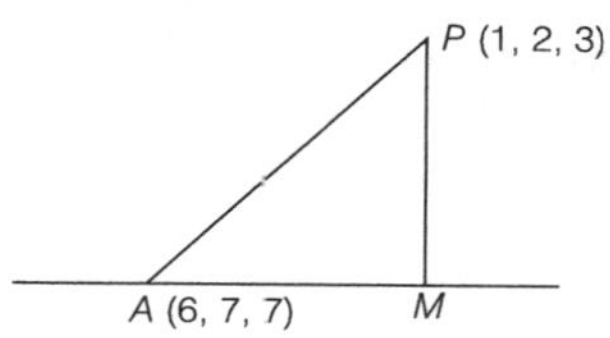

Let M be the foot of the perpendicular from $P(1, 2, 3)$ on the given line.

We have, $\quad \mathbf{AP} = -5\hat{i} - 5\hat{j} - 4\hat{k}$

$$|\mathbf{AP}| = \sqrt{25 + 25 + 16} = \sqrt{66}$$

Clearly, AM = Projection of **AP** on **b**

i.e. $\quad AM = \left|\dfrac{\mathbf{AP} \cdot \mathbf{b}}{|\mathbf{b}|}\right|$

$$= \left|\frac{(5\hat{i} + 5\hat{j} + 4\hat{k})\,(3\hat{i} + 2\hat{j} - 2\hat{k})}{|3\hat{i} + 2\hat{j} - 2\hat{k}|}\right|$$

$$= \left|\frac{15 + 10 - 8}{\sqrt{9 + 4 + 4}}\right|$$

$$= \sqrt{17}$$

$\therefore \quad PM = \sqrt{AP^2 - AM^2}$

$$= \sqrt{66 - 17} = 7$$

4. (b) Let the equation of line AB be

$$\frac{x - 0}{1} = \frac{y + a}{1} = \frac{z - 0}{1} = k \quad \text{[say]}$$

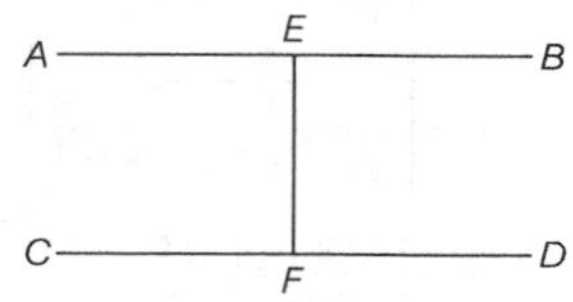

Any point on the line is $E(k, k - a, k)$.

Also, the equation of other line CD is

$$\frac{x + a}{2} = \frac{y - 0}{1} = \frac{z - 0}{1} = \lambda \quad \text{[say]}$$

Any point on the line $F(2\lambda - a, \lambda, \lambda)$.

Direction ratios of EF are

$$[(k - 2\lambda + a), (k - a - \lambda), (k - \lambda)].$$

Since, it is given that direction ratios of EF are proportional to 2, 1, 2.

$\therefore \quad \dfrac{k - 2\lambda + a}{2} = \dfrac{k - \lambda - a}{1} = \dfrac{k - \lambda}{2}$

On taking first two fractions, we get

$$k - 2\lambda + a = 2k - 2\lambda - 2a$$

$\Rightarrow \quad k = 3a \quad \ldots(i)$

taking second and third fraction, we get

$$2k - 2\lambda - 2a = k - \lambda$$

$\Rightarrow \quad k - \lambda = 2a$

$\Rightarrow \quad \lambda = 3a - 2a \quad$ [from Eq. (i)]

$\Rightarrow \quad \lambda = a$

Hence, coordinates of E are $(3a, 2a, 3a)$ and coordinates of F are (a, a, a).

Alternate Solution

We observe that, the points $E(3a, 2a, 3a)$ and $F(a, a, a)$ satisfy the equations of the lines $x = y + a = z$ and $x + a = 2y = 2z$, respectively. Also, direction ratios of EF are proportional to $3a - a, 2a - a, 3a - a$, i.e. 2, 1, 2.

5. (a) Given equation can be rewritten as

$$\frac{x - b}{a} = \frac{y - 0}{1} = \frac{z - d}{c}$$

and

$$\frac{x - b'}{a'} = \frac{y - 0}{1} = \frac{z - d'}{c'}$$

These lines will be perpendicular, if

$$aa' + cc' + 1 = 0 \quad [\because l_1 l_2 + m_1 m_2 + n_1 n_2 = 0]$$

6. (b) We know that the lines $\frac{x-x_1}{l_1}=\frac{y-y_1}{m_1}=\frac{z-z_1}{n_1}$ and $\frac{x-x_2}{l_2}=\frac{y-y_2}{m_2}=\frac{z-z_2}{n_2}$ intersect, if

$$\begin{vmatrix} x_2-x_1 & y_2-y_1 & z_2-z_1 \\ l_1 & m_1 & n_1 \\ l_2 & m_2 & n_2 \end{vmatrix}=0$$

So, given lines will intersect, if

$$\begin{vmatrix} 3-1 & k+1 & 0-1 \\ 2 & 3 & 4 \\ 1 & 2 & 1 \end{vmatrix}=0$$

$\Rightarrow \quad 2(3-8)-(k+1)(2-4)-(4-3)=0$

$\Rightarrow \quad -10+2k+2-1=0$

$\therefore \quad k=9/2$

7. (a) The equation of plane containing the line $\frac{x-3}{1}=\frac{y-6}{5}=\frac{z-4}{4}$ is

$a(x-3)+b(y-6)+c(z-4)=0$ ...(i)

where, $a+5b+4c=0$ ...(ii)

This plane will pass through (3, 2, 0), if

$a(0)+b(-4)+c(-4)=0$ ...(iii)

Solving Eqs. (i) and (iii), we get

$$\frac{a}{-20+16}=\frac{b}{0+4}=\frac{c}{-4+0}$$

$\Rightarrow \quad \frac{a}{1}=\frac{b}{-1}=\frac{c}{1}$

Putting the values of a, b, c in Eq. (i), we get

$$x-y+z-1=0$$

which is the required equation of plane.

8. (d) Given, equation of sphere

$x^2+y^2+z^2+4x-2y-6z-155=0$ ...(i)

Equation of plane $12x+4y+3z=327$ ...(ii)

Centre of sphere is (−2, 1, 3) and radius

$=\sqrt{4+1+9+155}$

$=\sqrt{169}=13$

Length of perpendicular from the centre of sphere to the plane

$$=\left|\frac{-2\times 12+1\times 4+3\times 3-327}{\sqrt{(12)^2+(4)^2+3^2}}\right| \quad \left(\because p=\frac{ax_1+by_1+cz_1+d}{\sqrt{a^2+b^2+c^2}}\right)$$

$$=\left|\frac{-11-327}{\sqrt{169}}\right|=\frac{338}{13}=26$$

$\therefore$ Shortest distance between the plane and the sphere

= 26 − Radius of sphere

= 26 − 13 = 13

9. (a) The equation of the plane with reference to the two systems of rectangular axes are

$\frac{x}{a}+\frac{y}{b}+\frac{z}{c}=1$ ...(i)

$\frac{x}{a'}+\frac{y}{b'}+\frac{z}{c'}=1$ ...(ii)

Since, the origin is same.

$\therefore$ Length of perpendicular from (0, 0, 0) to the plane (i)

= Length of perpendicular from (0, 0, 0) to plane (ii).

$$\Rightarrow \left|\frac{1}{\sqrt{\frac{1}{a^2}+\frac{1}{b^2}+\frac{1}{c^2}}}\right|=\left|\frac{1}{\sqrt{\frac{1}{a'^2}+\frac{1}{b'^2}+\frac{1}{c'^2}}}\right|$$

$$\Rightarrow \frac{1}{a^2}+\frac{1}{b^2}+\frac{1}{c^2}=\frac{1}{a'^2}+\frac{1}{b'^2}+\frac{1}{c'^2}$$

$$\Rightarrow \left(\frac{1}{a^2}+\frac{1}{b^2}+\frac{1}{c^2}\right)-\left(\frac{1}{a'^2}+\frac{1}{b'^2}+\frac{1}{c'^2}\right)=0$$

10. (c) Since, (0, 0, 0) does not satisfy the equation of the plane. So, it does not pass through the origin. Other options are true because the plane is in standard normal form.

11. (a) The equation of XY-plane is $z=0$, i.e.

$$0\cdot x+0\cdot y+z=0$$

Clearly, the given plane is perpendicular to this plane.

12. (b) Let the equation of required plane be

$$\frac{x}{a}+\frac{y}{b}+\frac{z}{c}=1$$

This meets the coordinate axes at

$A(a, 0, 0)$, $B(0, b, 0)$ and $C(0, 0, c)$.

The coordinates of centroid of ΔABC are $\left(\frac{a}{3},\frac{b}{3},\frac{c}{3}\right)$.

$\therefore \quad \frac{a}{3}=1, \frac{b}{3}=2$, and $\frac{c}{3}=3$

$\Rightarrow \quad a=3, b=6$ and $c=9$

Hence, the equation of plane is $\frac{x}{3}+\frac{y}{6}+\frac{z}{9}=1$.

13. (b) Clearly, **a** is perpendicular to the normals to the two planes determined by the given pair of vectors.

We have,

$\mathbf{n}_1$ = Normal vector to the plane determined by $\hat{\mathbf{i}}$ and $(\hat{\mathbf{i}}+\hat{\mathbf{j}})$

$\Rightarrow \quad \mathbf{n}_1=\hat{\mathbf{i}}\times(\hat{\mathbf{i}}+\hat{\mathbf{j}})=\hat{\mathbf{k}}$

and $\mathbf{n}_2$ = Normal vector to the plane determined by $\hat{\mathbf{i}}-\hat{\mathbf{j}}$ and $\hat{\mathbf{i}}+\hat{\mathbf{k}}$

$\Rightarrow \quad \mathbf{n}_2=(\hat{\mathbf{i}}-\hat{\mathbf{j}})\times(\hat{\mathbf{i}}+\hat{\mathbf{k}})=-\hat{\mathbf{i}}-\hat{\mathbf{j}}+\hat{\mathbf{k}}$

Since, **a** is perpendicular to $\mathbf{n}_1$ and $\mathbf{n}_2$.

$\therefore \quad \mathbf{a}=\lambda(\mathbf{n}_1\times\mathbf{n}_2)$

$=\lambda\{\hat{\mathbf{k}}\times(\hat{\mathbf{i}}-\hat{\mathbf{j}}+\hat{\mathbf{k}})\}$

$=\lambda(-\hat{\mathbf{j}}+\hat{\mathbf{i}})$

Let θ be the angle between **a** and $\hat{\mathbf{i}}-2\hat{\mathbf{j}}+2\hat{\mathbf{k}}$.

Then, $\cos\theta=\frac{\lambda(1+2+0)}{\lambda\sqrt{2}\sqrt{1+4+4}}=\frac{1}{\sqrt{2}}$

$\therefore \quad \theta=\frac{\pi}{4}$

14. (b) Clearly, the given line passes through the point

$$\mathbf{a}=2\hat{\mathbf{i}}-2\hat{\mathbf{j}}+3\hat{\mathbf{k}}$$

and is parallel to the vector $\mathbf{a}=\hat{\mathbf{i}}-\hat{\mathbf{j}}+4\hat{\mathbf{k}}$

The plane is normal to the vector

$$\mathbf{n}=\hat{\mathbf{i}}+5\hat{\mathbf{j}}+\hat{\mathbf{k}}$$

We have, $\mathbf{b}\cdot\mathbf{n}=1-5+4=0$

Thus, the line is parallel to the plane.

$\therefore$ Required distance

= Length perpendicular from a point on the line to the given plane

= Length of the perpendicular from $(2\hat{\mathbf{i}}-2\hat{\mathbf{j}}+3\hat{\mathbf{k}})$ to the given plane

$$=\left|\frac{(2\hat{\mathbf{i}}-2\hat{\mathbf{j}}+3\hat{\mathbf{k}})\cdot(\hat{\mathbf{i}}+5\hat{\mathbf{j}}+\hat{\mathbf{k}})-5}{\sqrt{1+25+1}}\right|$$

$$=\left|\frac{2-10+3-5}{3\sqrt{3}}\right|=\frac{10}{3\sqrt{3}}$$

15. (a)

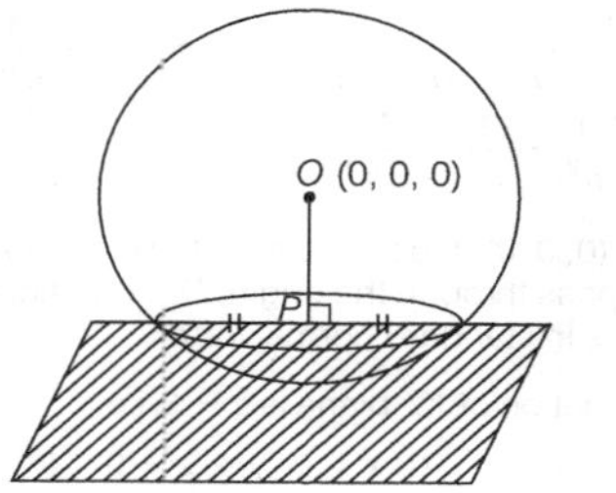

Let coordinate of P be (α, β, γ)

To find foot of perpendicular from $O(0, 0, 0)$ to

$x + y + z - 3\sqrt{3} = 0$

$$\Rightarrow \quad \frac{\alpha - 0}{1} = \frac{\beta - 0}{1} = \frac{\gamma - 0}{1} = \frac{-(-3\sqrt{3})}{1^2 + 1^2 + 1^2}$$

$$\Rightarrow \quad \alpha = \sqrt{3}, \beta = \sqrt{3}, \gamma = \sqrt{3}$$

So, position vector of centre of circle, is $\sqrt{3}(\hat{\mathbf{i}} + \hat{\mathbf{j}} + \hat{\mathbf{k}})$.

16. (b) We know that, a line lies in a plane if every point on the line is a point on the plane.

The coordinates of any point on the line L are

$$(3\lambda - 1, 2\lambda - 1, -\lambda + 3)$$

Substituting the coordinates of this point in the equation of plane. We have

$$x - 2y - z = 0$$

$$3\lambda + 1 - 4\lambda + 2 + \lambda - 3 = 0$$

Clearly, this point satisfy equation of plane. So, this point lies on the plane P.

Hence, L lies in P.

17. (a) Given equation of lines are

$$\mathbf{r} = (\hat{\mathbf{i}} + \hat{\mathbf{j}}) + \lambda(\hat{\mathbf{i}} + 2\hat{\mathbf{j}} - \hat{\mathbf{k}})$$

and $$\mathbf{r} = (\hat{\mathbf{i}} + \hat{\mathbf{j}}) + \mu(-\hat{\mathbf{i}} + \hat{\mathbf{j}} - 2\hat{\mathbf{k}})$$

The lines are parallel to the vectors

$$\mathbf{b}_1 = \hat{\mathbf{i}} + 2\hat{\mathbf{j}} - \hat{\mathbf{k}}$$

and $$\mathbf{b}_2 = -\hat{\mathbf{i}} + \hat{\mathbf{j}} - 2\hat{\mathbf{k}}$$

$\therefore$ The plane is normal to the vector

$$\mathbf{n} = \mathbf{b}_1 \times \mathbf{b}_2 = \begin{vmatrix} \hat{\mathbf{i}} & \hat{\mathbf{j}} & \hat{\mathbf{k}} \\ 1 & 2 & -1 \\ -1 & 1 & -2 \end{vmatrix} = -3\hat{\mathbf{i}} + 3\hat{\mathbf{j}} + 3\hat{\mathbf{k}}$$

The required plane passes through $(\hat{\mathbf{i}} + \hat{\mathbf{j}})$ and is normal to the vector $\mathbf{n}$.

Therefore, its equation is $\mathbf{r} \cdot \mathbf{n} = \mathbf{a} \cdot \mathbf{n}$

$$\Rightarrow \quad r \cdot (-3\hat{\mathbf{i}} + 3\hat{\mathbf{j}} + 3\hat{\mathbf{k}}) = (\hat{\mathbf{i}} + \hat{\mathbf{j}})(-3\hat{\mathbf{i}} + 3\hat{\mathbf{j}} + 3\hat{\mathbf{k}})$$

$$\Rightarrow \quad r \cdot (-3\hat{\mathbf{i}} + 3\hat{\mathbf{j}} + 3\hat{\mathbf{k}}) = -3 + 3 = 0$$

18. (b) Let r be the radius, therefore centre (r, r, r)

The sphere touches $x + 2y + 2z = 4$

$$\Rightarrow \quad \left|\frac{r + 2r + 2r - 4}{\sqrt{1 + 4 + 4}}\right| = r$$

$$\Rightarrow \quad |5r - 4| = 3r$$

$$\Rightarrow \quad r = 2 \text{ or } \frac{1}{2}$$

$\therefore$ Equation of sphere is $(x - 2)^2 + (y - 2)^2 + (z - 2)^2 = 2^2$

and $$\left(x - \frac{1}{2}\right)^2 + \left(y - \frac{1}{2}\right)^2 + \left(z - \frac{1}{2}\right)^2 = \left(\frac{1}{2}\right)^2$$

$$\Rightarrow \quad x^2 + y^2 + z^2 - 4x - 4y - 4z + 8 = 0$$

and $$x^2 + y^2 + z^2 - x - y - z + \frac{1}{2} = 0$$

19. (a) The equation of a line passing through the centre of the sphere and normal to the plane is

$$\mathbf{r} = (\hat{\mathbf{i}} + 2\hat{\mathbf{j}} - \hat{\mathbf{k}}) + \lambda(2\hat{\mathbf{i}} - 2\hat{\mathbf{j}} + \hat{\mathbf{k}})$$

Suppose it intersects the plane

$$\mathbf{r} \cdot (2\hat{\mathbf{i}} - 2\hat{\mathbf{j}} + \hat{\mathbf{k}}) = 6 \text{ at the point}$$

$$(\hat{\mathbf{i}} + 2\hat{\mathbf{j}} - \hat{\mathbf{k}}) + \lambda(2\hat{\mathbf{i}} - 2\hat{\mathbf{j}} + \hat{\mathbf{k}})$$

$$\therefore \quad [(\hat{\mathbf{i}} + 2\hat{\mathbf{j}} - \hat{\mathbf{k}}) + \lambda(2\hat{\mathbf{i}} - 2\hat{\mathbf{j}} + \hat{\mathbf{k}})] \cdot (2\hat{\mathbf{i}} - 2\hat{\mathbf{j}} + \hat{\mathbf{k}}) = 6 \Rightarrow \lambda = 1$$

$\therefore$ Required point is $(\hat{\mathbf{i}} + 2\hat{\mathbf{j}} - \hat{\mathbf{k}}) + (2\hat{\mathbf{i}} - 2\hat{\mathbf{j}} + \hat{\mathbf{k}}) = 3\hat{\mathbf{i}}$

20. (b) A vector normal to the plane containing **b** and **c** is $\mathbf{n} = \mathbf{b} \times \mathbf{c}$.

We have, BN = Projection of **a** on $\mathbf{n} = \mathbf{a} \cdot \hat{\mathbf{n}}$

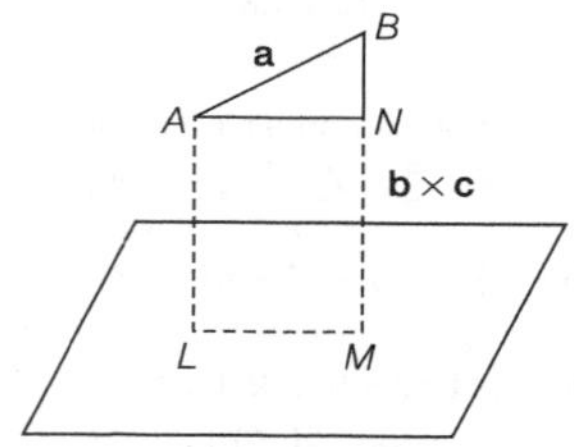

$\therefore$ Projection of **a** on the plane containing **b** and **c**

$$= LM = AN = \sqrt{AB^2 - BN^2}$$

$$= \sqrt{|\mathbf{a}|^2 - (\mathbf{a} \cdot \hat{\mathbf{n}})^2} = \sqrt{|\mathbf{a}|^2 |\hat{\mathbf{n}}|^2 - (\mathbf{a}^{-1} \cdot \hat{\mathbf{n}})^2}$$

$$= \sqrt{|\mathbf{a} \times \hat{\mathbf{n}}|^2} = |\mathbf{a} \times \hat{\mathbf{n}}| = \frac{|\mathbf{a} \times (\mathbf{b} \times \mathbf{c})|}{|\mathbf{b} \times \mathbf{c}|}$$

21. (c) The equation of the plane passes through (1, 1, 1) and direction ratios of normal to the plane in b, c, a.

$$b(x - 1) + c(y - 1) + a(z - 1) = 0 \quad \text{...(i)}$$

Now, $2001 = 3 \times 23 \times 29$

$\therefore \quad a < b < c$

$\Rightarrow \quad a = 3, b = 23, c = 29.$

Putting the values of a, b and c in d, we get

$$23x + 29y + 3z = 55$$

which is the equation of required plane.

22. (b) Let the direction ratio of the normal to the plane be proportional to a, b, c. Then, the equation of the plane is

$$a(x - 1) + b(y - 0) + c(z - 0) = 0 \quad \text{...(i)}$$

It passes through (0, 1, 0)

$$\therefore \quad a(-1) + b(1) + c(0) = 0$$

$$\Rightarrow \quad a = b \quad \text{...(ii)}$$

It is given that plane (i) makes an angle $\frac{\pi}{4}$ with the plane $x + y = 3$.

$$\therefore \quad \cos\frac{\pi}{4} = \frac{a \times 1 + b \times 1 + c \times 0}{\sqrt{a^2 + b^2 + c^2}\sqrt{1 + 1}}$$

$$\left[\because \cos\theta = \frac{a_1a_2 + b_1b_2 + c_1c_2}{\sqrt{a_1^2 + b_1^2 + c_1^2}\sqrt{a_2^2 + b_2^2 + c_2^2}}\right]$$

$$\Rightarrow \quad \frac{1}{\sqrt{2}} = \frac{a + b}{\sqrt{a^2 + b^2 + c^2}\sqrt{2}}$$

$$\Rightarrow \quad \sqrt{a^2 + b^2 + c^2} = a + b$$

$$\Rightarrow \quad a^2 + b^2 + c^2 = (a + b)^2$$

$$\Rightarrow \quad a^2 + b^2 + c^2 = a^2 + b^2 + 2ab$$

$$\Rightarrow \quad c^2 = 2ab \quad \text{...(iii)}$$

From Eqs. (ii) and (iii), we get

$$a : b : c = a : a : \sqrt{2}a = 1 : 1 : \sqrt{2}.$$

23. (d) We know that the lines

$$\frac{x - x_1}{l_1} = \frac{y - y_1}{m_1} = \frac{z - z_1}{n_1}$$

and $$\frac{x - x_2}{l_2} = \frac{y - y_2}{m_2} = \frac{z - z_2}{n_2}$$

are coplanar if,

$$\begin{vmatrix} x_2 - x_1 & y_2 - y_1 & z_2 - z_1 \\ l_1 & m_1 & n_1 \\ l_2 & m_2 & n_2 \end{vmatrix} = 0$$

So, the given lines will be coplanar, if

$$\begin{vmatrix} 1 - 2 & 4 - 3 & 5 - 4 \\ 1 & 1 & -k \\ k & 2 & 1 \end{vmatrix} = 0$$

$\Rightarrow$ $$k^2 + 3k = 0$$

$\Rightarrow$ $$k(k + 3) = 0$$

$\therefore$ $$k = 0, -3$$

24. (b) Let $\mathbf{n}_1$ and $\mathbf{n}_2$ be the vectors normal to the faces OAB and ABC. Then,

$$\mathbf{n}_1 = \mathbf{OA} \times \mathbf{OB} = \begin{vmatrix} \hat{\mathbf{i}} & \hat{\mathbf{j}} & \hat{\mathbf{k}} \\ 1 & 2 & 1 \\ 2 & 1 & 3 \end{vmatrix} = 5\hat{\mathbf{i}} + \hat{\mathbf{j}} - 3\hat{\mathbf{k}}$$

and $$\mathbf{n}_2 = \mathbf{AB} \times \mathbf{AC} = \begin{vmatrix} \hat{\mathbf{i}} & \hat{\mathbf{j}} & \hat{\mathbf{k}} \\ 1 & -1 & 2 \\ -2 & -1 & 1 \end{vmatrix} = \hat{\mathbf{i}} - 5\hat{\mathbf{j}} - 3\hat{\mathbf{k}}$$

If θ is the angle between the faces OAB and ABC, then

$$\cos\theta = \frac{\mathbf{n}_1 \cdot \mathbf{n}_2}{|\mathbf{n}_1||\mathbf{n}_2|}$$

$\Rightarrow$ $$\cos\theta = \frac{5 + 5 + 9}{\sqrt{25 + 1 + 9}\sqrt{1 + 25 + 9}} = \frac{19}{35}$$

$\therefore$ $$\theta = \cos^{-1}\left(\frac{19}{35}\right)$$

25. (a) We have,

$$\frac{x - 4}{1} = \frac{y - 2}{1} = \frac{z - k}{2} \quad \text{...(i)}$$

and $$2x - 4y + z = 7 \quad \text{...(ii)}$$

The given line passes through $(4, 2, k)$.

The line will lie in the plane

$$2x - 4y + z = 7$$

If the point $(4, 2, k)$ lies on the plane

$\therefore$ $$2 \times 4 - 4 \times 2 + k = 7$$

$\Rightarrow$ $$8 - 8 + k = 7$$

$\Rightarrow$ $$k = 7$$

26. (a) Given equation of spheres

$$S \equiv x^2 + y^2 + z^2 + 7x - 2y - z - 13 = 0$$

$$S' \equiv x^2 + y^2 + z^2 - 3x + 3y + 4z - 8 = 0$$

If these spheres intersect, then

$$S - S' = 0$$

represents the equation of common plane of intersection.

$\therefore$ $(x^2 + y^2 + z^2 + 7x - 2y - z - 13)$

$$-(x^2 + y^2 + z^2 - 3x + 3y + 4z - 8) = 0$$

$\Rightarrow$ $$10x - 5y - 5z - 5 = 0$$

$\Rightarrow$ $$2x - y - z = 1$$

27. (d) The equation of plane passing through $(1, -2, 3)$, is

$$a(x - 1) + b(y + 2) + c(z - 3) = 0 \quad \text{...(i)}$$

It passes through $(-1, 2, -1)$ and is parallel to the given line.

$\therefore$ $$a(-2) + b(4) + c(-4) = 0$$

and $$2a + 3b + 4c = 0$$

$\Rightarrow$ $$\frac{a}{28} = \frac{b}{0} = \frac{c}{-14}$$

$\therefore$ $$a : b : c = 2 : 0 : -1$$

28. (d) The given straight lines can be rewritten as

$$\frac{x - 1}{1} = \frac{y + 3}{-\lambda} = \frac{z - 1}{\lambda} = s$$

and $$\frac{x - 0}{1} = \frac{y - 1}{2} = \frac{z - 2}{-2} = t$$

These two lines are coplanar, if

$$\begin{vmatrix} x_1 - x_2 & y_1 - y_2 & z_1 - z_2 \\ l_1 & m_1 & n_1 \\ l_2 & m_2 & n_2 \end{vmatrix} = 0$$

$\Rightarrow$ $$\begin{vmatrix} 1 - 0 & -3 - 1 & 1 - 2 \\ 1 & -\lambda & \lambda \\ 1 & 2 & -2 \end{vmatrix} = 0$$

$\Rightarrow$ $$1(2\lambda - 2\lambda) + 4(-2 - \lambda) - 1(2 + \lambda) = 0$$

$\Rightarrow$ $$-8 - 4\lambda - 2 - \lambda = 0$$

$\Rightarrow$ $$5\lambda = -10$$

$\therefore$ $$\lambda = -2$$

29. (c) Let the equation of plane be

$$\frac{x}{a} + \frac{y}{b} + \frac{z}{c} = 1$$

It makes intercepts 3 and 4 on coordinate axes

$\Rightarrow$ $$a = 3, c = 4$$

Parallel to Y-axis

$$\frac{x}{0} = \frac{y}{1} = \frac{z}{0}$$

$\therefore$ $$0 \times \frac{1}{a} - 1 \times \frac{1}{b} + 0 \times \frac{1}{c} = 0$$

$\Rightarrow$ $$\frac{1}{b} = 0$$

$\Rightarrow$ $$\frac{x}{3} + \frac{z}{4} = 1$$

$\Rightarrow$ $$4x + 3z = 12$$

30. (b) Given,

$$12x^2 - 2y^2 - 6z^2 - 2xy - 8yz + 6zx = 0$$

The given equation can be factorised as

$$(3x + y + 3z)(4x - 2y - 2z) = 0$$

$\Rightarrow$ $$3x + y + z = 0 \text{ and } 4x - 2y - 2z = 0$$

Clearly, these two equations represent planes passing through the origin.

31. (a) Since,

$$\cos^2\alpha + \cos^2\beta + \cos^2\gamma = 1$$

[i.e. $l^2 + m^2 + n^2 = 1$]

$\Rightarrow$ $$1 - \sin^2\alpha + 1 - \sin^2\beta + 1 - \sin^2\gamma = 1$$

or $$\sin^2\alpha + \sin^2\beta + \sin^2\gamma = 2$$

Hence, both Statements are correct and Statement II is correct explanation of Statement I.

32. (c) $$xy + yz = 0$$

$\Rightarrow$ $$y(x + z) = 0$$

i.e. pair of perpendicular planes if

$$a_1x + b_1y + c_1z + d_1 = 0$$

and $$a_2x + b_2y + c_2z + d_2 = 0$$

are perpendicular

$\Rightarrow a_1a_2 + b_1b_2 + c_1c_2 = 0$

So, Statement II is Incorrect.

33. (d) As, $2x^2 - 6y^2 + 4z^2 + 18yz + 2zx + xy = 0 \Rightarrow \Delta \neq 0$

It does not represent pair of planes.

So, Statement I is incorrect.

34. (b) Shortest distance

$$= \left|\frac{(\mathbf{a}_2 - \mathbf{a}_2)\cdot(\mathbf{b}_1 \times \mathbf{b}_2)}{|\mathbf{b}_1 \times \mathbf{b}_2|}\right| = \left|\frac{\begin{vmatrix} -3+2 & 6-0 & 0-7 \\ -4 & 3 & 2 \\ -4 & 1 & 1 \end{vmatrix}}{\begin{vmatrix} \hat{\mathbf{i}} & \hat{\mathbf{j}} & \hat{\mathbf{k}} \\ -4 & 3 & 2 \\ -4 & 1 & 1 \end{vmatrix}}\right| = a$$

Hence, both Statements are correct, but Statement II is not correct explanation of Statement I.

35. (d) Dr's of straight line passing through two planes, is

$$= \begin{vmatrix} \hat{\mathbf{i}} & \hat{\mathbf{j}} & \hat{\mathbf{k}} \\ 3 & -6 & -2 \\ 2 & 1 & -2 \end{vmatrix}$$

$$= 14\,\hat{\mathbf{i}} + 2\,\hat{\mathbf{j}} + 15\,\hat{\mathbf{k}}$$

Let $z = 0$

$\Rightarrow \quad 3x - 6y = 15$

and $\quad 2x + y = 5$

On solving $x = 3, y = -1$

$\therefore$ Required equation of straight line is

$$\frac{x-3}{14} = \frac{y+1}{2} = \frac{z-0}{15}$$

So, Statement I is incorrect.

Revisal Problems (JEE Advanced)

1. (a) Given that P, Q, R are the images of a point $A(a, b, c)$ in, the XY, YZ and ZX-plane, respectively.

$\therefore$ Points P, Q, R are $(a, b, -c), (-a, b, c)$ and $(a, -b, c)$, respectively.

$\therefore$ Centroid of ΔPQR is $\left(\frac{a}{3}, \frac{b}{3}, \frac{c}{3}\right)$ i.e. $G \equiv \left(\frac{a}{3}, \frac{b}{3}, \frac{c}{3}\right)$

$\Rightarrow A, O, G$ are collinear.

So, area of ΔAOG is zero.

2. (d) Given,

$$(b+c)(y+z) - ax = b - c \quad \text{...(i)}$$
$$(c+a)(z+x) - by = c - a \quad \text{...(ii)}$$
$$(a+b)(x+y) - cz = a - b \quad \text{...(iii)}$$

$$\therefore \quad \Delta = \begin{vmatrix} -a & b+c & b+c \\ c+a & -b & c+a \\ a+b & a+b & -c \end{vmatrix}$$

On applying $R_1 \to R_1 + R_2 + R_3$

$$= \begin{vmatrix} a+b+c & a+b+c & a+b+c \\ c+a & -b & c+a \\ a+b & a+b & -c \end{vmatrix}$$

$$= (a+b+c)\begin{vmatrix} 1 & 1 & 1 \\ c+a & -b & c+a \\ a+b & a+b & -c \end{vmatrix}$$

On applying $C_1 \to C_1 - C_2$

$$= (a+b+c)\begin{vmatrix} 0 & 1 & 1 \\ 0 & -b & c+a \\ a+b+c & a+b & -c \end{vmatrix}$$

$$= (a+b+c)[(a+b+c)(c+a+b)]$$
$$= (a+b+c)^2(c+a+b)$$
$$= (a+b+c)^3$$

$$\Delta_x = \begin{vmatrix} b-c & b+c & b+c \\ c-a & -b & c+a \\ a-b & a+b & -c \end{vmatrix}$$

On applying $R_1 \to R_1 + R_2 + R_3$

$$= \begin{vmatrix} 0 & a+b+c & a+b+c \\ c-a & -b & c+a \\ a-b & a+b & -c \end{vmatrix}$$

$$= (a+b+c)\begin{vmatrix} 0 & 1 & 1 \\ c-a & -b & c+a \\ a-b & a+b & -c \end{vmatrix}$$

On applying $C_3 \to C_2 - C_3$

$$= (a+b+c)\begin{vmatrix} 0 & 0 & 1 \\ c-a & -(a+b+c) & c+a \\ a-b & a+b+c & -c \end{vmatrix}$$

$$= (a+b+c)^2\begin{vmatrix} 0 & 0 & 1 \\ c-a & -1 & c+a \\ a-b & 1 & -c \end{vmatrix}$$

$$= (a+b+c)^2|c-a+a-b|$$
$$= (c-b)(a+b+c)^2$$

Now, $\quad x = \dfrac{\Delta_x}{\Delta}$

$$= \frac{(c-b)(a+b+c)^2}{(a+b+c)^3}$$
$$= \frac{c-b}{a+b+c}$$

Similarly, we can find

$$y = \frac{\Delta_y}{\Delta} = \frac{(a-c)(a+b+c)^2}{(a+b+c)^3}$$
$$= \frac{a-c}{a+b+c}$$

and $\quad z = \dfrac{\Delta_z}{\Delta}$

$$= \frac{(b-a)(a+b+c)^2}{(a+b+c)^3}$$
$$= \frac{b-a}{a+b+c}$$

$$\therefore \quad x:y:z = \frac{c-b}{a+b+c} : \frac{a-c}{a+b+c} : \frac{b-a}{a+b+c}$$
$$= b-c : c-a : a-b$$

3. (b) Given system of linear equations

$$x - 2y + z = -4 \quad \text{...(i)}$$
$$2x - y + 2z = 2 \quad \text{...(ii)}$$
and $$x + y + \lambda z = 4 \quad \text{...(iii)}$$

$$\therefore \quad \Delta = \begin{vmatrix} 1 & -2 & 1 \\ 2 & -1 & 2 \\ 1 & 1 & \lambda \end{vmatrix}$$

$$= 1(-\lambda - 2) - 2(2\lambda - 2) + 1(2+1)$$
$$= -\lambda - 2 - 4\lambda + 4 + 3$$
$$= -5\lambda + 5$$
$$\Delta = 5(1-\lambda)$$

For no solution,

$$\Delta = 0$$

$$\Rightarrow \quad 5(1-\lambda) = 0$$

$$\Rightarrow \quad 1-\lambda = 0$$

$$\therefore \quad \lambda = 1$$

4. (b) Let line joining AB meet plane $2x + 3y + 5z = 1$ at P.

and $\quad P\left(\frac{\lambda+1}{\lambda+1}, \frac{-5\lambda}{\lambda+1}, \frac{7\lambda-3}{l+1}\right) \quad \left[\because \frac{AP}{PB} = \lambda\right]$

$$\Rightarrow \quad 2\left(\frac{\lambda+1}{\lambda+1}\right) + 3\left(\frac{-5\lambda}{\lambda+1}\right) + 5\left(\frac{7\lambda-3}{\lambda+1}\right) = 1$$

$$\Rightarrow \quad 2 - 15\lambda + 35\lambda - 15 = \lambda + 1$$

$$\Rightarrow \quad \lambda = 2/3$$

$$\Rightarrow \quad P(1, -2, 1)$$

$$\Rightarrow \quad AP = 2\sqrt{5}$$

5. (b) Let $Q(r, r, -r)$

PQ parallel

$$\frac{x}{1} = \frac{y}{1} = \frac{z}{-1}$$

$$\therefore \quad (\alpha - r)\cdot 1 + (\beta - r)\cdot 1 + (\gamma + r)(-1) = 0$$

$$\Rightarrow \quad r = \frac{\alpha + \beta - \gamma}{3}$$

$$\therefore \quad PQ^2 = \frac{2}{3}\{\alpha^2 + \beta^2 + \gamma^2 - \alpha\beta + \beta\gamma + \gamma\alpha\}$$

But $\quad PQ = 2\{$parallel distance from $P(\alpha, \beta, \gamma)$ to plane $x + y + z = 0\}$

$$\Rightarrow \quad \frac{2}{3}(\alpha^2 + \beta^2 + \gamma^2 - \alpha\beta + \beta\gamma + \gamma\alpha)$$

$$= 4\left\{\frac{\alpha + \beta + \gamma}{\sqrt{3}}\right\}^2$$

$$\therefore \quad \alpha^2 + \beta^2 + \gamma^2 + 5\,\alpha\beta + 3\beta\alpha + 3\gamma\alpha = 0$$

or $\quad x^2 + y^2 + z^2 + 5xy + 3yz + 3zx = 0$

6. (b) Equation of given plane $\frac{x}{a} + \frac{y}{b} + \frac{z}{c} = 1$ and given $a^2 + \beta^2 + c^2 = 9$.

Dr's of normal to the plane $\frac{1}{a}, \frac{1}{b}, \frac{1}{c}$.

Let P be the foot of perpendicular, then DR's of $OP = (x_1, y_1, z_1)$

OP is parallel to normal

$$\Rightarrow \quad \frac{x_1}{1/a} = \frac{b_1}{1/b} = \frac{z_1}{1/c} = k.$$

$$\Rightarrow \quad a = \frac{k}{x_1}, b = \frac{k}{y_1}, c = \frac{k}{z_1}$$

$$\Rightarrow \quad a^2 + b^2 + c^2 = 9$$

$$\Rightarrow \quad k^2\left(\frac{1}{x_1^2} + \frac{1}{y_1^2} + \frac{1}{z_1^2}\right) = 9 \qquad \ldots\text{(i)}$$

On substituting $a = \frac{k}{x_1}$, $b = \frac{k}{y_1}$, $c = \frac{k}{z_1}$ in $\frac{x_1}{a} + \frac{y_1}{b} + \frac{z_1}{c} = 1$,

we get

$$x_1^2 + y_1^2 + z_1^2 = k$$

$\therefore$ From Eq. (i),

$$(x_1^2 + y_1^2 + z_1^2)^2(x_1^{-2} + y_1^{-2} + z_1^{-2}) = 9$$

7. (c) Suppose M divides AC in the ratio of $\lambda : 1$. Then $M\left(\frac{-4\lambda+1}{\lambda+1}, \frac{\lambda-2}{\lambda+1}, \frac{\lambda+2}{\lambda+1}\right)$

$$\mathbf{BM} = \left(\frac{-5\lambda}{\lambda+1}, \frac{-3\lambda-6}{\lambda+1}, \frac{\lambda+2}{\lambda+1}\right)$$

and $\quad \mathbf{AC} = (-5, 3, -1)$

Now, $\quad \mathbf{BM} \perp \mathbf{AC}$

$$\Rightarrow \quad 25\lambda - 9\lambda - 18 - \lambda - 2 = 0$$

$$\Rightarrow \quad 15\lambda = 20$$

$$\Rightarrow \quad \lambda = 4/3$$

$$\therefore \quad \mathbf{BM} = \left(-\frac{20}{7}, -\frac{30}{7}, \frac{10}{7}\right)$$

8. (a) Dr's of line $= \begin{vmatrix} \hat{i} & \hat{j} & \hat{k} \\ 3 & 2 & 1 \\ 1 & 1 & -2 \end{vmatrix} = -5\hat{i} + 7\hat{j} + \hat{k}$

and $\quad \begin{vmatrix} \hat{i} & \hat{j} & \hat{k} \\ 1 & -1 & 4 \\ 1 & 1 & -4 \end{vmatrix} = 8\hat{j} + 2\hat{k}$

$$\therefore \quad \cos\theta = \frac{0 + 56 + 2}{\sqrt{25 + 44 + 1}\sqrt{64 + 4}} = \frac{29}{5\sqrt{51}}$$

9. (a) The cut $x = y$ separates the cube into points with $x < y$ and those with $x > y$.

So, number of pieces equals to the number of ways of arrangements of x, y and z, which is $(3)! = 6$

10. (d) If P is a variable point on the required plane the equation of plane is

$$[\mathbf{AP} \;\; \mathbf{AB} \;\; \mathbf{AB} \times \mathbf{AC}] = 0$$

$$\Rightarrow \quad 2x + y + z - 1 = 0$$

11. (c) Here, P and Q lie on the same side of XY-plane.

Image of P (1, 2, 3) on the XY-plane is $P'(1, 2, -3)$.

Reflected ray is

$$\frac{x-3}{2} = \frac{y-2}{0} = \frac{z-5}{8}$$

$$\Rightarrow \quad \frac{x-3}{1} = \frac{y-2}{0} = \frac{z-5}{4}$$

12. (b) Given, planes $y + z = 0$, $z + x = 0$, $x + y = 0$ meet at $O(0, 0, 0)$.

Let the tetrahedron be $OABC$ and the equation of one of the pair of opposite edges OA and BC be

$$y + z = 0, x + z = 0 \qquad \ldots\text{(i)}$$

and $\quad x + y = 0, x + y + z = a \qquad \ldots\text{(ii)}$

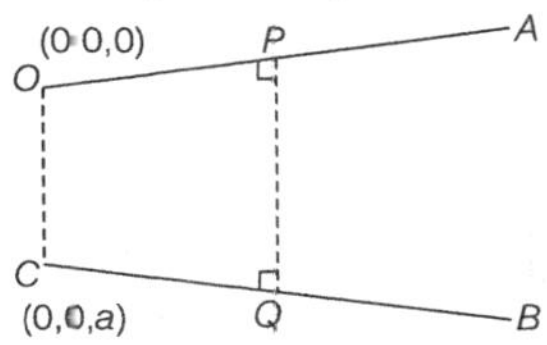

Eqs. (i) and (ii) can be expressed in symmetrical form as

$$\frac{x-0}{1} = \frac{y-0}{1} = \frac{z-0}{-1} \qquad \ldots\text{(iii)}$$

and $\quad \frac{x-1}{1} = \frac{y-0}{-1} = \frac{z-a}{0} \qquad \ldots\text{(iv)}$

DR's of OA and BC are $(1, 1, -1)$ and $(1, -1, 0)$.

Let PQ be the shortest distance between OA and BC having direction cosine (l, m, n).

$\therefore$ PQ is perpendicular to both OA and BC,

$$l + m + n = 0 \qquad \ldots\text{(v)}$$

$$l - m = 0 \qquad \ldots\text{(vi)}$$

On solving Eqs. (v) and (vi), we get

$$\frac{l}{1} = \frac{m}{1} = \frac{n}{2} = k$$

$$l^2 + m^2 + n^2 = 1$$

$$k^2 + k^2 + 4k^2 = 1$$

$$\Rightarrow \quad k = \frac{1}{\sqrt{6}}$$

$$\therefore \quad l = \frac{1}{\sqrt{6}} = m \text{ and } n = \frac{2}{\sqrt{6}}$$

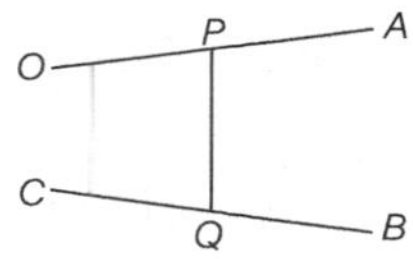

Shortest distance between OA and BC, i.e.

$$PQ = \text{Length of projection of } OC \text{ on } PQ$$

$$= |(x_2 - x_1)\,l + (y_2 - y_1)\,m + (z_2 - z_1)\,n|$$

$$= \left| 0 \cdot \frac{1}{\sqrt{6}} + 0 \cdot \frac{1}{\sqrt{6}} + a \cdot \frac{2}{\sqrt{6}} \right| = \frac{2a}{\sqrt{6}}$$

13. (a) $y(x + y) + z(x + y) = 0$

$$\Rightarrow \begin{cases} x + y = 0 \text{ having DR's } (1, 1, 0) \\ y + z = 0 \text{ having DR's } (0, 1, 1) \end{cases}$$

$$\therefore \quad \text{DR's line} = \begin{vmatrix} \hat{i} & \hat{j} & \hat{k} \\ 1 & 1 & 0 \\ 0 & 1 & 1 \end{vmatrix} = \hat{i} - \hat{j} + \hat{k} \text{ or } (1, -1, 1)$$

Point $O(0, 0, 0)$

$\therefore$ Shortest distance $= \sqrt{\dfrac{8}{3}}$

14. (a) Let $\quad \mathbf{a} = l_1\hat{i} + m_1\hat{j} + n_1\hat{k}$,

$$\mathbf{b} = l_2\hat{i} + m_2\hat{j} + n_2\hat{k}$$

and $\quad \mathbf{c} = l_3\hat{i} + m_3\hat{j} + n_3\hat{k}$.

Given that, **a**, **b**, **c** are three mutually perpendicular units Then,

$$p_1\hat{i} + q_1\hat{j} + r_1\hat{k} = \mathbf{b} \times \mathbf{c} = \mathbf{a}$$

Similarly, $\quad p_2\hat{i} + q_2\hat{j} + r_2\hat{k} = \mathbf{c} \times \mathbf{a} = \mathbf{b}$

and $\quad p_3\hat{i} + q_3\hat{j} + r_3\hat{k} = \mathbf{a} \times \mathbf{b} = \mathbf{c}$

These vectors are mutually perpendicular unit vectors.

15. (b,c) Observe that the lines L_1, L_2 and L_3 are parallel to the vector $(1, -1, -1)$.

Also, $\Delta = 0 = \Delta_1$ and $b_1c_2 - b_2c_1 \neq 0$

16. (a, c) DR's of AB are $\dfrac{x_2 - x_1}{AB}, \dfrac{y_2 - y_1}{AB}, \dfrac{z_2 - z_1}{AB}$

DRs of L are $\dfrac{l}{\sqrt{l^2 + m^2 + n^2}}, \dfrac{m}{\sqrt{l^2 + m^2 + n^2}}, \dfrac{n}{\sqrt{l^2 + m^2 + n^2}}$

$$\cos\theta = \frac{(x_2 - x_1)l + (y_2 - y_1)m + (z_2 - z_1)n}{|AB|\sqrt{l^2 + m^2 + n^2}}$$

Also, $\quad \mathbf{PQ} = AB\cos\theta \quad$ ...(i)

or $\quad \mathbf{PQ} = \dfrac{(x_2 - x_1)\,l + (y_2 - y_1)m + (z_2 - z_1)n}{\sqrt{l^2 + m^2 + n^2}}$

17. (a,b,c) Intersection of line with both the planes are the same

$$\Rightarrow \quad \frac{3}{3\beta^2 + 6(1 - 2\alpha) + 3} = \frac{-6}{6\alpha^2 + 6(1 - 2\beta) + 6}$$

$$\Rightarrow \quad 2(\beta - 1)^2 + 3(\alpha - 2)^2 = 0$$

$$\Rightarrow \quad \alpha = 2, \beta = 1$$

$\therefore (\alpha, \beta, \gamma) = (2, 1, 1)$ lie on (a, b, c).

18. (a,b,c) Let P be (x, y, z), then from the figure

$x = r\sin\theta\cos\phi$, $y = r\sin\theta\sin\phi$ and $z = r\cos\theta$

$1 = r\sin\theta\cos\phi$, $2 = r\sin\theta\sin\phi$ and $3 = r\cos\theta$

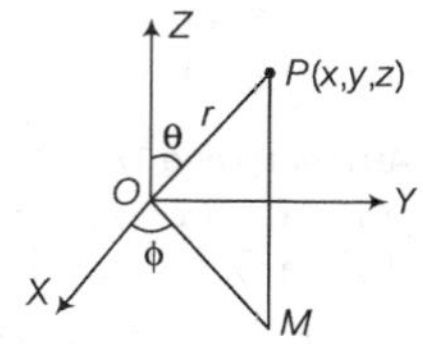

$$\Rightarrow \quad 1^2 + 2^2 + 3^2 = r^2$$

$$\Rightarrow \quad r = \pm\sqrt{14}$$

$$\therefore \quad \sin\theta\cos\phi = \frac{1}{\sqrt{14}}, \sin\theta\sin\phi = \frac{2}{\sqrt{14}}$$

and $\quad \cos\theta = \dfrac{3}{\sqrt{14}}$

[neglecting negative sign as θ and ϕ are acute]

$$\therefore \quad \frac{\sin\theta\sin\phi}{\sin\theta\cos\phi} = \frac{2}{1} \Rightarrow \tan\phi = 2$$

Also, $\quad \tan\theta = \dfrac{\sqrt{5}}{3}$

19. (a) Let the coordinates of A, B, C be respectively $(a, 0, 0)$, $(0, b, 0)$ and $(0, 0, c)$, so that the equation of plane is

$$\frac{x}{a} + \frac{y}{b} + \frac{z}{c} = 1 \qquad \text{...(i)}$$

Let (α, β, γ) be the coordinates of the centroid of tetrahedron $OABC$, then

$$\alpha = \frac{0 + a + 0 + 0}{4} \Rightarrow \alpha = \frac{a}{4} \qquad \text{...(ii)}$$

$$\beta = \frac{0 + 0 + b + 0}{4} \Rightarrow \beta = \frac{b}{4} \qquad \text{...(iii)}$$

$$\gamma = \frac{0 + 0 + 0 + c}{4} \Rightarrow \gamma = \frac{c}{4} \qquad \text{...(iv)}$$

Now, $\quad p = \left| \dfrac{0 + 0 + 0 - 1}{\sqrt{\dfrac{1}{a^2} + \dfrac{1}{b^2} + \dfrac{1}{c^2}}} \right|$

$$p = \frac{\pm 1}{\sqrt{\dfrac{1}{a^2} + \dfrac{1}{b^2} + \dfrac{1}{c^2}}}$$

$$\Rightarrow \quad \frac{1}{a^2} + \frac{1}{b^2} + \frac{1}{c^2} = \frac{1}{p^2}$$

On substituting the values of a, b, c from Eqs. (ii), (iii) and (iv), we get

$$\frac{1}{(4\alpha)^2} + \frac{1}{(4\beta)^2} + \frac{1}{(4\gamma)^2} = \frac{1}{p^2}$$

$$\Rightarrow \quad \frac{1}{x^2} + \frac{1}{y^2} + \frac{1}{z^2} = \frac{16}{p^2}$$

$$\Rightarrow \quad \frac{y^2z^2 + x^2z^2 + x^2y^2}{x^2y^2z^2} = \frac{16}{p^2}$$

$$\Rightarrow \quad y^2z^2 + x^2z^2 + x^2y^2 = \frac{16}{p^2}(x^2y^2z^2)$$

20. (b) Given, $\quad l + m + n = 0 \qquad$ ...(i)

and $\quad 4l = \dfrac{2mn}{m + n}$

$$\Rightarrow \quad l = \frac{mn}{2(m + n)} \qquad \text{...(ii)}$$

On eliminating l from Eqs. (i) and (ii), we get

$$-(m+n) = \frac{mn}{2(m+n)}$$

$$\Rightarrow \quad -2(m+n)^2 = mn$$

$$\Rightarrow \quad -2(m^2+n^2+2mn) = mn$$

$$\Rightarrow \quad -2m^2-2n^2-4mn = mn$$

$$\Rightarrow \quad 2m^2+2n^2+4mn = -mn$$

$$\Rightarrow \quad 2m^2+2n^2+5mn = 0$$

$$\Rightarrow \quad (m+2n)(2m+n) = 0$$

$\therefore$ Either $m+2n=0$ or $2m+n=0$ ...(iii)

Let the two lines whose direction cosines satisfy the given relations have their direction cosines (l_1, m_1, n_1) and (l_2, m_2, n_2). Then, from Eq. (iii)

$$m_1 + 2n_1 = 0$$

$$\Rightarrow \quad m_1 = -2n_1$$

where, $l_1 + m_1 + n_1 = 0$

$$l_1 = -(m_1+n_1)$$

$$l_1 = n_1$$

$$l_1 = -\frac{m_1}{2} = n_1$$

$$\Rightarrow \quad \frac{l_1}{1} = \frac{m_1}{-2} = \frac{n_1}{1} = \frac{\sqrt{l_1^2+m_1^2+n_1^2}}{\sqrt{1+4+1}}$$

$$\Rightarrow \quad l_1 = \frac{1}{\sqrt6}, m_1 = -\frac{2}{\sqrt6}, n_1 = \frac{1}{\sqrt6} \quad ...(iv)$$

Again, from Eq. (i), we get

$$2m_2 + n_2 = 0$$

and $l_2 = -m_2 - n_2 = m_2$

Now, $\frac{l_2}{1} = \frac{m_2}{1} = \frac{-n_2}{2}$

$$= \frac{\sqrt{l_2^2+m_2^2+n_2^2}}{\sqrt{1+1+4}} = \frac{1}{\sqrt6}$$

$$\Rightarrow \quad l_2 = \frac{1}{\sqrt6}, m_2 = \frac{1}{\sqrt6}, n_2 = -\frac{2}{\sqrt6}$$

$\therefore$ Then,

(i) $\frac{l_1}{l_2} + \frac{m_1}{m_2} + \frac{n_1}{n_2}$

$$= \frac{1/\sqrt6}{1/\sqrt6} + \frac{-2/\sqrt6}{1/\sqrt6} + \frac{1/\sqrt6}{-2/\sqrt6}$$

$$= 1 - 2 - \frac12 = -1 - \frac12 = -\frac32$$

(ii) $l_1l_2 + m_1m_2 + n_1n_2$

$$= \left(\frac{1}{\sqrt6}\right)\left(\frac{1}{\sqrt6}\right) + \left(-\frac{2}{\sqrt6}\right)\left(\frac{1}{\sqrt6}\right) + \left(\frac{1}{\sqrt6}\right)\left(-\frac{2}{\sqrt6}\right)$$

$$= \frac16 - \frac26 - \frac26 = -\frac12$$

(iii) $l_1m_1n_1 + l_2m_2n_2$

$$= \frac{1}{\sqrt6} \times \left(-\frac{2}{\sqrt6}\right) \times \left(\frac{1}{\sqrt6}\right) + \left(\frac{1}{\sqrt6} \times \frac{1}{\sqrt6}\right)\left(-\frac{2}{\sqrt6}\right)$$

$$= -\frac{2}{6\sqrt6} - \frac{2}{6\sqrt6}$$

$$= -\frac{4}{6\sqrt6} = -\frac{2}{3\sqrt6} = -\frac{\sqrt6}{9}$$

(iv) $(l_1 + l_2)(m_1 + m_2)(n_1 + n_2)$

$$= \left(\frac{1}{\sqrt6} + \frac{1}{\sqrt6}\right)\left(-\frac{2}{\sqrt6} + \frac{1}{\sqrt6}\right)\left(\frac{1}{\sqrt6} - \frac{2}{\sqrt6}\right)$$

$$= \frac{2}{\sqrt6} \times \frac{-1}{\sqrt6} \times \frac{-1}{\sqrt6} = \frac{2}{6\sqrt6} = \frac{\sqrt6}{18}$$

Hence, (i) → (s); (ii) → (p); (iii) → (q); (iv) → (r)

21. (c) Here, we have the determinant of the coefficient matrix of given equation as

$$\Delta = \begin{vmatrix} a & b & c \\ b & c & a \\ c & a & b \end{vmatrix}$$

$$= -(a+b+c)(a^2+b^2+c^2-ab-bc-ca)$$

$$= -\frac12 (a+b+c)[(a-b)^2+(b-c)^2+(c-a)^2]$$

(i) $a+b+c \neq 0$

and $a^2+b^2+c^2-ab-bc-ca = 0$

$$\Rightarrow \quad (a-b)^2+(b-c)^2+(c-a)^2 = 0$$

$$\Rightarrow \quad a = b = c$$

Therefore, this equation represents identical planes.

(ii) $a+b+c = 0$

and $a^2+b^2+c^2-ab-bc-ca \neq 0$

$$\Rightarrow \quad \Delta = 0$$

and a, b, c are not all equal.

Therefore, all equations are not identical but have infinite solution.

Hence, $ax + by = (a+b)z$ [using $a+b+c=0$]

and $bx + cy = (b+c)z$

$$\Rightarrow \quad (b^2-ac)y = (b^2-ac)z$$

$$\Rightarrow \quad y = z$$

$$\Rightarrow \quad ax + by + cy = 0$$

$$\Rightarrow \quad ax = ay$$

$$\Rightarrow \quad x = y = z$$

Therefore, this equation represents the line $x = y = z$.

(iii) $a+b+c \neq 0$

and $a^2+b^2+c^2-ab-bc-ca \neq 0$

$$\Rightarrow \quad \Delta \neq 0$$

The equations have only trivial solution,

i.e. $x = y = z = 0$

Therefore, the equations represent the plane meeting at a single point, namely origin.

(iv) $a+b+c = 0$

and $a^2+b^2+c^2-ab-bc-ca = 0$

$$\Rightarrow \quad a = b = c$$

and $\Delta = 0$

$$\Rightarrow \quad a = b = c = 0$$

⇒ All equations are satisfied by all x, y, z.

$\therefore$ The equations represent the whole of the three dimensional space.

Hence, (i) → (r), (ii) → (q), (iii) → (p), (iv) → (s)

22. (d)

(i) Let $P(x, y, z)$ be the point on the line segment joining (3, 3, 5) and (4, 6, 7) and divide it in the ratio $k : 1$.

K : 1

(3, 3, 5) $P(x, y, z)$ (4, 6, 7)

$$\therefore \quad x = \frac{4k+3}{k+1}$$

$$y = \frac{6k+3}{k+1}$$

$$z = \frac{7k+5}{k+1}$$

$\therefore$ Coordinates of point P are

$$\left(\frac{4k+3}{k+1}, \frac{6k+3}{k+1}, \frac{7k+5}{k+1}\right)$$

Also, it is given that projection of point P on coordinate axes are $\frac{32}{9}, \frac{42}{9}, \frac{55}{9}$.

$\therefore \quad \frac{4k+3}{k+1} = \frac{32}{9}; \frac{6k+3}{k+1} = \frac{42}{9}; \frac{7k+5}{k+1} = \frac{55}{9}$

$\Rightarrow \quad k = \frac{5}{4}$

(ii) Mid-point of BC is $\left(\frac{0+a}{2}, \frac{4+6}{2}, \frac{3+b}{2}\right)$

i.e. $\quad \left(\frac{a}{2}, 5, \frac{3+b}{2}\right)$

It is given that median AD is equally inclined to the coordinate axes.

$\therefore$ DR's of AD is $\pm\frac{1}{\sqrt{3}}, \pm\frac{1}{\sqrt{3}}, \pm\frac{1}{\sqrt{3}}$.

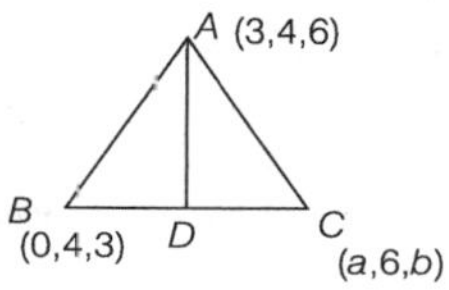

$\therefore$ Equation of AD

$$\frac{x-3}{1/\sqrt{3}} = \frac{y-4}{1/\sqrt{3}} = \frac{z-6}{1/\sqrt{3}} = \lambda \quad \text{...(i)}$$

Any point on the line (i) is

$$(\sqrt{3}\lambda + 3, \sqrt{3}\lambda + 4, \sqrt{3}\lambda + 6) \quad \text{...(ii)}$$

Point $D\left(\frac{a}{2}, 5, \frac{3+b}{2}\right)$ also lies on AD.

$\therefore \quad \sqrt{3}\lambda + 4 = 5$

$\Rightarrow \quad \lambda = \frac{1}{\sqrt{3}}$

From Eq. (ii), the coordinates of point are (4, 5, 7).

$\therefore \quad \left(\frac{a}{2}, 5, \frac{3+b}{2}\right) \equiv (4, 5, 7)$

$\Rightarrow \quad a = 8, \frac{3+b}{2} = 7$

$\Rightarrow \quad b = 11$

Now, $\quad a + b = 19$

(iii) Given equation of line

$$\frac{x+1}{-1} = \frac{y-12}{5} = \frac{z-7}{2} \quad \text{...(i)}$$

Put $x = 0$ in Eq. (i), we get

$$\frac{y-12}{5} = -1 \text{ and } \frac{z-7}{2} = -1$$

$\Rightarrow \quad y = 7$ and $z = 5$

Since, line (i) intersects the curve $yz = k$.

$\therefore \quad 7 \times 5 = k$

$\Rightarrow \quad k = 35$

(iv) Given,

$$|\mathbf{p}| = 2\sqrt{2}, |\mathbf{q}| = 3$$

Angle between p and q is $\frac{\pi}{4}$.

$\therefore \quad \cos\theta = \frac{\mathbf{p}\cdot\mathbf{q}}{|\mathbf{p}|\cdot|\mathbf{q}|}$

$\Rightarrow \quad \cos\frac{\pi}{4} = \frac{\mathbf{p}\cdot\mathbf{q}}{2\sqrt{2}\times 3}$

$\Rightarrow \quad \frac{1}{\sqrt{2}} = \frac{\mathbf{p}\cdot\mathbf{q}}{6\sqrt{2}}$

$\Rightarrow \quad \mathbf{p}\cdot\mathbf{q} = 6$

Now, $\quad (\text{Length})^2 = |4p + 5q|^2 = 16\times 8 + 25\times 9 + 40\times 6$

$= 593$

23. (b)

(i) Given, $\quad l^2 + m^2 - n^2 = 0 \quad$ ...(i)

$l + m + n = 0 \quad$ ...(ii)

Also, $\quad l^2 + m^2 + n^2 = 1 \quad$ ...(iii)

On solving Eqs. (i), (ii) and (iii), we get

$$m = \pm\frac{1}{\sqrt{2}}, n = \pm\frac{1}{\sqrt{2}} \text{ and } l = 0$$

$\Rightarrow \quad \theta = \frac{\pi}{3}$ or $\frac{\pi}{2}$

$\therefore \quad 4\sqrt{3}\sin\theta$, when $\theta = \frac{\pi}{3}$

$= 4\sqrt{3}\times\frac{\sqrt{3}}{2} = 6$

(ii) The equation of line joining points (1, 1, 1) and (1, – 1, 2) is

$$\frac{x-1}{0} = \frac{y-1}{-2} = \frac{z-1}{1} \quad \text{...(i)}$$

So, the equation of plane containing the line (i) is

$$a(x-1) + b(y-1) + c(z-1) = 0 \quad \text{...(ii)}$$

where, $\quad 0\cdot a - 2b + c = 0 \quad$ ...(iii)

Also, plane (ii) passes through (–2, – 2, 2).

$\therefore \quad -3a - 3b + c = 0 \quad$ ...(iv)

From Eqs. (iii) and (iv), we get

$$\frac{a}{(-2)-(-3)} = \frac{b}{(-3)-0} = \frac{c}{0-6}$$

$\Rightarrow \quad \frac{a}{1} = \frac{b}{-3} = \frac{c}{-6}$

$\therefore$ Required plane is

$$(x-1) - 3(y-1) - 6(z-1) = 0$$

$\Rightarrow \quad x - 1 - 3y + 3 - 6z + 6 = 0$

$\Rightarrow \quad x - 3y - 6z + 8 = 0$

$\Rightarrow \quad \frac{x}{-8} + \frac{y}{8/3} + \frac{z}{8/6} = 1$

$\therefore$ Sum of intercepts $= 8 + \frac{8}{3} + \frac{8}{6}$

$= \frac{48 + 16 + 8}{6} = \frac{72}{6} = 12$

(iii) $l = a\,\hat{\mathbf{i}} + b\,\hat{\mathbf{j}} + c\,\hat{\mathbf{k}}$

$\Rightarrow \quad |\hat{\mathbf{i}}| = 63$

$\Rightarrow \quad a^2 + b^2 + c^2 = (63)^2$

and $\quad a = 3\lambda, b = -2\lambda, c = 6\lambda$

$\Rightarrow \quad \lambda = \pm 9$

As, $\quad a < 0$

$\Rightarrow \quad a = -27, b = 18, c = -54$

$\therefore \quad 2a + b - c = 18$

(iv) Given planes are

$$x = cy + bz, y = az + cx \text{ and } z = bx + ay$$

The given planes are concurrent

$$\therefore \quad \begin{vmatrix} -1 & c & b \\ c & -1 & a \\ b & a & -1 \end{vmatrix} = 0$$

$\Rightarrow \quad a^2 + b^2 + c^2 + 2abc = 1$

24. (a) (i) Given equation of plane

$$ax - by + cz = 0 \quad \text{...(i)}$$

and line $\quad \frac{x-a}{a} = \frac{y-2d}{b} = \frac{z-c}{c} \quad$ ...(ii)

If line (ii) lies in plane (i), then

$$a\cdot a - b\cdot b + c\cdot c = 0$$

$\Rightarrow \quad a^2 - b^2 + c^2 = 0 \quad$ ...(iii)

and $\quad a\cdot a + (-b)(2d) + c(c) = 0$

$\Rightarrow \quad a^2 - 2bd + c^2 = 0 \quad \text{...(iv)}$

$\Rightarrow \quad a^2 + c^2 = 2bd$

$\Rightarrow \quad b^2 = 2bd \quad$ [from Eq. (iii)]

$\therefore \quad \dfrac{b}{d} = 2$

(ii) The given plane is

$$x - y + z - 5 = 0 \quad \text{...(i)}$$

The given line AB is

$$\frac{x-2}{2} = \frac{y+2}{3} = \frac{z-6}{-6} \quad \text{...(ii)}$$

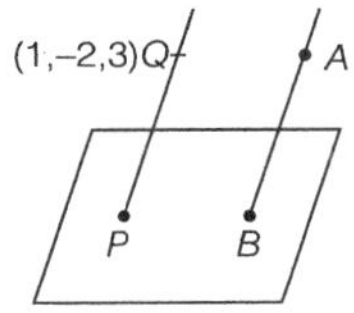

The equation of line passes through (1, – 2, 3) and parallel to line (ii) is

$$\frac{x-1}{2} = \frac{y+2}{3} = \frac{z-3}{-6} = \lambda \quad \text{...(iii)}$$

Coordinate of any point as the line (iii) is

$$P(2\lambda + 1, 3\lambda - 2, -6\lambda + 3)$$

If P is the point of intersection of Eqs. (i) and (iii), then it must lies on plane (i).

$\therefore \quad (2\lambda + 1) - (3\lambda - 2) + (-6\lambda + 3) - 5 = 0$

$\Rightarrow \quad 2\lambda - 3\lambda - 6\lambda + 1 + 2 + 3 - 5 = 0$

$\Rightarrow \quad -7\lambda + 1 = 0$

$\Rightarrow \quad \lambda = \dfrac{1}{7}$

$\therefore$ Coordinates of P are $\left(\dfrac{9}{7}, \dfrac{11}{7}, \dfrac{15}{7}\right)$.

$\therefore$ Distance between P and Q is

$$PQ = \sqrt{\left(\frac{9}{7} - 1\right)^2 + \left(\frac{11}{7} + 2\right)^2 + \left(\frac{15}{7} - 3\right)^2}$$

$$= \sqrt{\left(\frac{2}{7}\right)^2 + \left(\frac{25}{7}\right)^2 + \left(-\frac{6}{7}\right)^2}$$

$$= \sqrt{\frac{4}{49} + \frac{625}{49} + \frac{36}{49}} = \sqrt{\frac{665}{49}} = \frac{\sqrt{665}}{7}$$

(iii) Given equation of line

$$\frac{x-2}{1} = \frac{y-3}{1} = \frac{4-z}{k} \quad \text{...(i)}$$

and $\quad \dfrac{x-1}{k} = \dfrac{y-4}{2} = \dfrac{z-5}{1} \quad \text{...(ii)}$

Any point on the line (i) is $P(\lambda + 2, \lambda + 3, -k\lambda + 4)$.

If the lines (i) and (ii) intersect the point P must lies on line (ii).

$\therefore \quad \dfrac{\lambda + 2 - 1}{k} = \dfrac{\lambda + 3 - 4}{2} = \dfrac{-k\lambda + 4 - 5}{1}$

On taking first two terms of above equations

$$\frac{\lambda + 1}{k} = \frac{\lambda - 1}{2}$$

$\Rightarrow \quad k\lambda - k = 2\lambda + 2$

$\Rightarrow \quad \lambda(k - 2) = 2 + k$

$\Rightarrow \quad \lambda = \dfrac{k+2}{k-2} \quad \text{...(iii)}$

On taking last two terms, we get

$$\frac{\lambda - 1}{2} = \frac{-k\lambda - 1}{1} \Rightarrow \lambda - 1 = -2k\lambda - 2$$

$\Rightarrow \quad \lambda(1 + 2k) = -1$

$\Rightarrow \quad \lambda = -\dfrac{1}{(2k+1)} \quad \text{...(iv)}$

From Eqs. (iii) and (iv), we get

$$\frac{k+2}{k-2} = -\frac{1}{2k+1}$$

$\Rightarrow \quad (k + 2)(2k + 1) = -(k - 2)$

$\Rightarrow \quad 2k^2 + k + 4k + 2 = -k + 2$

$\Rightarrow \quad 2k^2 + 6k = 0$

$\Rightarrow \quad k(2k + 6) = 0$

$\Rightarrow \quad k = 0, -3$

(iv) Given equation of planes

$$kx + 4y + z = 0 \quad \text{...(i)}$$

$$4x + ky + 2z = 0 \quad \text{...(ii)}$$

$$2x + 2y + z = 0 \quad \text{...(iii)}$$

Now, equation of plane containing line of intersection of planes (i) and (ii), we get

$$(kx + 4y + z) + \lambda(4x + ky + 2z) = 0$$

$\Rightarrow \quad (k + 4\lambda)x + (4 + k\lambda)y + (1 + 2\lambda)z = 0 \quad \text{...(iv)}$

Since, plane (iii) also contains the line of intersection of planes (i) and (ii).

$\because$ Eqs. (iii) and (iv) are identical.

$\therefore \quad k + 4\lambda = 2 \quad \text{...(v)}$

$4 + k\lambda = 2 \quad \text{...(vi)}$

$1 + 2\lambda = 1 \quad \text{...(vii)}$

From Eq. (vii), $\quad 2\lambda = 0$

$\lambda = 0$

$\therefore \quad k = 2$

Hence, (i) $\to$ (r), (ii) $\to$ (q), (iii) $\to$ (p), (iv) $\to$ (r)

Explanations for Q. Nos. (25-27)

P_1 is the one bisecting plane of AB.

$\Rightarrow \quad 3x - 3y + z - 2 = 0$

and P_2 is $\quad x - y + \lambda z = 1$

P_1 is parallel to P_2

$\Rightarrow \quad \dfrac{3}{1} = \dfrac{1}{\lambda} \Rightarrow \lambda = \dfrac{1}{3}$

Let $\mathbf{x}_1$, $\mathbf{x}_2$ are the normals to P_1, P_2 respectively

$\therefore \quad \mathbf{x}_1 = 3\hat{\mathbf{i}} - 3\hat{\mathbf{j}} + \hat{\mathbf{k}}$

$\mathbf{x}_2 = \hat{\mathbf{i}} - \hat{\mathbf{j}} + \lambda\hat{\mathbf{k}}$

$\therefore \quad \cos\dfrac{\pi}{4} = \left|\dfrac{\mathbf{x}_1 \cdot \mathbf{x}_2}{|\mathbf{x}_1||\mathbf{x}_2|}\right| = \left|\dfrac{3 + 3 + \lambda}{\sqrt{19}\sqrt{2 + \lambda^2}}\right|$

$\Rightarrow \quad 17\lambda^2 - 24\lambda - 34 = 0$

$\therefore \quad \lambda_1\lambda_2 = -2$

Let $\dfrac{x-1}{1} = \dfrac{y+2}{3} = \dfrac{z-7}{-1} = k$

$\Rightarrow \quad x = k + 1, y = 3k - 2, z = -k + 7$

If it lies on P_1, then $k = 2$

$\therefore \alpha = 3, \beta = 4, r = 5$

25. (b) **26.** (d) **27.** (c)

Explanations for Q. Nos. (28 and 29)

If $D = (a_1, a_2, a_3)$, then $a_1 + 1 = 3$, $a_2 + 1 = 3$, $a_3 - 1 = 7$

$$D = (2, 1, 8)$$

Now, distance between AB and $CD = d = \dfrac{|\mathbf{AB} \times \mathbf{AD}|}{|\mathbf{AB}|}$

The vector normal to the plane of parallelogram is $\mathbf{AD} \times \mathbf{AB}$ and $\mathbf{AD} \times \mathbf{AB} = 2(7\hat{\mathbf{i}} + 2\hat{\mathbf{j}} - \hat{\mathbf{k}})$

Hence, projection of XY-plane is 2, YZ-plane is 14 and ZX-plane is 4.

28. (c) **29.** (d)

30. (d) The line $\frac{x-1}{3}=\frac{y-2}{-1}=\frac{z-3}{4}=r$

Any point say $B(3r+1, 2-r, 3+4r)$ (on line L)

$$\mathbf{AB} = 3r, -r, 4r+6$$

Hence, **AB** is parallel to $x+y+z=1$

$$\Rightarrow \quad 3r - r - 4r - 6 = 0$$

$$\Rightarrow \quad r = -3$$

$\therefore$ The coordinates of B is $(-8, 5, -9)$.

31. (b) The equation of plane containing the line L is

$$A(x-1)+B(y-2)+C(z-3)=0. \quad \ldots(\text{i})$$

where $3A - B + 4C = 0$

Plane (i) also contains point $A(1, 2, -3)$.

Hence, $C = 0$ and $3A = B$

The equation of plane $x-1+3(y-2)=0$

or $$x+3y-7=0$$

32. (a) The distance of point from the plane is

$$\frac{|1+3r+2-r-3-4r|}{\sqrt{1+1+1}}=\frac{|2r|}{\sqrt{3}}=\frac{4}{\sqrt{3}}$$

$$\Rightarrow \quad r = \pm 2$$

Hence, the points are $A(7, 0, 11)$ and $B(-5, 4, -5)$.

$$\Rightarrow \quad AB = \sqrt{416} = 4\sqrt{26}$$

33. (6) Let normal to the plane be

$$l\hat{\mathbf{i}} + m\hat{\mathbf{j}} + n\hat{\mathbf{k}}$$

$$2l + 3m + 4n = 0$$

and $$3l + 4m + 5n = 0$$

$$\Rightarrow \quad \frac{l}{-1}=\frac{m}{2}=\frac{n}{-1}$$

Equation of plane will be

$$a(x-1)+b(y-2)+c(z-3)=0$$

$$\Rightarrow \quad -1(x-1)+2(y-2)-1(z-3)=0$$

$$\Rightarrow \quad -x+1+2y-4-z+3=0$$

$$\Rightarrow \quad -x+2y-z=0$$

$$\Rightarrow \quad x-2y+z=0$$

$$\Rightarrow \quad \frac{|d|}{\sqrt{6}}=\sqrt{6}$$

$$\therefore \quad |d| = 6$$

34. (9) Here, $\frac{3\lambda-2\mu+6}{2}=\frac{2\lambda-4\mu+8}{2}=\frac{\lambda-3\mu+4}{1}$

On solving, we get

$$\mu = 0, \lambda = 2$$

Hence, $A(13, 11, 5)$; $B(1, -1, -1)$.

$$\Rightarrow \quad AB = \sqrt{144+144+36}$$

$$= \sqrt{324} = 18$$

$$\therefore \quad \frac{AB}{2} = 9$$

35. (0) Equation of the plane through $P(\alpha, \beta, \gamma)$ at right angles to OP is,

$$\alpha(x-\alpha)+\beta(y-\beta)+\gamma(z-\gamma)=0$$

$$\Rightarrow \quad x\alpha + y\beta + z\gamma = \alpha^2+\beta^2+\gamma^2$$

$$\Rightarrow \quad x\alpha + y\beta + z\gamma = p^2$$

This plane cuts axes at

$A\left(\frac{p^2}{\alpha}, 0, 0\right), B\left(\frac{p^2}{\beta}, 0\right)$ and $C\left(0, 0\, \frac{p^2}{r}\right)$.

Let $\Delta_x, \Delta_y, \Delta_z$ be the projection of ΔABC on YZ, ZX, XY-plane, respectively.

$$\therefore \quad \Delta_x = \frac{p^4}{2\beta\gamma}$$

Then, $$\Delta_y = \frac{p^4}{2\alpha\cdot\gamma}$$

$$\Delta_z = \frac{p^4}{2\alpha\gamma}$$

$$\therefore \quad \text{Area of } \Delta ABC = \sqrt{\Delta_x^2+\Delta_y^2+\Delta_z^2}$$

$$= \frac{p^4}{2}\sqrt{\frac{1}{\beta^2\gamma^2}+\frac{1}{\gamma^2\alpha^2}+\frac{1}{\alpha^2\gamma^2}}$$

$$= \frac{p^4}{2\alpha\beta\gamma}\sqrt{\alpha^2+\beta^2+\gamma^2}$$

$$= \frac{p^5}{2\alpha\beta\gamma} = A$$

$$\Rightarrow \quad \frac{(\alpha\beta\gamma)A}{p^5} = \frac{1}{2}$$

$$\therefore \quad \left[\frac{(\alpha\beta\gamma)A}{p^5}\right] = 0$$

36. (4) Here, $\begin{Vmatrix} 1-1 & 1 & 1 \\ \lambda & 3 & 2 & -3 \\ 3-\lambda & 1 & -2 \end{Vmatrix} = 0$

$$\therefore \quad \Delta_4 = \begin{vmatrix} 1 & -1 & 1 \\ \lambda & 3 & 2 \\ 3 & \lambda & 1 \end{vmatrix} = (\lambda-4)(\lambda+3) \quad \ldots(\text{i})$$

Also, $$\Delta_1 = \begin{vmatrix} -1 & 1 & 1 \\ 3 & 2 & -3 \\ \lambda & 1 & -2 \end{vmatrix} = 5(\lambda-2) \quad \ldots(\text{ii})$$

$$\Delta_2 = \begin{vmatrix} 1 & 1 & 1 \\ \lambda & 2 & -3 \\ 3 & 1 & -2 \end{vmatrix} = (3\lambda-16) \quad \ldots(\text{iii})$$

$$\Delta_3 = \begin{vmatrix} 1 & -1 & 1 \\ \lambda & 3 & -3 \\ 3 & \lambda & -2 \end{vmatrix} = (\lambda+3)(\lambda-2) \quad \ldots(\text{iv})$$

If the given planes form a triangular prism, then we know $\Delta_4 = 0$ and none of $\Delta_1, \Delta_2, \Delta_3$ is zero. Here from Eqs. (i), (ii), (iii) and (iv) we find that if $\lambda = 4$, then $\Delta_4 = 0$ and none of $\Delta_1, \Delta_2, \Delta_3$ is zero.

37. (4) $7x + 6y + 2z = 272$

and $$x - y + z = 16$$

$$\Rightarrow \quad 5x + 8y = 240$$

$$\Rightarrow \quad x = 48 - \frac{8}{5}y$$

Let $y = 5\lambda$, $\lambda \in I$

$$\Rightarrow \quad x = 48 - 8\lambda$$

and $$z = 16 + y - x$$

$$= 13\lambda - 32$$

But $x > 0$, $y > 0$ and $z > 0$

$$\Rightarrow \quad 48 - 8\lambda > 0$$

$$\Rightarrow \quad \lambda < \frac{48}{8}$$

$\Rightarrow \lambda \le 5$ and $13\lambda - 32 > 0$

$$\Rightarrow \quad \lambda > \frac{32}{13}$$

$$\Rightarrow \quad \lambda \ge 3$$

$$\therefore \quad \lambda \in [3, 5]$$

$$\therefore \quad Z_{min} = 39 - 32 = 7$$

$$\Rightarrow \quad x = 24, y = 15$$

$$\Rightarrow \quad x + y + z - 42 = 4$$

38. (6) The above lines in the symmetric form are

$$\frac{x-1}{2}=\frac{y-2}{2}=\frac{z}{4}$$

and $$\frac{x-2}{1}=\frac{y+2}{2}=\frac{z-3}{3}$$

$\therefore$ $$d=\frac{|[\mathbf{a}-\mathbf{c}\ \mathbf{b}\ \mathbf{d}]|}{|\mathbf{b}\times\mathbf{d}|}$$

Here, $$[\mathbf{a}-\mathbf{c}\ \mathbf{b}\ \mathbf{d}]=\begin{vmatrix}1 & -4 & 3\\ 2 & 3 & 4\\ 1 & 2 & 3\end{vmatrix}=12$$

$\therefore$ $$d=\frac{12}{\sqrt{1+4+1}}=2\sqrt{6}$$

$\Rightarrow$ $$\frac{d^2}{4}=6$$

39. (2) Given equation of line

$$\frac{x-2}{3}=\frac{y+1}{2}=\frac{z-1}{-1} \quad \text{...(i)}$$

Since, line (i) intersects the curve

$$xy=c^2 \quad \text{...(ii)}$$

In XY-plane, put $z=0$ in Eq. (i), we get

$$x=5,\ y=1$$

Now, substituting the values of x and y in Eq. (ii), we get

$$5\times 1=c^2 \Rightarrow c^2=5$$

$\Rightarrow$ $c=\pm\sqrt{5}$

$\Rightarrow$ $|c|=\sqrt{5}$

$\therefore$ $[|c|]=[\sqrt{5}]=2$

40. (2) Let the edges OA, OB and OC of the unit cube be along OX, OY and OZ, respectively.

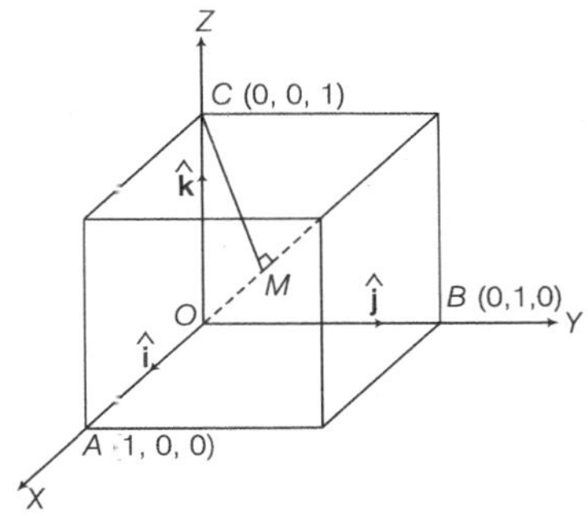

Since, $OA=OB=OC=1$ unit

$\Rightarrow$ $\mathbf{OA}=\hat{\mathbf{i}},\ \mathbf{OB}=\hat{\mathbf{j}},\ \mathbf{OC}=\hat{\mathbf{k}}$

Let CM be the perpendicular from the centre C on the diagonal OP. The vector equation of OP is $\mathbf{r}=\lambda(\hat{\mathbf{i}}+\hat{\mathbf{j}}+\hat{\mathbf{i}})$

$\therefore$ $\mathbf{OM}=$ projection of $\mathbf{OC}$ on $\mathbf{OP}=\mathbf{OC}\cdot\dfrac{\mathbf{OP}}{|\mathbf{OP}|}$

$$=\hat{\mathbf{k}}\cdot\frac{(\hat{\mathbf{i}}+\hat{\mathbf{j}}+\hat{\mathbf{k}})}{\sqrt{3}}=\frac{1}{\sqrt{3}}$$

Now, $|\mathbf{OC}|^2=|\mathbf{OM}|^2+|\mathbf{CM}|^2$

$\Rightarrow$ $|\mathbf{CM}|^2=|\mathbf{OC}|^2-|\mathbf{OM}|^2=1-\dfrac{1}{3}$

$\therefore$ $d^2=\dfrac{2}{3}\Rightarrow 3d^2=2$

JEE Main & AIEEE Archive

1. (d) Given equation of line is

$$\frac{x-2}{3}=\frac{y+1}{4}=\frac{z-2}{12}=\lambda \quad \text{(say) ...(i)}$$

and equation of plane is

$$x-y+z=16 \quad \text{...(ii)}$$

Any point on the line (i) is $(3\lambda+2,\ 4\lambda-1,\ 12\lambda+2)$

Let this point be point of intersection of the line and plane.

$\therefore$ $(3\lambda+2)-(4\lambda-1)+(12\lambda+2)=16$

$\Rightarrow$ $11\lambda+5=16$

$\Rightarrow$ $11\lambda=11$

$\Rightarrow$ $\lambda=1$

$\therefore$ Point of intersection is (5, 3, 14).

Now, distance between the points (1, 0, 2) and (5, 3, 14).

$$=\sqrt{(5-1)^2+(3-0)^2+(14-2)^2}$$

$$=\sqrt{16+9+144}=\sqrt{169}=13$$

2. (c) Let equation of plane containing the lines

$$2x-5y+z=3$$

and $$x+y+4z=5$$

be $(2x-5y+z-3)+\lambda(x+y+4z-5)=0$

$\Rightarrow$ $(2+\lambda)x+(\lambda-5)y+(4\lambda+1)z-3-5\lambda=0 \quad \text{...(i)}$

This plane is parallel to the plane $x+3y+6z=1$.

$\therefore$ $$\frac{2+\lambda}{1}=\frac{\lambda-5}{3}=\frac{4\lambda+1}{6}$$

On taking first two equalities, we get

$$6+3\lambda=\lambda-5$$

$\Rightarrow$ $2\lambda=-11$

$\Rightarrow$ $\lambda=-\dfrac{11}{2}$

On taking last two equalities, we get

$$6\lambda-30=3+12\lambda$$

$\Rightarrow$ $-6\lambda=33$

$\Rightarrow$ $\lambda=-\dfrac{11}{2}$

So, the equation of required plane is

$$\left(2-\frac{11}{2}\right)x+\left(\frac{-11}{2}-5\right)y+\left(-\frac{44}{2}+1\right)z-3+5\times\frac{11}{2}=0$$

$\Rightarrow$ $-\dfrac{7}{2}x-\dfrac{21}{2}y-\dfrac{42}{2}z+\dfrac{49}{2}=0$

$\Rightarrow$ $x+3y+6z-7=0$

3. (a) Given, $l+m+n=0$

$\Rightarrow$ $l=-(m+n)$

$\Rightarrow$ $(m+n)^2=l^2$

$\Rightarrow$ $m^2+n^2+2mn=m^2+n^2$ $\quad [\because l^2=m^2+n^2]$

$\Rightarrow$ $2mn=0$

Case I When $m=0$, then

$$l=-n$$

Hence, (l, m, n) is $(1, 0, -1)$.

Case II When $n=0$, then

$$l=-m$$

Hence, (l, m, n) is $(1, -1, 0)$.

$\therefore$ $$\cos\theta=\frac{1+0+0}{\sqrt{2}\times\sqrt{2}}=\frac{1}{2}$$

$\Rightarrow$ $\theta=\dfrac{\pi}{3}$

4. (a) Plane and line are parallel to each other. Equation of normal to the plane through the point (1, 3, 4) is

$$\frac{x-1}{2}=\frac{y-3}{-1}=\frac{z-4}{1}=k \quad \text{[say]}$$

Any point in this normal is $(2k+1, -k+3, 4+k)$.

$$\Rightarrow \left(\frac{2k+1+1}{2}, \frac{3-k+3}{2}, \frac{4+k+4}{2}\right) \text{ lies on}$$

$$\Rightarrow \quad 2(k+1)-\left(\frac{6-k}{2}\right)+\left(\frac{8+k}{2}\right)+3=0$$

$$\Rightarrow \quad k=-2$$

Hence, the point through which this image pass is

$$(2k+1, 3-k, 4+k)$$

i.e. $[2(-2)+1, 3+2, 4, -2]=(-3, 5, 2)$

Hence, equation of image line is

$$\frac{x+3}{3}=\frac{y-5}{1}=\frac{z-2}{-5}.$$

5. (c) Given planes are

$$2x+y+2z-8=0$$

and

$$2x+y+2z+\frac{5}{2}=0$$

Distance between two planes

$$=\frac{|c_1-c_2|}{\sqrt{a^2+b^2+c^2}}$$

$$=\left|\frac{-8-\frac{5}{2}}{\sqrt{2^2+1^2+2^2}}\right|$$

$$=\frac{\frac{21}{2}}{3}=\frac{7}{2}$$

6. (c) The given lines are

$$\frac{x-2}{1}=\frac{y-3}{1}=\frac{z-4}{-k} \quad \ldots\text{(i)}$$

and

$$\frac{x-1}{k}=\frac{y-4}{2}=\frac{z-5}{1} \quad \ldots\text{(ii)}$$

Condition for two lines are coplanar.

$$\begin{vmatrix} x_1-x_2 & y_1-y_2 & z_1-z_2 \\ l_1 & m_1 & n_1 \\ l_2 & m_2 & n_2 \end{vmatrix}=0$$

where, (x_1, y_1, z_1) and (x_2, y_2, z_2) are any points on the lines (i) and (ii), respectively and $< l_1, m_1, n_1 >$ and $< l_2, m_2, n_2 >$ are direction cosines of lines (i) and (ii), respectively.

$$\therefore \quad \begin{vmatrix} 2-1 & 3-4 & 4-5 \\ 1 & 1 & -k \\ k & 2 & 1 \end{vmatrix}=0$$

$$\Rightarrow \quad \begin{vmatrix} 1 & -1 & -1 \\ 1 & 1 & -k \\ k & 2 & 1 \end{vmatrix}=0$$

$$\Rightarrow \quad 1(1+2k)+(1+k^2)-(2-k)=0$$

$$\Rightarrow \quad k^2+2k+k=0$$

$$\Rightarrow \quad k^2+3k=0$$

$$\Rightarrow \quad k=0, -3$$

Note *If 0 appears in the denominator, then the correct way of representing the equation of straight line is*

$$\frac{x-2}{1}=\frac{y-3}{1}; z=4$$

and

$$x=1; \frac{y-4}{2}=\frac{z-5}{1}.$$

7. (a) **Given** A plane $P: x-2y+2z-5=0$

To find The equation of a plane parallel to given plane P and at a distance of 1 unit from origin. Equation of family of plane parallel to the given plane P is

$$Q: x-2y+2z+d=0$$

Also, perpendicular distance of Q from origin is 1 unit.

$$\Rightarrow \quad \left|\frac{0-2(0)+2(0)+d}{\sqrt{1^2+2^2+2^2}}\right|=1$$

$$\Rightarrow \quad \left|\frac{d}{3}\right|=1$$

$$\Rightarrow \quad d=\pm 3$$

Hence, the required equation of the plane parallel to P and at unit distance from origin is

$$x-2y+2z\pm 3=0$$

Hence, out of the given equations, option (a) is the only correct.

8. (c) **Given** Two lines $L_1: \frac{x-1}{2}=\frac{y+1}{3}=\frac{z-1}{4}$

and $L_2: \frac{x-3}{1}=\frac{y-k}{2}=\frac{z-0}{1}$

To find The value of k of the given lines L_1 and L_2 are intersecting each other.

Let $L_1: \frac{x-1}{2}=\frac{y+1}{3}=\frac{z-1}{4}=p$

and $L_2: \frac{x-3}{1}=\frac{y-k}{2}=\frac{z-0}{1}=q$

$\Rightarrow$ Any point P on line L_1 is of type $P(2p+1, 3p-1, 4p+1)$ and any point Q on line L_2 is of type $Q(q+3, 2q+k, q)$.

Since, L_1 and L_2 are intersecting each other, hence both points P and Q should coincide at the point of intersection, i.e. corresponding coordinates of P and Q should be same.

$2p+1=q+3$, $3p-1=2q+k$ and $4p+1=q$

On solving $2p+1=q+3$ and $4p+1=q$, we get the values of p and q as

$$p=\frac{-3}{2}$$

and

$$q=-5$$

On substituting the values of p and q in the third equation $3p-1=2q+k$, we get

$$3\left(\frac{-3}{2}\right)-1=2(-5)+k$$

$$\therefore \quad k=\frac{9}{2}$$

9. (d) Angle between straight line $\mathbf{r}=\mathbf{a}+\lambda\mathbf{b}$ and plane $\mathbf{r}\cdot\hat{\mathbf{n}}=\mathbf{d}$ is

$$\sin\theta=\frac{\mathbf{b}\cdot\hat{\mathbf{n}}}{|\mathbf{b}|\,|\hat{\mathbf{n}}|}$$

$$\therefore \quad \sin\theta=\frac{(\hat{\mathbf{i}}+2\hat{\mathbf{j}}+\lambda\hat{\mathbf{k}})\cdot(\hat{\mathbf{i}}+2\hat{\mathbf{j}}+3\hat{\mathbf{k}})}{\sqrt{1+4+\lambda^2}\sqrt{1+4+9}}$$

$$\Rightarrow \quad \sin\theta=\frac{5+3\lambda}{\sqrt{\lambda^2+5}\cdot\sqrt{14}}$$

Given, $\cos\theta=\sqrt{\frac{5}{14}}$

$$\therefore \quad \sin\theta=\frac{3}{\sqrt{14}} \Rightarrow \frac{3}{\sqrt{14}}=\frac{5+3\lambda}{\sqrt{\lambda^2+5}\cdot\sqrt{14}}$$

$$\Rightarrow \quad 9(\lambda^2+5)=9\lambda^2+30\lambda+25$$

$$\Rightarrow \quad 9\lambda^2+45=9\lambda^2+30\lambda+25 \Rightarrow 30\lambda=20$$

$$\therefore \quad \lambda=\frac{2}{3}$$

10. (a) Mid-point of AB is M (1, 3, 5)

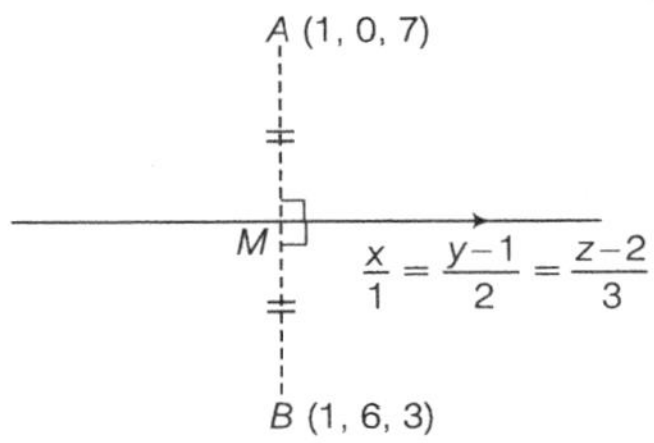

which lies on $\quad \dfrac{x}{1} = \dfrac{y-1}{2} = \dfrac{z-2}{3}$

as $\quad \dfrac{1}{1} = \dfrac{3-1}{2} = \dfrac{5-2}{3} \Rightarrow 1 = 1 = 1$

Hence, Statement II is correct.

Also, direction ratios of AB is given by $(1-1, 6-0, 3-7)$

i.e. $\quad (0, 6, -4) \quad \ldots$(i)

And direction ratios of straight line is $(1, 2, 3) \quad \ldots$(ii)

The two lines are perpendicular, if

$$0\,(1) + 6\,(2) - 4\,(3) = 12 - 12 = 0$$

Hence, Statement I is correct and Statement II is a correct explanation of Statement I.

11. (d) Let the coordinates of Q be $(2\lambda, 3\lambda + 2, 4\lambda + 3)$, which is any point on the straight line AB.

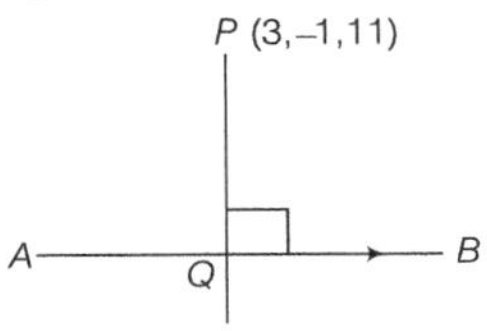

$\therefore$ DR's of PQ is $(2\lambda - 3, 3\lambda + 3, 4\lambda - 8)$.

Also, perpendicular to straight line AB

$$\frac{x}{2} = \frac{y-2}{3} = \frac{z-3}{4} = \lambda$$

having DR's (2, 3, 4).

Thus, $\quad 2\,(2\lambda - 3) + 3\,(3\lambda + 3) + 4\,(4\lambda - 8) = 0$

$\Rightarrow \quad 4\lambda - 6 + 9\lambda + 9 + 16\lambda - 32 = 0$

$\Rightarrow \quad 29\lambda - 29 = 0$

$\therefore \quad \lambda = 1$

Hence, coordinates of Q are (2, 5, 7).

$\therefore \quad |PQ| = \sqrt{(3-2)^2 + (-1-5)^2 + (7-11)^2}$

$\quad = \sqrt{1 + 36 + 16} = \sqrt{53}$

12. (b) Let Q be any point on the plane.

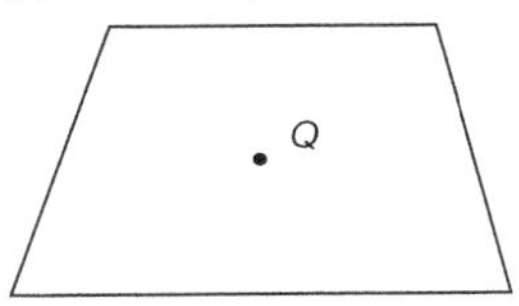

Then, equation of PQ is

$$\frac{x-1}{1} = \frac{y+5}{1} = \frac{z-9}{1} = \lambda,$$

where $P = (1, -5, 9)$.

$\therefore \quad x = \lambda + 1,\ y = \lambda - 5,\ z = \lambda + 9$ lies on the plane

$$x - y + z = 5$$

$\Rightarrow \lambda + 1 - \lambda + 5 + \lambda + 9 = 5$

$\therefore \quad \lambda = -10$

Hence, coordinates of Q are $Q\,(-9, -15, -1)$.

$\therefore \quad |PQ| = \sqrt{(10)^2 + (10)^2 + (10)^2} = 10\sqrt{3}$

13. (c) We know that,

$$\cos^2 45° + \cos^2 120° + \cos^2\theta = 1$$

$\Rightarrow \quad \dfrac{1}{2} + \dfrac{1}{4} + \cos^2\theta = 1$

$\Rightarrow \quad \cos^2\theta = \dfrac{1}{4} \Rightarrow \cos\theta = \pm\dfrac{1}{2}$

$\therefore \quad \theta = 60°$ or $120°$

14. (a) The image of the point (3, 1, 6) with respect to the plane $x - y + z = 5$ is

$$\frac{x-3}{1} = \frac{y-1}{-1} = \frac{z-6}{1} = \frac{-2\,(3-1+6-5)}{1+1+1}$$

$$\left[\because \frac{x-x_1}{a} = \frac{y-y_1}{b} = \frac{z-z_1}{c} = \frac{-2(ax_1 + by_1 + cz_1 + d)}{a^2 + b^2 + c^2}\right]$$

$\Rightarrow \quad \dfrac{x-3}{1} = \dfrac{y-1}{-1} = \dfrac{z-6}{1} = -2$

$\Rightarrow \quad x = 3 - 2 = 1,$

$\quad y = 1 + 2 = 3$

and $\quad z = 6 - 2 = 4,$

which shows that Statement I is correct.

We observe that the line segment joining the points $A(3, 1, 6)$ and $B(1, 3, 4)$ has direction ratios $2, -2, 2$ which are proportional to $1, -1, 1$, the direction ratios of the normal to the plane.

Hence, Statement I is correct. Thus, both Statements are correct and Statement II is a correct explanation of Statement I.

15. (b) DR's of given line are $(3, -5, 2)$.

DR's of normal to the plane $= (1, 3, -\alpha)$

$\therefore$ Line is perpendicular to the normal

$$3(1) - 5(3) + 2(-\alpha) = 0$$

$\Rightarrow \quad 3 - 15 - 2\alpha = 0$

$\Rightarrow \quad 2\alpha = -12$

$\Rightarrow \quad \alpha = -6$

Also, point $(2, 1, -2)$ lies on the plane.

$\therefore \quad 2 + 3 + 6(-2) + \beta = 0$

$\Rightarrow \quad \beta = 7$

$\Rightarrow \quad (\alpha, \beta) = (-6, 7)$

16. (c) Projection of a vector on coordinate axes are

$$x_2 - x_1,\ y_2 - y_1,\ z_2 - z_1$$

$\Rightarrow \quad x_2 - x_1 = 6,\ y_2 - y_1 = -3,\ z_2 - z_1 = 2$

Now, $\sqrt{(x_2 - x_1)^2 + (y_2 - y_1)^2 + (z_2 - z_1)^2} = \sqrt{36 + 9 + 4} = 7$

So, the DC's of the vector are $\dfrac{6}{7}, -\dfrac{3}{7}$ and $\dfrac{2}{7}$.

17. (d) Equation of line passing through $(5, 1, a)$ and $(3, b, 1)$ is

$$\frac{x-3}{5-3} = \frac{y-b}{1-b} = \frac{z-1}{a-1} \quad \ldots\text{(i)}$$

$$\left[\because \frac{x-x_1}{x_2-x_1} = \frac{y-y_1}{y_2-y_1} = \frac{z-z_1}{z_2-z_1}\right]$$

Point $\left(0, \dfrac{17}{2}, -\dfrac{13}{2}\right)$ satisfies Eq. (i), we get

$$-\frac{3}{2} = \frac{\frac{17}{2} - b}{1-b} = \frac{-\frac{13}{2} - 1}{a-1}$$

$\Rightarrow \quad a - 1 = \dfrac{\left(-\dfrac{15}{2}\right)}{\left(-\dfrac{3}{2}\right)} = 5 \Rightarrow a = 6$

Also, $\quad -3(1-b) = 2\left(\dfrac{17}{2} - b\right)$

$\Rightarrow \quad 3b - 3 = 17 - 2b \Rightarrow 5b = 20$

$\therefore \quad b = 4$

18. (b) Given, $\dfrac{x-1}{k} = \dfrac{y-2}{2} = \dfrac{z-3}{3}$ ...(i)

and $\dfrac{x-2}{3} = \dfrac{y-3}{k} = \dfrac{z-1}{2}$ ...(ii)

Since, lines intersect at a point. Then, shortest distance between them is zero.

$$\therefore \quad \begin{vmatrix} k & 2 & 3 \\ 3 & k & 2 \\ 1 & 1 & -2 \end{vmatrix} = 0$$

$\Rightarrow \quad k(-2k-2) - 2(-6-2) + 3(3-k) = 0$

$\Rightarrow \quad -2k^2 - 5k + 25 = 0$

$\Rightarrow \quad 2k^2 + 5k - 25 = 0$

$\Rightarrow \quad 2k^2 + 10k - 5k - 25 = 0$

$\Rightarrow \quad 2k(k+5) - 5(k+5) = 0$

$\Rightarrow \quad k = \dfrac{5}{2}, -5$

Hence, integer value of k is -5.

19. (a) Let the direction cosines of line L be l, m and n. Since, the line intersect the given planes, then the normal to the planes are perpendicular to the line L.

$\therefore \quad 2l + 3m + n = 0$ ...(i)

and $\quad l + 3m + 2n = 0$ ...(ii)

From Eqs. (i) and (ii), we get

$$\frac{l}{3} = \frac{m}{-3} = \frac{n}{3} = k \quad \text{[say]}$$

We know that, $l^2 + m^2 + n^2 = 1$

$\therefore \quad (3k)^2 + (-3k)^2 + (3k)^2 = 1$

$\Rightarrow \quad 27k^2 = 1$

$\Rightarrow \quad k = \dfrac{1}{3\sqrt{3}}$

$\therefore \quad l = \dfrac{1}{\sqrt{3}}$

$\Rightarrow \quad \cos \alpha = \dfrac{1}{\sqrt{3}}$

20. (d) Since, a line makes an angle of $\dfrac{\pi}{4}$ with positive directions of each of X-axis and Y-axis, therefore

$$\alpha = \frac{\pi}{4}, \quad \beta = \frac{\pi}{4}$$

We know that,

$$\cos^2 \alpha + \cos^2 \beta + \cos^2 \gamma = 1$$

$\therefore \quad \cos^2 \dfrac{\pi}{4} + \cos^2 \dfrac{\pi}{4} + \cos^2 \gamma = 1$

$\Rightarrow \quad \dfrac{1}{2} + \dfrac{1}{2} + \cos^2 \gamma = 1$

$\Rightarrow \quad \cos^2 \gamma = 0$

$\therefore \quad \gamma = \dfrac{\pi}{2}$

21. (a) Given equation of sphere

$$x^2 + y^2 + z^2 - 6x - 12y - 2z + 20 = 0$$

whose coordinates of centre are (3, 6, 1).

Since, one end of diameter is (2, 3, 5) and let the other end of diameter be (α, β, γ), then

$$\frac{\alpha+2}{2} = 3, \quad \frac{\beta+3}{2} = 6, \quad \frac{\gamma+5}{2} = 1$$

$\Rightarrow \quad \alpha = 4, \quad \beta = 9 \quad \text{and} \quad \gamma = -3$

Hence, the coordinates of other points is (4, 9, – 3).

22. (d) Given equations can be rewritten as

$$\frac{x-b}{a} = \frac{y-0}{1} = \frac{z-d}{c}$$

and $\dfrac{x-b'}{a'} = \dfrac{y-0}{1} = \dfrac{z-d'}{c'}$

These lines will be perpendicular, if

$$aa' + cc' + 1 = 0 \quad [\because l_1l_2 + m_1m_2 + n_1n_2 = 0]$$

23. (d) We know that, the image (x, y, z) of a point (x_1, y_1, z_1) in a plane $ax + by + cz + d = 0$ is given by

$$\frac{x-x_1}{a} = \frac{y-y_1}{b} = \frac{z-z_1}{c} = \frac{-2(ax_1 + by_1 + cz_1 + d)}{a^2+b^2+c^2}$$

Thus, the image of point (–1, 3, 4) in a plane $x - 2y = 0$ is given by

$$\frac{x+1}{1} = \frac{y-3}{-2} = \frac{z-4}{0} = \frac{-2[1\times(-1) + (-2)\times 3 + 0 \times 4]}{1+4}$$

$\Rightarrow \quad \dfrac{x+1}{1} = \dfrac{y-3}{-2} = \dfrac{z-4}{0} = \dfrac{-2(-7)}{5}$

$\Rightarrow \quad x = \dfrac{14}{5} - 1 = \dfrac{9}{5}, \; y = -\dfrac{28}{5} + 3 = -\dfrac{13}{5}$ and $z = 4$

Hence, the image of point (–1, 3, 4) is $\left(\dfrac{9}{5}, -\dfrac{13}{5}, 4\right)$.

24. (b) Equation of given spheres are

$$x^2 + y^2 + z^2 + 6x - 8y - 2z = 13 \quad \text{...(i)}$$

and $x^2 + y^2 + z^2 - 10x + 4y - 2z = 8$ ...(ii)

whose centres are (– 3, 4, 1) and (5, – 2, 1).

Mid-point of (– 3, 4, 1) and (5, – 2, 1) is (1, 1, 1).

Since, the given plane $2ax - 3ay + 4az + 6 = 0$ passes through (1, 1, 1).

$\therefore \quad 2a - 3a + 4a + 6 = 0$

$\Rightarrow \quad 3a = -6$

$\Rightarrow \quad a = -2$

25. (d) Direction ratios of line are

$$(a_1, b_1, c_1) = (1, 2, 2)$$

And direction ratios of a plane are

$$(a_2, b_2, c_2) = (2, -1, \sqrt{\lambda})$$

$$\therefore \quad \sin\theta = \frac{a_1a_2 + b_1b_2 + c_1c_2}{\sqrt{a_1^2 + b_1^2 + c_1^2}\sqrt{a_2^2 + b_2^2 + c_2^2}}$$

$$= \frac{1\times 2 + 2(-1) + 2\times\sqrt{\lambda}}{\sqrt{1^2+2^2+2^2}\sqrt{2^2+1^2+(\sqrt{\lambda})^2}}$$

$\Rightarrow \quad \dfrac{1}{3} = \dfrac{2\sqrt{\lambda}}{3\sqrt{5+\lambda}} = 5 + \lambda = 4\lambda$

$\Rightarrow \quad \lambda = \dfrac{5}{3}$

26. (c) The given equations of lines can be rewritten as

$$\frac{x}{3} = \frac{y}{2} = \frac{z}{-6} \quad \text{and} \quad \frac{x}{2} = \frac{y}{-12} = \frac{z}{-3}$$

$\therefore$ Angle between the lines is

$$\theta = \cos^{-1}\left(\frac{3\times 2 + 2(-12) - 6(-3)}{\sqrt{3^2+2^2+(-6)^2}\sqrt{(2)^2+(-12)^2+(-3)^2}}\right)$$

$$\left[\because \cos\theta = \frac{a_1\cdot a_2 + b_2\cdot b_2 + c_1\cdot c_2}{\sqrt{a_1^2+b_1^2+c_1^2}\sqrt{a_2^2+b_2^{\,2}+c_2^2}}\right]$$

$$= \cos^{-1}(0) = 90^\circ$$

27. (c) Since, the centre of sphere $x^2 + y^2 + z^2 - x + z - 2 = 0$ is $\left(\frac{1}{2}, 0, -\frac{1}{2}\right)$ and radius $= \sqrt{\frac{1}{4} + \frac{1}{4} + 2} = \frac{\sqrt{10}}{2}$

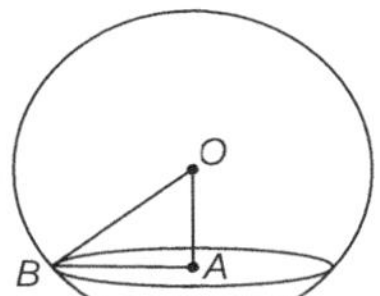

Distance from centre of sphere to the given plane

$$OA = \left|\frac{\frac{1}{2} + \frac{1}{2} - 4}{\sqrt{1 + 4 + 1}}\right| = \frac{3}{\sqrt{6}}$$

So, radius of circle,

$$AB = \sqrt{OB^2 - OA^2} = \sqrt{\frac{10}{4} - \frac{9}{6}} = \sqrt{\frac{30 - 18}{12}} = \sqrt{\frac{12}{12}} = 1$$

28. (c) A line makes angle θ with X-axis and Z-axis and β with Y-axis.

$\therefore \quad l = \cos\theta, \quad m = \cos\beta, \quad n = \cos\theta$

$\because \quad l^2 + m^2 + n^2 = 1$

$\therefore \quad \cos^2\theta + \cos^2\beta + \cos^2\theta = 1$

$\Rightarrow \quad 2\cos^2\theta = 1 - \cos^2\beta$

$\Rightarrow \quad 2\cos^2\theta = \sin^2\beta \quad \ldots(i)$

But it is given that, $\sin^2\beta = 3\sin^2\theta \quad \ldots(ii)$

From Eqs. (i) and (ii), we get

$$3\sin^2\theta = 2\cos^2\theta$$

$\Rightarrow \quad 3(1 - \cos^2\theta) = 2\cos^2\theta$

$\Rightarrow \quad 3 = 5\cos^2\theta$

$\therefore \quad \cos^2\theta = \frac{3}{5}$

29. (c) We know that, the distance between two parallel planes $ax + by + cz + d_1 = 0$ and $ax + by + cz + d_2 = 0$ is

$$\left|\frac{d_2 - d_1}{\sqrt{a^2 + b^2 + c^2}}\right|$$

Therefore, the distance between $4x + 2y + 4z - 16 = 0$ and $4x + 2y + 4z + 5 = 0$ is

$$\left|\frac{5 + 16}{\sqrt{16 + 4 + 16}}\right| = \left|\frac{21}{\sqrt{36}}\right| = \frac{21}{6} = \frac{7}{2}.$$

30. (b) Let the equation of line AB be

$$\frac{x - 0}{1} = \frac{y + a}{1} = \frac{z = 0}{1} = k \quad \text{[say]}$$

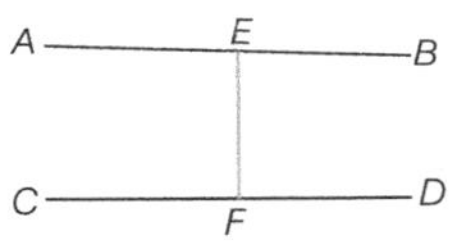

Any point on the line is $F(k, k - a, k)$.

Also, the equation of other line CD is

$$\frac{x + a}{2} = \frac{y - 0}{1} = \frac{z - 0}{1} = \lambda \quad \text{[say]}$$

Any point on the line is $E(2\lambda - a, \lambda, \lambda)$.

Direction ratios of EF are $[(k - 2\lambda + a), (k - a - \lambda), (k - \lambda)]$.

Since, it is given that direction ratios of EF are proportional to 2, 1, 2.

$$\therefore \quad \frac{k - 2\lambda + a}{2} = \frac{k - \lambda - a}{1} = \frac{k - \lambda}{2}$$

On solving first and second fractions, we get

$$k - 2\lambda + a = 2k - 2\lambda - 2a$$

$\Rightarrow \quad k = 3a \quad \ldots(i)$

On solving second and third fractions, we get

$$2k - 2\lambda - 2a = k - \lambda$$

$\Rightarrow \quad k - \lambda = 2a$

$\Rightarrow \quad \lambda = 3a - 2a \quad$ [from Eq. (i)]

$\therefore \quad \lambda = a$

Hence, coordinates of E are $(3a, 2a, 3a)$ and coordinates of F are (a, a, a).

31. (a) The given straight line can be rewritten as

$$\frac{x - 1}{1} = \frac{y + 3}{-\lambda} = \frac{z - 1}{\lambda} = s$$

and $\quad \frac{x - 0}{1} = \frac{y - 1}{2} = \frac{z - 2}{-2} = t$

These two lines are coplanar, if

$$\begin{vmatrix} x_1 - x_2 & y_1 - y_2 & z_1 - z_2 \\ l_1 & m_1 & n_1 \\ l_2 & m_2 & n_2 \end{vmatrix} = 0$$

$$\Rightarrow \quad \begin{vmatrix} 1 - 0 & -3 - 1 & 1 - 2 \\ 1 & -\lambda & \lambda \\ 1 & 2 & -2 \end{vmatrix} = 0$$

$$\Rightarrow \quad \begin{vmatrix} 1 & -4 & -1 \\ 1 & -\lambda & \lambda \\ 1 & 2 & -2 \end{vmatrix} = 0$$

$\Rightarrow \quad 1(2\lambda - 2\lambda) + 4(-2 - \lambda) - 1(2 + \lambda) = 0$

$\Rightarrow \quad -8 - 4\lambda - 2 - \lambda = 0$

$\Rightarrow \quad -10 = 5\lambda \Rightarrow \lambda = -2$

32. (d) Given equation of two spheres are

$$S \equiv x^2 + y^2 + z^2 + 7x - 2y - z - 13 = 0$$

and $S' \equiv x^2 + y^2 + z^2 - 3x + 3y + 4z - 8 = 0$

If these spheres intersect, then $S - S' = 0$ represents the equation of common plane of intersection.

$\therefore \quad (x^2 + y^2 + z^2 + 7x - 2y - z - 13)$

$\quad - (x^2 + y^2 + z^2 - 3x + 3y + 4z - 8) = 0$

$\Rightarrow \quad 10x - 5y - 5z - 5 = 0$

$\Rightarrow \quad 2x - y - z = 1$

33. (c) Given equation of sphere is

$$x^2 + y^2 + z^2 - 2x - 2y - 4z - 19 = 0$$

whose centre is $(-1, 1, 2)$

and radius $= \sqrt{(-1)^2 + (1)^2 + (2)^2 + 19} = \sqrt{25} = 5$

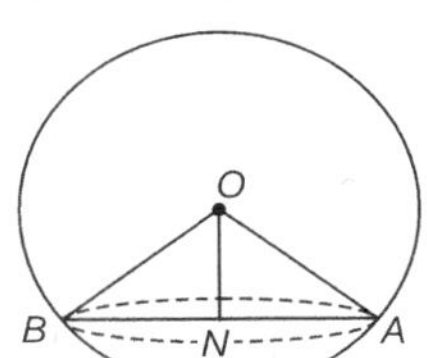

Also, equation of plane is

$$x + 2y + 2z + 7 = 0$$

Length of the perpendicular from centre O on the plane is

$$ON = \frac{-1 \times 1 + 1 \times 2 + 2 \times 2 + 7}{\sqrt{1^2 + 2^2 + 2^2}} = \frac{12}{3} = 4$$

In ΔOAN, $5^2 = 4^2 + AN^2$ [by Pythagoras theorem]

$\Rightarrow \quad AN^2 = 25 - 16 \Rightarrow AN^2 = 9$

$\therefore \quad AN = 3$

34. (c) Given lines $\frac{x-2}{1}=\frac{y-3}{1}=\frac{z-4}{-k}$ ...(i)

and $\frac{x-1}{k}=\frac{y-4}{2}=\frac{z-5}{1}$ ...(ii)

are coplanar.

$$\therefore \begin{vmatrix} -1 & 1 & 1 \\ 1 & 1 & -k \\ k & 2 & 1 \end{vmatrix} = 0 \quad \left[\because \begin{vmatrix} x_2-x_1 & y_2-y_1 & z_2-z_1 \\ l_1 & m_1 & n_1 \\ l_2 & m_2 & n_2 \end{vmatrix} = 0\right]$$

$$\Rightarrow -1(1+2k)-1(1+k^2)+1(2-k)=0$$

$$\Rightarrow -2k-1-1-k^2+2-k=0$$

$$\Rightarrow -k^2-3k=0$$

$$\Rightarrow k=0 \text{ or } -3$$

35. (c) Given equations of sphere and plane are

$$x^2+y^2+z^2+4x-2y-6z-155=0$$

and $$12x+4y+3z-327=0$$

Centre of the sphere is $(-2, 1, 3)$ and radius

$$=\sqrt{4+1+9+155}=\sqrt{169}=13$$

Length of the perpendicular from centre to the plane

$$=\left|\frac{-2\times 12+1\times 4+3\times 3-327}{\sqrt{(12)^2+(4)^2+3^2}}\right|$$

$$\left[\because P=\left|\frac{ax_1+by_1+cz_1+d}{\sqrt{a^2+b^2+c^2}}\right|\right]$$

$$=\left|\frac{-11-327}{\sqrt{169}}\right|=\frac{338}{13}=26$$

$\therefore$ Shortest distance between the plane and sphere

= 26 – Radius of sphere

= 26 – 13 = 13

36. (d) Consider OX, OY, OZ and Ox, Oy, Oz are two systems of rectangular axes.

Let their corresponding equations of plane be

$$\frac{x}{a}+\frac{y}{b}+\frac{z}{c}=1 \quad \text{...(i)}$$

and $$\frac{x}{a'}+\frac{y}{b'}+\frac{z}{c'}=1 \quad \text{...(ii)}$$

Length of perpendicular from origin to Eqs. (i) and (ii) must be same.

$$\therefore \frac{1}{\sqrt{\frac{1}{a^2}+\frac{1}{b^2}+\frac{1}{c^2}}}=\frac{1}{\sqrt{\frac{1}{a'^2}+\frac{1}{b'^2}+\frac{1}{c'^2}}}$$

$$\Rightarrow \frac{1}{a^2}+\frac{1}{b^2}+\frac{1}{c^2}=\frac{1}{a'^2}+\frac{1}{b'^2}+\frac{1}{c'^2}$$

$$\Rightarrow \frac{1}{a^2}+\frac{1}{b^2}+\frac{1}{c^2}-\frac{1}{a'^2}-\frac{1}{b'^2}-\frac{1}{c'^2}=0$$

37. (a) A parallelopiped is formed by planes drawn through the points (2, 3, 5) and (5, 9, 7), parallel to the coordinate planes.

Let a, b and c be the lengths of edges, then

$$a=5-2=3$$

$$b=9-3=6$$

and $$c=7-5=2$$

So, the length of diagonal of a parallelopiped

$$=\sqrt{a^2+b^2+c^2}=\sqrt{9+36+4}=\sqrt{49}=7 \text{ units}$$

38. (b) The equation of plane containing the line $\frac{x-x_1}{l}=\frac{y-y_1}{m}=\frac{z-z_1}{n}$ is

$$a(x-x_1)+b(y-y_1)+c(z-z_1)=0$$

Since, the normal to the plane is perpendicular to the given line.

$$\therefore al+bm+cn=0$$

39. (b) The equation of a line through the centre $\hat{\mathbf{j}}+2\hat{\mathbf{k}}$ and normal to the given plane is

$$\mathbf{r}=\hat{\mathbf{j}}+2\hat{\mathbf{k}}+\lambda(\hat{\mathbf{i}}+2\hat{\mathbf{j}}+2\hat{\mathbf{k}}) \quad \text{...(i)}$$

This meets the plane at a point for which we must have

$$[(\hat{\mathbf{j}}+2\hat{\mathbf{k}})+\lambda(\hat{\mathbf{i}}+2\hat{\mathbf{j}}+2\hat{\mathbf{k}})]\cdot(\hat{\mathbf{i}}+2\hat{\mathbf{j}}+2\hat{\mathbf{k}})=15$$

$$\Rightarrow 2(1+2\lambda)+\lambda+2(2+2\lambda)=15$$

$$\Rightarrow 6+9\lambda=15$$

$$\therefore \lambda=1$$

On putting $\lambda=1$ in Eq. (i), we get

$$\mathbf{r}=\hat{\mathbf{i}}+3\hat{\mathbf{j}}+4\hat{\mathbf{k}}$$

Hence, centre of circle is (1, 3, 4).

JEE Advanced & IIT JEE Archive

1. (b,d) Here, $P_3: (x+z-1)+\lambda y=0$

i.e. $P_3: x+\lambda y+z-1=0$ ...(i)

whose distance from (0, 1, 0) is 1.

$$\therefore \frac{|0+\lambda+0-1|}{\sqrt{1+\lambda^2+1}}=1$$

$$\Rightarrow |\lambda-1|=\sqrt{\lambda^2+2}$$

$$\Rightarrow \lambda^2-2\lambda+1=\lambda^2+2$$

$$\Rightarrow \lambda=-\frac{1}{2}$$

$\therefore$ Equation of P_3 is $2x-y+2z-2=0$.

$\because$ Distance from (α, β, γ) is 2.

$$\therefore \frac{|2\alpha-\beta+2\gamma-2|}{\sqrt{4+1+4}}=2$$

$$\Rightarrow 2\alpha-\beta+2\gamma-2=\pm 6$$

$$\Rightarrow 2\alpha-\beta+2\gamma=8$$

and $$2\alpha-\beta+2\gamma=-4$$

2. (a,b) Since, L is at constant distance from two planes P_1 and P_2.

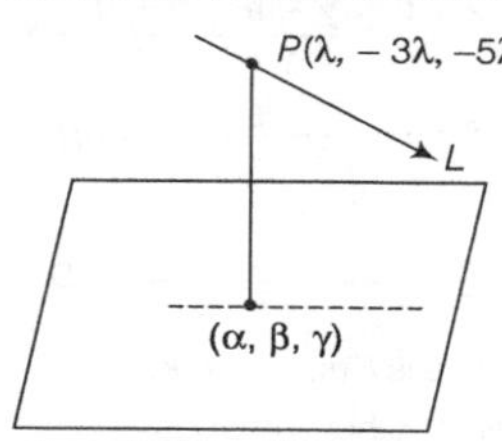

$P_1: x+2y-z+1=0$

Therefore, L is parallel to the line through intersection of P_1 and P_2.

$$\therefore \text{DR's of } L=\begin{vmatrix} \hat{\mathbf{i}} & \hat{\mathbf{j}} & \hat{\mathbf{k}} \\ 1 & 2 & -1 \\ 2 & -1 & 1 \end{vmatrix}$$

$$=\hat{\mathbf{i}}(2-1)-\hat{\mathbf{j}}(1+2)+\hat{\mathbf{k}}(-1-4)$$

$$=\hat{\mathbf{i}}-3\hat{\mathbf{j}}-5\hat{\mathbf{k}}$$

$\therefore$ DR's of L are (1, –3, –5) passing through (0, 0, 0).

Now, equation of L is

$$\frac{x-0}{1}=\frac{y-0}{-3}=\frac{z-0}{-5}$$

For any point on L,

$$\frac{x}{1}=\frac{y}{-3}=\frac{z}{-5}=\lambda \qquad \text{[say]}$$

i.e. $P(\lambda, -3\lambda, -5\lambda)$

If (α, β, γ) is foot of perpendicular from P on P_1, then

$$\frac{\alpha-\lambda}{1}=\frac{\beta+3\lambda}{2}=\frac{\gamma+5\lambda}{-1}=k \qquad \text{[say]}$$

$$\Rightarrow \quad \alpha=\lambda+k$$
$$\beta=2k-3\lambda$$
$$\gamma=-k-5\lambda$$

which satisfy $P_1 : x+2y-z+1=0$

$$\Rightarrow (\lambda+k)+2(2k-3\lambda)-(-k-5\lambda)+1=0 \Rightarrow k=-\frac{1}{6}$$

$$\therefore \quad x=-\frac{1}{6}+\lambda$$
$$Zy=-\frac{1}{3}-3\lambda$$
$$z=\frac{1}{6}-5\lambda$$

which satisfy options (a) and (b).

3. (a) Given, $\mathbf{OQ}=(1-3\mu)\hat{\mathbf{i}}+(\mu-1)\hat{\mathbf{j}}+(5\mu+2)\hat{\mathbf{k}}$,

$\mathbf{OP}=3\hat{\mathbf{i}}+2\hat{\mathbf{j}}+6\hat{\mathbf{k}}$ [where, O is origin]

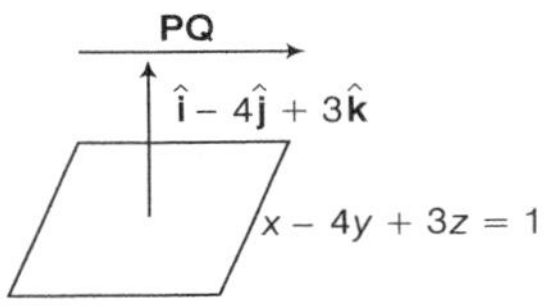

Now, $\mathbf{PQ}=(1-3\mu-3)\hat{\mathbf{i}}+(\mu-1-2)\hat{\mathbf{j}}+(5\mu+2-6)\hat{\mathbf{k}}$

$=(-2-3\mu)\hat{\mathbf{i}}+(\mu-3)\hat{\mathbf{j}}+(5\mu-4)\hat{\mathbf{k}}$

$\because$ **PQ** is parallel to the plane $x-4y+3z=1$

$$\therefore \quad -2-3\mu-4\mu+12+15\mu-12=0$$
$$\Rightarrow \quad 8\mu=2$$
$$\Rightarrow \quad \mu=\frac{1}{4}$$

4. (c) Since, $l=m=n=\frac{1}{\sqrt{3}}$

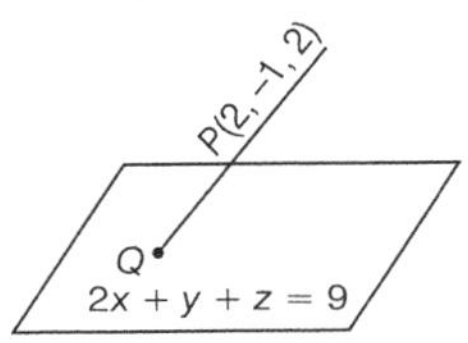

$\therefore$ Equation of line is

$$\frac{x-2}{1/\sqrt{3}}=\frac{y+1}{1/\sqrt{3}}=\frac{z-2}{1/\sqrt{3}}$$

$$\Rightarrow \quad x-2=y+1=z-2=r \qquad \text{[say]}$$

$\therefore$ Any point on the line is

$$Q \equiv (r+2, r-1, r+2)$$

$\because$ Q lies on the plane $2x+y+z=9$

$$\therefore \quad 2(r+2)+(r-1)+(r+2)=9$$
$$\Rightarrow \quad 4r+5=9$$
$$\Rightarrow \quad r=1$$

$\therefore Q(3, 0, 3)$

$$\therefore \quad PQ=\sqrt{(3-2)^2+(0+1)^2+(3-2)^2}=\sqrt{3}$$

5. (d) Given three planes are

$$P_1 : x-y+z=1 \qquad \text{...(i)}$$
$$P_2 : x+y-z=-1 \qquad \text{...(ii)}$$
and $$P_3 : x-3y+3z=2 \qquad \text{...(iii)}$$

Solving Eqs. (i) and (ii), we have

$$x=0$$
$$z=1+y$$

which does not satisfy Eq. (iii)

As, $$x-3y+3z=0-3y+3(1+y)$$
$$=3\ (\neq 2)$$

So, Statement II is correct.

Next, since we know that direction ratios of line of intersection of planes $a_1x+b_1y+c_1z+d_1=0$

and $a_2x+b_2y+c_2z+d_2=0$ is

$$b_1c_2-b_2c_1, c_1a_2-a_1c_2, a_1b_2-a_2b_1$$

Using above result

Direction ratios of lines L_1, L_2 and L_3 are

0, 2, 2 ; 0, – 4, – 4 ; 0, – 2, – 2 respectively.

$\Rightarrow$ All the three lines L_1, L_2 and L_3 are parallel pairwise.

Thus, Statement I is incorrect.

6. (b) The equations of given lines in vector form may be written as

$$L_1 : \mathbf{r}=(-\hat{\mathbf{i}}-2\hat{\mathbf{j}}-\hat{\mathbf{k}})+\lambda(3\hat{\mathbf{i}}+\hat{\mathbf{j}}+2\hat{\mathbf{k}})$$

and $$L_2 : \mathbf{r}=(2\hat{\mathbf{i}}-2\hat{\mathbf{j}}+3\hat{\mathbf{k}})+\mu(\hat{\mathbf{i}}+2\hat{\mathbf{j}}+3\hat{\mathbf{k}})$$

Since, the vector perpendicular to both L_1 and L_2.

$$\therefore \quad \begin{vmatrix} \hat{\mathbf{i}} & \hat{\mathbf{j}} & \hat{\mathbf{k}} \\ 3 & 1 & 2 \\ 1 & 2 & 3 \end{vmatrix}=-\hat{\mathbf{i}}-7\hat{\mathbf{j}}+5\hat{\mathbf{k}}$$

$\therefore$ Required unit vector

$$=\frac{(-\hat{\mathbf{i}}-7\hat{\mathbf{j}}+5\hat{\mathbf{k}})}{\sqrt{(-1)^2+(-7)^2+(5)^2}}$$
$$=\frac{1}{5\sqrt{3}}(-\hat{\mathbf{i}}-7\hat{\mathbf{j}}+5\hat{\mathbf{k}})$$

7. (d) The shortest distance between L_1 and L_2 is

$$\left|\frac{\{2-(-1)\,\hat{\mathbf{i}}+(2-2)\,\hat{\mathbf{j}}+3-(-1)\,\hat{\mathbf{k}}\}\cdot(-\hat{\mathbf{i}}-7\hat{\mathbf{j}}+5\hat{\mathbf{k}})}{5\sqrt{3}}\right|$$
$$=\left|\frac{(3\hat{\mathbf{i}}+4\hat{\mathbf{k}})\cdot(-\hat{\mathbf{i}}-7\hat{\mathbf{j}}+5\hat{\mathbf{k}})}{5\sqrt{3}}\right|$$
$$=\frac{17}{5\sqrt{3}} \text{ units}$$

8. (c) The equation of the plane passing through the point (–1, – 2, – 1) and whose normal is perpendicular to both the given lines L_1 and L_2 may be written as

$$(x+1)+7(y+2)-5(z+1)=0$$
$$\Rightarrow \quad x+7y-5z+10=0$$

The distance of the point (1, 1, 1) from the plane

$$=\left|\frac{1+7-5+10}{\sqrt{1+49+25}}\right|$$
$$=\frac{13}{\sqrt{75}} \text{ unit}$$

9. (b) Let $\Delta=\begin{vmatrix} a & b & c \\ b & c & a \\ c & a & b \end{vmatrix}$

$$=-\frac{1}{2}(a+b+c)[(a-b)^2+(b-c)^2+(c-a)^2]$$

(i) If $a+b+c\neq 0$ and $a^2+b^2+c^2=ab+bc+ca$

$\Rightarrow \Delta=0$ and $a=b=c\neq 0$

$\Rightarrow$ The equations represent identical planes.

(ii) $a+b+c=0$ and $a^2+b^2+c^2 \neq ab+bc+ca$

$\Rightarrow \Delta = 0$

$\Rightarrow$ The equations have infinitely many solutions.

$$ax + by = (a+b)z$$
$$bx + cy = (b+c)z$$
$$\Rightarrow \quad (b^2 - ac)y = (b^2 - ac)z$$
$$\Rightarrow \quad ax + by + cy = 0$$
$$\Rightarrow \quad y = z \Rightarrow ax = ay$$
$$\Rightarrow \quad x = y = z.$$

(iii) $a+b+c \neq 0$ and $a^2+b^2+c^2 \neq ab+bc+ca$

$$\Rightarrow \quad \Delta \neq 0$$

$\Rightarrow$ The equation represents planes meeting at only one point.

(iv) $a+b+c=0$ and $a^2+b^2+c^2 = ab+bc+ca$

$$\Rightarrow \quad a = b = c = 0$$

$\Rightarrow$ The equations represent whole of the three dimensional space.

10. (d) Given planes are $3x - 6y - 2z = 15$

and $2x + y - 2z = 5$

For $z = 0$, we get $x = 3, y = -1$

Direction ratios of planes are $< 3, -6, -2 >$ and $< 2, 1, -2 >$ then the DR's of line of intersection of planes is $< 14, 2, 15 >$ and line is

$$\frac{x-3}{14} = \frac{y+1}{2} = \frac{z-0}{15} = \lambda \quad \text{(say)}$$

$\Rightarrow$ $x = 14\lambda + 3, \quad y = 2\lambda - 1, \quad z = 15\lambda$

Hence, Statement I is incorrect but Statement II is correct.

11. (d) Let the equation of plane be

$$a(x-1) + b(y+2) + c(z-1) = 0$$

which is perpendicular to $2x - 2y + z = 0$

and $x - y + 2z = 4$

$$\Rightarrow \quad 2a - 2b + c = 0$$

and $a - b + 2c = 0$

$$\Rightarrow \quad \frac{a}{-3} = \frac{b}{-3} = \frac{c}{0}$$

$$\Rightarrow \quad \frac{a}{1} = \frac{b}{1} = \frac{c}{0}$$

So, the equation of plane is

$$x - 1 + y + 2 = 0 \text{ or } x + y + 1 = 0$$

Its distance from the point (1, 2, 2) is

$$\frac{|1+2+1|}{\sqrt{2}} = 2\sqrt{2}$$

12. (a) Since, $\frac{x}{a} + \frac{y}{b} + \frac{z}{c} = 1$ cuts the coordinate axes at $A(a, 0, 0)$, $B(0, b, 0)$, $C(0, 0, c)$

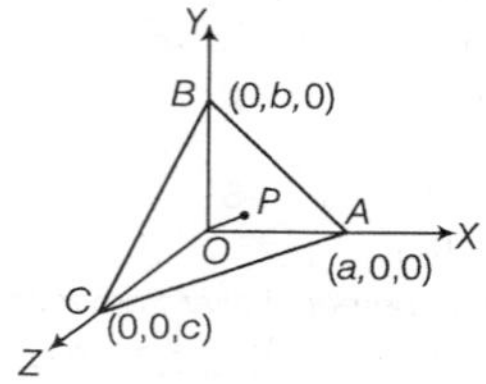

and its distance from origin = 1

$$\therefore \quad \frac{1}{\sqrt{\frac{1}{a^2} + \frac{1}{b^2} + \frac{1}{c^2}}} = 1$$

or $$\frac{1}{a^2} + \frac{1}{b^2} + \frac{1}{c^2} = 1 \quad \text{...(i)}$$

where P is centroid of triangle

$$\therefore \quad P(x, y, z) = \left(\frac{a+0+0}{3}, \frac{0+b+0}{3}, \frac{0+0+c}{3}\right)$$

$$\Rightarrow \quad x = \frac{a}{3}, y = \frac{b}{3}, z = \frac{c}{3} \quad \text{...(ii)}$$

$\therefore$ From Eqs. (i) and (ii), we get

$$\frac{1}{9x^2} + \frac{1}{9y^2} + \frac{1}{9z^2} = 1$$

or $$\frac{1}{x^2} + \frac{1}{y^2} + \frac{1}{z^2} = 9 = K$$

$$\therefore \quad K = 9$$

13. (0) Equation of plane containing the lines $2x - y + z - 3 = 0$ and $3x + y + z = 5$ is

$$(2x - y + z - 3) + \lambda(3x + y + z - 5) = 0$$

$$\Rightarrow \quad (2 + 3\lambda)x + (\lambda - 1)y + (\lambda + 1)z - 3 - 5\lambda = 0$$

Since, distance of plane from (2, 1, – 1) to above plane is $1/\sqrt{6}$.

$$\therefore \quad \left|\frac{6\lambda + 4 + \lambda - 1 - \lambda - 1 - 3 - 5\lambda}{\sqrt{(3\lambda+2)^2 + (\lambda-1)^2 + (\lambda+1)^2}}\right| = \frac{1}{\sqrt{6}}$$

$$\Rightarrow \quad 6(\lambda - 1)^2 = 11\lambda^2 + 12\lambda + 6$$

$$\Rightarrow \quad \lambda = 0, -\frac{24}{5}$$

$\therefore$ Equations of planes are

$$2x - y + z - 3 = 0$$

and $$62x + 29y + 19z - 105 = 0$$

Last 3 Years' Questions
JEE Main & Advanced
(2016-2018)

JEE Main

2016

1. The eccentricity of the hyperbola whose length of the latusrectum is equal to 8 and the length of its conjugate axis is equal to half of the distance between its foci, is **[Hyperbola]**

(a) $\frac{4}{3}$ (b) $\frac{4}{\sqrt{3}}$ (c) $\frac{2}{\sqrt{3}}$ (d) $\sqrt{3}$

2. The distance of the point $(1, -5, 9)$ from the plane $x - y + z = 5$ measured along the line $x = y = z$ is **[3D Geometry]**

(a) $3\sqrt{10}$ (b) $10\sqrt{3}$

(c) $\frac{10}{\sqrt{3}}$ (d) $\frac{20}{3}$

3. If the line, $\frac{x-3}{2} = \frac{y+2}{-1} = \frac{z+4}{3}$ lies in the plane, $lx + my - z = 9$, then $l^2 + m^2$ is equal to **[3D Geometry]**

(a) 26 (b) 18 (c) 5 (d) 2

4. Let $\hat{\mathbf{a}}$, $\hat{\mathbf{b}}$ and $\hat{\mathbf{c}}$ be three unit vectors such that $\hat{\mathbf{a}} \times (\hat{\mathbf{b}} \times \hat{\mathbf{c}}) = \frac{\sqrt{3}}{2}(\hat{\mathbf{b}} + \hat{\mathbf{c}})$. If $\hat{\mathbf{b}}$ is not parallel to $\hat{\mathbf{c}}$, then the angle between $\hat{\mathbf{a}}$ and $\hat{\mathbf{b}}$ is **[Vectors]**

(a) $\frac{3\pi}{4}$ (b) $\frac{\pi}{2}$ (c) $\frac{2\pi}{3}$ (d) $\frac{5\pi}{6}$

2017

5. If the image of the point $P(1, -2, 3)$ in the plane $2x + 3y - 4z + 22 = 0$ measured parallel to the line $\frac{x}{1} = \frac{y}{4} = \frac{z}{5}$ is Q, then PQ is equal to **[3D Geometry]**

(a) $3\sqrt{5}$ (b) $2\sqrt{42}$ (c) $\sqrt{42}$ (d) $6\sqrt{5}$

6. The distance of the point $(1, 3, -7)$ from the plane passing through the point $(1, -1, -1)$ having normal perpendicular to both the lines **[3D Geometry]**
$\frac{x-1}{1} = \frac{y+2}{-2} = \frac{z-4}{3}$ and $\frac{x-2}{2} = \frac{y+1}{-1} = \frac{z+7}{-1}$, is

(a) $\frac{20}{\sqrt{74}}$ units (b) $\frac{10}{\sqrt{83}}$ units (c) $\frac{5}{\sqrt{83}}$ units (d) $\frac{10}{\sqrt{74}}$ units

7. The eccentricity of an ellipse whose centre is at the origin is 1/2. If one of its directrices is $x = -4$, then the equation of the normal to it at $\left(1, \frac{3}{2}\right)$ is **[Ellipse]**

(a) $2y - x = 2$ (b) $4x - 2y = 1$

(c) $4x + 2y = 7$ (d) $x + 2y = 4$

8. If a hyperbola passes through the point $P(\sqrt{2}, \sqrt{3})$ and has foci at $(\pm 2, 0)$, then the tangent to this hyperbola at P also passes through the point **[Hyperbola]**

(a) $(3\sqrt{2}, 2\sqrt{3})$ (b) $(2\sqrt{2}, 3\sqrt{3})$

(c) $(\sqrt{3}, \sqrt{2})$ (d) $(-\sqrt{2}, -\sqrt{3})$

9. Let $\mathbf{a} = 2\hat{\mathbf{i}} + \hat{\mathbf{j}} - 2\hat{\mathbf{k}}$, $\mathbf{b} = \hat{\mathbf{i}} + \hat{\mathbf{j}}$ and $\mathbf{c}$ be a vector such that $|\mathbf{c} - \mathbf{a}| = 3$, $|(\mathbf{a} \times \mathbf{b}) \times \mathbf{c}| = 3$ and the angle between $\mathbf{c}$ and $\mathbf{a} \times \mathbf{b}$ is 30°. Then, $\mathbf{a} \cdot \mathbf{c}$ is equal to **[Vectors]**

(a) $\frac{25}{8}$ (b) 2 (c) 5 (d) $\frac{1}{8}$

10. The radius of a circle having minimum area, which touches the curve $y = 4 - x^2$ and the lines $y = |x|$, is **[Parabola]**

(a) $2(\sqrt{2} + 1)$ (b) $2(\sqrt{2} - 1)$ (c) $4(\sqrt{2} - 1)$ (d) $4(\sqrt{2} + 1)$

2018

11. If the curves $y^2 = 6x$, $9x^2 + by^2 = 16$ intersect each other at right angles, then the value of b is **[Ellipse]**

(a) 6 (b) $\frac{7}{2}$ (c) 4 (d) $\frac{9}{2}$

12. Tangent and normal are drawn at $P(16, 16)$ on the parabola $y^2 = 16x$, which intersect the axis of the parabola at A and B, respectively. If C is the centre of the circle through the points P, A and B and $\angle CPB = \theta$, then a value of $\tan\theta$ is **[Parabola]**

(a) $\frac{1}{2}$ (b) 2 (c) 3 (d) $\frac{4}{3}$

13. Tangents are drawn to the hyperbola $4x^2 - y^2 = 36$ at the points P and Q. If these tangents intersect at the point $T(0, 3)$, then the area (in sq units) of ΔPTQ is **[Hyperbola]**

(a) $45\sqrt{5}$ (b) $54\sqrt{3}$ (c) $60\sqrt{3}$ (d) $36\sqrt{5}$

14. If L_1 is the line of intersection of the planes $2x - 2y + 3z - 2 = 0$, $x - y + z + 1 = 0$ and L_2 is the line of intersection of the planes $x + 2y - z - 3 = 0$, $3x - y + 2z - 1 = 0$, then the distance of the origin from the plane, containing the lines L_1 and L_2 is **[3D Geometry]**

(a) $\frac{1}{4\sqrt{2}}$ (b) $\frac{1}{3\sqrt{2}}$

(c) $\frac{1}{2\sqrt{2}}$ (d) $\frac{1}{\sqrt{2}}$

15. The length of the projection of the line segment joining the points $(5, -1, 4)$ and $(4, -1, 3)$ on the plane, $x + y + z = 7$ is **[3D Geometry]**

(a) $\frac{2}{\sqrt{3}}$ (b) $\frac{2}{3}$ (c) $\frac{1}{3}$ (d) $\sqrt{\frac{2}{3}}$

16. Let $\mathbf{u}$ be a vector coplanar with the vectors $\mathbf{a} = 2\hat{\mathbf{i}} + 3\hat{\mathbf{j}} - \hat{\mathbf{k}}$ and $\mathbf{b} = \hat{\mathbf{j}} + \hat{\mathbf{k}}$. If $\mathbf{u}$ is perpendicular to $\mathbf{a}$ and $\mathbf{u} \cdot \mathbf{b} = 24$, then $|\mathbf{u}|^2$ is equal to **[Vectors]**

(a) 336 (b) 315 (c) 256 (d) 84

JEE Advanced

2016

1. In a ΔXYZ, let x, y, z be the lengths of sides opposite to the angles X, Y, Z respectively and $2s = x + y + z$. If $\frac{s-x}{4} = \frac{s-y}{3} = \frac{s-z}{2}$ and area of incircle of the ΔXYZ is $\frac{8\pi}{3}$ then **[3D Geometry Type 2]**

(a) area of the ΔXYZ is $6\sqrt{6}$

(b) the radius of circumcircle of the ΔXYZ is $\frac{35}{6}\sqrt{6}$

(c) $\sin\frac{X}{2}\sin\frac{Y}{2}\sin\frac{Z}{2} = \frac{4}{35}$

(d) $\sin^2\left(\frac{X+Y}{2}\right) = \frac{3}{5}$

Note ***Type-1*** *Only one correct option,* ***Type-2*** *More then one correct option,* ***Type-3*** *Single digit Answer,* ***Type-4*** *Numerical value based*

2. The circle $C_1 : x^2 + y^2 = 3$ with centre at O intersects the parabola $x^2 = 2y$ at the point P in the first quadrant. Let the tangent to the circle C_1 at P touches other two circles C_2 and C_3 at R_2 and R_3, respectively. Suppose C_2 and C_3 have equal radii $2\sqrt{3}$ and centres Q_2 and Q_3, respectively. If Q_2 and Q_3 lie on the Y-axis, then **[Parabola, Type 2]**

(a) $Q_2Q_3 = 12$ (b) $R_2R_3 = 4\sqrt{6}$

(c) area of the ΔOR_2R_3 is $6\sqrt{2}$ (d) area of the ΔPQ_2Q_3 is $4\sqrt{2}$

3. Let P be the image of the point (3, 1, 7) with respect to the plane $x - y + z = 3$. Then, the equation of the plane passing through P and containing the straight line $\frac{x}{1} = \frac{y}{2} = \frac{z}{1}$ is

[3D Geometry, Type 1]

(a) $x + y - 3z = 0$ (b) $3x + z = 0$ (c) $x - 4y + 7z = 0$ (d) $2x - y = 0$

4. Let $\hat{u} = u_1\hat{i} + u_2\hat{j} + u_3\hat{k}$ be a unit vector in R^3 and $\hat{w} = \frac{1}{\sqrt{6}}(\hat{i} + \hat{j} + 2\hat{k})$. Given that there exists a vector $\vec{v}$ in R^3, such that $|\hat{u} + \vec{v}| = 1$ and $\hat{w} \cdot (\hat{u} + \vec{v}) = 1$.

Which of the following statement(s) is/are correct? **[Vectors, Type 2]**

(a) There is exactly one choice for such $\vec{v}$ (b) There are infinitely many choices for such $\vec{v}$

(c) If $\hat{u}$ lies in the XY-plane, then $|u_1| = |u_2|$ (d) If $\hat{u}$ lies in the XY-plane, then $2|u_1| = |u_3|$

5. Let P be the point on the parabola $y^2 = 4x$, which is at the shortest distance from the centre S of the circle $x^2 + y^2 - 4x - 16y + 64 = 0$. Let Q be the point on the circle dividing the line segment SP internally. Then, **[Parabola, Type 2]**

(a) $SP = 2\sqrt{5}$

(b) $SQ : QP = (\sqrt{5} + 1) : 2$

(c) the x-intercept of the normal to the parabola at P is 6

(d) the slope of the tangent to the circle at Q is $\frac{1}{2}$

Paragraph (Q. 6-7)

Let $F_1\ (x_1, 0)$ and $F_2\ (x_2, 0)$, for $x_1 < 0$ and $x_2 > 0$, be the foci of the ellipse $\frac{x^2}{9} + \frac{y^2}{8} = 1$. Suppose a parabola having vertex at the origin and focus at F_2 intersects the ellipse at point M in the first quadrant and at point N in the fourth quadrant. **[Ellipse]**

6. The orthocentre of ΔF_1MN is

(a) $\left(-\frac{9}{10}, 0\right)$ (b) $\left(\frac{2}{3}, 0\right)$

(c) $\left(\frac{9}{10}, 0\right)$ (d) $\left(\frac{2}{3}, \sqrt{6}\right)$

7. If the tangents to the ellipse at M and N meet at R and the normal to the parabola at M meets the X-axis at Q, then the ratio of area of ΔMQR to area of the quadrilateral MF_1NF_2 is

(a) 3 : 4 (b) 4 : 5 (c) 5 : 8 (d) 2 : 3

2017

8. If $2x - y + 1 = 0$ is a tangent to the hyperbola $\frac{x^2}{a^2} - \frac{y^2}{16} = 1$ then which of the following CANNOT be sides of a right angled triangle? **[Hyperbola, Type 2]**

(a) a, 4, 1 (b) $2a$, 4, 1 (c) a, 4, 2 (d) $2a$, 8, .

9. If a chord, which is not a tangent, of the parabola $y^2 = 16x$ has the equation $2x + y = p$, and mid-point (h, k), then which of the following is(are) possible value(s) of p, h and k?

[Parabola, Type 2]

(a) $p = -1, h = 1, k = -3$ (b) $p = 2, h = 3, k = -4$

(c) $p = -2, h = 2, k = -4$ (d) $p = 5, h = 4, k = -3$

Directions (Q.Nos. 10-12) by appropriately matching the information given in the three columns of the following table.

Columns 1, 2 and 3 contain conics, equations of tangents to the conics and points of contact, respectively.

	Column-1		Column-2		Column-3
(I)	$x^2 + y^2 = a^2$	(i)	$my = m^2 x + a$	(P)	$\left(\frac{a}{m^2}, \frac{2a}{m}\right)$
(II)	$x^2 + a^2 y^2 = a^2]$	(ii)	$y = mx + a\sqrt{m^2 + 1}$	(Q)	$\left(\frac{-ma}{\sqrt{m^2+1}}, \frac{a}{\sqrt{m^2+1}}\right)$
(III)	$y^2 = 4ax$	(iii)	$y = mx + \sqrt{a^2 m^2 - 1}$	(R)	$\left(\frac{-a^2 m}{\sqrt{a^2 m^2+1}}, \frac{1}{\sqrt{a^2 m^2+1}}\right)$
(IV)	$x^2 - a^2 y^2 = a^2$	(iv)	$y = mx + \sqrt{a^2 m^2 + 1}$	(S)	$\left(\frac{-a^2 m}{\sqrt{a^2 m^2-1}}, \frac{-1}{\sqrt{a^2 m^2-1}}\right)$

[Hyperbola, Type 1]

10. For $a = \sqrt{2}$, if a tangent is drawn to a suitable conic (Column 1) at the point of contact $(-1, 1)$, then which of the following options is the only CORRECT combination for obtaining its equation?

(a) (I) (ii) (Q)
(b) (I) (i) (P)
(c) (III) (i) (P)
(d) (II) (ii) (Q)

11. The tangent to a suitable conic (Column 1) at $\left(\sqrt{3}, \frac{1}{2}\right)$ is found to be $\sqrt{3}x + 2y = 4$, then which of the following options is the only CORRECT combination?

(a) (IV) (iv) (S)
(b) (II) (iv) (R)
(c) (IV) (iii) (S)
(d) (II) (iii) (R)

12. If a tangent to a suitable conic (Column 1) is found to be $y = x + 8$ and its point of contact is (8, 16), then which of the following options is the only CORRECT combination?

(a) (III) (i) (P)
(b) (I) (ii) (Q)
(c) (II) (iv) (R)
(d) (III) (ii) (Q)

13. Let O be the origin and let PQR be an arbitrary triangle. The point S is such that

$$\mathbf{OP} \cdot \mathbf{OQ} + \mathbf{OR} \cdot \mathbf{OS} = \mathbf{OR} \cdot \mathbf{OP} + \mathbf{OQ} \cdot \mathbf{OS} = \mathbf{OQ} \cdot \mathbf{OR} + \mathbf{OP} \cdot \mathbf{OS}$$

Then the triangle PQR has S as its **[Vectors, Type 1]**

(a) centroid
(b) orthocentre
(c) incentre
(d) circumcentre

14. The equation of the plane passing through the point (1, 1, 1) and perpendicular to the planes $2x + y - 2z = 5$ and $3x - 6y - 2z = 7$ is **[3D Geometry, Type 1]**

(a) $14x + 2y - 15z = 1$ (b) $-14x + 2y + 15z = 3$ (c) $14x - 2y + 15z = 27$ (d) $14x + 2y + 15z = 31$

Paragraph (Q. 15-16)

Let O be the origin and **OX, OY, OZ** be three unit vectors in the directions of the sides **QR, RP, PQ** respectively, of a triangle PQR. **[Vectors]**

15. If the triangle PQR varies, then the minimum value of $\cos(P + Q) + \cos(Q + R) + \cos(R + P)$ is

(a) $-\frac{3}{2}$
(b) $\frac{3}{2}$
(c) $\frac{5}{3}$
(d) $-\frac{5}{3}$

16. $|\mathbf{OX} \times \mathbf{OY}| =$ **[Vectors]**

(a) $\sin(P + Q)$ (b) $\sin(P + R)$ (c) $\sin(Q + R)$ (d) $\sin 2R$

2018

17. Let $P_1 : 2x + y - z = 3$ and $P_2 : x + 2y + z = 2$ be two planes. Then, which of the following statement(s) is (are) TRUE? **[3D Geometry, Type 2]**

(a) The line of intersection of P_1 and P_2 has direction ratios 1, 2, −1

(b) The line $\frac{3x-4}{9} = \frac{1-3y}{9} = \frac{z}{3}$ is perpendicular to the line of intersection of P_1 and P_2

(c) The acute angle between P_1 and P_2 is 60°

(d) If P_3 is the plane passing through the point $(4, 2, -2)$ and perpendicular to the line of intersection of P_1 and P_2, then the distance of the point (2, 1, 1) from the plane P_3 is $\frac{2}{\sqrt{3}}$

18. Let **a** and **b** be two unit vectors such that $\mathbf{a} \cdot \mathbf{b} = 0$. For some $x, y \in R$, let $\mathbf{c} = x\mathbf{a} - y\mathbf{b} + (\mathbf{a} \times \mathbf{b})$. If $|\mathbf{c}| = 2$ and the vector **c** is inclined at the same angle α to both **a** and **b**, then the value of $8\cos^2 \alpha$ is .... **[Vectors, Type 4]**

19. Consider two straight lines, each of which is tangent to both the circle $x^2 + y^2 = (1/2)$ and the parabola $y^2 = 4x$. Let these lines intersect at the point Q. Consider the ellipse whose centre is at the origin $O(0, 0)$ and whose semi-major axis is OQ. If the length of the minor axis of this ellipse is $\sqrt{2}$, then which of the following statement(s) is (are) TRUE? **[Ellipse, Type 2]**

(a) For the ellipse, the eccentricity is $1/\sqrt{2}$ and the length of the latus rectum is 1

(b) For the ellipse, the eccentricity is 1/2 and the length of the latus rectum is 1/2

(c) The area of the region bounded by the ellipse between the lines $x = \frac{1}{\sqrt{2}}$ and $x = 1$ is $\frac{1}{4\sqrt{2}}(\pi - 2)$

(d) The area of the region bounded by the ellipse between the lines $x = \frac{1}{\sqrt{2}}$ and $x = 1$ is $\frac{1}{16}(\pi - 2)$

20. Let P be a point in the first octant, whose image Q in the plane $x + y = 3$ (that is, the line segment PQ is perpendicular to the plane $x + y = 3$ and the mid-point of PQ lies in the plane $x + y = 3$) lies on the Z-axis. Let the distance of P from the X-axis be 5. If R is the image of P in the XY-plane, then the length of PR is .... **[3D Geometry, Type 4]**

21. Consider the cube in the first octant with sides OP, OQ and OR of length 1, along the X-axis, Y-axis and Z-axis, respectively, where $O(0, 0, 0)$ is the origin. Let $S\left(\frac{1}{2}, \frac{1}{2}, \frac{1}{2}\right)$ be the centre of the cube and T be the vertex of the cube opposite to the origin O such that S lies on the diagonal OT. If $\mathbf{p} = \mathbf{SP}$, $\mathbf{q} = \mathbf{SQ}$, $\mathbf{r} = \mathbf{SR}$ and $\mathbf{t} = \mathbf{ST}$, then the value of $|(\mathbf{p} \times \mathbf{q}) \times (\mathbf{r} \times \mathbf{t})|$ is ..... . **[Vectors, Type 4]**

22. Let $H : \frac{x^2}{a^2} - \frac{y^2}{b^2} = 1$, where $a > b > 0$, be a hyperbola in the XY-plane whose conjugate axis LM subtends an angle of 60° at one of its vertices N. Let the area of the ΔLMN be $4\sqrt{3}$. **[Hyperbola]**

	List-I		List-II
P.	The length of the conjugate axis of H is	1.	8
Q.	The eccentricity of H is	2.	$\frac{4}{\sqrt{3}}$
R.	The distance between the foci of H is	3.	$\frac{2}{\sqrt{3}}$
S.	The length of the latus rectum of H is	4.	4

The correct option is

(a) $P \to 4; Q \to 2; R \to 1; S \to 3$ (b) $P \to 4; Q \to 3; R \to 1; S \to 2$

(c) $P \to 4; Q \to 1; R \to 3; S \to 2$ (d) $P \to 3; Q \to 4; R \to 2; S \to 1$

Answer *with* Explanations

JEE Main

1. (c) We have, $\frac{2b^2}{a} = 8$ and $2b = ae$

$\Rightarrow \quad b^2 = 4a$ and $2b = ae$

Consider $\quad 2b = ae$

$\Rightarrow \quad 4b^2 = a^2e^2$

$\Rightarrow \quad 4a^2(e^2 - 1) = a^2e^2$

$\Rightarrow \quad 4e^2 - 4 = e^2 \quad [\because a \neq 0]$

$\Rightarrow \quad 3e^2 = 4$

$\Rightarrow \quad e = \frac{2}{\sqrt{3}} \quad [\because e > 0]$

2. (b) Equation of line passing through the point (1, – 5, 9) and parallel to $x = y = z$ is

$$\frac{x-1}{1} = \frac{y+5}{1} = \frac{z-9}{1} = \lambda \text{ (say)}$$

Thus, any point on this line is of the form $(\lambda + 1, \lambda - 5, \lambda + 9)$.
Now, if $P(\lambda + 1, \lambda - 5, \lambda + 9)$ is the point of intersection of line and plane, then

$(\lambda + 1) - (\lambda - 5) + \lambda + 9 = 5$

$\Rightarrow \quad \lambda + 15 = 5 \Rightarrow \lambda = -10$

$\therefore$ Coordinates of point P are$(-9, -15, -1)$.
Hence, required distance

$$= \sqrt{(1+9)^2 + (-5+15)^2 + (9+1)^2}$$

$$= \sqrt{10^2 + 10^2 + 10^2} = 10\sqrt{3}$$

3. (d) Since, the line $\frac{x-3}{2} = \frac{y+2}{-1} = \frac{z+4}{3}$ lies in the plane $lx + my - z = 9$, therefore we have $2l - m - 3 = 0$

[$\because$ normal will be perpendicular to the line]

$\Rightarrow \quad 2l - m = 3 \quad$ ...(i)

and $\quad 3l - 2m + 4 = 9$

[$\because$ point (3, – 2, – 4) lies on the plane]

$\Rightarrow \quad 3l - 2m = 5 \quad$ ...(ii)

On solving Eqs. (i) and (ii), we get

$l = 1$ and $m = -1$

$\therefore \quad l^2 + m^2 = 2$

4. (d) Given, $|\hat{\mathbf{a}}| = |\hat{\mathbf{b}}| = |\hat{\mathbf{c}}| = 1$

and $\quad \hat{\mathbf{a}} \times (\hat{\mathbf{b}} \times \hat{\mathbf{c}}) = \frac{\sqrt{3}}{2}(\hat{\mathbf{b}} + \hat{\mathbf{c}})$

Now, consider $\hat{\mathbf{a}} \times (\hat{\mathbf{b}} \times \hat{\mathbf{c}}) = \frac{\sqrt{3}}{2}(\hat{\mathbf{b}} + \hat{\mathbf{c}})$

$\Rightarrow \quad (\hat{\mathbf{a}} \cdot \hat{\mathbf{c}})\,\hat{\mathbf{b}} - (\hat{\mathbf{a}} \cdot \hat{\mathbf{b}})\,\hat{\mathbf{c}} = \frac{\sqrt{3}}{2}\hat{\mathbf{b}} + \frac{\sqrt{3}}{2}\hat{\mathbf{c}}$

On comparing, we get

$$\hat{\mathbf{a}} \cdot \hat{\mathbf{b}} = -\frac{\sqrt{3}}{2}$$

$\Rightarrow \quad |\hat{\mathbf{a}}|\,|\hat{\mathbf{b}}| \cos\theta = -\frac{\sqrt{3}}{2}$

$\Rightarrow \quad \cos\theta = -\frac{\sqrt{3}}{2} \quad [\because |\hat{\mathbf{a}}| = |\hat{\mathbf{b}}| = 1]$

$\Rightarrow \quad \cos\theta = \cos\left(\pi - \frac{\pi}{6}\right)$

$\Rightarrow \quad \theta = \frac{5\pi}{6}$

5. (b) Any line parallel to $\frac{x}{1} = \frac{y}{4} = \frac{z}{5}$ and passing through $P(1, -2, 3)$ is

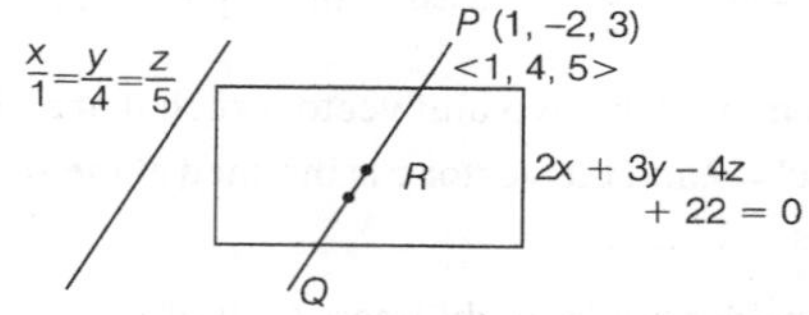

$$\frac{x-1}{1} = \frac{y+2}{4} = \frac{z-3}{5} = \lambda \quad \text{(say)}$$

Any point on above line can be written as $(\lambda + 1, 4\lambda - 2, 5\lambda + 3)$.

$\therefore$ Coordinates of R are $(\lambda + 1, 4\lambda - 2, 5\lambda + 3)$.

Since, point R lies on the above plane.

$\therefore \quad 2(\lambda + 1) + 3(4\lambda - 2) - 4(5\lambda + 3) + 22 = 0 \Rightarrow \lambda = 1$

So, point R is (2, 2, 8).

Now, $PR = \sqrt{(2-1)^2 + (2+2)^2 + (8-3)^2} = \sqrt{42}$

$\therefore \quad PQ = 2PR = 2\sqrt{42}$

6. (b) Given, equations of lines are

$$\frac{x-1}{1} = \frac{y+2}{-2} = \frac{z-4}{3}$$

and $$\frac{x-2}{2} = \frac{y+1}{-1} = \frac{z+7}{-1}$$

Let $\mathbf{n}_1 = \hat{\mathbf{i}} - 2\hat{\mathbf{j}} + 3\hat{\mathbf{k}}$ and $\mathbf{n}_2 = 2\hat{\mathbf{i}} - \hat{\mathbf{j}} - \hat{\mathbf{k}}$

$\therefore$ Any vector $\mathbf{n}$ perpendicular to both $\mathbf{n}_1, \mathbf{n}_2$ is given by

$$\mathbf{n} = \mathbf{n}_1 \times \mathbf{n}_2$$

$$\Rightarrow \quad \mathbf{n} = \begin{vmatrix} \hat{\mathbf{i}} & \hat{\mathbf{j}} & \hat{\mathbf{k}} \\ 1 & -2 & 3 \\ 2 & -1 & -1 \end{vmatrix} = 5\hat{\mathbf{i}} + 7\hat{\mathbf{j}} + 3\hat{\mathbf{k}}$$

$\therefore$ Equation of a plane passing through (1, – 1, – 1) and perpendicular to $\mathbf{n}$ is given by

$5(x-1) + 7(y+1) + 3(z+1) = 0$

$\Rightarrow \quad 5x + 7y + 3z + 5 = 0$

$\therefore$ Required distance $= \left|\frac{5 + 21 - 21 + 5}{\sqrt{5^2 + 7^2 + 3^2}}\right| = \frac{10}{\sqrt{83}}$ units

7. (b) We have, $e = \frac{1}{2}$ and $\frac{a}{e} = 4$

$\therefore \quad a = 2$

Now, $\quad b^2 = a^2(1 - e^2) = (2)^2\left[1 - \left(\frac{1}{2}\right)^2\right]$

$= 4\left(1 - \frac{1}{4}\right) = 3 \Rightarrow b = \sqrt{3}$

∴ Equation of the ellipse is

$$\frac{x^2}{(2)^2} + \frac{y^2}{(\sqrt{3})^2} = 1$$

$$\Rightarrow \quad \frac{x^2}{4} + \frac{y^2}{3} = 1$$

Now, the equation of normal at $\left(1, \frac{3}{2}\right)$ is

$$\frac{a^2x}{x_1} - \frac{b^2y}{y_1} = a^2 - b^2$$

$$\Rightarrow \quad \frac{4x}{1} - \frac{3y}{(3/2)} = 4 - 3$$

$$\Rightarrow \quad 4x - 2y = 1$$

8. (b) Let the equation of hyperbola be $\frac{x^2}{a^2} - \frac{y^2}{b^2} = 1.$

$$\therefore \quad ae = 2 \Rightarrow a^2e^2 = 4$$

$$\Rightarrow \quad a^2 + b^2 = 4$$

$$\Rightarrow \quad b^2 = 4 - a^2$$

$$\therefore \quad \frac{x^2}{a^2} - \frac{y^2}{4 - a^2} = 1$$

Since, $(\sqrt{2}, \sqrt{3})$ lie on hyperbola.

$$\therefore \quad \frac{2}{a^2} - \frac{3}{4 - a^2} = 1$$

$$\Rightarrow \quad 8 - 2a^2 - 3a^2 = a^2(4 - a^2)$$

$$\Rightarrow \quad 8 - 5a^2 = 4a^2 - a^4$$

$$\Rightarrow \quad a^4 - 9a^2 + 8 = 0$$

$$\Rightarrow \quad (a^4 - 8)(a^4 - 1) = 0$$

$$\Rightarrow \quad a^4 = 8, a^4 = 1$$

$$\therefore \quad a = 1$$

Now, equation of hyperbola is $\frac{x^2}{1} - \frac{y^2}{3} = 1.$

∴ Equation of tangent at $(\sqrt{2}, \sqrt{3})$ is given by

$$\sqrt{2}x - \frac{\sqrt{3}y}{3} = 1$$

$$\Rightarrow \quad \sqrt{2}x - \frac{y}{\sqrt{3}} = 1$$

which passes through the point $(2\sqrt{2}, 3\sqrt{3})$.

9. (b) We have, $\mathbf{a} = 2\hat{\mathbf{i}} + \hat{\mathbf{j}} - 2\hat{\mathbf{k}}$

$$\Rightarrow \quad |\mathbf{a}| = \sqrt{4 + 1 + 4} = 3$$

and $\mathbf{b} = \hat{\mathbf{i}} + \hat{\mathbf{j}} \Rightarrow |\mathbf{b}| = \sqrt{1 + 1} = \sqrt{2}$

Now, $|\mathbf{c} - \mathbf{a}| = 3 \Rightarrow |\mathbf{c} - \mathbf{a}|^2 = 9$

$$\Rightarrow \quad (\mathbf{c} - \mathbf{a}) \cdot (\mathbf{c} - \mathbf{a}) = 9$$

$$\Rightarrow \quad |\mathbf{c}|^2 + |\mathbf{a}|^2 - 2\mathbf{c} \cdot \mathbf{a} = 9 \quad \ldots \text{(i)}$$

Again, $|(\mathbf{a} \times \mathbf{b}) \times \mathbf{c}| = 3$

$$\Rightarrow \quad |\mathbf{a} \times \mathbf{b}||\mathbf{c}| \sin 30^\circ = 3 \Rightarrow |\mathbf{c}| = \frac{6}{|\mathbf{a} \times \mathbf{b}|}$$

$$\text{But } \mathbf{a} \times \mathbf{b} = \begin{vmatrix} \hat{\mathbf{i}} & \hat{\mathbf{j}} & \hat{\mathbf{k}} \\ 2 & 1 & -2 \\ 1 & 1 & 0 \end{vmatrix} = 2\hat{\mathbf{i}} - 2\hat{\mathbf{j}} + \hat{\mathbf{k}}$$

$$\therefore \quad |\mathbf{c}| = \frac{6}{\sqrt{4 + 4 + 1}} = 2 \quad \ldots \text{(ii)}$$

From Eqs. (i) and (ii), we get

$$(2)^2 + (3)^2 - 2\mathbf{c} \cdot \mathbf{a} = 9$$

$$\Rightarrow \quad 4 + 9 - 2\mathbf{c} \cdot \mathbf{a} = 9$$

$$\Rightarrow \quad \mathbf{c} \cdot \mathbf{a} = 2$$

10. (c) Let the radius of circle with least area be r.

Then, then coordinates of centre = $(0, 4 - r)$.

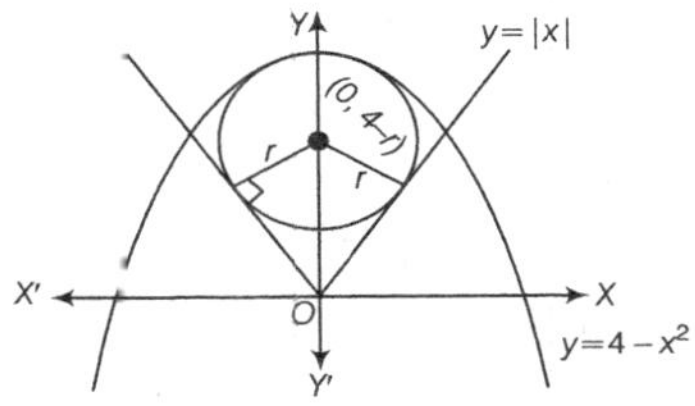

Since, circle touches the line $y = x$ in first quadrant.

$$\therefore \quad \left|\frac{0 - (4 - r)}{\sqrt{2}}\right| = r$$

$$\Rightarrow \quad r - 4 = \pm r\sqrt{2}$$

$$\Rightarrow \quad r = \frac{4}{\sqrt{2} + 1} \text{ or } \frac{4}{1 - \sqrt{2}}$$

$$\text{But} \quad r \neq \frac{4}{1 - \sqrt{2}} \quad \left[\because \frac{4}{1 - \sqrt{2}} < 0\right]$$

$$\therefore \quad r = \frac{4}{\sqrt{2} + 1} = 4(\sqrt{2} - 1)$$

11. (d) We have, $y^2 = 6x$

$$\Rightarrow \quad 2y\frac{dy}{dx} = 6 \Rightarrow \frac{dy}{dx} = \frac{3}{y}$$

Slope of tangent at (x_1, y_1) is $m_1 = \frac{3}{y_1}$

Also, $9x^2 + by^2 = 16$

$$\Rightarrow \quad 18x + 2by\frac{dy}{dx} = 0 \Rightarrow \frac{dy}{dx} = \frac{-9x}{by}$$

Slope of tangent at (x_1, y_1) is $m_2 = \frac{-9x_1}{by_1}$

Since, these are intersection at right angle.

$$\therefore \quad m_1m_2 = -1 \Rightarrow \frac{27x_1}{by_1^2} = 1$$

$$\Rightarrow \quad \frac{27x_1}{6bx_1} = 1 \quad [\because y_1^2 = 6x_1]$$

$$\Rightarrow \quad b = \frac{9}{2}$$

12. (b) Equation of tangent and normal to the curve $y^2 = 16x$ at (16, 16) is $x - 2y + 16 = 0$ and $2x + y - 48 = 0$, respectively.

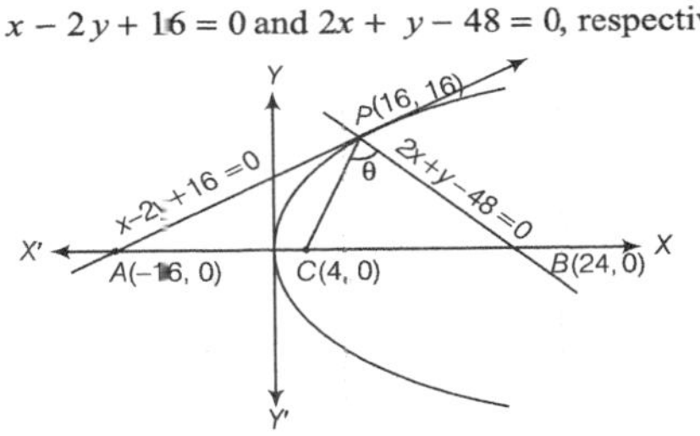

$A = (-16, 0)$; $B = (24, 0)$

C is the centre of circle passing through PAB

i.e. $\quad C = (4, 0)$

$$\text{Slope of } PC = \frac{16-0}{16-4} = \frac{16}{12} = \frac{4}{3} = m_1$$

$$\text{Slope of } PB = \frac{16-0}{16-24} = \frac{16}{-8} = -2 = m_2$$

$$\tan\theta = \left|\frac{m_1 - m_2}{1 + m_1 m_2}\right|$$

$$\Rightarrow \quad \tan\theta = \left|\frac{\frac{4}{3} + 2}{1 - \left(\frac{4}{3}\right)(2)}\right| \quad \Rightarrow \quad \tan\theta = 2$$

13. *(a)* Tangents are drawn to the hyperbola $4x^2 - y^2 = 36$ at the point P and Q.

Tangent intersects at point $T(0, 3)$

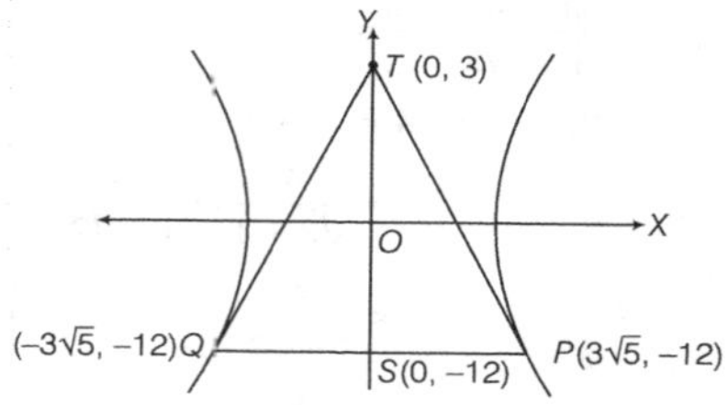

Clearly, PQ is chord of contact.

$\therefore$ Equation of PQ is $-3y = 36 \Rightarrow y = -12$

Solving the curve $4x^2 - y^2 = 36$ and $y = -12$,

we get $\quad x = \pm 3\sqrt{5}$

$$\text{Area of } \Delta PQT = \frac{1}{2} \times PQ \times ST = \frac{1}{2}(6\sqrt{5} \times 15) = 45\sqrt{5}$$

14. *(b)* L_1 is the line of intersection of the plane $2x - 2y + 3z - 2 = 0$ and $x - y + z + 1 = 0$ and L_2 is the line of intersection of the plane $x + 2y - z - 3 = 0$ and $3x - y + 2z - 1 = 0$

$$\text{Since } L_i \text{ is parallel to } \begin{vmatrix} \hat{i} & \hat{j} & \hat{k} \\ 2 & -2 & 3 \\ 1 & -1 & 1 \end{vmatrix} = \hat{i} + \hat{j}$$

$$L_2 \text{ is parallel to } \begin{vmatrix} \hat{i} & \hat{j} & \hat{k} \\ 1 & 2 & -1 \\ 3 & -1 & 2 \end{vmatrix} = 3\hat{i} - 5\hat{j} - 7\hat{k}$$

Also, L_2 passes through $\left(\frac{5}{7}, \frac{8}{7}, 0\right)$. [put $z = 0$ in last two planes]

So, equation of plane is

$$\begin{vmatrix} x - \frac{5}{7} & y - \frac{8}{7} & z \\ 1 & 1 & 0 \\ 3 & -5 & -7 \end{vmatrix} = 0$$

$$\Rightarrow \quad 7x - 7y + 8z + 3 = 0$$

Now, perpendicular distance from origin is

$$\left|\frac{3}{\sqrt{7^2 + 7^2 + 8^2}}\right| = \frac{3}{\sqrt{162}} = \frac{1}{3\sqrt{2}}$$

15. *(d)* **Key idea** *length of projection of the line segment joining* $\mathbf{a}_1$ *and* $\mathbf{a}_2$ *on the plane* $\mathbf{r} \cdot \mathbf{n} = d$ *is* $\left|\frac{(\mathbf{a}_2 - \mathbf{a}_1) \times \mathbf{n}}{|\mathbf{n}|}\right|$

Length of projection the line segment joining the points (5, −1, 4) and (4, −1, 3) on the plane $x + y + z = 7$ is

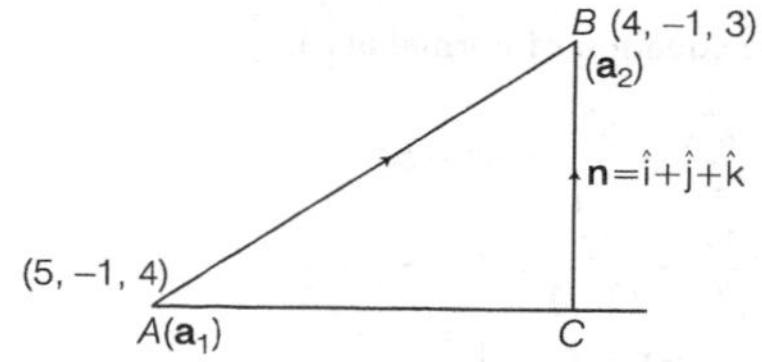

$$AC = \left|\frac{(\mathbf{a}_2 - \mathbf{a}_1) \times \mathbf{n}}{|\mathbf{n}|}\right| = \frac{|(-\hat{\mathbf{i}} - \hat{\mathbf{k}}) \times (\hat{\mathbf{i}} + \hat{\mathbf{j}} + \hat{\mathbf{k}})|}{|i + j + k|}$$

$$AC = \frac{|\hat{\mathbf{i}} - \hat{\mathbf{k}}|}{\sqrt{3}} \Rightarrow AC = \frac{\sqrt{2}}{\sqrt{3}} = \sqrt{\frac{2}{3}}$$

Alternative Method

Clearly, DR's of AB are $4 - 5, -1 + 1, 3 - 4$, i.e. $-1, 0, -1$ and DR's of normal to plane are 1, 1, 1.

Now, let θ be the angle between the line and plane, then θ is given by

$$\sin\theta = \frac{|-1 + 0 - 1|}{\sqrt{(-1)^2 + (-1)^2}\sqrt{1^2 + 1^2 + 1^2}} = \frac{2}{\sqrt{2}\sqrt{3}} = \sqrt{\frac{2}{3}}$$

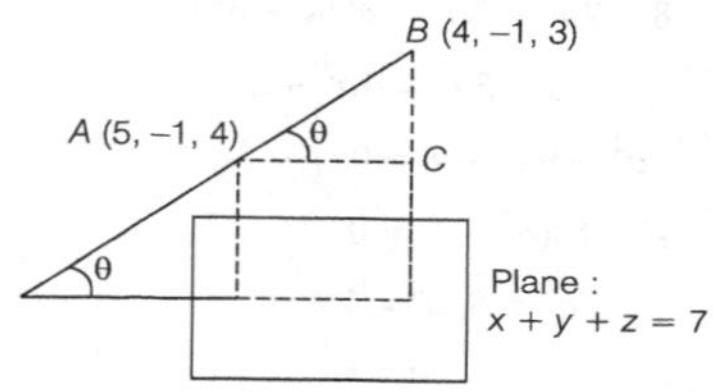

$$\Rightarrow \quad \sin\theta = \sqrt{\frac{2}{3}} \Rightarrow \cos\theta = \sqrt{1 - \sin^2\theta} = \sqrt{1 - \frac{2}{3}} = \frac{1}{\sqrt{3}}$$

Clearly, length of projection

$$= AB\cos\theta = \sqrt{2}\frac{1}{\sqrt{3}} \qquad [\because AB = \sqrt{2}]$$

$$= \sqrt{\frac{2}{3}}$$

16. *(a)* **Key idea** *If any vector* $\mathbf{x}$ *is coplanar with the vector* $\mathbf{y}$ *and* $\mathbf{z}$, *then* $\mathbf{x} = \lambda\mathbf{y} + \mu\mathbf{z}$

Here, $\mathbf{u}$ is coplanar with $\mathbf{a}$ and $\mathbf{b}$.

$$\therefore \quad \mathbf{u} = \lambda\mathbf{a} + \mu\mathbf{b}$$

Dot product with $\mathbf{a}$, we get

$$\mathbf{u} \cdot \mathbf{a} = \lambda(\mathbf{a} \cdot \mathbf{a}) + \mu(\mathbf{b} \cdot \mathbf{a})$$

$$\Rightarrow \quad 0 = 14\lambda + 2\mu \qquad \ldots\text{(i)}$$

$$[\because \mathbf{a} = 2\hat{\mathbf{i}} + 3\hat{\mathbf{j}} - \hat{\mathbf{k}}, \mathbf{b} = \hat{\mathbf{j}} + \hat{\mathbf{k}}, \mathbf{u} \cdot \mathbf{a} = 0]$$

Dot product with $\mathbf{b}$, we get

$$\mathbf{u} \cdot \mathbf{b} = \lambda(\mathbf{a} \cdot \mathbf{b}) + \mu(\mathbf{b} \cdot \mathbf{b})$$

$$24 = 2\lambda + 2\mu \qquad \ldots\text{(ii)} \quad [\because \mathbf{u} \cdot \mathbf{b} = 24]$$

Solving Eqs. (i) and (ii), we get

$$\lambda = -2, \mu = 14$$

Dot product with $\mathbf{u}$, we get

$$|\mathbf{u}|^2 = \lambda(\mathbf{u} \cdot \mathbf{a}) + \mu(\mathbf{u} \cdot \mathbf{b})$$

$$|\mathbf{u}|^2 = -2(0) + 14(24) \Rightarrow |\mathbf{u}|^2 = 336$$

JEE Advanced

1. (a, c, d) Given a ΔXYZ, where

$$2s = x + y + z$$

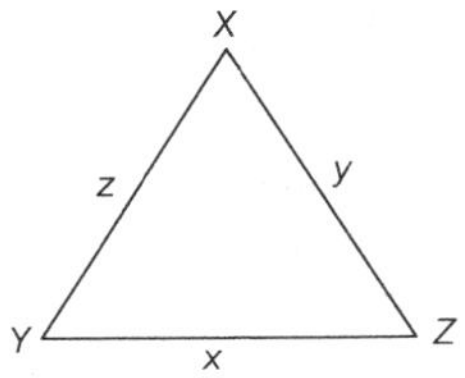

and $\dfrac{s-x}{4} = \dfrac{s-y}{3} = \dfrac{s-z}{2}$

$\therefore \quad \dfrac{s-x}{4} = \dfrac{s-y}{3} = \dfrac{s-z}{2}$

$$= \frac{3s-(x+y+z)}{4+3+2} = \frac{s}{9}$$

or $\dfrac{s-x}{4} = \dfrac{s-y}{3} = \dfrac{s-z}{2} = \dfrac{s}{9} = \lambda$ (let)

$\Rightarrow \quad s = 9\lambda, s = 4\lambda + x, s = 3\lambda + y$

and $s = 2\lambda + z$

$\therefore \quad s = 9\lambda, x = 5\lambda,$

$y = 6\lambda, \ z = 7\lambda$

Now, $\Delta = \sqrt{s(s-x)(s-y)(s-z)}$ [heron's formula]

$= \sqrt{9\lambda \cdot 4\lambda \cdot 3\lambda \cdot 2\lambda} = 6\sqrt{6}\lambda^2$ ...(i)

Also, $\pi r^2 = \dfrac{8\pi}{3}$

$\Rightarrow \quad r^2 = \dfrac{8}{3}$ ...(ii)

and $R = \dfrac{xyz}{4\Delta} = \dfrac{(5\lambda)(6\lambda)(7\lambda)}{4 \cdot 6\sqrt{6}\lambda^2} = \dfrac{35\lambda}{4\sqrt{6}}$ ...(iii)

Now, $r^2 = \dfrac{8}{3} = \dfrac{\Delta^2}{S^2} = \dfrac{216\lambda^4}{81\lambda^2}$

$\Rightarrow \quad \dfrac{8}{3} = \dfrac{8}{3}\lambda^2$ [from Eq. (ii)]

$\Rightarrow \quad \lambda = 1$

(a) $\Delta XYZ = 6\sqrt{6}\lambda^2 = 6\sqrt{6}$

$\therefore$ Option (a) is correct.

(b) Radius of circumcircle

$$(R) = \frac{35}{4\sqrt{6}}\lambda = \frac{35}{4\sqrt{6}}$$

$\therefore$ Option (b) is incorrect.

(c) Since,

$$r = 4R \sin\frac{X}{2} \cdot \sin\frac{Y}{2} \cdot \sin\frac{Z}{2}$$

$\Rightarrow \quad \dfrac{2\sqrt{2}}{\sqrt{3}} = 4 \cdot \dfrac{35}{4\sqrt{6}} \sin\dfrac{X}{2} \cdot \sin\dfrac{Y}{2} \cdot \sin\dfrac{Z}{2}$

$\Rightarrow \quad \dfrac{4}{35} = \sin\dfrac{X}{2} \cdot \sin\dfrac{Y}{2} \cdot \sin\dfrac{Z}{2}$

$\therefore$ Option (c) is correct.

(d) $\sin^2\left(\dfrac{X+Y}{2}\right) = \cos^2\left(\dfrac{Z}{2}\right)$, as

$$\frac{X+Y}{2} = 90^\circ - \frac{Z}{2}$$

$$= \frac{s(s-z)}{xy} = \frac{9 \times 2}{5 \times 6} = \frac{3}{5}$$

$\therefore$ Option (d) is correct.

2. (a, b, c) Given,

$C_1 : x^2 + y^2 = 3$ intersects the parabola $x^2 = 2y$.

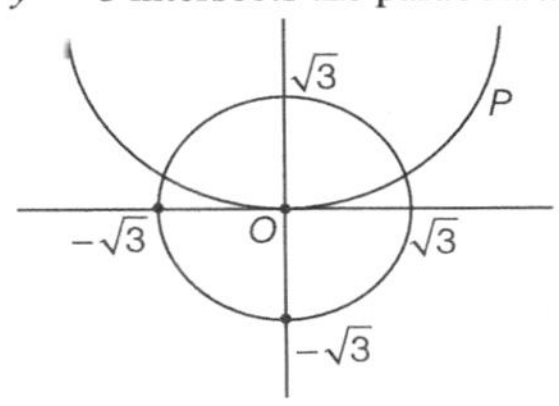

On solving $x^2 + y^2 = 3$ and $x^2 = 2y$, we get

$y^2 + 2y = 3 \Rightarrow y^2 + 2y - 3 = 0$

$\Rightarrow \quad (y+3)(y-1) = 0$

$\therefore \quad y = 1, -3$

[neglecting $y = -3$, as $-\sqrt{3} \le y \le \sqrt{3}$]

$\therefore \quad y = 1 \Rightarrow x = \pm\sqrt{2}$

$\Rightarrow \quad P(\sqrt{2}, 1) \in$ I quadrant

of tangent at $P(\sqrt{2}, 1)$ to $C_1 : x^2 + y^2 = 3$ is

$\sqrt{2}x + 1 \cdot y = 3$ ...(i)

Now, let the centres of C_2 and C_3 be Q_2 and Q_3, and tangent at P touches C_2 and C_3 at R_2 and R_3 shown as below

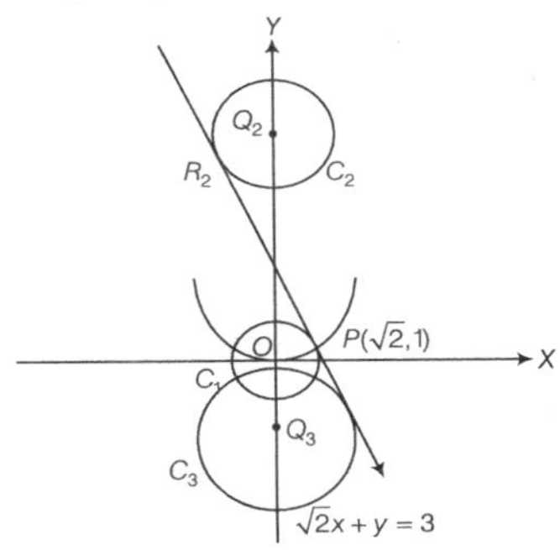

Let Q_2 be $(0, k)$ and radius is $2\sqrt{3}$.

$\therefore \quad \dfrac{|0+k-3|}{\sqrt{2+1}} = 2\sqrt{3} \Rightarrow |k-3| = 6 \Rightarrow k = 9, -3$

$\therefore Q_2(0, 9)$ and $Q_3(0, -3)$

Hence, $Q_2Q_3 = 12$

$\therefore$ Option (a) is correct.

Also, R_2R_3 is common internal tangent to C_2 and C_3,

and $r_2 = r_3 = 2\sqrt{3}$

$\therefore \quad R_2R_3 = \sqrt{d^2 - (r_1 + r_2)^2}$

$= \sqrt{12^2 - (4\sqrt{3})^2}$

$= \sqrt{144 - 48} = \sqrt{96} = 4\sqrt{6}$

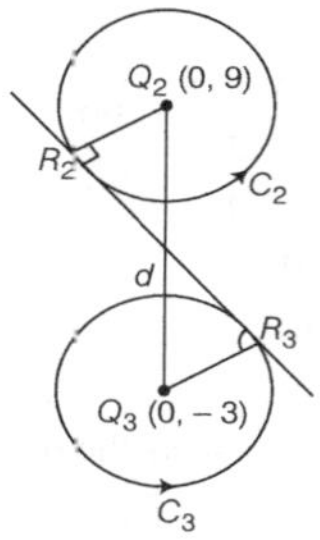

$\therefore$ Option (b) is correct.

$\because$ Length of perpendicular from $O(0, 0)$ to R_2R_3 is equal to radius of $C_1 = \sqrt{3}$.

$$\therefore \text{ Area of } \Delta OR_2R_3 = \frac{1}{2} \times R_2R_3 \times \sqrt{3}$$

$$= \frac{1}{2} \times 4\sqrt{6} \times \sqrt{3} = 6\sqrt{2}$$

$\therefore$ Option (c) is correct.

$$\text{Also, area of } \Delta PQ_2Q_3 = \frac{1}{2} Q_2Q_3 \times \sqrt{2}$$

$$= \frac{\sqrt{2}}{2} \times 12 = 6\sqrt{2}$$

$\therefore$ Option (d) is incorrect.

3. **(c)** Let image of $Q(3, 1, 7)$ w.r.t. $x - y + z = 3$ be $P(\alpha, \beta, \gamma)$.

$$\therefore \quad \frac{\alpha - 3}{1} = \frac{\beta - 1}{-1} = \frac{\gamma - 7}{1}$$

$$= \frac{-2(3 - 1 + 7 - 3)}{1^2 + (-1)^2 + (1)^2}$$

$$\Rightarrow \alpha - 3 = 1 - \beta = \gamma - 7 = -4$$

$$\therefore \quad \alpha = -1, \beta = 5, \gamma = 3$$

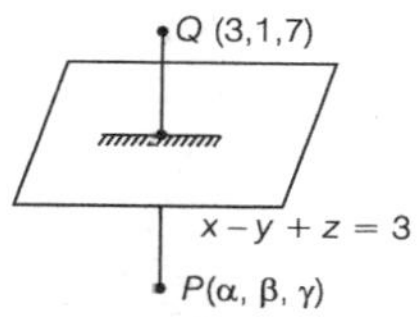

Hence, the image of $Q(3, 1, 7)$ is $P(-1, 5, 3)$.

To find equation of plane passing through $P(-1, 5, 3)$ and containing $\frac{x}{1} = \frac{y}{2} = \frac{z}{1}$

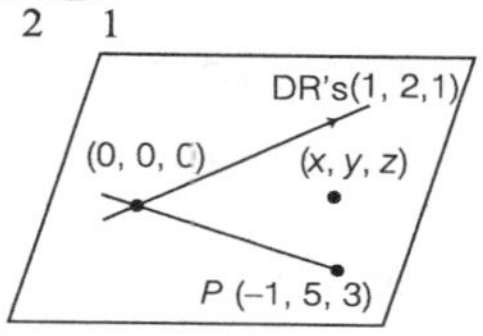

$$\Rightarrow \begin{vmatrix} x - 0 & y - 0 & z - 0 \\ 1 - 0 & 2 - 0 & 1 - 0 \\ -1 - 0 & 5 - 0 & 3 - 0 \end{vmatrix} = 0$$

$$\Rightarrow x(6 - 5) - y(3 + 1) + z(5 + 2) = 0$$

$$\therefore \quad x - 4y + 7z = 0$$

4. **(b, c)** Let θ be the angle between $\hat{u}$ and $\vec{v}$.

$$\therefore |\hat{u} \times \vec{v}| = 1 \Rightarrow |\hat{u}||\vec{v}| \sin\theta = 1$$

$$\therefore \quad |\vec{v}| \sin\theta = 1 \qquad [\because |\hat{u}| = 1] \ldots \text{(i)}$$

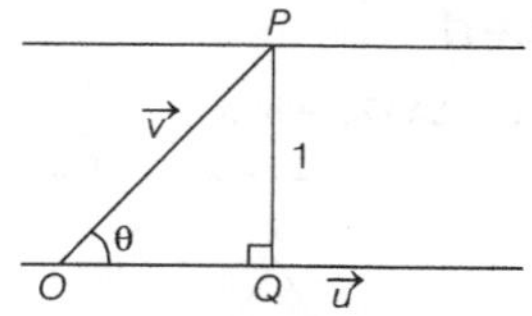

Clearly, there may be infinite vectors $\overrightarrow{OP} = \vec{v}$, such that P is always 1 unit distance from $\hat{u}$.

$\therefore$ Option (b) is correct.

Again, let ϕ be the angle between $\hat{w}$ and $\hat{u} \times \vec{v}$.

$$\therefore \hat{w} \cdot (\hat{u} \times \vec{v}) = 1 \Rightarrow |\hat{w}||\hat{u} \times \vec{v}| \cos\phi = 1$$

$$\Rightarrow \quad \cos\phi = 1 \Rightarrow \phi = 0$$

$$\text{Thus,} \quad \hat{w} = \hat{u} \times \vec{v}$$

Now, if $\hat{u}$ lies in XY-plane, then

$$\hat{u} \times \vec{v} = \begin{vmatrix} \hat{i} & \hat{j} & \hat{k} \\ u_1 & u_2 & 0 \\ v_1 & v_2 & v_3 \end{vmatrix}$$

$$\text{or} \quad \hat{u} = u\hat{i} + u_2\hat{j}$$

$$\therefore \quad \hat{w} = (u_2v_3)\hat{i} - (u_1v_3)\hat{j} + (u_1v_2 - v_1u_2)\hat{k}$$

$$= \frac{1}{\sqrt{6}}(\hat{i} + \hat{j} + 2\hat{k})$$

$$\therefore \quad u_2v_3 = \frac{1}{\sqrt{6}} \Rightarrow u_1v_3 = \frac{-1}{\sqrt{6}}$$

$$\Rightarrow \quad \frac{u_2v_3}{u_1v_3} = -1 \quad \text{or} \quad |u_1| = |u_2|$$

$\therefore$ Option (c) is correct.

Now, if $\hat{u}$ lies in XZ-plane, then $\hat{u} = u_1\hat{i} + u_3\hat{k}$

$$\therefore \quad \hat{u} \times \vec{v} = \begin{vmatrix} \hat{i} & \hat{j} & \hat{k} \\ u_1 & 0 & u_3 \\ v_1 & v_2 & v_3 \end{vmatrix}$$

$$\Rightarrow \quad \hat{w} = (-v_2u_3)\hat{i} - (u_1v_3 - u_3v_1)\hat{j} + (u_1v_2)\hat{k}$$

$$\Rightarrow \quad \hat{w} = \frac{1}{\sqrt{6}}(\hat{i} + \hat{j} + 2\hat{k})$$

$$\Rightarrow \quad -v_2u_3 = \frac{1}{\sqrt{6}} \text{ and } u_1v_2 = \frac{2}{\sqrt{6}}$$

$$\therefore \quad |u_2| = 2|u_3|$$

$\therefore$ Option (d) is incorrect

5. **(a, c, d)** Tangent to $y^2 = 4x$ at $(t^2, 2t)$ is

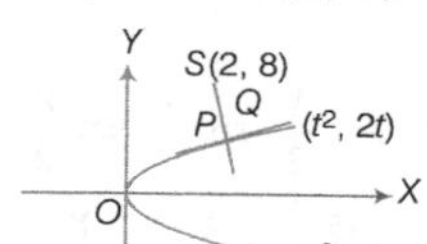

$$y(2t) = 2(x + t^2) \Rightarrow yt = x + t^2 \qquad \ldots\text{(i)}$$

Equation of normal at $P(t^2, 2t)$ is

$$y + tx = 2t + t^3$$

Since, normal at P passes through centre of circle $S(2, 8)$.

$\therefore 8 + 2t = 2t + t^3 \Rightarrow t = 2$, i.e. $P(4, 4)$

[since, shortest distance between two curves lie along their common normal and the common normal will pass through the centre of circle]

$\therefore \quad SP = \sqrt{(4-2)^2 + (4-8)^2} = 2\sqrt{5}$

$\therefore$ Option (a) is correct.

Also, $\quad SQ = 2$

$\therefore \quad PQ = SP - SQ = 2\sqrt{5} - 2$

Thus, $\quad \dfrac{SQ}{QP} = \dfrac{1}{\sqrt{5}-1} = \dfrac{\sqrt{5}+1}{4}$

$\therefore$ Option (b) is incorrect.

Now, x-intercept of normal is $x = 2 + 2^2 = 6$

$\therefore$ Option (c) is correct.

Slope of tangent $= \dfrac{1}{t} = \dfrac{1}{2}$

$\therefore$ Option (d) is incorrect

6. (a) Here, $\quad \dfrac{x^2}{9} + \dfrac{y^2}{8} = 1 \quad \ldots(i)$

has foci $(\pm ae, 0)$

where, $a^2e^2 = a^2 - b^2 \Rightarrow a^2e^2 = 9 - 8$

$\Rightarrow ae = \pm 1$ i.e. $F_1, F_2 = (\pm 1, 0)$

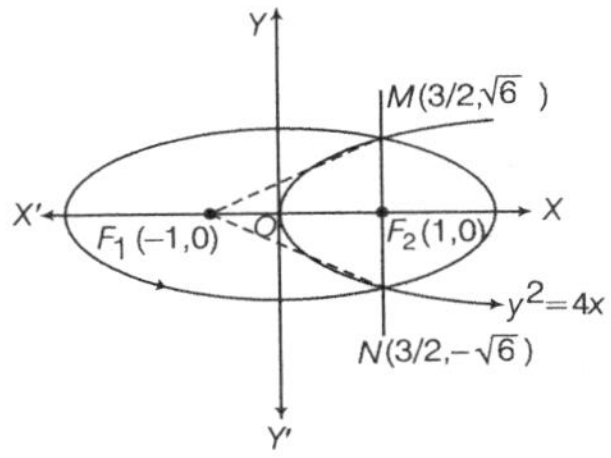

Equation of parabola having vertex $O(0,0)$ and $F_2(1, 0)$ (as, $x_2 > 0$)

$$y^2 = 4x \quad \ldots(ii)$$

On solving $\dfrac{x^2}{9} + \dfrac{y^2}{8} = 1$ and $y^2 = 4x$, we get

$$x = 3/2 \text{ and } y = \pm\sqrt{6}$$

Equation of altitude through M on NF_1 is

$$\frac{y-\sqrt{6}}{x-3/2} = \frac{5}{2\sqrt{6}}$$

$$\Rightarrow \quad (y-\sqrt{6}) = \frac{5}{2\sqrt{6}}(x - 3/2) \quad \ldots(iii)$$

and equation of altitude through

$$F_1 \text{ is } y = 0 \quad \ldots(iv)$$

On solving Eqs. (iii) and (iv), we get

$\left(-\dfrac{9}{10}, 0\right)$ as orthocentre.

7. (c) Equation of tangent at $M(3/2, \sqrt{6})$ to $\dfrac{x^2}{9} + \dfrac{y^2}{8} = 1$ is

$$\frac{3}{2}\cdot\frac{x}{9} + \sqrt{6}\cdot\frac{y}{8} = 1 \quad \ldots(i)$$

which intersect X-axis at (6, 0).

Also, equation of tangent at

$N(3/2, -\sqrt{6})$ is

$$\frac{3}{2}\cdot\frac{x}{9} - \sqrt{6}\,\frac{y}{8} = 1 \quad \ldots(ii)$$

Eqs. (i) and (ii) intersect on X-axis at $R(6, 0)$. $\quad \ldots(iii)$

Also, normal at $M(3/2, \sqrt{6})$ is

$$y - \sqrt{6} = \frac{-\sqrt{6}}{2}\left(x - \frac{3}{2}\right)$$

On solving with $y = 0$,

we get $\quad Q(7/2, 0) \quad \ldots(iv)$

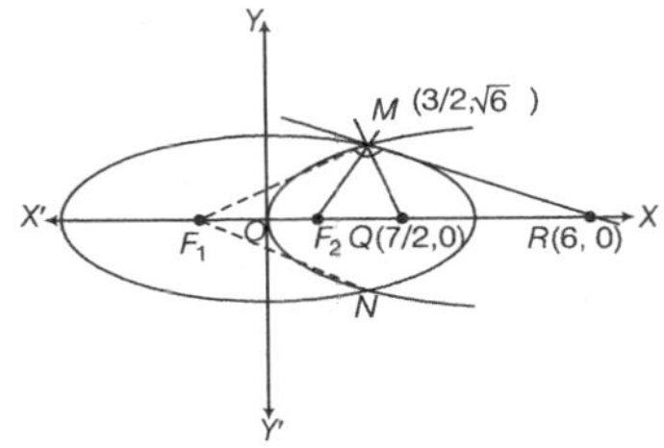

$\therefore$ Area of $\Delta MQR = \dfrac{1}{2}\left(6 - \dfrac{7}{2}\right)\sqrt{6} = \dfrac{5\sqrt{6}}{4}$ sq units

and area of quadrilateral MF_1NF_2

$= 2 \times \dfrac{1}{2}\{1 - (-1)\}\sqrt{6} = 2\sqrt{6}$ sq units

$\therefore \quad \dfrac{\text{Area of } \Delta MQR}{\text{Area of quadrilateral } MF_1NF_2} = \dfrac{5}{8}$

8. (a,b,c) Tangent $\equiv 2x - y + 1 = 0$

Hyperbola $\equiv \dfrac{x^2}{a^2} - \dfrac{y^2}{16} = 1$

It point $\equiv (a\sec\theta, 4\tan\theta)$,

tangent $\equiv \dfrac{x\sec\theta}{a} - \dfrac{y\tan\theta}{4} = 1$

On comparing, we get

$$\sec\theta = -2a$$

$$\tan\theta = -4$$

$$\Rightarrow \quad 4a^2 - 16 = 1$$

$$\therefore \quad a = \frac{\sqrt{17}}{2}$$

Substitute the value of a in option (a), (b), (c) and (d).

9. (b) Equation of chord with mid-point (h, k).

$$T = S_1$$

$$yk - 8x - 8h = k^2 - 16h$$

$$2x - \frac{yk}{4} = 2h - \frac{k^2}{4}$$

$\because \quad 2x + y = p$

$\therefore \quad k = -4$ and $p = 2h - 4$

Solutions (Q. Ncs. 13-15)

Using standard equations

I-ii-Q $\quad$ II-iv-R

III-i-P $\quad$ IV-iii-S

Cinoarubg wutg eqyatuibs ub guveb qyestuibsm we get

10. (a) Either option (a) or (c) is correct. But $a = \sqrt{2}$ and (–1, 1) satisfy $x^2 + y^2 = a^2$.

So, option (a) is correct.

11. *(b)* Either option (b) or (c) is correct.

Satisfying the point $\left(\sqrt{3}, \frac{1}{2}\right)$ in (II), we get

$$a^2 = 4.$$

$\therefore$ The conic is $x^2 + 4y^2 = 4$

Now, equation of tangent at $(\sqrt{3}, \frac{1}{2})$ is $\sqrt{3}x + 2y = 4$, which is the given equation.

So, option (b) is correct.

12. *(a)* Either option (a), (b) or (c) is correct.

Satisfying the point (8, 16) in (III), we get

$\therefore$ The conic is $y^2 = 32x = 16(2x)$

Now, equation of tangent at (8, 16) is $y \cdot 16 = 16(x + 8)$

$\Rightarrow y = x + 8$, which is the given in equation.

So, option (a) is correct.

Solutions. (Q.Nos. 16-18)

$$f(x) = x + \ln x - x \ln x$$

$$f(1) = 1 > 0$$

$$f(e^2) = e^2 + 2 - 2e^2 = 2 - e^2 < 0$$

$\Rightarrow f(x) = 0$ for some $x \in (1, e^2)$

$\therefore$ I is correct

$$f'(x) = 1 + \frac{1}{x} - \ln x - 1$$

$$= \frac{1}{x} - \ln x$$

$f'(x) > 0$ for $(0, 1)$

$f'(x) < 0$ for (e, ∞)

$\therefore$ P and Q are correct, II is correct, III is incorrect.

$$f''(x) = \frac{-1}{x^2} - \frac{1}{x}$$

$f''(x) < 0$ for $(0, \infty)$

$\therefore$ S, is correct, R is incorrect. IV is incorrect.

$$\lim_{x\to\infty} f(x) = -\infty$$

$$\lim_{x\to\infty} f'(x) = -\infty$$

$$\lim_{x\to\infty} f''(x) = 0$$

$\therefore$ ii, iii, iv are correct.

13. *(b)* $\mathbf{OP} \cdot \mathbf{OQ} + \mathbf{OR} \cdot \mathbf{OS} = \mathbf{OR} \cdot \mathbf{OP} + \mathbf{OQ} \cdot \mathbf{OS}$

$\Rightarrow \mathbf{OP}(\mathbf{OQ} - \mathbf{OR}) + \mathbf{OS}(\mathbf{OR} - \mathbf{OQ}) = 0$

$\Rightarrow (\mathbf{OP} - \mathbf{OS})(\mathbf{OQ} - \mathbf{OR}) = 0$

$\Rightarrow \mathbf{SP} \cdot \mathbf{RQ} = 0$

Similarly $\mathbf{SR} \cdot \mathbf{PQ} = 0$ and $\mathbf{SQ} \cdot \mathbf{PR} = 0$

$\therefore$ S is orthocentre.

14. *(d)* Let the equation of plane be $ax + by + cz = 1$. Then

$$a + b + c = 1$$

$$2a + b - 2c = 0$$

$$3a - 6b - 2c = 0 \Rightarrow a = 7b$$

$$c = \frac{15b}{2}$$

$$b = \frac{2}{31}, a = \frac{14}{31}, c = \frac{15}{31}$$

$\therefore \quad 14x + 2y + 15z = 31$

15. *(a)* $\cos(P + Q) + \cos(Q + R) + \cos(R + P)$

$$= -(\cos R + \cos P + \cos Q)$$

Max. of $\cos P + \cos Q + \cos R = \frac{3}{2}$

Min. of $\cos(P + Q) + \cos(Q + R) + \cos(R + P)$ is $= -\frac{3}{2}$

16. *(a)* $\sin R = \sin(P + Q)$

17. *(c, d)* We have,

$P_1 : 2x + y - z = 3$

and $P_2 : x + 2y + z = 2$

Here, $\vec{n_1} = 2\hat{i} + \hat{j} - \hat{k}$

and $\vec{n_2} = \hat{i} + 2\hat{j} + \hat{k}$

(a) Direction ratio of the line of intersection of P_1 and P_2 is θ $\vec{n_1} \times \vec{n_2}$

i.e. $\begin{vmatrix} \hat{i} & \hat{j} & \hat{k} \\ 2 & 1 & -1 \\ 1 & 2 & 1 \end{vmatrix} = (1 + 2)\hat{i} - (2 + 1)\hat{j} + (4 - 1)\hat{k}$

$$= 3(\hat{i} - \hat{j} + \hat{k})$$

Hence, statement a is false.

(b) We have,

$$\frac{3x - 4}{9} = \frac{1 - 3y}{9} = \frac{z}{3}$$

$$\Rightarrow \frac{x - \frac{4}{3}}{3} = \frac{\left(y - \frac{1}{3}\right)}{-3} = \frac{z}{3}$$

This line is parallel to the line of intersection of P_1 and P_2. Hence, statement (b) is false.

(c) Let acute angle between P_1 and P_2 be θ.

We know that, $\cos\theta = \frac{\vec{n_1} \cdot \vec{n_2}}{|\vec{n_1}||\vec{n_2}|}$

$$= \frac{(2\hat{i} + \hat{j} - \hat{k}) \cdot (\hat{i} + 2\hat{j} + \hat{k})}{|2\hat{i} + \hat{j} - \hat{k}||\hat{i} + 2\hat{j} + \hat{k}|}$$

$$= \frac{2 + 2 - 1}{\sqrt{6} \times \sqrt{6}} = \frac{1}{2}$$

$$\theta = 60°$$

Hence, statement (c) is true.

(d) Equation of plane passing through the point $(4, 2, -2)$ and perpendicular to the line of intersection of P_1 and P_2 is

$$3(x - 4) - 3(y - 2) + 3(z + 2) = 0$$

$$\Rightarrow 3x - 3y + 3z - 12 + 6 + 6 = 0$$

$$\Rightarrow x - y + z = 0$$

Now, distance of the point (2, 1, 1) from the plane $x - y + z = 0$ is

$$D = \left|\frac{2 - 1 + 1}{\sqrt{1 + 1 + 1}}\right| = \frac{2}{\sqrt{3}}$$

Hence, statement (d) is true.

18. *(3)* We have,

$$\vec{c} = x\vec{a} + y\vec{b} + \vec{a} \times \vec{b} \text{ and } \vec{a} \cdot \vec{b} = 0$$

$$|\vec{a}| = |\vec{b}| = 1 \text{ and } |\vec{c}| = 2$$

Also, given $\vec{c}$ is inclined on $\vec{a}$ and $\vec{b}$ with same angle α.

$\therefore \quad \vec{a} \cdot \vec{c} = x|\vec{a}|^2 + y(\vec{a} \cdot \vec{b}) + \vec{a} \cdot (\vec{a} \times \vec{b})$

$$|\vec{a}||\vec{c}|\cos\alpha = x + 0 + 0$$

$$x = 2\cos\alpha$$

Similarly,

$$|\vec{b}||\vec{c}|\cos\alpha = 0 + y + 0$$

$\Rightarrow \quad y = 2\cos\alpha$

$$|\vec{c}|^2 = x^2 + y^2 + |\vec{a} \times \vec{b}|^2$$

$$4 = 8\cos^2\alpha + |a|^2|b|^2\sin^2 90^\circ$$

$$4 = 8\cos^2\alpha + 1$$

$\Rightarrow \quad 8\cos^2\alpha = 3$

19. *(a, c)* We have, Equations of circle

$$x^2 + y^2 = \frac{1}{2}$$

and Equations of parabola

$$y^2 = 4x$$

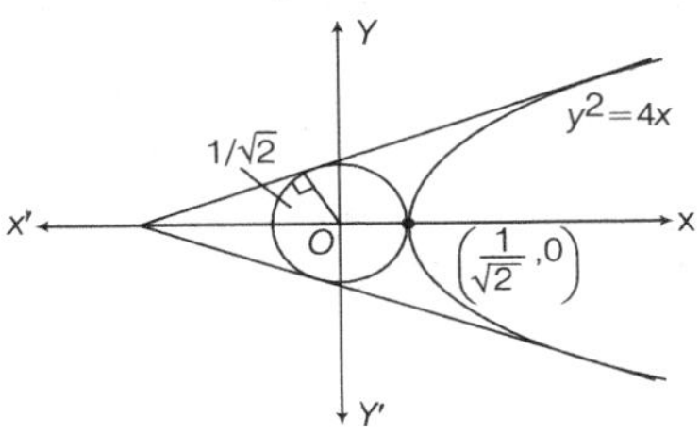

Let the equation of common tangent of parabola and circle is

$$y = mx + \frac{1}{m}$$

Since, radius of circle $= \dfrac{1}{\sqrt{2}}$

$\therefore \quad \dfrac{1}{\sqrt{2}} = \left|\dfrac{0 + 0 + \frac{1}{m}}{\sqrt{1+m^2}}\right|$

$\Rightarrow \quad m^4 + m^2 - 2 = 0$

$\Rightarrow \quad m = \pm 1$

$\therefore$ Equation of common tangents are

$$y = x + 1$$

and $\quad y = -x - 1$

Intesection point of common tangent at $Q(-1, 0)$

$\therefore$ Equation of ellipse

$$\frac{x^2}{1} + \frac{y^2}{1/2} = 1$$

where, $a^2 = 1, b^2 = 1/2$

Now, eccentricity $(e) = \sqrt{1 - \dfrac{b^2}{a^2}} = \sqrt{1 - \dfrac{1}{2}} = \dfrac{1}{\sqrt{2}}$

and length of latusrectum $= \dfrac{2b^2}{a} = \dfrac{2\left(\frac{1}{2}\right)}{1} = 1$

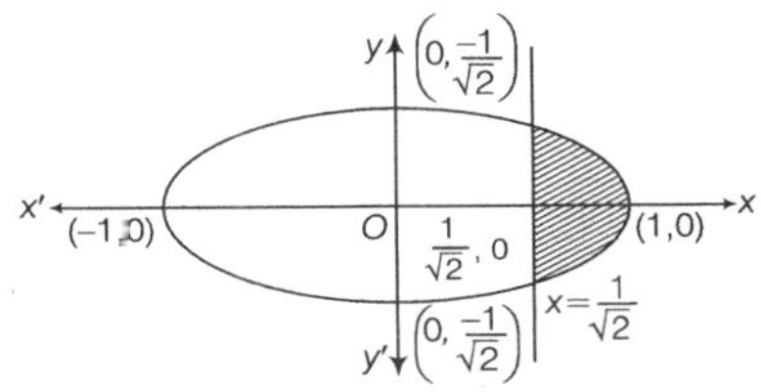

$\therefore \quad$ Area of shaded region $= 2\displaystyle\int_{1/\sqrt{2}}^{1} \frac{1}{\sqrt{2}}\sqrt{1-x^2}\,dx$

$$= \sqrt{2}\left[\frac{x}{2}\sqrt{1-x^2} + \frac{1}{2}\sin^{-1}x\right]_{1/\sqrt{2}}^{1}$$

$$= \sqrt{2}\left[\left(0 + \frac{\pi}{4}\right) - \left(\frac{1}{4} + \frac{\pi}{8}\right)\right]$$

$$= \sqrt{2}\left(\frac{\pi}{8} - \frac{1}{4}\right) = \frac{\pi - 2}{4\sqrt{2}}$$

20. *(8)* Let $P(\alpha, \beta, \gamma)$ and R is image of P in the XY-plane.

$\therefore \; R(\alpha, \beta, -\gamma)$

Also, Q is the image of P in the plane $x + y = 3$

$\therefore \quad \dfrac{x-\alpha}{1} = \dfrac{y-\beta}{1} = \dfrac{z-\gamma}{0}$

$$= \frac{-2(\alpha+\beta-3)}{2}$$

$x = 3-\beta,\; y = 3-\alpha,\; z = \gamma$

Since, Q is lies on Z-axis

$\therefore \quad \beta = 3, \alpha = 3, z = \gamma$

$\therefore \quad P(3, 3, \gamma)$

Given, distance of P from X-axis be 5

$\therefore \quad 5 = \sqrt{3^2 + \gamma^2}$

$$25 - 9 = \gamma^2$$

$\Rightarrow \quad \gamma = \pm 4$

Then, $\quad PR = |2\gamma| = |2 \times 4| = 8$

21. *(0.5)* Here, $F(1, 0, 0)$, $Q(0, 1, 0)$, $R(0, 0, 1)$, $T = (1, 1, 1)$ and $S = \left(\dfrac{1}{2}, \dfrac{1}{2}, \dfrac{1}{2}\right)$.

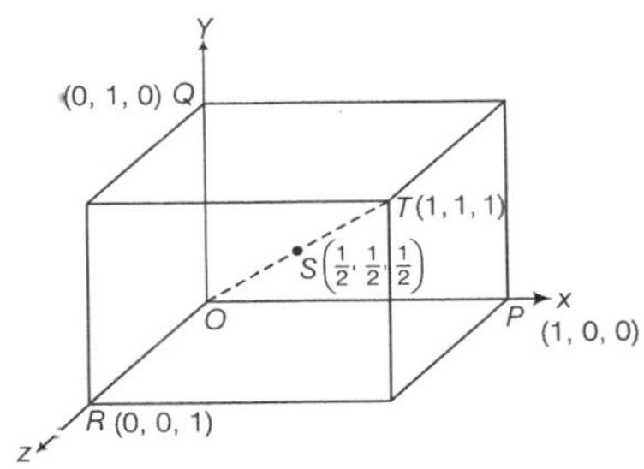

Now, $\quad \vec{p} = \overrightarrow{SP} = \overrightarrow{OP} - \overrightarrow{OS}$

$$= \left(\frac{1}{2}\hat{i} - \frac{1}{2}\hat{j} - \frac{1}{2}\hat{k}\right) = \frac{1}{2}(\hat{i} - \hat{j} - \hat{k})$$

$$\vec{q} = \overrightarrow{SQ} = \frac{1}{2}(-\hat{i} + \hat{j} - \hat{k})$$

$$\vec{r} = \overrightarrow{SR} = \frac{1}{2}(-\hat{i} - \hat{j} + \hat{k})$$

and $\qquad \vec{t} = \overrightarrow{ST} = \frac{1}{2}(\hat{i} + \hat{j} + \hat{k})$

$$\vec{p} \times \vec{q} = \frac{1}{4}\begin{vmatrix} \hat{i} & \hat{j} & \hat{k} \\ 1 & -1 & -1 \\ -1 & 1 & -1 \end{vmatrix} = \frac{1}{4}(2\hat{i} + 2\hat{j})$$

and $\qquad \vec{r} \times \vec{t} = \frac{1}{4}\begin{vmatrix} \hat{i} & \hat{j} & \hat{k} \\ -1 & -1 & 1 \\ 1 & 1 & 1 \end{vmatrix} = \frac{1}{4}(-2\hat{i} + 2\hat{j})$

Now, $(\vec{p} \times \vec{q}) \times (\vec{r} \times \vec{t}) = \frac{1}{16}\begin{vmatrix} \hat{i} & \hat{j} & \hat{k} \\ 2 & 2 & 0 \\ -2 & 2 & 0 \end{vmatrix}$

$$= \frac{1}{16}(8\hat{k}) = \frac{1}{2}\hat{k}$$

$\therefore \qquad |(p \times q) \times (\vec{r} \times \vec{t})| = \left|\frac{1}{2}\hat{k}\right| = \frac{1}{2} = 0.5$

22. *(b)* We have,

Equation of hyperbola

$$\frac{x^2}{a^2} - \frac{y^2}{b^2} = 1$$

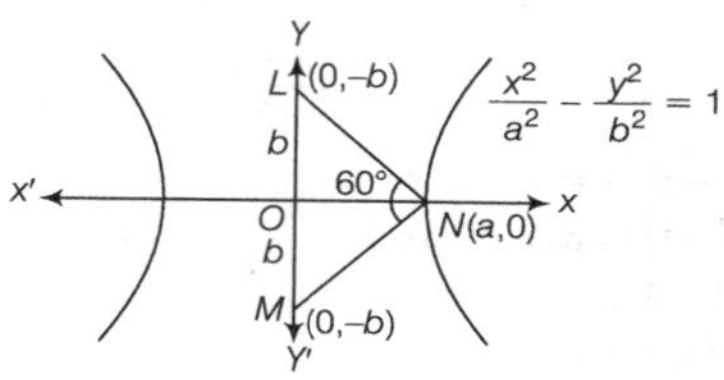

It is given,

$$\angle LNM = 60^\circ$$

and Area of $\Delta LMN = 4\sqrt{3}$

Now, ΔLNM is an equilateral triangle whose sides is $2b$

$[\because \Delta LON \cong \Delta MOL; \quad \therefore \quad \angle NLO = \angle NMO = 60^\circ]$

$\therefore \quad$ Area of $\Delta LMN = \frac{\sqrt{3}}{4}(2b)^2$

$\Rightarrow \qquad 4\sqrt{3} = \sqrt{3}b^2 \Rightarrow b = 2$

Also, area of $\Delta LMN = \frac{1}{2}a(2b) = ab$

$\Rightarrow \qquad 4\sqrt{3} = a(2)$

$\Rightarrow \qquad a = 2\sqrt{3}$

(P) Length of conjugate axis $= 2b = 2(2) = 4$

(Q) Eccentricity $(e) = \sqrt{1 + \frac{b^2}{a^2}} = \sqrt{1 + \frac{4}{12}}$

$$= \frac{4}{2\sqrt{3}} = \frac{2}{\sqrt{3}}$$

(R) Distance between the foci $= 2ae = 2 \times 2\sqrt{3} \times \frac{2}{\sqrt{3}} = 8$

(S) The length of latusrectum $= \frac{2b^2}{a} = \frac{2(4)}{2\sqrt{3}} = \frac{4}{\sqrt{3}}$

$P \to 4$; $Q \to 3$; $R \to 1$; $S \to 2$

Hence, option (b) is correct.